Advances in East Asian Agricultural Origins Studies: The Pleistocene to Holocene Transition

Advances in East Asian Agricultural Origins Studies: The Pleistocene to Holocene Transition

Editors

Pei-Lin Yu
Ikeya Kazunobu
Meng Zhang

MDPI • Basel • Beijing • Wuhan • Barcelona • Belgrade • Manchester • Tokyo • Cluj • Tianjin

Editors
Pei-Lin Yu
Boise State University
USA

Ikeya Kazunobu
National Museum of Ethnology
Japan

Meng Zhang
Fudan University
China

Editorial Office
MDPI
St. Alban-Anlage 66
4052 Basel, Switzerland

This is a reprint of articles from the Special Issue published online in the open access journal *Quaternary* (ISSN 2571-550X) (available at: https://www.mdpi.com/journal/quaternary/special_issues/asian_agricultural_origins).

For citation purposes, cite each article independently as indicated on the article page online and as indicated below:

LastName, A.A.; LastName, B.B.; LastName, C.C. Article Title. *Journal Name* **Year**, *Volume Number*, Page Range.

ISBN 978-3-0365-3407-7 (Hbk)
ISBN 978-3-0365-3408-4 (PDF)

Contents

About the Editors . . . . **vii**

Pei-Lin Yu, Kazunobu Ikeya and Meng Zhang
Introduction: New Discoveries and Theoretical Implications for the Last Foraging and First Farming in East Asia
Reprinted from: *Quaternary* **2021**, *4*, 37, doi:10.3390/quat4040037 . . . . **1**

Kazunobu Ikeya
Ethnoarchaeology of Introducing Agriculture and Social Continuity among Sedentarised Hunter–Gatherers: The Transition from the Jomon to the Yayoi Period
Reprinted from: *Quaternary* **2021**, *4*, 28, doi:10.3390/quat4030028 . . . . **5**

Shin'ichiro Fujio
Early Grain Cultivation and Starting Processes in the Japanese Archipelago
Reprinted from: *Quaternary* **2021**, *4*, 3, doi:10.3390/quat4010003 . . . . **21**

Ruizhe Liu, Hui Liu and Shengqian Chen
Alternative Adaptation Strategy during the Paleolithic–Neolithic Transition: Potential Use of Aquatic Resources in the Western Middle Yangtze Valley, China
Reprinted from: *Quaternary* **2020**, *3*, 28, doi:10.3390/quat3030028 . . . . **37**

Pei-Lin Yu
Modeling Incipient Use of Neolithic Cultigens by Taiwanese Foragers: Perspectives from Niche Variation Theory, the Prey Choice Model, and the Ideal Free Distribution
Reprinted from: *Quaternary* **2020**, *3*, 26, doi:10.3390/quat3030026 . . . . **55**

Chao Zhao
The Climate Fluctuation of the 8.2 ka BP Cooling Event and the Transition into Neolithic Lifeways in North China
Reprinted from: *Quaternary* **2020**, *3*, 23, doi:10.3390/quat3030023 . . . . **79**

Meng Zhang
Microblade–Based Societies in North China at the End of the Ice Age
Reprinted from: *Quaternary* **2020**, *3*, 20, doi:10.3390/quat3030020 . . . . **105**

Jun Takakura
Rethinking the Disappearance of Microblade Technology in the Terminal Pleistocene of Hokkaido, Northern Japan: Looking at Archaeological and Palaeoenvironmental Evidence
Reprinted from: *Quaternary* **2020**, *3*, 21, doi:10.3390/quat3030021 . . . . **137**

Robert L. Kelly
New Discoveries and Theoretical Implications for the Last Foraging and First Farming in East Asia
Reprinted from: *Quaternary* **2021**, *4*, 40, doi:10.3390/quat4040040 . . . . **157**

About the Editors

Pei-Lin Yu is Affiliate Professor of Anthropology at Boise State University and Staff Archaeologist at the Corps of Engineers. She received her B.S. at the University of New Mexico, and M.A. and Ph.D. at Southern Methodist University. Yu is a Fulbright Senior Scholar and author of one monograph, editor of two Special Issues and four edited volumes, and numerous peer reviewed articles on hunting and gathering lifeways, the transition to agriculture, cultural heritage preservation, and Indigenous participation in archaeology and anthropology.

Kazunobu Ikeya is a Professor at the National Museum of Ethnography in Osaka, Japan. He received his B.A. at Tohoku University, M.A. at the Univ. of Tsukuba, and Ph.D. at Tohoku University in 2003. He has published a wide-ranging variety of articles in peer-reviewed publications that focus on environmental anthropology, hunting and gathering societies, cultural geography, and ethnoarchaeology of southern Africa and Japan.

Meng Zhang is a Junior Research Fellow (Assistant Professor) at the Department of Cultural Heritage and Museology, and the Institute of Archaeological Science, Fudan University. He was awarded his B.A. and M.A. at Jilin University, and his Ph.D. at the Department of Anthropology, University of New Mexico. His monograph entitled *Late Pleistocene and Early Holocene Microblade-based Industries in Northeastern Asia: A Macroecological Approach to Foraging Societies* (2021) has been published by the British Archaeological Reports (BAR) Publishing.

quaternary

Editorial

Introduction: New Discoveries and Theoretical Implications for the Last Foraging and First Farming in East Asia

Pei-Lin Yu [1], Kazunobu Ikeya [2] and Meng Zhang [3,4,*]

1 Department of Anthropology, Boise State University, Boise, ID 83725, USA; pei-linyu@boisestate.edu
2 Department of Modern Society and Civilization, National Museum of Ethnology, Osaka 565-8511, Japan; ikeya@minpaku.ac.jp
3 Department of Cultural Heritage and Museology, Fudan University, Shanghai 200433, China
4 Institute of Archaeological Science, Fudan University, Shanghai 200433, China
* Correspondence: zhangmengwb@fudan.edu.cn

Worldwide, scientific understanding about domestication and the origins of food production is undergoing rapid change based on new data from discoveries in paleoclimates and environments, paleobiology, and archaeology. Two major periods of transition include the Pleistocene to Holocene of about 12,000–9000 BP, and another in the Middle Holocene (7000–4000 BP) [1]. The larger narrative of the transition from foraging to farming has been mainly built based on the archaeological record from Near East and Mesoamerica; in contrast, East Asia has been less integrated in worldwide archaeological research, in part because of language boundaries. Obviously, this region, well-known for cultivation of millet, rice, and other cultigens, cannot be ignored in order to do empirical generalization and theory building for this anthropologically oriented research question. In this Special Issue of *Quaternary* we focus on East Asia because of the broad array of habitats and deep time horizons that enable explorations of variability in agricultural origins and adoptions. The regions we now call northern and southern China, Inner Mongolia, Japan, and the Taiwan island have seen steady growth in the number of archaeological and other scientific investigations that are germane to the foraging to agriculture transition in both key periods, due to the rapid pace of salvage archaeology resulting from development projects and the strength of academic and museum programs that foster archaeological research [2–6]. Additionally and importantly, critical and productive re-examination of important legacy research documents, samples, and collections is facilitating the growth of new ways to investigate legacy data from prior explorations [3,4,7–9]. The multi-lingual publication rate is growing, as evidenced by the birth of new English language journals devoted exclusively to Asian archaeology and special issues (like this one) in multi-disciplinary publications, and the availability of open-source access is accelerating the growth of international readership, comparative approaches, and collaborations [3,9–11].

In the field of East Asian archaeology, methodological advances in chronometric analysis have fueled an increase in the sample of securely dated contexts and clarified the timing of change as well as allowed for critical comparison of regional sequences. Detailed evidence for paleoclimatic deteriorations and ameliorations allow us to posit testable correlations between climatic fluctuations and adaptive responses in subsistence, settlement, and social organization. Environmental analysis brings habitat and niche into focus, so that overlaps and differences can be identified in landscapes preferred by foragers and farmers. Detailed analysis of biomorphology, isotopic signatures, and genetic information can reveal domestication thresholds for a wide variety of crop species. Examining technological systems from the standpoint of social organization, such as implications of microblade-based societies, can sharpen distinctions between groups that at first glance appear to be homogeneous, or find commonalities between apparently different systems. Use wear and residue analysis are contributing to better knowledge about subsistence of the transition, and actualistic and experimental methods are important middle range approaches to open

Citation: Yu, P.-L.; Ikeya, K.; Zhang, M. Introduction: New Discoveries and Theoretical Implications for the Last Foraging and First Farming in East Asia. *Quaternary* **2021**, *4*, 37. https://doi.org/10.3390/quat4040037

Received: 15 June 2020
Accepted: 5 August 2020
Published: 18 November 2021

up new lines of hypothesis development regarding testable expectations for the contexts and evolutionary pathways of subsistence and technology. Finally, reference information from global and local ethnographies of foragers and farmer-gardeners, and from the sister discipline of human behavioral ecology, are providing rich insights into testable factors that influence decision-making in the rapidly changing adaptive theater of the Pleistocene to Holocene transition, and play an increasingly important role in our understanding of the origins of East Asian food production.

The origin and spread of agriculture in East Asia during the transitional period occurred within a varied theater of habitats and foraging adaptations. The mountain chains, plateaus, hilly flanks, and river-crossed plains of China offer a fascinating "laboratory" that includes centers of domestication as well as a variable pace and process of agricultural adoptions by foraging neighbors. In the north, the vast distances and high connectivity appear to facilitate the development of mobile socio-technological systems such as microblade-based societies that were maintained and evolved through frequent aggregations [12] and a supple adaptive capacity for climatic shifts [13] that offered both 'push and pull' for experimentation with cultivation. In the Middle Yangtze River valley and other complex riverine habitats, hunter-gatherers mapped onto abundant resource patches and in some cases semi-sedentized societally complex foraging remained a viable lifeway compared to rice agriculture [14]. The roles of women in procuring abundant wild aquatic resources through the use of standardized technological systems have been under-studied and are worthy of further investigation.

Moving offshore, the abundant littoral, estuarine, and oceanic resources of neighboring islets offered Paleolithic hunting and gathering societies a means of broadening the subsistence niche with aquatic resources, with later specialization through development of technology like boats, lines, and net systems to acquire pelagic foods of the deep ocean [15]. In the case of Taiwan, the arrival of Neolithic farmer-fishers from SE China with their millet and rice initially added competitive pressure for land and resources and narrowed foraging options; but ultimately this phenomenon also provided a means for some hunter-gatherers to broaden their niche by adding low-effort and mobility friendly cultigens to aquatic and hunted foods.

By comparison, in the Japanese archipelago the characteristics of many habitats fostered the continued adaptive value and maintenance of hunting and gathering in some areas. This is in contrast to the basins of Yangtze River and Yellow River, which were centers of agriculture. In this respect, it is also necessary to conduct comparisons with surrounding areas including northeastern and southern parts of China. Of the various cultivated plants of the Japanese archipelago, including soybeans, foxtail millet (*Setaria italica, awa* in Japanese), pearl millet (*Echinochloa esculenta, hie* in Japanese), and rice, the latter is the most important crop that contributed to the naissance of farming. Because of the prevailing culture of indigenous hunter-gatherers and fishermen in the Japanese archipelago, even if cultivated plants other than rice had been introduced the shift of peoples' livelihoods to agriculture would have been complex. Paddy rice cultivation became established in northern Kyushu by immigrants from the Korean Peninsula [16].

Two questions relate to the shifting of subsistence in the Japanese Archipelago: first, why did the culture of hunter-gatherers persist? The transition period from the Pleistocene to the Holocene at ca. 11,500 BP occurred at the early phase of the Jomon period (16,000–3000 BP, about 13,000 years total) in Japan. Jomon hunter-gatherers lived on various islands from Hokkaido to the main island of Okinawa over a range of 3000 km north and south. Recent studies have elucidated that azuki beans, soybeans, millet, and others were cultivated in some parts of the country during the middle Jomon period [17], but cultivation was just one of many activities of subsistence, along with hunting, gathering, and fishing. In fact, cultivated plants would have been less important than gathering forest nuts (walnuts, acorns, conker, chestnuts, beech, etc.). During the transitional period from the Pleistocene to the Holocene, it is possible that modes of hunting, gathering, and fishing were influenced by a warming climate, but the introduction of agriculture and its social effects have yet to be recognized. This may have been the case because the diverse natural resources of forests, rivers and sea in the Japanese archipelago were abundant, secure, and stable. In fact, it has been revealed that human cultures changed from the late Pleistocene to the early Holocene in Hokkaido [18]. In the eastern part of Hokkaido the discovery of many early Holocene pit dwellings are evidence that the number of settlements was increasing, suggestive of higher population densities, or growing sedentism, or both.

The second question concerns the adoption of agriculture (rice farming) in the Japanese Archipelago? Why was it not adopted in some cases, and what are the salient differences? Much later than in the rest of East Asia, in the tenth century BCE immigrants from the Korean Peninsula introduced rice cultivation to northern Kyushu. Thus the first farming cultures of the Japanese Archipelago were born, and varied relationships developed between farmers and hunter-gatherers. It is estimated that their relationships endured for different periods in the Fukuoka Plain, Osaka Plain, Kanto Plain, and Tohoku Region [19,20]. If examined at the level of individual archeological sites on Fukuoka Plain [20], clearly their relationship changed over time. For example, hunter-gatherers still lived in Hokkaido and Okinawa during the late period [21]. About 500 years elapsed between the introduction of rice farming in Kyushu and its spread to the Tohoku region of Japan. Moreover, rice farming was then limited to Kyushu, Honshu, and Shikoku, which indicates not only limitations to conduct rice cultivation in cold regions, but also the reluctance of indigenous hunter-gatherers—who were well-adapted to abundant natural resources—to convert to the farming lifestyle.

Taken individually, the new methods and discoveries in these case studies are provocative. Approaches of macroecology, human behavioral ecology, ethnoarchaeology, experimental archaeology, and others are applied to provide explanatory and referential frameworks on the foraging to farming transitions in East Asia. Results indicate very early experimentation in food production; growing evidence for the persistence of foraging; new perspectives on the roles of aggregations and gendered food procurement; and the surprisingly intertwined relationship between wild aquatic foods, niche, migration, and the agricultural experiment. Taken as a whole, the emerging picture of Paleolithic to Neolithic transitions in East Asia is revolutionary, overturning prior assumptions about both hunter-gatherers and farmers and revealing new aspects of the tempo and mode of the change. From forests, plateaus, grassy plains, and mountains to coasts, islands, and archipelagos, we are now positioned to take an unprecedented opportunity to understand and predict variability in the tempo and mode of the Paleolithic to Neolithic transition across one of the biggest landmasses on the planet. Clearly, the Russian Far East, Mongolia, Korea, Sakhalin, and Island Southeast Asia are important transitional regions experiencing rapid research growth, and these regions merit robust attention in their own right.

This Special Issue of *Quaternary*, "Advances in East Asian Agricultural Origins Studies: The Pleistocene to Holocene Transition", aims to present advances in research from established and upcoming researchers working in mainland China, Taiwan Island, and the Japanese Archipelago, with the purpose of evaluating the significance of Paleolithic cultural influences on the transition to Neolithic adaptations by comparing cultural evolutionary scenarios through time and across space. The array of approaches is multidisciplinary, including new archaeological excavation data, revisiting legacy data, fresh perspectives on paleoclimate and environments, and reference information from hunting and gathering peoples and small scale farmers. Quantitative, qualitative, and integrated data and methodologies are featured. A critical synthetic review chapter provides a global perspective on this rapidly growing field including theoretical and methodological considerations for the case studies.

The co-editors hope that growing our understanding of the East Asian transition from foraging to Neolithic agriculture—a hallmark of the Quaternary and among the most dramatic and influential processes in the history of modern humans—will help to deepen our understanding of variability of human adaption and decision-making against the backdrop of dramatic and unpredictable climate change that marks the transition from terminal Pleistocene to incipient Holocene. We also hope these articles help to inform the general trajectory of scientific inquiry into the foundational characteristics of Late Quaternary growth and migrations of human populations, societal complexity, subsistence and economic transformations of landscape, niche construction, migration, boundary defense, and other density-dependent phenomena that have had profound consequences for the planet's ecosystems and contemporary human societies.

Author Contributions: This article was mainly written by P.-L.Y. and edited by K.I. and M.Z. This Special Issue was designed and organized by P.-L.Y., K.I. and M.Z., with equal contributions. P.-L.Y. provided most English editing of the articles. All authors have read and agreed to the published version of the manuscript.

Funding: This article was not supported by external funding.

Acknowledgments: The co-editors thank the editorial staff of *Quaternary* and anonymous reviewers for their thoughtful contributions to improve this article and the Special Issue. Individual contributors are also thanked for extraordinary scholarship and efforts to publish in a venue that will reach an international audience.

Conflicts of Interest: The authors declare no conflict of interest.

References

1. Larson, G.; Piperno, D.R.; Allaby, R.G.; Purugganan, M.D.; Andersson, L.; Arroyo-Kalin, M.; Barton, L.; Vigueira, C.C.; Denham, T.; Dobney, K.; et al. Current perspectives and the future of domestication studies. *Proc. Natl. Acad. Sci. USA* **2014**, *111*, 6139–6146. [CrossRef]
2. Chen, S.Q.; Yu, P.L. Early "Neolithics" of China: Variation and Evolutionary Implications. *J. Anthropol. Res.* **2017**, *73*, 149–180. [CrossRef]
3. Habu, J.; Lape, P.V.; Olsen, J.W. *Handbook of East and Southeast Asian Archaeology*; Springer: New York, NY, USA, 2017; ISBN 9781493965199.
4. Ikeya, K.; Ogawa, H.; Mitchell, P. *Interactions between Hunter–Gatherers and Farmers: From Prehistory to Present*; National Museum of Ethnology: Osaka, Japan, 2009; ISBN 9784901906654.
5. Kuo, S.C. *New Frontiers in the Neolithic Archaeology of Taiwan (5600–1800 BP)*; Springer: New York, NY, USA, 2019; ISBN 9789813292628.
6. Wu, X.H.; Zhang, C.; Goldberg, P.; Cohen, D.; Pan, Y.; Arpin, T.; Bar-Yosef, O. Early Pottery at 20,000 Years Ago in Xianrendong Cave, China. *Science* **2012**, *336*, 1696–1700. [CrossRef] [PubMed]
7. Bettinger, R.L.; Barton, L.; Morgan, C. The origins of food production in north China: A different kind of agricultural revolution. *Evol. Anthropol. Issues News Rev.* **2010**, *19*, 9–21. [CrossRef]
8. Chen, S.Q. *Prehistoric Modernization: A Cultural Ecological Approach of the Origins of Agriculture in China*; Science Press: Beijing, China, 2014; ISBN 9787030397263. (In Chinese)
9. Chen, S.Q.; Yu, P.-L. Intensified Foraging and the Roots of First Farming in China. *J. Anthropol. Res.* **2017**, *73*, 381–411. [CrossRef]
10. Fuller, D.Q.; Qin, L.; Zheng, Y.F.; Zhao, Z.J.; Chen, X.G.; Hosoya, L.A.; Sun, G.-P. The Domestication Process and Domestication Rate in Rice: Spikelet Bases from the Lower Yangtze. *Science* **2009**, *323*, 1607–1610. [CrossRef] [PubMed]
11. Gross, B.L.; Zhao, Z. Archaeological and genetic insights into the origins of domesticated rice. *Proc. Natl. Acad. Sci. USA* **2014**, *111*, 6190–6197. [CrossRef] [PubMed]
12. Zhang, M. Microblade–based societies in North China at the end of the Ice Age. *Quaternary* **2020**, *3*, 20. [CrossRef]
13. Zhao, C. The climate fluctuation of the 8.2 ka BP cooling event and the transition into Neolithic lifeways in North China. *Quaternary* **2020**, *3*, 23. [CrossRef]
14. Liu, R.Z.; Liu, H.; Chen, S.Q. Alternative Adaptation Strategy during the Paleolithic-Neolithic Transition: Potential Use of Aquatic Resources in the Western Middle Yangtze Valley, China. *Quaternary* **2020**, *3*, 28. [CrossRef]
15. Yu, P.-L. Modeling Incipient Use of Neolithic Cultigens by Taiwanese Foragers: Perspectives from Niche Variation Theory, the Prey Choice Model, and the Ideal Free Distribution. *Quaternary* **2020**, *3*, 26. [CrossRef]
16. Kaner, S.; Yano, K. Early Agriculture in Japan. In *The Cambridge World History*; Barker, G., Ed.; Cambridge University Press: Cambridge, UK, 2015; Volume 14, pp. 353–386, ISBN 9780511978807.
17. Fujio, S. Early Grain Cultivation and Starting Processes in the Japanese Archipelago. *Quaternary* **2021**, *4*, 3. [CrossRef]
18. Takakura, J. Rethinking the Disappearance of Microblade Technology in the Terminal Pleistocene of Hokkaido, Northern Japan: Looking at Archaeological and Palaeoenvironmental Evidence. *Quaternary* **2020**, *3*, 21. [CrossRef]
19. Nasu, H.; Momohara, A. The beginnings of rice and millet agriculture in prehistoric Japan. *Quat. Int.* **2015**, *397*, 504–512. [CrossRef]
20. Ikeya, K. Ethnoarchaeology of Introducing Agriculture and Social Continuity among Sedentarised Hunter–Gatherers: The Transition from the Jomon to the Yayoi Period. *Quaternary* **2021**, *4*, 28. [CrossRef]
21. Crawford, G.W. Advances in Understanding Early Agriculture in Japan. *Curr. Anthropol.* **2011**, *52*, S331–S345. [CrossRef]

quaternary

MDPI

Article

Ethnoarchaeology of Introducing Agriculture and Social Continuity among Sedentarised Hunter–Gatherers: The Transition from the Jomon to the Yayoi Period

Kazunobu Ikeya

National Museum of Ethnology, Osaka 565-8511, Japan; ikeya@minpaku.ac.jp

Abstract: This study was conducted to elucidate the introduction of agriculture and social continuity from the Jomon to the Yayoi period, from an ethnoarchaeological perspective. The Yayoi period has been divided into two types: a broad spectrum economy that relied on many kinds of resources, such as rice, millet, and nuts, and a selective economy that specialised in rice and wild boar. However, it is not clear how the livelihoods shifted from the Jomon to the Yayoi period. In this study, ethnohistorical materials were examined first. Ethnohistorical reference materials gathered worldwide have revealed three relationships between hunter–gatherers and farmers: coexistence, fusion, and assimilation. Focusing on fusion, this study examined situations of hunting, gathering, and fishing, as inferred from ruins of the Late and Final Jomon period, and assessed their relationships with agriculture using ethnohistorical reference materials of the Early Edo period. There were not many social changes caused by the introduction of field farming; however, the introduction of paddy rice cultivation had different effects on society depending on the level of investment in obtaining water from streams and springs and creating irrigation features.

Keywords: dry-field farming; first farmers; Jomon; paddy rice farming; sedentarised hunter-gatherers; Yayoi

Citation: Ikeya, K. Ethnoarchaeology of Introducing Agriculture and Social Continuity among Sedentarised Hunter–Gatherers: The Transition from the Jomon to the Yayoi Period. *Quaternary* **2021**, *4*, 28. https://doi.org/10.3390/quat4030028

Academic Editor: José Javier Baena Preysler

Received: 6 July 2021
Accepted: 30 August 2021
Published: 14 September 2021

1. Introduction and Study Location

Prehistoric hunter–gatherers who became sedentarised in western Asia are said to have begun farming and raising livestock more than 10,000 years ago [1]. Later, after farmers populated this area, hunter–gatherers and farmers coexisted [2]. The earliest clear evidence for the use of rice in the lower Yangtze region of Zhejian, China, is between 6900 and 6600 years ago [3], but this was not yet cultivated. At that time, indigenous hunter–gatherers formed a diverse array of relationships when they encountered migrant rice farmers.

In the case of the Japanese archipelago, a large shift is known to have occurred from prehistoric hunter–gatherers (Jomon people using Jomon earthenware, [4,5]) to prehistoric agricultural farmers (Yayoi people using Yayoi earthenware) except in Hokkaido. Regional differences are apparent, as the landmass extends around 2000 km from Hokkaido to Kyushu. People from the Korean Peninsula with knowledge of rice farming first settled in northern Kyushu (Figure 1). In Japan, rice cultivation was first adopted by new immigrant populations at the Nabatake archaeological site in the Fukuoka Plain of northern Kyushu during the Early Yayoi period, although this was not irrigated but rain-fed rice cultivation. Millet farming was also conducted there [6,7] The floodplain of the valley was wetland; the hills behind were evergreen forests. Under such circumstances, rice farming in the form of dry-field farming began on the lower slopes of the plain or at the edges of lowland areas. Later, in the Early Yayoi period, the Itazuke site village was established with pit dwellings and a circumjacent ring moat 5–6 m deep. Irrigated paddy rice cultivation was combined with millet fields producing *awa* (foxtail millet, *Setaria italica*) and *hie* (a kind of Japanese barnyard millet, *E. esculenta)*.

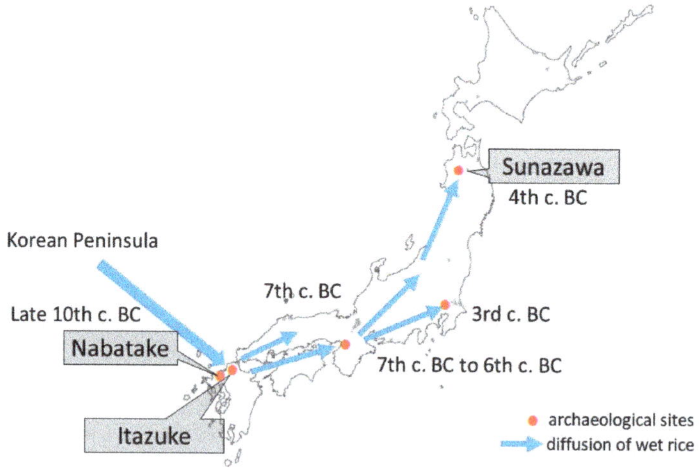

Figure 1. Diffusion of wet rice farming and archaeological sites in Japan ([8]. Used by permission).

On the other hand, according to Shitara [9], the economy in the Yayoi period can be divided into two types: an exhaustive economy that relied on many kinds of resources, such as rice, millet, and nuts, and a selective economy that specialised in rice and wild boar. However, the transition between livelihoods from the Jomon to the Yayoi period is not clear. The wet rice cultivation technique was accepted by hunter-gatherers from Honshu, Shikoku, and Kyushu over an approximately 700-year span [10]. For example, on the Fukuoka Plain, archaeological remains of paddy field farms from the latter half of the 10th century BC have been found in the downstream basin of northern Kyushu's plains. The transition process from hunter–gatherer to farmer can be classified into the following three categories [11]. In the first category, native hunter–gatherers and migrant farmers established a new settlement together in a downstream basin. Residents specialised together in paddy rice cultivation. In this case, sometimes a ring moat was built around the settlement, as in the Itazuke ruins of the Fukuoka Plain, and the settlement endured to become a local main village. In the second case, the two groups established a settlement together but later abandoned the settlement and moved to another place. The last case occurs in the context of food stress caused by a population increase, in which farmers who lived in a downstream basin moved upstream into areas populated by hunter–gatherers and started food production. Apparently, hunter–gatherers began farming after being influenced by immigrant farmers from the Korean Peninsula; that is, they were transformed gradually into farmers.

Herein, a dynamic model of the relationship between hunter–gatherers and farmers is presented based on the question of how sedentarised hunter–gatherers began farming during the transition period from hunter–gatherer to farmer in East Asia. It is said that field agriculture such as soybean and azuki bean cultivation had already been introduced in the Jomon period by the time of rice introduction [12,13], but the influence of legume cultivation on hunter–gatherer societies is not yet known. Similarly, the impact of the introduction of paddy rice cultivation on hunter–gatherer societies in the Yayoi period is not well understood. It is the objective of this paper to examine the sustainability and transformation of livelihoods and society after the introduction of agriculture, using ethnoarchaeological reference data from similar sites that are relevant to the ways that agriculture and other livelihoods were combined.

The author has selected sites from the Late and Final Jomon periods in the Japanese archipelago where the use of resources at the time of interest can be reconstructed. Sites were also selected from early modern *Matagi* settlements, which are not from the Early Yayoi period but show the influence of the introduction of paddy rice cultivation on a hunter–gatherer society. This is due to the fact that there is abundant information on livelihoods such as hunting, gathering, and farming in the *Matagi* settlements, which are located in almost the same locations as the Jomon sites.

The author conducted an ethnoarchaeological study on the use of people's natural resources in a remote mountain village called a *Matagi* settlement [14], which possesses both historical sites of the Late and Final Jomon periods, and archival materials on paddy rice cultivation or swidden agriculture in the 17th century. The *Matagi*, who believe in the mountain god named Yamano-kami, are hunters who combine gathering, fishing, small farming, and trading. The historical relationship between the *Matagi* and Jomon people is unclear. Actual practices and methods of resource control related to cave hunting and trapping of Japanese black bears have already been reported from ethnohistorical perspectives [15].

The study site is about 30 km upstream of the Miomote River in central Japan, which empties into the Sea of Japan. In the study site, a small flat river terrace surrounds the deep slope of the mountains in the upper Miomote River. The Motoyashiki archaeological site (Figures 1 and 2) dates to the Late and Final Jomon periods. The entire settlement (approximately 180 m east–west, and 90 m north–south) was investigated at the time of the Okumiomote Dam's construction. It consists of excavated pillar buildings, pit dwellings, buried earthenware (200), stone graves (89), and a processing facility for *tochi* (horse chestnut, *Aesculus turbinata*) including water features for leaching or removing lye from chestnuts. Large amounts of wood and wood products, bone and horn products, plant-dependent substances, and animal-dependent substances had already been excavated there, including Jomon pottery and stone tools [16].

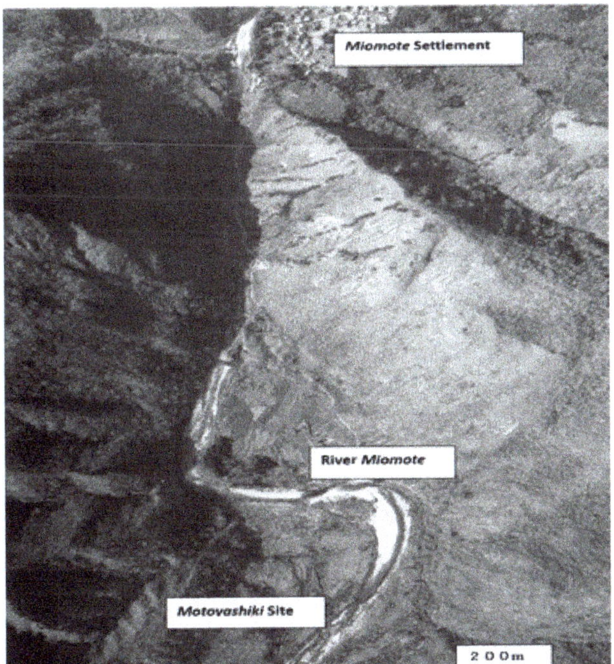

Figure 2. Study Area: Motoyashiki archaeological site at the late and final Jomon periods and Miomote settlement at the Early Edo period [16].

2. Theoretical Framework (Presumption): Transition from Jomon to Yayoi Period and Group Relationships

Studies of sedentarised hunter–gatherers have clarified two types of responses to introduced agriculture [17]. In the first, hunter–gatherers introduced agriculture into their life: they hunted and gathered while farming. In the second, hunter–gatherers maintained a coexistent relationship with farmers. In the second adaptive response, hunter–gatherers sometimes helped with farmers' activities (e.g., harvesting) and received food in return. Both types of incorporation of agriculture have been identified among the San people of the Kalahari Desert [18]. The Aka and Mbuti people in the Congo Basin, the Aeta of the Philippines, and the Mlabri of Thailand have established the latter type [19–23]. Nevertheless, it remains unknown why such a difference has arisen among hunter–gatherers.

Figure 3 presents a dynamic model of the relationship between hunter–gatherers and farmers by examining the ethnohistories of present-day hunter–gatherers (Figure 3). Earlier studies, mainly archaeological or ethnological, revealed characteristics of several individual cases which can be sorted into three types of relationships between hunter–gatherers and farmers—that is, coexistence and symbiosis, fusion, and assimilation. First, assimilation is the situation in which all cultural elements or habits of one group disappear after contact. In such a case, the //Gana San would marry into farming societies and lose their own language and traditions. Next, coexistence and symbiosis are conditions under which the hunting and gathering group was maintained even after contacting another group. For example, very few Tsila San hunter–gatherers intermarried with Kgalagadi farmers; rather, they worked in the farmers' fields and received food. Lastly, fusion is the condition by which external elements are added after contacting another group. In this case, trace characteristics of the existing group remain. For example, a //Gana San hunter–gatherer would marry a farmer, combine hunting, gathering, and farming, and use both languages. Why did such differences arise? Based on the example of hunter–gatherers in the Kalahari Desert, group size and population caused this variation, because Kgalagadi farmers migrated inside the hunter–gatherers' territory in the 18th century [18].

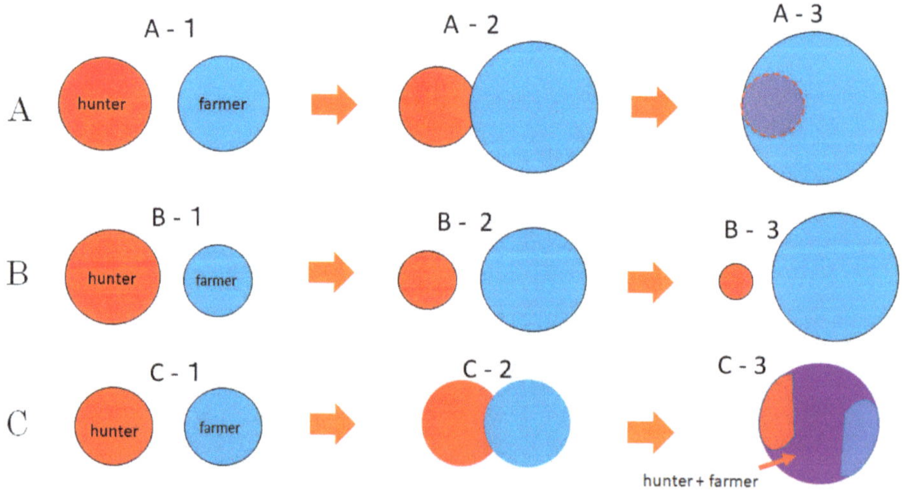

Figure 3. Dynamic model of relationships between hunter-gatherers and farmers: (**A**) assimilation; (**B**) coexistence/symbiosis; (**C**) fusion. Circle size represents the population size (made by K. Ikeya).

Coexistence and symbiosis occur when the respective population sizes of the two groups are similar. However, although the population of the Mlabri hunter–gatherers remained smaller than that of the farmers, they maintained their culture and identity [22]. Fusion occurred when the farmer group became larger than the hunter-gatherer group, although hunter–gatherers continued to maintain their cultural traditions strongly. Assimilation also occurred when the farmer group was larger than the hunter–gatherer group.

We may now ask, what type of relationship did hunter–gatherers and farmers establish during the transition from the Jomon to the Yayoi period? First, one must consider the time during which hunter–gatherers met farmers in northern Kyushu, approximately 600 BC. It is unknown whether the two groups coexisted with no mutual relationship, or whether they shared a symbiotic condition with some kind of trade or exchange. As already described, an assimilation type of relationship occurred in the Fukuoka Plain after hunter–gatherers encountered farmers. At the same time, coexistence or symbiotic relationships also formed. Finally, a fusion type of relationship might have arisen between them after the end of their existence as hunters and farmers. This suggests that the relationships between the same groups could change over time based on circumstances.

In the example of the Osaka Plain, hunter–gatherers and farmers made contact in the sixth century BC, after which the two groups formed a coexistent relationship. Although we found a wooden farming tool at the farming site in the plain, which used Yayoi earthenware of the 'Ongagawa' type, remains of wood materials were to be found [24]. We surmise that the farmer would have taken the wooden farming tool from the hunter–gatherer site at the foot of the mountains. In the fifth century BC, they mingled increasingly. Subsequently, a group of people identifiable as Yayoi farmers emerged in the fourth century BC [25]. This sequence of events corresponds with the assimilation model. The transition from coexistence to assimilation evidently took approximately 200 years.

The farmers then spread to the eastern region of the main island (Honshu) of Japan by the fifth century BC. Indigenous hunter–gatherer groups gradually mingled with the farmers. The subsequent development varied among regions; it stopped at the fusion stage in some regions, while in other regions, hunter–gatherers assimilated with farmers. Paddy fields continued to spread to eastern Japan and the Tohoku region along the plain, including the Niigata region, except in the mountainous areas. The northernmost area is the Sunazawa ruins in the Tsugaru Plain of Aomori Prefecture. Paddy field farmers had long been unable to live in this area, a situation that may be explained using the coexistence model. In this habitat, survival was possible with only hunting and gathering. The Yayoi farming society was established throughout Honshu and Kyushu by the first century BC.

This context partly explains how hunter–gatherers became involved in farming. First, the mutual relationship between coexistence and symbiosis appeared during this period. Activities such as exchange, trade, intermarriage, and entrusting of labour, such as farming, presumably arose during this period. Recovering archaeological evidence of entrusted labour being practised in prehistory is difficult, although it might be inferred ethnographically. Another part of the context is the discovery of the responses of hunter–gatherers that are related to their foraging occupation while being influenced by farmers. Understanding the responses of these societies to the introduction of agriculture is especially important.

3. From Sedentarised Hunter-Gatherers to Early Farmers: Transition and Changes of Fusion Types C-1, C-2, and C-3

An examination of the transition from Jomon to Yayoi shows that the timing of the introduction of paddy fields varies considerably within the Japanese archipelago. Paddy villages developed during the Yayoi period in some areas such as the Fukuoka Plain, but paddy field farming did not begin on a full scale until the Edo period in the mountainous region near the Sea of Japan. Documents from the Early Edo period about paddy fields are available for research in that region. Although the habitat and socioeconomic contexts are different, certain characteristics about the onset of early paddy field farming can be examined using Edo-period reference information. Documents about rice and vegetable farming of the time are expected to have utility for framing factors that influenced Japan's first farmers while taking into account the effects of the Tokugawa Japan domain called *han* in the Edo period (1603–1868).

The study area is a village upstream of the Miomote River in the northern region of Niigata Prefecture. This place is known for the Motoyashiki ruins from the latter half and final phase of the Jomon period, and 19 other sites of Jomon ruins sitting on river terraces (C-1). At the time, groups of hunter–gatherers maintained coexistence and symbiosis with each other. It remains unknown when and how upland farming and paddy farming were subsequently introduced into the area because no Yayoi ruins have been found. It is likely that farmers practising small paddy farming immigrated to Motoyashiki during the mediaeval or Early Edo periods. In other words, although the C-2 stage is assumed to have occurred, informative archaeological materials to verify this fact are lacking. This suggests that both dry-field farming and paddy field farming were being practised by the Early Edo period, which shows the C-3 phase. This shows the possibilities of the transition from the C-1 to the C-3 fusion types.

3.1. Combined Occupations of Sedentarised Hunter-Gatherers (C-1): Latter Half and Final Phase of the Jomon Period

The 19 Jomon sites used in this study are located in three areas: Motoyashiki, along the Miomote River (Figure 2), Shimokubo, near the confluence of the Miomote and Suezawa Rivers, and Doromatasawa, along the Doromata River. Rock shelters are located in the mountains upstream of Iwaisawa, a tributary of the Miomote River [26]. Many pieces of earthenware, stone slabs for crushing nuts (hereafter, grinding stones), animal bones, charcoal, and other artifacts have been excavated from these 19 sites. By the Late Jomon period, the Shimokubo and Maeda sites from the Middle-Jomon period had been abandoned, although the rock shelters continued to be used. The Motoyashiki site, covering an area of 100 m east–west and 70 m north–south, replaced them. Artifacts such as stone spear points, monuments, stone plates, whetstones, clay figures, chipped stone axes, and ground stone axes have been excavated from Motoyashiki.

Plant remains representing 15 species were discovered at the Motoyashiki site. These included edible plants (nine types) and inedible plants (six types) [27]. Among the edible plants, horse chestnuts were excavated in the greatest quantities. They comprised approximately 28% of the total (3039 pieces) plant remains. Horse chestnuts were followed by beech, which comprised approximately 20% of the total amount (2212 pieces). Other plants discovered include Japanese walnuts, chestnuts, Japanese pepper, acorns, *Vitis* (grape) spp., Leguminae, and dogwood.

The following animal remains (calcined/whitened from high heat) were excavated at the Motoyashiki site: Japanese serow, Japanese deer, boars/Suidae (only one), badgers, Japanese raccoon dogs, Japanese red foxes, rabbits, flying squirrels, Japanese black bears, and other mammals [28]. Among the mammals, the relative weights of the excavated remains of Japanese serows and Japanese black bears were higher than that of Sika deer (Figure 4). Among the fish remains, Salmonidae and *Salvelinus* accounted for 44.4%. Fishing tools made of bones and horns were excavated as well. In addition, large amounts of frog remains were unearthed.

The Choja-Iwaya Rock Shelter Site is located approximately 8 km northeast of the Motoyashiki site. The small amount of grinding stones discovered at the site suggests that rock shelters are unlikely to have been used for long-term residence [26]. Although the grinding stones reflect a combination of types during the Jomon period, they are not time diagnostic. Analysis of animal teeth identified a large number of Japanese serows, along with Japanese black bears and Japanese macaques. In the case of the Japanese serows, the growth lines on the cementum of the lower jaws facilitated estimation of the animals' ages; estimation of the time of death indicated hunting activity from the end of winter to early spring. These findings indicate that this rock shelter was most likely a hunting camp.

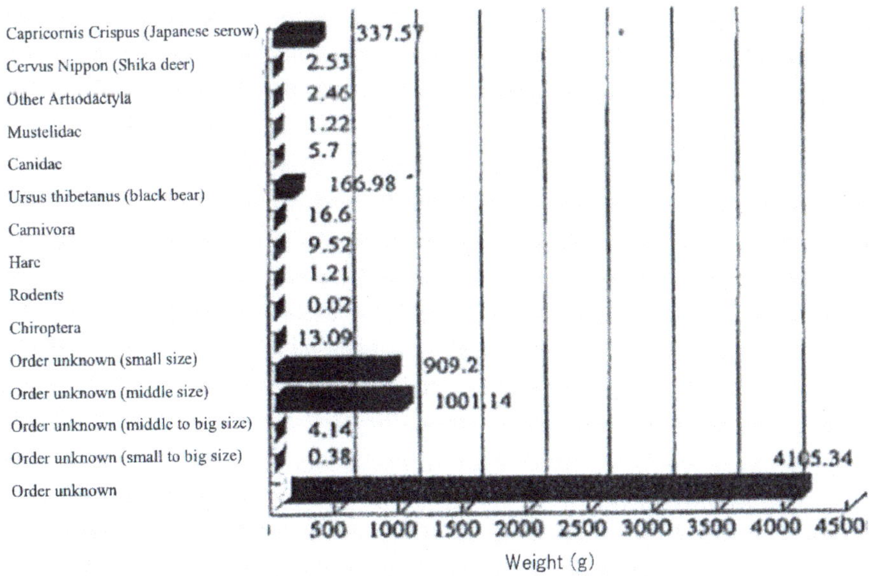

Figure 4. Weights of excavated animal remains [29].

Large quantities of chestnuts, walnuts, and horse chestnut endocarps were excavated at the Motoyashiki site, which was actively settled approximately 3000 years ago. The genetic diversity score obtained from DNA analyses of the excavated chestnut fruits was 1.517, which is similar to those of the Sannai-Maruyama Special Historical Site in Aomori Prefecture and the Sakuramachi site in Toyama Prefecture [30]. These results imply that the chestnut–people relationship termed "plant husbandry" or woodland management [31] had developed to a considerable degree. Although chestnuts are larger in western Japan and smaller in eastern Japan, those found at Motoyashiki were small. The excavated walnuts were identified from DNA analysis results as Japanese walnuts and heartnuts.

Artifacts brought from outside the local area have been discovered at the Motoyashiki site. An example is shark teeth; no vertebrae have been discovered. This suggests that people likely used the teeth as accessories and status symbols, rather than consuming shark meat as food [28]—although preservation issues might skew the findings regarding teeth. At the same time, the absence of the canine teeth of Asian black bears in the excavation implies that they were removed from this site. The discovery of a large number of unfinished ground stone axes suggests that the axes were being exported from this region and that there was a group specialising in their production (Sato et al. 2002). Furthermore, the percentage of excavated obsidian ore, which was used as material for various objects, was as follows: 20% (10 pieces) were produced in Itayama in the city of Shibata, 20% (10 pieces) were produced in Mt. Gassan in Yamagata prefecture, and 6% were produced in Mt. Kirigamine in Nagano Prefecture (Figure 5). This discovery implies that these were traded goods brought in from the outside. Additionally, appraisal of the gems excavated from the site indicates that the jade was light green to green in colour and was produced in the Oumi area of Itoigawa [32].

Figure 5. Obsidian from the Motoyashiki site [29].

Based on the discussion presented above, the following combination of occupations can be inferred from the archaeological resources of the Motoyashiki site (latter half and final phase of the Jomon period) (Table 1). The first is the addition of exchange and trade activities to hunting, fishing, and nut and fruit gathering. With regard to hunting, the animal bones found at the Choja-Iwaya Rock Shelter Site suggest that the favoured prey were Japanese serows, bears, and monkeys, rather than wild boars and deer, which had been commonly hunted during the Jomon period. As the furthest upstream area for salmon migration in the Miomote River did not reach the Motoyashiki site, people more likely fished for trout rather than salmon. It is unclear whether there is a relationship between the Choja-Iwaya site in the mountain and the Motoyashiki site near the river. It is assumed that there was a social network during the same period.

Table 1. Occupation calendar at the Motoyashiki site (latter half and final phase of the Jomon period).

	Month											
	4	5	6	7	8	9	10	11	12	1	2	3
Hunting												
bear	......											
serow											.....	
Gathering												
royal fern (*zenmai*)	....											
bracken (*warabi*)		....										
horse chestnut (*tochi*)						......						
walnuts						......						
chestnuts						....						
maitake mushroom						....						
Fishing												
trout				....								
char				....								

Source: Information about the timing of hunting, gathering, and fishing was obtained from studies of animal and plant remains [26,28,30]. (Illustration by K. Ikeya).

3.2. Combined Occupations among Early Farmers (C-3): Early Edo Period (1655)

Archival information about the hunting, gathering, fishing, and trading conditions during the Edo period (1603–1868) is described below. People hunted commercially by hunting hibernating bears in early spring and trapping bears in the fall. The commercial value of bear gall bladders was high at the time. Notably, each household was assigned its own bear trapping area, according to documents from the Edo period [15]. Hunting of bears in early spring and serows in the coldest part of winter was conducted in groups. It was a ritual practice; hunters were required to use specific mountain words when in the mountains. The practice of serow hunting is evidenced by the fact that a villager by the name of Kanshichi presented serow fur as a year-end gift to the daimyo's councillor in charge of his master's castle. A man named Sagoemon from an area called Kiriba, who is mentioned in a cadastral register of the time, trapped bears.

Regarding gathering, the commercial gathering of sedge and chestnuts was conducted. Sedge was valued as a material for mats. Regarding chestnuts, people were granted certain rights to use shared chestnut forests. To explain further, the mountainous area on the opposite side of a river (termed foothills) consisted of chestnut forests. The area of forests shared by the village was approximately 10 ha [33]. Gathering activities were conducted based on village rules. During the fall harvesting season, the chief of the district would determine the gathering day and then notify each household of the date and time. The gathering period would be approximately two weeks, during which the gathering would be conducted six or seven times. On the morning of the specified day, one person from each household would collect all the chestnuts gathered by all the participants at the house of the district chief, who would then distribute them based on the number of household members.

Moving on to fishing, there was a productive salmon fishery. A skilled person could sometimes catch more than 10 salmon per day. Trout was the most important fish, and August through September was the best fishing season. There were two types of hunting methods: one was to dive and stab trout gathering in the depths using a double-headed spear, and the other was to place a fishing trap in the shallows.

The distribution of farmland during the Early Edo period can be estimated from a cadastral register of 1655. Figure 6 depicts the distribution of arable land for rice crops and shifting cultivation in the Early Edo period (1655). Rice fields were distributed near the Miomote settlement (Table 2 and Figure 6), with shifting cultivation fields in Motoyashiki (where a Jomon archaeological site is located). Incidentally, Kamimukai and Nakadouri in this region were used for shifting cultivation until the 1950s (Figure 6). This diagram shows that rice farming was concentrated in most of the area around the Miomote village, whereas both slash-and-burn farming and rice farming were conducted in the area around the Motoyashiki site. The two types of farming were also distributed in the area near the Suezawa River. As slash-and-burn farming continued into the 1950s, the following information about the condition of slash-and-burn farming is useful as a frame of reference.

Slash-and-burn farming, called *Kano*, was conducted on shared lands. Around 20 July, during the hottest part of summer, land users would go to a site together and assign households to farming areas by drawing lots. This farmland assignment was called *Kappatsuke*. Slash-and-burn farming was conducted by mowing grass in each person's assigned lot, leaving green grass only at the boundaries, drying the lot for three to five days, then burning the area. Swiddens were also visible on a steep hill with a 30° slope. As a crop, buckwheat (soba) was planted on a flat field in the first year. Buckwheat can grow in a cool climate, is unaffected by soil conditions, and can be grown easily without much labour. The growth period is short: seeds planted in summer mature in slightly over two months. During the second and third years, people would make ridges and plant soy, adzuki beans, and foxtail millet (awa). They would subsequently move the swidden field elsewhere on a rotating basis.

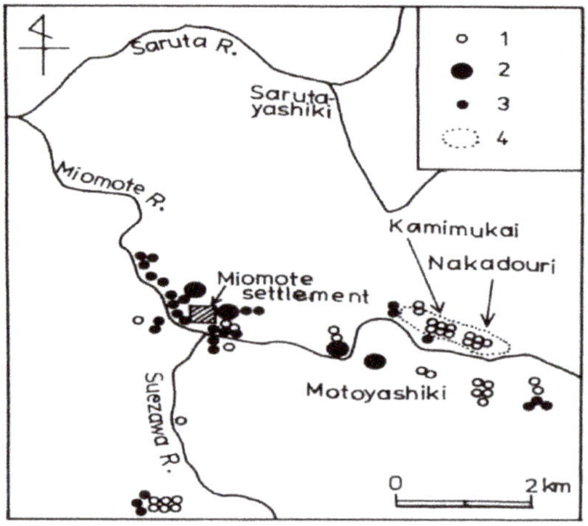

Figure 6. Distribution of shifting cultivation and rice fields in 1655: 1, one shifting cultivation field; 2, ten rice crop fields; 3, one rice crop field; 4, distribution of shifting cultivation later, in approximately 1955 [34].

Table 2. Occupation calendar of the Miomote settlement (Early Edo period).

	Month											
	4	5	6	7	8	9	10	11	12	1	2	3
Hunting												
bear	....						....					
Serow												......
Gathering												
royal fern (*zenmai*)		......										
bracken (*warabi*)		....										
horse chestnut (*tochi*)							....					
walnuts							....					
chestnuts							....					
maitake mushroom							....					
Fishing												
trout					....							
char					....							
Farming												
rice			....				....					
Japanese millet			....				....					
millet												
azuki beans		.....				.....						
Soybeans		.....				.....						
buck wheat					......		......					

Source: Information about the timing of hunting, gathering, fishing, and farming was obtained from studies of animal and plant remains [33,35,36]. (Illustration by K. Ikeya).

Similarly, the following reference information is useful for paddy field farming. Tasks in a year would include, in order, ploughing in May, rice planting in June, paddy field management from July to September, and harvesting in October. Short-handled hoes were used for ploughing before the period of high economic growth. Rice planting was conducted through mutual assistance of a group, with the head family and branch families. Water management was necessary after rice planting. The water would be cold because of the mountainous location, and therefore, a method called *Nurume* was used to raise the water temperature. Long, winding irrigation channels were built to expose water to sunlight for long hours. Subsequently, the people would weed the fields. Harvesting was conducted through a mutual assistance relationship.

A registry of the registered rice paddies, new paddies, and wet and dry fields was established in the Miomote village by 1655. All 16 households owned paddy fields, permanent fields, and swiddens, despite some differences among households in terms of upper and middle paddy fields. The swidden areas of households ranged considerably, from 3 to 100 terraces, with an average of 42 ridges [34]. Disparities among paddy field areas of households were also evident. New rice fields had not been developed at the time.

Irrigation technology is particularly important for paddy field farming; Figure 7 depicts the distribution of irrigation channels in Miomote. The dates of construction for most of the channels are unknown, but they were clearly constructed in the Edo Era because they are listed in the land register ('Tochi-daicho') of the Early Meiji period. This canal network has four sources of water: mountain streams, springs, marsh and pond water, and rivers. The canals extended throughout the flatlands of the settlement.

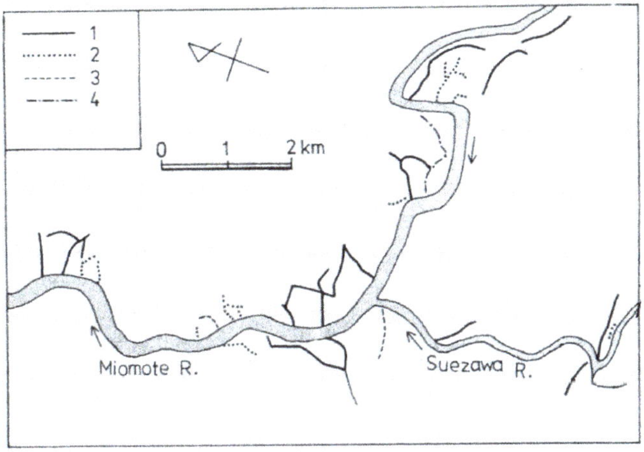

Figure 7. Distribution of irrigation channels: 1, mountain stream water; 2, spring water; 3, marsh and pond water; 4, river water [34].

This information suggests that the increase in rice yield, achieved during the Edo period, was accomplished by virtue of changes in irrigation technology. Irrigation using water drawn from a river, pond, or wetlands presumably existed before the Edo period (1655). During the Edo period, long-distance irrigation from the foot of mountains and using water drawn from rivers was likely developed. In the case of one canal for the mountain streams, water from the Tochidaira and Umahiki streams enters at the foot of the mountain and flows into the rice field along the Suezawa River. The canal length is more than one kilometre. The elders of the village say that the rice field was for common use. It is likely that the villagers originally dug the canal at the foot of the mountain.

The discussion presented above suggests that people of the Edo period combined their occupations, which included hunting, gathering, fishing, slash-and-burn farming, paddy field farming, and trade. The emphasis here, however, is on the position of farming in society. As the study region was mountainous, the amount of rice production was unstable, and no village had been developed. Products obtained from hunting and gathering were commercialised, and the villagers were thus actively engaged in hunting and gathering. Consequently, although the hunter–gatherers' valued practice of equally distributing shared resources was maintained strongly in hunting and gathering activities, disparities in terms of land ownership among households were evident in slash-and-burn farming and paddy field farming. In other words, the organisation of early farmers in this period entailed strong ideas of egalitarianism in the economies surrounding natural resources, partly because of the economic importance of hunting and gathering. This finding clearly reflects the characteristics of the fusion model.

4. Results and Discussion: Establishment of Farming by Sedentarised Hunter-Gatherers and Social Continuity

Based on current knowledge regarding the initial transition of sedentarised hunter–gatherers to farming between Japan's Jomon period and Yayoi period, this study has presented dynamic models of intergroup relationships between hunter–gatherers and farmers. Ethnoarchaeological reference data regarding the ways people combined farming and other occupations, and how they maintained and changed their occupations and society after introducing farming, were considered in terms of the dynamic fusion models, to refine expectations for modes of the transition. More specifically, the study specifically addressed the upstream mountainous area of the Miomote River area, a unique location where it is possible to assess detailed information about the archaeological site from the latter half and final phase of the Jomon period and ancient writings about slash-and-burn farming and paddy fields from the early Edo period at the same location in the Japanese archipelago. This study's results suggest the following three findings:

(1) A dynamic model of intergroup relationships between hunters and farmers.

Three relationship types are found in ethnohistorical resources for hunter–gatherers and farmers: coexistence and symbiosis, fusion, and assimilation. Coexistence and symbiosis are states in which the integrity of groups' lifeways is maintained after contact between the two groups. Fusion occurs in cases in which other factors are mixed in the contact between the groups. Assimilation occurs when the cultural factors and customs of one group disappear because of intergroup contact. Factors such as the population size of each group, the presence of intermarriage relationships, and the level of power to maintain a group culture are likely to be related to differences among these three relationships.

(2) How farming and other occupations were combined for the Jomon Period: Support for the fusion model.

Occupations in the latter half and final phase of the Jomon period included exchange and trade in addition to hunting, fishing, and gathering of nuts and fruits. These were combined depending on the season.

With regard to hunting, the animal bones found at the Choja-Iwaya Rock Shelter Site suggest that people primarily hunted Japanese serows, Asian black bears, and Japanese macaques rather than wild boars and deer, which were commonly hunted during the Jomon period [31,37]. Regarding gathering, large amounts of horse chestnuts and beech were found in excavations. Chestnuts were also used, and are distinguished by mechanisms of social control. Cereals have not been found. The spatial aspects of these activities indicate that gathering was conducted around the villages, and hunting was conducted around the gathering areas. In addition, the space for exchange and trade extended outside the hunting areas. This finding is presented as a three-layered concentric circle. People of the Edo period combined the occupations of hunting, gathering, fishing, slash-and-burn farming, paddy field farming, and trade. These activities also formed a three-layer structure consisting of farming and gathering around villages, hunting, and trade (Figure 8).

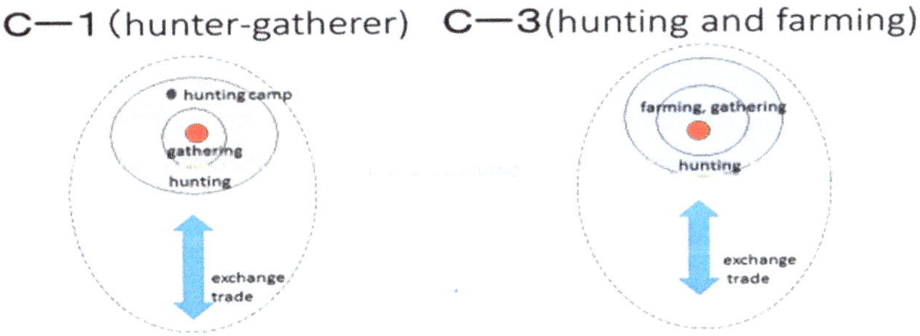

Type Period Subsistence Land use
C-1 late and final Jomon hunting, gathering, fishing, exchange commons
C-3 early Edo hunting, gathering, fishing, exchange commons
millet farming commons
paddy rice farming private land

Figure 8. Transition from hunter–gatherer (C-1) to early farmer (C-3) (C, fusion type). (Illustration by K. Ikeya).

Changes in combinations of occupations due to the introduction of farming are apparent in the comparison of the two periods. At the research site, the combination of farming activities, gathering edible wild plants in the spring, fishing in the summer, and gathering mushrooms and fruits in the fall would not cause an overlap of the periods of the various associated occupations (refer to Tables 1 and 2). In the Edo period, however, hunting methods and resource distribution for more concentrated use of resources had developed due to the commercialisation of forest products and social control methods of managing such activities. This particularly affected the trapping of bears and gathering chestnuts in the fall.

(3) Maintenance and changes in occupations and society after the introduction of farming: validation of the fusion model.

In this case, study, hunting, gathering, and fishing were maintained even after the introduction of farming. At the time, equal distribution of the shared resources of local residents, a method characteristic of hunter–gatherers, was practised during the hunting of bears and the gathering of sedge and chestnuts. Farmland, however, was divided into paddy fields, dry fields, and slash-and-burn farming. Disparities among households were evident in their areas of land for slash-and-burn farming and paddy field farming because irrigation systems and water management are fundamentally important for paddy fields. An ethos of land ownership had presumably developed by this time.

Therefore, hunter–gatherers most likely did not become rice farmers (assimilation type) despite the customs related to rice farming [35]. In other words, these findings suggest that although hunter–gatherers combined occupations and engaged in hunting as well as gathering and farming, the idea of egalitarian resource allocation that is characteristic of hunter–gatherers would be strong in the local community (characteristics of the fusion type). Paddy rice farming depended heavily on a high degree of social cooperation, community cohesion, and reciprocity, and thus resisted the development of social hierarchies [37].

A similar case is the San people, hunter–gatherers living in the southern region of Africa, who combined hunting and gathering with farming. A new combination of occupations was created by combining farming with hunting and gathering, even after small-scale agriculture was introduced [38,39]. Rather than becoming farmers, they maintained the values of the hunter–gatherer society. For instance, when certain households had harvested cultivated watermelons, the household that harvested a large amount temporarily stored the watermelons [39]. Subsequently, the watermelons harvested from the household's farm were distributed to other households.

In the Japanese archipelago, the modes of introduction for paddy field farming vary among regions. The common occupation in rice farming villages of the plains is the paddy field method. Although fishing in paddy fields was allegedly combined in some cases [40], the percentage of hunting and gathering was not large. This is a case of the assimilation type. In contrast, in mountainous regions such as the area studied here, it is expected that a lack of land suitable for paddy fields and the small volumes of rice production encouraged the combination of slash-and-burn farming and paddy field farming with hunting, gathering, and fishing to maintain subsistence. This case illustrates the fusion type.

As described in the examples presented above, agriculture can be combined with other pre-existing occupations by conducting it during different seasons. Moreover, cereal farming and paddy field farming (water management and land ownership) relate to the land very differently. The values of the hunter–gatherer society are maintained if the ratio of paddy field farming is low. If the ratio is high, then it is expected that the shift to the assimilation type would be based on how closely people are related to the land. There were not many social changes caused by the introduction of field farming; however, the introduction of paddy rice cultivation had different effects on society depending on the level of investment in channelling water from streams and springs and creating waterways. Based on these findings, regional diversity among the first farmers during the transition from the Jomon period to the Yayoi period can likely be explained by the types of mutual interaction between hunter–gatherers and farmers, the extent of investment in paddy field farming, and limiting characteristics of the habitat.

Funding: This study is a part of the Cultural History of PaleoAsia-Integrative research on the formative processes of modern human cultures in Asia project and is supported by the Japan Society for the Promotion of Science (Grant Number: 16H06409).

Institutional Review Board Statement: Not applicable.

Informed Consent Statement: Not applicable.

Data Availability Statement: Not applicable.

Acknowledgments: I am very grateful to Shin'ichiro Fujio and Jun Takakura, who provided information and advice. Additionally, I am grateful for the funding support from the Cultural History of PaleoAsia-Integrative research on the formative processes of modern human cultures in Asia project. I am also grateful for the support from the Center for Northeast Asia Area Studies at the National Museum of Ethnology.

Conflicts of Interest: The authors declare no conflict of interest.

References

1. Miyake, Y. Hunter-Gatherer Societies in the Prehistoric West Asia. In *Global Environmental History from the Perspective of Hunter-Gatherers: Coexistence with Nature, Neighbors, and Civilization (Shuryo Saishumin kara mita Chikyu Kankyo-shi)*; Ikeya, K., Ed.; University of Tokyo Press: Tokyo, Japan, 2017; pp. 58–73; ISBN 978-4-13-060317-1. (In Japanese)
2. Bar-Yosef, O.; Meadow, R.H. The Origins of Agriculture in the near East. In *Last Hunters-First Farmers: New Perspective on the Prehistoric Transition to Agriculture*; Price, T.D., Gebauer, B., Eds.; School of Advanced Research Press: Santa Fe, NM, USA, 1995; pp. 39–94.
3. Fuller, D.Q.; Qin, L.; Zheng, Y.; Zhao, Z.; Chen, X.; Hosoya, L.A.; Sun, G.-P. The Domestication Process and Domestication Rate in Rice. *Science* **2009**, *323*, 1607–1610. [CrossRef] [PubMed]
4. Habu, J. *Ancient Jomon of Japan*; Cambridge University Press: Cambridge, UK, 2004; ISBN 978-0521776707.
5. Habu, J. Resource Use by Jomon People and Long-Term Changes of Their Culture. In *Hunting-Gathering Culture in Japan (Nihon no Shuryo Saishu Bunka)*; Ikeya, K., Ed.; Sekaishiso-Sha: Kyoto, Japan, 2005; pp. 45–72; ISBN 978-4790711285. (In Japanese)
6. Nasu, H. The Initial Form of Rice and Millet Cultivation during the Final Jomon-Yayoi Transition Era from the View of Archaeobotanical Weed Assemblages. *Bull. Natl. Mus. Jpn. Hist.* **2014**, *187*, 95–110. (In Japanese)
7. Nasu, H.; Momohara, A. The Beginnings of Rice and Millet Agriculture in Prehistoric Japan. *Quat. Int.* **2016**, *397*, 504–512. [CrossRef]
8. Fujio, S. *History of Yayoi Period (Yayoi Jidai no Rekishi)*; Kodansha: Tokyo, Japan, 2015; ISBN 978-4062883306. (In Japanese)

9. Shitara, H. *From the Jomon Period to the Yayoi Period, Iwanami Lecture, Japanese History*; Iwanami Shoten: Tokyo, Japan, 2013; Volume 1. (In Japanese)
10. Fujio, S. Early Grain Cultivation and Starting Processes in the Japanese Archipelago. *Quaternary* **2021**, *4*, 3. [CrossRef]
11. Fujio, S. The Formation of Yayoi Culture in Fukuoka Plain. *Bull. Natl. Mus. Jpn. Hist.* **1999**, *77*, 51–84. (In Japanese)
12. Nakayama, S. *Plant Archaeology and Origins of Japanese Agriculture (Shokubutsu Kokogaku to Nihon no Noko no Rekishi)*; Doseisha: Minato City, Tokyo, 2010. (In Japanese)
13. Obata, H. *Northeast Asian Paleoethnobotany and Jomon Agriculture (Touhoku Ajia Kominzoku Shokubutsugaku to Jomon Noko)*, Doseisha: Minato City, Tokyo, 2011. (In Japanese)
14. Ikeya, K. Bear Ritual of the Matagi and the Ainu in Northeastern Japan. In *Circumpolar Animism and Shamanism*; Yamada, T., Irimoto, T., Eds.; Hokkaido University Press: Sapporo, Japan, 1997; pp. 55–63.
15. Ikeya, K. Mobility and Territoriality among Hunting-Farming-Trading Societies: The Case Study of Bear Hunting in Mountain Environments of Northeastern Japan. In *Beyond Affluent Foragers: Rethinking Hunter-Gatherer Complexity*; Grier, C., Kim, J., Uchiyama, J., Eds.; Oxbow Books: Oxford, UK, 2006; pp. 34–44.
16. Education Board of Asahi Village. *Excavation Report of the Archeological Sites related to Okumiomote Dam Construction, IV (Okumiomote Dam kanren Iseki Hakkutsu Chosa Hokokusho 4), Motoyashiki Ruins*; Education Board of Asahi Village: Murakami, Japan, 1995. (In Japanese)
17. Yatsuka, H.; Ikeya, K. Farming Practices among African Hunter-Gatherers: Diversifying without Loss of the Past. In *Rethinking African Agriculture: How Non-Agrarian Factors Shape Peasant Livelihoods*; Hyden, G., Sugimura, K., Tsuruta, T., Eds.; Routledge, Taylor and Francis Group: London, UK, 2020; pp. 49–63.
18. Ikeya, K. The Historical Dynamics of the Socioeconomic Relationships between the Nomadic San and the Rural Kgalagadi, Botswana. *Notes Rec.* **1999**, *31*, 19–32.
19. Turnbull, C. *The Forest People*; Simon & Schuster: New York, NY, USA, 1961.
20. Terashima, H. Economic Exchange and the Symbiotic Relationship between the Mbuti (Efe)Pygmies and the Neighbouring Farmers. *Sprache Gesch. Afr.* **1986**, *7*, 391–405.
21. Headland, T.N.; Reid, L.A.; Bicchieri, M.G.; Bishop, C.A.; Blust, R.; Flanders, N.E.; Gardner, P.M.; Hutterer, K.L.; Marciniak, A.; Schroeder, R.F.; et al. Hunter-Gatherers and Their Neighbors from Prehistory to the Present [and Comments and Replies]. *Curr. Anthropol.* **1989**, *30*, 43–66. [CrossRef]
22. Ikeya, K.; Nakai, S. Historical and Contemporary Relations between Mlabri and Hmong in Northern Thailand. In *Senri Ethnological Studies 73 Interactions between Hunter-Gatherers and Farmers: From Prehistory to Present*; Ikeya, K., Nakai, S., Eds.; National Museum of Ethnology: Osaka, Japan, 2009; pp. 247–261.
23. Takeuchi, K. Interethnic Relationships between Pygmies and Farmers. In *Hunter-Gatherers of the Congo Basin: Cultures, Histories, and Biology of African Pygmies*; Hewlett, B.S., Ed.; Transaction Publishers: Piscataway Township, NJ, USA, 2014; pp. 299–320. ISBN 1412853613.
24. Harunari, H. *The Beginning of the Yayoi Period*; Tokyo University Press: Tokyo, Japan, 1990; ISBN 978-4130241113.
25. Fujio, S. Interaction between the Jomon Farmer and the Yayoi Farmer in the Beginning of the Yayoi Period along the Old Kawachi Lake. *Bull. Natl. Mus. Jpn. Hist.* **2009**, *152*, 373–400. (In Japanese)
26. Ono, A. (Ed.) *Choja-Iwaya Rock Shelter Site*; Education Board of Asahi Village: Niigata Prefecture, Japan, 1993. (In Japanese)
27. Education Board of Asahi Village. *Excavation Report of the Archeological Sites Related to Okumiomote Dam Construction (Okumiomote Dam kanren Iseki Hakkutsu Chosa Hokokusho) XIV, Motoyashiki Site II Upper*; Education Board of Asahi Village: Murakami, Japan, 2002. (In Japanese)
28. Tomioka, N.; Asari, H. Analysis of Animal Remains Excavated from Motoyashiki Site in Niigata Prefecture. In *Excavation Report of the Archeological Sites Related to Okumiomote Dam Construction (Okumiomote Dam kanren Iseki Hakkutsu Chosa Hokokusho) XIV, Motoyashiki Site II Upper*; Education Board of Asahi Village: Murakami, Japan, 2002; pp. 334–340. (In Japanese)
29. Education Board of Asahi Village. *Excavation Report of the Archeological Sites related to Okumiomote Dam Construction, I(Okumiomote Dam kanren Iseki Hakkutsu Chosa Hokokusho 1), Shimozori Ruins*; Education Board of Asahi Village: Niigata Prefecture, Japan, 1990. (In Japanese)
30. Sato, Y. *Castanea crenata and Its DNA Analysis Excavated from Motoyashiki Site in Niigata Prefecture Excavation Report of the Archeological Sites Related to Okumiomote Dam Construction (Okumiomote Dam kanren Iseki Hakkutsu Chosa Hokokusho)XIV, Motoyashiki Site II Upper*; Education Board of Asahi Village: Niigata Prefecture, Japan, 2002; pp. 357–365. (In Japanese)
31. Crawford, G.W. Advances in Understanding Early Agriculture in Japan. *Curr. Anthropol.* **2011**, *52* (Suppl. 4), S331–S345. [CrossRef]
32. Miyajima, H. Identification of Beads Excavated from the Motoyashiki Site. In *Excavation Report of the Archeological Sites Related to Okumiomote Dam Construction (Okumiomote Dam kanren Iseki Hakkutsu Chosa Hokokusho) XIV, Motoyashiki Site II Upper*; Education Board of Asahi Village: Murakami, Japan, 2002; pp. 414–418. (In Japanese)
33. Watanabe, S. (Ed.) *Okumiomote: A Mountain Village of Uetsu Boundary (Uetsu Kokkyo No Sanson Okumiomote)*; Yamagata Chiri-Danwakai: Yamagata, Japan, 1979. (In Japanese)
34. Ikeya, K. Rice Crops and Shifting Cultivation in Miomote, Murakamihan in the Edo Era. *Sci Rep. Tohoku Univ.* **1987**, *37*, 41–51.
35. Education Board of Asahi Village (Ed.) *Folklore of Asahi Village (Asahi Mura No Minzoku) II*; Education Board of Asahi Village: Murakami, Japan, 1984. (In Japanese)

36. Ikeya, K. *Social Monograph among Wild Plant Gatherers: Natural Resource Use and Territoriality (Sansai Torino Shakaishi-Shigen Riyo to Teritori)*; Tohoku University Press: Miyagi, Japan, 2003; ISBN 4-925085-75-1. (In Japanese)
37. Kaner, S.; Yano, K. Early Agriculture in Japan. In *The Cambridge World History*; Barker, G., Ed.; Cambridge University Press: Cambridge, UK, 2015; pp. 353–386.
38. Ikeya, K. Dry Farming among the San in the Central Kalahari. *Afr. Study Monogr. Suppl.* **1996**, *22*, 85–100.
39. Ikeya, K. Environment and Resource Management among the San in Botswana. *Nomadic Peoples* **2001**, *4*, 67–82. [CrossRef]
40. Yasumuro, S. *Japanese Livelihood (Nihon Minzoku Nariwai Ron)*; Keiyusha: Tokyo, Japan, 2012; ISBN 978-4-87449-093-8. (In Japanese)

quaternary

MDPI

Article

Early Grain Cultivation and Starting Processes in the Japanese Archipelago

Shin'ichiro Fujio [1,2,3,4]

1 Department of Archaeology, National Museum of Japanese History, Chiba 285-8502, Japan; shin-f@rekihaku.ac.jp
2 Department of Anthropology, Boise State University, Boise, ID 83525, USA
3 National Museum of Japanese History, 117 Jonaicho, Sakura, Chiba 285-0017, Japan
4 Institute of Archaeological Science, Fudan University, Shanghai 200433, China

Abstract: This paper presents a specific examination of the introduction of grain cultivation and the processes of development in the Japanese Archipelago. In fact, no definitive archaeological evidence has been found that Jomon hunter–gatherers cultivated grain in the Japanese Archipelago; the earliest potential evidence of grain is a stamp mark of rice on the surface of a final late-Jomon, in about 11th century BC, pottery found at the Itaya 3 site in Shimane Prefecture. Current evidence indicates that the first grain cultivation was started by Jomon people who adopted irrigated wet rice cultivation that had arrived from the Korean Peninsula to northern parts of Kyushu, and gradually spread eastward thereafter. This study specifically examines four regions, including northern Kyushu, Kinki, southern Kanto, and northern Tohoku, in order to investigate the processes of grain cultivation initiation and spread. First, the years during which wet rice cultivation started in each region are estimated based on carbon-14 dating of earthenware types used during that period. Secondly, the timing of the spread of wet rice cultivation has been estimated based on carbon-14 dating of earthenware. Subsequently, differences in the periods between the initiation and dissemination of wet rice cultivation were estimated. Results suggest that dissemination took place over approximately 250 years in northern Kyushu, where wet rice cultivation first started. The time required for adoption decreased gradually as the trend moved eastward. It was estimated to have taken approximately 150 years in Kinki and 20–30 years in southern Kanto, taking place at about the same time. A factor, significantly contributing to such differences in timing and development processes among regions, was likely the relationship between the first farmers who introduced wet rice farming and the indigenous hunter–gatherers who lived there.

Keywords: bronze age; hunter gatherers; interaction; irrigation system; Jomon people; Korean Peninsula; wet rice cultivation

Citation: Fujio, S. Early Grain Cultivation and Starting Processes in the Japanese Archipelago. *Quaternary* **2021**, *4*, 3. https://doi.org/10.3390/quat4010003

Received: 19 June 2020
Accepted: 5 January 2021
Published: 27 January 2021

1. Introduction

Since the last glacial maximum, about 22,000 years ago, rising temperatures in the Japanese archipelago saw the emergence of pottery about 16,000 years ago. Earthenware appeared during the late glacial period [1]. In terms of Japanese archaeology, the Old Stone Age to the Jomon era is indicated by the appearance of earthenware [2].

The temperature rose only slightly during the Younger Dryas period of about 14,000 years ago, then warmed drastically from about 11,000 years ago [3]. This change marked the beginning of the post-glacial period in the Japanese archipelago. In this region, no edible wild grains grew during the post-glacial period. Major foods sought by Jomon people included nuts and other forest plants, forest animals, and seafood [4]. At the Sannai Maruyama site in Aomori Prefecture, a chestnut forest (*Castanea crenata*) existed that was as dense as today's chestnut gardens [5]. Stable sustenance using chestnuts as a staple food for subsistence is known to have continued for more than 500 years. Archaeological evidence suggests that Jomon people were able to grill and boil chestnuts.

Keiji Imamura defined the Neolithic culture of the Japanese archipelago as a "forest Neolithic culture" that used forest plants as a major food, in contrast to the "grassland Neolithic culture", which used grains and livestock as major foods in western Asia [6]. About 4200 years ago, the climate cooled suddenly and the Sannai Maruyama site was abandoned. Jomon people switched their subsistence emphasis from chestnuts to more cold-resistant horse chestnuts (*Aesculus turbinata*). Jomon people were compelled to go through complicated processes, such as heating and immersion, to make horse chestnuts edible [5]. Therefore, such water-exposed remains are being investigated continually at various locations.

In addition to gathering, the Jomon people might have conducted domestication of some plants. Soybeans and azuki beans (*Glycine soja* and *Vigna angularis*, respectively) are grown on Honshu Island and were collected by Jomon people as early as 10,000 years ago. Seiji Nakayama and Hiroki Obata confirmed from replica analysis that the wild bean size increased gradually over time, approaching the size of modern soybeans and adzuki beans (Figure 1). Nakayama and Obata designated these as Jomon beans, which are smaller than modern soybeans and adzuki beans, but suggest that they might constitute evidence of plant domestication by Jomon people [7].

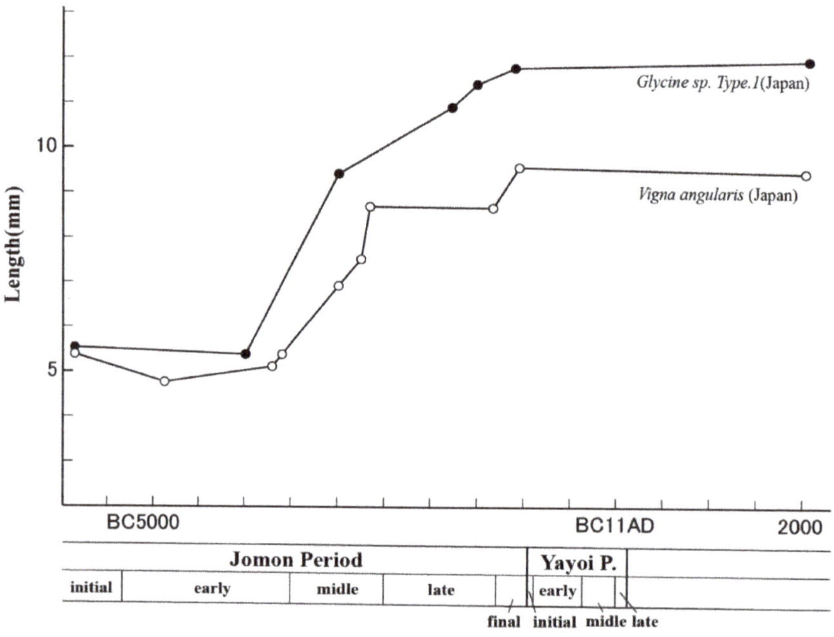

Figure 1. Changes in legume size during Japanese archipelago prehistory (Modified from Figure 101; referred from Hiroki Obata [7]).

From that experience with beans, Jomon people might have learned that managing plants would produce delicious and larger nuts or beans. Therefore, grain cultivation arrived at a time when the Jomon people had already accumulated selective cultivation knowledge and experience over thousands of years.

2. Who Brought Grain Cultivation to the Japanese Archipelago?

2.1. Diffusion from the Korean Peninsula to the Japanese Archipelago

Grain cultivation was introduced to the Japanese Archipelago from outside cultures in the late 10th century BC. Ancient Korean people of Bronze-age culture brought wet rice cultivation to northern parts of Kyushu [8]. The wet rice cultivation technique, which incorporated irrigation systems, was accepted by hunter–gatherers from Honshu, Shikoku, and Kyushu over an approximately 700-year span (Figure 2).

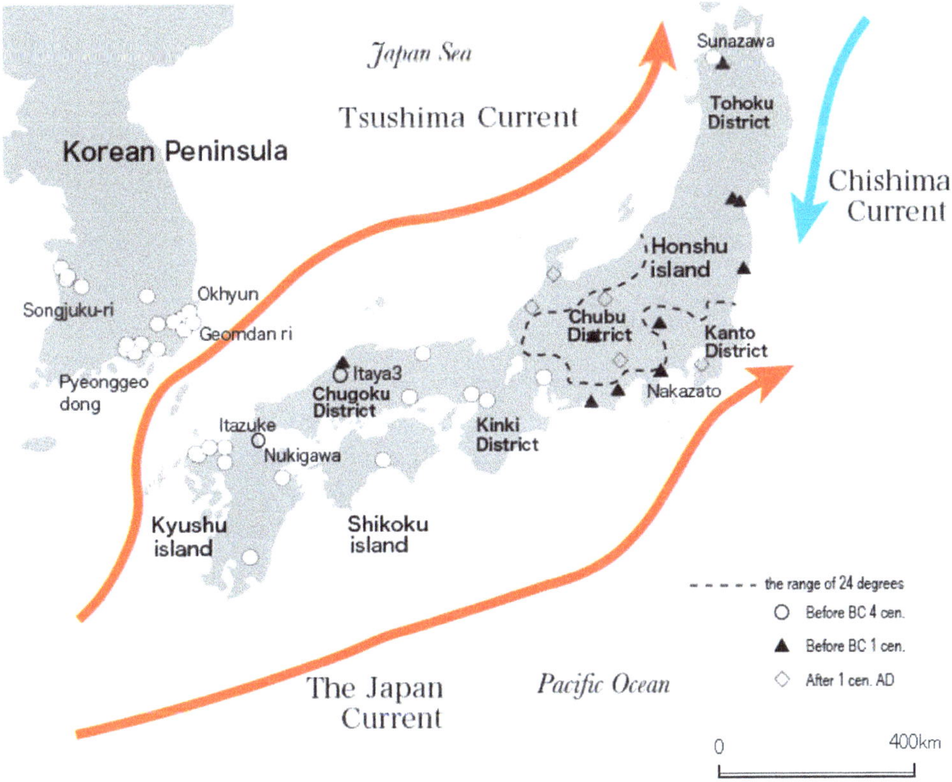

Figure 2. Geography, ocean currents, and prehistoric sites of the Korean Peninsula and the Japanese archipelago. (by Author).

The Tsushima warm current flows along the western Japanese archipelago; the Chishima cold current and the Japanese warm current flow along the eastern side. Therefore, during summer on the western Sea of Japan side, the Tohoku region is warm, but the eastern or Pacific side remains cold because of oceanic currents. The distribution of early rice farming sites is in areas with about 24 °C average temperature in August.

2.2. Introduction of Rice Cultivation

It seems that two occurrences supported the spread of grain cultivation to western Japan during the Jomon period (Figure 3). The first was the spread of rice and millet in the 11th century BC from the southeastern Korean Peninsula to eastern Kyushu and the Chugoku region [9]. The second was the spread of rice cultivation incorporating irrigation systems from the southern Korean Peninsula to northwestern Kyushu in the second half of the 10th century BC [10].

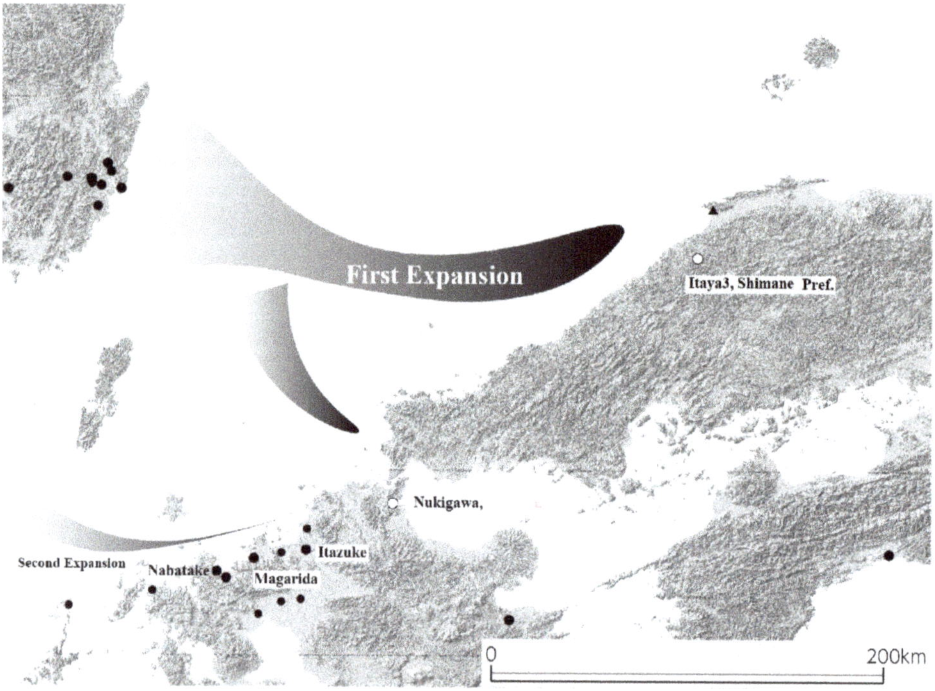

Figure 3. Two opportunities are apparent for grain cultivation to have spread from the Korean Peninsula to northern Kyushu and the Chugoku region. (by Author).

2.3. First Expansion

Evidence for Japan's oldest rice is the impression of rice plants on pottery excavated from the Itaya III site in Shimane Prefecture (Figure 4) [11]. This evidence has been dated to the 11th century BC, at the end of the Jomon period. Contemporaneous harvesting tools (stone reaping knives) for gathering rice ears have also been found at the Nukigawa site in Kitakyushu-shi, Fukuoka Prefecture (Figure 5) [12]. This evidence indicates that rice was present, but it is not clear that rice was being cultivated. However, even if Jomon people grew rice, they did not subsist mainly on rice production. Since then, no evidence of the spread of rice cultivation has been found in this area. Rice cultivation in this area was not generalized until the early Yayoi period, about 400 years later.

The following archaeological evidence confirms that Kyushu's interaction with the Korean Peninsula took place approximately 7000 years ago. The first is that obsidian from Koshidake Mountain, Saga Prefecture was found in the southern part of the Korean Peninsula, and the joint fishing hooks and stone harpoons for catching large fishes are widely distributed in the area from the east coast of the Korean peninsula to the west coast of Kyushu. In particular, the presence of common fishing gear in both areas indicates the possibility that there were people engaged in fishing activities in both areas, and there is no doubt that exchanges of both areas were common. Takakura Hiroaki calls this type of exchange the "fishermen's exchange" [13]. The first expansions of the 11th century BC may have been brought about through such exchanges.

2.4. Second Expansion

Paddy field cultivation began during the second half of the 10th century BC, in the initial Yayoi period, along the shores of the Genkai Nada in northwestern Kyushu. The paddy field excavated at the Itazuke site in Fukuoka Prefecture was full-scale, including irrigation facilities (Figure 6) [14]. Tools for rice cultivation were also present, such as stylized wooden farming tools, continental polished stoneware to make them, and stone reaping knives used for gathering rice ears.

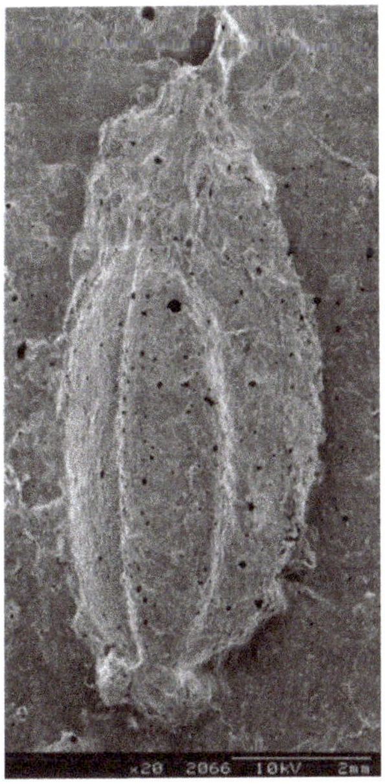

Figure 4. Evidence of oldest rice in Japanese Archipelago revealed by the replica method provided by Tsuyoshi Ushino. The replica method investigates stamp marks by embedding a resin in a fine hole indented on the pottery surface, removal after hardening, and observation of the surface using an electron microscope.

Figure 5. Stone reaping knives and a series of pottery items excavated from the Nukigawa site, Fukuoka Prefecture. Provided by Kitakyushu City Buried Cultural Property Research Office.

Figure 6. Reconstructed picture of paddy fields and wells and drainage channels during the latter 10th century BC excavated at Itazuke Site, Fukuoka Prefecture, provided by Fukuoka City Buried Cultural Property Center.

The paddy field found in the Itazuke site is a large-scale paddy field plot with a section of more than 300 square meters. Weirs installed at 50-m intervals in the main waterways supplied paddy fields with water.

2.5. Each Homeland

In the southern part of the Korean Peninsula, paddy field cultivation began during the 11th century BC. Small-plot paddy fields have been found at the Ulsan Okhyun site (Figure 7) [15]. The common pattern of pottery excavated at Itaya III Site indicates that the Yeungnam area might be its home region [9]. At that time, material culture of the southern Korean Peninsula in 11th century BC included bronze swords, which had not been used before the formation of agricultural society. The social situation was unstable. Moreover, the cultural elements of the southern Korean peninsula found in western Japan are fragmentary. Therefore, it is unlikely that rice was transmitted by large-scale movements of people from the southern Korean Peninsula.

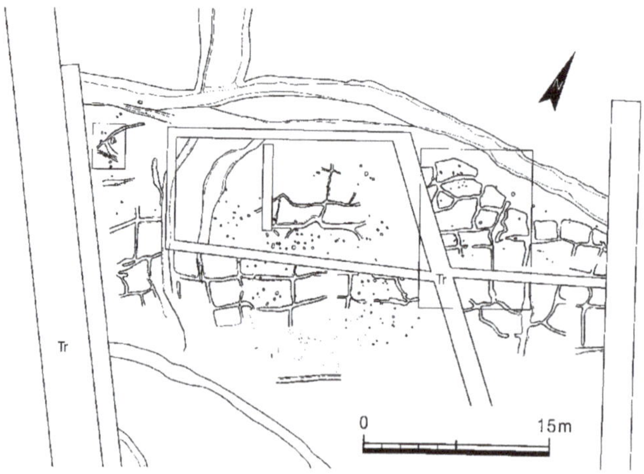

Figure 7. Oldest paddy field on the Korean Peninsula, Okhyun site, Ulsan, Korea. Provided by Busan University Museum.

During the 10th century BC, ditch-enclosed settlements began to appear in the Yeungnam area; an agricultural society had become established. Consequently, social contradictions became apparent. Some people who escaped from the society have been identified, who would have crossed the Korean strait to Kyushu [16].

One theory holds that, given the characteristics of small pots excavated from both areas, the origin of persons arriving from across the Korean strait was the homeland of the Nakdong River Basin. This second spread of paddy field cultivation occurred along with the spread of bronze age culture based on paddy rice cultivation, such as tools and techniques for paddy field cultivation, as well as a festival to pray for fertility. Such practices might have been imported along with the movement of large groups of people.

3. Interaction between Native Hunter–Gatherers and Newcomer Farmers

3.1. Introduction

Jomon people from about 16,000 to 3000 years ago depended on gathering, fishing, hunting, and some farming for subsistence. They cultivated gourds, soybeans, and adzuki beans but not grains such as foxtail millet (awa, *Setaria italica*), or rice. Consequently, farming was only a component of Jomon subsistence. Farmers who came from the southern Korean Peninsula began wet rice cultivation with irrigation systems 3000 years ago in cooperation with original inhabitants of the Fukuoka Plain in northern Kyushu. Their interaction had just begun to resemble similar situations found in prehistoric Europe [17].

Research of the time period of interactions among these peoples has included AMS-^{14}C analyses of the Fukuoka Plain, Osaka Plain, Ashigara Plain, and Hirosaki Plain. Based on those results, the times of interaction are known to be shorter for east and north areas. The reason for this phenomenon must be discussed.

3.2. Analysis

In Figure 8, the left panel portrays a set of pottery used by a farmer who began working on the Fukuoka Plain in the 10th century BC (Figure 8 left). The right panel depicts a set of pottery used by a farmer 200 years later in the 8th century BC (Figure 8 right). Cooking pots came to show a new form and style through cultural interaction. Moreover, one can ascertain the time of interaction and when such jars were used through ^{14}C analysis of soot adhering to the cooking pot surface, as shown in the left and right panels of Figure 9.

Figure 8. (**Left**), Set of pottery used by a farmer during earliest farming on the Fukuoka Plain in the 10th century BC; (**Right**), Set of pottery used by a farmer 200 years later during the 8th century BC. (photograph by Fujio).

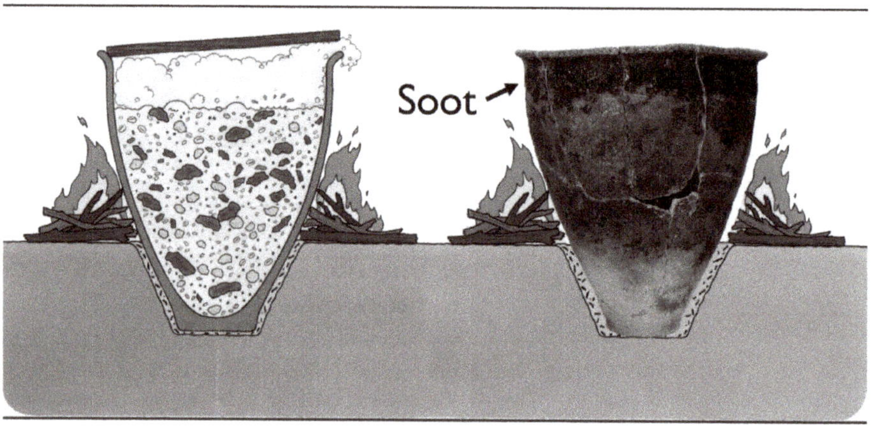

Figure 9. Schematic diagram of cooking in the early Yayoi period and the attachment of soot. (referred from Fujio) By carbon-14 dating of the soot which collected on the surface of the pottery during cooking, one can ascertain the age during which the pottery was used [18].

3.3. Northern Kyushu

Figure 10 portrays the distribution of sites on the Sawara Plain and Fukuoka Plain at the end of the final Jomon period in the 11th century BC [19]. Sites are rarely found in downstream basins on the plain. They are distributed in the middle and upper reaches of rivers. Native hunters and gatherers lived at sites in middle and upstream areas suitable for collecting nuts, fishing, and hunting, preferring these areas to the downstream basins.

Figure 11 presents the distribution of sites in the same region in the second half of the latter of 10th BC, when paddy field cultivation began [19]. The archaeological remains of paddy field farmers appeared in the downstream basin on the plain, where few archaeological sites had been identified to that time. For farmers, the downstream basin on the plains was a convenient place for paddy rice cultivation. *Sumiwake*, so-called habitat segregation, had begun.

In the Sawara Plain, numerous rice paddy farmers' settlements appear in the downstream region. Even in the Fukuoka Plain, numerous rice paddy farmers' settlements have appeared.

Circumstances leading to this situation are explained below [20].

The first farmers, as newcomers, developed paddy fields in downstream basins on the plains that had been used only slightly by the indigenous hunter–gatherers. Because the latter cultivated beans and other crops, collected nuts, and hunted in middle and upstream areas, the former settled there peacefully with few conflicts of interest. However, they did have mutual relations. Apparently, farmers were unable to cultivate wet rice independently. For example, without help from indigenous hunter–gatherers familiar with the land, they would have been unable to gain knowledge about rare raw materials: hard stone that might be processed for use as axes to cut down trees and hard oak trees with wood suitable for crafting wooden farming tools. The indigenous hunter–gatherers might have also been valuable for marriage and as laborers necessary to grow wet rice.

The farmers were also valuable for indigenous hunter–gatherers. Farmers offered food such as rice in exchange for information about local rare materials, as well as cultivation knowledge and social opportunities. These groups might therefore be regarded as interdependent. About 250 years passed from the start of farming to their interaction, which ended when hunter–gatherers themselves became farmers.

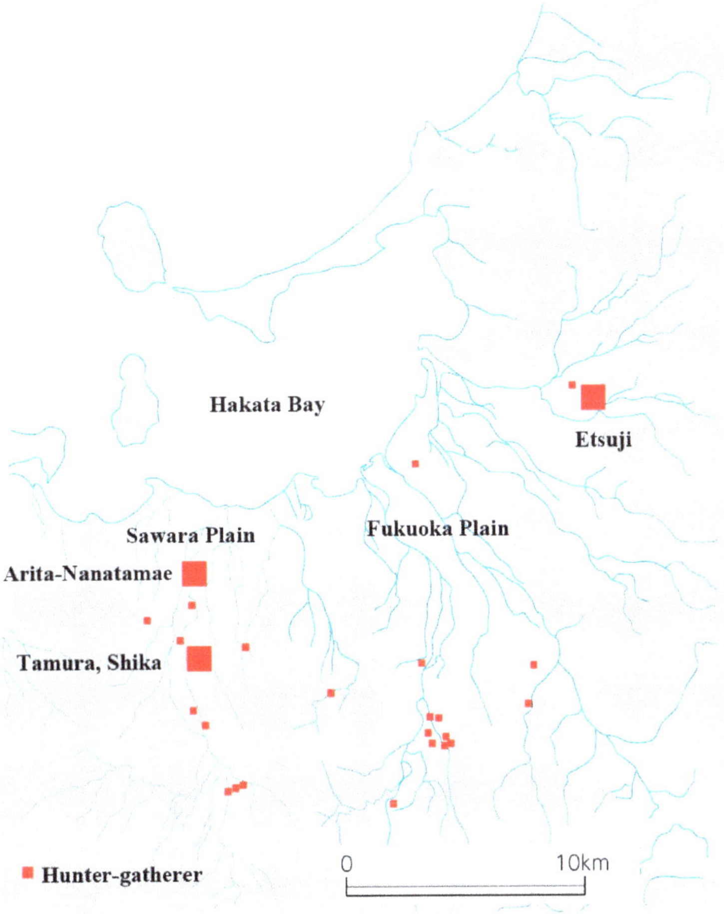

Figure 10. Distribution of sites at the end of final Jomon period in the Sawara Plain and Fukuoka Plain [19] (by Author). The Tamura and Shika sites are indigenous hunter–gatherer villages that have persisted since the late Jomon period. They are located in the middle reaches of the Sawara Plain. The Arita Nanatamae Site, which appears at the end of the final Jomon period, is located in the downstream area. It persisted in the initial Yayoi period. No site was found in the lower reaches of the Fukuoka Plain. The sites are located only in the upper and middle reaches.

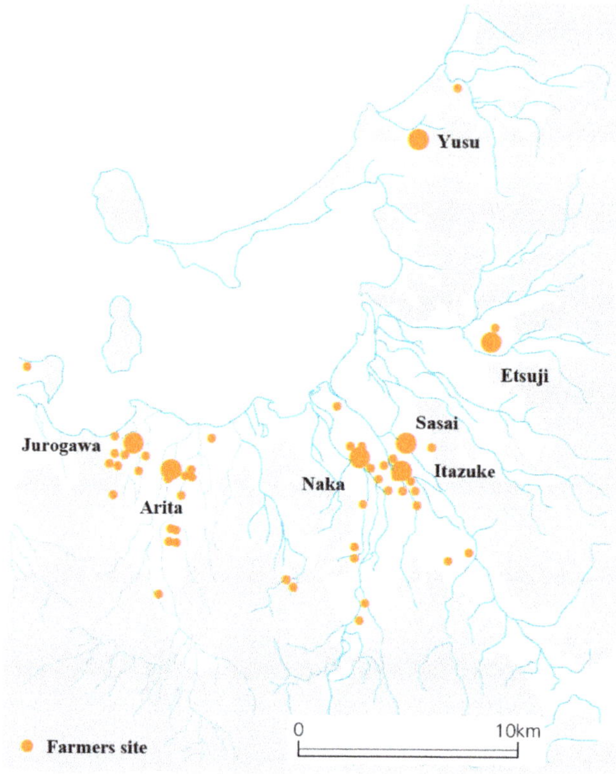

Figure 11. Distribution of initial Yayoi period sites in the Sawara Plain and Fukuoka Plain [19] (by Author).

3.4. Osaka Plain

On the Osaka Plain, farmer settlement occurred on the banks of Old Kawachi lagoon, whereas hunter–gatherer settlements were located on the upper and middle plains and hill-sides in the 6th century BC to 5th century BC [21]. Their mutual interaction for 150 years has been ascertained from the character of pottery used by farmers and hunter–gatherers [22].

Reportedly, the Sanuki region in the northeastern part of Shikoku brought paddy field cultivation to coastal areas of Osaka Bay. Large amounts of Sanukite from the Sanuki region have been excavated from sites on the coast of Osaka Bay at the beginning of the Yayoi period [23].

3.5. Ashigara Plain in Kanagawa Prefecture, West of Kanto District

At Ashigara Plain in Kanagawa Prefecture, west of Kanto District, Jomon people had been cultivating foxtail millet, etc., in the upper and lower reaches of the river for about 500 years. Around 250 BC, paddy field cultivation suddenly began along the lower reaches of the river. Paddy field farmers set up paddy fields in the lower reaches of the plains, where no hunter–gatherers lived [19]. The farmers created ditch-enclosed settlements and built burial mounds surrounded by square moats (Figure 12). Because their pottery has been found in western Japan, it is believed that paddy field farmers immigrated from the west. Their interaction continued for 30–50 years [24]. The *Sumiwake* is shown between at the lower plain and hillside.

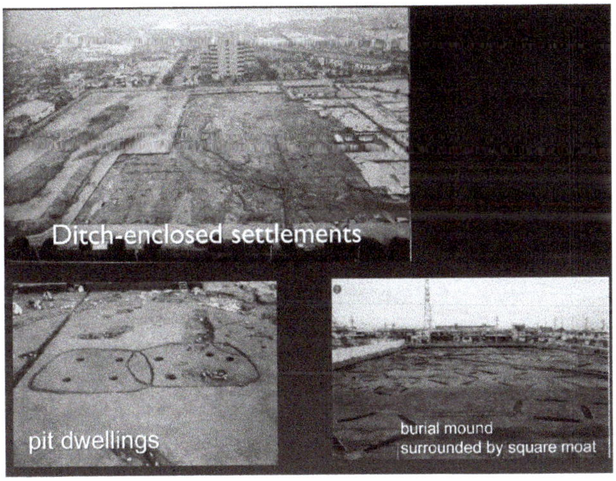

Figure 12. Ditch-enclosed settlement with a pit dwelling and a burial mound surrounded by a square moat excavated at the Nakazato site, which appeared during the third century BC. Provided by Tamagawa Cultural Property Research Institute.

3.6. Northern Tohoku District

On the Tsugaru Plain in Aomori prefecture, in the northern part of Tohoku district, the northernmost paddy field, at 40° N latitude, was discovered at the Sunazawa site, Hirosaki-shi, Aomori Prefecture (Figure 13).

Figure 13. Oldest paddy field in the Tohoku region: Sunazawa site. Provided by the Hirosaki City Board of Education. There were seven large-scale paddy fields discovered at the Sunazawa site, which has no irrigation facility. The water flows naturally downward. The long axis of No. 2 paddy field is about 13.6 m.

Jomon people began wet rice cultivation with independently operated irrigation systems in the 4th century BC. Sunazawa people fundamentally used paddy field cultivation with traditional stone tools from the final Jomon period. They also continued traditional festivals using clay figurines, stone rods, and other materials. Paddy field farmers using these Jomon-origin festival tools resided only in Aomori prefecture.

Phytolith analysis results suggest that wet rice cultivation was conducted at the Sunazawa site for 12–13 years. Wet rice farming was practiced in this region for approximately 300 years thereafter. However, the paddy fields were eventually inundated and ruined because of the falling temperatures and flooding which occurred approximately in the 1st century BC [25]. Farmers subsequently returned to hunting and gathering without resuming wet rice cultivation. Not until ancient, but more recent, times did grain cultivation recommence in this region [19]. The Epi–Jomon culture would therefore continue without establishing agricultural society, let alone a transition to the Kofun period.

3.7. Discussion

As described above for these four regions, the processes extending from the beginning of paddy field cultivation to the establishment of a fully agricultural society are readily apparent [23]. The duration of the transition was shorter in eastern regions. The analytical results also revealed that, likely as a result of climatic changes, some people resumed their original hunting and gathering life after ceasing the wet rice farming practices that they had used continuously for approximately 300 years. The models presented below illustrate the durations of interaction (Table 1) [24].

(A) 250 years, e.g., Fukuoka Plain.
(B) 150 years, e.g., Osaka Plain, probably western Japan.
(C) 20–50 years, e.g., Kanto Plain, probably eastern Japan.
(D) Native hunter–gatherers started wet rice cultivation independently without any new arrival of plant farmers. They passed down Jomon festivals. No mutual exchange occurred. They resumed hunting and gathering after approximately 300 years.

Table 1. Durations of interaction.

Model	Duration of Interaction	Newcomers	Example
A	250 years	Korean Peninsula	Northern Kyushu
B	150	e.g., Sanuki	Osaka Bay area
C	30–50	e.g., Harima, Mikawa	Eastern Japan
D	0	0	Northern Tohoku

Carbon-14 dating of soot on the surface of cooking utensils found at the oldest paddy field site in the four regions supports the following hypothesis: Although immigrants spread from Kyushu to Honshu, excluding northern Tohoku, the period of interactions between natives and immigrants became shorter as they advanced to the east. Native hunter–gatherers came to dominate smaller areas as the introduction of wet rice farming moved eastward. People from the Korean Peninsula brought wet rice farming into northern Kyushu, not only as an occupation and method of sustenance, but as a cultural complex including social and ritual systems with wet rice cultivation as the productive base.

Establishment took 250 years presumably because of the considerable time necessary for adaptation of wet rice cultivation, which was the production base for the Bronze Age culture of the Korean Peninsula, to environmental conditions prevailing in northern Kyushu. In other words, most hunter–gatherers in northern Kyushu might have taken time to convert wet rice cultivation from the Korean Peninsula into the Yayoi rice farming appropriate for the ecosystems and daytime lengths of northern Kyushu. This adaptation represents the creation of the Yayoi wet rice farming.

Why, then, did the period of interaction decrease as the trend moved to the eastern regions? First, acceptance of the system based on wet rice cultivation, of which the conversion to the Yayoi system had been completed in northern Kyushu, was likely to have been easier for people of Osaka Bay coastal areas, where the natural environment, ecosystems, and culture resembled those of northern Kyushu. The reason for the shorter periods required by people in the regions other than northern Kyushu might have been that they had accepted the Japanese archipelago version of Yayoi farming that had been created in northern Kyushu [26].

The first people to bring wet rice cultivation into the Ashigara Plain in southern Kanto were purportedly wet rice farmers who had migrated from western Japan, as suggested by the introduction of cooking earthenware from the western region. In other words, wet rice cultivation was started by migrants from western Japan, rather than by indigenous hunter–gatherers. Moated settlements and square-moated burial precincts suggest rapid formation of agricultural society, which was achieved through the efforts of such migrants.

The last region is northern Tohoku. The local people resumed their original hunting and gathering life after relinquishing agriculture, which had served them for approximately 300 years from the start. That reversion occurred because they accepted wet-rice farming into their comprehensive work structure maintained since the Jomon period, but only as a part-time livelihood. If paddy fields were buried by severe flooding, such as that which occurred before the 1st century BC, then the people would not have had such severe difficulty in abandoning wet rice farming for alternative resources: it was merely one of several occupations on which they relied for subsistence. It was fortunate that they had adopted wet rice cultivation while maintaining labor organization and clay figure festivals and other rituals continuously since the Jomon period.

4. Conclusions: Early Grain Cultivation and the Starting Process

Two groups of farmers resided on Honshu, Shikoku, and Kyushu islands: those who started wet rice cultivation and those who started millet cultivation [19]. Hunter–gatherers of northern Kyushu and northern Tohoku first began wet rice cultivation (Figure 14). However, hunter–gatherers in other areas first began millet cultivation and subsequently switched to wet rice cultivation several hundred years later. Hunter–gatherers of the latter areas started paddy rice cultivation after cultivating foxtail millet and broomcorn millet (*Panicum* spp.) for about 300–500 years (Figure 14) as a means of subsistence that complemented wild foods.

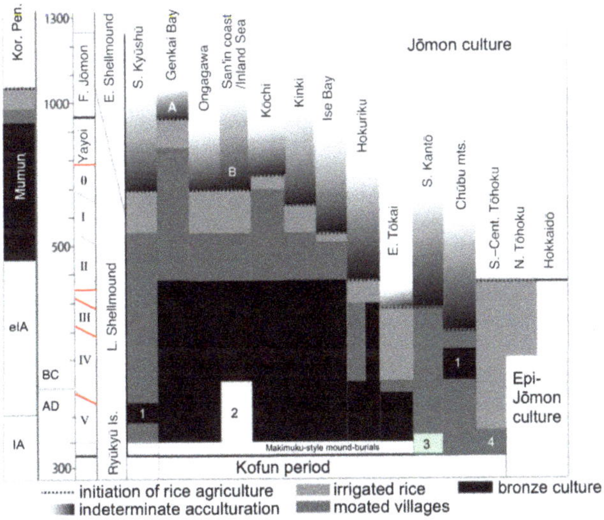

Figure 14. Regional distribution of the three elements of Yayoi culture: 1, bronze cache (Cache is designed to hold a bronze sword or a bronze bell-shaped vessel in soil until it is time to be spent next.); 2, withdrawal from bronze culture; 3, small bronze bell; and 4, burial mound surrounded by square moat. Japanese version provided by [19] (Fujio 2013: Figure 80) English version provided by Gina Barnes [27].

Komoto designated such a subsistence structure as comprehensive and regarded it as a generalized subsistence strategy of Jomon society [28]. However, paddy field farmers in northern Kyushu also collected, hunted, and fished, yet subsisted mainly on paddy field cultivation. Komoto designated this as a selective subsistence structure and regarded it as the representative subsistence structure of Yayoi society.

People who grew foxtail millet (awa) and broomcorn millet in the Jomon-like occupational structure crafted Jomon culture-specific festival tools such as clay figurines and stone rods. By contrast, paddy field farmers practicing selective subsistence had no clay figurines or stone rods. This important difference signifies that Jomon-like rituals are compatible with millet cultivation in the comprehensive occupation structure, but not with paddy field cultivation in the selective occupation structure. The rice prayer festival and the clay figurines and stone rod rituals might be completely different. Wet rice farmers in Aomori Prefecture used clay figurines probably because they used paddy field cultivation under a comprehensive subsistence structure.

The period from the start of millet cultivation to the start of irrigated paddy rice cultivation is designated as one of indeterminate acculturation. At this stage, neither Jomon nor Yayoi culture predominated. The transition period was short in areas where paddy field farmers might have migrated, as in northern Kyushu, and longer in areas where native hunter–gatherer and hunting people started with millet, as in western and eastern Japan. The length of the transition period increases with distance to the east, with that in Chubu and southern Kanto region lasting as long as 500 years. Figure 14 illustrates that the transitional period length varies among regions.

The next subject is the process that took place after adopting wet rice cultivation. There are four patterns that have been developed from the lengths of period between the introduction and dissemination of wet-rice farming in the regions. The analysis revealed that the period became shorter as the introduction of wet rice farming moved eastward, i.e., approximately 250 years in northern Kyushu, approximately 150 years in Kinki, and 20–30 years in southern Kanto. Only the hunter–gatherers in northern Tohoku resumed hunting and gathering after practicing wet rice farming for approximately 300 years. Thus, pre-existing conditions of subsistence structure are demonstrated to have a clear and predictable influence upon the adoption, retention, and intensification of wet paddy agriculture in the Japanese Archipelago.

Funding: This research received no external funding.

Institutional Review Board Statement: Not applicable.

Informed Consent Statement: Not applicable.

Data Availability Statement: Not applicable.

Acknowledgments: This article is a textual version of contents presented at the session organized by K. Ikeya "Interactions between Prehistoric Hunter-Gatherers and Neighbors in Asia "of the 8th World Archaeological Congress 2016 in Kyoto. In writing this article, I extend my gratitude to Kazunobu Ikeya of the National Museum of Ethnology.

Conflicts of Interest: The author declares no conflict of interest.

References

1. Taniguchi, Y. The beginning of the Jomon period: Rethinking the incipient Jomon. In *The Archaeology of the Jomon Period*; Doseisha: Tokyo, Japan, 2010; Volume 1, pp. 79–97. ISBN 978-4886215086. (In Japanese)
2. Kobayashi, T. The Origin of the Jomon pottery. *Archaeol. J.* **1962**, *100*, 26–30. (In Japanese)
3. Kudo, Y. *Environment and Culture History of the Upper Paleolithic and the Jomon Period*; Shinsensha: Tokyo, Japan, 2012; ISBN 978-4787712035. (In Japanese)
4. Tsuji, S. A Land Ecosystem in the Transition to the Jomon Age. *Quat. Res.* **1997**, *36*, 309–318. (In Japanese) [CrossRef]
5. Goto, K.; Tsuji, S. Vegetation history since the Early Jomon Period at Ooyazawa, Aomori, in the southern part of the Aomori Plain. *Quat. Int.* **2000**, *9*, 43–53. (In Japanese)
6. Imamura, K. Chestnut and walnut trees were cultivated plants in Europe. *Archaeol. J.* **2009**, *594*, 31–34. (In Japanese)

7. Obata, H.; Sasaki, Y.; Senba, Y. Impressions on pottery revealed cultivation of Glycine max subsp. max (soybean) in the late to latest Jomon periods in Kyushu Island. *Jpn. J. Hist. Bot.* **2007**, *15*, 97–114. (In Japanese)
8. Ahn, J.-H. Rethinking on the origin of Songgungni culture. *Yongnam Archaeol.* **2019**, *83*, 91–125. (In Korean)
9. Cheon, S.-H. The Transion of regional relationship between Korean peninsula and the western Japan in the Mumon Period. *Korean Antiq.* **2009**, *73*, 33–55. (In Korean)
10. Fujio, S. When did the wet rice cultivation with the irrigation system begin in the Western Japan. *Bull. Natl. Mus. Jpn. Hist.* **2014**, *183*, 113–143. (In Japanese)
11. Nakazawa, M. Continental grain in the early Yayoi period as seen from the analysis of earthenware indentation by the replica method. *Archaeol. J.* **2019**, *729*, 14–19. (In Japanese)
12. Maeda, Y.; Takesue, J. The reaping knife in final Jomon period discovered from Nukigawa site, Kitakyushu-shi. *Bull. Inst. Res. Kyushu Cult. Hist.* **1994**, *39*, 65–90. (In Japanese)
13. Takakura, H. *The Age of Golden Seal States*; Aoki shoten: Tokyo, Japan, 1985. (In Japanese)
14. Yamasaki, S. Paddy field in Northern Kyushu. In *The Appearance of the Yayoi Culture in Japan*; Bunken Shuppan: Tokyo, Japan, 1991; pp. 350–394. (In Japanese)
15. Kyungnam University Museum & Pusan National University Museum. *Okhyeon*; Kyungnam University Museum & Pusan National University Museum: Pusan, Korea, 2015. (In Korean)
16. Ahn, J.-H. Yeungnam society and Yayoi culture during the establishment of Songjuku-ri culture. In *Archaeology of the Yayoi Period*; Doseisha: Tokyo, Japan, 2009; Volume 2, pp. 73–89. (In Japanese)
17. Dennell, R.W. The hunter-gatherer/agricultural frontier in Prehistoric temperate Europe. In *The Archaeology of Frontier and Boundaries*; Green, S.W., Perlman, S.M., Eds.; Academic Press, Inc.: Cambridge, MA, USA, 1985; pp. 113–139.
18. Fujio, S.; Imamura, I.; Nishimoto, T. When did the wet-rice cultivation begin In Japanese Archipelago? *SOKENDAI Rev. Cult. Soc. Stud.* **2005**, *1*, 73–96. (In Japanese)
19. Fujio, S. *Reconstruction of the Yayoi Culture Image*; Yoshikawa-Kobunkan: Tokyo, Japan, 2013; ISBN 978-4642093293. (In Japanese)
20. Fujio, S. The Formation of Yayoi Culture in Fukuoka Plain. *Bull. Natl. Mus. Jpn. Hist.* **1999**, *77*, 51–84. (In Japanese)
21. Kobayashi, K.; Harunari, H.; Akiyama, K. ^{14}C Dating of Yayoi Period in Kawachi Districts. *Bull. Natl. Mus. Jpn. Hist.* **2008**, *139*, 17–51. (In Japanese)
22. Fujio, S. Interaction between the Jomon Farmer and the Yayoi Farmer at the beginning of the Yayoi Period along the Old Kawachi Lake. *Bull. Natl. Mus. Jpn. Hist.* **2009**, *152*, 373–400. (In Japanese)
23. Harunari, H. *The Beginning of the Yayoi Period*; UP Selection Book; Tokyo University Press: Tokyo, Japan, 1990; ISBN 978-4130241113. (In Japanese)
24. Fujio, S. Interaction between Garden Culture People and Farmer. In Proceedings of the T06-Q Interactions between Prehistoric Hunter-Gatherers and Neighbors in Asia the 8th World Archaeological Congress, Kyoto, Japan, 28 August–2 September 2016.
25. Nakatsuka, T.; Sano, M.; Li, Z.; Xu, C.; Tsushima, A.; Shigeoka, Y.; Sho, K.; Ohnishi, K.; Sakamoto, M.; Ozaki, H.; et al. A 2600-year summer climate reconstruction in central Japan by integrating tree-ring stable oxygen and hydrogen isotopes. *Clim. Past* **2020**, *16*, 2153–2172. [CrossRef]
26. Fujio, S. Inter-regional differences in adaptation to rice cultivation. *Q. Archaeol. Stud.* **1991**, *38*, 30–54. (In Japanese)
27. Barnes, G.L. The Jomon-Yayoi transition in eastern Japan: Enquiries from the Kanto region. *Jpn. J. Archaeol.* **2019**, *7*, 1–51.
28. Komoto, M. Early Agricultural Culture in Northeast Asia: Focusing on the analysis of natural relics. In *The Formation of Early Yayoi Culture in Japan*; Bunken Shuppan: Tokyo, Japan, 1991; pp. 553–613. (In Japanese)

quaternary

MDPI

Article

Alternative Adaptation Strategy during the Paleolithic–Neolithic Transition: Potential Use of Aquatic Resources in the Western Middle Yangtze Valley, China

Ruizhe Liu [1], Hui Liu [2] and Shengqian Chen [3,*]

[1] Department of Anthropology, University of New Mexico, Albuquerque, NM 87131, USA; rzliu@unm.edu
[2] Hubei Provincial Institute of Cultural Relics and Archaeology, Wuhan 430077, China; njliuhui1998@sina.com
[3] Department of History, Renmin University of China, Beijing 100872, China
* Correspondence: csq@ruc.edu.cn

Received: 15 June 2020; Accepted: 14 September 2020; Published: 21 September 2020

Abstract: The middle Yangtze Valley is an important region for studying the origin of rice agriculture. Archaeological remains of rice have been found in sites such as Pengtoushan and Bashidang as early as 8000 years ago. However, we know little about the adaptive diversity in this region as research has mostly focused on rice cultivation. With the help of new discoveries, this paper explores another adaptation behavior pattern that emphasizes the utilization of aquatic resources in the western part of the Jianghan Plain during the Paleolithic–Neolithic transition. Although the sea level was continuously rising with the warming process of early Holocene, the lakes that later became widely distributed were still in formation, thus not available for human utilization in the middle Yangtze Valley. However, most notably, the amelioration of the environment was producing a sort of new niche: utilizing aquatic resources became possible at least in parts of this region. A case study from the Guanzhou Site, based on lithics, suggests there were increasing demands for the utilization of aquatic resources. The study proposes that at least two different adaptation strategy changes occurred in the western Jianghan Plain during the Paleolithic–Neolithic transition, i.e., rice agriculture or complex hunting–gathering. These strategies represent adaptations to the different ecological conditions at the crossroads of habitat types.

Keywords: complex hunting–gathering; Paleolithic–Neolithic transition; aquatic utilization; China

1. Introduction

About ten thousand years ago, the arrival of the Holocene was marked by a stable and mild climate after the last ice age terminated. Several dramatic changes in behaviors of prehistoric hunter-gatherers happened during the Paleolithic–Neolithic transition (PNT), such as the appearance of food production, complex social systems, complicated human behaviors, ideology and sedentism [1]. Meanwhile, in a general trajectory, the human adaptation based on hunting and gathering was replaced by reliance on food production, a process termed the origin of agriculture, also described by V. Gordon Childe as the "Neolithic Revolution" [2].The global impact of the agricultural revolution on humankind has given rise to "hot" topics including population growth, environmental degradation, the appearance of sedentism, social complexity, and other changes to adaptation strategies. Although the process was prolonged for several thousand years and there are various cultural responses aside from food production, the origin of agriculture has been a major research focus for decades [1,3–5], in other words, agriculture, as an important adaptation during the PNT have been explored much, especially in West Asia, which addresses the influence of environment and population growth [6–8]. However, in East Asia, further research is not only needed on the origin of agriculture, but also research relating to nonagricultural adaptations, which have received much less attention.

At present, evidence is fairly clear on the origin of agriculture in China, which includes two centers: North China (the east edge of the Loess Plateau) and South China (the plains of the middle and lower reaches of the Yangtze River). Yet the PNT of noncentral (noncenter regions: except the Loess Plateau and the plain of middle and lower reaches of the Yangtze River, the rest of the places around China are noncenter regions, such as Northeast China, South of the Five Ridges, Qinghai–Tibet Plateau, the mountain areas of the Yangtze reaches (Xijiang region) and so on) regions is still unclear. Since we have studied the question of the PNT for many years, in this paper we investigate the archaeological records in the Xiajiang area (Xiajiang area: the mountain and gorge area along the Yangtze River, including several important terrains, the Three Gorges, mountain area in Southwest China, the edge of Yunnan–Guizhou Plateau) and its adjacent regions, especially first-hand unearthed lithics from the Guanzhou site in Hubei Province. A special pattern of cultural adaptation can be recognized, which is characterized by a unique lithic technology: the "ridged hammer bipolar flaking" (RHBF) technique (see below), a representative technique in the Xiajiang area. Until now, very little research has focused on the question of the PNT process in this region, and only a few studies have been involved (e.g., [9,10]). Therefore, the present paper based on an analysis of lithic assemblages in this region can fill in gaps in the question to a certain extent and enrich our overall understanding the diversity about the PNT.

In general, the topography of China consists of three parts: the Tibetan Plateau, the floodplains and hills in the east, and the middle zone between these two parts. The region we discuss, the western part of middle Yangtze valley, is located in the eastern margin of the middle zone where rivers flow into floodplains. The power of rivers to carry sediment rapidly recedes with decreased flow rates, which leads to two consequences that could have impacted lives of prehistoric hunter-gatherers. First, the deposition of river boulders near the reach where a river enters floodplain. This zone can provide abundant raw material, for stone tool manufacturing, of adequate size. In contrast, in the lower reach, it is difficult to find lithic raw material of equivalent size. The other impact is the formation of wetlands in the floodplain, including lakes and seasonal wetlands. Effectively exploiting these special ecological zones would require more complicated technologies: for example, boat and paddle. Therefore, along a river there is an optimal zone for prehistoric hunter-gatherers, where they may find rich lithic raw materials, diverse food resources and a landscape suited for relatively easier mobility. The region discussed above is the transitional zone between the plateaus and hilly floodplains. Notably, the lithic source outcrops are located in more eastern areas, since the sea level in the Terminal Pleistocene did not reach the present level, and lakes along the middle Yangtze valley had not taken their current shape until about 6000 BP.

According to the latest archaeological evidence, the earliest sites with Neolithic features such as domestication, pottery, and sedentism are all located in the hilly flanks of the middle Yangtze valley. Carbonized rice remains suggesting domestication at the site of Xianrendong and Diaotonghuan in Wannian County of Jiangxi Province [11–14] and Yuchanyan in Dao County of Hunan Province, correspondingly [15,16]. The Yangtze valley in the subtropical zone is a marginal habitat for the growth of wild rice. Up to the Neolithic, there is a continuous sequence from the Shangshan Culture [17], the Kuahuqiao Culture [18–20] to the Hemudu Culture [21] in the lower Yangtze valley; a sequence similar to that from the Pengtoushan Culture [22,23] and the Chenbeixi Culture [24,25], the Lower Zaoshi Culture [26] to the Daxi Culture [27] in the middle Yangtze valley. It is noted that the PNT can be traced back to about 20,000 years ago, when pottery, polished stone tools and intensive utilization of wild rice appeared in several sites, and this process can be extended to around 8000 BP. This shows the process from the origin of agriculture to the formation of full-developed Neolithic societies is gradual. After the PNT, the Neolithic sites diffused from hilly flanks to floodplains. This tendency is attested by the site distribution from the Pengtoushan Culture to the Daxi Culture. At the same time, agriculture diffuses westward to mountainous zones, including the Sichuan Basin, in which the pottery assemblage of the Daxi Culture has been found. Agriculture in the other parts, confined by natural conditions, was adopted much later. For example, rice farming in Guizhou appeared in the second millennium BC.

The above is the overall situation that we have known up to this time, but the PNT is not clear for the marginal regions of agricultural origin. This transition, which happened worldwide, not only relates to environmental changes at the end of Pleistocene, but also more commonly implies the change in human adaptation—that is, origins of agriculture in various regions of the world. This change brought challenges and opportunities for the hunter-gatherers of noncentral regions in at least three aspects: (1) formation of a new niche with climate amelioration; (2) agriculture or food production as a new technology of cultural adaptation; (3) released constraints from ecological conditions and the inertia of the foraging lifeway inherited from their ancestors. It is generally assumed that hunter-gatherers in the marginal region were not influenced by the changes of the PNT including those from environmental and cultural backgrounds, but continuously kept foraging—their existing lifeways—and thus were forced to accept agriculture later during processes of immigration or other stressors. This assumption has been partially rejected by the results of recently analyzed archaeological records. In fact, foragers could have changed as early as in the central region for agricultural origins, but the form of change could have been different.

2. Background of the RHBF Technique

The ridged hammer bipolar flaking (RHBF) technique has been recognized in the eastern region of the upper reaches of the Yangtze River since the discovery and excavation of a series of sites from the Upper Paleolithic to the early Neolithic in recent years. These new archaeological records provide a precious opportunity to learn about the cultural diversity of this region.

The research relating to the RHBF technique began in the 1970s in China. At that time, RHBF flakes were first found at a prehistoric cave site in Guizhou Province [28]. Cao [28] did a preliminary study on these lithic artifacts, and then he named this technique after creating simple experiments. In the report, he described the process in detail: a flat and round boulder used for raw material was brought into contact with the anvil held at a slant, the experimenter firmly held the boulder with one hand and also grasped a hammer in other hand, and then smashed one end of the boulder with the sharp edge of the hammer. Generally, by hitting four or five times, a flake would fall off the boulder. Furthermore, in the following decades, the RHBF technique was reported in many other provinces (Figure 1).

Although the number of sites found in China is not small (Figure 1), due to insufficient understanding of RHBF technology and dating problems in previous studies, a systematic study on this lithic technology has not been conducted. Firstly, the definition is still not clear; the early definition was made based on flakes discovered in the cave site and the experiments we mentioned above. However, some following studies reported this type of technique or flakes as the Yangtze technique [9] or zero-platform flakes [29]. Secondly, although there are some experiments for the RHBF technique in previous studies, they almost all emphasize the special characteristics of the RHBF technique or other artifacts unearthed along with RHBF flakes [10]. However, the technical process and function of RHBF technology, particularly, the meaning of cultural adaptation, have not been deeply explored. Thirdly, chronology of the RHBF technique is still a puzzle. On the one hand, some typical sites located in the wilderness area of South China are difficult to get reliable absolute date, since it is impossible to correlate lithic findings to stratigraphy. On the other hand, some sites with RHBF flakes are rescue excavations, thus the dating obtained in Southwest China is speculative and unreliable.

Figure 1. The distribution of Ridged-hammer Bipolar Flaking (RHBF) artifacts, China (1:32000000). 1. Ganlanba; 2. Tangzigou; 3. Laolongdong; 4. Baiwei; 5. Gexinqiao; 6. Banang; 7. Bailan; 8. Dongjian; 9. Nalao; 10. Balie; 11. Dingbang; 12. Weibo; 13. Donglong; 14. Yandongpo; 15. Yajiao; 16. FuyuanDahe; 17. Latuo; 18. QujingDahe; 19. Zhangkoudong; 20. Nayong Zhangkoudong; 21. Baiyanjiao; 22. Xiaohuidong; 23. Maomaodong; 24. Chuandong; 25. Ma'anshan; 26. Xinhuang Paleolithic site; 27. Xianrenqiao; 28. Yandunbao; 29. Chibaling; 30. Ranjialukou; 31. Zhongbazi; 32. Qiaojiayuanzi; 33. Ou'tang Neolithic site; 34. Honghuatao; 35. Zhujiatai; 36. Pengtoushan; 37. Xiaomaicheng; 38. Heilongtan; 39. Nanyangwucheng; 40. Zhaojiahuagougoukou; 41. Sifangdong; 42. Chuanfandong; 43. Lingfengdong; 44. Huangditong; 45. Changbin Culture.

The RHBF technique is long-standing in prehistory. The earliest existence of this technique could date back to the early Paleolithic, at a cave site called Wanshouyan in Fujian Province, Southeast China [30]. Additionally, sites in southwest China supposedly have evidence that the RHBF technique lasted until the Shang (c. 1570–1045 BC) and Zhou (c. 1045–256 BC) dynasties [31]. Archaeological sites associated with the RHBF technique are also widely distributed—Southeastern, Northern, and Southwestern China (Figure 1). However, we argue that this specific technique is mainly distributed in Southwest China, especially in the region from the Xiajiang region to the Yunnan–Guizhou Plateau (part of which extends into the territory of Guangxi) (Figure 1). The present paper argues that most sites in other regions of China cannot be classified into RHBF technique sites for three reasons:

(1) Some artifacts lack the important features of RHBF flakes, e.g., greater width than length, and a linearly notched striking point. For example, although the site in Southeast China, Wanshouyan [30] consists of some RHBF flakes, most of them are only in a similar shape of RHBF flakes and lack the diagnostic features of this technique.

(2) The sample size of RHBF artifacts is small, including only a few flakes discovered from sites of North China.

(3) Sporadic discoveries in Southeast and North China have not reported such technique in detail, which makes it difficult to figure out whether they are truly representative of the RHBF technique. Compared with sites in North and Southeast China, sites in the southwest that report RHBF techniques are numerous and mostly cluster in the Xiajiang region. Some of them are also much more clearly described in reports, which provides more compelling contextual evidence to prove the primary existence of RHBF techniques in Southwest China.

3. Materials and Methods

In this study, the Guanzhou site (Figure 2) was examined based on the observations of lithics and archaeological experiments conducted in 2018. The Guanzhou site, a typical RHBF site, was excavated by the Hubei Provincial Institute of Cultural Relics and Archaeology in 2016. The site is now located on an eyot or small island in the Yangtze River (Figure 2), which was once connected to the shore about the time before the Ming and Qing dynasties (Ming dynasty tombs were also found on the island, but no later tombs were found). This landform was then cut off from the right bank of the river by a flood and became an eyot. The 2016 excavation unearthed more than 8000 pieces of lithic, of which more than 2000 pieces were RHBF flakes. Among the most exciting and uncommon artifacts were hammers, anvils, and lithic cores—from the RHBF technique—were also discovered, which have provided us with valuable materials for reconstructing this unique lithic technology. Carbon-14 dating indicates this site could date back to around 8500 years ago [32], the early Neolithic. Similarly, sites with rich RHBF flakes are also known as the Cibingzhou site (the report of this site has not published yet, however, we have visited this site during summer, 2019 and observed the unearthed artifacts) and Maomaodong site in Guizhou [33,34], as well as the Tangzigou site in Yunnan [35]. During the PNT, the popularity of RHBF technology seems to reach its peak, and was widely distributed in China, especially in the Xiajiang region [36]. In 2019, we inspected the excavated materials at relevant sites in Guizhou, Yunnan, Guangxi, Hunan and other provinces, and confirmed the existence of RHBF techniques in these areas.

(a) (b)

Figure 2. The location of Guanzhou site. (**a**), 1:8,000,000; (**b**) 1:500,000.

The Paleolithic–Neolithic transition occurred worldwide, at least in Eurasia, but different regions do not share the common pattern as can be seen in representative local materials. The eastern region of the upper reaches of the Yangtze River in this study are represented by RHBF technology. In other words, the rise and fall of RHBF technology can better reflect the process of the PNT period in this area, and not only can this show changes in livelihoods, to some extent it, can also show changes in social organization. The method adopted in this article was used to combine the sample analysis and archaeological experiments with ethnoarchaeology to reconstruct the RHBF technology and the function of RHBF flakes, and on this basis, to further understand their meaning in terms of division of labor. Furthermore, we will also draw on the evidence of ethnoarchaeology to support our reasoning.

In the past ten years, we have carried out a series of experimental archaeological studies of stone tools [36–38], which helped us to analyze lithic production and form a set of effective methodologies. Our study can be divided into four levels as follows (Figure 3).

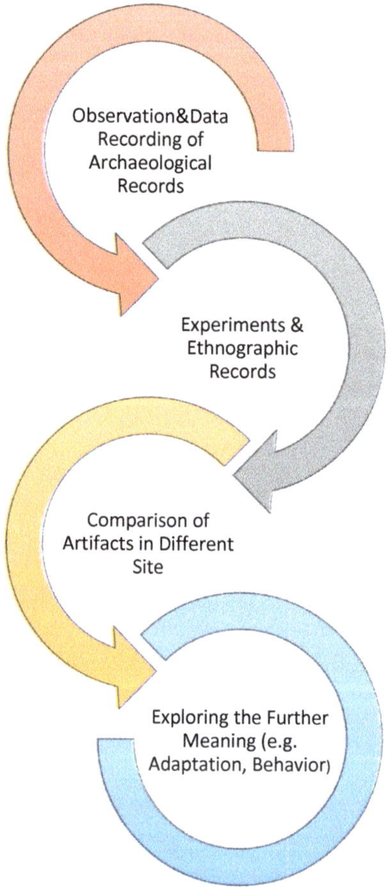

Figure 3. Four steps in lithic analysis.

Firstly, observations of the characteristics of the archaeological artifacts, including the classification and description of artifacts, the observation of the use-wear, the analysis of the technology process, and the preliminary assumption of the artifact's function. Secondly, experimental archaeology and looking for other reference information to support our assumption, such as ethnographic reportage. Thirdly, comparison of artifacts from different sites by considering the local ecology to further verify our judgment. Finally, further explore the meaning of cultural adaptation behind the artifacts or technique. This methodology has been used to analyze the relationship between the prehistoric stone industry in western Liaoning and primitive agriculture; since stone tools are usually the most complete archaeological materials, the analysis of lithics can provide an important reference to understand cultural as well as evolutionary significance in different periods. Therefore, the following section will use this methodology to explore the significance of RHBF technology during the PNT. In this study, the reasoning and analysis will be developed in the following steps:

(1) The observation and data recording of lithic artifacts (c. 8000 pieces) from the Guanzhou site, including length, width, height, weight, types of use-wear, position and level, breakage and so on.
(2) Replication experiments conducted on the bank of the Yangtze River adjacent to the Guanzhou site. Seven experimenters participate: 2 young male students, 2 young female students, 2 experienced males who have conducted lithic experiments for several years, and 1 less experienced male.
(3) Comparative experiments with other lithic techniques, such as throwing, anvil and hammer techniques.
(4) A functional experiment aimed to examine the usage of RHBF flakes.

Among the four steps, two key issues need to be addressed. On the one hand, the experimental location is important. Before the flood separated the Guanzhou site from the bank, the experiment location was connected to the site at about 250 m. The close distance between the experimental location and the site means that they share a similar ecological environment, both locales are covered with gravel, and the raw materials are the same as those found in the site so that equivalent ecology and raw material as the Guanzhou site could be easily accessible. On the other hand, comparative experimentation is significant because it is a way to figure out if other lithic techniques can produce the same flakes—although the RHBF technique is regarded as a throwing or anvil technique in previous studies. Since the comparative experiment is not closely related to topic of this study, the result of the comparative experiment will be not shown in the following section.

4. Result

4.1. Observation

Our investigation found that the most distinctive product of RHBF technology is the RHBF flake (Figure 4). From morphological observations, the ventral surface is very flat (Figure 5), with almost no bulb of percussion or striking platform, so it was also termed a zero-platform flake by Li [29]. The striking points are usually in a linearly concave shape, which is obviously different from the striking points of other techniques, which usually produce obvious bulbs of percussion. According to the measurement statistics of 2167 RHBF specimens unearthed from the Guanzhou site, such flakes often have a greater width than length (Figure 6), with an average length of 76.2 mm and a median of 73.2 mm, an average width of 96.2 mm as well as a median of 93.2 mm. The average thickness was 15.8 mm, and the median was 15.8 mm (Figure 6). If specimens with a thickness greater than 30mm are excluded, which are also extremely large specimens, then the thickness variation range of the stone pieces is very consistent [36]. In other words, RHBF flakes are the products of a highly consistent operation. Furthermore, by collecting data from 14 sites' reports, we found that the average length of RHBF flakes is 82.1 mm, the average width is 95.6 mm, and the average thickness is 19.1 mm, which is similar to what we get in Guanzhou site, although the amount of data from other sites is not large: only data from 22 pieces' flakes are accessible [36].

Based on these observations, some assumptions regarding the function of RHBF flakes can be developed. The edges of these flakes are thin, sharp and the design and the shape are very suitable for cutting activities, so we assume such flakes can be directly used as scrapers or knives. However, due to the obvious feature of thickness, RHBF flakes are not suitable for cutting large or tough items, such as thicker branches of trees. In addition, because larger RHBF flakes also have advantages in terms of weight, they can also be used directly for chopping activities—an alternative hypothesis relating to function. Additionally, the coexisting lithic tools found at the Guanzhou site may rationalize our assumptions. More than 400 scrapers and 200 chopping tools were found in the site. The majority of scrapers are made from RHBF flakes, the so-called repairing wear of which are mostly caused by use. Additionally, most scrapers have no obvious breakage: the edge is smooth, which also indicates the processing object should be not so hard or even soft. As to chopping tools, most of them are also made from RHBF flakes, but are much larger and heavier when compared with the RHBF flakes of scrapers. Furthermore, RHBF cores are important materials for making chopping tools.

Figure 4. RHBF flakes in Guanzhou site. 1. 2016SGT1818⑤11; 2. 2016SGT1818⑥415; 3. 2016SGT1818⑥505 4. 2016SGT1818⑥657; 5. 2016SGT1818⑥751; 6. 2016SGT1819⑦82; 7. 2016SGT1919⑤7; 8. 2016SGT2019⑤5; 9. 2016SGT2019⑤11; 10. 2016SGT2119④60-1.

(**a**) (**b**)

Figure 5. RHBF flakes in Guanzhou site: (**a**) ventral side, (**b**) dorsal side.

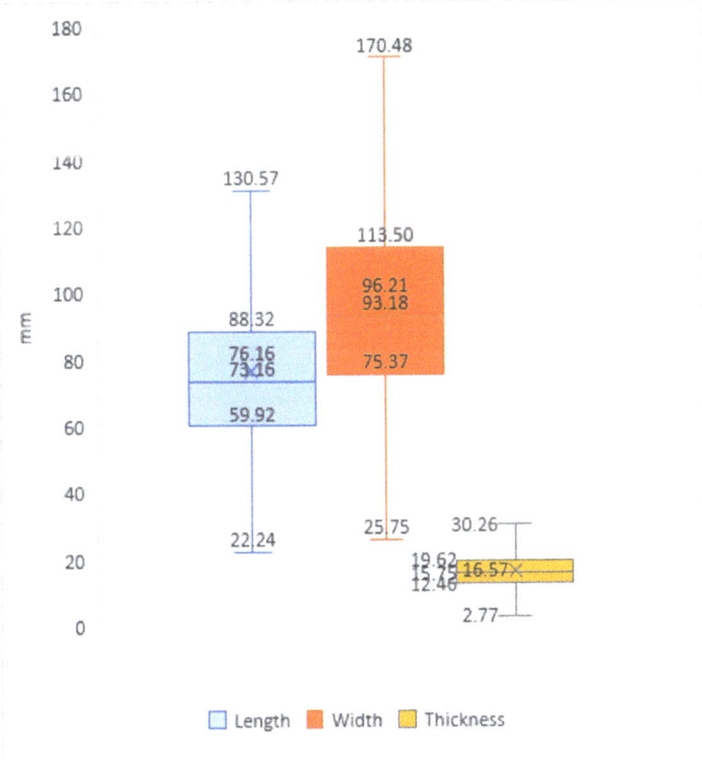

Figure 6. Length, width and thickness of RHBF flakes.

The observation of lithic artifacts in the Guanzhou site also sheds light on the process of the RHBF technique. For example, several hammers found with RHBF flakes and tools show significantly different features from hammers found in other archaeological sites. Hammers unearthed in Guanzhou site are cobbles in oval shape with a great size (c. 238 mm) and weight (c.1408 g) that cannot be easily held by one hand. In addition, striped collision marks are mainly concentrated in the middle of flat side (Figure 7) which is unique and different from the point-like use-wear of other hammer techniques. Because there is no use-wear at both ends of the hammer, we may assume the way to use it is holding it with both hands to produce flakes. Besides, the striped use-wear (Figure 7) in the middle of RHBF hammers also indicates that hammers struck an extremely narrow and linear platform, in other words, the raw material was once placed at an angle so that only the edge of the raw material could be struck with the hammer.

4.2. Replication Experiment

In the replication experiment, the key was how to place the raw material securely. Several methods have been tried, including hand-held, stone-placement and mud-placement methods. The hand-held method is extremely dangerous and unlikely to be chosen because it is easy to get injured and holding a large hammer in one hand also makes it difficult to produce enough percussion power. Besides, it is also impossible to use other small stones to support the raw material, since the raw material cannot be steadily fixed in such a situation. Later, wet mud was considered for the location of the site, as this is easily accessible at the riverbank. The wet mud experiment is efficient in terms of supporting the raw material and tilting it slightly on the flat anvil (Figure 8).

Figure 7. RHBF hammer (2016SGT2019⑦109).

Figure 8. Wet mud fixing.

After exploring a method for setting up raw material, the replication process becomes much easier to figure out. Firstly, in order to reduce the counterforce from the anvil, the anvil is buried in the sand. Secondly, the operator lifts the stone hammer with both hands over his or her head (Figure 9), and then strikes the raw material with the help of gravitational potential energy. Generally, a flake can be produced after two or three times of striking. The characteristics of the experimental flakes are exactly the same as those in the site, suggesting the application of this method in the experiment might be the one employed by the person who produced the flakes during the Pleistocene. Experiments revealed two unique features of the RHBF technique. One shows that the RHBF technique is highly reliant on raw material. In the experiment, each boulder can usually only produce one flake, and the reason is because of a lack of a perfect flat form with regard to the boulder; in other words, once the core is split the position of the striking point becomes too thin to produce more flakes. Therefore, such technique is wasteful unless the raw materials are abundant, and such an uneconomical technique also indicates a local raw material source. The other feature of the RHBF technique concerns gender and labor division. In the experiment, female operators also participated and found that such technique is simple enough that even students who have no experience in lithic production can successfully produce RHBF flakes. Thus, we may assume that since such technique is not restricted to one gender, and so labor division could be further explored.

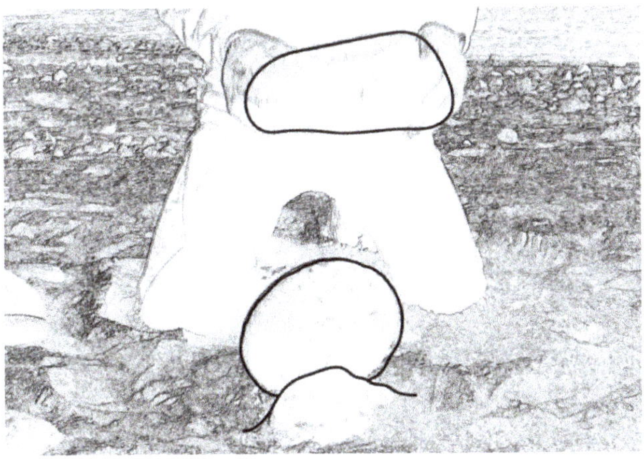

Figure 9. RHBF technique process.

4.3. Functional Experiments

Along with RHBF flakes and tools, a high number of fish bones were also discovered in the Guanzhou site. Taking the site location into consideration, an assumption relating to fish processing was examined in functional experiments. Experimental research shows that RHBF flakes are extremely effective for scaling (Figure 10). On the one hand, RHBF flakes are knife-like, and are thin enough to be scrapers. On the other hand, the edge of RHBF flakes is not sharp as knife, so when scaling, it cannot cut into the flesh. Furthermore, RHBF flakes are also efficient in cutting off parts of fish such as the belly, tail and head. We processed four grass carp of about 2000 g in total, which took an average of 10 min. After a period of practice, and with some supplemental tools with sharp points that could be used to pierce the belly, the processing time can be greatly reduced. Moreover, the site faunal remains show that fish caught here are mainly herring, with large individuals that can reach tens of kilograms. Thus, effective scrapers must be used to handle such large catches. The RHBF flakes are usually regular in shape (Figure 4), knife-like, and the sharpness of the blade edge is also appropriate for fish processing. Interestingly, in the Guanzhou site, fish bones are concentrated in the earlier stratum which contains a lower amount of RHBF flakes compared to the upper stratum. The inconsistency between the amount of fish bones and flakes in the strata suggests changes in fish processing that could be linked with mobility patterns: in the earlier period, people tended to conduct fishing, processing, and consumption on site; at the later stage, because of a greater need of food, people carried their catch to the settlement to process and consume, which may reflect logistical mobility, a possibly more specialized society, and labor division.

Figure 10. RHBF flake used for fish scaling.

5. Discussion

5.1. Aquatic Adaptation in East Asia and the RHBF Technique

Although aquatic resources are abundant and relatively accessible, most models of human evolution have all but ignored the role of aquatic or maritime adaptations during the earlier stage of human history [39]. Besides farming, the use of aquatic resources is also an important adaptation during the PNT, especially in East Asia. By research on the organic residue of Neolithic pottery from Sakhalin Island in the Russian Far East, Gibbs [40] state that early pottery on Sakhalin was used for the processing of aquatic species, and that its adoption formed part of a wider Neolithic transition involving the reorientation of local lifeways towards the exploitation of marine resources, and other evidence in Russian far east also supports the aquatic use in hunter-gatherer groups [39]. In Japan and Korean, there is also evidence showing that people in the upper Paleolithic and PNT have access to the aquatic resources [41–43]. In China, sites in the PNT indicate there is a potential adaptation strategy relating to aquatic resources, and based on what we mentioned above, the RHBF technique has a potential in aquatic utilization.

The RHBF technique is mainly distributed in the Xiajiang region, where a mountain and gorge area are located in the middle reaches of the Yangtze River. Unlike the plains area of the Yangtze River, the Xiajiang region is not a center for agricultural origins. However, the rise of the RHBF technique may indicate there is an alternative pathway in the cultural evolution during the PNT. Cibingzhou, another site found recently in Guizhou Province, could also support our hypothesis of an alternative strategy in the PNT. Interestingly, the Cibingzhou site shares many similarities with the Guanzhou site. First, the artifacts found were from the same period, the PNT. Secondly, the Cibingzhou site is also located on an eyot that was originally connected to the shore. Additionally, the river was diverted recently, and erosion cut off the eyot. Finally, and most importantly, a large number of RHBF artifacts have been unearthed in the Cibingzhou site. Thus, judging from current records, sites with many RHBF flakes are mostly found near rivers or other water resources, and appeared around 8000 years ago. Because there is a possibility that the RHBF technique is suitable to process fish, based on the functional experiment, it is reasonable to assume that sites near water and containing RHBF artifacts may reflect dependence upon aquatic resources such as fishing. The use of aquatic resources could be an option for people living in areas that are marginal for agriculture. Moreover, fishing, as an adaptive way of enhanced utilization for hunter-gatherers in nonagricultural areas can relieve subsistence pressure on growing populations. Socially complex societies can potentially form if aquatic resources are rich enough, such as those known on the northwest coast of North America.

However, most results of aquatic utilization are based on pottery or the residue on pottery, and so a lithic analysis may not be the direct way to draw a conclusion on whether the RHBF technique is used for aquatic resources. Thus, three important things need to be addressed in future research. Firstly, a use-wear lithic analysis should be conducted to test if RHBF flakes are used to process fish. Secondly, pottery analysis and residue research should be taken into consideration. Thirdly, complete research on the fish bones discovered in the Guanzhou site should provide more information relating to aquatic utilization, which cannot be ignored.

5.2. The Adaptation Patterns in Different Regions during the PNT Period

During the PNT, there were several patterns in China for responding to the ecological and social changes: wheat agriculture in North China and rice agriculture that originated on the plains of the middle and lower reaches of the Yangtze River. In the region south of the Five Ridges, this change manifested as "low-level food production" [44], which is a mixture of rhizome planting, aquatic resource utilization, and hunting and gathering. In the northeast region, the end of the Pleistocene saw an increase in effective precipitation, and the use of aquatic resources began around 12,000 years ago. A series of sites appeared along rivers and lakes, forming the so-called "fishing and hunting of the Neolithic Age" [45]. The Yanshan Great Wall zone in North China is located in an ecologically interlaced zone, and local livelihood patterns throughout the Neolithic era fluctuated between hunting and gathering and agriculture along with climate change [46]. There has been a long-term lack of investigation regarding adaptation strategies in the grassland area in the PNT. However, recent investigations and excavations of a series of sites of Yumin culture (the materials and reports on the Yumin cuture have not been published, the information here is based on some of our investigation and observations of artifacts of Yumin culture) have explored the changing processes of the grassland area: an adaptive method that maintains obvious seasonal utilization, which can be distinguished between winter and summer. In the southwestern region, there is a partial overlap with the area discussed in this article: the continuous hunting and gathering and maintaining usage of stone tools, which is also called "post-Paleolithic era". However, what our studies have addressed here is the diversity of this post-Paleolithic era, since at least three adaptation strategies can be distinguished. With the advent of the Holocene, RHBF technology brought a significant differentiation of cultural adaptation. From the current materials, at least three different cultural adaptation methods include the following. The first occured in the upper and middle reaches of the Yangtze River transition zone where some population groups moved toward the center of the origin of agriculture, then these groups continued to spread into the plains and reclaim new farmland. Such spreading trends can be seen from the distribution features of the sites from the PNT to the late Neolithic period. Later agriculture dispersed from the edge of the basin to the hilly area in the early Neolithic period, and then entered the Jianghan Plain and Dongting Lake Plain in the middle and late stages. The second adaptation strategy was found in both the transition zone and some areas of in eastern upper reaches of the Yangtze River, where it is more convenient to use aquatic resources. During that time, hunter-gatherers began to emphasize the use of aquatic resources, a strategy which produces results similar to agricultural production in that it can also result in a higher population density and social complexity. The third adaptation strategy took place in areas that are neither suitable for early agriculture nor for the use of aquatic resources, where population groups continued their previous hunting and gathering living style. However, there is a possibility that other methods were used to control population growth, which could ensure that the population density was maintained at lower levels than the local carrying capacity if the population groups here only relied on hunting–gathering for living.

In the second strategy, groups took full advantage of the RHBF technique, which endured for a prolonged time and reached a peak in these areas; even as late as the Shang and Zhou dynasties, this technique could still be found. However, since the development of agriculture played the main role in the first strategy, RHBF technology was eventually replaced by the development of tools for processing crops, such as grinding tools. In the third strategy, RHBF technology was used to a certain extent, but not though a monopolistic technique, as other bipolar or percussion techniques still existed concurrently.

The PNT happened in Asia. For example, pottery has been discovered from 10,000 years ago in most Asian areas. Since pottery is a product of intensification [47], to some extent representing the changing trends of residency, we can be sure that the transition was a universal event, although the transition varied in time, space and strategies. Particularly, the adaptation in Southwest China, especially from the perspective of lithics, calls for further exploration since it shares some strategies with all of Asia. For example, lithic assemblages from southern Yunnan are similar to the Hoa Binh culture, an important archaeological assemblage found in Northern Vietnam [48–50]. This reflects a special adaption strategy that occurred in Southeast Asia.

5.3. *The Possibility of Labor Division*

Similar lithic artifacts are found around the world, especially in North America. The Shoshone of western interior North America used similar tools called teshoa [11]. Such records provide a clue about the process of the RHBF manufacturing technique as well as tool function. A teshoa is kind of stone knife used by Shoshone women to process hide and meat. "When a woman needed a knife for butchering and skin dressing, she selected two quartzite cobblestones from the nearest stream; she then used one as an anvil to knock a large spall from the other. When the edge of the teshoa had done its job, it could be discarded, and a new one could quickly be made whenever it was needed" [51]. Such ethnology records show not only the process of teshoa manufacturing but are also consistent with our assumption regarding labor division. Among our experiments, results show that the RHBF technique is compatible with producing large amounts of flakes in a short time: students without much experience in lithic production can make RHBF flakes effectively because such techniques do not require percussion strength and experiences; thus people of all ages and genders can accomplish it easily.

However, the ethnographic records relating to labor division in China, especially the Xiajiang region, are still unclear, and thus, research on teshoa can only provide us with a possible hypothesis regarding labor division for the RHBF technique. In addition, based on our experiments, the results indicate that the female students and/or young people could conduct such a technique, and all were efficient in using flakes to process fish. Nevertheless, further research is needed to test such a hypothesis.

6. Conclusions and Further Suggestions

The RHBF technique is representative of the Xiajiang region during the PNT, which indicates a possible emphasis upon aquatic resources. This hypothesis is also supported by our experiments and ethnographic records. Additionally, the RHBF technique also shows diversity in the adaptation behaviors of the PNT in terms of not being restricted to agriculture—people living in marginal areas for agriculture adopted much more diverse and complex hunting and gathering strategies to support their subsistence. However, such an alternative adaptation and RHBF technique still need further exploration, since there is no systemic analysis of fish bones unearthed in the sites of the Xiajiang region. The lack of local ethnographic records calls for further consideration of the gendered division of labor. Furthermore, the compelling topic of complex hunter-gatherers also needs to be further addressed: is intensification through the use of aquatic resources a pathway to a complex society in China [52], and is division of labor a possible signal? More archaeological materials, especially evidence for sedentism, will be discovered and studied in marginal areas for agricultural origins, particularly in the Xiajiang region.

Author Contributions: R.L.: data collection and analysis, writing—original draft preparation; H.L.: resources; S.C.: supervision, funding acquisition. All authors have read and agreed to the published version of the manuscript.

Funding: This research was funded by the Fundamental Research Funds for the Central Universities, and the Research Funds of Renmin University of China, grant number 19XNL010.

Acknowledgments: Thanks to the Hubei Provincial Institute of Cultural Relics and Archaeology, who provided the material and resources for us. Additionally, thanks for the funding support from the Fundamental Research Funds for the Central Universities, and the Research Funds of Renmin University of China. In the experiment section, thanks to people who helped us, Yixin Zhou, Bing Dong, Jun Zhao, Xiaohan Wang, Chanyang Ye, Jiyuan He and Siqi Fan, we could not have finished this project in time without your attendance and help. Additionally, thanks for the English editing from Pei-lin Yu.

Conflicts of Interest. The authors declare no conflict of interest

References

1. Winterhalder, B.; Kennett, D.J. Behavioral ecology and the transition from hunting and gathering to agriculture. In *Behavioral Ecology and the Transition to Agriculture*; University of California Press: Berkeley, CA, USA, 2006; Volume 1.
2. Childe, V.G. *Man Makes Himself*; Watts & Co.: London, UK, 1936.
3. Bellwood, P.; Gamble, C.; Le Blanc, S.A.; Pluciennik, M.; Richards, M.; Terrell, J.E. First Farmers: The Origins of Agricultural Societies, by Peter Bellwood. Malden (MA): Blackwell, 2005; ISBN 0-631-20565-9 hardback£ 60; ISBN 0-631-20566-7 paperback£ 17.99, xix+ 360 pp., 59 figs., 3 tables. *Camb. Archaeol. J.* **2007**, *17*, 87. [CrossRef]
4. Crawford, G.W.; Shen, C. The origins of rice agriculture: Recent progress in East Asia. *Antiquity* **1998**, *72*, 858–866. [CrossRef]
5. Eshed, V.; Gopher, A.; Pinhasi, R.; Hershkovitz, I. Paleopathology and the origin of agriculture in the Levant. *Am. J. Phys. Anthropol.* **2010**, *143*, 121–133. [CrossRef] [PubMed]
6. Dalou, A.Y.A.; ElSerogy, A.M.; Al-Shorman, A.A.; Alrousan, M.; Khwaileh, A. Bioarchaeology, conservation and display a 16k-human skeleton, Jordan. *Mediterr. Archaeol. Archaeom.* **2017**, *17*, 251–263.
7. Shqiarat, M. History and archaeology of water management in Jordan through ages. *Sci. Cult.* **2019**, *5*, 41–54.
8. Peroschi, M.E.; Mailland, F.; Mailland, I.; Anati, E. Nahal Karkom, a pre-pottery neolithic B site in the southern negev, Israel: Archaeological report and archaeometric analyses. *Mediterr. Archaeol. Archaeom.* **2018**, *18*, 169–193.
9. Gao, X.; Wei, Q.; Li, G. A report of the third excavation of the Ranjialukou Paleolithic site. *Acta Anthr. Sin* **2008**, *27*, 1–11. (In Chinese)
10. Zhang, C.; Lin, C. The analysis of lithic artfacts from Honghuatao Site, Neolithic site. *Relics South* **2008**, 68–77. (In Chinese)
11. Wan, Z.; Ma, Z.; Yang, X.; Zhang, C.; Zhou, G. Starch residuals from shell tools from sites of Xianrendong and Diaotonghuan and its implication for paleoclimate. *Si Ji Yan Jiu Quat. Sci.* **2012**, *32*, 256–263. (In Chinese)
12. Peng, S. A breakthrough in prehistoric archaeology, Jiangxi Province—A discussion of the discory of Xianrendong and Diaotonghuan site. *Nong Ye Kao Gu Agric. Archaeol.* **1998**, *1*, 389–392. (In Chinese)
13. Peng, S.; Zhou, G. Xianrendong and Diaotonguan site-a case study of transition pattern from Paleolithic to Neolithic. *Agric. Archaeol.* **2004**, 29–39. (In Chinese)
14. Li, J. The second report of Xianrendong site, Jiangxi Province. *Cult. Relics* **1976**, *12*. (In Chinese)
15. Liu, Z. The great historic implication of the discovery of Yunchanyan site. *Agric. Archaeol.* **1996**, *3*, 95–98. (In Chinese)
16. Zhang, W.; Yuan, J. The preliminary study on the ancient rice excavated from Yuanchanyan, Daoxian, Hunan Province. *Acta Agron. Sin.* **1998**, *4*, 416–420. (In Chinese)
17. Zheng, Y.; Jiang, L. Rice remains in Shangshan site and its implication. *Archaeology* **2007**, *9*, 19–25. (In Chinese)
18. Yang, X.; Jiang, L. Starch analysis to reveal the prehistoric diet of people from Kuahuqiao site, Zhejiang Province. *Chin. Sci. Bull.* **2010**, *55*, 596–602. (In Chinese)
19. Wang, X. The discussion of Kuhuqiao Culture. *Sichuan Cult. Relics* **2006**, *4*, 28–35. (In Chinese)
20. Zheng, Y.; Jiang, L.; Zheng, J. Study on the remains of ancient rice from Kuohuqiao site in Zhejiang Province. *Chin. J. Rice Sci.* **2004**, *18*, 119–124. (In Chinese)
21. Wang, H. Hemudu site and Hemudu Culture. *Southeast Cult.* **2000**, *7*, 15–22. (In Chinese)
22. Pei, A. The discussion of rice remains in Pentoushan culture and prehistoric rice agriculture in China. *Agric. Archaeol.* **1989**, *2*, 102–108. (In Chinese)
23. Pei, A. The second discussion of rice remains in Pentoushan culture and prehistoric rice agriculture in China. *Agric. Archaeol.* **1998**, *1*, 193–202. (In Chinese)

24. Yang, Q. The discussion of Chengbeixi Culture. *Southeast Cult.* **1991**, 206–212. (In Chinese)
25. Lin, C.; Hu, H. Chengbeixi, Pengtoushan culture and the rice agriculture in China. *Agric. Archaeol.* **1993**, 116–122. (In Chinese)
26. He, J. Agriculture of earlier stage in the Dongting lake area. *Huaxia Archaeol.* **1997**, 25–29. (In Chinese)
27. Zhang, Z. The analysis of Daxi culture. *Jianghan Archaeol.* **1982**, 66–71. (In Chinese)
28. Cao, Z.; Zhou, G.; Cheng, S.; Hui, X.; Jiu, D.; Qi, S.; Hua, W.; Zhi, Y. Xiaohuidong site, a Paleolithic site in Guizhou Province. *Vertebr Palasiat* **1978**, *16*, 67–74. (In Chinese)
29. Li, Y. Note on the classification of the platform of the flake. *Acta Anthr. Sin.* **1984**, 253–258. (In Chinese)
30. Li, J.; Chen, Z.; Yu, S. Lingfengdong-the first lower paleolithic site found in Fujian Province. *Acta Anthr. Sin.* **2001**, *20*, 247–255. (In Chinese)
31. Li, Y.; Yu, X.; Hou, Y. On the ridged-hammer bipolar flaking and its products in the Three Gorges region. In Proceedings of the Tenth Annual Meeting of the Chinese Society of Vertebrate Paleontology, Xiamen, China, 20–22 November 2006; Wei, D., Ed.; China Ocean Press: Beijing, China, 2006; pp. 261–272. (In Chinese).
32. Neolithic Site in Hubei Province: Guanzhou. In *The Important Archaeological Discoveries in China, 2017*; Cultural Relics Publishing House: Beijing, China, 2018; ISBN 978-7-5010-5565-4. (In Chinese)
33. Cao, Z. On the Paleolithc artifacts from Maomaodong-the rock shelter site, Guizhou Province. *Vertebr. Palasiat.* **1982**, *2*, 155–168. (In Chinese)
34. Cao, Z. The preliminary study of bone tools and antler spades from the rock shelter site of Maomaodong. *Acta Anthr. Sin.* **1982**, *1*, 36–42. (In Chinese)
35. Yang, F. Several questions regarding Tangzigou Culture. *J. Yunnan Univ. Natl.* **1990**, 82–88. (In Chinese)
36. Liu, R. Study on Ridged-Hammer Bipolar Flaking Technique. Master's Thesis, Renmin University of China, Beijing, China, 2019. (In Chinese).
37. Chen, S. The questions regarding "seeing humans through materials" in archaeogical studies. *Archaeology* **2014**, *10*, 61–67. (In Chinese)
38. Chen, S.; Yang, K.; Li, B.; Zhu, Y.; Ji, P. A study of the stone tools from the Haminmanga. *Acta Anthr. Sin.* **2016**, 522–536. (In Chinese)
39. Erlandson, J.M. The archaeology of aquatic adaptations: Paradigms for a new millennium. *J. Archaeol. Res.* **2001**, *9*, 287–350. [CrossRef]
40. Gibbs, K.; Isaksson, S.; Craig, O.E.; Lucquin, A.; Grishchenko, V.A.; Farrell, T.F.; Thompson, A.; Kato, H.; Vasilevski, A.A.; Jordan, P.D. Exploring the emergence of an 'Aquatic' Neolithic in the Russian Far East: Organic residue analysis of early hunter-gatherer pottery from Sakhalin Island. *Antiquity* **2017**, *91*, 1484–1500. [CrossRef]
41. Shoda, S.; Lucquin, A.; Yanshina, O.; Kuzmin, Y.; Shevkomud, I.; Medvedev, V.; Derevianko, E.; Lapshina, Z.; Craig, O.E.; Jordan, P. Late Glacial hunter-gatherer pottery in the Russian Far East: Indications of diversity in origins and use. *Quat. Sci. Rev.* **2020**, *229*, 106124. [CrossRef]
42. Otsuka, Y. The background of transitions in microblade industries in Hokkaido, northern Japan. *Quat. Int.* **2017**, *442*, 33–42. [CrossRef]
43. Jun, C.P.; Yi, S.; Lee, S.J. Palynological implication of Holocene vegetation and environment in Pyeongtaek wetland, Korea. *Quat. Int.* **2010**, *227*, 68–74. [CrossRef]
44. Smith, B.D. Low-level food production. *J. Archaeol. Res.* **2001**, *9*, 1–43. [CrossRef]
45. Zhao, B. Mode of subsistence in Neolithic in Nengjiang River area. *Archaeology* **2007**, *11*, 55–61.
46. Chen, S. *Prehistoric Modernization: Studies on the Agricultural Origin from the Perspective of Cultural Ecology*; Science Press of China: Beijing, China, 2013. (In Chinese)
47. Rice, P.M. On the origins of pottery. *J. Archaeol. Method Theory* **1999**, *6*, 1–54. [CrossRef]
48. Benecke, N. Seasonal dating of fish remains from the Hoabinhian Site Can-Cave (Ky Son District, Hoa Binh Province, Vietnam). *Archaeofauna* **1995**.
49. Chung, T.N. Technological and typological characteristics of the chipped tools in the Hoa Binh culture. *Vietnam Soc. Sci.* **1994**, *5*, 46–55.
50. Phūkhačhōn, S. *Archaeological Research of the Hoabinhian Culture, or, Technocomplex and Its Comparison with Ethnoarchaeology of the Phi Tong Luang, a Hunter-Gatherer Group of Thailand*; Archaeologica Venatoria; Institut für Urgeschichte der Universität Tübingen: Tübingen, Germany, 1988; ISBN 978-3-921618-27-1.

51. Eyman, F. *The Teshoa, a Shoshonean Woman's Knife, a Study of American Indian Chopper Industries*; Milton, P., Ed.; Pennsylvania Archaeologist Society: Casper, WY, USA, 1968; p. 34.
52. Zangrando, A.F. Is fishing intensification a direct route to hunter-gatherer complexity? A case study from the Beagle Channel region (Tierra del Fuego, southern South America). *World Archaeol.* **2009**, *41*, 589–608. [CrossRef]

quaternary

MDPI

Article

Modeling Incipient Use of Neolithic Cultigens by Taiwanese Foragers: Perspectives from Niche Variation Theory, the Prey Choice Model, and the Ideal Free Distribution

Pei-Lin Yu

Department of Anthropology, Boise State University, 1910 University Dr., Boise, ID 83725, USA; pei-linyu@boisestate.edu

Received: 3 June 2020; Accepted: 14 August 2020; Published: 7 September 2020

Abstract: The earliest evidence for agriculture in Taiwan dates to about 6000 years BP and indicates that farmer-gardeners from Southeast China migrated across the Taiwan Strait. However, little is known about the adaptive interactions between Taiwanese foragers and Neolithic Chinese farmers during the transition. This paper considers theoretical expectations from human behavioral ecology based models and macroecological patterning from Binford's hunter-gatherer database to scope the range of responses of native populations to invasive dispersal. Niche variation theory and invasion theory predict that the foraging niche breadths will narrow for native populations and morphologically similar dispersing populations. The encounter contingent prey choice model indicates that groups under resource depression from depleted high-ranked resources will increasingly take low-ranked resources upon encounter. The ideal free distribution with Allee effects categorizes settlement into highly ranked habitats selected on the basis of encounter rates with preferred prey, with niche construction potentially contributing to an upswing in some highly ranked prey species. In coastal plain habitats preferred by farming immigrants, interactions and competition either reduced encounter rates with high ranked prey or were offset by benefits to habitat from the creation of a mosaic of succession ecozones by cultivation. Aquatic-focused foragers were eventually constrained to broaden subsistence by increasing the harvest of low ranked resources, then mobility-compatible Neolithic cultigens were added as a niche-broadening tactic. In locations less suitable for farming, fishing and hunting continued as primary foraging tactics for centuries after Neolithic arrivals. The paper concludes with a set of evidence-based archaeological expectations derived from these models.

Keywords: Paleolithic Taiwan; aquatic-focused foraging; Neolithic Taiwan; agricultural adoption; niche variation theory; invasion theory; prey choice model

---

1. Introduction

" ... these vast and largely uncharted [transitional] regions are not just uninhabited territory crossed on the way to an anticipated agricultural destination by evolutionary interstates without exits. They are, to the contrary, regions occupied by diverse, vibrant, and successful human societies that have developed stable, long-term economic solutions that combine low-level reliance on domesticates with continued use and management of wild species" [1].

Research into the encounters, interactions, and knowledge transfer between hunting and gathering peoples and agriculturalists began with collection of field data in the 1960s and 70s. Hunting and gathering societies were observed using a range of interaction strategies from unilinear agricultural adoption to sustained symbiosis founded on forest product exchange [2]. Interaction studies continued with cultural history investigations and pattern recognition in the 1980s and 90s, and moved toward the synthesis of the ways that variability in relationships can lead to greater understanding of human social change [2,3]. This observed variability is germane to understanding evolutionary relationships, and framing archaeological expectations for major transitions such as the Neolithic.

The topic of agricultural adoptions by island hunter-gatherers offers perspectives on Neolithic onsets, particularly regarding the influences of aquatic resources, immigration, demographic pressure, constraints on mobility, and migrations [4,5]. Taiwan's biogeographic position at the east–west nexus between Southeast China and the islands of the Pacific Ocean and the north–south nexus of the Japanese archipelago and SE Asia places it at the intersection of diverse climates and ecosystems and the center of a cultural and adaptive interaction sphere rooted in the Paleolithic. As such, the island is an important case study with relevance for the Neolithicization of coastal and island Southeast Asia and eventual Austronesian expansions across Oceania [6–13]. The Neolithic onset of Taiwan is notable for its late date of circa (c.) 6000 BP compared to mainland China [10–14]. Neolithic cultures proceeded to expand, become regionally differentiated, and complex at a rapid rate: by 4500 BP, agriculture-based societies had radiated to most of the island, growing millet, rice, tubers, and tree crops; Middle Neolithic settlements are marked by social ranking, megaliths, production of prestige goods, and sizeable structures and features [15–17].

The nature of the initial interaction with Paleolithic groups, which established the foundations for the decline of foraging and replacement by agriculture, is not yet well understood, although most Taiwanese researchers agree that Neolithic cultures are not directly descended from Upper Paleolithic peoples [16]. Where might we find reference information about the adaptive responses of native Paleolithic foraging peoples to invasive dispersals of Neolithic immigrants? A promising area is the evolutionary study of various non-human species in order to better understand the process and its outcomes. With specific reference to subsistence changes associated with invasive dispersals, behavioral ecologists have investigated and described important parameters that influence the decision making of migrating individuals as well as the attendant costs and benefits to dispersers [18,19]. The adaptive response of native populations—who are 'dispersed upon'—is less well studied, but of particular evolutionary interest in the case of Neolithic transitions. The decision-making process for groups facing unfamiliar crop foods and cultivation techniques, their assessment of opportunity costs to mobility needed for food, technological materials, social networks, and other resources [2,20–22], and the ensuing societal and environmental feedback loops make up the localized and evolutionarily significant initial conditions of food producing societies.

In order to derive a working hypothesis about the adoption of agricultural practices by Paleolithic hunter-gatherers experiencing invasive migrations, this paper describes Taiwan's unique geographic position and biodiversity, summarizes current environmental and archaeological knowledge about the transitional Paleolithic to Neolithic transitional period, models Late Paleolithic foraging niche breadth using Lewis Binford's database of hunting and gathering peoples, and assesses the implications of niche variation theory, the prey choice model, and the ideal free distribution for mutually influential adaptive responses of Paleolithic Taiwanese foragers and arriving Neolithic Chinese farmer-fishers. Predicting variations in this process has important implications for settlement, the adoption of cultivation, and social integration by native foragers. The paper concludes with proposed lines of evidence for hypothesis testing with the archaeological record.

A note on spelling: for Chinese words, the Pinyin system of romanization is used except in the case where the Wade–Giles system is used for longstanding and familiar names. For Taiwanese indigenous words, Romanized versions that are commonly accepted in the scholarly literature are used. For site dates, BP or BC are used as presented in cited sources.

2. Human Behavioral Ecology and Subsistence Strategies of Invasive Dispersals

Human behavioral ecology (HBE) has developed an array of intellectual tools to explore the dynamic relationships between organisms and the environment using an evolutionary framework [23–28]. In comparison and complementary with macroecological models that explore broad patterns in extant data, concept models are derived from the principals of natural selection and general theory of behavior. These models seek to link behavior to expected material outcomes [28]. The development of testable hypotheses regarding the cumulative effects of decisions and broader patterns of evolutionary change aspire to bridge these two scales of analysis [26,29].

From early studies in optimal foraging, HBE is now exploring the socioecological contexts influencing why individuals modify their environments, move to new places, begin to produce food, and enter into cooperative or coercive social relations [26]. The ideal free distribution and its corollary, the ideal despotic variant [30] provide an explanatory framework to predict when individuals will disperse or migrate to a new habitat, based on density-dependent changes in the suitability of the habitats available to them [31]. This model ranks habitats by their quality, assessed by the fitness of the occupants. The Ideal Free Distribution (IFD) assumes that a dispersing organism will decide to settle in the preferred option of several habitats that differ in suitability (e.g., availability of resources, exposure to hazards, and other characteristics). Desirable habitats are occupied chronologically in rank order, presuming no competition from others. The ideal despotic variant (IDV) highlights immigration and differential access to resources, in which groups seeking to settle in already-occupied highly ranked habitats are deterred from entry by defense and are thereby forced to settle in lower-ranked locations. The habitat implications for this model can be provocative: the Allee effect [32] can result in positive feedback effects of population density in contrast to density dependent resource depression. Fruitful assessments of the Allee effect include North America and Australia, where certain prey types predicted by the Prey Choice Model (PCM) would be pursued on encounter and increase population size with increasing land use intensity [33–35]. The ideal free distribution has also been examined in island habitats, immigration, and settlement in Polynesia [31], California [36], and will be the focus of a future paper concerning Taiwan [37].

Returning to the impacts of dispersing farmers on foraging, the niche concept is foundational to both behavioral ecology and macroevolutionary approaches. Binford's [20] definition of niche is the dynamic articulation between the capabilities of a group of 'actors' and the organizational properties of their habitat: the types and quantities of resources as well as how, when, and why actors use them. A foraging niche is positioned along a continuum from specialized to generalized, in which generalized actors feed on a greater proportion of the species available in the habitat, and specialized actors feed on a smaller proportion [20,38]. Foraging niches are in a constant state of flux through adaptive responses to changes in prey or resource type, frequency, and distribution. Organizational properties of the habitat influence the distribution and frequency of prey. The process by which individual foragers take into account energetic costs and returns when encountering potential prey (including plants) is described in the encounter-contingent prey choice model (PCM) [39,40]. With a stipulated goal of maximizing net energy returns, the order of ranking for prey is determined by the evaluation of which items to take upon encounter [26,41–45]. The prey choice model predicts that, all other things being equal, high ranked resources will always be taken upon encounter, but lower ranked resources will only be taken if they will increase the overall return rate (expressed as energy/time). Under conditions of resource stress, the reduced encounter rate for highly ranked resources increases the likelihood that lower ranked resources will be taken. The implication is that scarcity of high ranked resources tends to encourage a broader dietary niche as energetic costs of the search eventually overwhelm the advantages of being selective.

During invasive dispersals of morphologically and functionally similar populations into occupied territories, the overlap between niches that results from similar prey choices increases competition, depresses prey, and in turn narrows the niche breadth of both dispersers and natives [46]. With regard to invasive immigrations, understanding the initial contexts and processes of competitive interactions provides an important benchmark for predicting the long-term consequences of invasion. Foraging niche breadth is one such factor: a broad or generalized foraging niche is considered to be key in determining the success of dispersers [46], but less is known about the effects upon pre-existing niche breadths of native species. Generalized native foraging niches would provide a wider range of prey options under competitive pressure; therefore identifying conditions for niche generalization versus specialization is helpful in characterizing niche breadth.

Niche variation theory describes the process of 'ecological release' by which foraging niche breadth broadens and diversifies when inter-species competition is reduced [18,47–50]. This mode of niche expansion increases foraging variation among individuals (or groups) rather than individual niche breadth. Subsequently, within-population variation is a key requirement for frequency-dependent interactions that may drive evolutionary diversification and influence the population dynamics and ecological interactions of species [18] (Figure 1).

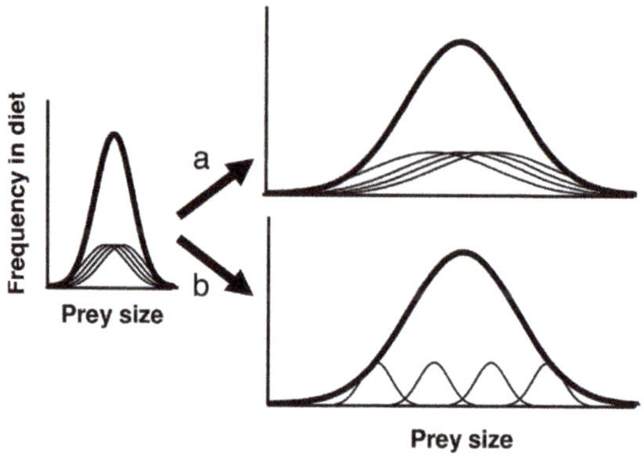

Figure 1. Two alternative patterns of population niche expansion. A population that uses a narrow range of prey sizes can increase its population niche breadth (bold lines) in two ways. (**a**) Inter-group niche breadths (thin lines) can expand, resulting in no increased niche variation among individuals. (**b**) Individual group niche breadths can remain limited, whereas groups diverge from each other to increase among-group variation (after [18], used by permission of lead author).

The obverse of ecological release—competitive pressure—is predicted by "classic invasion theory" (*sensu* [38]). A successful invading group may displace similar native species by overlapping their niche through direct competition for preferred prey. A generalist diet is a common trait of successful invaders, allowing them to exploit diverse resources in novel environments, competitively exclude native species, and breed successfully [51]. Under this theory, the decline of native species following an invasion may be a consequence of direct interspecific competition for prey e.g., [52,53]. If species occupy a larger niche in the absence of interspecific competition [50,54] under conditions of competition, both dispersing and native species are expected to occupy smaller niches than their allopatric equivalents [55]. In a potential variant, increased competition for resources could necessitate an increase in dietary niche breadth in order to maintain energy requirements [56] (also see below). Empirical evidence for different taxa indicates that interspecific competition can cause the niche widths of consumers to increase or decrease, depending on the context [57], with generalists having more capacity to increase dietary niche breadth. Niche breadths of both dispersing and native competitors is an important contextual factor in determining invasion success and predicting the impacts of invaders on biodiversity and foraging niche (e.g., [46,58]). The human foraging niche is highly flexible and may expand, contract, or shift; under some circumstances, the diets of different groups may functionally be as disparate as separate species. All other things being equal, in the early phases of invasion, native populations with a broad dietary niche are expected to have more adaptive options when facing competition.

Behavioral ecology applications have largely focused on in situ domestication or local adoptions of plant foods by landlocked human groups including intensified use of wild species and the domestication process for early cultigens [3,41–45,59]. Gremillion and Piperno [42] employed the prey choice concept to predict the manipulation of low-ranked plants that eventually became fully domesticated seed crops. In the case of invasive competition, the prey choice model predicts that dispersing competitors' appropriation of a higher proportion of available resources would incentivize natives to consider including unfamiliar, and (at least initially) low-ranked prey in the diet. Of course, there are limits to the expansion of niche by incorporating lower ranked prey items; eventually low ranked options are exhausted. The niche construction theory [59,60] employs Elton's [38] concept of an organism or population that modifies habitats to enhance energetic rates of return. With regard to plant domestication, there is room for debate regarding the onset of niche construction as a maximizing strategy during times of resource abundance (*sensu* [60,61]) or a Malthusian response to resource depression [59]. The latter approach is germane to resource depressions such as those we expect to follow invasive dispersals. A niche construction tactic like agriculture could initially offset direct competition for forager resources by providing an additional food source, but as foraging complements most traditional farming diets, it is expected that highly preferred wild resources would become depleted within a certain radius of sedentized farming communities. Therefore, a correlation to the working hypothesis for the Taiwanese Neolithic is that niche construction practices of immigrating farmers reduced mobility and limited foraging to the surrounding area. Preferred wild prey would rapidly become depleted in these areas, reducing encounter rates for native foragers. The harvest of lower ranked prey would broaden the subsistence niche for both the forager and farmer in these locations, and when the limits of low ranked prey are reached, foragers would be faced with two alternatives: adopt agricultural practices or migrate. This process is predicted for areas where farmers settled initially in direct overlap with forager territories.

A hypothesis may now be developed for the Neolithic transition of Taiwan: *If the assumptions of PCM hold*, with increasing land use intensification, high ranked wild prey should rapidly become depleted in areas of high suitability, reducing encounter rates for native foragers. The harvest of lower ranked prey should broaden the subsistence niche for both foragers and farmers in these locations, and when the limits of low ranked prey are reached, foragers would be faced with two alternatives: adopt agricultural practices or migrate. This process is predicted for areas where farmers directly overlapped with forager territories.

If assumptions of an IFD with Allee effects hold, we expect to see an initial increase in habitat suitability associated with habitat modification resulting from farming and/or intensified foraging, and associated increases in encounter rates for certain prey types that do better with heterogeneous vegetation and successional communities. In this modified expectation of niche variation theory, ecological release and niche variation resulted not from decreased population densities and resource demand, but manipulation of habitat characteristics that increase encounter rates.

The utility of model-based approaches for formulating explanations about evolutionary transitions of human subsistence is mediated by the scientists' ability to make testable inferences about the foraging niches of individuals and societies that we cannot observe directly. The archaeological remains of past dietary behaviors serve as the basis—albeit imperfectly preserved—for inferences regarding prey choice and variations in niche breadth (Figure 2; see also [26]).

Assembling reference knowledge about processes that are germane to, but independent of, past phenomena of interest can complement limited evidentiary data to help guide archaeological research questions [20,29,62–64]. Behavioral ecology concept models can be informative about the dynamics at the level of the individual, and assist in the development of hypotheses about systems that can be assessed with archaeological data. The case of the Taiwan Neolithic offers an opportunity to examine reference information about foraging niches from a mountainous island habitat.

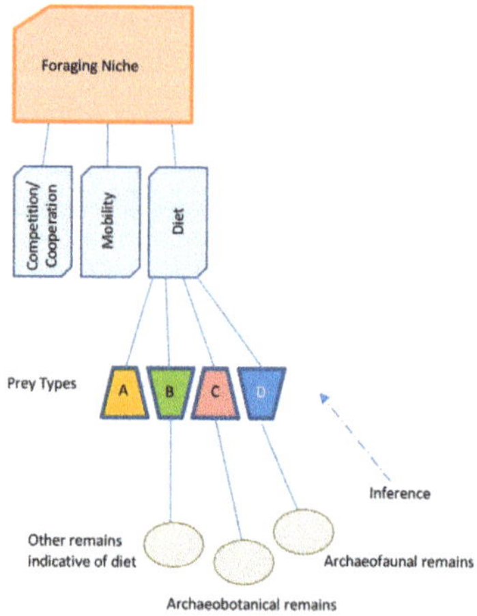

Figure 2. Relationship of niche to diet, prey types, and archaeological inference.

3. Materials and Methods

To scope a test for the effectiveness of the prey choice model versus the ideal free distribution with Allee effects, the paper first modeled Paleolithic foraging to create a ranking of wild prey types preferred by foragers. This is followed by a summary of the salient characteristics of Taiwan's first farmers to estimate the influence of farming combined with prey choice on habitat preferences.

3.1. Modelling Taiwan's Paleolithic Niche Breadth: Paleoenvironment and Subsistence

The unique geographic position, climate, and landforms of Taiwan have had a profound influence on ecosystem characteristics and contexts since the Last Glacial Maximum. The island lies at the intersection of multiple climatic zones and ecosystems, conditioning for highly diverse habitats and access by boat from the west, north, and south. Along the eastern margin of the Asian continental shelf and the western edge of the Pacific Ocean and between longitude 120° E and 122° E and latitude 21° N and 25° N, Taiwan's geographic area is about 36,000 km^2, with a north–south distance of c. 394 km and east–west distance of about 140 km. Okinawa lies to the north, China lies c. 130 km to the west, Luzon Island about 250 km to the south, and the Pacific ocean to the east. Dozens of surrounding islands include the Penghu (Pescadores) group, Ludao (Green Island), and Lanyu (Orchid Island). The coastline is about 1566 km around the perimeter, and together with 129 sizeable rivers and more than 100 wetlands and estuaries, provides abundant aquatic habitats. Taiwan's western coastal plain is crossed by large meandering rivers, grading upward through dense deciduous forest in mountain footslopes to sub-tropical cypress and other evergreens at mid altitudes to an alpine oak scrub and steppe. Before urban development, estuaries and extensive low gradient wetlands lined the northern and western coasts, with a large lake basin where the Taipei metropolis now sits. On the east coast, most piedmonts are narrowly constricted and limited flat areas are mostly alluvial outflow deposits from steep mountain rivers.

Taiwan's landmass is more than 80% mountainous with tectonic activity manifested by volcanism, earthquakes, and frequent landslides. These steep mountains create diverse microhabitats and influence the mobility and foraging patterns of game species. The central mountain range forms the spine of the island, with more than 100 peaks higher than 3000 m above sea level, although the island is only c. 144 km at its widest point. The smaller coastal mountain range to the east is about 140 km long with peaks of about 1000–1500 m. A long, narrow inland valley runs north–south between the central and coastal mountain ranges, and three major lake basins are located in the north, west, and west-center (as Figure 3).

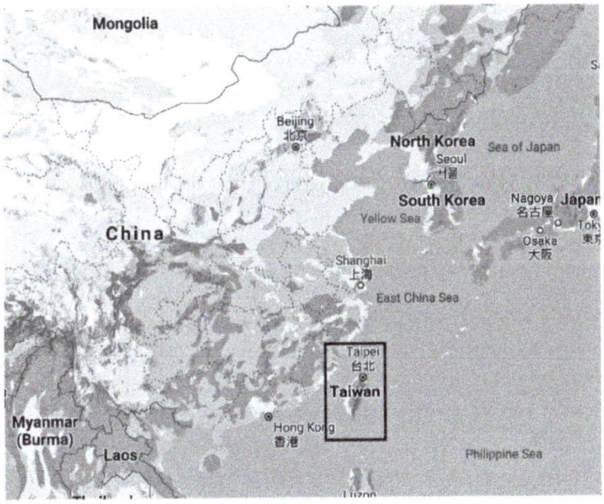

Figure 3. Location map of Taiwan.

Currently, the Tropic of Cancer runs through Taiwan, creating a sub-tropical climate merging into tropical conditions in the south. Sea breezes, typhoons, and monsoons keep modern temperatures warm, with annual averages between 21.5 and 24.5 Centigrade. Annual humidity is high with average rainfall between 1820 and 2720 mm. High seasonal variability in rainfall comes from the rainshadow effect and seasonal variation in wind direction; the central west coast is driest and the northeast coast, the wettest.

The wild vegetation of Taiwan is adapted to the temperature and moisture gradient [65], and the two main forest ecoregions currently include subtropical evergreen forests across most of the island and tropical monsoon rainforests on the southern tip. Temperate and sub-alpine communities exist in high mountain settings. During the middle Holocene optimum, pollen records indicate that during the time period around 6100 BP, warm climates allowed tropical evergreen forests to grow across most of the island [66]. Overall, climates have undergone major fluctuations in the past 12,000 years, and during the Tainan Paleoenvironmental Period (early to middle Holocene, 10–5 kyr BP), temperatures were warmer overall. A variety of geological indicators show that melting glaciers and local tectonics led to a rise in regional sea levels, reaching a maximum at around 6500 BP [67] and inundating the margins of Taiwan's western coastal plain, the Penghu Islands and others [68], and the Southeast Chinese coast across the strait [69]. Environmental fluctuations during the subsequent Nankuanli Geological Period (c. 5000–3000 BP) included actively fluctuating sea levels [67,70]. Marine transgressions and retreats created a dynamic estuarine and littoral environment, particularly in the low relief western and southern plains of Taiwan and the SE China coast. Chen et al. [68] noted localized major marine transgressions at c. 4700 BP, and at the same time, uplifts in the southwest central mountains caused aggradation and drainage formation and movements. Estuarine and marshland resources were abundant, albeit shifting, during this period of environmental upheaval on both sides of the Taiwan Strait.

These warmer climates likely created wind patterns, rainfall, seasonal variance, and ecosystems that are different from today's. Rising sea levels would have altered land availability, and vegetation communities would have resembled those currently on the southern tip of the island. However, ecosystems likely varied according to elevation, slope, and aspect, much like today.

Largely isolated from mainland influences since the late Pleistocene, Taiwan's highly diverse flora and fauna include unique island forms such as the Taiwanese cypress (*Taiwania cryptomerioides*), bay tree (*Machilus kusanoi*), Formosan bear (*Ursus thibetanus formosanus*), clouded leopard (*Neofelis nebulosa brachyura*), boar (*Sus scrofa taivanus*), Sika deer (*Cervus nippon taioanensis*), sambar (*Rusa unicolor swinhoei*), serow (*Capricornis swinhoei*), muntjac (*Muntiacus reevesi*), pangolin (*Manis pentadactyla pentadactyla*), rock macaque (*Macaca cyclopis*), and giant flying squirrel (*Petaurista alborufus*). Many large bodied species now limited to forested upland reserved areas were broadly distributed in the piedmont and forested interface with the coastal plain [71]. The island is home to a wide variety of endemic and migratory birds, and the marine fishery includes diverse and abundant shellfish, crab, and shallow water fish populations on the west and south coasts. A productive pelagic fishery lies off the east coast, and edible kelp and algae are common. The pre-20th century freshwater fishery was also diverse including shellfish, crayfish, and catadromous (sea-spawning migratory) and landlocked species including Formosan salmon (*Oncorhynchus masou formosanus*).

3.2. Paleolithic Environment and Cultural Adaptations

Archaeological preservation of the Paleolithic to Neolithic transitional period is complicated by marine transgressions, mountain building, and sediment aggradation during the significant period between 14,000 and 6000 BP [5,67]. Currently, Taiwan's earlier Paleolithic cultural sites, termed Changbinian, have been found primarily in caves along the rugged east coast and date to c. 20,000–6000 BP [72–77]. The type site of Baxiandong (Pah-Hsien-Tung) is made up of a beach caves located in rugged terrain that was somewhat inland during the early period of occupation (Figure 4). The lithic assemblage of Baxiandong lower levels includes unifacially flaked choppers and cobble flake tools (c. 20,000–25,000 BP) [75] made of sandstone pebbles sourced from nearby beaches [77] that were knapped on-site based on refit data [72]. Changbin core/flake pebble technological systems show affinities with the Paleolithic cultures of the Ryuku Islands and the Philippines [77]. Faunal remains mostly represent fish, shellfish, and cervids [75,77].

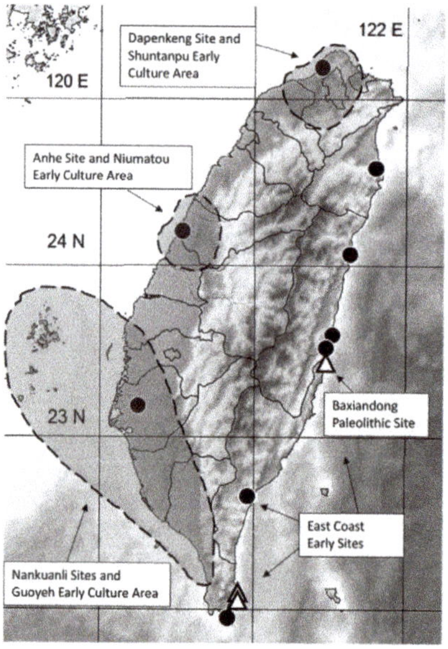

Figure 4. Map of Taiwan's Paleolithic and early Neolithic culture areas and sites (Adapted from [78]).

From Changbinian origins, regional foraging variants emerged at about 15,000–5000 BP. These cultures are termed as the Persistent Upper Paleolithic because they retain Paleolithic technological systems and lack ceramics [79]. The upper Paleolithic levels of Baxiandong include mostly unretouched flakes, but some pieces show evidence of faceted striking platforms. Tool types include side scrapers, points, knives, and notched scrapers made of higher quality raw material such as quartzite and chert [72]. Hearth-like ash lenses have also been found [72,77]. Other Persistent Upper Paleolithic sites are located in the northwest near Taipei [73,74,80] including the northwest Wangxing culture, which terminates at c. 6000 BP and is described as an adaptation particular to this region [81,82].

Persistent Upper Paleolithic sites are also found the central west near Miaoli and Taichung [80–83], in the southwest near Tainan [74], on the southern tip at O-luan-pi, Xiaoma, and Longkeng [84], the southeast near Pingtung [85,86] and southeast near Taitung [87]. In particular, the time period of 6500–5000 BP is well-represented in the south. Organic preservation is rare at most Paleolithic sites, mostly consisting of fragmented animal and fish bone as well as shell [ibid], which suggests a mixed subsistence strategy. The presence of horn implements and needles from large body size game at Baxiandong indicates that deer and other large prey were taken regularly [75–77]. Overall, archaeological evidence indicates that Taiwan's Paleolithic cultures were evolutionarily dynamic and adapted to diverse habitats, with some emphasis on aquatic resources. From Asian cobble technologies after the Pleistocene to Holocene transition, Persistent Upper Paleolithic cultures shifted to smaller lithics of higher quality raw materials, retouch techniques, and an increase in bone tools as well as regional variation in subsistence, technology, and settlement.

3.3. *Reference Information about Foraging from the Binford Hunter-Gatherer Database*

Lewis R. Binford's database of hunting and gathering peoples [20,88] includes environmental data for climate, topography, soils, and primary (plant) and secondary (animal) biomass from individual weather stations around the globe. Johnson [29] demonstrates that habitat data can be used to project hunter-gatherer subsistence, social organization, and demography where foraging societies no longer reside, based on regressions of variables of climate data and living hunter-gatherer societies. The Binford database is used to calculate projections for foraging behaviors based on a global sample of data from 339 ethnographically documented hunting and gathering societies that is geo-referenced to climatic and environmental parameters. This allows for informed estimates of foraging behaviors where detailed ethnographic information is not available such as Taiwan. The projection for Taiwan's anticipated foraging niche breadth is based on modeled foraging subsistence diversity and climate and environmental data from 27 Taiwanese weather stations (Figure 5).

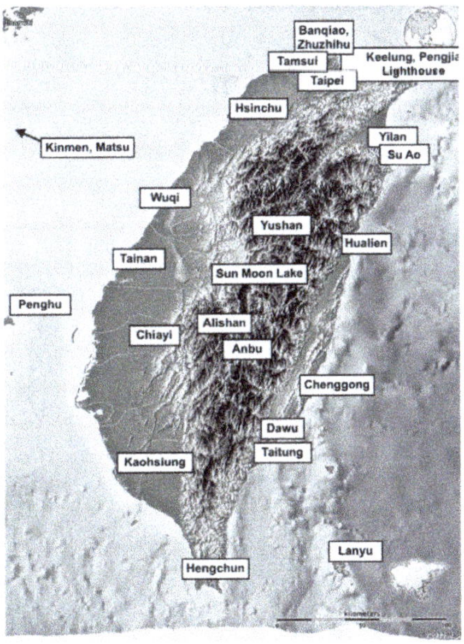

Figure 5. Taiwanese weather stations where temperature, rainfall, soil, and vegetation data were used to derive environmental and ethnographic projections [20,88].

Environmental calculations were derived from average temperature and rainfall data from the early 2000s as well as soil type, vegetation, modeled ungulate biomass, elevation, and proximity to the coast. Thus, the Binford projections for Taiwan do not represent accurate environmental conditions of the middle Holocene. Rather, this information offers a frame of reference for foraging lifeways that are plausible for a warm, mountainous island setting and can be assessed using more direct lines of evidence such as paleoclimatic records and archaeological information. This paper selected the following subsistence related variables from the Binford database [88].

a. WHUNTP/WGATHP/WFISHP: Expected percentage of hunting, gathering, and fishing/aquatic resources derived from ethnographically known hunter-gatherer groups who reside in habitats with similar environmental characteristics.
b. SUBSPE: Ordinal classification of projected foraging subsistence specialty. 1 = primarily hunting or dependence on terrestrial animals; 2 = gathering or dependence on terrestrial plants; 3 = primarily fishing or dependence on aquatic prey.
c. SUBDIV: An index of foraging diversity or 'evenness' that is calculated using Simpson's diversity index (see below for details of calculation). Expressed as a decimal value between 0 and 1 in which values approaching 1 indicate less evenness or more specialization.

4. Results

4.1. *Foraging Subsistence Projections and Preferred Wild Prey in Taiwan*

To estimate foraging subsistence focus, SUBSPE, an ordinal variable is calculated for predominant expected dependence upon foraging modes available in the habitat. Options include terrestrial animal focus (1 = hunting), terrestrial plants (2 = gathering), and aquatic species (3 = fishing/shellfish). For example, the SUBSPE for the weather station located near the Tamsui estuary in the north estimated more than 50% dependence on aquatic food resources. Therefore, SUBSPE was designated as "3" for Tamsui. Overall, the Binford projections for Taiwan foraging subsistence indicate that, if Paleolithic hunter gatherers were behaving as they did in similar habitats during the ethnographic present, most groups would be focused on fishing/aquatic resources (Figure 6). The only exceptions are in the central or coastal mountains, which are predicted to rely mostly on hunting.

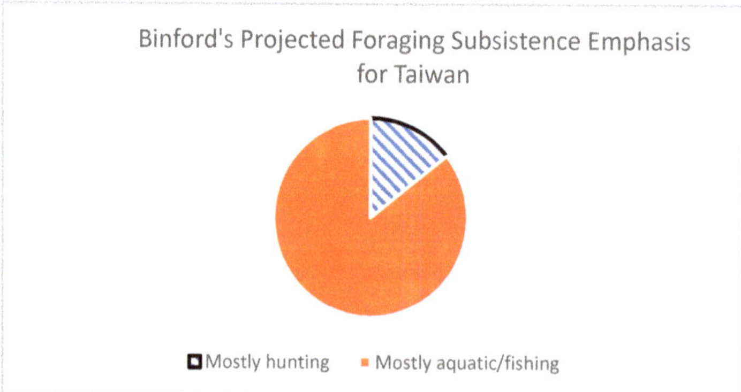

Figure 6. Binford's expected subsistence emphasis (SUBSPE) for 27 Taiwanese weather stations.

In his worldwide analysis of hunter gatherer subsistence patterns, Binford [20] noted several interesting generalizations about aquatic-focused foraging. Usually, early stage niche broadening is accomplished by including more terrestrial plant foods (often those requiring additional processing). Binford notes that the majority of aquatically focused hunter-gatherer groups worldwide are situated in forest plant communities that generally offer little accessible plant food for hunter-gatherers [20]:

"...Human dependence upon aquatic resources occurs either as a supplement to a plant-based strategy or as the primary strategy in environments that prohibit plant-based subsistence options. The shift to a highly aquatic focus happens only in cases where edible plants are rare". The low frequency of edible endemic Taiwanese forest species [89] and corresponding predominance of non-native plant food species in traditional diets of Taiwan's indigenous Amis and Paiwan peoples [90,91] are consistent with Binford's generalization: specialization in aquatic foraging occurred early in Taiwan due to the lack of terrestrial plant food species. Most edible plant foods endemic to the island that are known today include ferns, fungi, fruit, and algae or kelp.

Binford's projections for Taiwan's foraging subsistence focus by percentage (Figure 7) indicate that the distance to the coast is a predictor of aquatic specialization, with mountain sites such as Alishan, Sun Moon Lake, and Yushan comprising the highest terrestrial hunting focus. The outlier Chiayi is distant from the coast, but the high aquatic focus could be influenced by a local river. Notably, no Taiwanese site was modeled to have greater than 37% dependence on terrestrial plant foods.

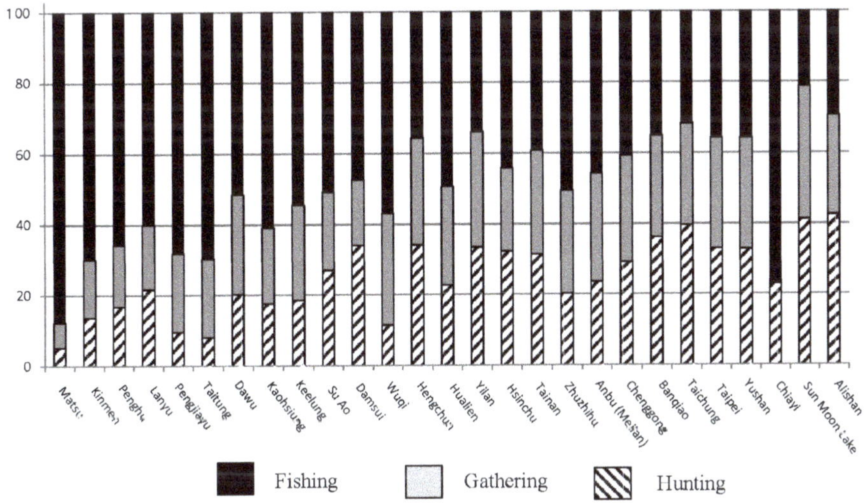

Figure 7. Binford's estimated percent contribution to foraging subsistence by distance to coast.

Foraging niche breadth can be characterized as the degree of evenness between relative reliance on terrestrial animals, plants, and aquatic foods: more evenness indicates a generalized and broad niche. To make an estimate for Taiwan foraging, Simpson's diversity index [92] (originally calculated for biodiversity) is used to measure evenness, defined as the relative degree of reliance of hunting, gathering, and aquatic subsistence for each location. This calculation is made using variables that reflect 'packed' population conditions that are referable to those of the ethnographic present, >9.1 persons/km^2 (Binford's WHUNTP, WGATHP, and WFISHP variables) [20,88]. The diversity index is expressed as $D = \Sigma[n/N]^2$ in which n = the total number of organisms of a particular species and N = the total number of organisms of all species. The equation used in this calculation is: Foraging diversity [D] = $(\text{WHUNTP} + \text{WGATHP} + \text{WFISHP})^2/10{,}000$. The value D ranges between 0 and 1, where 0 represents infinite diversity and 1, no diversity. Thus a value closer to 1 indicates un-evenness and a high degree of specialization. Thus, the percentage dependence on either hunting, gathering, or fishing is substituted for the number of organisms in the diet, and the result is divided by 10,000 for ease of viewing.

Figure 8 indicates that foraging is specialized upon fish/aquatics near the coast, then shifts inland to mountain-oriented hunting. Again, the outlier is Chiayi, possibly influenced by a large freshwater river with migratory catadromous fish. Overall, the Taiwan foraging diversity projection reflects Binford's expectation that niche specialization may occur in groups that are primarily dependent upon aquatic resources under nonpacked conditions, where there are few or no wealth or rank differentials [20]. Furthermore, Paleolithic population densities are likely to have increased over time as a consequence of the predictability of aquatic resources: Binford's database of known foragers indicates that aquatic-dependent groups have higher population densities (17.9 persons per 100 square kilometers) than plant gatherers (10.5 persons) or hunters (3.4 persons) [20].

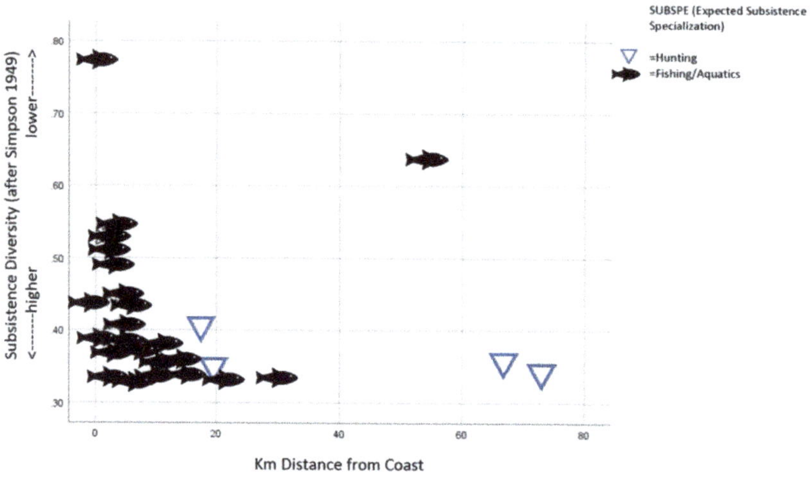

Figure 8. Binford's [20,88] estimated diversity of foraging subsistence for Taiwan, assessed by distance from the coast, calculated using Simpson's diversity index [92].

Binford's projections from global hunting and gathering data have interesting implications for the evolutionary pathway of foraging in Paleolithic Taiwan. During the early period of c. 20–10,000 BP, foraging groups moving to the island, most likely over land, added aquatic resources as a niche broadening tactic. With the onset of Holocene conditions and population infilling from c. 10,000–6000 BP, foragers became specialized in aquatic foods. Shellfish and other nearshore/wetland species required little effort to harvest and could be taken by children and the elderly. Through specialized technology (boats, nets, spears, harpoons), large bodied offshore species could be pursued. With growing population densities and competition for coastal resources, the aquatic focused niche became filled and foraging shifted inland toward montane terrestrial resources, with ungulates preferred. Taiwan's rugged mountain uplands typically require arduous foot travel from the coast, which would likely necessitate a near full-time commitment to hunting.

Projections from reference information are consistent with post-ecological release described by the niche variation hypothesis: early Paleolithic human migration to the island allowed for niche-broadening over time and increasing among-group variation toward either mountain hunting or coastal fishing. These niches eventually became specialized to form the regionally diverse Persistent Upper Paleolithic foraging adaptation of Taiwan. These projections allow us to estimate rank order of the foragers' preferred wild prey as follows.

If the prey rankings in Table 1 are correct, then high ranked habitats for foragers were coastal, lake, and wetland, followed by coastal plain, piedmont, and forested uplands. The last habitats to be occupied were predicted as high mountain regions that are sub-alpine today and likely forested during the warmer mid-Holocene. Projections from reference information are consistent with post-ecological release described by the niche variation hypothesis: early Paleolithic human migration to the island allowed for niche-broadening over time and increasing among-group variation toward either mountain hunting or coastal fishing. These niches eventually became specialized to form the regionally diverse Persistent Upper Paleolithic foraging adaptation of Taiwan.

Table 1. Estimated ranking of preferred prey and habitat, Upper Paleolithic foragers.

Prey Type Rank Order	Habitat Type
1. Shellfish, nearshore fish including migratory catadromous species, waterfowl	Littoral/estuary/delta wetland/lake basin
2. artiodactyls (deer, sambar, muntjac, serow), perissodactyls (boar)	Coastal plain, forested mountains, valleys
3. Deep water fish, turtles, marine mammals	Pelagic/offshore, islets
4. Arboreal and burrowing prey (macaque, pangolin, flying squirrel, and birds)	Forested valleys, piedmont and uplands
5. Resident river fish, catadromous species, turtles	Riverine/inland
6. Wild plants (geophytes, ferns, fungus, fruit, algae)	Forested valleys, piedmont and uplands

4.2. *Transitional Neolithic Paleoenvironments and Archaeology*

Against this heterogeneous and dynamic environmental and cultural backdrop, the first evidence of sedentized agriculture appears in Taiwan between 6000 and 5000 BP during what is termed the Dapenkeng (previously Tap'enkeng) period [4,10,11,15–17,67,78,93]. Saltwater inundation and the loss of low lying arable lands and productive wetlands on China's southeast margin could have created conditions favoring sea-faring over further investment in local agriculture, leading to migrations across the Taiwan Strait [5]. Regional Neolithic variants emerged early (see Figure 4, above), as evidenced by the Shuntanpu Early Culture in the north, Niumatou Early Culture in the central region, and Guoye in the south. At the time of writing, the influence of Paleolithic regionalization on early Neolithic cultures is not known.

Dapenkeng period sites are mostly distributed on the southern, northern, and eastern coasts around the island [16]. Hung and Carson [4] noted that today's alluvial plains accumulated mostly after 3000 BC, especially along the western coastline. Thus, most early Dapenkeng communities overlooked swampy nearshore environments, which have since became in-filled with alluvium [4,67]. Archaeological materials from the Dapenkeng type site (c. 5600 BP) near Tamsui in northern Taiwan include cord-marked pottery, polished stone adzes and harvesting knives, drilled slate projectile points, and baked clay spindle whorls [10–12]. Pottery is thick walled, sand-tempered and cord marked; stone adzes are quadrangular in cross-section and usually polished. Pecked pebbles were possibly used as net sinkers, and distinctive bark-cloth beaters were used to create raw material for clothing and other purposes [93]. Other cultural traits of Shuntanpu Early Culture include site placement on coastal and stream terraces, and planned semisedentary communities with extensive ditch and well construction and ash pits [4,10,11,93–95]. Faunal remains from the Dapenkeng type site include fish, shellfish, and mammal remains [10,89]. Macrobotanical evidence of millet and rice has also been found at multiple Shuntanpu sites, along with irrigation features [78,93].

Niumatou Early Culture sites (c. 5600–4200 BP) in the Taichung Basin of central Taiwan include the large Anhe site. Niumatou culture is remarkable for extensive cemeteries that indicate a wide range of ages and sexes, and tooth extraction practices [96]. Well-crafted burial objects include sandy paste ceramics (both red and occasionally black wares) such as *dou* double and triple cups, tapa bark beaters, and nephrite adzes and jewelry [ibid]. The ceramics have strong affinities to Southeast Chinese wares across the strait, indicating continued cultural interchange [78]. Sika deer, muntjac, small fish, shark, and shellfish remains indicate a mix of terrestrial and aquatic foods, along with macrobotanical evidence for rice [78,96].

In the south, recent excavations near Tainan have revealed a major Neolithic site complex at Nan-kuan-li and Nan-kuan-li East. The Dapenkeng period/Guoye phase dates to between 5000 and 4300 BP [16,67,97]. Site placement was preferentially on plains, terraces, and low hills facing wetlands [67,78]. Material evidence includes a robust assemblage of soft pottery of red and brown ware, rounded bottoms and feet, and decorations from cord-marking or comb-incised or shell-impression [67,97]. The toolkit is dominated by wood working and farming implements such as polished stone hoes, axes, adzes, and chisels. A small number of hoes are fashioned of thick shell. Drilled stone knives, stone projectile points, fishnet sinkers, and tapa bark beaters are common [67,97]. The consistent presence of olivine basalt from Penghu off the southwest coast indicates a strong trade connection with people of that island complex throughout the Guoye period [67]. Bone, tooth, and shell were used for drills, needles, jewelry, and cutting implements [ibid]. Faunal remains were well preserved, with shellfish predominant as well as a variety of sharks and other cartilaginous fish like stingray as well as bony fish from estuarine and pelagic habitats [16,97]. Reptiles include tortoise, soft shell turtle, and snake, and wild mammals include rat, sika deer, muntjac, Formosan badger, Formosan wild boar, and a small species of wild cat. Domesticated dog bones are found in midden settings as well as buried intentionally [67]. Plant remains include diverse wild species, dominated by chinaberry tree [Melia azedarach] as well as nigaki shrub (*Picrasma quassioides*) and hackberry tree (*Celtis sinensis*) [16]. Seeds and fruits may have been used for food, similar to Donghulin and other transitional Neolithic sites in China [98–100]. Both carbonized rice and millet remains have been found, with millet somewhat earlier in the sequence [67]. The local landrace of rice may have been domesticated locally based on seed morphology [67], whereas foxtail millet (*Setaria* spp.) likely arrived from China.

On the east coast, Dapenkeng period sites are smaller, consisting of either nephrite quarries or concentrations of cord-marked soft pottery below later Neolithic settlements. No settlements of the size and complexity of the western and southern regions have yet to be discovered [78], thus subsistence evidence is not sufficient for detailed discussion yet.

Overall, archaeological evidence suggests that the subsistence niche during the transitional Neolithic was broadly based. Evidence for fishing and shellfish collection predominates over farming, hunting, and collecting [67]. Taiwan's first farmers employed diverse cultivation techniques for two crop types: dryland (millet) and wet paddy (rice), which suggests influences both from the northern and eastern regions of mainland China [67,100]. Other crops such as geophytes and tree crops also may have been cultivated. Sizeable communities, burial grounds, and irrigation features along with distinctive vessel shapes and tapa bark beaters and genetic affinities of millet and rice suggest ongoing influences and interchanges with the cultures of what are now the Fujian coast and Guangdong/Pearl River delta regions in southeast China [78].

These data suggest that rats, mustelids, reptiles, and cats were part of the earliest Neolithic diet. Although differences in the preservational environment could allow for better species identification in Neolithic depositional contexts, this listing is broader than the Paleolithic and includes more genera and smaller body size prey. This, in combination with crop cultivation, allows us to estimate the rank order of the farmer-fisher's preferred prey as follows (the term 'prey' is used generally here, including cultigens).

If the prey rankings in Table 2 are correct, then high ranked habitats for Taiwan's first farmers were initially coastal, lake, and wetland adjacent to coastal plains. These habitats required to practice cultivation and access wild aquatic prey are distributed on the northwestern, west, and southern areas of the island. Piedmont, forested uplands, and the steep east coast would have been occupied after preferred habitats became infilled. The least desirable habitats to be used consistently by early farmers have been predicted as high mountain regions that are sub-alpine today and likely forested during the warmer mid-Holocene.

Table 2. Estimated ranking of preferred prey and habitat, Incipient Neolithic farmer-fishers.

Prey Type Rank Order	Habitat Type
1. Shellfish, nearshore fish including migratory catadromous species, waterfowl	Littoral/estuary/delta wetland/lake basin
2. artiodactyls (deer, sambar, muntjac, serow), perissodactyls (boar)	Coastal plain, forested mountains, valleys
3. Cultigens and edible weeds	Coastal plain, valleys
4. Arboreal and burrowing prey (macaque, pangolin, flying squirrel, and birds)	Forested valleys, piedmont and uplands
5. Resident river fish, catadromous species, turtles	Riverine/inland
6. Deep water fish, turtles, marine mammals	Pelagic/offshore, islets
7. Wild plants (geophytes, ferns, fungus, fruit, algae)	Forested valleys, piedmont and uplands

5. Discussion

The estimated prey rankings indicate that the overlap in habitat preference of foragers and farmer-fishers includes coastal plains, lakes, and other flat wetland areas with easy access to the coast or major bodies of water. These are located in the northwest, west, and south. More mobile foragers would have had an edge when exploiting the rugged east coast or the mountain uplands. The working hypothesis prediction regarding the earliest Neolithic in Taiwan can now be refined as follows.

If the assumptions of the PCM hold, high ranked aquatic prey (e.g., large shellfish and other nearshore species) and ungulates should rapidly become depleted in highly ranked habitats. The harvest of lower ranked prey such as small body size mammals and reptiles, and wild plants, should become more common. This process would end when foragers decide to adopt agricultural practices, or migrate to the east coast or mountain habitats.

If assumptions of an IFD with Allee effects hold, the flatlands of the northwest, west, and south adjacent to the coast should see increases in encounter rates for sika deer, pigs, and muntjac, which do well with more heterogeneous vegetation and successional communities associated with farming [71]. It is less clear if farming and/or intensified foraging would improve encounter rates with near shore aquatic prey, although in upland lake habitats, indigenous Taiwanese create fish islands of bamboo and thatching to improve the survival of young fry. The antiquity of this practice is not known, but it is reasonable. Given the scarcity of edible wild plant options in Taiwan, the adoption of crops would have provided a niche construction tactic to increase energetic returns per area. Edible commensal weeds that volunteer in the fields likely provided a dietary complement [90].

This process has important implications for predicting the ways that feedbacks in the consequences of resource/land use may shape interactions in diverse subsistence strategies, settlement, and social interactions. If, as modeled by Binford's database, the Persistent Upper Paleolithic foraging niche was specialized in either coastal or mountain resources, foragers of agriculturally desirable regions such as the northwestern, western, and southern coastal floodplains, flat, well-drained areas near river confluences, and terraces would have been affected the earliest by immigrating farmers. Competition for wild resources needed by foragers would have been direct, continuous, and increasing as farmers settled and expanded. The prey choice model predicts immediate and unidirectional drops in preferred prey, which would result in a 'push' for foragers in these regions to move, or adopt unfamiliar crops and cultivation techniques. The IFD with Allee effects predicts an initial increase in encounters with preferred prey due to beneficial niche construction activities, which might delay—though not for long—the thresholds for foragers to cultivate or move. Another consideration is the 'pull' associated with forager-farmer social interactions that rapidly reduced language barriers and other obstacles to knowledge exchange.

A more nuanced process is expected for areas not initially settled by early Neolithic agriculturalists in the mountain uplands, islets, and rugged east coast. Here, processes predicted by the PCM and the IFD with Allee effects were delayed. Competitive pressure on wild resources needed by foragers would initially have been indirect, gradual, and sporadic, most likely driven by the movements of foragers in response to immigrating farmers rather than by the farmers themselves. The 'pull' to expand niche breadth by adopting crops and cultivation knowledge would be weaker, with crop types adopted in ascending order of cost and risk to existing foraging lifeways (Yu, in press). Figure 9 shows areas preferred by farmers for settlement and variability in expected modes of competitive pressure on wild resources needed by foragers.

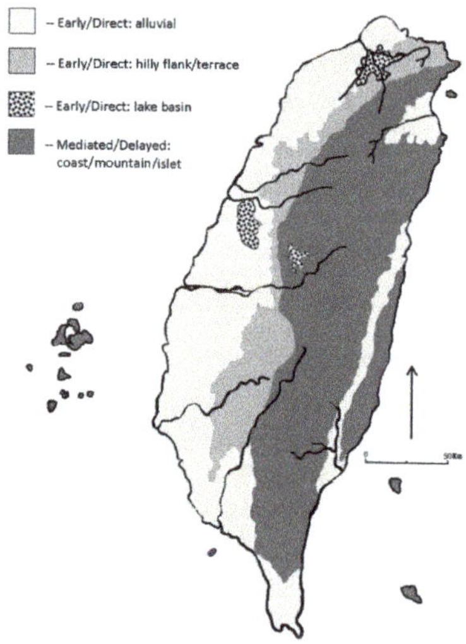

Figure 9. Geographic representation of Taiwan's modes of variable invasive pressure on wild resources associated with Neolithic onset, based on Binford's subsistence projections [37] used by permission.

In sum, niche variation theory, the prey choice model, and the IFD with Allee characteristics are useful heuristic tools for the latest Paleolithic and earliest Neolithic of Taiwan when combined with macroecological reference information about late-stage foraging. However, is an important difference in extrapolating theoretical predictions derived from non-humans for human groups: our flexible and constructible niche. Neolithic immigrants initially enhanced vegetative diversity through cultivation, potentially increasing encounter rates with high ranking prey. Farmers also brought along novel food species such as seed and possibly tuber crops, thus foragers under resource pressure could offset narrowing niche options by adopting crops that they deemed the least costly to mobility needed for acquiring preferred wild resources [37]. A working chronological reconstruction of Taiwan's Neolithic transition follows:

1. Early Paleolithic foragers migrated to Taiwan from the mainland c. 20,000 years ago, moving into new territory under conditions of ecological release.
2. Niche variation theory predicts an expansion of foraging niche breadth, increasing among-group variation within the niche. By the Persistent Upper Paleolithic, foraging niches divided into two specialized modes: either aquatic resources or mountain hunting.
3. The arrival and first dispersal of Neolithic Chinese farmers caused farmer-forager competition for wild resources, especially aquatic, in flat areas near coasts and wetlands. The prey choice model predicts that the niche overlap would lead to reductions in preferred prey and resource depression, whereas the IFD with Allee effects predicts that niche constriction could have increased encounter rates with preferred prey, at least in the beginning.
4. Foragers choosing to remain in colonized areas responded adaptively by taking low-ranked prey, then including low-ranked, low-cost cultigens in the diet, expanding the foraging niche temporarily (Figure 10). The PCM predicts this would have occurred rapidly; the IFD predicts a delay due to the beneficial effects of niche construction.
5. This phenomenon occurred early and rapidly in coastal areas and neighboring flat regions that were favorable to arriving farmers, then dispersed into more distant mountain uplands and the east coast as demographic packing continued.
6. The transition was more gradual in the central mountains and east coast. Foragers who opted to move away from farmer interactions may have encountered forager-forager competition and

resource depression in those areas, and demographic growth fueled a cycle that ultimately ended full-time hunting and gathering across Taiwan.

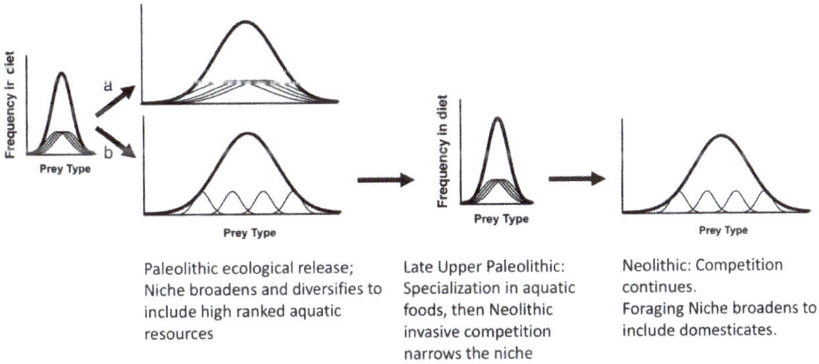

Figure 10. Schematic of the Niche Variation Theory applied to Neolithic impacts on foraging niche (adapted from [18], used by permission). (**a**) Inter-group niche breadths expand, resulting in no increased niche variation among individuals. (**b**) Individual group niche breadths remain limited, then groups diverge to increase among-group variation.

This qualitative assessment of the PCM and IFD has utility for making predictive statements about timing and intensity of the Taiwan Neolithic transition, particularly from the foraging perspective. In fact, this finding explains why foraging persisted in the mountains and along the east coast, even into the Middle Neolithic. The eventual demise of the Paleolithic lifeway can be predicted once the foraging niche could no longer be expanded through the addition of low cost options, or prolonged through niche construction [37]. Even after the full commitment to cultivation was made, selective forces actively favored a prolonged stage of mixed economies that included wild prey, commensal species, and vegeculture resources alongside cereal cultigens. This moderate risk/moderate yield mixed strategy with Neolithic roots is seen in Japan, Southeast China, and other areas worldwide, and played a role in oceanic migrations. The mixed strategy maintains diverse and nutritious diets, sustains crop diversity and wild resource viability, reduces travel and processing costs, and hedges against risk and loss [1].

The limitations of testing this approach for the Taiwan Neolithic transition include the imperfect preservation of material evidence for foraging subsistence behaviors, and therefore small sample sizes and lack of a basis for strong inferential arguments. However, as the rate of discovery of well-dated early sites increases, the likelihood of new Persistent Upper Paleolithic and Dapenkeng sites—and therefore the sample size—is increasing. The PCM and IFD working scenarios could be assessed using various classes of archaeological evidence as new data become available (Table 3).

Table 3. Potential lines of evidence to assess Prey Choice Model and Ideal Free Distribution expectations.

Model/Prediction	Categories of Evidence (Incipient NEOLITHIC, c. 6000–5000 BP)
1. Prey Choice Model/Resource Depression	Subsistence: Decreases in frequency, diversity, age, and size of high ranked prey species relative to lower ranked prey. Technology: Increased processing tools and features such as ovens and graters. Settlement: Temporal and geographic overlap between Paleolithic and Neolithic sites in flatlands near the coast. Low frequency of earliest Neolithic sites in mountain uplands, islets, and east coast.
2. IFD/Allee Effects/Niche Construction and Resource Benefits	Subsistence: Initially, increases in frequency, diversity, age, and size of highly ranked prey types relative to lower ranked prey. Could decrease over time after population rises. Technology: Few processing tools and features such as ovens and graters; more procurement tools (e.g., fishing and hunting equipment). Settlement: Temporal and geographic overlap between Paleolithic and Neolithic sites in flatlands near the coast; initial increase in Paleolithic sites close to Neolithic settlements, then replacement by agricultural sites. Low frequency of earliest Neolithic sites in mountain uplands, islets, and east coast.

6. Conclusions

A major challenge in assessing hypotheses for this time period in Taiwan is the lack of means to distinguish early Neolithic sites from sites in which Paleolithic foragers were experimenting with cultivation. Although the presence of ceramics is one indicator of farming, some of Taiwan's 'last foragers' may have been able to obtain ceramics or ceramic manufacturing techniques while maintaining a mostly foraging lifestyle. Across the strait in SE China, a prolonged Neolithic transition included coastally-adapted hunter-gatherers who used ceramics, lived in semi-sedentary villages, and created cemeteries as late as 3000 BP [17,101–103].

Nevertheless, predictions derived from the niche variation theory, the prey choice model, and the IFD with Allee effects allow us to explore alternative processes for the incipient Neolithic. Taiwan's geographic position and diverse habitats and foraging opportunities ensured that the transition to farming was neither simple nor linear, but may be predicted. With the increase in chronometrically dated sites from this time period, archeofauna and other lines of evidence can be used to determine alternative explanations for the ways that Taiwan's Upper Paleolithic foragers and Chinese Neolithic farmer-fishers evaluated the costs and benefits and made strategic adjustments to their subsistence niche as they encountered new food types and each other to create a mosaic of potential habitat choices. If the PCM is supported, the ability to adjust prey choice and niche breadth ensured that the Persistent Upper Paleolithic endured in areas remote from farmers as late as 5000 BP, and likely even later. The IFD with Allee effects also accommodates prolonged foraging in areas where increased diversity of habitat succession and plant communities arose from the mosaic created by cultivation.

On the Neolithic side, farmers continued to include wild and semi-domesticated species alongside crops, with aquatic resources as a major, if not the most important dietary staple. The role of prey choice, niche, and habitat preferences in invasive dispersal and native adaptations is central to understanding variability in crop adoption by foragers, helping to frame the process in an evolutionary framework that places the foraging legacy in a key position for influencing evolutionary transitions to agriculture and later pathways to complexity.

Funding: This research was funded by the J. William Fulbright Foreign Scholarship Program and the Taiwan Foundation for Scholarly Exchange.

Acknowledgments: Research associated with this study was generously supported by a Fulbright Senior Researcher Fellowship under the robust support of Fulbright Taiwan. Maa-ling CHEN (National Taiwan University) shared archaeological wisdom and sources, helped me to navigate the ins and outs of living and traveling in Taiwan, generously invited me to archaeology classes, and facilitated the translation of reports. Chen, Chia-yu HU, and Hsiu-fang TSAI arranged for a viewing of the Dapenkeng type site artifacts and records at the National Taiwan Museum of Anthropology in Taipei. Fu CHUN, Mei-Zhen YE, and Kun-Hsiu LEE, all from the National Taiwan Museum of Prehistory in Taitung, facilitated a tour of the Baxiandong Paleolithic site and provided useful technical reports. Yu-Mei CHEN and Yi-Chang LIU of Academia Sinica's Institute of History and Philology kindly hosted a lecture and provided invaluable technical materials and advice regarding Taiwanese prehistory. Thanks to Kazunobu IKEYA for permission to cite from his important works on Asian hunting and gathering societies, and Daniel Bolnick for permission to adapt figures from his important work on evolutionary aspects of niche and invasions. My appreciation goes to the peer reviewers whose comments strengthened and clarified the theoretical concepts in the paper. My father, J. S. YU, translated the literature sources and during scholarly discussions, assisted with reverse translation and transcription of recordings, and provided important insights about Taiwanese traditional crops from personal and family experience. I am profoundly grateful for his help, his tireless and humorous companionship, and his belief in the value of scientific anthropological research for the good of all human societies.

Conflicts of Interest: The author declares no conflict of interest.

References

1. Smith, B.D. Low-level food production. *J. Archaeolog. Res.* **2001**, *9*, 1–43. [CrossRef]
2. Ikeya, K.; Ogawa, H.; Mitchell, P. (Eds.) *Interactions between Hunter-Gatherers and Farmers: From Prehistory to Present*; National Museum of Ethnology: Osaka, Japan, 2009; ISBN 9784901906654.

3. Winterhalder, B.; Kennett, D.H. Behavioral ecology and the transition from hunting and gathering to agriculture. In *Behavioral Ecology and the Transition to Agriculture*; Kennett, D., Winterhalder, B., Eds.; University of California Press: Berkeley, CA, USA, 2006; pp. 1–21, ISBN 9780520246478.
4. Hung, H.C.; Carson, M.T. Foragers, fishers and farmers: Origins of the Taiwanese Neolithic. *Antiquity* **2014**, *88*, 1115–1131. Available online: https://www.cambridge.org/core/journals/antiquity/article/foragers-fishers-and-farmers-origins-of-the-taiwanese-neolithic/7B50C25B6298EB3AEABE1E8EEFD030C8 (accessed on 15 April 2020). [CrossRef]
5. Peltier, W.R. On eustatic sea level history: Last Glacial Maximum to Holocene. *Quat. Sci. Rev.* **2002**, *21*, 377–396. Available online: https://www.sciencedirect.com/science/article/pii/S0277379101000841 (accessed on 15 April 2020). [CrossRef]
6. Rollett, B.V.; Zheng, Z.; Yue, Y. Holocene sea-level change and the emergence of Neolithic seafaring in the Fuzhou Basin (Fujian, China). *Quat. Sci. Rev.* **2011**, *30*, 788–797. Available online: https://www.sciencedirect.com/science/article/pii/S0277379111000308 (accessed on 15 April 2020). [CrossRef]
7. Bulbeck, D. An integrated perspective on the Austronesian diaspora: The switch from cereal agriculture to maritime foraging in the colonisation of Island Southeast Asia. *Aust. Archaeol.* **2008**, *67*, 31–51. Available online: https://www.tandfonline.com/doi/abs/10.1080/03122417.2008.11681877 (accessed on 15 April 2020). [CrossRef]
8. Bellwood, P. Southeast China and the Prehistory of the Austronesians. In *Lost Maritime Cultures: China and the Pacific*; Jiao, T.L., Ed.; Bishop Museum Press: Honolulu, HI, USA, 2007; pp. 36–53, ISBN 9781581780635.
9. Bellwood, P. Formosan prehistory and Austronesian dispersal. In *Austronesian Taiwan: Linguistics, History, Ethnology, Prehistory*; Blundell, D., Ed.; N. W. Lin Foundation for Culture and Educational Endowment: Berkeley, CA, USA, 2009; pp. 336–364, ISBN 986-8537819.
10. Blust, R. *The Austronesian Languages*; Research School of Pacific and Asian Studies: Canberra, Australia, 2009; ISBN 9780858836020.
11. Chang, K.C. The Neolithic Taiwan Strait. *Kaogu* **1989**, *6*, 569.
12. Chang, K.C. *Fengbitou, Tapenkeng, and the Prehistory of Taiwan*; Yale University Publications in Anthropology: New Haven, CT, USA, 1969; ISBN 9781444335293.
13. Chang, K.C.; Goodenough, W.H. Archaeology of southeastern China and its bearing on the Austronesian homeland. In *Prehistoric Settlement of the Pacific*; Goodenough, W.H., Ed.; American Philosophical Society: Philadelphia, PA, USA, 1996; pp. 28–35, ISBN 9780871698650.
14. Pawley, A. The Austronesian dispersal: Languages, technologies and people. In *Examining the Farming/language Dispersal Hypothesis*; Bellwood, P., Renfrew, C., Eds.; MacDonald Institute for Archaeological Research: Cambridge, UK, 2002; pp. 251–273, ISBN 9781902937205.
15. Tsang, C.H. Recent discoveries at the Tapenkeng Culture Sites in Taiwan: Implications for the Problem of Austronesian Origins. In *The Peopling of East Asia: Putting Together Archaeology, Linguistics, and Genetics*; Sagart, L., Blench, R., Sanchez-Mazas, A., Eds.; Routledge Curzon: New York, NY, USA, 2005; pp. 63–73, ISBN 9781138862234.
16. Bellwood, P. Taiwan and the Prehistory of the Austronesian-Speaking Peoples. In *Ethnos, Geography and Development: An Interdisciplinary Approach to Human-Environment Relations*; Kuan, D.W., Ed.; Shung Ye Museum of Formosan Aborigines: Taipei, Taiwan, 2017; pp. 3–33, ISBN 9789869239639.
17. Li, K.T. First Farmers and Their Coastal Adaptation in Prehistoric Taiwan. In *A Companion to Chinese Archaeology*; Underhill, A.P., Ed.; Blackwell: Oxford, UK, 2013; pp. 612–633, ISBN 9781444335294.
18. Liu, Y.C. Prehistory and Austronesians in Taiwan: An archaeological perspective. In *Austronesian Taiwan: Linguistics, History, Ethnology, Prehistory*; Blundell, D., Ed.; N. W. Lin Foundation for Culture and Educational Endowment: Berkeley, CA, USA, 2009; pp. 365–400, ISBN 9868537800.
19. Bolnick, D.I.; Svanbäck, R.; Araújo, M.S.; Persson, L. Comparative support for the niche variation hypothesis that more generalized populations also are more heterogeneous. *Proc. Natl. Acad. Sci. USA* **2007**, *104*, 10075–10079. Available online: https://www.pnas.org/content/104/24/10075 (accessed on 15 April 2020). [CrossRef]
20. Fix, A.G.; Shepherd, J.C. *Migration and Colonization in Human Microevolution*; Cambridge University Press: Cambridge, MA, USA, 1999; Volume 24, ISBN 9780521592062.

21. Binford, L.R. *Constructing Frames of Reference: An Analytical Method for Archaeological Theory Building Using Ethnographic and Environmental Data Sets*; University of California Press: Berkeley, CA, USA, 2001; ISBN 0520223934.
22. Svizzero, S. Persistent controversies about the Neolithic Revolution. *J. Hist. Archaeol. Anthr. Sci.* **2017**, *1*, 00013. [CrossRef]
23. Yu, P. Ethnoarchaeology of foraging and the case of the vanishing agriculturalists in the Amazon Basin. *J. Anthr. Archaeol.* **2015**, *38*, 59–66. Available online: https://www.sciencedirect.com/science/article/pii/S0278416514000683 (accessed on 15 April 2020). [CrossRef]
24. Bird, D.W.; O'Connell, J.F. Behavioral ecology and archaeology. *J. Archaeol. Res.* **2006**, *14*, 143–188. [CrossRef]
25. Bird, D.W.; O'Connell, J.F. Human behavioral ecology. In *Archaeological Theory Today*; Hodder, I., Ed.; Polity Press: Cambridge, MA, USA, 2012; pp. 37–61, ISBN 9780745653075.
26. Broughton, J.M.; Cannon, M.D. (Eds.) *Evolutionary Ecology and Archaeology*; University of Utah Press: Salt Lake City, UT, USA, 2010; ISBN 9780874809350.
27. Codding, B.F.; Bird, D.W. Behavioral ecology and the future of archaeological science. *J. Archaeol. Sci.* **2015**, *56*, 9–20. Available online: https://www.sciencedirect.com/science/article/pii/S0305440315000655?casa_token=yVm483vDuhgAAAAA:tPRWrRljxrWNR0_FAJj0Bx03La8Z_ickOJAQI_53Skf_p7T3chXy8ogpRHbpOmcHromv9ZhaPw (accessed on 5 July 2020). [CrossRef]
28. Lupo, K.D. Evolutionary foraging models in zooarchaeological analysis: Recent applications and future challenges. *J. Archaeol. Res.* **2007**, *15*, 143–189. Available online: https://link.springer.com/article/10.1007/s10814-007-9011-1 (accessed on 5 July 2020). [CrossRef]
29. O'Connell, J.F. Ethnoarchaeology needs a general theory of behavior. *J. Archaeol. Res.* **1995**, *3*, 205–255. [CrossRef]
30. Johnson, A.L. Exploring adaptive variation among hunter-gatherers with Binford's frames of reference. *J. Archaeol. Res.* **2014**, *22*, 1–42. Available online: https://link.springer.com/article/10.1007/s10814-013-9068-y (accessed on 15 April 2020). [CrossRef]
31. Fretwell, S.D.; Lucas, H.L. On territorial behavior and other factors influencing habitat distribution in birds. *Acta Biotheor.* **1969**, *19*, 16–36. Available online: https://link.springer.com/article/10.1007/BF01601953 (accessed on 5 July 2020). [CrossRef]
32. Kennett, D.J.; Anderson, A.; Winterhalder, B. The Ideal Free Distribution, food production, and the colonization of Oceania. In *Behavioral Ecology and the Transition to Agriculture*; Kennett, D.J., Winterhalder, B., Eds.; University of California Press: Berkeley, CA, USA, 2009; pp. 265–288, ISBN 9780520246478.
33. Allee, W.C.; Bowen, E. Studies in animal aggregations: Mass protection against colloidal silver among goldfishes. *J. Exp. Zool.* **1932**, *61*, 185–207. [CrossRef]
34. Codding, B.F.; Parker, A.K.; Jones, T.L. Territorial behavior among Western North American foragers: Allee effects, within group cooperation, and between group conflict. *Quat. Int.* **2019**, *518*, 31–40. Available online: https://www.researchgate.net/publication/321270258_Territorial_behavior_among_Western_North_American_foragers_Allee_effects_within_group_cooperation_and_between_group_conflict (accessed on 18 July 2020). [CrossRef]
35. Bliege Bird, R.B.; McGuire, C.; Bird, D.W.; Price, M.H.; Zeanah, D.; Nimmo, D.G. Fire mosaics and habitat choice in nomadic foragers. *Proc. Natl. Acad. Sci. USA* **2020**, *117*, 12904–12914. Available online: https://www.pnas.org/content/117/23/12904.short (accessed on 5 July 2020). [CrossRef]
36. Bliege Bird, R.B.; Bird, D.W.; Fernandez, L.E.; Taylor, N.; Taylor, W.; Nimmo, D. Aboriginal burning promotes fine-scale pyrodiversity and native predators in Australia's Western Desert. *Biol. Conserv.* **2018**, *219*, 110–118. Available online: https://www.sciencedirect.com/science/article/abs/pii/S0006320717317962 (accessed on 18 July 2020). [CrossRef]
37. Winterhalder, B.; Kennett, D.J.; Grote, M.N.; Bartruff, J. Ideal free settlement of California's northern Channel Islands. *J. Anthropol. Archaeol.* **2010**, *29*, 469–490. Available online: https://www.sciencedirect.com/science/article/abs/pii/S0278416510000425 (accessed on 5 July 2020). [CrossRef]
38. Yu, P. Tempo and mode of Neolithic crop adoption by Paleolithic hunter-gatherers of Taiwan: Ethnoarchaeological and behavioral ecology perspectives. In *Hunter-Gatherers in Asia: From Prehistory to Present*; Ikeya, K., Nishiaki, Y., Eds.; Senri Ethnological Studies (SES)/National Museum of Ethnography: Osaka, Japan, 2021. (in press)

39. Elton, C.S. *The Ecology of Invasions by Plants and Animals*; University of Chicago Press: Chicago, IL, USA, 1958; ISBN 9780521592062.
40. MacArthur, R.H.; Pianka, E.R. On optimal use of a patchy environment. *Am. Nat.* **1966**, *100*, 603–609. Available online: https://www.jstor.org/stable/2459298?seq=1 (accessed on 5 July 2020). [CrossRef]
41. Stephens, D.W.; Krebs, J.R. *Foraging Theory*; Princeton University Press: Princeton, NJ, USA, 1987; Volume 1, ISBN 9780691084428.
42. Gremillion, K.J. Seed processing and the origins of food production in eastern North America. *Am. Antiq.* **2004**, *69*, 215–233. Available online: https://www.jstor.org/stable/4128417?seq=1#metadata_info_tab_contents (accessed on 10 June 2018). [CrossRef]
43. Gremillion, K.J.; Piperno, D.R. Human behavioral ecology, phenotypic (developmental) plasticity, and agricultural origins: Insights from the emerging evolutionary synthesis. *Curr. Anthropol.* **2009**, *50*, 615–619. Available online: https://www.journals.uchicago.edu/doi/abs/10.1086/605360 (accessed on 15 April 2020). [CrossRef] [PubMed]
44. Miller, D.S. *From Colonization to Domestication: Population, Environment, and the Origins of Agriculture in Eastern North America*; University of Utah Press: Salt Lake City, UT, USA, 2018; ISBN 1607816164.
45. Pearsall, D.M. Investigating the Transition to Agriculture. *Curr. Anthropol.* **2009**, *50*, 609–613. Available online: https://www.jstor.org/stable/10.1086/605406#metadata_info_tab_contents (accessed on 15 April 2020). [CrossRef]
46. Winterhalder, B.; Kennett, D.H. Four neglected concepts with a role to play in explaining the origins of agriculture. *Curr. Anthropol.* **2009**, *50*, 645–648. Available online: https://www.jstor.org/stable/10.1086/605355?seq=1#metadata_info_tab_contents (accessed on 15 April 2020). [CrossRef] [PubMed]
47. Jackson, M.C.; Britton, J.R. Stable isotope analyses indicate trophic niche overlap of invasive Pseudorasbora parva and sympatric cyprinid fishes. *Ecol. Freshw. Fish* **2013**, *22*, 654–657. Available online: https://onlinelibrary.wiley.com/doi/pdf/10.1111/eff.12063 (accessed on 15 April 2020). [CrossRef]
48. Hamilton, S.; Johnston, R.F. Evolution in the House Sparrow; VI. variability and niche width. *Auk* **1978**, *95*, 313–323. Available online: https://academic.oup.com/auk/article-abstract/95/2/313/5208845?redirectedFrom=fulltext (accessed on 15 April 2020).
49. Soule, M.; Stewart, B.R. The "niche-variation" hypothesis: A test and alternatives. *Am. Nat.* **1970**, *104*, 85–97. Available online: https://www.jstor.org/stable/2459075?seq=1#metadata_info_tab_contents (accessed on 15 April 2020). [CrossRef]
50. Werner, T.K.; Sherry, T.W. Behavioral feeding specialization in *Pinaroloxias inornata*, the "Darwin's finch" of Cocos Island, Costa Rica. *Proc. Natl. Acad. Sci. USA* **1987**, *84*, 5506–5510. Available online: https://www.pnas.org/content/84/15/5506 (accessed on 15 April 2020). [CrossRef]
51. Van Valen, L. Morphological variation and width of ecological niche. *Am. Nat.* **1965**, *99*, 377–390. Available online: https://www.jstor.org/stable/2459179?seq=1#metadata_info_tab_contents (accessed on 15 April 2020). [CrossRef]
52. Snyder, W.E.; Evans, E.W. Ecological effects of invasive arthropod generalist predators. *Annu. Rev. Ecol. Evol. Syst.* **2006**, *37*, 95–122. Available online: https://www.annualreviews.org/doi/full/10.1146/annurev.ecolsys.37.091305.110107 (accessed on 15 April 2020). [CrossRef]
53. Olsson, K.; Stenroth, P.; Nyström, P.; Graneli, W. Invasions and niche width: Does niche width of an introduced crayfish differ from a native crayfish? *Freshw. Biol.* **2009**, *54*, 1731–1740. Available online: https://portal.research.lu.se/portal/en/publications/invasions-and-niche-width-does-niche-width-of-an-introduced-crayfish-differ-from-a-native-crayfish(e78fca3d-d88c-40fc-9628-292b33506e33).html (accessed on 15 April 2020). [CrossRef]
54. Porter, S.D.; Savignano, D.A. Invasion of polygyne fire ants decimates native ants and disrupts arthropod community. *Ecology* **1990**, *71*, 2095–2106. Available online: https://esajournals.onlinelibrary.wiley.com/doi/10.2307/1938623 (accessed on 15 April 2020). [CrossRef]
55. Hutchinson, G.E. Concluding Remarks. *Cold Spring Harb. Symp. Quant. Biol.* **1957**, *22*, 415–421. Available online: http://symposium.cshlp.org/content/22/415 (accessed on 1 September 2020). [CrossRef]
56. Bolnick, D.I.; Ingram, T.; Stutz, W.E.; Snowberg, L.K.; Lau, O.L.; Paull, J.S. Ecological release from interspecific competition leads to decoupled changes in population and individual niche width. *Proc. R. Soc. B Biol. Sci.* **2010**, *277*, 1789–1797. [CrossRef] [PubMed]

57. Svanbäck, R.; Bolnick, D.I. Intraspecific competition affects the strength of individual specialization: An optimal diet theory method. *Evol. Ecol. Res.* **2005**, *7*, 993–1012. [CrossRef]
58. Araújo, M.S.; Bolnick, D.I.; Layman, C.A. The ecological causes of individual specialisation. *Ecol. Lett.* **2011**, *14*, 948–958. [CrossRef] [PubMed]
59. Penk, M.; Irvine, K.; Donohue, I. Ecosystem-level effects of a globally-spreading invertebrate invader are not moderated by a functionally similar native. *J. Anim. Ecol.* **2015**, *84*, 1628–1636. [CrossRef]
60. Smith, B.D. A comparison of niche construction theory and diet breadth models as explanatory frameworks for the initial domestication of plants and animals. *J. Archaeol. Res.* **2015**, *23*, 215–262. Available online: https://link.springer.com/article/10.1007/s10814-015-9081-4 (accessed on 15 April 2020). [CrossRef]
61. Boserup, E. *The Conditions of Agricultural Growth: The Economics of Agrarian Change under Population Pressure*; Transaction Publishers: New Brunswick, NJ, USA, 2011; ISBN 9781138537187.
62. Zeanah, D.W. Foraging models, niche construction, and the Eastern Agricultural Complex. *Am. Antiq.* **2017**, *82*, 3–24. Available online: https://www.cambridge.org/core/journals/american-antiquity/article/foraging-models-niche-construction-and-the-eastern-agricultural-complex/7481D608E03F7864C6C6FB5990F1B9F1 (accessed on 15 April 2020). [CrossRef]
63. David, N.; Kramer, C. *Ethnoarchaeology in Action*; Cambridge University Press: New York, NY, USA, 2001; ISBN 9780521667791.
64. Gifford-Gonzales, D. Ethnoarchaeology—Looking back, looking forward. *SAA Archaeol. Rec.* **2010**, *10*, 22–25. Available online: http://digital.ipcprintservices.com/publication/?i=30669&article_id=308302&view=articleBrowser (accessed on 15 April 2020).
65. Yu, P. Ethnoarchaeology as a strategy for building frames of reference for research problems. In *Hunter Gatherer and Mid-Range Societies, Encyclopedia of Global Archaeology*; Prentiss, A., Ed.; Springer Publishing: New York, NY, USA, 2014; ISBN 9781441904263.
66. Song, Y.C.; Xu, G.S. A Scheme of Vegetation Classification of Taiwan, China. *Acta Bot. Sin.* **2003**, *45*, 883–895. Available online: https://europepmc.org/article/cba/364797 (accessed on 15 April 2020).
67. Lee, C.Y.; Liew, P.M. Late Quaternary vegetation and climate changes inferred from a pollen record of Dongyuan Lake in southern Taiwan. *Palaeogeogr. Palaeoclimatol. Palaeoecol.* **2010**, *287*, 58–66. Available online: https://www.sciencedirect.com/science/article/pii/S0031018210000167?casa_token=3oPH7oCmZY8AAAAA:sbce5Q82gnmOkAUpF9GWZFF-YE2S-XwRbanPffYAc6cEC9J-DQgcq6xbiK9G6unkjBbAIXOnqg (accessed on 5 July 2020). [CrossRef]
68. Tsang, C.H.; Li, K.T. *Archaeological Heritage in the Tainan Science Park of Taiwan*; National Museum of Prehistory: Taitung City, Taiwan, 2018; ISBN 9789860474138.
69. Chen, C.T.A.; Ruo, R.; Paid, S.C.; Liu, C.T.; Wong, G.T.F. Exchange of water masses between the East China Sea and the Kuroshio off northeastern Taiwan. *Cont. Shelf Res.* **1995**, *15*, 19–39. Available online: https://www.sciencedirect.com/science/article/pii/0278434393E0001O (accessed on 22 April 2020). [CrossRef]
70. Lin, J.X.; Dai, L.P.; Yuan, W.; Min, L. Quaternary marine transgressions in eastern China. *J. Palaeogeogr.* **2012**, *1*, 105–125. Available online: https://www.sciencedirect.com/science/article/pii/S209538361530119X (accessed on 22 April 2020).
71. McCullough, D.R.; Takatsuki, S.; Kaji, K. (Eds.) *Sika deer: Biology and Management of Native and Introduced Populations*; Springer Science & Business Media: Berlin, Germany, 2008.
72. Lien, C.M. Chang-pin Culture of Taiwan and Characteristics of Its Lithic Industry. In *Emergence and Diversity of Modern Human Behavior in Paleolithic Asia*; Kaifu, Y., Izuho, M., Goebel, T., Sato, H., Ono, A., Eds.; Texas A & M Press: College Station, TX, USA, 2014; pp. 239–248, ISBN 9781623492762.
73. Sung, W.S. The archaeology of Taiwan. In *Taiwan of China*; Chen, C.L., Ed.; Zhongyang Wenwu Gongyingshe: Taipei, Taiwan, 1980; pp. 93–220. (In Chinese)
74. Sung, W.S. *Changpinian: A Newly Discovered Pre-Ceramic Culture from the Agglomerate Caves on the East Coast of Taiwan*; Newsletter of Ethnological Society of Taiwan: Taipei, Taiwan, 1969; Volume 9, pp. 1–9.
75. Tsang, C.H.; Chen, W.S.; Li, K.T. *A Progress Report on First Year Results of the Ba Xian Dong Archaeological Survey Research Project*; Academia Sinica: Taipei, Taiwan, 2009. (In Chinese)
76. Tsang, C.H.; Chen, W.S.; Li, K.T. (Eds.) *Tai Dong Chang Bin Ba Xian Dong Yi Zhi Diao Cha Yan Jiu Ji Hua (Di Er Nian)*; Academia Sinica: Taipei, Taiwan, 2011. (In Chinese)

77. Tsang, C.H. Lun Changbin Wenhua de Niandai yu Leiyuan. In *Baxiandong Guodeng Yizhi Baohu yu Yanjiu Guoji Xueshu Yantaohui Luwenji*; Wenhuabu Wenhua Zichanju, Taidongxian Zhengfu, Zhongyang Yanjiuyan, Lishi Yuyan yankuisuo Zhixing: Taipei, Taiwan, 2013; pp. 1–23. (In Chinese)
78. Kuo, S.C. *New Frontiers in the Neolithic Archaeology of Taiwan (5600–1800 BP)*; Springer: New York, NY, USA, 2019; ISBN 9789813292628.
79. Chen, W.C. The Early Occupation of Taiwan. In *The Handbook of East and Southeast Asian Archaeology*; Habu, J., Lape, P.V., Olsen, J.W., Eds.; Springer: New York, NY, USA, 2017; pp. 277–291, ISBN 9781493965212.
80. Liu, Y.; Guo, S.; Lu, R. *Taimin Diqu Kaogu Yihi Pucha Yanjiu Jihua (Diqiqi)*; Taipeixian, Jilongshi, Taipeishi, Neizhengbu Weituo, Zhongyang Yanjiuyan, Lishi Yuyan yankuisuo Zhixing: Taipei, Taiwan, 2004.
81. Liu, Y. 2011 Zhuminzhi Kaogupian. In *Taiwan Quanzhi*; Guoshiguan Taiwan Wenxianguan: Nantou, Taiwan, 2011; Volume 3.
82. Liu, Y.; Chen, J.; Zheng, H.; Li, J. *Taizhongxian Kaogu Yizhi Pucha yu Yanjiu Jihua Yanjiu Baogao*; Taizhongixian Wenhuaju Weituo, Zhongyan Yanjiuyuan Renwen Shehui Kexue Zongxin, Kaoguxue Yanjiu Zhuanti Zhongxin Zhixing: Taichung, Taiwan, 2007.
83. Liu, Y. Disanzhang: Shiqian Yizhi. In *Taizhong Xianzhi (Juanyi) Tudizhi*; Taizhongxian Zhengfu: Taichung, Taiwan, 1989; pp. 773–849.
84. Li, K.T. *1985 Duiyu Taiwan Kaogu Yanjiu de Ryogan Renshi*; Taiwan Wenxian: Taipei, Taiwan, 1985; Volume 36, pp. 15–23.
85. Huang, S.; Chen, Y.; Yan, X. *Kending Guojia Gongyuan Kaogu Minzhu Diaocha Baogao*; Neizhengbu Yingjianshu Kending Guojia Gongyuan Guanlichu Weituo, Guoli Taiwan Daxue Renlei Xuexi Zhixing: Taipei, Taiwan, 1987.
86. Li, K.T.; Liu, Y.; Chang, C. *Eluanbi Gongyuan Kaogu Diaocha Baogao*; Jiaotongbu Guangguangju Kending Fenjing Tedingqu Guanlingchu Weituo, Guoli Taiwan Daxue Renlei Xuexi Zhixing: Taipei, Taiwan, 1983.
87. Huang, S.; Chen, Y. *Donghe Diqu Yizhi Shijue di Shiqian Wenhua Chongjian*; Xingzhengyuan Wenhua Jianshe Weiyuanhui Weituo, Guoli Taiwan Daxue Renlei Xuexi Zhixing: Taipei, Taiwan, 1990.
88. Binford, L.R.; Johnson, A.L. Program for Calculating Environmental and Hunter-Gatherer Frames of Reference (ENVCALC2.1). Updated Java Version. August 2014. Available online: http://ajohnson.sites.truman.edu/data-and-program/ (accessed on 25 June 2018).
89. Lin, C.; Taiwan Forestry Research Institute, Taipei; Yu, P.; Boise State University, Boise, ID, USA. Personal communication, 2017.
90. Butal, A.; Tung, G.S.; Miaraw, P. *The Ethnobotany of Amis in Eastern Formosa*; Jen Teh Yen, Council of Agriculture Executive Yuan: Taipei, Taiwan, 2009; ISBN 9789860223156.
91. Lu, T.H.; Ke, Y.N.; Lin, S.F.; Lu, S.Y. *Plants Used by Paiwan in Taiwan*; Report prepared for Taiwan Forestry Bureau, Council of Agriculture, Executive Yuan: Taipei, Taiwan, 2011.
92. Simpson, E.H. The Measurement of Diversity. *Nature* **1949**, *163*, 688. Available online: https://www.nature.com/articles/163688a0 (accessed on 15 April 2020). [CrossRef]
93. Chu, C.Y. *Report on the restoration, excavation, construction, and supervision of the Talungtung Site (Dalongdong Yizhi Qiangjiu Fajue Ji Shigong Jiankan Jihuau Chengguo Baogao)*; Tree Valley Foundation, Taipei City Government Department of Cultural Affairs: Taipei, Taiwan, 2012. (In Chinese)
94. Huang, S.C. *Report on Archaeological Investigation at Pa-chia Village, Kui-jun Township*; Bulletin of the Department of Archaeology and Anthropology: Tainan, Taiwan, 1974; pp. 62–68. (In Chinese)
95. Tsang, C.H. *Archaeology of the Penghu Islands*; Academia Sinica: Taipei, Taiwan, 1992; ISBN 9789576710483. (In Chinese)
96. National Museum of Natural Science. *Report on the Salvage Excavation Project at the Construction Site at the "Anho Road Site"*; Report commissioned by Ching He Construction Co., Ltd.; National Museum of Natural Science: Taichung City, Taiwan, 2016. (In Chinese)
97. Tsang, C.H. (Ed.) *Final Report on the Tainan Science Park Archaeological Rescue and Monitoring Project*; Taiwan National Museum of Prehistory: Taitung City, Taiwan, 2006. (In Chinese)
98. Tsang, C.H. *Final Report on the Archaeological Project of Restoring Areas of the Taoyeh Site Excluded from the Preservation Area, Tainan Science Park*; Report prepared for the Southern Taiwan Science Bureau; Institute of History and Philology, Academia Sinica: Taipei, Taiwan, 2004. (In Chinese)

99. Chen, S.Q.; Yu, P. Early 'Neolithics' of China: Variation and Evolutionary Implications. *J. Anthropol. Res.* **2017**, *73*, 149–180. Available online: https://www.journals.uchicago.edu/doi/abs/10.1086/692104?af=R (accessed on 22 April 2020). [CrossRef]
100. Liu, L.; Field, J.; Fullagar, R.; Zhao, C.; Chen, X.; Yu, J. A functional analysis of grinding stones from an early Holocene site at Donghulin, North China. *J. Archaeol. Sci.* **2010**, *37*, 2630–2639. Available online: https://www.sciencedirect.com/science/article/pii/S0305440310001858 (accessed on 22 April 2020). [CrossRef]
101. Sagart, L.; Hsu, T.F.; Tsai, Y.C.; Wu, C.C.; Huang, L.T.; Chen, Y.C.; Chen, Y.F.; Tseng, Y.C.; Lin, H.Y.; Hsing, Y.I.C. A northern Chinese origin of Austronesian agriculture: New evidence on traditional Formosan cereals. *Rice* **2018**, *11*, 57–73. Available online: https://thericejournal.springeropen.com/articles/10.1186/s12284-018-0247-9 (accessed on 15 April 2020). [CrossRef]
102. Jiao, T.L. Toward an alternative perspective on the foraging and low-level food production on the coast of China. *Quat. Int.* **2016**, *419*, 54–61. Available online: https://www.sciencedirect.com/science/article/pii/S1040618215006606 (accessed on 15 April 2020). [CrossRef]
103. Zhang, C.; Hung, H.C. Later hunter-gatherers in southern China, 18,000–3000 BC. *Antiquity* **2012**, *86*, 11–29. Available online: https://www.cambridge.org/core/journals/antiquity/article/later-huntergatherers-in-southern-china-18-0003000-bc/AE598987DF32F318EB18B4CA16B814E0 (accessed on 15 April 2020).

quaternary

Article

The Climate Fluctuation of the 8.2 ka BP Cooling Event and the Transition into Neolithic Lifeways in North China

Chao Zhao [1,2,3]

1 School of Cultural Heritage, Northwest University, Xi'an 710069, China; chz63@pitt.edu
2 Department of Anthropology, University of Pittsburgh, Pittsburgh, PA 15260, USA
3 The Louis Frieberg Center for East Asian Studies, Hebrew University of Jerusalem, Jerusalem 91905, Israel

Received: 22 May 2020; Accepted: 13 July 2020; Published: 4 August 2020

Abstract: Early Neolithic lifeways in North China, which are marked by a low-level food production economy, population aggregation, and sedentism, thrived just after the period of a climatic cooling event at 8.2 ka. Instead of simply regarding this climate fluctuation as a cause for the significant socio-economic transition, this paper attempts to explore the interplay between people's choices of coping strategies with climate change as a perspective to learn how people respond to this climate fluctuation and how such responses generated the interlocked socio-economic transitions. This analysis indicates that pre-existing changes in human adaptive behaviors prior to the cooling events were sufficient to enable people in certain areas to apply the intensification of food procurement in circumscribed territories as a strategy to cope with the climate fluctuations of the 8.2 ka BP cooling event. The application of such a coping strategy facilitated the economic and sociopolitical transition into Neolithic lifeways and led to the flourishing development of Neolithic cultures after 8 ka BP in North China.

Keywords: climate fluctuation; 8.2 ka BP cooling event; transitional sites; Early Neolithic; adaptive strategy; North China

1. Introduction

The development of Neolithic culture occurred right after the eighth millennium BP in North China, which is marked as one of the most dramatic economic, social, and cultural changes in Chinese history. For hundreds of thousands of years in North China, human beings lived in mobile hunting–gathering lifeways in simple and small groups. Then, during the early Holocene (approx.: 12–7 ka BP), significant changes happened and facilitated the multi-faceted transformations of people's lifeways in terms of economic practice, sociopolitical organization, and cultural identity. By contrast with Paleolithic lifeways, people in the Neolithic period were integrated into larger social communities, lived in sedentary or semi-sedentary villages, developed new types of art forms and beliefs, and carefully treated the dead to increase inter-personal social and cultural connections [1]. Rather than seasonally exploiting food resources in a large range of territories, people began to intensively use the local resources in the circumscribed environment, firstly by the combined reliance on intensified hunting–gathering and low-level primitive farming and then by further developed agriculture [2,3]. Such lifeways stimulated the booming of the population and fundamentally changed the pattern of interactions between humans and the environment [4].

It is widely assumed the Holocene climatic optimum facilitated the prevalence of Neolithic lifeways since it provided an affluent environment of food resources and provided suitable climate conditions for farming [5]. However, a finer-scale examination of the climatic change suggests that there are multiple rounds of climate fluctuations in post-glacial periods, reflected as the oscillations between the warmer–wetter and colder–drier periods [6–8]. The 8.2 ka BP cooling event is such a climate fluctuation and was expressed as a drop in temperature during the general trend towards the warmer and wetter climate in Early Holocene. This cooling event occurred at around 8.2 ka BP and lasted for two to four centuries (some records detected that the decrease in temperature was initiated as early as 8.4 ka BP) [9,10]. The archaeological records in North China indicate a dramatic increase in site numbers and the thriving development of Neolithic culture after 8–7.8 ka BP, implying that the all-round and widespread transformation into Neolithic lifeways in North China happened soon after the 8.2 ka BP cooling event. Such a temporal correlation implies that the 8.2 ka BP cooling event might have played an important role in stimulating the significant cultural transition. However, rather than simply regarding this climate fluctuation as a cause of the cultural transition, I present how the interplay between humans and the environment across the 8.2 ka BP cooling event drove the cultural change.

Accumulative cultural preparation is an important focus of this study, since how people respond to climate change deeply relies upon their knowledge derived from long-term experience. One aspect of such cultural preparation is a suite of technological packages used to intensively exploit resources and massively store them for delayed consumption since sedentary lifeways and large group aggregation require sufficient food supplies in a confined area [11–13]. Another aspect of cultural preparation is organizational innovation, since the need to maintain large group aggregation and strengthen the sense of territoriality required more sophisticated sociopolitical organization and intensified social connections than ever before [14–16]. Such cultural preparations were formed and developed along with human coping strategies to deal with environmental changes. Some traits of such cultural preparations sporadically occurred as early as the Late Paleolithic but were significantly developed and deeply converged with other traits in the Early Neolithic period. Therefore, learning how people are technically and sociopolitically organized either to cope with challenges or take advantage of the favorable conditions of the environment over a long time span would be an insightful approach to explore how the human–environment interaction during the 8.2 ka BP cooling event generated the profound social and economic transition into the Neolithic lifeways.

2. Materials and Methods

2.1. The Study Area and the Research Approach

North China (about 33°–43° N, 100°–125° E) is a vast geographic region in East Asia that features a temperate semi-humid monsoon climate. It is separated from Central and South China by the Qinling mountains and Huai River to the south, bordered by the Mongolian Plateau and the Yinshan and Yanshan Mountains to the north, and bounded by the Yellow Sea and Bohai Sea to the east and the neighboring Gobi Desert region in the west. The Yanshan and Taihang Mountains divide North China into two regions. The area to the west and north is the Loess Plateau, which is comprised by loess-covered highland, mountains, hills, and basins. The area to the east is the North China plain, a flat landscape with large water bodies and sporadically distributed hills and low mountains. Influenced by the alternations of summer and winter monsoon, North China has hot and wet summers and cold dry winters. As for the intra-environmental variations across the sub-regions, the weather becomes drier moving from the southeast to the north and west, where the strength of the summer monsoon declines gradually. The climate becomes colder moving from lower elevation plains to the mountains and highlands [17].

North China is one of a few centers in the world where complex agricultural systems emerged independently. The most widely consumed staple grains in the Neolithic period of North China were two types of millet, namely foxtail millet (*Setaria* spp.) and broomcorn millet (*Panicum* spp.) (foxtail and broomcorn, *Setaria* and *Panicum* spp., respectively) [3], while people in the southern and eastern parts of North China also consumed domesticated rice (*Oryza* spp.) as a supplemental food resource [18]. The onset of flourishing Neolithic cultures in North China was once thought as the outcome of fully fledged millet agriculture [19]. However, recent studies have revealed that multiple factors of Neolithic lifeways, such as sedentism, large group aggregation, and cultural and technological sophistication, all came before the full establishment of the agricultural economy [5,20]. Thus, early Neolithic cultures were based upon a low-level food production economy characterized by the broad-spectrum use of wild resources and the incorporation of domesticates as auxiliary resources [2].

North China is an ideal area to investigate how the 8.2 ka BP cooling event impacted the patterns of human–environment interactions and facilitated the significant cultural transitions. This climatic event is well evident in the paleoclimate records of China. Furthermore, the growing body of archaeological discoveries provides sufficient evidence to learn how people respond culturally during climate fluctuation. In the following section, the manifestation of the 8.2 ka BP cooling event in China is presented under the general paleoclimate background during the Paleolithic–Neolithic transition in North China. It provides a better understanding of the nature of this climate fluctuation at the suitable regional scale focused on by this paper. Then, the range of the sites used for the studies are mapped out in a sequential chronological frame, and the charts of the sum probability of their radiocarbon dating results are compared with the charts of the paleoclimatic proxies to show the concurrence pattern between the climate fluctuation and the significant cultural transition. Finally, the possible modes of coping strategies under climate fluctuation are summarized according the previous anthropological studies and how they are reflected by archaeological evidence is discussed, shedding light on the range of evidence I collected and discuss in the following sections to learn what specific coping strategies people have applied, how their application was based on the previous cultural preparation, and how the application of such strategies impacted the subsequent cultural development.

2.2. *Paleoclimate Fluctuation and the Paleolithic–Neolithic Transition in North China*

The early and mid-Holocene are the key periods of significant social transformation into Neolithic lifeways. On average, increased humidity and temperature distinguish the early-mid Holocene "Anathermal" from the preceding Younger Dryas. At the peak of the wet/warm period, the average temperature was 2~3 °C higher than today's standard and the annual precipitation was 50~300 mm more than it is currently in North China [21,22]. However, the wet and warm periods were not consistent, but punctuated by periodical aridity and temperature decreases [23]. The 8.2 ka cooling event is an example of one period with noteworthy magnitude [24].

The 8.2 ka BP event is apparent in many climate records, particularly from the Northern Hemisphere. The event itself was most likely caused by meltwater escaping from Lake Agassiz-Ojibway into the Atlantic Ocean via the Hudson Bay, which altered thermohaline circulation [9,25,26]. The evidence of such a climate anomaly around 8.2 ka BP is clearly shown in ice cores from Greenland, where air temperatures dropped by 3 to 6 ± 2 °C [27]. The reinforced Northern Hemisphere cooling would have increased the temperature gradient between high and low latitudes, which caused the migration of the inter-tropical convergence zone southward. Such an effect would have led to the weakening of the East Asian monsoons and increased aridity [28].

The high resolution, well-dated paleoclimate proxies from East Asia can reflect the impact of this cooling event. The pollen records from Bigeum Island near the Korean Peninsula revealed a rapid drop in arboreal pollen frequency and a corresponding increase in fern spores, which implies that an abrupt dry and/or cold event significantly impacted the distribution of vegetation on the island [29]. The changing $\delta^{18}O$ values of the Guliya ice core in the Tibetan Plateau and stalagmites from Lianhua Cave in the Loess Plateau of North China, Heshang Cave in Central China, and Dongge Cave in South China all reflect an interval of weakening Asian monsoons during a certain period between 8.4 and 8.0 ka BP, coinciding within error with the 8.2 ka BP event reflected by the Greenland ice cores [30–33]. Furthermore, a 10-year moving average annual rainfall record in southwest China during the 8.2 ka BP event was reconstructed based on a central-scale model and the comparison of two high-resolution stalagmite $\delta^{18}O$ records from Dongge Cave and Heshang Cave. This reconstructed record reveals that the mean annual precipitation in southwest China during the central 8.2 ka BP event was less than that of the present (1950–1990) by ~200 mm and decreased by ~350 mm in ~70 years [34]. The stalagmite records of Shihua Cave and Nuanhe Cave in Northeast China indicate that the summer monsoon-dominated $\delta^{18}O$ record only weakly express the 8.2 ka event. Nonetheless, the variations of the winter-dominated proxies of $\delta^{13}C$ and Ba/Ca reflect a colder and drier climate initiated at 8.42 ka BP and centered at ~8.2 ka BP [35]. A clear drop in $\delta^{13}C$ at 8.6–8.1 ka BP can also be detected from the peat bog record of Hongyuan in the eastern Tibetan Plateau, indicating an obvious dry/cold period with decreased humidification [36].

A sharp increase in precipitation and temperature can also be detected by these paleoclimatic proxies after the 8.2 ka BP cooling event, which marks the onset of the mid-Holocene climatic optimum. The stalagmite record of Lianhua Cave shows a 2.5‰ $\delta^{18}O$ isotope depletion at 8.1 ka [33]. Such a magnitude of change is roughly similar to the transition from the Younger Dryas to the Early Holocene at 11.5 ka [5].

The cultural development and site distributions throughout the period from 11,500 to 7000 BP are presented in Figure 1. The comparison of climate fluctuation and cultural development (particularly referring to the occupation intensity of different types of sites) throughout this period is presented in Figure 2 (Supplementary Table S1). These two figures reflect the patterns of concurrence between climate and cultural changes, suggesting that the transitional sites occurred along with climate change into warmer and wetter conditions initiated from the beginning of Holocene (after 12 ka BP). These sites still share many features in common with the Upper Paleolithic sites in North China (45–12 ka BP) but, meanwhile, have shown the buds of the Neolithic cultural traits. During or even preceding the 8.2 ka BP cooling event, the Early Neolithic sites, characterized by settled inhabitation, large group aggregation, and intensive site use, emerged but were very rare and distributed only in limited areas of North China. With the onset of the climax of the Holocene climatic optimum (7.8–7 ka BP), the numbers of Early Neolithic sites soared and they were widely distributed across North China with sharply increased occupation intensity. In the following sections, the specific interplay between humans and the environment will be analyzed and discussed to learn the underlying mechanisms of this significant social change under climate fluctuations.

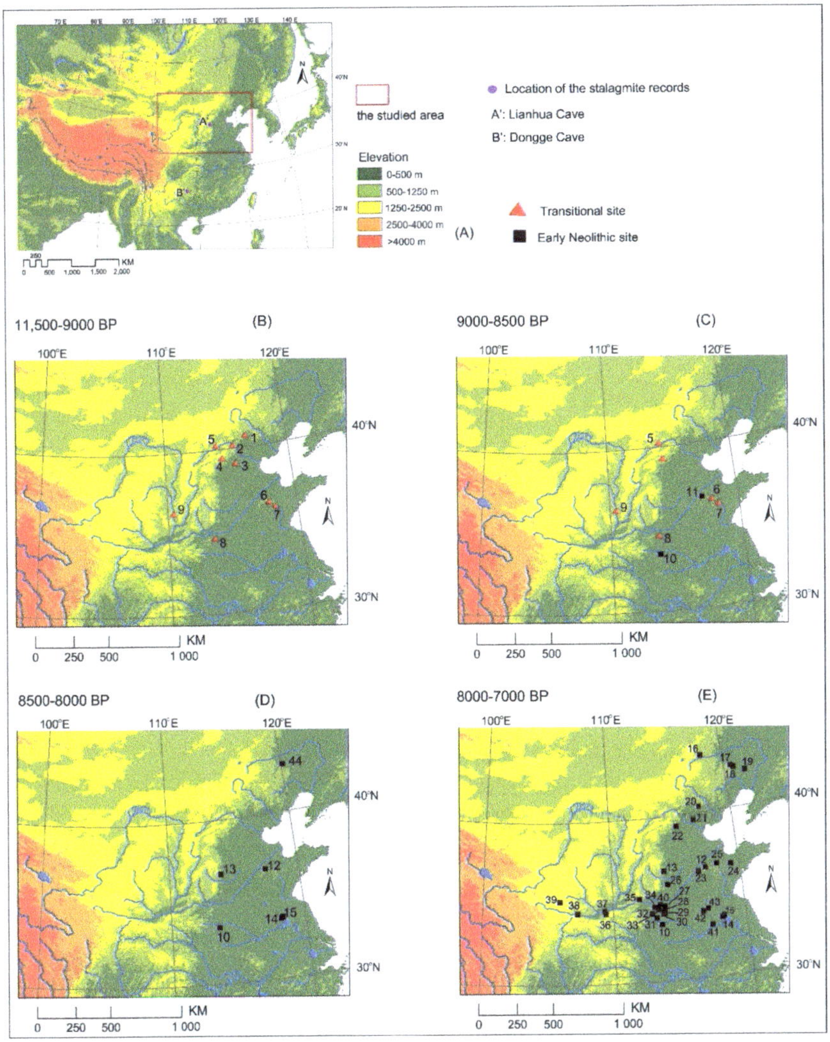

Figure 1. The distributions of the archaeological sites from 11,500 to 7000 cal BP. (**A**). Studied area; (**B**). The distributions of the archaeological sites from 11,500 to 9000 cal BP; (**C**). The distributions of the archaeological sites from 9000 to 8500 cal BP; (**D**). The distributions of the archaeological sites from 8500 to 8000 cal BP; (**E**). The distributions of the archaeological sites from 8000 to 7000 cal BP (The name of the sites: 1: Zhuannian [37]; 2: Donghulin [38]; 3: Nanzhuangtou [39]; 4: Ximiao [40]; 5: Yujiagou [41]; 6: Huangya [42]; 7: Bianbiandong [43]; 8: Lijiagou [44]; 9: Shizitan S9 and 12G [45,46]; 10: Jiahu [47]; 11: Zhangmatun [48]; 12: Xihe [49]; 13: Cishan [50]; 14: Shunshanji [51]; 15: Hanjing [52]; 16: Baiyinchanghan [53]; 17: Xinglonggou [54]; 18: Xinglongwa [55]; 19: Chahai [56]; 20: Shangzhai [57]; 21: Beiwang [58]; 22: Beifudi [59]; 23: Yuezhuang [60]; 24: Qianbuxia [61]; 25: Houli [61]; 26: Huawo [62]; 27: Shawoli [63]; 28: Peiligang [64]; 29: Tanghu [65]; 30: Shigu [66]; 31: Shuiquan [67]; 32: Zhongshanzhai [68]; 33: Egou [69]; 34: Tieshenggou [70]; 35: Bancun [71]; 36: Beiliu [72]; 37: Baijia [73]; 38: Beishouling [74]; 39: Dadiwan [75]; 40: Maliangou [76]; 41: Shuangdun [77]; 42: Shishanzi [78]; 43: Xiaoshankou [79]; 44: Xiaohexi [80]).

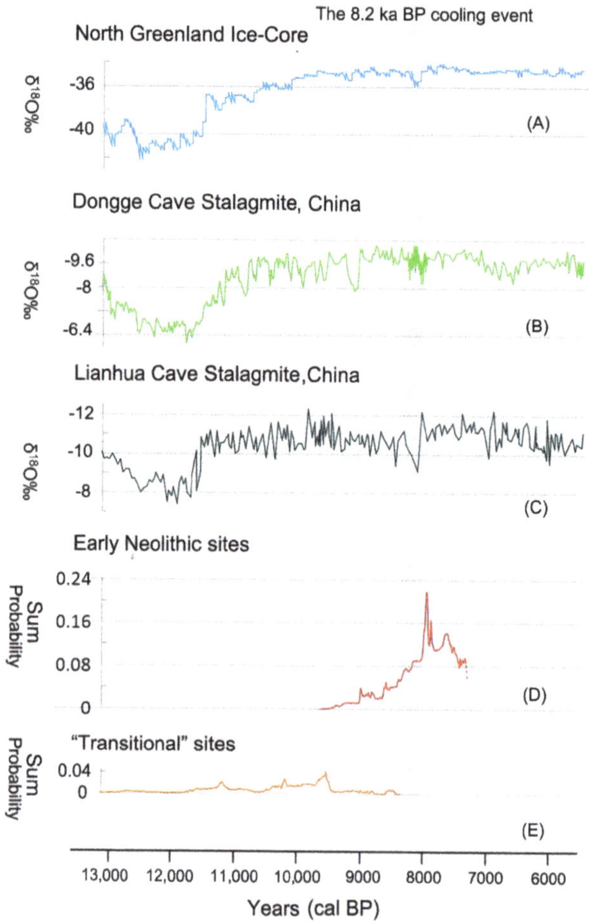

Figure 2. Graphic comparison of human occupation intensity and climate fluctuation. The shaded gray bar shows the time span of the 8.2 ka BP cooling event. The curve (**A**) represents hemispheric changes in precipitation and temperature recorded in the North Greenland ice core [81]. Higher $\delta^{18}O$ values indicate warmer temperatures and higher humidity. The curve (**B**) represents a proxy for monsoon intensity in South China from the Dongge Cave stalagmite records [31]. Higher $\delta^{18}O$ values indicate strong winter monsoons (cold/dry) and lower $\delta^{18}O$ values indicate stronger summer monsoons (warm/wet). The curve (**C**) represents a proxy for monsoon intensity in North China from the Lianhua Cave stalagmite records [33,82]. Higher $\delta^{18}O$ values indicate strong winter monsoons (cold/dry) and lower $\delta^{18}O$ values indicate stronger summer monsoons (warm/wet). The curves in (**D**,**E**) respectively indicate the relative occupation intensity of the Early Neolithic and the transitional sites across North China based on the sum probability of radiocarbon age estimates. The basic information of the radiocarbon dates (including the dataset of the radiocarbon dating and the explanation of how the sum probability distribution model was built) is presented in the supplementary data.

2.3. *Human Responses to the Impacts of Climatic Change*

How climate change impacts on the development of human society has long been a hot topic attracting broad academic attention. Nowadays, environmental determinism is already obsolete and few scholars would regard environmental change as the decisive force leading to social changes.

The influences of climate change on human society are now evaluated under the complex patterns of the interplay between humans and the environment since we realized that human beings are able to flexibly handle the situations caused by environmental changes in various ways [83]. Facing similar natural challenges caused by environmental deterioration, different human communities might come up with varied coping strategies to deal with them. Such divergent strategies would have different impacts on the trajectory of local social and cultural development [84–86]. The choice of coping strategy is usually the result of previously held beliefs, social–political organization, technological bases, and historical interaction with neighboring societies [87,88]. Moreover, the potential advantageous conditions due to climatic amelioration may only benefit people and stimulate cultural fluorescence when people are fully prepared in cultural and social dimensions to sufficiently make use of such conditions [11,89].

The change into a colder and drier climate in historic and present day North China is often considered as climate deterioration because: (1) it would lead to the reduction of total biomass and therefore cause a decrease in resource affluency and the sizes of rich resource patches [90]; (2) it will also shorten the growing season, leading to the extension of the lean season for food resources [91,92]; (3) it will lead to unfavorable conditions for farming activity, at least in some parts of North China, due to the threat of disasters like drought and frost [93,94]. The sharp climatic change from a warm/wet period into a cold/dry period could also increase the risks of subsistence practice and force people to adjust their coping strategies, since the past experience of subsistence strategy might no longer fit well with the changing environment situation [95].

Based on an anthropological view, the ancient foragers might have had several options to handle the situation of such climate deterioration, as listed below:

(1) To increase mobility: This can be achieved as either expanding the foray distance or increasing the frequency of movement, or both [96]. Such a coping strategy could have enabled foragers to use a larger range of territory to procure resources and make up the deficiency of food resources in any confined areas, or reduce the resource pressures of any circumscribed regions [90,96].

(2) To actively intensify the yields of food resources in the circumscribed area where the resources are relatively abundant or have a suitable environment for food production [87]: The surplus of the food resources usually went into storage and could be consumed later in lean seasons [97,98]. A substantial amount of effort and/or prerequisites are required to fulfill such a strategy, such as technological improvements, new inventions, part-time cultivation, and improved methods and investments for resource management [13,99,100]. Such requirements call for intensified investment in the circumscribed area and therefore people must have settled down and developed effective ways to defend the resources within their immediate territory [14].

(3) To make long-distance migrations: Since environmental deterioration exerts impacts to varying degrees in different places, some areas might be less severely affected by climate change and might serve as destinations for migration [101]. Differing from the two strategies mentioned above, the migration strategy might have not necessarily required people to modify their adaption strategies since people could have still relied on past coping strategy by moving into the places where their past experience of subsistence strategy still fulfilled their needs [102]. However, since these destinations are usually not "empty", the long-distance migration might have caused changes in social and cultural interaction, either reflected by inter-group conflict or amalgamation [103].

(4) To develop social alliances and increase the intensity of social exchanges: This can increase reciprocal ties and buffer against subsistence risks [104]. This strategy is especially suitable in more heterogenous environments where the neighboring people may have relied on different subsistence bases and could have mutually complemented each other's strategies through exchange [11].

Different coping strategies can be reflected by a range of archaeological evidence listed in Table 1. In the following section, the multiple lines of evidence from the sites will be analyzed to learn the specific coping strategies people relied on to deal with the climate fluctuation in North China. The sites include both the ones during the time span of the 8.2 ka BP cooling event and prior to this climate fluctuation. Since the choice of coping strategy is closely based on the cultural preparedness that evolved from prior subsistence practices, the incorporation of the earlier sites strategies provides us an important cultural context to evaluate how and why people chose certain strategies to cope with the climate fluctuation of the 8.2 ka BP cooling event.

Table 1. Different coping strategies under climatic deterioration and their corresponding material implications reflected by archaeological evidence.

Coping strategy under climatic deterioration	Archaeological manifestation
Increase mobility	Scarcity of the sites; lack of the permanent site structure; thin cultural deposits; portable toolkits.
Intensify the resource exploitation of the circumscribed affluent area	Toolkits used for intensive resource extraction; traces of intensified site occupation (thick cultural layer, permanent site structure, etc.); faunal and floral evidence showing the intensified resource use; traces of increased social integration (enlargement of the site, non-utilitarian goods); traces of territoriality (cemetery, defense structure).
Long-distance migration	Evidence of cultural transmission; the sharp decline in site numbers contrasted by the dramatic increase in site numbers in another area.
Alliance and trans-regional exchange	The presence of exotic goods, mutual cultural influences; stylish cultural markers showing the presence of reciprocal ties.

3. Results

3.1. The "Transitional Remains" of the Early Holocene (11.5–8.5 ka BP)

In China, "Paleolithic" and "Neolithic" are short-hand terms for Pleistocene foragers and primitive Holocene farmers [4]. The differences between the two temporal/cultural entities are so pronounced that their shared heritage and genealogy are largely unexplored [4]. In past studies, some scholars have drawn attention to the lack of information connecting terminal Pleistocene foragers to the Early Neolithic farmers (or low-level food producers) and cited a "gap" in the archaeological record of North China from about 11,500 to 9000 BP [105,106]. However, although the evidence is still scarce, transitional sites that emerged in the Early Holocene prior to the full appearance of "Neolithic packages" have been found in different regions of North China. The names of the distributions of these sites are listed in maps "B" and "C" in Figure 1.

People were probably not sedentary and were organized as relatively small groups in this period. This inference is made from the evidence that: (1) no permanent dwelling structures have ever been found across the sites; (2) the site size is usually smaller than that of village-like settlements, except for the Nanzhuangtou site (which is estimated to be as large as 2 ha, but the precision of the estimation is still debated) [39]. The open-air site Donghulin is as large as 0.3 ha, smaller than most of the Neolithic sites found in a later period [38]. The Bianbiandong and Huangya sites are rock shelter sites, with only narrow spaces for daily activities [42]. The Lijiagou and Shizitan Locality 9/Locality12 G sites are deeply buried under the earth, so it is hard to make a reliable estimation of the site area. However, according to the exposure of the lithic remains on the profile, the site areas are probably not large [45,46,107].

Nonetheless, other lines of evidence suggest that people might have put more investment in site construction than in earlier periods. Burned earthen surfaces have been found in both Donghulin and Bianbiandong sites [38,43]. These are in an irregular oval shape and are harder than the surrounding depositions, which is inferred as living floors [38,43]. Decayed wooden poles have been found in the Nanzhuangtou site, which are suggested to have been used for constructing simple dwellings [39]. The hearths identified in the Donghulin site are in a more complex structure than the ones found in the Upper Paleolithic period (45–12 ka BP). The lower level of the hearths was arranged neatly by placing rocks in a circle. They are about 0.5–1 m in diameter and 0.2–0.3 m in depth, which is thicker than most of those found in the earlier period, showing the more intensive and perhaps persistent use of them [38]. Piles of rocks in semi-circular shapes have been found in the Lijiagou site. Such rock crescents might be correlated with dwelling construction and indicate more labor investment for inhabitation [107].

The occurrence of secondary burial in the Donghulin and Bianbiandong sites also implies that people had relatively permanent settlements at this time, to which the dead were brought back for ritual burials [108].

The faunal and floral remains found during this period indicate that people relied on a wide range of wild resources for their subsistence base. Some r-selected species (referring to the ones that quickly and massively produce short-lived "cheap" offspring), which are abundant in nature and hard to deplete in a short time, were intensively exploited, such as acorns in the mountainous regions and water caltrop in wetland environments [109]. Meanwhile, the earliest evidence of millet domestication was identified from the flotation results of the Donghulin site and starch analysis from the Zhuannian and Nanzhuangtou sites. The morphological studies of these millet remains show a combination of both wild and domesticated features, indicating that millet was still undergoing domestication [110,111]. Even though flotation results show that the broomcorn and foxtail millet make up only a tiny proportion of the charred seeds, morphological changes from the wild versions suggest that people had begun to carry out intensive intervention and management of the growth and reproduction of millet [111].

Corresponding technologies related to resource intensification are widely found in this period. Except for Shizitan Loc 9/Loc 12 G, pottery sherds have been discovered in all other sites [112]. Pottery provided people with innovative cooking techniques, which facilitated a more thorough intake of nutrients from food resources, like the grease from animal bones and hard-to-digest plant resources such as millet [92,113]. However, pottery is breakable and not suited transport over long distances, therefore it does not fit well in highly mobile lifeways [114]. The pottery found in this period is simple in form, crude, coarse-grained, and fired at low temperatures. The quantities are small in comparison with later period sites [115]. Such evidence indicates that even though the pottery and accompanying new cooking methods were widely adopted by foragers, vessels were not as intensively used as by the subsequent Neolithic farmers. The mobile lifeways and the limited levels of resource intensification might have confined the extent of people's reliance on pottery use [113].

Like pottery, grinding stone tools are also used for the intensive exploitation of food resources. They can process a wide range of foods, like nuts, tubers, bulbs, roots, and grass seeds, grinding them into powders that facilitate storage and consumption [116]. The grinding stones can be found in all of the abovementioned sites, except for Shizitan S12G [112]. However, most of the grinding stones found in this period are less regular in shape, with narrower grinding surfaces in comparison with those found in Early Neolithic sites after 8 ka BP [117] (the grinding slabs shown in Figure 3 and Figure 5 can provide an intuitive comparison). Moreover, grinding slabs make up only a small proportion of the lithic tool assemblage in each site except Nanzhuangtou, where only very limited number of lithic remains have been found [39]. In the Donghulin site, the grinding tools only comprise 11.05% of the tool assemblage [118]. In the Lijiagou site, they are only 13% of the tool assemblage in its upper cultural layer (approx. ~10.5–10 ka BP) [44]. Chipped stone tools still dominate the lithic assemblages in these sites. Such a pattern indicates that, even though grinding stones were widely adopted by people as auxiliary tools for food processing, they were not yet expected to play an important role in massively processing food resources at the unprecedented levels seen in the Early Neolithic period (about 8.5–7 ka BP).

To sum up, the features of the transitional sites, on the one hand, more closely resemble the preceding Upper Paleolithic sites in terms of the lack of permanent dwelling structures, small site size, chipped stone-dominated tool assemblage, and thin cultural depositional layers. Such similarities indicate that people in the Early Holocene at least partially inherited the socio-economic organization and land-use strategies from terminal Pleistocene foragers. However, the techniques necessary for the intensive exploitation of resources that were prevalent in the subsequent Neolithic period had already appeared and were widely adopted during this period [112]. The presence of domesticated millet also indicates that the knowledge of intensive resource management for certain species, and even cultivation, was developed in this period [111].

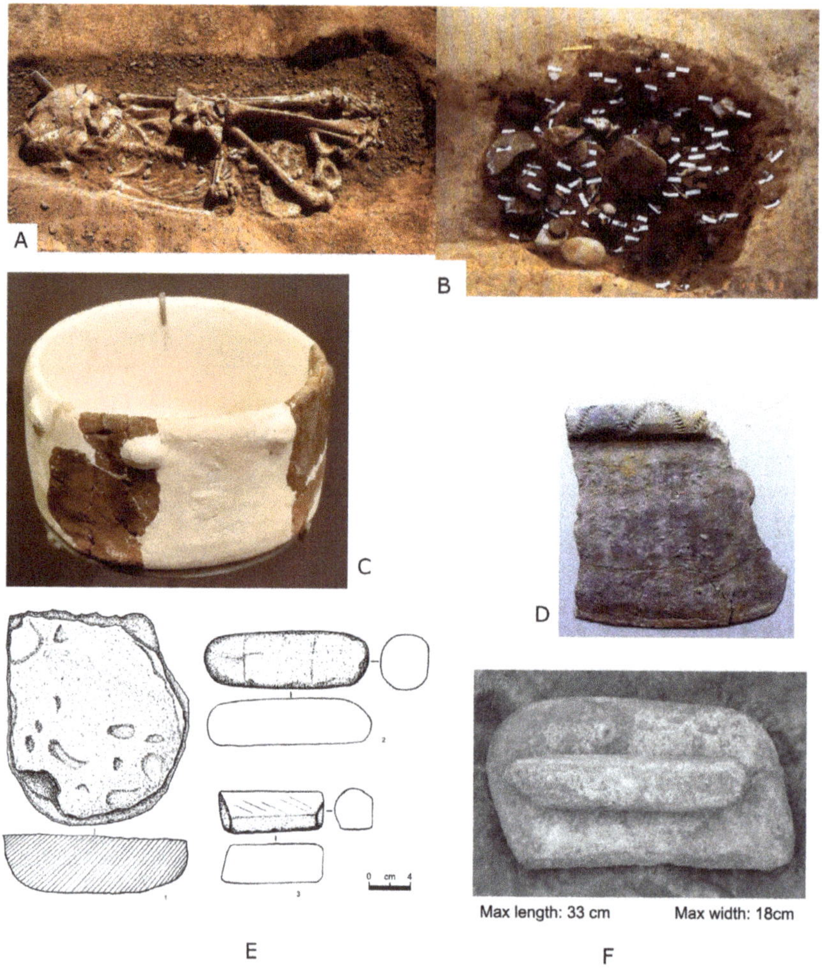

Figure 3. Features and artifacts found from transitional sites (11.5–8.5 ka BP) which show the innovative cultural traits that herald the "Neolithic package" and differ from typical Paleolithic remains ((**A**): sedentary burial, Donghulin site (2 in Figure 1); (**B**): stone-lined hearth, Donghulin site (2 in Figure 1); (**C**): Pottery bowl, Zhuannian site (1 in Figure 1); (**D**): pottery sherd with decorations on the rim, Donghulin site (2 in Figure 1); (**E**): stone slab and rollers Bianbiandong site (7 in Figure 1); (**F**): stone slab and roller, Donghulin site (2 in Figure 1), the maximum length and width are marked in the graph, after [37,38,43]).

Though these techniques and knowledge were still in their primary stage of development and had not yet became prevalent, they did stimulate subtle behavioral modifications and provided cultural preparedness for further social and economic transformation. For instance, such techniques and knowledge enabled people to fully exploit certain kinds of food resources that thrived due to climate amelioration at the beginning of Early Holocene, such as acorns and water caltrop. The paleoclimate data indicate that the beginning of the Early Holocene (which started at around 12 ka BP) was a warmer and wetter period in comparison with the Terminate Pleistocene period [31,82]. The warm and wet climate facilitated the recovery of forests in the mountainous region of North China and may have increased the distribution of nut-bearing trees, like oaks. The wet climate also expanded wetland areas and facilitated the growth of highly productive wetland food resources [109]. Even though sedentary lifeways had not yet been formed, the enhanced food resource supplies or the need to intensively exploit food resources in certain areas might have led people to stay for longer durations, making more intensive and regular use of certain sites within their land-use system.

3.2. Pioneering Changes in Coping Strategies Across the 8.2 ka BP Cooling Event in North China

Dating results for the transitional sites in the Early Holocene reveal that most of them are concentrated between 11.5 and 8.5 ka BP, suggesting that the main occupational phase of these sites is before the 8.2 ka BP cooling event. There are large "empty" areas where no archaeological sites have been discovered during the 8.2 ka BP cooling event except for a few "pioneering" sites found in specific regions which provided rich and diversified food resources. These sites show similar features to those thriving Early Neolithic settlements that appeared after 8 ka BP. This pattern implies significant modifications to human coping strategies in the period of the 8.2 ka BP cooling event.

The appearance of the "pioneering sites", characterized by larger-scale population amalgamation, sedentary lifeways, intensive site use, and rich material remains, suggests that people in some areas adopted the strategy of staying put in resource-rich areas and investing in ways to increase the local food supply as a means to adapt to climatic deterioration. Several sites have reflected the adoption of such a strategy, as discussed below.

3.2.1. The Jiahu Site

The Jiahu site (shown as 10 on maps "D" and "E" in Figure 1), is located in the Huai River Valley on the southeast edge of the Huanghuai alluvial plain in Wuyang County, Henan Province. Based on radiocarbon dates and cultural features, the Jiahu site was occupied from 9000 to 7600 BP, and can be divided into three phases: (1) 9000–8500 BP; (2) 8500–8000 BP; (3) 8000–7600 BP [119]. The total area of the site is about 5.5 ha and more than 2900 m^2 have been excavated over eight seasons of excavation from 1983 to 2013 [120,121]. Phase 1 of the Jiahu site was before the appearance of the 8.2 ka BP cooling event and was contemporary with some "transitional remains", such as the upper layer of the Lijiagou site and the lower layer of Shizitan Locality 9. Certain features of cultural remains in Jiahu Phase 1 differ pronouncedly from the "transitional remains" for various aspects. One crucial difference is the complexity of the site structures. Fifteen houses have been discovered, which are widely distributed from the west to east zone of the site and can be divided into five clusters. Fifty-one burials are laid close to the pit houses. In addition, two kilns and numerous pit structures have been found within the excavation zone. The pit structures were used as storage facilities or midden deposits [47]. Such complexity of site structure implies that people were sociopolitically organized in different ways and occupied the sites with more stable residency. The clustered distribution of houses probably reveals the amalgamation of different groups (which might have family ties) rather than the natural growth of a small band [122]. More investment for dwelling construction, storage facilities, and garbage disposal indicate a pattern of relatively stable habitation rather than ephemeral site use [123]. Moreover, pottery was used far more frequently in the Jiahu site [47]. The forms of pottery were more diversified than those found in the transitional sites, and they were made in more regularized shapes and fired at higher temperatures. The grinding stones were also made in more regular shapes [124].

Phase 2 of Jiahu had temporal overlap with the 8.2 ka BP cooling event. In this period, the settlement was further developed and became prosperous. The increasing number of artifacts, as well as houses, burials, and pit structures, indicates that people made more intensified use of the site during this period [47]. The houses were still distributed in clusters (the Jiahu report divides them into six clusters). The increase in burial numbers from 51 to 348 between Phase 1 and Phase 2 suggests a sharp increase in population size [119,121]. Unlike Phase 1, when burials and houses were mixed together, concentrated cemetery zones appeared. The distributional pattern indicates that, while they were spatially separated from the houses, each of the cemetery zones roughly corresponded with one house cluster. Analysis of the strontium isotopes of human bones from the cemetery indicates that, beginning in Phase 2, a few people from other regions migrated into the Jiahu site [125]. Some special items, such as turtle shells with gravel, a bone flute, and other carved bone "folk-shaped" items, have been found in a few of the burials [47]. Such items (seen as Figure 4) probably served as ritual paraphernalia and the appearance of them implies the elaboration of ritual activity during this period.

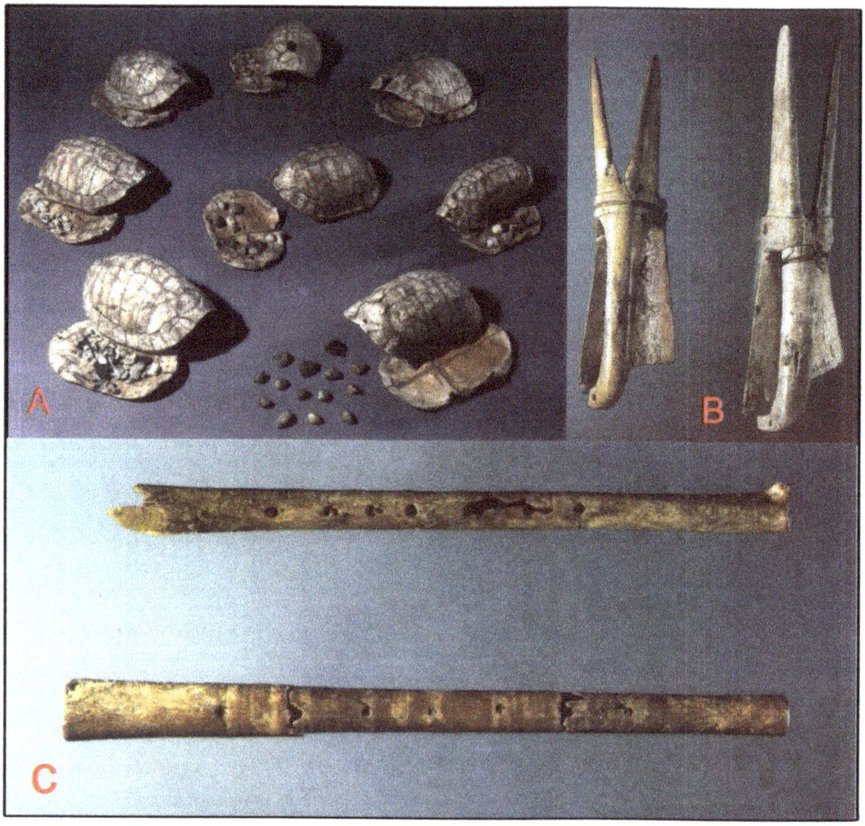

Figure 4. The items possibly used as ritual paraphernalia in the Jiahu site ((**A**) turtle shell with gravel, found in burial M363; (**B**) bone folk-shaped item; **1**: found in burial M395, **2**: found in burial M363. (**C**) bone flute; **1**: found in burial M57, **2**: found in burial M68; after [47]).

How could the Jiahu community have become prosperous during the general background of climatic deterioration? This could well be closely related to people's strategical exploitation of rich resource patches during the period of climate deterioration. The Jiahu site is located on the low-lying alluvial plain and near the confluence area of the ancient Hui and Sha rivers. Based on the study of geomorphology and sedimentology, Jiahu is situated on a hillock slightly higher than the surrounding area and faces large marshes and waterbodies to the east. Such a location offered easy access to different ecozones for procuring food resources within a circumscribed area. The marsh to the east of the site provides highly productive wetland food resources and is an ideal place for rice cultivation. To the east, a zone with rich aquatic resources provides habitat for freshwater fishes and shellfish. To the west of the site is a vast plain with terrestrial resources like deer, tree nuts, and fruit [47]. The Jiahu tool assemblages (seen as Figure 5) indicate that, through all three occupational phases, people were equipped with complex techniques to procure various resources from different ecozones: specially designed bone harpoons in varying sizes were used for gathering fish, and bone arrowheads were used for hunting. The grinding tools were enlarged and produced in more formal shapes, making up higher proportions within the lithic tool assemblages (30% of the lithic tools in Jiahu Phase 2) in comparison with those of "transitional remains" [125]. Such changes reflect that specialized technologies were used intensively for the bulk processing of plant resources. The stone hoes and sickles may have been used as tilling and harvesting tools relevant to farming activity [124]. The flotation results and faunal remains also indicate that people extracted a broad spectrum of wild plant and animal species from different ecozones and incorporated domesticated rice and pigs as supplemental food resources [126,127].

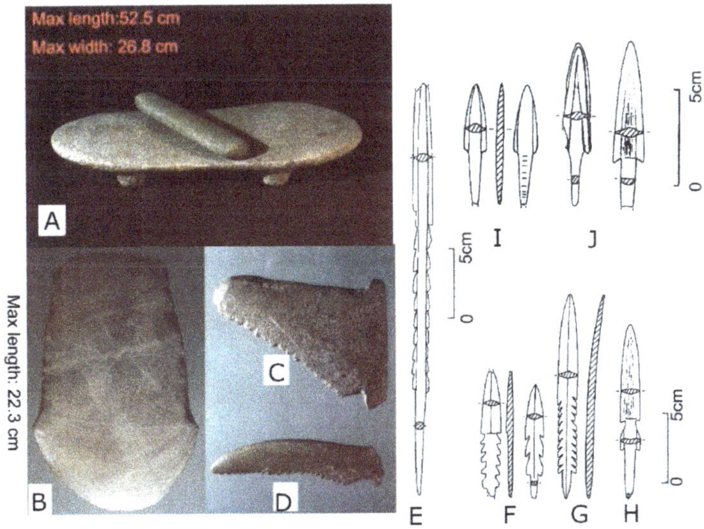

Figure 5. Toolkits of the Jiahu site which show the intensity of the techniques of food acquisition ((**A**): stone slab and roller, the size of the stone slab is marked in the figure; (**B**): stone hoe; (**C**,**D**): stone sickle; (**E**–**H**): bone harpoon; (**I**,**J**): arrowhead, after [47]).

The climatic deterioration and sophistication of the techniques for resource acquisition increased the value of specific rich resource patches, like the area around the Jiahu site. For one thing, climate deterioration resulted in a "patchy" distribution of freshwater resources that effectively constrained the distribution of the plants and animals that depended on them. This would concentrate the distributions of rich resource patches into smaller areas and increase the inter-patch distance [90]. Additionally, food resources in such rich patches could be fully exploited by human beings with the assistance of complex techniques. Under such circumstances, people would likely be tethered to such constrained areas instead of making a long-distance movement. This would facilitate the growth of local populations and even attract people from less affluent areas. The enlargement of the community size required more food supplies and therefore stimulated further resource intensification. Besides, increasing population density would have led to scalar stress among community members and might even have cause social conflict, which called for innovative sociopolitical solutions to cope [128]. The elaboration of ritual activities, the division and management of cemeteries, and perhaps even the construction of ditches at the boundary of the settlement at Phase 2 were likely used as a cultural means for the Jiahu community to enhance social integration. The intensification of ritual activities, as ethnographic evidence indicates, could also potentially have been used as a means to deal with subsistence stresses and risks, which often appear when factors like population growth, social interaction, environmental change, and climate fluctuation have broken up the pre-existing balance between humans and the environment [86,129].

3.2.2. Shunshanji and Hanjing

Shunshanji and Hanjing are two adjacent sites located in the low-lying hills of the lower Huai River drainage area in the eastern part of Huanghuai Plain, which is the transitional zone between North and Central China (respectively shown as 14 and 15 in maps "D" and "E" of Figure 1).

The cultural remains of Shunshanji can be divided into three phases. Radiocarbon dates indicate that Phase 1 and Phase 2 lasted from 8.5–8 ka BP (the boundary between Phases 1 and 2 is estimated to be at 8.3~8.2 ka BP), while Phase 3 is estimated as the period of 8–7 ka BP. The size of Shunshanji is 17.5 ha, which is the largest ever found before 8 ka BP in North China. An area of 2750 m^2 has been excavated in total over three field seasons. The settlement has a ditch that encircles an area of nearly 7.5 ha. The cultural remains were found both within and outside of the ditch, including houses, pit structures, and burials [51].

Phytoliths from the ditch profile reflect the impact of the 8.2 ka BP cooling event at the beginning of Phase 2, with types indicating the cold climate significantly increasing during this period [130]. Yet like the Jiahu site, the Shunshanji cultural remains suggest that the settlement became prosperous during Phase 2. During this period, the ditch was constructed and a cemetery appeared to the northwest of the ditch. The distribution of cultural remains expanded and a richer amount of artifacts and relics was deposited [51]. Such a pattern indicates more intensified occupation of the site. According to the presence of a large quantity of deer, boar, fish, and turtle bones and the features of lithic types [51,131], the subsistence base of the settlement was still based on the intensive exploitation of wild resources. Rice phytoliths have also been found, and interestingly, the amount of them in Phase 2 is higher than those from both Phases 1 and 3, which may suggest that people intensified the cultivation of rice to some extent during the cooling period and decreased the level of intensification when the climate returned to a warmer and wetter pattern [130]. Whether or not people applied institutionalized rituals to enhance social integration at Shunshanji is still not clear. But the placement of small animal- or human-shaped pottery figurines in the ditch might be correlated with specific rituals or sacrifices [132].

The Hanjing site is 4 km west of Shuishanji and radiocarbon dates indicate that the main occupational phase was contemporaneous with Shuishanji [52]. The Hanjing site is about 5 ha and a 3 ha area of the site was encircled by a ditch. The ditch was infilled with numerous types of garbage, like broken ceramics and animal bones. The evidence shows that the ditch was filled and the roads and activity zones were formed on top of it during the late occupational phase, suggesting the expansion of the site [52]. Besides the ditches, houses, and pit structure, a section of paddy field has been found, directly indicating intentional rice cultivation [131].

3.2.3. Zhangmatun and Xihe

Zhangmatun and Xihe are two adjacent sites (slightly more than 20 km apart) located in the transitional zone between the hills/low mountains of the Taiyi Mountain range and the North China alluvial plain in Shandong Province in eastern North China (respectively shown as 11 and 12 on maps "C", "D", and "E" of Figure 1). The Zhangmatun site is estimated to be as large as 0.7 ha, and 600 m^2 have been excavated. One pit house and one pit structure have been identified, along with a rich amount of faunal and floral remains. Dating results show that this site was occupied from 9 to 8.5 ka BP [48]. Although the limited excavation area does not allow us to explore the site's layout features, the appearance of a pit house and associated rich food remains resemble the site features of Jiahu Phase 1. Though the site size is smaller than Jiahu, it is larger than most of the sites classified as "transitional remains". Such evidence suggests that the transition into Neolithic lifeways, as indicated by growing residential stability, increasing levels of resource intensification, and the enlargement of group size, appeared prior to the 8.2 ka BP cooling event in this region.

The Xihe site dates to between 8.4 and 7.7 ka BP. The early occupational period of the site was contemporary with the 8.2 ka BP cooling event. More than 2400 m^2 of the site have been excavated, revealing over 30 houses and numerous surrounding pit structures. Houses were more densely distributed than in the Zhangmatun site. The largest house is more than 70 m^2 and most are as large as 25~50 m² [49]. The living floors and walls are more elaborately treated than the house found in Zhangmatun. The density of pottery in Xihe is also higher than in Zhangmatun, indicating that people used pottery far more frequently [133]. All these lines of evidence suggest that during the period of the cooling event, people were more stably settled in the circumscribed area, perhaps with a larger population aggregation and more intensified social integration (evidenced by the more compact distribution of houses and pit structures). However, the Xihe site is devoid of the elaborate expressions of ritual and symbolic affairs seen in the Jiahu site, indicating a less complex sociopolitical integration than at the Jiahu site.

The faunal and floral remains of Zhangmatun and Xihe indicate that people relied heavily on a wide range of wild resources and also incorporated domesticated crops and pigs as supplemental foods [20,134]. The domesticated crops in Zhangmatun were two types of millet (broomcorn and foxtail) [135]. The crops in Xihe were rice and foxtail millet [136]. The practice of rice cultivation, as well as a large amount of fish and bird bones found in both the Zhangmatun and Xihe sites, suggest that in addition to the exploitation of terrestrial resources, wetland and freshwater resources were intensively used [20].

3.2.4. Cishan

The Cishan site is located in the transitional zone between the Taihang Mountains and the North China plain (shown as 13 on maps "D" and "E" of Figure 1). Cishan has been dated between 8.4 and 7.6 ka BP. Three pit houses and 889 pit structures have been found within the 5519 m² excavation zone [137]. The site sits on a loess platform near a large river, and is not far from the forested piedmont zone. Such a location provides good opportunities for people to get resources from different environmental zones near the site [138].

The nature of the site is highly disputed. Given the facts that (1) millet and hackberry (*Celtis sinensis*) seeds have been found in some of the pit structures and (2) the majority of the remains are identified as pit structures instead of houses, some scholars have considered that the Cishan site was used as a specific place to store food resources and even hold sacrifice activities based on stored foods [139,140]. However, some scholars still insist that the Cishan was a residential site since it was excavated during the early period when the archaeologists still lacked the experience to distinguish large pit structures and small and simply built pit houses without interior hearths [141–143]. I agree with the inference that the Cishan site was a residential settlement for the following reasons: (1) some simply constructed houses could be mis-identified as large pit structures (i.e., middens or storage pits) and (2) the diversity of tools suggests that people were engaged in various activities besides the storage behaviors at Cishan, like woodworking, food processing, hunting, fishing, and cultivation [50]. The highly diversified faunal remains, such as migratory birds and deer with both dropped and undropped antlers, suggest multi-seasonal hunting [144]. Therefore, the Cishan community also shows a coping strategy of intensively using resources around the site and massively storing the food for delayed consumption in lean seasons. The large site area (estimated as roughly 10 ha) probably implies a large group aggregation as well.

However, the in situ distribution of about 140 suites of tool compounds reflects the behavior of intentionally preparing and leaving the tools for later use. This implies that people might have left the site temporarily and anticipated a later return [123]. Such combined evidence suggests that, on one the hand, Cishan communities stayed for a longer period of time in the site and repetitively used it based on a broad range of resources exploited from the surrounding environment. For another thing, the characteristics indicating sedentism were not yet stable.

3.2.5. Xiaohexi Remains

Prior to the flourishing development of Xinglongwa culture around 8~7.2 ka BP, a couple of newly discovered sites have been attributed to "Xiaohexi culture" in the west Liao River Valley in the northeast part of North China [145] (the type site, Xiaohexi, is marked as 16 on the maps "D" and

"E" of Figure 1). The Xiaohexi sites are usually located on slopes or high loess platforms along river valleys [146]. The age of these remains is mainly inferred from the stratigraphic evidence. The previous excavation of the Baiyinchanghan site revealed that a pit house of Xiaohexi culture was stratigraphically overlain by a pit house of early Xinglongwa culture [147], suggesting that at least part of the Xiaohexi remains might be even earlier than Xinglongwa culture, potentially with a time overlap with the 8.2 ka BP cooling event.

The Xiohexi sites are organized as hamlets or small villages. The appearance of pit houses and diversified toolkits indicates more settled lifeways with the intensive use of a wide range of surrounding resources near the site [80]. However, the Xiaohexi pottery is crude and simple in form [148]. The cultural deposits are thin and there are no complex overlaps of houses and pit structures built at different times, as seen in the Jiahu site [145]. Such patterns indicate that the degree of residential stability was lower than that of Jiahu.

3.2.6. Nucleated Sedentism and Resource Intensification: An Innovative Coping Strategy Occurred in Multiple Sites but in Different Expressions

All the sites I discussed above in this section (Section 3.2) point to increasing sedentism and resource intensification as a way to cope with environmental change—But expressed to varying degrees. The sedentism of the Jiahu site seems more stable since there is a complex pattern in which later houses or pit structures overlaid those built in earlier times. Corresponding sociopolitical practices to support the operation of the large and sedentary community of Jiahu were also more elaborate. The Shuishanji site is the largest site in this period, probably indicating the largest scale of population aggregation. The appearance of the cemetery and possible evidence of ritual or sacrifice activities also indicate intensified social integration. The fine-scale evaluation of the residential stability of Shuishanji is difficult since an only a very limited number of houses has been excavated. Yet according to the period of site occupation and relatively thick cultural deposition, we can at least infer that people inhabited and used this settlement with high levels of intensity.

As for the other sites, although they show tendencies of technological organization that are similar to Jiahu or Shunshanji/Hanjing, such as the increasing use of pottery and more sophisticated lithic and bone tool assemblages, they were less developed both in terms of the stability of sedentism or group aggregation and the elaboration of sociopolitical integration like those of Jiahu or Shuishanji. Such a pattern suggests that, even though people have attempted to rely on the increase in residential stability and resource intensification as a way to deal with environmental changes, the extent to which such a strategy can be fulfilled is still heavily influenced by the constraints of the local environment. The scale of the intensive use of wetland and aquatic resources in Jiahu, Shunshanji, and Hanjing seems larger than that of other sites mentioned above, indicating the importance of the exploitation of waterfront environments for the thriving development of Neolithic communities during the period of the 8.2 ka BP cooling event.

3.2.7. The Scarcity of the Archaeological Remains across North China and Its Implications

The distribution of the above-mentioned sites across multiple regions of North China suggest that the new coping strategy, applied to deal with the 8.2 ka BP cooling event, was not an occasionally occurring phenomenon confined to any specific region. Nonetheless, the site numbers are still scarce compared to those after 8 ka BP. There are vast gap areas distributed "in between" these sites which are devoid of any detachable sites. Though the absence of archaeological remains in these gap areas does not necessarily mean that people migrated out of such regions and left large portions of North China unoccupied, it is highly possible that people increased their mobility and left few visible remains for archaeologists since, as discussed in Section 2.3 of this article, increasing mobility could be an efficient coping strategy when used in an area with few or less rich resources to alleviate resource pressure and expand the area for resource exploitation [90]. Hence, the limited number of intensively occupied sites accompanied by vast "in between" areas devoid of any visible archaeological remains likely indicates that people might have adopted divergent ways to cope with environmental changes (nucleated sedentism, intensified resource exploitation vs. increase in mobility) across the area with varying degrees of resource affluency.

4. Discussion

4.1. Preexisting Cultural Preparations and the Path to the Neolithic Lifeway

The transitional remains of the Paleolithic to Neolithic disappeared during the period of the 8.2 ka BP cooling event. Such a change indicates that the land-use pattern, as reflected by the transitional sites, did not work well as an adaptive strategy for this climate fluctuation. People were either forced to return to more mobile lifeways to buffer against subsistence risks caused by the cooling event, or develop new forms of socioeconomic coping solutions.

The scarcity of archaeological remains probably indicates that people increased residential mobility in many regions of North China. However, a few sites, discussed in Section 3.2, like Jiahu, Shunshanji, Hanjing, Zhangmatun, Xihe, Cishan, and Xiaohexi, are quite distinctive since they show evidence for the initial formation of a "Neolithic package", indicating that people were economically and sociopolitically organized in new ways to pursue subsistence needs and interact with the environment. These groups were aggregated in larger communities and settled down in confined places based on the intensive resource exploitation of the surrounding area, combined with intensified hunting–gathering, part-time farming, and even small-scale animal husbandry. The locations of such settlements are usually in rich resource patches like the transitional zone between mountainous areas and plains, or between dryland and marshes and large water bodies. For one thing, some of these ecozones have certain types of highly productive food resources. For another, easy access to different ecozones provided people with diversified ranges of plant and animal resources to use [149].

People were able to fully exploit such "rich areas" partly based on the cultural preparations developed prior to the 8.2 ka BP cooling event. The onset of climate amelioration in the Early Holocene (12–8.5 ka BP) changed the resource distributional pattern and provided opportunities for people to experiment with new ways of resource exploitation and to modify their previous subsistence practices. Grinding stones and pottery became more widely adopted as part of the resource intensification techniques. Domesticated millet was incorporated for the first time into the diet, implying accumulated knowledge about plant domestication and cultivation. Such changes provided important technical and knowledge support for the further development of resource intensification. Accompanied by the modifications of subsistence practices, people began to stay for longer periods at certain sites and their mobility gradually decreased. Some even attempted to settle down in a few resource-rich areas with larger group sizes during 9~8.5 ka BP, as reflected by Jiahu Phase 1 and the Zhangmatun site.

Therefore, before the coming of the 8.2 ka BP cooling event, people in some regions of North China were fully prepared to intensify resource exploitation and increase sedentism as coping strategies to deal with climatic deterioration. However, since people's capabilities for food production were still low, and their subsistence basis was still heavily dependent on wild resources, the application of such a coping strategy only happened in the areas with rich resource patches, where people could take full advantage of extracting highly productive natural resources.

4.2. The Impact of Climate Change Coping Strategies on the Thriving Development of Neolithic Culture

Before the period of the 8.2 ka BP cooling event, sedentary settlements were only found sparsely in a few resource-affluent areas in North China. However, after the end of the 8.2 ka BP cooling event, sedentary settlements that embody distinctive Neolithic features thrived and became widely spread across different regions of North China (shown in Figure 1). Such a pattern indicates that the period around the 8.2 ka BP cooling event was a key time for the formation of Neolithic lifeways.

The specific interactive pattern between environmental change and the practice of coping strategies accounts for such significant social and cultural transitions. The drier and colder climate of the 8.2 ka BP cooling event reduced general resource affluence in North China and pushed the population in certain regions to aggregate in the constrained resource patches of relatively richer resource distributions than the surrounding areas. Under the general background of climatic deterioration, the values of such resource patches would have increased, which facilitated people to maintain long-term access to these places through more stable residency to ensure more persistent occupation and exclusive exploitation of such "sweet spots". As a result, the land-use pattern became transformed from relatively extensive resource acquisition across broad regions into intensive resource exploitation in a circumscribed environment. Such a change created favorable conditions, as well as the stimulus, for the development and prevalence of techniques relevant to intensive resource exploitation, which contributed to the subsequent thriving development of Neolithic lifeways in two main aspects: first, it increased the levels of resource intensification and enabled people to acquire wild resources effectively and in substantial quantity in the circumscribed region during affluent seasons and store them for use during the lean season. On the other hand, as an important part of the resource intensification technique, the knowledge and skills of farming became further developed and people were able to use the benefits of the climatic amelioration after the end of the 8.2 ka BP cooling event to expand farming activity and increase the importance of domesticates as supplement food resources.

Besides consolidated economic bases, the maintenance of large group aggregation in the settled area also required sociopolitical innovations; otherwise, the population was inclined to split away and disperse when faced with increasing scalar stresses resulting from local population increases and densities [128]. Cemeteries, symbolic items, and ritual activities could have been effectively used to strengthen group identity and increase social integration [150,151]. The need to apply such sociopolitical innovations might at first have arisen in large aggregated communities that appeared around 8.5–8 ka BP, since the increased population was inclined to nucleate rather than split away under conditions when the environment outside the resource patches was less productive and there was a need for people to stay put in larger community sizes to facilitate perimeter defense. Among the sites found from around 8.5–8 ka BP, Jiahu, Shunshanji, and Hanjing show relatively large population aggregation and the most stable form of sedentism. The appearance of a cemetery in Jiahu and Shuishanji, as well as the emergence of ritual paraphernalia in Jiahu Phase 2, are ahead of the prevalent applications of such sociopolitical innovations during the prosperity of Neolithic culture after 8 ka BP [152]. Such a leading development of sociopolitical innovations might have provided the basis for future wide adoption by later societies.

Therefore, the application of resource intensification and increasing sedentism to deal with the climate fluctuation caused by the 8.2 ka BP cooling event provided favorable conditions for the convergence of the multiple "Neolithic traits" together, such as the sophistication of resource intensification techniques, sedentary inhabitation, the enlargement of group size, and the elaboration of social interaction. This combination generated new full-fledged forms of socioeconomic solutions to meet subsistence needs. Due to the environmental constraints around 8.2 ka BP, the appliance of such socioeconomic solutions was only possible in a few rich resource patches and was developed at different levels of prosperity in different regions. Nonetheless, after the 8.2 ka BP cooling event, the climate began to ameliorate and the warmer/wetter climate increased general resource affluence, creating more rich resource patches and decreasing the risks to food production. As a result, the practice of such new socioeconomic lifeways was less confined by environment constraints and was more widely adopted by people in different regions of North China.

5. Conclusions

As indicated by the material implications of the archaeological remains, certain types of "Neolithic traits" budded from the terminal Pleistocene to the early Holocene, accompanied by and partly as a consequence of climate amelioration. However, the dispersal of fully-fledged Neolithic lifeways across different regions of North China did not appear until the onset of the Holocene climate after 8–7.8 ka BP. The climate fluctuation caused by the 8.2 ka BP cooling event, which happened just before the coming of the climate optimum, exerted a significant impact on the development of the Neolithization process.

The occurrence of "pioneering" sites, with comprehensive Neolithic features during the period of 8.5–8 ka BP, suggests that changing environmental conditions required people to settle in relatively rich resource patches and form intensive interpersonal relationships, as well as between humans and resources. Such an intensive relationship required people to provide sufficient food supplies in the confined environment and develop more sophisticated socioeconomic mechanisms to reconcile social conflict and enhance the cohesion of the community. Therefore, as indicated by archaeological evidence, intensified techniques for resource acquisition were further developed based on the previous cultural accumulations. The innovative sociopolitical ways used to meet the social demand of nucleated sedentary societies also emerged during this period. The interplay of these two ever-evolving factors has fundamentally changed humans' land-use strategies and formed a new full-fledged sociopolitical practice for people to integrate their societies and interact with the environment. Taking favorable opportunities brought by climate amelioration after the period of the 8.2 ka BP cooling event, this Neolithic lifeway was widely adopted by people, it expanded into broader regions and was further developed, which was reflected in the thriving development of Early Neolithic cultures in North China.

Supplementary Materials: The following are available online at http://www.mdpi.com/2571-550X/3/3/23/s1, Table S1: The radiocarbon dates of the sites used for constructing the cumulative probability model of Figure 2 in the paper.

Funding: This research received no external funding.

Acknowledgments: Pei-Lin Yu from Boise State University contributed to the revision of the manuscript. Pu Yang from Nanjing University of Information Science and Technology provided important suggestions for the analysis of the Paleoclimate data. Yonaton Goldsmith from the Hebrew University of Jerusalem provided the important paleoclimatic dataset.

Conflicts of Interest: The author declares no conflict of interest.

References

1. Shelach, G. The Development of Agriculture and Sedentary Life in North China. In *The Archaeology of Early China: From Prehistory to the Han Dynasty*; Shelach-Lavi, G., Ed.; Cambridge University Press: Cambridge, UK, 2015; pp. 68–102. [CrossRef]
2. Chen, S.; Yu, P.-L. Early "Neolithics" of China: Variation and evolutionary implications. *J. Anthropol. Res.* **2017**, *73*, 149–180. [CrossRef]
3. Zhao, Z. New Archaeobotanic Data for the Study of the Origins of Agriculture in China. *Curr. Anthropol.* **2011**, *52*, S295–S306. [CrossRef]
4. Barton, L.W. *Early Food Production in China's Western Loess Plateau*; University of California Davis: Davis, CA, USA, 2009.
5. Shelach-Lavi, G.; Teng, M.; Goldsmith, Y.; Wachtel, I.; Stevens, C.J.; Marder, O.; Wan, X.; Wu, X.; Tu, D.; Shavit, R.; et al. Sedentism and plant cultivation in northeast China emerged during affluent conditions. *PLoS ONE* **2019**, *14*, e0218751. [CrossRef] [PubMed]
6. Schettler, G.; Liu, Q.; Mingram, J.; Stebich, M.; Dulski, P. East-Asian monsoon variability between 15,000 and 2000 cal. yr BP recorded in varved sediments of Lake Sihailongwan (northeastern China, Long Gang volcanic field). *Holocene* **2006**, *16*, 1043–1057. [CrossRef]
7. Xu, Q.; Xiao, J.; Li, Y.; Tian, F.; Takeshi, N. Pollen-based quantitative reconstruction of Holocene climate changes in the Daihai Lake Area, Inner Mongolia, China. *J. Clim.* **2010**, *23*, 2856–2868. [CrossRef]
8. Wen, R.; Xiao, J.; Chang, Z.; Zhai, D.; Xu, Q.; Li, Y.; Itoh, S.; Lomtatidze, Z. Holocene climate changes in the mid-high-latitude-monsoon margin reflected by the pollen record from Hulun Lake, northeastern Inner Mongolia. *Quat. Res.* **2010**, *73*, 293–303. [CrossRef]
9. Alley, R.B.; Mayewski, P.A.; Sowers, T.; Stuiver, M.; Taylor, K.C.; Clark, P.U. Holocene climatic instability: A prominent, widespread event 8200 yr ago. *Geology* **1997**, *25*, 483. [CrossRef]
10. Rohling, E.J.; Pälike, H. Centennial-scale climate cooling with a sudden cold event around 8200 years ago. *Nature (London)* **2005**, *434*, 975–979. [CrossRef]

11. Chen, S. *The Prehistoric Modernization—An Ecological-based Exploration of the Process of Origin of Agriculture in China*; China Science Publishing: Beijing, China, 2013. (In Chinese)
12. Kelly, R.L. Mobility/Sedentism: Concepts, Archaeological Measures, and Effects. *Annu. Rev. Anthropol.* **1992**, *21*, 43–66. [CrossRef]
13. Cleland, C.E. The Focal-Diffuse model: An evoluntionary perspective on the prehistorical cultural adaptions of the eastern United States. *Midcont. J. Archaeol.* **1976**, *1*, 59–76.
14. Bar-Yosef, O.; Belfer-Cohen, A. The origins of sedentism and farming communities in the Levant. *J. World Prehistory* **1989**, *3*, 447–498. [CrossRef]
15. Bender, B. Gatherer-hunter to farmer: A social perspective. *World Archaeol.* **1978**, *10*, 204–222. [CrossRef]
16. Bandy, M.S.; Fox, J.R. Becoming Villagers: The Evolution of Early Village. In *Becoming Villagers: Comparing Early Village Societies*; Bandy, M.S., Fox, J.R., Eds.; The University of Arizona Press: Tucson, AZ, USA, 2010; pp. 1–16.
17. Ren, M. *The Compendium of Chinese Natural Geography*; The Commercial Press: Beijing, China, 1999; ISBN 9787100026277. (In Chinese)
18. Cohen, D.J. The beginnings of agriculture in China: A multiregional view. *Curr. Anthropol.* **2011**, *52*, S273–S293. [CrossRef]
19. Chen, W. The origin and development of the primitive agriculture. *Agricultural Archaeology* **2005**, *1*, 8–15. (In Chinese)
20. Wu, W. *Research on the Subsistence of Peiligang Period in North China*; Shandong University: Jinan, China, 2014. (In Chinese)
21. Fang, X.; Liu, C.; Hou, G. Reconstruction of precipitation pattern of China in the Holocene Megathermal. *Sci. Geogr. Sin.* **2011**, *31*, 1287–1292. (In Chinese)
22. Shi, Y.; Kong, Z.; Wang, S.; Tang, L.; Wang, F.; Yao, T.; Zhao, X.; Zhang, P.; Shi, S. Climate and environment of the Holocene megathermal maximum in China. *Sci. China Ser. B* **1993**, *27*, 481–493.
23. Shi, Y.; Kong, Z.; Wang, S.; Tang, L.; Wang, F.; Yao, T.; Zhao, X.; Zhang, P.; Shi, S. The climatic fluctuation and important event during the Holocene Megathermal in China. *Sci. China Ser. B* **1994**, *37*, 353–365.
24. Hai, C.; Fleitmann, D.; Edwards, L.R.; Wang, X.; Cruz, F.W.; Auler, A.S.; Mangini, A.; Wang, Y.; Kong, X.; Burns, S.J.; et al. Timing and structure of the 8.2 kyr B.P. event inferred from $\delta^{18}O$ records of stalagmites from China, Oman, and Brazil. *Geology* **2009**, *37*, 1007–1010.
25. Teller, J.T.; Leverington, D.W.; Mann, J.D. Freshwater outbursts to the oceans from glacial Lake Agassiz and their role in climate change during the last deglaciation. *Quat. Sci. Rev.* **2002**, *21*, 879–887. [CrossRef]
26. Cronin, T.M.; Vogt, P.R.; Willard, D.A.; Thunell, R.; Halka, J.; Berke, M.; Pohlman, J. Rapid sea level rise and ice sheet response to 8200-year climate event. *Geophys. Res. Lett.* **2007**, *34*. [CrossRef]
27. Kobashi, T.; Severinghaus, J.P.; Brook, E.J.; Barnola, J.-M.; Grachev, A.M. Precise timing and characterization of abrupt climate change 8200 years ago from air trapped in polar ice. *Quat. Sci. Rev.* **2007**, *26*, 1212–1222. [CrossRef]
28. Chiang, J.C.H.; Bitz, C.M. Influence of high latitude ice cover on the marine Intertropical Convergence Zone. *Clim. Dyn.* **2005**, *25*, 477–496. [CrossRef]
29. Park, J.; Park, J.; Yi, S.; Kim, J.C.; Lee, E.; Jin, Q. The 8.2 ka cooling event in coastal East Asia: High-resolution pollen evidence from southwestern Korea. *Sci. Rep.* **2018**, *8*, 12423. [CrossRef] [PubMed]
30. Wang, N.; Yao, T.; Thompson, L.G.; Henderson, K.A.; Davis, M.E. Evidence for cold events in the early Holocene from the Guliya ice core, Tibetan Plateau, China. *Chin. Sci. Bull.* **2002**, *47*, 1422–1427. [CrossRef]
31. Dykoski, C.A.; Edwards, R.L.; Cheng, H.; Yuan, D.; Cai, Y.; Zhang, M.; Lin, Y.; Qing, J.; An, Z.; Revenaugh, J. A high-resolution, absolute-dated Holocene and deglacial Asian monsoon record from Dongge Cave, China. *Earth Planet. Sci. Lett.* **2005**, *233*, 71–86. [CrossRef]
32. Li, X.; Hu, C.; Huang, J.; Xie, S.; Baker, A. A 9000-year carbon isotopic record of acid-soluble organic matter in a stalagmite from Heshang Cave, central China: Paleoclimate implications. *Chem. Geol.* **2014**, *388*, 71–77. [CrossRef]
33. Dong, J.; Shen, C.-C.; Kong, X.; Wang, H.-C.; Jiang, X. Reconciliation of hydroclimate sequences from the Chinese Loess Plateau and low-latitude East Asian Summer Monsoon regions over the past 14,500 years. *Palaeogeogr. Palaeoclimatol. Palaeoecol.* **2015**, *435*, 127–135. [CrossRef]
34. Liu, Y.; Hu, C. Quantification of southwest China rainfall during the 8.2 ka BP event with response to North Atlantic cooling. *Clim. Past* **2016**, *12*, 1583–1590. [CrossRef]

35. Wu, J.Y.; Wang, Y.J.; Cheng, H.; Kong, X.G.; Liu, D. Stable isotope and trace element investigation of two contemporaneous annually-laminated stalagmites from northeastern China surrounding the "8.2 ka event". *Clim. Past* **2012**, *8*, 1497–1507. [CrossRef]
36. Wang, H.; Hong, Y.; Lin, Q.; Hong, B.; Zhu, Y.; Wang, Y.; Xu, H. Response of humification degree to monsoon climate during the Holocene from the Hongyuan peat bog, eastern Tibetan Plateau. *Palaeogeogr. Palaeoclimatol. Palaeoecol.* **2010**, *286*, 171–177. [CrossRef]
37. Yu, J. The discovery of the Early Neolithic site Zhuannian, Beijing. *Beijing Wenbo* **1998**, *3*, 2–4. (In Chinese)
38. Zhao, C. The prehistoric Donghulin site in Mentougou district, Bejing City. *Kaogu* **2006**, *7*, 3–8. (In Chinese)
39. Li, J.; Qiao, Q.; Ren, X. The report of 1997's excavation of Nanzhuangtou site in Xushui, Hebei. *Acta Archaeol. Sin.* **2010**, *3*, 362–391. (In Chinese)
40. Yuan, S. AMS radiocarbon dating of Xinglong carved antler, Shiyu and Ximiao sites. *Acta Anthropol. Sin.* **1993**, *1*, 92–95. (In Chinese)
41. Mei, H. *The Transition from Paleolithic to Neolithic age in Nihewan Basin—The Discovery and Research of Yujiagou Site in Yangyuan*; Peking University: Beijing, China, 2007. (In Chinese)
42. Sun, B.; Cui, S. An exploration of the Early Neolithic remains in Shandong region. *Cultral Relics Cent. China* **2008**, *3*, 23–28. (In Chinese)
43. Sun, B.; Wagner, M.; Zhao, Z.; Li, G.; Wu, X.; Tarasov, P.E. Archaeological discovery and research at Bianbiandong early Neolithic cave site, Shandong, China. *Quat. Int.* **2014**, *348*, 169–182. [CrossRef]
44. Wang, Y.; Zhang, S.; Gu, W.; Wang, S.; He, J.; Wu, X.; Qu, T.; Zhao, J.; Chen, Y.; Bar-Yosef, O. Lijiagou and the earliest pottery in Henan Province, China. *Antiquity* **2015**, *89*, 273–291. [CrossRef]
45. Shi, J.; Song, Y. The excavation to Locality S9 of the Shizitan Site in Ji County, Shanxi. *Kaogu* **2010**, *10*, 7–17. (In Chinese)
46. Shi, J.; Chen, H.; Song, Y.; Shi, X.; Wei, X.; Li, L. Preliminary report on the excavation at the Locality S12G of the Shizitan site in Ji County, Shanxi Province. *Kaogu Yu Wenwu* **2013**, *3*, 3–8. (In Chinese)
47. Henan Antique Archaeology Institute. *Wuyang Jiahu*; China Science Publishing: Beijing, China, 1999. (In Chinese)
48. Wang, F.; Li, M. The Early Neolithic remains of Zhangmatun Site in Jinan City. *Kaogu* **2018**, *2*, 116–120. (In Chinese)
49. Liu, Y.; Wang, Z.; Zhang, K.; Li, S.; Wang, Z.; Zang, Z.; Sun, T.; Zhang, Z. The excavation report of the Xihe Site in Zhangqiu, 2008. *Haidai Kaogu* **2012**, *5*, 67–138. (In Chinese)
50. Sun, D.; Liu, Y.; Chen, G. The Cishan site in Wuan, Hebei. *Acta Anthropol. Sin.* **1981**, *3*, 303–338. (In Chinese)
51. Lin, L.; Gan, H.; Yan, L. The excavation of the Shunshanji site of Neolithic Age in Sihong County, Jiangsu. *Acta Anthropol. Sin.* **2014**, *4*, 519–562. (In Chinese)
52. Dai, X.; Zhuang, L.; Yu, H.; Lu, H.; Sun, Y.; Lu, J.; Shi, Y.; Fan, J.; Jiang, M.; Fang, D.; et al. 2014 excavational report of the Hanjing Site in Sihong, Jiangsu Province. *Culture of Southeast China* **2018**, *1*, 20–27. (In Chinese)
53. Inner Mongolia Antique Archaeology Institute. *Baiyinchanghan—The Excavation Report of the Neolithic Settlement*; China Science Publishing: Beijing, China, 2004. (In Chinese)
54. Liu, G.; Jia, X.; Zhao, M.; Tian, G.; Shao, G. The excavation of the Xinglonggou settlement in 2002–2003 season of the Chifeng city, Inner Mongolia. *Kaogu* **2004**, *7*, 3–8. (In Chinese)
55. Yang, H.; Zhu, Y.; Kong, Z.; Du, N. The preliminary report of the Xinglongwa site in Aohan Banner, Inner Mongolia. *Kaogu* **1985**, *10*, 865–874. (In Chinese)
56. Liaoning Antique Archaeology Institute. *Chahai: The Excavation Report of a Neolithic Settlement*; Antique Publishing House: Beijing, China, 2012. (In Chinese)
57. Yu, J.; Wang, Y. The preliminary report of the Shangzhai Neolithic site in Pinggu, Beijing. *Wenwu* **1989**, *8*, 1–8. (In Chinese)
58. Chen, Z.; Liu, H.; Zhang, X.; Yang, G.; Zheng, S.; Zhang, Y.; Yuan, Q.; Fu, Y.; Liu, H.; Zhou, S.; et al. The archaeoloigcal report of the Beiwang site, Langfang. *Wenwu Chunqiu* **2010**, *1*, 17–29. (In Chinese)
59. Duan, H. *Beifudi: The Prehistoric Site of Yi River Valley*; Cutural Relics Press: Beijing, China, 2007. (In Chinese)
60. He, D.; Xu, X. A preliminary exploration of Houli culture in Jinan region. *Shiqian Yanjiu* **2013**, *30*, 115–124. (In Chinese)
61. Sun, Q. *A Study of Houli Culture*; Shandong University: Jinan, China, 2014. (In Chinese)
62. Geng, Q. The trial excavation of the Huawo site in Qi county, Henan. *Kaogu* **1981**, *3*, 279–281. (In Chinese)
63. Wang, J. The Shawoli Neolithic site in Xinzheng, Henan. *Kaogu* **1983**, *12*, 1057–1065. (In Chinese)

64. Ren, W.; Wang, J.; Zheng, N. The report of the 1979-year excavation of the Peiligang site. *Acta Anthropol. Sin.* **1984**, *1*, 23–52. (In Chinese)
65. Xin, Y.; Hu, Y.; Zhang, Y.; Liu, Q. Excavation to the remains of Peiligang culture of the Tanghu site at Xinzheng City, Henan in 2007. *Kaogu* **2010**, *5*, 3–23. (In Chinese)
66. Guo, T.; Chen, J. The Archaeological report of Shigu site, Changge. *Huaxia Archaeol.* **1987**, *1*, 3–125. (In Chinese)
67. Zheng, N. The preliminary report of the excavation of the Shuiquan Neolithic site of Jia county, Henan. *Kaogu* **1992**, *10*, 865–874. (In Chinese)
68. Zheng, N. The preliminary report of the excavation of the Zhongshanzhai site in Linru, Henan. *Kaogu* **1986**, *7*, 577–585. (In Chinese)
69. Yang, Z. The preliminary report of the Egou Beigang Neolithic site in Mi county, Henan. *Wenwu* **1979**, *5*, 14–19. (In Chinese)
70. Li, Y. The preliminary report of the trial excavation of the Tieshenggou Early Neolithic site in Gong county, Henan. *Wenwu* **1980**, *5*, 16–19. (In Chinese)
71. Wang, J.; Zhang, X. A discussion on chronology of Yangshao culture remains at the Bancun site and the related issues. *Kaogu yu Wenwu* **2001**, *3*, 41–50. (In Chinese)
72. Zhang, R. The preliminary report of the second and third excavation of the Beiliu site in Weinan. *Shiqian Yanjiu* **1986**, *4*, 111–128. (In Chinese)
73. Chinese Academy of Social Sciences. *The Excavation Report of Baijia Cun Site in Lintong*; Bashu Publishing House: Chengdu, China, 1994. (In Chinese)
74. Chen, Y. A re-exmination of the Beishouling Early Neolithic site. *Huaxia Archaeol.* **1990**, *3*, 70–85. (In Chinese)
75. Gansu Antique Archaeology Institute. *Qinan Dadiwan—The Archaeological Report of the Neolithic Site*; Cultural Relics Publishing: Beijing, China, 2006. (In Chinese)
76. Li, Y. The survey and trial excavation of the Maliangggou site in Mi county, Henan. *Kaogu* **1981**, *3*, 282–284. (In Chinese)
77. Anhui Antique Archaeology Institute; Bengbu Museum. *Bengbu Shuangdun—Neolithic Site Excavation Report*; China Science Publishing: Beijing, China, 2008. (In Chinese)
78. Jia, Q. Shishanzi Neolithic site in Suixi, Anhui. *Kaogu* **1992**, *3*, 193–203. (In Chinese)
79. Wang, J.; Wu, J.; Liang, Z. The preliminary report of the Xiaoshankou and Gutaisi sites in Su county, Anhui. *Kaogu* **1993**, *12*, 1062–1075. (In Chinese)
80. Yang, H.; Lin, X. The preliminary analysis of Xiaohexi Sites in Aohan Banner, Inner Mongolia. *Beifang Wenwu* **2009**, *2*, 3–6. (In Chinese)
81. Andersen, K.K.; Azuma, N.; Barnola, J.M.; Bigler, M.; Biscaye, P.; Caillon, N.; Chappellaz, J.; Clausen, H.B.; Dahl-Jensen, D.; Fischer, H.; et al. High-resolution record of Northern Hemisphere climate extending into the last interglacial period. *Nature* **2004**, *431*, 147–151. [CrossRef] [PubMed]
82. Dong, J.; Shen, C.-C.; Kong, X.; Wu, C.-C.; Hu, H.-M.; Ren, H.; Wang, Y. Rapid retreat of the East Asian summer monsoon in the middle Holocene and a millennial weak monsoon interval at 9 ka in northern China. *J. Asian Earth Sci.* **2018**, *151*, 31–39. [CrossRef]
83. Halstead, P.; O'Shea, J.O. Introduction: Cultural responses to risk and uncertainty. In *Bad Year Economics: Cultural Responses to Risk and Uncertainty*; Halstead, P., O'Shea, J.O., Eds.; Cambridge University Press: Cambridge, UK, 1989; pp. 1–7.
84. Rosen, A.M. *Civilizing Climate: Social Responses to Climate Change in the Ancient Near East*; Altamira: London, UK, 2007.
85. Steward, J.H. Ecological Aspects of Southwestern Society. *Anthropos* **1937**, *32*, 87–104.
86. O'Shea, J.O.; Halstead, P. *Conclusion: Bad year economics In Bad Year Economics: Cultural Responses to Risk and Uncertainty*; Halstead, P., O'Shea, J.O., Eds.; Cambridge University Press: Cambridge, UK, 1989; pp. 123–126.
87. Bar-Yosef, O. Climatic fluctuations and early farming in West and East Asia. *Curr. Anthropol.* **2011**, *52*, S175–S193. [CrossRef]
88. Minc, L.D.; Smith, K.P. The spirit of survival: Cultural responses to resource variability in North Alaska. In *Bad Year Economics: Cultural Responses to Risk and Uncertainty*; Halstead, P., O'Shea, J.O., Eds.; Cambridge University Press: Cambridge, UK, 1989; pp. 8–39.
89. Chen, S.; Yu, P.-L. Intensified foraging and the roots of farming in China. *J. Anthropol. Res.* **2017**, *73*, 381–412. [CrossRef]

90. Barton, L.; Brantingham, P.J.; Ji, D. Late Pleistocene climate change and Paleolithic cultural evolution in northern China: Implications from the Last Glacial Maximum. In *Developments in Quaternary Sciences*; Elsevier: Amsterdam, The Netherlands, 2007; Volume 9, pp. 105–128.
91. Ji, D.; Chen, F.; Bettinger, R.I.; Elston, R.; Geng, Z.; Barton, L.M.; Wang, H.; An, C.; Zhang, D. Human response to the Last Glacial Maximum: Evidence from North China. *Acta Anthropol. Sin.* **2005**, *4*, 270–282. (In Chinese)
92. Elston, R.G.; Guanghui, D.; Dongju, Z. Late Pleistocene intensification technologies in Northern China. *Quat. Int.* **2011**, *242*, 401–415. [CrossRef]
93. Fang, X.; Jiang, H.; Lian, P. Range and rate of abrupt change of precipitation around 3500 ka BP in the North China Farming-grazing transitional zone. *Earth Sci. Front.* **2002**, *1*, 163–167. (In Chinese)
94. Hou, G.; Liu, F.; Liu, C.; Fang, X. Prehistorical Cultural Transition Forced by Environmental Change in Mid-Holocene in Gansu-Qinghai Region. *Acta Anthropol. Sin.* **2009**, *1*, 53–58. (In Chinese)
95. Chen, S. The adaptative changes of the prehistoric cultures in the zones along Yanshan mountains and the Great Wall. *Acta Anthropol. Sin.* **2011**, *1*, 1–22. (In Chinese)
96. Kelly, R.L. Hunter-Gatherer mobility strategies. *J. Anthropol. Res.* **1983**, *39*, 277–306. [CrossRef]
97. Rowley-Conwy, P.A.; Zvelebil, M. Saving it for later: Storage by prehistoric hunter-gatherers in Europe. In *Bad Year Economics: Cultural Responses to Risk and Uncertainty*; Cambridge University Press: Cambridge, UK, 1989; pp. 40–56.
98. Testart, A.; Forbis, R.G.; Hayden, B.; Ingold, T.; Perlman, S.M.; Pokotylo, D.L.; Rowley-Conwy, P.; Stuart, D.E. The significance of food storage among Hunter-Gatherers: Residence patterns, population densities, and social inequalities. *Curr. Anthropol.* **1982**, *23*, 523–537. [CrossRef]
99. Hayden, B.; Bowdler, S.; Butzer, K.W.; Cohen, M.N.; Druss, M.; Dunnell, R.C.; Goodyear, A.C.; Hardesty, D.L.; Hassan, F.A.; Kamminga, J.; et al. Research and development in the stone age: Technological transitions among Hunter-Gatherers. *Curr. Anthropol.* **1981**, *22*, 519–548. [CrossRef]
100. Rowley-Conwy, P.; Layton, R. Foraging and farming as niche construction: Stable and unstable adaptations. *Philos. Trans. Biol. Sci.* **2011**, *366*, 849–862. [CrossRef]
101. Black, R.; Adger, W.N.; Arnell, N.W.; Dercon, S.; Geddes, A.; Thomas, D. The effect of environmental change on human migration. *Glob. Environ. Chang.* **2011**, *21*, S3–S11. [CrossRef]
102. Flohr, P.; Fleitmann, D.; Matthews, R.; Matthews, W.; Black, S. Evidence of resilience to past climate change in Southwest Asia: Early farming communities and the 9.2 and 8.2 ka events. *Quat. Sci. Rev.* **2016**, *136*, 23–39. [CrossRef]
103. Zvelebil, M. The mesolithic context of the transition to farming. In *Hunters in Transition: Mesolithic Societies of Temperate Eurasia and Their Transition to Farming*; Zvelebil, M., Ed.; Cambridge University Press: Cambridge, UK, 1986; pp. 5–15.
104. Weissner, P. Risk, Reciprocity and Social Influence in !Kung San Politics. In *Politics and History in Band Societies*; Leacock, E., Lee, R., Eds.; Cambridge University Press: Cambridge, UK, 1982; pp. 61–84.
105. Lu, T.L.D. *The Transition from Foraging to Farming and the Origin of Agriculture in China*; BAR International Series 774; British Archaeological Reports: Oxford, UK, 1999.
106. An, Z. Archaeological research on Neolithic China. *Curr. Anthropol.* **1988**, *29*, 753–759.
107. Wang, Y.; Zhang, S.; He, J.; Wang, S.; Zhao, J.; Qu, T.; Wang, J.; Gao, X. The excavation of the Lijiagou Site in Xinmi city, Henan. *Kaogu* **2011**, *4*, 3–9. (In Chinese)
108. Liu, L.; Chen, X. Neolithization: Sedentism and food production in the Early Neolithic (7000–5000 BC). In *The archaeology of China: From the Late Paleolithic to the Early Bronze Age*; Liu, L., Chen, X., Eds.; Cambridge University Press: Cambridge, UK, 2012; pp. 123–168.
109. Yang, X.; Ma, Z.; Li, J.; Yu, J.; Stevens, C.; Zhuang, Y. Comparing subsistence strategies in different landscapes of North China 10,000 years ago. *Holocene* **2015**, *25*, 1957–1964. [CrossRef]
110. Yang, X.; Wan, Z.; Perry, L.; Lu, H.; Wang, Q.; Zhao, C.; Li, J.; Xie, F.; Yu, J.; Cui, T.; et al. Early millet use in northern China. *Proc. Natl. Acad. Sci. USA* **2012**, *109*, 3726–3730. [CrossRef] [PubMed]
111. Zhao, Z. The process of origin of agriculture in China: Archaeological evidence from flotation results. *Quanternary Sci.* **2014**, *1*, 73–84. (In Chinese)
112. Chen, Y.; Qu, T. An Exploration of the complex phenomena of society in North China at around 10 ka B.P. *Zhongyuan Wenwu* **2012**, *3*, 20–26. (In Chinese)
113. Sturm, C.; Clark, J.K.; Barton, L. The Logic of Ceramic Technology in Marginal Environments: Implications for Mobile Life. *Am. Antiq.* **2016**, *81*, 645–663. [CrossRef]

114. Eerkens, J.W. Residential mobility and pottery sse in the western Great Basin. *Curr. Anthropol.* **2003**, *44*, 728–738. [CrossRef]
115. Shelach-Lavi, G.; Tu, D. Food, pots and socio-economic transformation: The beginning and intensification of pottery production in North China. *Archaeol. Res. Asia* **2017**, *12*, 1–10. [CrossRef]
116. Liu, L.; Field, J.; Fullagar, R.; Zhao, C.; Chen, X.; Yu, J. A functional analysis of grinding stones from an early holocene site at Donghulin, North China. *J. Archaeol. Sci.* **2010**, *37*, 2630–2639. [CrossRef]
117. Xiang, J. The Variations of the Origin of Grounding Stone between South and North China. *Nanfang Wenwu* **2014**, *2*, 101–109. (In Chinese)
118. Cui, T. *The Study of Donghulin Site's Lithic Assemblage—The Lithic Industry and Human Behavioral in the Process of Paleolithic to Neolithic Age*; Peking University: Beijing, China, 2010. (In Chinese)
119. Henan Antique Archaeology Institute; University of Science and Technology of China. *Wuyangjiahu II*; China Science Publishing: Beijing, China, 2015. (In Chinese)
120. Chi, Z.; Hung, H.-C. Jiahu 1: Earliest farmers beyond the Yangtze River. *Antiquity* **2013**, *87*, 46–63. [CrossRef]
121. Yang, Y.; Zhang, J.; Lan, W.; Cheng, Z.; Yuan, Z.; Zhu, Z. The Excavation of the Jiahu Site in Wuyang County, Henan in 2013. *Kaogu* **2017**, *12*, 3–20. (In Chinese)
122. Lin, Y. The key changes during the transition from Paleolithic to Neolithic Age in North China. *Wenwu Chunqiu* **2016**, *2*, 12–17. (In Chinese)
123. Li, B. *The Abandonment Process Research of Early Neolithic Sites in Northern China*; Jilin University: Changchun, China, 2018. (In Chinese)
124. Cui, Q. *The Lithics Analysis of Jiahu Site in Wuyang, Henan*; University of Science and Technology of China: Hefei, China, 2018. (In Chinese)
125. Lai, Y.; Zhang, J.; Yin, R. On instruments of production and economic structure of Jiahu site in Wuyang. *Zhongyuan Wenwu* **2009**, *2*, 22–28. (In Chinese)
126. Zhao, Z.; Zhang, J. Report on the analysis of the results of the 2001 flotation of the Jiahu site. *Kaogu* **2009**, *8*, 84–93. (In Chinese)
127. Cucchi, T.; Hulme-Beaman, A.; Yuan, J.; Dobney, K. Early Neolithic pig domestication at Jiahu, Henan Province, China: Clues from molar shape analyses using geometric morphometric approaches. *J. Archaeol. Sci.* **2011**, *38*, 11–22. [CrossRef]
128. Bandy, M.S. Fissioning, scalar stress, and social evolution in early village societies. *Am. Anthropol.* **2004**, *106*, 322–333. [CrossRef]
129. Bollig, M. *Risk Management in a Hazardous Environment—A Comparative Study of Two Pastoral Societies*; Springer: Berlin/Heidelberg, Germany, 2006.
130. Wu, W.; Lin, L.; Gan, H.; Jin, G. Environment and subsistence of the second phase of Shunshanji site: Evidence from phytolith analysis. *Zhongguo Nongshi* **2017**, *36*, 3–14. (In Chinese)
131. Zhuang, L.; Yu, H.; Qiu, Z.; Yan, L.; Liu, S.; Pan, M.; Ning, Z.; Jiang, M.; Fang, D.; Ma, G.; et al. 2015 and 2016 Excavational Report of the Hanjing Site in Sihong, Jiangsu Province. *Dongnan Wenhua* **2018**, *1*, 28–39. (In Chinese)
132. Lin, L.; Gan, H.; Yan, L.; Jiang, F. The Shunshanji site of the Neolithic age in Sihong County, Jiangsu. *Kaogu Xuebao* **2013**, *7*, 3–14. (In Chinese)
133. Liu, Y.; Lan, Y.; Tong, P. Excavation at the Neolithic Xihe site in Zhangqiu City, Shandong in 1997. *Kaogu* **2000**, *10*, 15–28. (In Chinese)
134. Song, Y. The comprehensive analysis of fauna remains of Houli cultural period in Jinan region. *Huaxia Kaogu* **2016**, *3*, 53–59. (In Chinese)
135. Wu, W.; Jin, G.; Wang, X. Plant cultivation and the subsistence of Houli Culture: Evidences from Zhangmatun site, Jinan city. *Zhongguo Nongshi* **2015**, *34*, 3–13. (In Chinese)
136. Wu, W.; Zhang, K.; Wang, Z.; Jin, G. The Analysis of the Plant Remains from Xihe Site, Zhangqiu (2008). *Dongfang Kaogu* **2013**, *10*, 373–390. (In Chinese)
137. Jin, J. A preliminary exploration of the "Tool Compounds" in late phase of the Cishan site. *Kaogu* **1995**, *3*, 231–237. (In Chinese)
138. Tong, W. The primary farming remains and the related issues of Cishan site. *Nongye Kaogu* **1984**, *1*, 194–207. (In Chinese)

139. Li, G. The transmission of millet and broomcorn millet around eight thousand years in North China and transfer characteristics of Cishan site in the east line of Taihang Mountain. *Nanfang Wenwu* **2018**, *1*, 229–251. (In Chinese)
140. Bu, G. The sacrificial site of Cishan and the related issues. *Wenwu* **1987**, *11*, 43–47. (In Chinese)
141. Luo, P. The houses of Cishan people. *Wenwu Chunqiu* **2006**, *1*, 1–2. (In Chinese)
142. Chen, S. *Adaptive Changes of Prehistoric Hunter—Gatherers during the Pleistocene-Holocene Transition in China*; Southern Methodist University: Dallas, TX, USA, 2004.
143. Yue, Q. An investigation of "Pit Structure" of the Cishan site. *Gujin Nongye* **1992**, *2*, 27–30. (In Chinese)
144. Zhou, B. The fauna remains of the Cishan site, Hebei. *Kaogu Xuebao* **1981**, *3*, 339–347. (In Chinese)
145. Suo, X. The preliminary analysis of Xiaohexi Culture. *Kaogu yu Wenwu* **2005**, *1*, 23–27. (In Chinese)
146. Wu, L. The micro-perspective analysis of the settlement of Xiaohexi culture in West Liao River Valley. *J. Chifeng Univ. Soc. Sci.* **2014**, *35*, 1–5. (In Chinese)
147. Suo, X.; Guo, Z. Xiaohexi Culture remains at the Baiyinchanghan Site. *Res. China's Front. Archaeol.* **2004**, *3*, 301–310. (In Chinese)
148. Shelach-Lavi, G.; Teng, M.; Goldsmith, Y.; Wachtel, I.; Ovadia, A.; Wan, X.; Marder, O. Human adaptation and socioeonomic change in northeast China: Results of the Fuxin regional survey. *J. Field Archaeol.* **2016**, *41*, 467–485. [CrossRef]
149. Janz, L. Fragmented landscapes and economies of abundance: The Broad-Spectrum revolution in Arid East Asia. *Curr. Anthropol.* **2016**, *57*, 537–564. [CrossRef]
150. Cohen, A.B. "Ritualization" in early village society: The case of the lake Titicaca Basin Formative. In *Becoming Villagers: Comparing Early Village Societies*; Bandy, M.S., Fox, J.R., Eds.; The University of Arizona Press: Tucson, AZ, USA, 2010; pp. 81–99.
151. Mantha, A. Territoriality, social boundaries and ancestor veneration in the central Andes of Peru. *J. Anthropol. Archaeol.* **2009**, *28*, 158–176. [CrossRef]
152. Chen, M. *A Study of the Cutlure and Society of Peiligang Period*; Fudan University: Shanghai, China, 2013. (In Chinese)

quaternary

Article

Microblade–Based Societies in North China at the End of the Ice Age

Meng Zhang [1,2]

[1] Department of Cultural Heritage and Museology, Fudan University, Shanghai 200433, China; zhangmengwb@fudan.edu.cn
[2] Institute of Archaeological Science, Fudan University, Shanghai 200433, China

Received: 23 May 2020; Accepted: 3 July 2020; Published: 9 July 2020

Abstract: One of the most prominent cultural changes during the end of Ice Age in northeastern Asia was the adoption of microblade technology by prehistoric hunter–gatherers to deal with the challenge brought by the climate deterioration and oscillation during and post the Last Glacial Maximum (LGM). The Pleistocene to Holocene transition in North China witnessed the rise of broader spectrum subsistence alongside a series of cultural changes, including adoption of food production, highly mobile lifeways being replaced by sedentism, and the formation of new social organization based on their agricultural land–use patterns. From the perspective of technological change, this project aims to build a socio–ecological framework to examine the cultural change of prehistoric microblade–based societies. In contrast to previous studies, the present research employs a macroecological approach based on Binford's *Constructing Frames of Reference* (2001) to reconstruct the behaviors and demography of prehistoric foraging groups, under both modern and LGM climate conditions. Three case studies are conducted to show cultural and technological changes among microblade–based societies in North China during the Pleistocene–Holocene transition.

Keywords: microblade technology; broad spectrum revolution; Pleistocene to Holocene transition; origin of food production; hunter–gatherers; macroecology; *Constructing Frames of Reference*

1. Introduction

During the Pleistocene to Holocene transition, northern China not only witnessed a broad spectrum revolution, intensification, and food production, but also saw the increasing use of durable stone tools linked to a terrestrial plant–dependent economy and the fading of microblade technology linked to highly mobile hunting–dominated lifeways. This process is generalized by Chen and Yu [1,2] as the Paleolithic to Neolithic Transition (PNT) (ca. 15,000–8500 Carbon 14 [C14] BP) in the period after the Late Upper Paleolithic (LUP) (ca. 24,000–15,000 C14 BP). Archaeological records suggest that the early Neolithic sites and sites with cultivation dated to the terminal Pleistocene are both located in relatively small habitats in basins, valleys and foothills of forest–steppe ecotones in the zone between the terrestrial plant threshold and the storage threshold in Lewis Binford's system (effective temperature [ET] = 12.75 − 15.25 °C) [3]. Agriculture appeared on the North China Plain after humans achieved sophisticated food production technologies during the Middle Holocene. Chen also attributes food production to collectors who practiced logistical strategies in Binford's forager–collector system [4,5]. Microblade–based societies cannot securely be classified as either foragers or collectors, but according to Chen's proposition, it seems that the foragers in northern China (if there were such) may have changed their subsistence–settlement system to collecting during the terminal Pleistocene. Current archaeological records are inadequate to fully test this hypothesis, but they can be investigated from a macroecological and analogical viewpoint. Recent research indicates that in northern China during the transition from the unpacked to the packed population condition, hunting-dominated subsistence is expected to become gathering-dominated subsistence in the west (the Loess Plateau), and fishing–dominated subsistence in the east (North China Plain) in which lakes and marshes formed at the end of the Last Ice Age [6]. Most early sites associated with cultivation, ground stone tools, and microblade assemblages are located along the boundary of the gathering–fishing–dominated subsistence line, suggesting that prehistoric foraging societies might have been practicing a mixed economy.

This project only focuses on the relationship between the rise of food production and the decline of microblade–based societies, within the time range from the Last Glacial Maximum to early Holocene. The decline of microblade–based societies cannot be explained as humans migrating out to the frontiers, since the archaeological record in Northern China tends to support an explanation of cultural transition rather than replacement [7]. In the following three sub–topics concerning Northern China, I apply the macroecological approach to study the development and decline of microblade–based societies under the background of the PNT, with the time range from the LGM to the early Holocene.

2. Methodology

The present project employs an analogical research strategy from the anthropological side to evaluate ideas generalized from the archaeological side in the previous studies which mainly followed a cultural–historical paradigm. The main methodology adopted here is macroecology, developed from the late Lewis Binford's book *Constructing Frames of Reference* [3], which has been fully standardized by his colleague Amber Johnson as a calculation program EnvCalc 2.1 [8]. This program makes possible the reconstruction of prehistoric hunter–gatherers' foraging strategies possible on a global scale [9] and the regional scale [7,10,11]. More than 300 variables derived from the program can be grouped as four datasets: (1) Climate, biomes, and habitat; (2) Minimalist terrestrial model; (3) Modelling density-dependent change in hunter-gatherer subsistence; and (4) Growth rate model and density–controlled subsistence (the variables referring to the present research are listed in Table 1). These variables provide an irreplaceable referential framework to help anthropologists predict/project hunter–gatherers' foraging strategies in a specific region even the foraging populations no longer exist due to cultural transformation or resettlement. For example, several variables, including duration of growing season (GROWC), net aboveground productivity (NAGP), and terrestrial model value for population density (TERMD2), are used to test the hypotheses of the transition from hunter–gathering to farming in the Near East by Binford [3] (pp. 434–464).

Table 1. Key variables used in this research project.

Topics and Models	Hunter-Gatherers	Variable	Definition [1]
Climate, Biomes, and Habitat	Climate	GROWC	Length of growing season: Length of growing season. Number of months with mean temperatures greater than 8 degrees Celsius
		ET	Effective temperature: a measure designed to examine biological implications of ambient warmth
	Biomes and habitat	NAGP	Net above–ground productivity
		BIO5	Primary biomass
		EXPREY	Expected moderate body-size ungulate biomass (kg/km^2)

Table 1. *Cont.*

Topics and Models	Hunter-Gatherers	Variable	Definition [1]
Minimalist Terrestrial Model	Population density	TERMH2	Number of persons per 100 km^2 unit who could be supported by the ungulate resources alone
		TERMG2	Number of persons per 100 km^2 unit who could be supported by the plant resources alone
		TERMD2	Population density expected at a particular location, expressed in terms of persons per 100 km^2
	Subsistence specialty	SUBSPX2	Terrestrial model expected subsistence bias for use with ethnographic cases h = hunting, g = gathering, m = mixed, u = uninhabited
Modelling Density–Dependent Change in Hunter–Gatherer Subsistence	Subsistence	SUBSPE	Ordinal classification of projected hunter–gatherer subsistence specialty 1 = hunting, 2 = gathering, 3 = fishing
	Population density	WDEN	Projected hunter-gatherer density
	Group size	EXGRP1	Projected mean size of smallest residential group, segmented by group pattern and subsistence specialization bias
		EXGRP2	Projected mean size of largest residential seasonal camps, segmented by group pattern and subsistence specialization bias
		EXGRP3	Projected mean size of periodic regional camps, segmented by group pattern and subsistence specialization bias
	Mobility	EXDMOV1	Projected total distance moved, scaled for subsistence type, for groups with year round camp to camp mobility pattern
		EXDMOV2	Projected total distance moved, scaled for subsistence type, for groups who move into and out of a central location or who are primarily sedentary
Growth Rate Model and Density–Controlled Subsistence	Subsistence specialty against the unpacked background	UPSUBSPE	Ordinal classification of projected unpacked HG subsistence specialty (packing multiplier = 0.5) 1 = hunting, 2 = gathering, 3 = fishing
	Subsistence specialty against the packed background	D1PSUBSPE	Ordinal classification of projected packed HG subsistence specialty (packing multiplier = 1) 1 = hunting, 2 = gathering, 3 = fishing

Note: [1] from "variables used or calculated in EnvCalc2.1 program (except UPSUBSPE and D1PSUBSPE) [8].

The author's dissertation combines macroecological approach of hunter–gatherers, modern and simulated LGM climatic data, and prehistoric technological organization associated with microblade technology to investigate the role of microblade technology in the development of human adaptations in NE Asia, especially northern China, during the closing millennia of the Upper Pleistocene and across the Pleistocene–Holocene transition [6]. Microblade–based societies as a new concept provides a fresh perspective to characterize a socio-technologically adaptive convergence and radiation of human groups before and after the Last Glacial Maximum (LGM) across NE Asia, in which people adopted microblade technology as the basis for the lithic elements of composite tools/weapons to help organize their lives in their territories, depending, of course, on the availability and quality of local lithic resources.

3. Archaeological Sites Associated with Microblade Technology in North China

North China, in the narrow sense, refers two subregions: the Loess Plateau in the west and the North China Plain in the east. According to archaeological record recovered, the southern limit of distribution of microblade technology during the late Pleistocene is northern side of the Qinling Mountain and the Huaihe River, which is also the north–south division of modern Chinese geography. The northern line of North China is along the Great Wall, which is also the steppe–forest ecotone during the early Holocene.

Since this paper mainly focuses on the period of origin of food production, using Shizitan site as an example to discuss the possibility of plant-intensified exploitation behaviors during the LGM, and at the end, using the Donghulin site to talk about the demise of microblade–based societies, data collected for this project are associated with those sites dated to the Pleistocene–Holocene transition (Table 2). They are mainly Accelerator Mass Spectrometry Carbon 14 (AMS C14) dates, as well as latitudes and longitudes of the sites. To further discuss issues of the broad spectrum revolution, archaeological data are categorized into artifact assemblage, site organization, and biological remains are also shown in Table 3.

Table 2. Chronometric dates of the sites in the Paleolithic to Neolithic Transition (PNT) group.

Site	Layer	Material	Method	Lab. No.	Dates	Cal. BC (2-sigma)	Reference
Nanzhuangtou	layer 5–6	wood	AMS C14	BK86120	9875 ± 160	10,028–9112 (87.2%) 9083–9041 (1.1%) 9026–8838 (7.1%)	Yuan et al. [12]
		wood	AMS C14	Bk86121	9690 ± 95	9298–8799 (95.4%)	Yuan et al. [12]
		wood	AMS C14	BK87093	9810 ± 100	9669–9116 (86.1%) 9076–9055 (0.6%) 9016–8843 (8.7%)	Yuan et al. [12]
		wood	AMS C14	BK89064	9850 ± 90	9752–9722 (1%) 9692–9146 (94.4%)	Yuan et al. [12]
	Layer 6	charcoal	AMS C14	BK87075	10,510 ± 100	10,731–10,149 (95.4%)	Yuan et al. [12]
	Layer 5	silt	AMS C14	BK87086	9980 ± 100	9983–9941 (1.6%) 9877–9266 (93.8%)	Yuan et al. [12]
		silt	AMS C14	BK87088	10,815 ± 140	11,096–10,576 (94%) 10,518–10,485 (1.4%)	Yuan et al. [12]
Zhuannian			AMS C14		9200 ± 100	8701–8676 (1.4%) 8646–8250 (94%)	Ren and Wu [13]
			AMS C14		c.9800		Yu [14]
Donghulin	M1 (Burial 1)	human bone	AMS C14		9570 ± 70	9216–8755 (95.4%)	Cui [15]
	Layer 8, T3	charcoal	AMS C14		9180 ± 80	8608–8273 (95.4%)	Cui [15]
	Layer 7, T3	charcoal	AMS C14		9155 ± 40	8528–8521 (0.8%) 8473–8282 (94.6%)	Cui [15]
	Layer 4, T3	charcoal	AMS C14		8780 ± 90	8205–8035 (20.6%) 8016–7606 (74.8%)	Cui [15]
	Layer 4, T9	charcoal	AMS C14		8805 ± 50	8204–8102 (13.8%) 8096–8036 (5.6%) 8015–7711 (75.6%) 7692–7685 (0.4%)	Cui [15]

Table 2. *Cont.*

Site	Layer	Material	Method	Lab. No.	Dates	Cal. BC (2–sigma)	Reference
Donghulin	Layer 3, T9	charcoal	AMS C14		8772 ± 40	8165–8138 (1.5%) 7973–7647 (93.9%)	Cui [15]
	Layer 2, T9	Pineapple	AMS C14		8885 ± 55	8240–7825 (95.4%)	Cui [15]
Lijiagou (north area)	Layer 6	charcoal	AMS C14	BA091494	8950 ± 40	8272–8166 (44.3%) 8131–7970 (51.1%)	He et al. [16]
	Layer 5	charcoal	AMS C14	BA091417	8015 ± 35	7064–6814 (95.4%)	He et al. [16]
	Layer 4	charcoal	AMS C14	BA091416	7740 ± 40	6642–6483 (95.4%)	He et al. [16]
Lijiagou (south area)	Layer 4	charcoal (3 samples)	AMS C14	10300–10500 cal. BP			Zhao et al. [17]
Bianbiandong		Human skull	AMS C14	BA04308	8675 ± 40	7786–7768 (2%) 7759–7592 (93.4%)	Sun et al. [18]
		Human skull	AMS C14	BA04309	8670 ± 30	7738–7597 (95.4%)	Sun et al. [18]
		Animal skull	AMS C14	BA04310	10,030 ± 40	9800–9786 (1.2%) 9773–9381 (94.2%)	Sun et al. [18]
		Human skull	AMS C14	BA04317	8730 ± 50	7941–7605 (95.4%)	Sun et al. [18]
	Area I Layer 3A	Animal bone	AMS C14	BA07226	8505 ± 45	7596–7507 (95.4%)	Sun et al. [18]
	Area I Layer 3B	Animal bone	AMS C14	BA07227	8585 ± 40	7704–7700 (0.5%) 7681–7541 (94.9%)	Sun et al. [18]
	Area I Layer 4	Animal bone	AMS C14	BA07228	8180 ± 45	7321–7067 (95.4%)	Sun et al. [18]
	Layer 2	Animal bone	AMS C14	BA07229	6120 ± 45	5211–4944 (95.4%)	Sun et al. [18]
	Collected	Animal bone	AMS C14	BA07230	6305 ± 40	5364–5214 (95.4%)	Sun et al. [18]
Kengnan	Layer 2, 3		C14	~10,000–9000 BP (calibrated?)			Li, Song [19]
Yujiagou			OSL		18 ka–14 ka		Mei [20]
Ma'anshan [1]					–		

Table 2. *Cont.*

Site	Layer	Material	Method	Lab. No.	Dates	Cal. BC (2–sigma)	Reference
Hutouliang	Terrace II	bone	Conv. C14	PV–0156	10,690 ± 120	10,879–10,428 (93.3%) 10,391–10,352 (0.8%) 10,335–10,287 (1.2%)	IA–CASS [21]
	Layer 6 (upper)	polymineral	OSL		9.2 ± 1.4 ka		Nagatomo et al. [22]
		polymineral	OSL		9.0 ± 1.3ka		Nagatomo et al. [22]
Mengjiaquan					–		
Jijitan [1]					–		
Xueguan		charcoal	Conv.C14	BK81016	13,170 ± 150	14,274–13,371 (95.4%)	IA–CASS [21]
Lingjing	Layer 5	charred material	AMS C14	IAAA–92123	11,360 ± 50	11,350–11,145 (95.4%)	Li et al. [23]
	Layer 5	charred material	AMS C14	IAAA–92124	11,490 ± 50	11,502–11,285 (95.4%)	Li et al. [23]
	Layer 5	charred material	AMS C14	IAAA–92125	11,930 ± 50	12,009–11,930 (8.6%) 11,915–11,625 (86.8%)	Li et al. [23]
	Layer 5	charcoal	AMS C14	IAAA–100080	28,610 ± 120	31,223–30,139 (95.4%)	Li et al. [23]
	Layer 5	burnt bone	AMS C14	IAAA–100082	11,520 ± 40	11,499–11,331 (95.4%)	Li et al. [23]
	Layer 5	burnt bone	AMS C14	IAAA–102634	11,400 ± 50	11,406–11,168 (95.4%)	Li et al. [23]
	Layer 5	charcoal	AMS C14	IAAA–102635	11,600 ± 50	11,601–11,365 (95.4%)	Li et al. [23]
	Layer 5	charred material on sherd	AMS C14	IAAA–102636	8610 ± 40	7723–7576 (95.4%)	Li et al. [23]
	Layer 5	charred material (charcoal?)	AMS C14	IAAA–102638	10,180 ± 40	10,106–9757 (95.4%)	Li et al. [23]
	Layer 5	charred material (charcoal?)	AMS C14	IAAA–102639	11,710 ± 50	11,763–11,721 (3.2%) 11,681–11,483 (92.2%)	Li et al. [23]
	Layer 5	charred material (charcoal?)	AMS C14	IAAA–102640	11,860 ± 50	11,822–11,613 (95.4%)	Li et al. [23]
	Layer 5	burnt bone	AMS C14	IAAA–102641	11,290 ± 50	11,307–11,115 (95.4%)	Li et al. [23]

Table 2. *Cont.*

Site	Layer	Material	Method	Lab. No.	Dates	Cal. BC (2–sigma)	Reference
Lingjing	Layer 5	charred material (charcoal?)	AMS C14	IAAA–102642	10,970 ± 50	11,032–10,772 (95.4%)	Li et al. [23]
	Layer 5	charred material (burnt bone?)	AMS C14	IAAA–102643	11,280 ± 50	11,302–11,109 (94.4%)	Li et al. [23]
	Layer 5	Charcoal (?)	AMS C14	IAAA–102644	11,550 ± 50	11,525–11,335 (95.4%)	Li et al. [23]
	Layer 5	charred material (?)	AMS C14	IAAA–102645	11,470 ± 50	11,489–11,251 (95.4%)	Li et al. [23]
	Layer 5	charcoal	AMS C14	IAAA–102647	11,220 ± 50	11,242–11,046 (95.4%)	Li et al. [23]
	Layer 5	charcoal	AMS C14	IAAA–102648	11,520 ± 50	11,511–11,321 (95.4%)	Li et al. [23]
	Layer 5	charred material (burnt bone?)	AMS C14	IAAA–102649	11,800 ± 50	11,801–11,587 (89%) 11,576–11,532 (6.4%)	Li et al. [23]
	Layer 5	charred material (charcoal?)	AMS C14	IAAA–102650	11,860 ± 50	11,822–11,613 (95.4%)	Li et al. [23]
Shizitan, Locality S29	Layer 1	bone	AMS C14	BA10129	11,175 ± 60	11,203–10,903 (95.4%)	Song [24]
	Layer 2 (203 cm deep)	Bone	AMS C14	BA101414	14,650 ± 70	16,077–15,674 (95.4%)	Song and Shi [25]
	Layer 2 (200–206 cm deep)	Bone	AMS C14	BA10132	15,725 ± 80	17,238–16,844 (95.4%)	Song and Shi [25]
	Layer 2 (248 cm deep)	Charcoal	AMS C14	BA10131	16,760 ± 65	18,494–18,058 (95.4%)	Song and Shi [25]
	Layer 2 (276 cm deep)	Bone	AMS C14	BA101416	15,390 ± 70	16,862–16,553 (95.4%)	Song and Shi [25]
	Layer 3 (282 cm deep)	Bone	AMS C14	BA101419	17,200 ± 50	18,987–18,615 (95.4%)	Song and Shi [25]
	Layer 3 (303 cm deep)	Bone	AMS C14	BA10133	17,360 ± 60	19,232–18,762 (95.4%)	Song and Shi [25]
	Layer 4 (465 cm deep)	Bone	AMS C14	BA101420	17,500 ± 70	19,449–18,936 (95.4%)	Song and Shi [25]

Table 2. *Cont.*

Site	Layer	Material	Method	Lab. No.	Dates	Cal. BC (2–sigma)	Reference
Shizitan, Locality S29	Layer 4 (469 cm deep)	Bone	AMS C14	BA10134	16,170 ± 50	17,754–17,367 (95.4%)	Song and Shi [25]
	Layer 4 (605 cm deep)	Bone	AMS C14	BA10135	16,930 ± 50	18,649–18,268 (95.4%)	Song and Shi [25]
	Layer 4 (620 cm deep)	Bone	AMS C14	BA101422	16,750 ± 80	18,511–18,030 (95.4%)	Song and Shi [25]
	Layer 4 (622 cm deep)	Bone	AMS C14	BA101421	18,570 ± 60	20,638–20,363 (95.4%)	Song and Shi [25]
	Layer 4 (624 cm deep)	Bone	AMS C14	BA101423	19,210 ± 80	21,495–20,943 (95.4%)	Song and Shi [25]
	Layer 4 (640 cm deep)	Tooth	AMS C14	BA10136	17,040 ± 60	18,804–18,391 (95.4%)	Song and Shi [25]
	Layer 5 (772 cm deep)	Bone	AMS C14	BA10137	18,360 ± 70	20,477–20,025 (95.4%)	Song and Shi [25]
	Layer 5 (787 cm deep)	Charcoal	AMS C14	BA10485	20,420 ± 80	22,973–22,319 (95.4%)	Song and Shi [25]
	Layer 5 (750.5 cm deep)	Bone	AMS C14	BA101426	19,650 ± 100	22,028–21,432 (95.4%)	Song and Shi [25]
	Layer 5 (751.5 cm deep)	Charcoal	AMS C14	BA101427	19,510 ± 70	21,815–21,244 (95.4%)	Song and Shi [25]
	Layer 5 (750.5 cm deep)	Charcoal	AMS C14	BA101428	19,940 ± 70	22,284–21,811 (95.4%)	Song and Shi [25]
	Layer 5 (804 cm deep)	Charcoal	AMS C14	BA101429	19,710 ± 80	22,050–21,535 (95.4%)	Song and Shi [25]
	Layer 5 (801.8 cm deep)	Charcoal	AMS C14	BA101430	19,860 ± 70	22,185–21,707 (95.4%)	Song and Shi [25]
	Layer 6 (968 cm deep)	Bone	AMS C14	BA101431	18,140 ± 80	20,326–19,812 (95.4%)	Song and Shi [25]

Table 2. *Cont.*

Site	Layer	Material	Method	Lab. No.	Dates	Cal. BC (2–sigma)	Reference
Shizitan, Locality S29	Layer 6 (964 cm deep)	Charcoal	AMS C14	BA101433	20,410 ± 80	22,961–22,306 (95.4%)	Song and Shi [25]
	Layer 6 (964 cm deep)	Bone	AMS C14	BA101434	19,850 ± 80	22,190–21,680 (95.4%)	Song and Shi [25]
	Layer 6 (961.5 cm deep)	Tooth	AMS C14	BA121954	20,155 ± 45	22,471–22,067 (95.4%)	Song and Shi [25]
	Layer 6 (1004 cm deep)	Charcoal	AMS C14	BA10487	20,500 ± 100	23,117–22,388 (95.4%)	Song and Shi [25]
	Layer 6 (1004 cm deep)	Bone	AMS C14	BA10488	18,090 ± 70	20,242–19,733 (95.4%)	Song and Shi [25]
	Layer 6 (1004 cm deep)	Bone	AMS C14	BA121951	18,280 ± 45	20,393–19,978 (95.4%)	Song and Shi [25]
	Layer 6 (1026 cm deep)	Charcoal	AMS C14	BA101438	20,350 ± 90	22,886–22,201 (95.4%)	Song and Shi [25]
	Layer 7 (1160 cm deep)	Bone	AMS C14	BA121960	21,690 ± 80	24,151–23,840 (95.4%)	Song and Shi [25]
	Layer 7 (1160 cm deep)	Bone	AMS C14	BA101439	19,650 ± 80	21,996–21,465 (95.4%)	Song and Shi [25]
	Layer 7 (1160 cm deep)	Charcoal	AMS C14	BA101442	20,010 ± 70	22,353–21,901 (95.4%)	Song and Shi [25]
	Layer 7–8 boundary (1355 cm deep)	Bone	AMS C14	BA101445	20,510 ± 90	23,107–22,414 (95.4%)	Song and Shi [25]
	Layer 8 (1425 cm deep)	Charcoal	AMS C14	BA101444	24,185 ± 90	26,575–25,976 (95.4%)	Song and Shi [25]
Shizitan, Locality S5	Layer 1	bone	AMS C14	BA101404	9220 ± 50	8565–8299 (95.4%)	Song and Shi [26]

Table 2. *Cont.*

Site	Layer	Material	Method	Lab. No.	Dates	Cal. BC (2–sigma)	Reference
Some other sites dated to PNT time range, but not related to origin of food production (data shown below, recorded as others in the maps)							
Shuidonggou Locality 12	Layer 11	charcoal	AMS C14	LUG06–54	9797 ± 91	9654–9580(2.5%) 9548–9478 (2.3%) 9465–9117 (80.9%) 9074–9056 (0.5%) 9013–8911 (7%) 8905–8845 (2.2%)	Liu et al. [27]
	Layer 11	quartz	OSL	IEE1110	11.6 ± 0.6 ka		Nagatomo et al. [22]
	L1–L5	cultural layers formed between 11,000 and 12,000 years ago, starting at the middle of the Younger Dryas					Yi [28]
Gezishan 53S/0W	Stratum E	–		Beta 97241	10,230 ± 50	10,189–9810 (95.4%)	Elston et al. [29]
	Stratum E/F	–		Beta 86731	11,620 ± 70	11,626–11,353 (95.4%)	Elston et al. [29]
	Stratum G2	–		Beta 97242	12,710 ± 70	13,398–12,857 (95.4%)	Elston et al. [29]
Gezishan 3N/3W	Stratum D	–		Beta 86732	10,020 ± 60	9818–9318 (95.4%)	Elston et al. [29]
	Stratum D	–		Beta 97346	10,130 ± 70	10,086–9452 (95.4%)	Elston et al. [29]
PY–04		charcoal	AMS C14	CAMS94202	10,670 ± 40	10,759–10,618 (95.4%)	Barton et al. [30]

Note: [1] no absolute date available.

Table 3. Presence of markers of the PNT in North China, by site [3] Table 2 modified from [2]. These sites are mapped in Figure 1.

Site	Type	Artifact Assemblage								Site Organization							Biological Remains			
		A	B	C	D	E	F	G	H	I	J	K	L	M	N	O	P	Q	R	S
Nanzhuangtou	PNT	0	1	1	1	0	0	0	0	1	1	0.5	0	1	1	1	0	0	1	1
Zhuannian	PNT	1	1	1	1	0	1	1	1	0	1	0	0	0	0	0	0	0	0	0
Donghulin [1]	PNT	1	1	1	1	0	1	1	1	1	1	1	1	1	1	1	1	0.5	0	0
Lijiagou	PNT	1	1	1	0	1	0	1	1	0.5	1	0.5	0	1	0	0	0	0	0	0
Bianbiandong	PNT	0	1	1	1	0	0	0	1	1	1	0.5	0.5	1	0	1	0	0	0	0
Kengnan	PNT	0	1	1	0	0	0	0	1	0	0	0	0	0	0	0	0	0	0	0
Yujiagou	PNT	1	1	0	0	1	0	0	0	0	0	0	0	0	0	0	0	0	0	0
Ma'anshan	PNT	1	1	0	0	1	0	0	1	1	1	0	0	1	0	0	0	0	0	0
Hutouliang	PNT	1	0.5	0	0	1	0	0	0	0	0	0	0	0	0	0	0	0	0	0
Mengjiaquan	PNT	1	0	0	0	0	0	1	1	0	0	0	0	0	0	0	0	0	0	0
Jijitan	PNT	1	0	0	0	1	0	0	0	0	0	0	0	0	0	0	0	0	0	0
Xueguan	PNT	1	0	0	0	0	0	0.5	0	0	0	0	0	0	0	0	0	0	0	0
Lingjing [1]	PNT	1	1	0	0	0	0	0	0	0	0	0	0	0	0	0	0	0	0	0
Shizitan [2]	LUP	1	0	1	1	0	0	0	1	1	1	0	0	0	0	0	1	0	0	0
Xiachuan	LUP	1	0	0	0	1	0	0	1	0	0	0	0	0	0	0	0	0	0	0
Longwangchan	LUP	1	0	1	1	0	0	1	1	0.5	1	0	0	0	0	0	0	0	0	0
1 = present			0.5 = possibly present											0 = not discovered yet.						

Notes: The rows: A = microblade, B = pottery, C = mortar (millstone), D = pestle (millstick), E = adze, F = stone container, G = polished tool, H = diversity of artifact types, I = hearth, J = diversity of site types, K = structure, L = grave, M = central camp, N = ditch, O = pit, P = foxtail millet, Q = broomcorn millet, R = dog, S = wild pig (boar). The columns: Other sites without data except microblades: Xishi (Late Upper Paleolithic, LUP); Shuidonggou Loc. 12; Dadiwan (Component 4); Gezishan (Pigeon Mountain); PY–04, all classified as "other" (see maps of Northern Chinanorthern China). [1] updated information according to Cui [15]; [2] new site added according to Li et al. [23]; [3] updated information of Locality 29 according to Liu et al. [31].

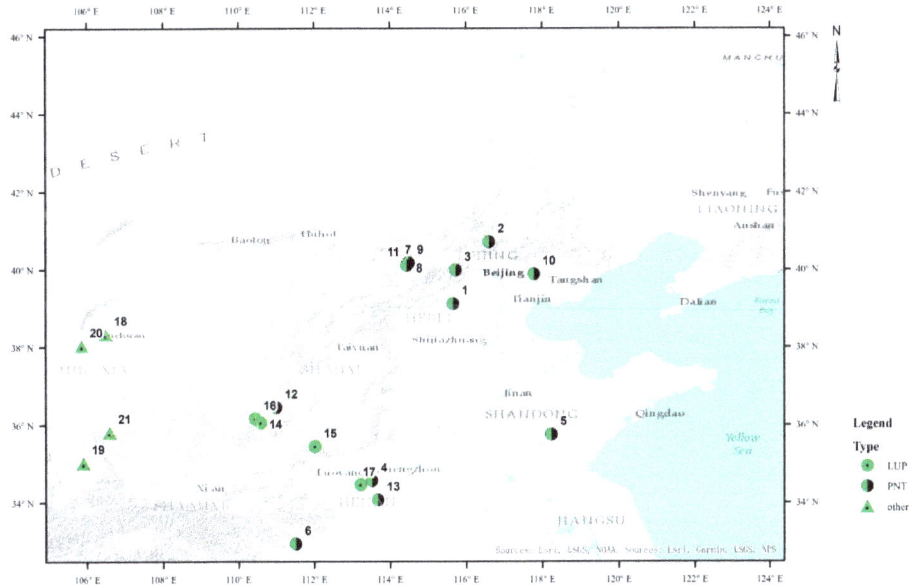

Figure 1. Archaeological sites dated to Pleistocene-Holocene transition in Northern China. PNT sites: 1 Nanzhuangtou; 2 Zhuannian; 3 Donghulin; 4 Lijiagou; 5 Bianbiandong; 6 Kengnan; 7 Yujiagou; 8 Ma'anshan; 9 Hutouliang; 10 Mengjiaquan; 11 Jijitan; 12 Xueguan; 13 Lingjing LUP sites: 14 Shizitan; 15 Xiachuan; 16 Longwangchan Other sites: 17 Xishi; 18 Shuidonggou Locality 12; 19 Dadiwan, Component 4; 20 Gezishan; 21 PY-04.

4. Microblade–Based Societies during the Last Glacial Maximum

Among archaeological sites in north China associated with microblade assemblages, the Shizitan site is best studied, especially in the terms of interdisciplinary approach focusing on human behaviors at a specific site. Shizitan is located at the Jixian County, SW Shanxi Province. Several localities are distributed along the Qingshuihe River, a tributary of the Yellow River. In recent years, Li Liu, in collaboration with local archaeologists, has been conducting a research project on plant use at this site, based on use–wear and residue analyses, suggesting that prehistoric plant use in northern China can be traced back to the immediately pre–LGM period, about 28 ky cal. BP [31–34]. The plant use, including harvesting and processing using stone tools, can be divided into 6 phases (Table 4). This work suggests near–continuous exploitation of various plants in the Middle Yellow River Valley (both the Xiachuan site and the Longwangchan site share many material features with the Shizitan site, including microblades, grinding stones, and hematite processing) [31]. Their investigation provides substantial evidence of an early broad–spectrum subsistence strategy in the Upper Paleolithic northern China.

Table 4. Phases of plant use at the Shizitan (SZT) site (data collected from [31]).

Phase	Site and Stratum	Material	Function for Plants	Plant Use
Phase I (pre-Last Glacial Maximum (LGM))	Stratum 8, SZT 29	2 FL 2 SL	FL: cutting, scraping SL: processing	Starchy plants, including Panicoideae (possibly Job's tears), Triticeae and yam No clear evidence for wild millet collection
Phase II (the early part of LGM)	Stratum 7, SZT 29	7 MB 3 FL 4 SL 1HS	MB: cutting FL: plant cutting SL: processing * HS: processing	Wild cereals (e.g., Triticeae, Job's tears and wild millets), tubers and other plants
Phase III (the late part of LGM)	Strata 6, 5, and 4, SZT 29	1 MB 7 FL 3 SL	MB: cutting FL: cutting SL: processing	Wild cereals, and especially Panicoideae
Phase IV (the terminal LGM)	Strata 3 and 2, SZT 29	4 FL 4 SL	FL: cutting SL: processing	Grasses, including Panicoideae and Triticeae. Tubers are much less well represented
Phase V (warmer conditions)	Stratum 1, SZT 29	2 FL	FL: cutting	Unknown, only one non–specific phytolith was recovered.
Phase VI (Early Holocene)	Stratum 1, SZT 5	2 FL	FL: cutting	Unknown, only starch

Notes: FL: flake, MB: microblade, SL: slab or slab fragment, HS: handstone. * Both slab fragments and handstones are used for processing plant food and minerals.

Liu et al. [33] propose the viewpoint, "The trajectory from intensified collection of a wide range of wild plants to domestication of a small number of species was a very long process in north China", and suggest that it "parallels the transition from the 'broad spectrum revolution' to agriculture in the Near East" (pp. 3524). Although they did not use the term "intensification" to generalize this process, plant processing could be seen as a means of food resource intensification, since people need to invest much labor to extract energy from plants, including tubers and seeds. Plant use might suggest a local population packing effect during the LGM in the Middle Yellow River Valley, possibly caused by southward movement of prehistoric peoples from the north or the west, which can be tested as a hypothesis.

The effective temperature (ET) value of the Shizitan site under LGM climatic conditions was below the terrestrial plant threshold, suggesting that the prehistoric foraging societies needed to depend on terrestrial animal resources during the LGM. However, a relatively longer growing season than that of Siberia provides this location higher net above–ground productivity and primary biomass. The Shizitan site is also located in the belt of high secondary biomass, making it a place that possessed plenty of both terrestrial plants and animals. The Qingshuihe River, a small tributary of the Yellow River, also provided people with enough water and aquatic resources for survival, although archaeologists have not found substantial evidence of fishing. The maps in Figures 2–6 support the following characterization of the environment and resource use at the Shizitan site:

> *"Around 23,000–18,000 years ago during the LGM, the Qingshui River valley appears to have been an area with a wide range of faunal and floral resources, which attracted small hunting-gathering groups. In addition to hunting, people collected and processed many types of plants, including grass seeds of Triticeae and Paniceae,* Vigna *beans,* D. opposita *yam, and* T. kirilowii *snakegourd roots, among others".* [32] (p. 5384)

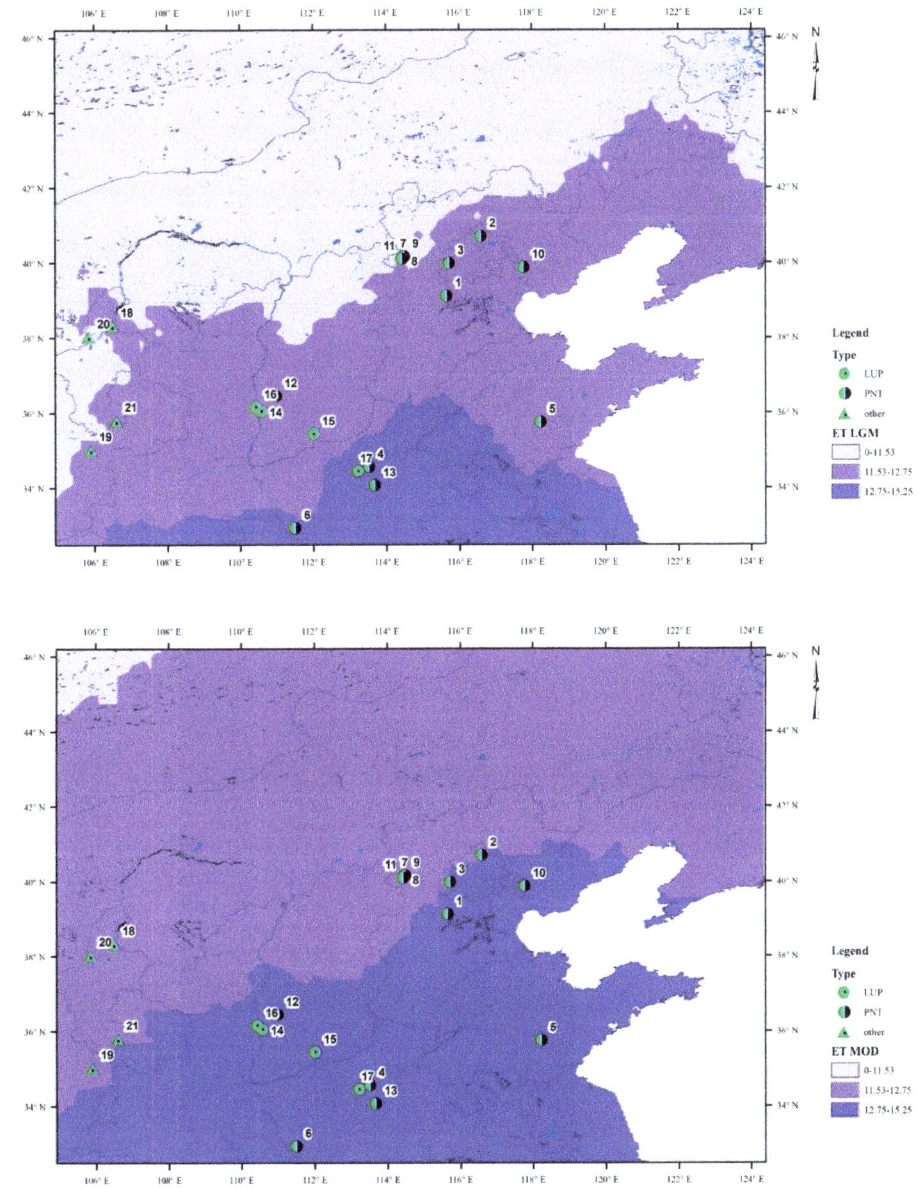

Figure 2. Archaeological sites dated to the Pleistocene to Holocene transition on the maps of effective temperature (ET) in Northern China under LGM and modern climatic conditions.

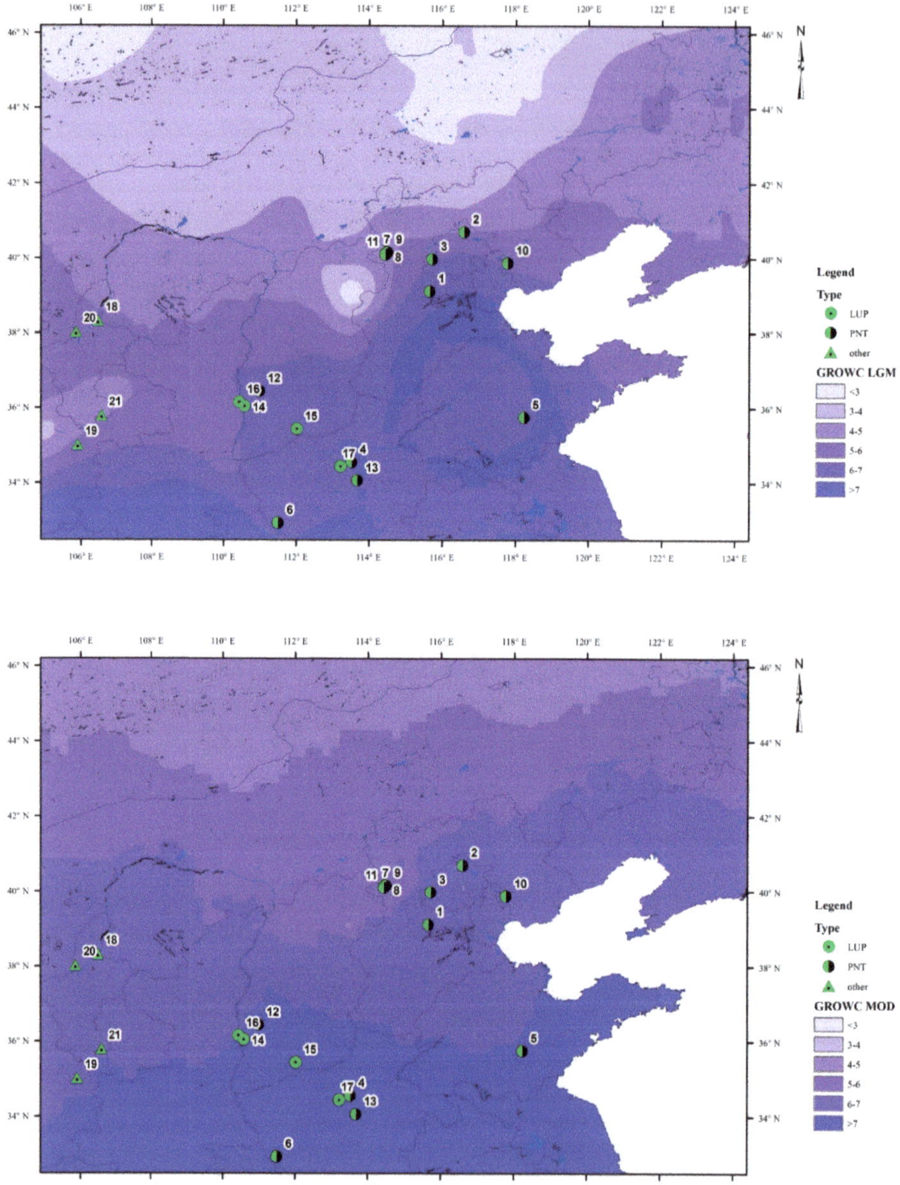

Figure 3. Archaeological sites dated to the Pleistocene to Holocene transition on the maps of growing season (GROWC) of Northern China under LGM and modern climatic conditions.

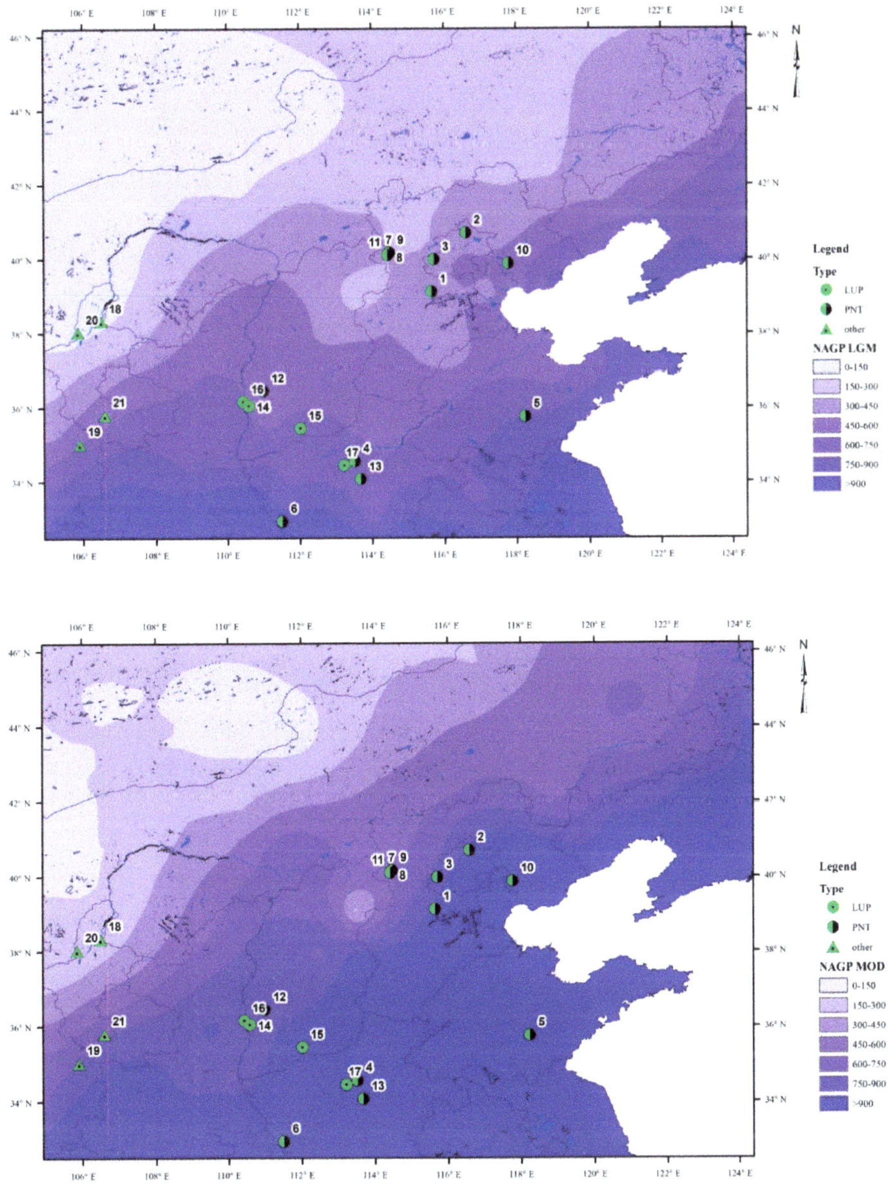

Figure 4. Archaeological sites dated to the Pleistocene to Holocene transition on the maps of net above-ground productivity (NAGP) in Northern China under LGM and modern climatic conditions.

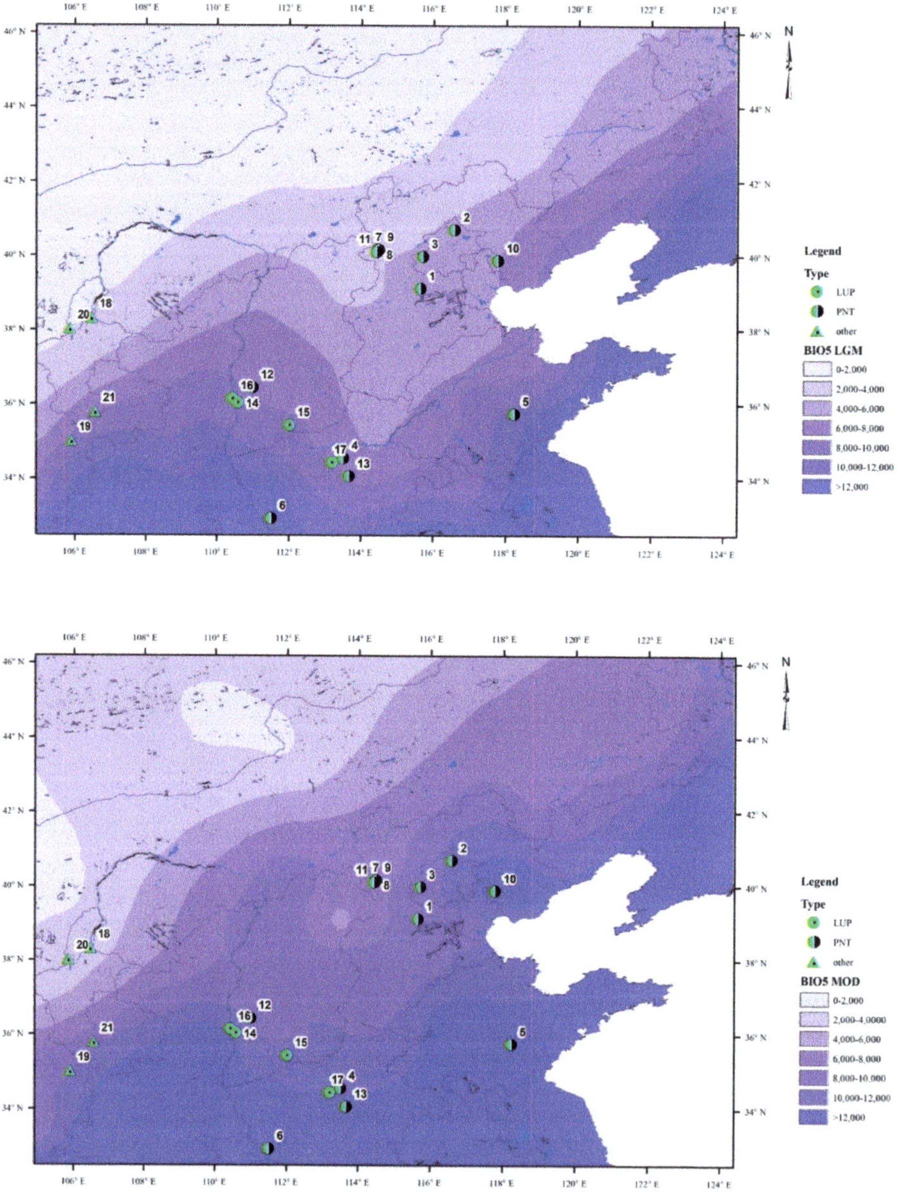

Figure 5. Archaeological sites dated to the Pleistocene to Holocene transition on the maps of primary biomass (BIO5) of Northern China under LGM and modern climatic conditions.

Thus, there is no question that prehistoric hunter–gatherers in this region adopted a mixed economy to satisfy their subsistence requirements, which might have been a general strategy adopted by the peoples during the LGM. The peoples in the lower latitudes or above the subpolar bottleneck (ET = 11.53 °C) have significantly more terrestrial plant resources to utilize.

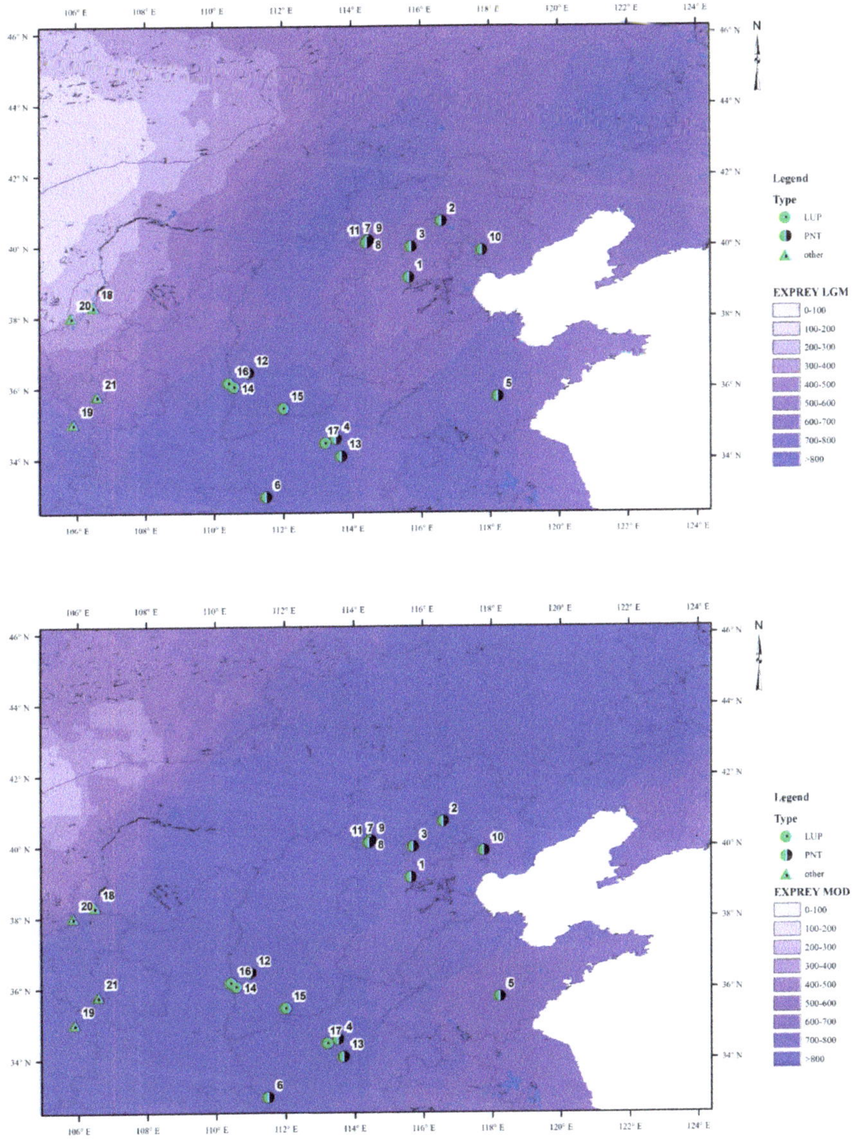

Figure 6. Archaeological sites dated to the Pleistocene to Holocene transition on the maps of secondary biomass (expected moderate body–size ungulate biomass, EXPREY) of northern China under LGM and modern climatic conditions.

The question is whether resource intensification on terrestrial plants was practiced during the LGM. According to the maps produced under the growth rate model, unpacked/packed maps show that the location of the Shizitan site is expected to have favored a gathering-dominated economy if population became packed, while it would have favored a hunting–dominated economy if population was unpacked (Figure 7). The ethnographic projected population density map under LGM climatic conditions (Figure 8) suggests that local resource cannot support a high population density in the region where the Shizitan site is located (2–3.033 persons per 100 km^2). The patterns resulting from these projections can only be used if hunter–gatherers in this location were organized like recent hunter–gatherers in similar contexts, rather than used to say what the archaeological record actually was. In addition, the packing threshold was developed using data for mobile plant-dependent hunter–gatherers who used no long-term storage—the density threshold at which highly mobile hunters might need to intensify would be lower. In other words, the occupants at the Shizitan site during the LGM might have intensified plant resources especially for storage. But more archaeological evidence is needed to support this hypothesis.

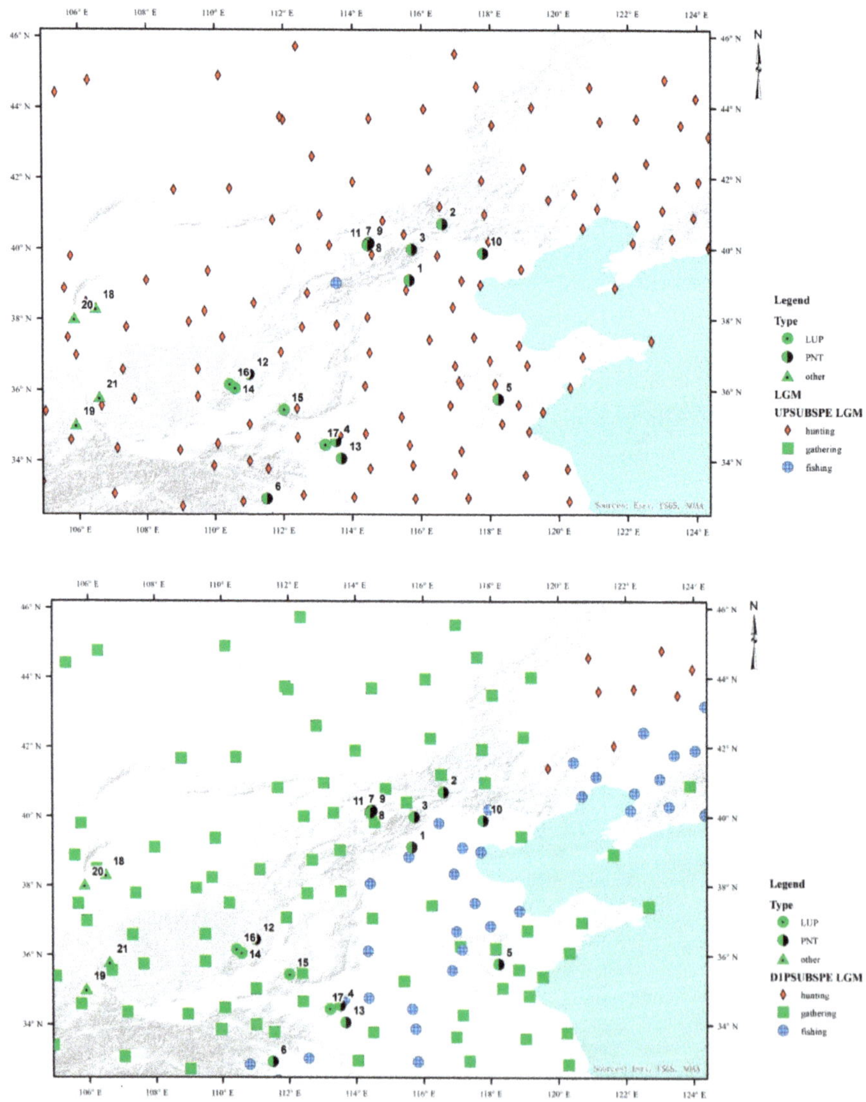

Figure 7. Archaeological sites dated to the Pleistocene to Holocene transition on the maps of projected unpacked and packed HG subsistence specialty (UPSUBSPE and D1PSUBSPE) in northern China under LGM climatic conditions.

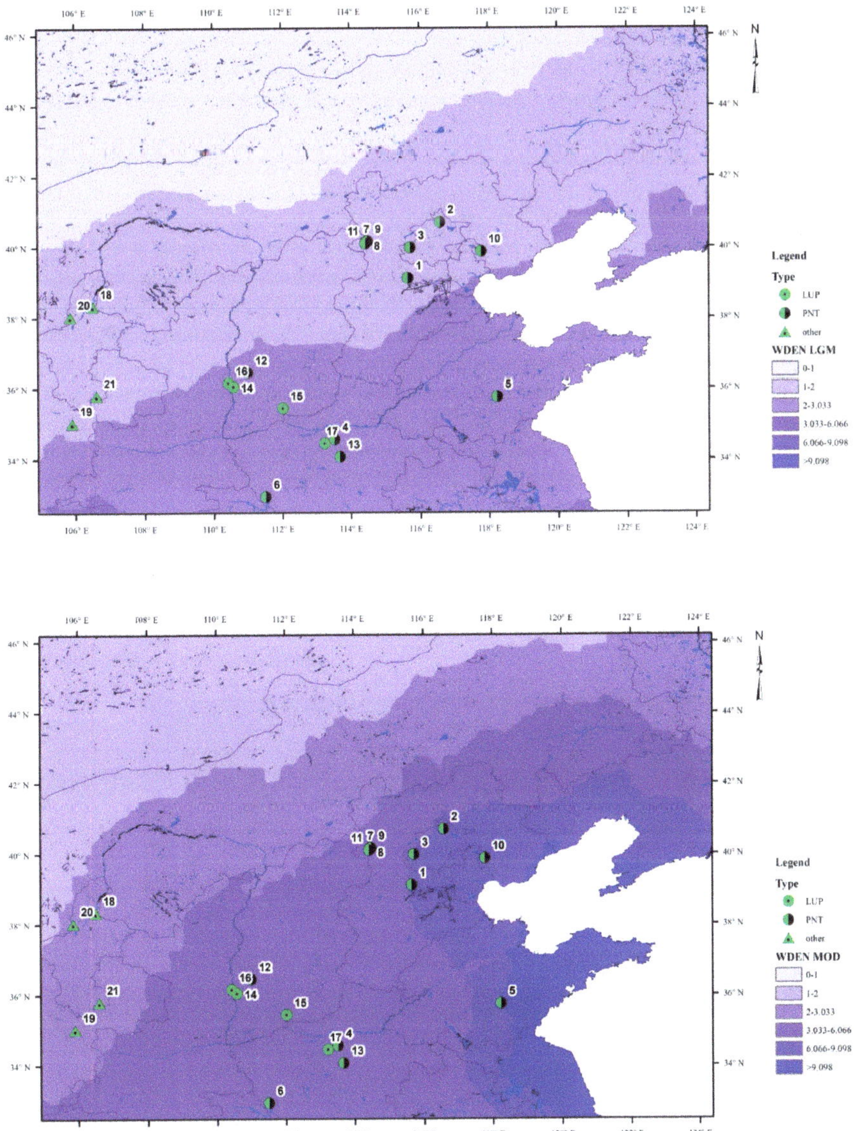

Figure 8. Archaeological sites dated to the Pleistocene to Holocene transition on the maps of projected HG population density (WDEN) of northern China under LGM and modern climatic conditions.

Is there any other possible explanation for the existence of intensive plant use at the Shizitan site? The map of EXDMOV1 (projected total distance moved, scaled for subsistence type, for groups with year round camp to camp mobility pattern) shows that foraging societies in northern China during the LGM had to adopt highly mobile lifeways (comparing with the values of EXDMOV1 under current climatic conditions) (Figure 9). Widespread microblade technology provided them with effective and light composite weaponry to hunt animals; some microblades were used as reaping knife inserts to harvest and process plant resources. Microblade–based societies successfully adapted to the deteriorated environment with relatively lower primary and secondary biomass. The region

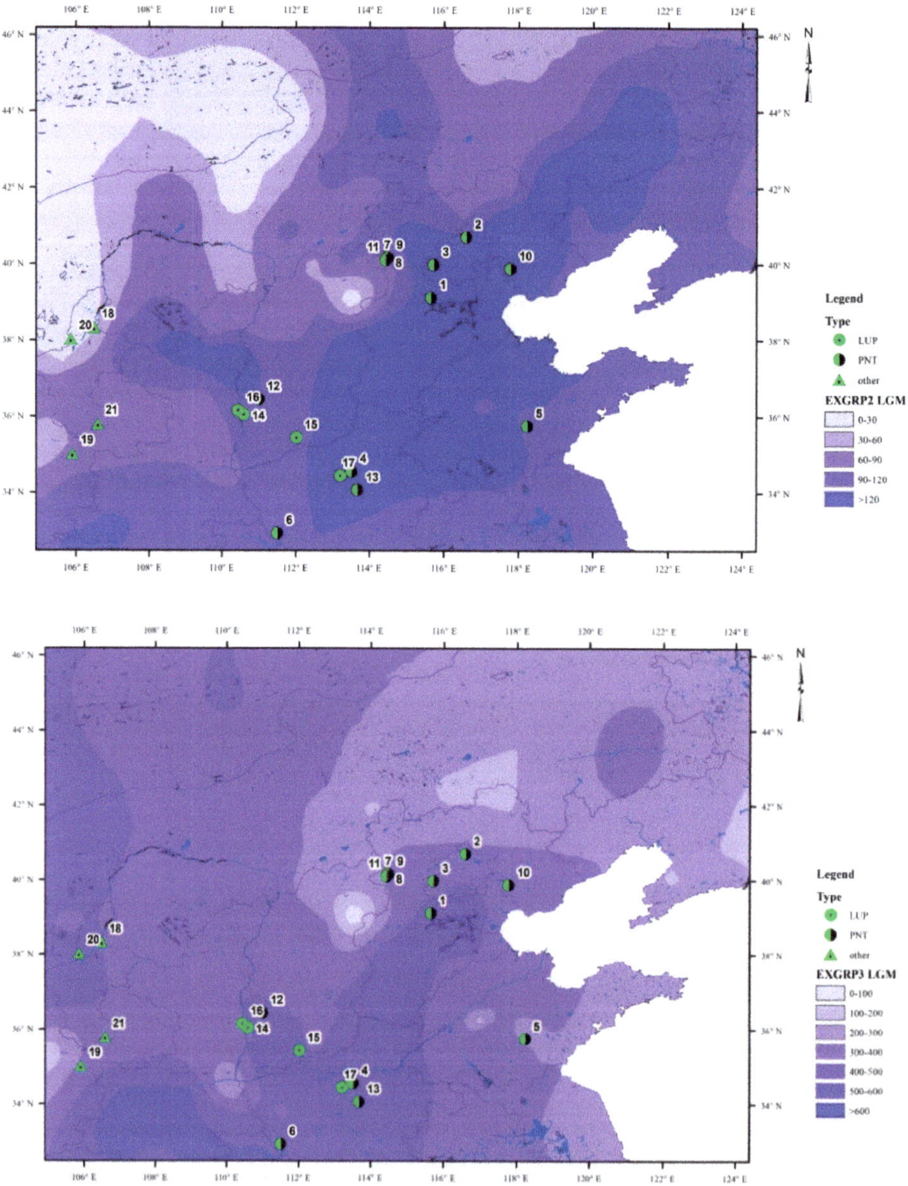

Figure 10. Archaeological sites dated to the Pleistocene to Holocene transition on the maps of expected hunter–gatherer group size in northern China under LGM climatic conditions.

Table 5. Hearths in the Shizitan site, Locality S29 [25].

Type of Hearth	Number	Percentage
Above–ground hearth	265	92.98%
Above–ground hearth with sandstone rocks	13	4.56%
Round–pit hearth	5	1.76%
Hearth with stone–floored pit	2	0.7%
Total	285	100%

To sum up, the macroecological approach suggests that intensification might not have existed at the Shizitan site during the LGM. The archaeological record of plant resource exploitation at the Shizitan site indicates that increasing population densities might have made intensification important. However, plants may have been used as supplementary food resources (tubers and wild millet) in daily life. Just as the maps suggested, the plant foods also might have been used to satisfy the regional band requirements of periodic aggregations at the Shizitan site (e.g., Localities S29 and Loc. 14). Unfortunately, because of a lack of lithic refitting research, the hypothesis proposed here remains speculative, although the limited available evidence is suggestive. Limited by the preliminarily published data, testing this hypothesis will require more evidence and question-oriented studies in the future. Plant use at the Shizitan site lasted until the early Holocene, when it was totally abandoned. This is a key site for archaeologists to investigate the process of a potential broad spectrum revolution in the Upper Paleolithic of the Middle Yellow River Basin, and/or to study subsistence-settlement systems and technological organization of microblade-based societies in northern China, since it possesses a complete record from Phase I to Phase III (and some in Phase IV).

5. Broad Spectrum Revolution and Intensification during the Pleistocene–Holocene Transition

The Late Pleistocene and Early Holocene in northern China witnessed the transformation from societies composed of mobile hunter-gatherers equipped with microblade technology to societies equipped with ground stone tools, pottery, food production, and durable dwellings. There have been many studies on these aspects of the transition. Although their appearances can be dated to different specific times in different regions, the formation of agricultural societies can be seen as an emergent event, rather than as the simple overlap of historical events. Chen and Yu [1,2] have termed this period the Paleolithic to Neolithic transition (PNT). They "synthesize a broad range of diagnostic artifacts, settlements, site structure, and biological remains to develop a working hypothesis that agriculture was differentially developed or adopted according to 'initial conditions' of habitat, resource structure, and cultural organization" [2]. The authors noted multiple and divergent evolutionary pathways, including northern China (divided at the Taihangshan Mts. into the North China Plain and Loess Plateau). Published data on the chronology and archaeological record of northern China linked to the PNT are presented, clearly showing basic information on diagnostic artifacts, settlements, and site structure associated with social organization, and potential species for domestication Tables 1 and 2 in ref [2], with updated information from recently published results. The two papers by Chen and Yu [1,2] clearly and successfully illustrate cultural change during the PNT of Northern and Southern China, but the authors did not fully apply a macroecological approach in their study.

Since I have shown a series of maps of Northern China for the Shizitan case under LGM conditions, I now only show the corresponding maps under modern climatic conditions, representing the early Holocene. Readers can compare the maps under the two sets of climatic conditions.

The Pleistocene–Holocene transition witnessed a dramatic change from glacial to interglacial, especially after the Younger Dryas. The Terminal Pleistocene after the LGM (starting at 15 ky uncalibrated BP), experienced several climatic phases of relatively warm and cold periods. The Younger Dryas is seen as a stimulus for food production in Northern China and the Near East [7]. In this project, the maps of LGM climatic conditions can be read similarly to those under Younger Dryas conditions, since paleoclimate data suggests that the Younger Dryas was associated with marine isotopic values as low as those of the LGM, but during a much shorter time (about 1300 years, 12.9–11.6 ky cal. BP, or about 600 years, 10.8–10.2 ky uncalibrated BP) [34,35].

Effective temperature dramatically increased and the isoline of the terrestrial plant threshold (ET = 12.75 °C) moved northward. Several sites with the best evidence of food production are located around this isoline (Nanzhuangtou, Zhuannian, and Donghulin) (Figure 2). In contrast to the more localized and fragmentary distribution of growing season isolines under LGM climatic conditions, the length of the growing season dramatically increased, causing more gradual change along latitudes (34° N–40° N) under interglacial conditions (Figure 3), which could help in the formation of belts favorable for stable plant growth and provide a good initial condition for cultivation. The increases of ET and length of growing season resulted in an increase of net above–ground productivity and primary biomass (Figures 4 and 5). Secondary biomass also increased and the central region moved northwest to the modern forest–steppe ecotone along the southwest–northeast China line Figures 5 and 7 in ref [6]. Under modern climatic conditions (early Holocene), the three sites—Nanzhuangtou, Zhuannian, and Donghulin—are all located at the edge of the zones with high values of both primary and secondary biomass, i.e., the transition zone (Figure 6). This phenomenon matches the basic hypothesis advanced by Binford [36,37], that early food production would appear in zones adjacent to areas with the most resources, as a cultural adaptation after mobility declined or ceased to be effective to solve the problems of population growth.

The growth rate model also shows the replacement of terrestrial hunting by gathering and aquatic resource use. Different from the maps under LGM climatic conditions, the subsistence replacement indeed happened during the Pleistocene to Holocene transition—Chen and Yu's Paleolithic to Neolithic Transition. For northern China (above 30° N), hunting in some places was replaced by the gathering of terrestrial plants (e.g., the Gezishan site in Upper Yellow River Basin), by aquatic resource use (some sites in NE China, such as the Houtaomuga and Shuangta sites), or food production (i.e., Nanzhuangtou, Zhuannian, Donghulin, and other PNT sites, as well as the early Neolithic sites) (Figure 11). The hunters north to the Yanshan Mts. (north to the area of Beijing) retained their hunting–dominated lifeways in a mixed economy with the aid of advanced microblade technology (pencil–shaped microcores) during the early and middle Holocene, perhaps as specialized hunters who exchanged wild meat with neighboring agriculturists, who were in turn finally replaced by pastoralism (about 1000 BC) [38]. The sites of Nanzhuangtou, Zhuannian, and Donghulin are all located in the boundary regions of combined hunting, gathering and fishing under the packed population condition, suggesting that they needed to manage a pluralistic subsistence style. This is supported by the coexistence of microblade–equipped hunting weaponry, pottery, mortars and pestles, and polished tools at the Zhuannian and Donghulin sites. The lack of any microblade assemblage at the Nanzhuangtou site perhaps resulted from the sampling representation of the excavated area (300 out of 2000 m^2), or a higher level of food production with a very low level of hunting activities. In addition, evidence of hunting activity calls for further research focus, since some sites are located near water systems and people are expected to have exploited some aquatic resources.

The increase of sedentary life can also be investigated from the viewpoint of a macroecological approach, in terms of projected population density and expected distance of residential movement per year. The Nanzhuangtou, Zhuannian, and Donghulin sites are located at the position of 2/3 packing threshold, and the Bianbiandong cave site is at the packing threshold (Figure 8). Thus, the people at Bianbiandong should have experienced population pressure, while the occupants at sites Nanzhuangtou, Zhuannian, and Donghulin might have experienced population packing if they emphasized hunting during the early Holocene. If they were mobile hunter–gatherers, their mobility should have dramatically decreased from 250/300–350 miles travelled per year during the Younger Dryas (see the EXDMOV1 under LGM climatic conditions), to 200–250 miles per year during the early Holocene (Figure 9). However, if they were sedentary hunter-gatherers their mobility only needed to be 20–40 miles per year during early Holocene (Figure 12). For people who practiced cultivation, the decrease in mobility could focus on a habitat where they could use low-energy food resources through dietary expansion and/or intensification. Therefore, zones with relatively lower primary and secondary biomass compared to neighboring regions associated with optimal resource distribution could serve as new locations in which 'daughter' groups could practice food production [37,38]. The lifeways built on sedentism (decreased mobility) favored durable ground/polished and expedient stone tools, rather than highly curated and formal chipped stone tools designed for mobile hunting, because the most important part of the economy became plant resource–use and even cultivation [39,40]. This may explain the disappearance of microblade assemblages in early Neolithic sites in northern China.

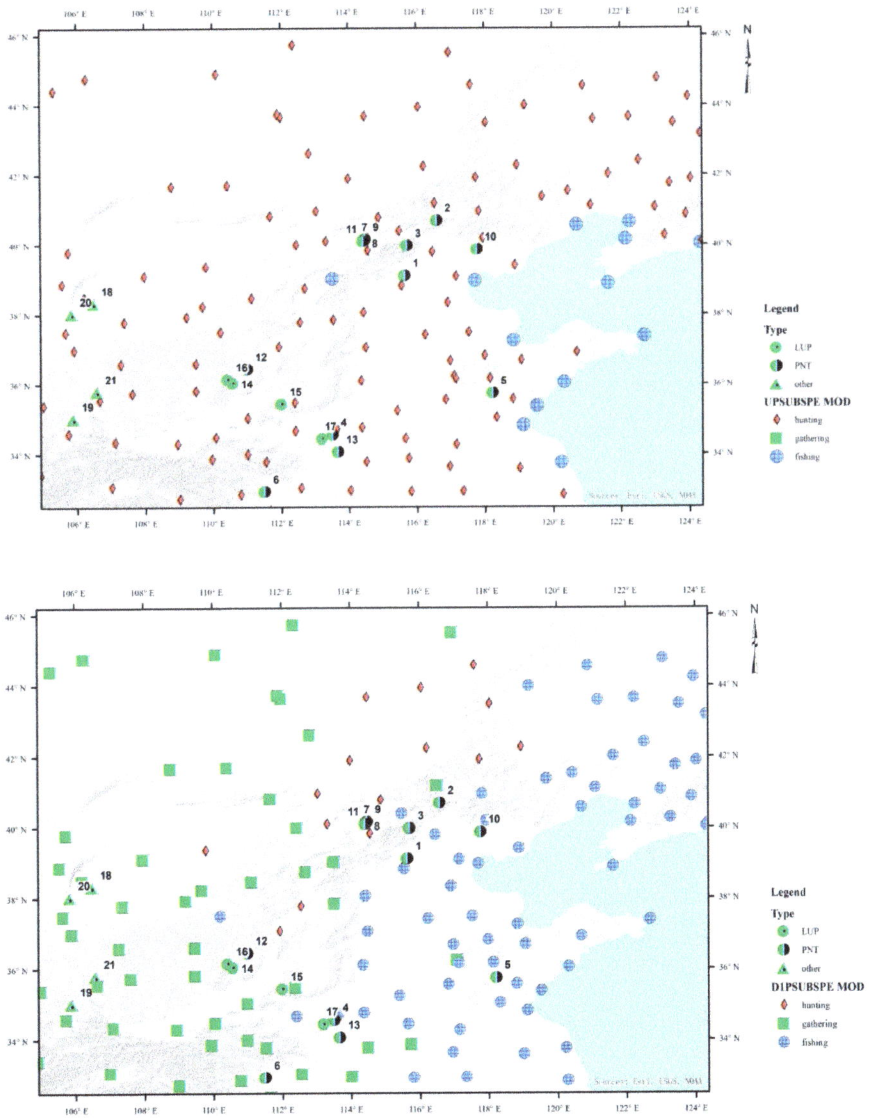

Figure 11. Archaeological sites dated to the Pleistocene to Holocene transition on the maps of projected unpacked and packed hunter–gatherer subsistence specialty (UPSUBSPE and D1PSUBSPE) in northern China under modern climatic conditions.

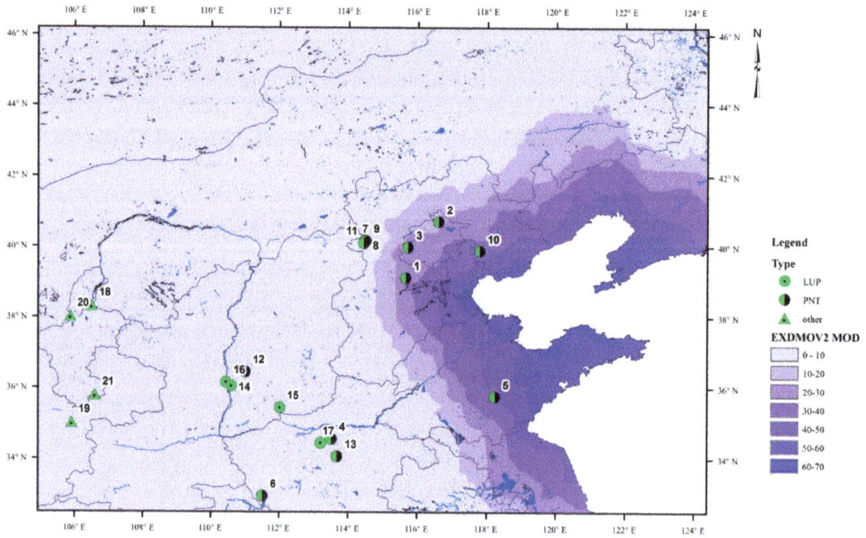

Figure 12. Archaeological sites dated to the Pleistocene–Holocene transition on the map of EXDMOV2 (expected distance of residential movement per year of hunter–gatherers with central location or who are primarily sedentary) in northern China under modern climatic conditions.

6. The Demise of Microblade–Based Societies

The Early Holocene witnessed the disappearance of microblade technologies in North China and the demise of microblade-based societies, although there were a few later sites associated with some of these technologies. However, investigating this cultural process is tough, since those specializing in Neolithic archaeology rarely report those data in the early stages. There are few published data on microblades or microcores in the Neolithic sites (8500–2000 BP) of the northern China cultural region. Although this can be explained as a reporting bias of Neolithic archaeologists, in fact there may be a real absence or very few microblades in the sites. In contrast, Holocene microblade assemblages have been reported in some sites of the Inner Mongolia Steppe, in NE and NW China, on the Tibetan Plateau, and in SW China, all of which are frontier regions of China. Details of the disappearance of microblade technology are little known, let alone the process of the decline and disappearance of microblade–based societies. Among the sites dated to the late PNT after the Younger Dryas event, only the Donghulin site has been researched in detail in Tian-Xing Cui's [15] dissertation. Microblade assemblages at this site co-exist with ground stone and other percussion tools. By contrast with the classic microblade technology at sites dated to terminal Pleistocene, such as Hutouliang [41], Lingjing [42], Yujiagou [20], and Shuidonggou Locality 12 [28,43], the microcores at the Donghulin site show great variety with unstandardized production methods (Cui used many subtypes to classify them), and some tools show expedient characteristics. The degeneration of microblade production methods might be a result of the transition to a patchy environment composed of forest and grassland. Forest animals dominate the fauna of the Donghulin site (Figure 13), which might impede the use of weaponry equipped with microblades, since open landscape was being replaced by forest and marshes that favor alternative hunting technologies. Paleoethnobotanical research suggests that the occupants at the Donghulin site intensified the use of plant products. Macrobotanical remains including Gramineae, Leguminosae, and *Celtis sinensis*, and microbotanical remains (starch) on mortars and pestles including Triticeae and wild foxtail millet (*Setaria italica*), imply a broad–spectrum economy and intensification. Sedentism had already been adopted by the late PNT occupants, which also reduced the use of microblade technology designed for highly mobile lifeways.

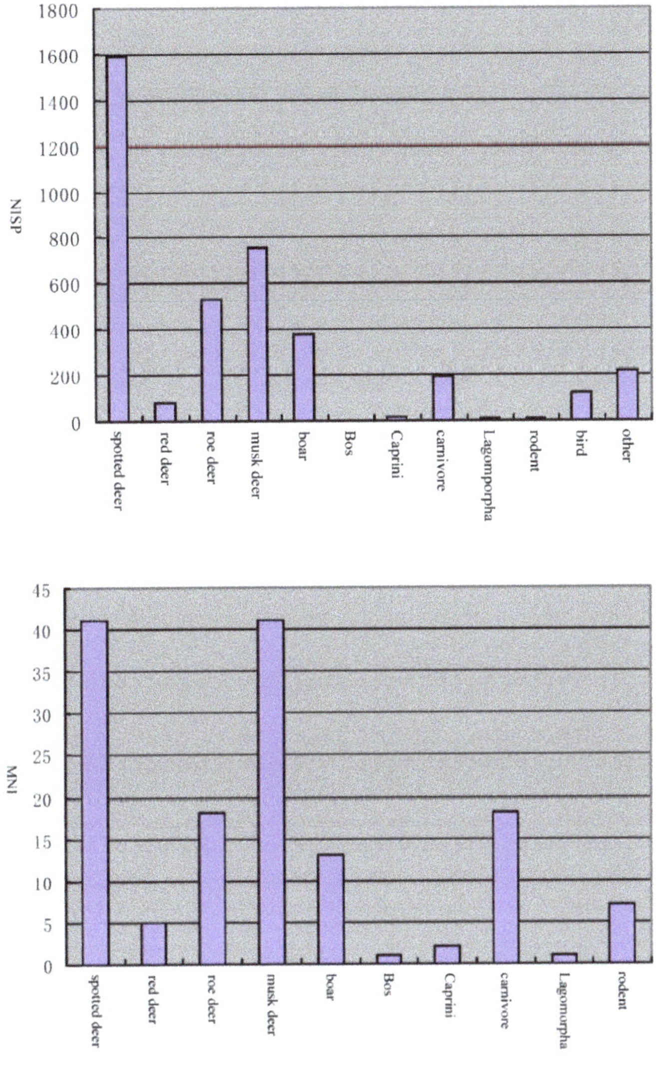

Figure 13. Number of identified specimens (NISPs) and minimal number of individuals (MNIs) of fauna at the Donghulin site (permission from Dr. Tian–Xing Cui).

To sum up, the current published archaeological record is inadequate to explain the cultural process of microblade technology disappearance in northern China. However, the correlation between the decline of microblade–based societies and the rise of agriculture is certain, and we need further detailed studies to build the links.

7. Conclusions

This paper does not aim to provide a synthesis of archaeological sites associated with microblade technology, but rather to explore the mechanisms of cultural change for prehistoric hunter–gatherers equipped with microblade products during and after the Last Glacial Maximum. Three sequential stages of microblade–based societies are addressed in this paper, and the conclusion can be summarized as follows:

(1) Based on the calculated results in the EnvCalc2.1 program under LGM climatic conditions, especially the variables BIO5 (primary biomass), WDEN (projected density) and EXGRP3 (annual or several-year aggregation size), the plant use at the Shizitan site might not be linked with resource intensification, but rather with social group aggregation in a region with relatively high productivity of plant resources. However, the equifinality in the archaeological explanation needs to be further studied to evaluate alternative hypotheses, including the scale and seasonality of human aggregation and the possibility of resource intensification.

(2) The sites dated to the Pleistocene–Holocene transition experienced Paleolithic to Neolithic technological and economic transitions, marked by use of pottery, appearance of permanent settlements, and widespread adoption of ground stone tools. The broad spectrum revolution and subsistence intensification can be explained within the framework of the macroecological approach; the replacement of the hunting-dominated economy by a gathering-and/or-fishing-dominated economy is linked with population growth during the interglacial or interstadial periods, matching the maps under the packed condition of regional population. Decreased mobility also provided the occupants an opportunity to use low-energy food resources through dietary expansion and/or intensification. Against this background, microblade-based weaponry became less significant in daily life due to the previously open landscape becoming more closed, leading to a decrease in the density of large herd ungulates, and durable ground/polished and expedient stone tools were favored.

(3) The disappearance of microblade technology accompanies the end of microblade-based societies in North China. At present, there are very few publications on this period but the microblade assemblages at the Donghulin site show much less standardization in microblade production. This might be linked with the full adaptation of sedentary lifeways and significant decreases of wild ungulates during the Neolithic.

The present research provides a case study using the macroecological approach for exploring the technological change of prehistoric foraging societies equipped with microblade technology in North China, facing dramatic climatic oscillation during the end of the Ice Age. Although a research framework has been built in this project, more substantial archaeological research is needed using systematic observation of interassemblage variability, a well–controlled chronological database, detailed studies on the function of targeted tools with the aid of use–wear analysis, and interdisciplinary projects on land–use patterns and resource exploitation to test hypotheses proposed here on the cultural dynamics of the microblade-based societies.

Funding: This research received no external funding

Acknowledgments: This paper is developed from the author's dissertation at the Department of Anthropology, University of New Mexico, under the supervision of Lawrence Straus, Bruce Huckell, Robert Hitchcock, Amber Johnson, and Pei–Lin Yu. The research was supported by the Binford Fellowship, TAships, CSC, Moore Research Grant, and the CCKF when I was in the PhD program at the UNM. This project is also supported by the Subject Fundamental Development Project of Fudan University. Yanhua Song and Fagang Wang as well as other archaeologists showed me Paleolithic collections in Shanxi and Hebei Provinces. Tianxing Cui gave me permission to use the archaeological profile as Figure 13 in the present paper. The reviewers provided valuable and constructive suggestions. In addition, this paper is dedicated to Lewis Binford and Ofer Bar–Yosef, who provided me with a toolkit and precious suggestions to deal with the microblade technology issue in prehistory.

Conflicts of Interest: The author declares no conflict of interest.

References

1. Chen, S.-Q.; Yu, P.-L. Early "Neolithics" of China: Variation and Evolutionary Implications. *J. Anthropol. Res.* **2017**, 73, 149–180. [CrossRef]
2. Chen, S.-Q.; Yu, P.-L. Intensified Foraging and the Roots of Farming in China. *Anthropol. Res.* **2017**, 73, 381–412. [CrossRef]

3. Binford, L.R. *Constructing Frames of Reference: An Analytical Method for Archaeological Theory Building Using Hunter-Gatherer and Environmental Data Sets*; University of California Press: Berkeley, CA, USA, 2001.
4. Chen, S.-Q. Adaptive Changes of Hunter-Gatherers during the Late Pleistocene-Early Holocene Transition in China. *Acta Anthropol. Sin.* **2006**, *25*, 195–207. (In Chinese)
5. Binford, L.R. Willow smoke and dogs' tails: Hunter-gatherer settlement systems and archaeological site formation. *Am. Antiq.* **1980**, *45*, 4–20. [CrossRef]
6. Zhang, M. Explaining Variation and Change among Late Pleistocene and Early Holocene Microblade–Based Societies in Northeastern Asia. Ph.D. Thesis, University of New Mexico, Albuquerque, NM, USA, May 2019.
7. Chen, S.-Q. *Prehistoric Modernization: A Cultural Ecological Approach of the Origins of Agriculture*; Science Press: Beijing, China, 2014. (In Chinese)
8. Binford's Hunter-Gatherer Data. Available online: http://ajohnson.sites.truman.edu/data-and-program/ (accessed on 20 June 2019).
9. Johnson, A.L. Exploring adaptive variation among hunter-gatherers with Binford's frames of reference. *J. Anthropol. Res.* **2014**, *22*, 1–42. [CrossRef]
10. Johnson, A.L.; Collins, C.; Turchin, P.; Cesaretti, R. *Global & Regional Frameworks for Comparing Agricultural Intensification & Productivity Across Cases*; SAA: Washington, DC, USA, 2018.
11. Yu, P.-L. Pit Cooking and Intensification of Subsistence in the American Southwest and Pacific Northwest. Ph.D. Thesis, Department of Anthropology, Southern Methodist University, Dallas, TX, USA, July 2006.
12. Yuan, S.-X.; Chen, T.-M.; Zhou, K.-S. Radiocarbon dates and pollen analysis of the Nanzhuangtou site (Appendix of Primary text excavation report of the Nazhuangtou site, Xushui County, Hebei Province). *Kaogu* **1992**, *11*, 967–970. (In Chinese)
13. Ren, S.-N.; Wu, Y.-L. *Chinese Archaeology: Neolithic*; China Social Science Press: Beijing, China, 2010. (In Chinese)
14. Yu, J.-C. Neolithic discoveries and studies of Beijing. In *Basheji*; Bingwen, Y., Ed.; Beijing Library Press: Beijing, China, 1998; pp. 39–44. (In Chinese)
15. Cui, T.-X. A Study on Lithic Assemblage of the Donghulin Site: Lithic Industry and Human Behaviors in the Paleolithic to Neolithic Transition, Dissertation, School of Archaeology and Museology. Ph.D. Thesis, Peking University, Beijing, China, 2010. (In Chinese).
16. He, J.-N.; Zhang, S.-L.; Wang, S.-Z.; Wang, J.-Y.; Zhao, J.-F.; Gao, X.-X.; Wang, Y.-P. Archaeological report of Lijiagou site (north area), Xinmi, Henan Province (2009). *Anc. Civiliz.* **2013**, *9*, 177–207. (In Chinese)
17. Zhao, J.-F.; Zhang, S.-L.; Wang, S.-Z.; He, J.-N.; Gao, X.-X.; Wang, J.-Y.; Wang, Y.-P. Archaeological report of Lijiagou site (south area), Xinmi, Henan Province (2009). *Anc. Civiliz.* **2013**, *9*, 208–239. (In Chinese)
18. Sun, B.; Wagner, M.; Zhao, Z.-J.; Li, G.; Wu, X.-H.; Tarasov, P.E. Archaeological discovery and research at Bianbiandong early Neolithic cave site, Shandong, China. *Quat. Int.* **2014**, *348*, 169–182. [CrossRef]
19. Li, W.-C.; Song, G.-D.; Wu, Y. Preliminary analysis of starch grains on the surface of stone artifacts from the Kengnan site. *Acta Anthr. Sin.* **2014**, *33*, 70–81. (In Chinese)
20. Mei, H.-J. Transition from Paleolithic to Neolithic in the Nihewan Basin: A Study of the Discoveries from the Yujiagou Site. Ph.D. Thesis, College of Archaeology and Museology, Peking University, Beijing, China, 2007. (In Chinese).
21. IA–CASS. *Radiocarbon Dates in Chinese Archaeology, 1965–1991*; Institute of Archaeology, Ed.; Wenwu Press (Cultural Relics Publishing House): Beijing, China, 1991. (In Chinese)
22. Nagatomo, T.; Shitaoka, Y.; Namioka, H.; Sagawa, M.; Wei, Q. OSL dating of the strata at Paleolithic sites in the Nihewan Basin, China. *Acta Anthr. Sin.* **2009**, *28*, 276–284. (In Chinese)
23. Li, Z.-Y.; Xing, Y.; Mu, J.-H.; Wu, X.-L.; Li, Y.-N.; Kato, S. Excavation report of microliths in Layer 5 at the Lingjing Xuchang Man site, 2008–2013. *Huaxia Archaeol.* **2018**, *2*, 3–33. (In Chinese)
24. Song, Y.-H. Study on the Quartzite Artifacts in Shizitan Site, Jixian, Shanxi Province. Ph.D. Thesis, University of Chinese Academy of Sciences, Beijing, China, 2011. (In Chinese).
25. Song, Y.-H.; Shi, J.-M. A preliminary excavation report of the Locality S29, Shizitan site, Ji County, Shanxi Province. *Kaogu* **2017**, *2*, 35–51. (In Chinese)
26. Song, Y.-H.; Shi, J.-M. Preliminary excavation report of Shizitan Site, Locality S5, Ji County, Shanxi Province. *Kaogu* **2016**, *4*, 3–15. (In Chinese)
27. Liu, D.-C.; Chen, F.-Y.; Zhang, X.-L.; Pei, S.-W.; Gao, X.; Xia, Z.-K. Preliminary comments on the paleoenvironment of the Shuidonggou Locality 12. *Acta Anthropol. Sin.* **2008**, *27*, 295–303. (In Chinese)

28. Yi, M.-J. Adaptive Strategies of Hunter-Gatherers during the Late Upper Paleolithic in Northern China: An Archaeological Research on the Shuidonggou 12. Ph.D. Thesis, Institute of Vertebrate Paleontology and Paleoanthropology, University of Chinese Academy of Sciences, Beijing, China, 2012. (In Chinese).
29. Elston, R.G.; Xu, C.; Madsen, D.B.; Zhong, K.; Bettinger, R.L.; Li, J.-Z.; Brantingham, P.J.; Wang, H.M.; Yu, J. New dates for the north China Mesolithic. *Antiquity* **1997**, *71*, 985–993. [CrossRef]
30. Barton, L.; Brantingham, P.J.; Ji, D.-X. Late Pleistocene climate change and Paleolithic cultural evolution in northern Chinanorthern China: Implications from the Last Glacial Maximum. *Dev. Quat. Sci.* **2007**, *9*, 105–128.
31. Liu, L.; Levin, M.J.; Bonomo, M.F.; Wang, J.-J.; Shi, J.-M.; Chen, X.-C.; Han, J.-Y.; Song, Y.-H. Harvesting and processing wild cereals in the Upper Palaeolithic Yellow River Valley, China. *Antiquity* **2018**, *92*, 603–619. [CrossRef]
32. Liu, L.; Bestel, S.; Shi, J.-M.; Song, Y.-H.; Chen, X.-C. Paleolithic human exploitation of plant foods during the last glacial maximum in North China. *Proc. Natl. Acad. Sci. USA* **2013**, *110*, 5380–5385. [CrossRef]
33. Liu, L.; Ge, W.; Bestel, S.; Jones, D.; Shi, J.-M.; Song, Y.-H.; Chen, X.-C. Plant exploitation of the last foragers at Shizitan in the Middle Yellow River Valley China: Evidence from grinding stones. *J. Archaeol. Sci.* **2011**, *38*, 3524–3532. [CrossRef]
34. Straus, L.G.; Goebel, T. Humans and Younger Dryas: Dead end, short detour, or open road to the Holocene? *Quat. Int.* **2011**, *242*, 259–261. [CrossRef]
35. Straus, L.G. The archaeology of the Pleistocene-Holocene transition in Southwest Europe. In *Humans at the End of the Ice Age: The Archaeology of the Pleistocene–Holocene Transition*; Straus, L.G., Eriksen, B.V., Erlandson, J.M., Yesner, D.R., Eds.; Plenum Press: New York, NY, USA; London, UK, 1996; pp. 83–99.
36. Binford, L.R. *In Pursuit of the Past: Decoding the Archaeological Record*; Thames and Hudson London: New York, NY, USA, 1983.
37. Binford, L.R. Post-Pleistocene adaptation. In *New Perspectives in Archeology*; Binford, S.R., Binford, L.R., Eds.; Aldine Publ. Co.: Chicago, IL, USA, 1968; pp. 313–341.
38. Zhang, M. Microblade-based Societies: A new perspective on microblade technology in northern Chinanorthern China after the Last Glacial Maximum. *Acta Archaeol. Sin.* **2020**. in press (In Chinese)
39. Andrefsky, W. Raw-material availability and the organization of technology. *Am. Antiq.* **1994**, *59*, 21–34. [CrossRef]
40. Nelson, M.C. The study of technological organization. *Archaeol. Method Theory* **1991**, *3*, 57–100.
41. Zhu, Z.-Y.; Gao, X. A study of wedge–shaped cores from Hutouliang site. *Acta Anthropol. Sin.* **2006**, *25*, 129–142. (In Chinese)
42. Li, Z.-Y.; Li, Y.-N.; Kato, S. Observations of microblade core technologies from Level 5 of the Xuchang Man site, Lingjing. *Acta Anthr. Sin.* **2014**, *33*, 285–303. (In Chinese)
43. Yi, M.-J.; Gao, X.; Wang, H.-M.; Pei, S.-W.; Chen, F.-Y. A Study on Cores Unearthed from the Shuidonggou Locality 12 in 2007. *Acta Anthr. Sin.* **2015**, *34*, 166–179. (In Chinese)

quaternary

Article

Rethinking the Disappearance of Microblade Technology in the Terminal Pleistocene of Hokkaido, Northern Japan: Looking at Archaeological and Palaeoenvironmental Evidence

Jun Takakura

Archaeological Research Center, Hokkaido University, Sapporo 0600811, Japan; jun-ta@let.hokudai.ac.jp

Received: 15 June 2020; Accepted: 17 July 2020; Published: 20 July 2020

Abstract: Archaeological research, for several decades, has shown that various microblade technologies using obsidian and hard shale appeared and developed from the Last Glacial Maximum to the terminal Pleistocene (Bølling–Allerød–Younger Dryas) in Hokkaido, Northern Japan. It is well accepted that microblade technology was closely related to the high mobility of foragers to adapt to harsh environments. Recent archaeological and palaeoenvironmental evidence from Hokkaido demonstrates that the disappearance of microblade technology occurred during the terminal Pleistocene, influenced by a wide range of factors, including changes in landscape, climate, subsistence and human populations. The goal of this paper is to provide an overview of the current state of research on the process and background of the disappearance of microblade technology and to discuss prospects for future research. This paper will (1) review palaeoenvironmental research in Hokkaido on changes in climate and biological composition from the terminal Pleistocene to the initial Holocene; (2) survey changes in the technological adaptations and resource use of humans based on the archaeological evidence; and (3) discuss how the abrupt fluctuations of climate that occurred in the terminal Pleistocene affected human behaviour and demographics in Hokkaido.

Keywords: microblade technology; palaeoenvironment; Hokkaido; terminal Pleistocene; initial Holocene

1. Introduction

The accumulation of Upper Palaeolithic studies in Hokkaido, for several decades, has revealed that there are numerous lithic assemblages characterised by various microblade technologies using raw materials such as obsidian and hard shale [1–3]. Detailed lithic technological analyses and chronological reconstructions in Hokkaido have made it clear that the various microblade technologies that appeared and developed were closely related to the behavioural strategies of mobile hunter-gatherers adapted to living in harsh environments from the Last Glacial Maximum (LGM; 26,500–19,500 years ago) [4] to the terminal Pleistocene (Bølling–Allerød–Younger Dryas) [1–3]. In this paper, 'microblade' refers to parallel-sided small artefacts possessing one or more ridges running parallel to their long axes, generally 15–60 mm long, 4–10 mm wide, and 1–2 mm thick [5–9]. Systematic observation of the techno-typological characteristics of microblade cores and lithic refitted pieces has demonstrated that highly standardised microblades were produced through complicated lithic reduction sequences from lithic raw materials [2,10,11]. Microblade technology enables the production and maintenance of compact and lightweight stone tools that can be used as part of composite tools. For humans adapted to a cold and dry environment in the Upper Palaeolithic, the manufacturing and use of microblades can be considered an important technological choice for behavioural adaptation in terms of efficiently using and transporting lithic raw materials collected during foraging [3,12,13].

It is widely known that microblade technology was distributed across regions of northeast Asia, such as North China, the Korean peninsula, the Russian Far East, Mongolia and Siberia [13–22]. Many researchers have made advances with comparing the techno-typological features of microblade assemblages and accumulating radiocarbon dates. As a result, we are currently in the process of getting a better understanding of the spatiotemporal patterns of microblade technology in northeast Asia. Understanding when and how microblade technology appeared and disappeared in each region has been a topic of debate in northeast Asian Upper Palaeolithic research. Although several hypotheses on the origin(s) of microblade technology in northeast Asia have been presented and debated until recently [6,13,15,16,22,23], the disappearance of microblade technology in each region has received less attention.

According to records from ice cores from Greenland, drastic and abrupt environmental changes occurred repeatedly from the terminal Pleistocene to the initial Holocene [24]. A warming period occurred from 14,700 years ago until 12,900 years ago called the Bølling–Allerød, followed by a cooling period called the Younger Dryas. The Younger Dryas is designated as lasting until 11,650 years ago. After the end of the Pleistocene, a period of global warming began. The Preboreal is designated as lasting from 11,650 years ago until 8500 years ago. Quaternary scientists have long demonstrated that episodes of glacial readvance and retreat, shifts in precipitation and forest destruction and re-growth occurred during the transition from the terminal Pleistocene to the initial Holocene [25].

Techno-typological comparisons and some radiocarbon dates show that microblade technology disappeared in Hokkaido during the terminal Pleistocene [2–4,26,27]. To explain the background of the disappearance of microblade technology in terms of relationship to palaeoenvironment change would offer valuable insights into the nature of the technological and behavioural characteristics of microblade technology in Hokkaido, as would identifying the differences with regions such as Siberia, Mongolia and North China, where microblade technology continued after the onset of the Holocene [8,21]. Furthermore, recent archaeological research in eastern Hokkaido shows that different techno-complexes, including not microblades but various bifacial points and burins associated with potteries, emerged during the terminal Pleistocene [27–29]. Did the appearance of these techno-complexes play a role in the end of microblade technology in Hokkaido? Or should we expect a significant overlap for some time? The archaeological sequence of Hokkaido in the transition from the terminal Pleistocene to the initial Holocene and its implication for the process of how the palaeoenvironment related to the changes in human adaptive systems are rarely a subject of major agreement among scholars working in the region [27,29]. Understanding the relationship between them will be beneficial to any final interpretation of past human adaptations. This will provide insights into the process through which hunter-gatherers who had adapted to the warming of the natural environment after the Preboreal adopted sedentary lifestyles and highly diversified subsistence, making use of marine-based and other stable food sources [28,30,31].

Recently, there have been some important archaeological achievements that suggest how hunter-gatherers in Hokkaido accomplished technological adaptations and resource use during the transitional period from the terminal Pleistocene to the initial Holocene. Based on these studies, this paper will survey the process through which microblade technology disappeared in Hokkaido and the background behind this process, influenced by a wide range of factors, including changes in landscape, climate, subsistence and human populations. This paper will (1) review research on the palaeoenvironment carried out in Hokkaido regarding changes in climate and biogeography from the terminal Pleistocene to the initial Holocene; (2) assess changes in the technological adaptations and resource use of humans by focusing on lithic production technology and patterns of land use; and (3) discuss the ways in which sudden cooling affected human behaviour and demographics, particularly during the Younger Dryas.

2. Palaeoenvironment

Hokkaido is a large island located in the northern tip of the Japanese archipelago (Figure 1). During the LGM, lower sea levels caused by a cooler climate meant that Hokkaido was connected to the island of Sakhalin via the Soya land bridge and to Siberia via the Tatar land bridge. Humans and the faunal community were able to disperse from the Eurasian continent to Hokkaido without any straits to interrupt their migration. The resulting palaeoenvironment of Hokkaido during the LGM was somewhat more continental in character than it is at present. On the contrary, there was no land bridge across the Tsugaru Strait dividing Hokkaido and Honshu, thus, creating a barrier for humans and the faunal community [32]. This brought about disparity between Hokkaido and Honshu in terms of various cultural features and the composition of the faunal community during the Upper Palaeolithic [33–35]. Some researchers suggest that by the onset of the Bølling–Allerød interstadial the bridge between Sakhalin and Hokkaido had disappeared and Hokkaido was surrounded by ocean [26].

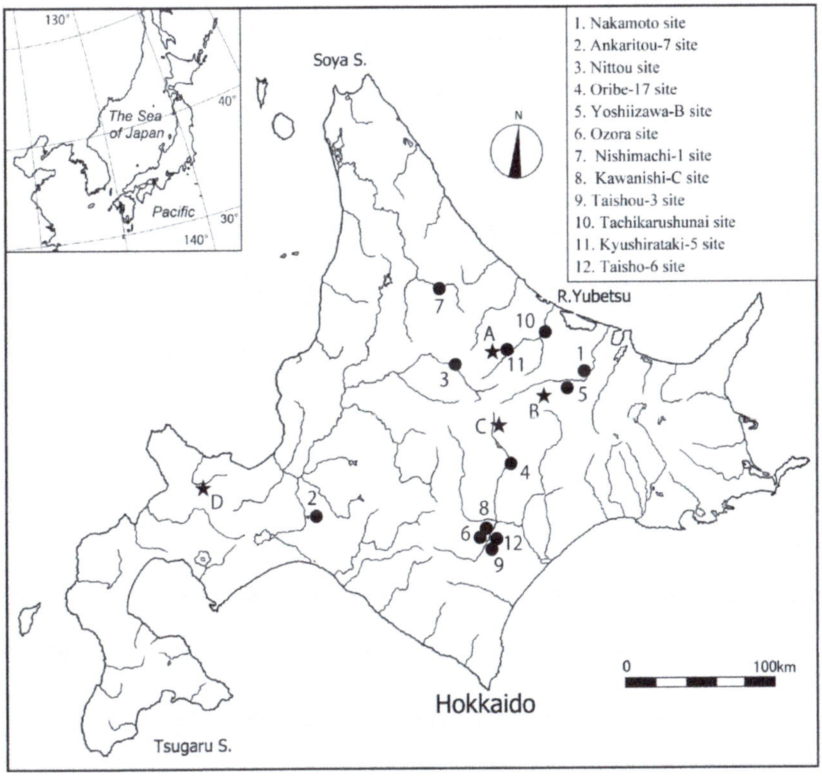

Figure 1. Locations of the sites referenced in this text. Major obsidian sources: (**A**) Shirataki; (**B**) Oketo; (**C**) Tokachi-Mitsumata; (**D**) Akaigawa.

The analysis of oxygen isotope ratios and dust in varved sediment has led to highly detailed explanations of climate changes from the Last Glacial, which experienced drastic fluctuations in temperature, to the Holocene. The analysis of varved sediment in Lake Tougetsu and Lake Suigetsu in western Japan shows that a cooling event occurred corresponding with the Younger Dryas [36–38]. The palaeoenvironment changes occurring on Hokkaido have been approached by many fields, including geology, topography, palaeopedology, palaeoecology and microfossils. In particular, there has seen plentiful research focusing on the changes in vegetation via pollen analysis since the 1960s [39,40].

The analysis of pollen deposits at Kenbuchi, northern Hokkaido, shows that during the latter stage of the Marine Isotope Stage (MIS) 3, about 30,000 years ago, *Picea jezoensis* and/or *Picea glehnii* were predominant and boreal forests mixed with *Pinus pumila* were widespread. Starting from around 25,000 years ago, cooling led to the spread of steppes, where *Pinus pumila* was mixed with *Larix gmelinii*. Beginning around 21,000 years ago, *Pinus pumila* suddenly decreased, while *Picea jezoensis* and/or *Picea glehnii* increased. From around 17,000 years ago, there was another cooling period, and the appearance of *Pinus pumila* and *Microcachrys tetragona* increased. Beginning around 14,000 years ago, the ratio of *Pinus pumila* pollen in the record suddenly declined; however, a temporary recovery has also been observed. This period is thought to be the Younger Dryas. The prevalence of the *Quercus* subgenus increased around 10,000 years ago [41–43]. Pollen analyses conducted at Nakafurano in central Hokkaido show that although *Pinus pumila* and *Microcachrys tetragona* declined when *Picea jezoensis* and/or *Picea glehnii* increased between 14,000 and 12,000 years ago, there was a temporary period of cooling from 12,000 to 11,500 years ago, during which *Pinus pumila* increased. After this, conifer forests declined, and *Betula* forests expanded. Around 10,000 years ago, they were joined by *Juglans*, and around 9000 years ago, temperate, broad-leaf, mostly *Quercus* forests were formed [41]. Thus, it was only from around 10,000 years ago that humans in Hokkaido could begin to make intensive use of nuts as a stable food source.

According to pollen analysis, a significant expansion of subarctic conifer forests occurred in the mountainous areas of northern Honshu and in Hokkaido during the Younger Dryas. However, we know significantly less about the mammals that humans hunted relative to our knowledge of contemporaneous flora, due to the lack of palaeontological and zooarchaeological records describing the faunal community of the terminal Pleistocene. Palaeontological records from Hokkaido show that mammoths and bison existed there during the LGM, around 25,000 to 17,000 years ago [33,34,44]. However, the woolly rhinoceros and other species of large ungulates, such as the *Equus* and *Gazella* genera, common to the steppe and frequently observed during the Last Glacial in northeast China, are not attested at all in Hokkaido and neither are the *Hyaena*. Hokkaido is, thus, thought to have been less cold and dry than the northeastern part of the Eurasian continent, placing it near the edge of the distribution area for mammoths, which were distributed mainly across the mammoth steppe [34,44]. It is not known exactly when mammoths, bison, and other large mammals disappeared from Hokkaido. Taking into account the abovementioned changes in vegetation, it seems that climate warming gradually shifted their habitat north after the end of the LGM [45,46]. Therefore, we can estimate that at the time of the terminal Pleistocene, after large fauna such as mammoths and bison declined and were no longer a stable and dependable source of human food, the faunal community was composed mainly of mammals that were small, or of medium size, such as deer [47]. This, then, was the faunal community that would continue into the Holocene and that Jomon hunter-gatherers would rely on for diet [26,48].

Regardless of when the large fauna disappeared from Hokkaido, it appears that there were major changes in the faunal communities targeted by hunter-gatherers with microblades before the onset of the terminal Pleistocene. Interestingly, recent zooarchaeological analysis at the Yujiagou site (Layer 4 and Layer 3b), part of the Hutouliang Site Complex in North China, shows that the terminal Pleistocene hunters with microblades preferred juvenile gazelles and horses and used bone marrow as a significant source of energy [49]. Therefore, this analysis highlights that there were differences between the animal resources available in Hokkaido and North China during the terminal Pleistocene, although human foraging behaviours using microblade technology were dispersed in both regions. It is necessary to note that the terminal Pleistocene hunter-gatherers with microblades in Hokkaido had to adapt differently to changing environments.

3. Microblade Technology in the Terminal Pleistocene

From the LGM to the terminal Pleistocene, microblade assemblages, including microblades and various types of microblade cores mainly made from obsidian and hard shale, continued to exist in Hokkaido. Researchers have long focused their attention on the construction of a chronological framework and a techno-typological classification system for microblade technology [10,50,51]. In the study of microblade technology in northeast Asia, the Yubetsu method is most commonly seen in which a microblade was produced from a bifacial core blank with symmetrical cross sections and forming platforms by removing spalls [52,53]. Moreover, the Rankoshi, Tougeshita, Pirika, Oshorroko, Hirosato, Horoka and Momijiyama methods are also defined in Hokkaido and relate to differences in the manufacturing processes of microblade core blanks [10]. These microblade reduction methods are used as indices to locate lithic assemblages chronologically. The Rankoshi, Tougeshita and Pirika methods appeared around 25,000 to 24,000 years ago during the LGM [1,2].

There are few reliable radiocarbon dates available to determine the period to which microblade assemblage continued, which was at some point in the terminal Pleistocene [54]. Lithic assemblages associated with the Hirosato method (the Hirosato techno-complex) (Figure 2) obtained from the Nakamoto site indicate radiocarbon dates of around 15,200 to 13,700 years ago [1]. These dates were obtained from charcoal associated with hearths. This definitively shows that the lithic assemblage associated with the Hirosato method continued until the period of warming known as Bølling–Allerød in the terminal Pleistocene. The Hirosato techno-complex is confirmed in the Ankaritou-7 site [55], the Nakamoto site [1], the Nittou site [56], etc. Lithic artefacts, such as bifacial stemmed points, bifacial points and stone axes, associated with the Hirosato method have been found to show commonalities with artefacts associated with the Oshorroko method (Figure 3). As a result, lithic assemblages associated with the Oshorroko method (the Oshorroko techno-complex) have also been chronologically located during the same warm period (the Bølling–Allerød) [1,3,47,57]. The Oshorroko techno-complex has been confirmed in the Oribe-17 site [58], the Yoshiizawa-B site [57,59,60], the Ozora site [61], etc. Furthermore, lithic assemblages with small boat-shaped tools (the boat-shaped tool techno-complex) (Figure 4) have also been chronologically located during the terminal Pleistocene [47]. The boat-shaped tool techno-complex has been confirmed in the Nishimachi-1 site [62], the lithic concentration No.17 at the Kawanishi-C site [63], etc. Microblade technology is not seen in the lithic assemblages with the boat-shaped tools. However, the techno-typological features of burins, end-scrapers, bifacial points and bifacial stemmed points included in the lithic assemblage, as well as the technical features of the reduction sequence of the blades that serve as their materials, have recognised some commonalities with both the Hirosato and Oshorroko techno-complexes. Thus, the assertion that they should be placed at the same chronological stage is convincing [64].

Researchers have come to no agreement regarding the chronological relationship of these different techno-complexes [54]. Recently, it has been proposed that they appeared together and co-existed for some time. Some researchers have focused on the presence of newly invented tools, such as bifacial stemmed points and stone axes, that have been recognised in these techno-complexes [1,47,65]. A further accumulation of reliable radiocarbon dates showing clear relationships with lithic assemblages is urgently needed to resolve this issue of chronology. However, such additional research is hampered by the lack of any features, such as hearths, and complex post-depositional disturbances that are characteristic of the terminal Pleistocene sites in Hokkaido.

It is possible to identify flaking techniques from the assessment of crack velocity through the analysis of fracture wings observed on flaking surfaces [66]. In a large proportion of cases in Hokkaido, microblades were systematically detached by pressure flaking, with the exception of some cases [67–71]. The results of use-wear analyses based on materials from the Yubetsu method dated to around 18,000 to 17,000 years ago illustrate that standard and formal microblades were used as hunting implements, with many of them generally hafted on the side of bone/antler tools, according to the parallel striations identified along the sides of the microblades [72,73]. Low discovery rates and weak development of the use-wear observed on the microblades can be interpreted as indicating that they were not used as processing tools, such as burins and end-scrapers [73]. It allows us to infer that the microblades from the terminal Pleistocene sites were probably used as weapon insets, although further functional research will be needed.

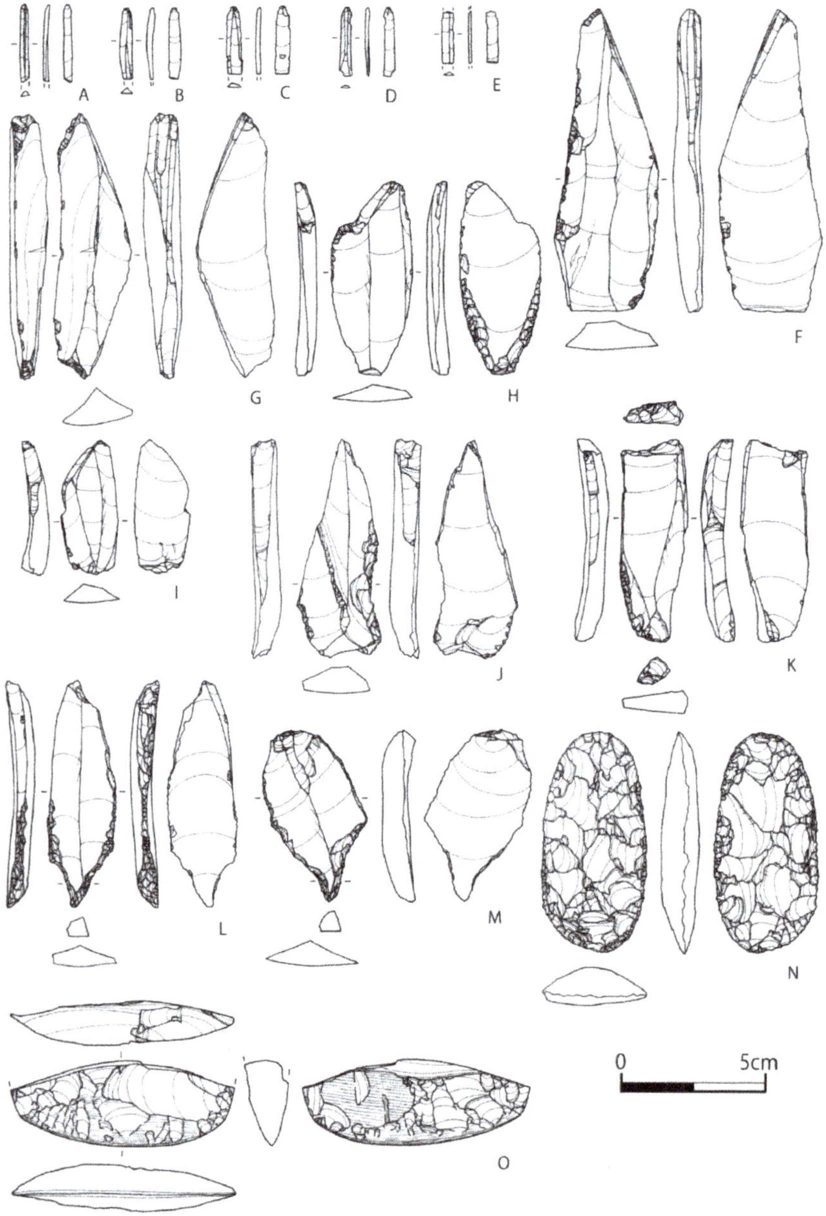

Figure 2. Stone artefacts in the Hirosato techno-complex. Examples from the Ankaritou-7 site in central Hokkaido: (**A–E**) microblades; (**F,G**) Hirosato-type microblade cores; (**H–K**) burins; (**L,M**) drills; (**N**) bifacial tool; (**O**) broken edge-ground stone axe. Reproduced from [55].

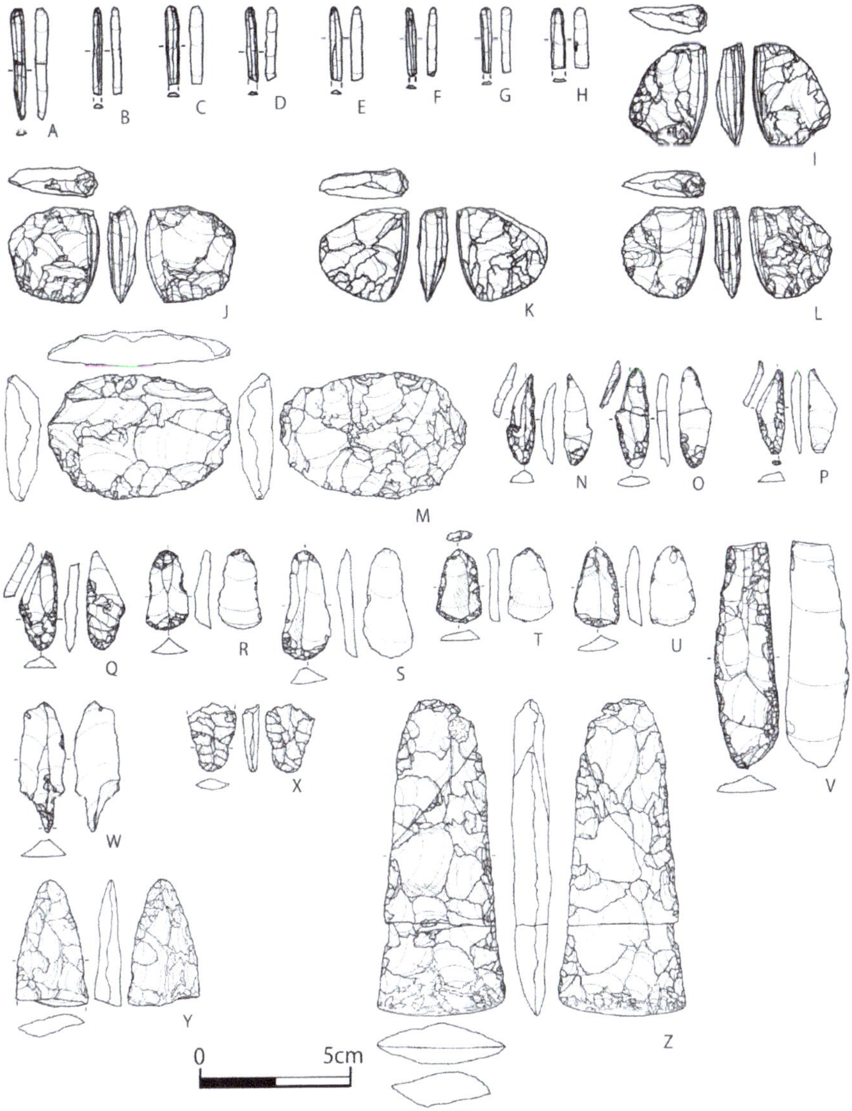

Figure 3. Stone artefacts in the Oshorroko techno-complex. Examples from the Oribe-17 site in eastern Hokkaido: (**A–H**) microblades; (**I–L**) Oshorroko-type microblade cores; (**M**) bifacial blank; (**N–Q**) burins; (**R–U**) end-scrapers; (**V**) side-scraper; (**W**) drill; (**X**) broken bifacial stemmed point; (**Y**) broken bifacial point; (**Z**) edge-ground stone axe. Reproduced from [58].

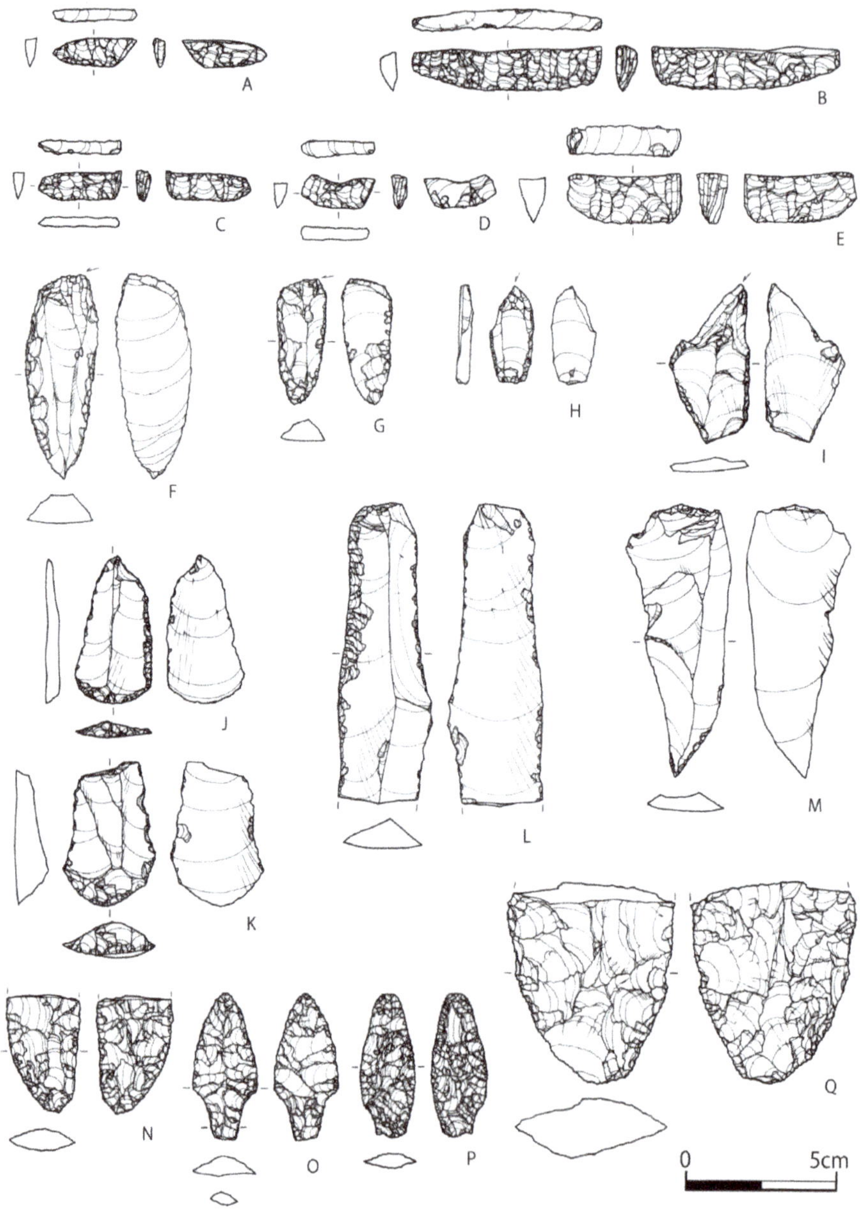

Figure 4. Stone artefacts in the boat-shaped tool techno-complex. Examples from the Nishimachi-1 site in northern Hokkaido: (**A–E**) boat-shaped tools; (**F–I**) burins; (**J**,**K**) end-scrapers; (**L**) side-scraper; (**M**) drill; (**N**,**Q**) broken bifacial points; (**O**,**P**) bifacial stemmed points. Reproduced from [62].

Microblade assemblages in the terminal Pleistocene are frequently accompanied by standard and abundant burins and end-scrapers made of blades. High proportions of formally shaped tools indicate the preparation of curated tool kits. The use-wear analysis of these burins shows that use-related traces are mainly located on burin facet edges and that they were often used in the manufacture and maintenance of bone/antler tools [65,74]. Burins have often been observed in lithic assemblages before the appearance of microblade assemblages in Hokkaido. However, use-wear analysis has not detected any signs of the manufacture and maintenance of bone/antler tools from these assemblages [75]. Accordingly, the beginning of intensive manufacture and the use of bone/antler tools is thought to be closely related to the emergence of microblade assemblages [76,77].

The use-wear analysis of end-scrapers from the microblade assemblages in the terminal Pleistocene suggests that they were intensively used in hide processing [65]. In addition, the traces of intensive rejuvenations of edges on end-scrapers can often be observed in these microblade assemblages [78,79]. Such evidence enables us to infer that foragers using microblade technology in the terminal Pleistocene conducted the production and maintenance of a diversified set of relatively specialised tools, reflecting the occurrence of task-specific occupation. It seems that this behavioural strategy brought about the inter- and intra-site variability of stone toolkits among the microblade assemblages in the terminal Pleistocene of Hokkaido [1,47,65,80].

It is widely known that many obsidian artefacts have been recovered from the Upper Palaeolithic sites in Hokkaido. There are twenty-one obsidian sources on Hokkaido [81]. In particular, four major sources, namely Shirataki, Oketo, Tokachi-Mitsumata and Akaigawa (Figure 1), had been mainly exploited for acquiring lithic raw materials throughout the Upper Palaeolithic. The provenance studies of Shirataki obsidian, including the results from the Sokol site in Sakhalin, show that the extra-regional transportations associated with the Yubetsu method extended throughout the Hokkaido and Sakhalin islands [82–88]. The long-distance transport of complete toolkits made from a single obsidian source can inform us of the long-distance movements (about 400 km) that the hunter-gatherers with microblade technology accomplished before the onset of the terminal Pleistocene. Several authors have attempted to discuss the diachronic change of the obsidian usage pattern during the Upper Palaeolithic based on the compiled data from which obsidian provenance studies have been carried out thus far [82,86–88]. The results of analyses from the Oshorroko and the boat-shaped tool techno-complexes demonstrate that these were dominated by obsidian from the nearest and various major sources, typically procured within 100 km. Debitage analyses and studies of technological activities among these techno-complexes indicate that preforms and finished tools were often transported a relatively limited distance, and secondary reduction and tool refurbishing were the most common activities observed in a number of sites [57,60,64,78,79]. In contrast, the results of analyses from the Hirosato techno-complex show that obsidian and/or artefacts from Oketo were frequently transported and used for producing various stone tools including large blades (up to 30 cm long) [87]. Extra-regional raw materials in the terminal Pleistocene microblade assemblages occur primarily as finished points and/or other tools that most likely moved through social networks. These trends of obsidian usage may be consistent with changes in the faunal community and subsistence during the terminal Pleistocene.

4. Techno-Complex with Potteries in the Terminal Pleistocene

In recent years, it has been shown that there were hunter-gatherers in Hokkaido during the Bølling–Allerød who produced and used potteries but did not use microblades [27,89,90]. From the excavation at the Taisho-3 site in eastern Hokkaido, a lithic assemblage composed mainly of small obsidian bifacial points associated with potteries has been discovered (Figure 5). Based on such techno-typological features, the Taisho-3 techno-complex can be defined. There are no lithic tools made from blades in the Taisho-3 techno-complex. This clearly differs technically and typologically from the microblade assemblages described earlier, showing that two distinct cultural groups existed. Although this may be somewhat influenced by the marine reservoir effect, we know from radiocarbon dating of charred residues from potteries at the Taisho-3 site that the Taisho-3 techno-complex dates between 15,000 and 13,700 years ago (Table 1).

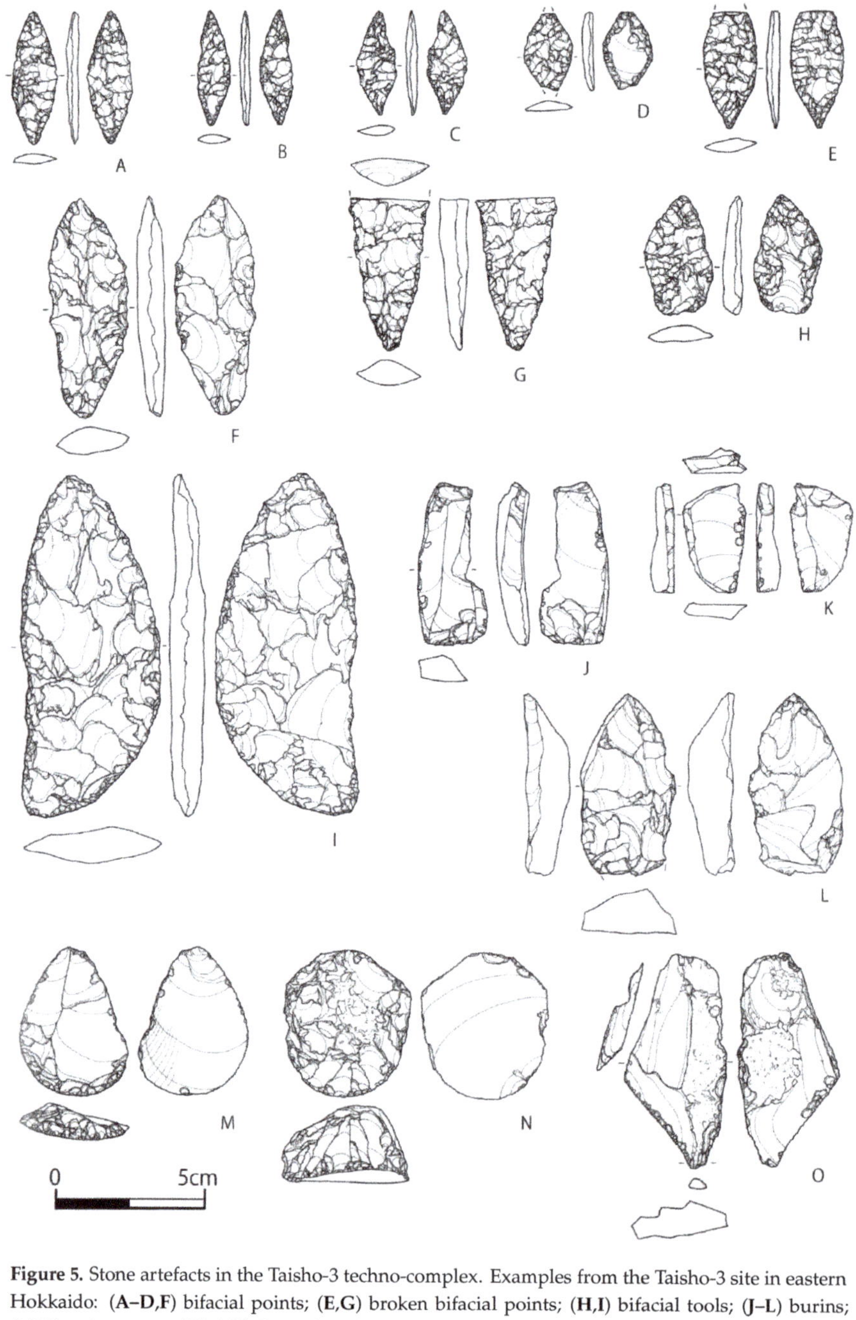

Figure 5. Stone artefacts in the Taisho-3 techno-complex. Examples from the Taisho-3 site in eastern Hokkaido: (**A–D,F**) bifacial points; (**E,G**) broken bifacial points; (**H,I**) bifacial tools; (**J–L**) burins; (**M,N**) end-scrapers; (**O**) drill. Reproduced from [90].

Table 1. AMS radiocarbon dates from the Taisho-3 site (calibrated using IntCal 13).

Lab Number	Material	14C BP	1σ	δ13C	Calibrated date, cal BP (2σ) upper	Calibrated Date, cal BP (2σ) Lower	References
Beta-194626	Charred remains inside pottery	12,400	40	−23.5	14,780	14,170	[90]
Beta-194627	Charred remains inside pottery	12,220	40	−24.0	14,260	13,980	[90]
Beta-194628	Charred remains inside pottery	12,350	40	−23.7	14,675	14,120	[90]
Beta-194629	Charred remains inside pottery	12,460	40	−22.6	14,960	14,265	[90]
Beta-194630	Charred remains inside pottery	12,210	40	−23.4	14,240	13,970	[90]
Beta-194631	Charred remains inside pottery	12,130	40	−23.3	14,140	13,830	[90]
IAAA-41603	Charred remains inside pottery	12,290	60	−21.6	14,630	14,025	[90]
IAAA-41604	Charred remains inside pottery	12,330	70	−23.2	14,745	14,060	[90]
IAAA-41605	Charred remains inside pottery	12,120	60	−22.1	14,145	13,780	[90]
IAAA-41606	Charred remains inside pottery	12,470	60	−21.7	15,025	14,250	[90]
IAAA-41607	Charred remains inside pottery	12,160	60	−22.5	14,210	13,815	[90]

The Taisho-3 site is located on a slightly elevated terrace that forms a natural embankment adjacent to a river (Figure 6). Just below the stratum containing the cultural horizon during the terminal Pleistocene is a gravel layer that was formed by river flooding; thus, human activity was conducted in a riverbed-like environment [90]. Unlike the Taisho-3 site, almost all other sites where microblade assemblages have been discovered are located on river terraces at a distance from rivers. The environmental setting of the Taisho-3 site demonstrates that aquatic resources were important for the daily activities of the hunter-gatherers who lived there. Stable carbon and nitrogen isotope analysis and gas chromatography–mass spectrometry analysis using charred residues from the potteries at the Taisho-3 site have shown that fish and other aquatic resources were exploited by the terminal Pleistocene hunter-gatherers [28,91]. These results are consistent with the patterns of land use inferred from the location of the Taisho-3 site.

In recent years, the Taisho-3 techno-complex has also been discovered from the excavation at the location M–I of the Tachikarushunai site, located in the Yubetsu River basin [92]. Nevertheless, it is apparent that sites with a Taisho-3 techno-complex are extremely rare compared to those with microblade assemblages in the terminal Pleistocene, although a taphonomic bias cannot be excluded for this record. The potteries discovered at the Taisho-3 site include the so-called nail-impressed wares (Figure 7). This pottery type has been found in several terminal Pleistocene sites in Honshu [93]. Some researchers argue that the technological characteristics of bifacial points have also been found to show commonalities with lithic artefacts among the Incipient Jomon from Honshu [94]. Accordingly, there is a strong possibility that the archaeological materials from the Taisho-3 site and the location M–I of the Tachikarushunai site indicate the results of small-scaled migrations from Honshu and adaptations to Hokkaido [27,29,89,94].

Lithic assemblages, composed mainly of leaf-shaped, foliate denticulate bifacial points (the Kyushirataki-5 techno-complex), have also recently been discovered [54,95]. Techno-typological comparisons of bifacial points have shown that they have significant commonalities with lithic assemblages from the first half of the terminal Pleistocene in Honshu [29,95]. However, there has been no radiocarbon dating of samples that clearly correspond to the Kyushirataki-5 techno-complex. In addition, it is necessary to note that there are also very few sites confirmed to be of this techno-complex. This can be explained by the reflection that group migration from Honshu most likely occurred on a small scale. It is apparent that we need to obtain reliable radiocarbon dates from future excavations to determine the chronological position of this techno-complex.

Figure 6. Distant view of excavation at the Taisho-3 site. Reproduced from [90], with the permission of the Obihiro Centennial City Museum.

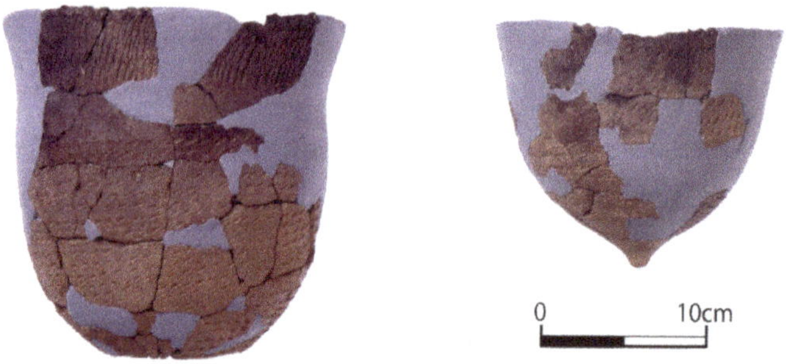

Figure 7. Potteries from the Taisho-3 site. These are the so-called nail-impressed wares. Reproduced from [90], with the permission of the Obihiro Centennial City Museum.

Based on radiocarbon dates obtained from the Taisho-3 techno-complex and those from the microblade assemblages in the terminal Pleistocene, it is difficult to assume that there was a temporal gap between the two. However, there are contrary views that they were not parallel in time [96]. As mentioned, if we are to consider the disparity in the patterns of land use, it is quite possible that they coexisted in Hokkaido during the same period of time [29,89]. Migratory communities new to the area might have developed a niche that had not been used among the hunter-gatherers in the area, who had until then, relied upon microblade technology. In other words, the investigation of the Taisho-3 site suggests that these communities migrated into eastern Hokkaido and made use of the aquatic resources they found there. The small size of the community, in addition to the fact that they explored new resources for their subsistence, could explain their ability to coexist with indigenous groups in the same region who produced and used microblades. It is noteworthy that Hokkaido provided a setting for multiple groups, with different cultural lineages developing different niches, to coexist during the warming phase in the terminal Pleistocene.

5. Disappearance of Microblade Technology

We have no reliable radiocarbon dates from archaeological sites to fill the gaps beginning from around 13,700 years ago in Hokkaido to the time when the Holocene began [26,54,97,98]. Using techno-typology, some argue that the lithic assemblages accompanying the small boat-shaped tools can be chronologically located to the period from around 13,700 years ago to 11,650 years ago [54]. At the present stage of knowledge, this possibility cannot be completely excluded. Regardless of its validation, there are apparently very few sites where lithic assemblages have been discovered that support this hypothesis [54]. Remarkable differences clearly exist in comparison to the number of sites, until the warming phase in the terminal Pleistocene.

Based on these trends, we can conclude that there was most likely a large change in the population of humans residing in Hokkaido during the Younger Dryas. The intense cooling of the Younger Dryas made adaptation to the environments in Hokkaido difficult for groups that had survived throughout the long period of the Last Glacial using microblade technology. As suggested above, it seems that mammoths and other large fauna declined or no longer existed in Hokkaido during this period. The large animal resources that the hunter-gatherers using microblade technology had stably exploited became scarce before the onset of the terminal Pleistocene. It seems that the re-cooling of the Younger Dryas period brought about the abrupt decrease of resources, such as small- or medium-sized mammals, including deer, that the hunter-gatherers maintaining microblade technology during the Bølling–Allerød had adapted to. This may have been the trigger for the end of microblade technology in Hokkaido. As suggested from the palaeoenvironment and archaeological evidence, the exploitations of large mammals and long-distance foraging could be maintained during the Younger Dryas in North China and Siberia [21,49,99], but not in Hokkaido.

Here, it is interesting to note that the clear failure of the traditions that were brought by the groups that migrated from Honshu reflects the Taisho-3 and Kyushirataki-5 techno-complexes during the terminal Pleistocene. Beyond one temporal gap, archaeological sites with potteries did not appear in eastern Hokkaido until the Holocene [98]. After the second half of the Preboreal, the number of sites suddenly increased in Hokkaido [27,89,100,101]. It is presumed that the abrupt cooling of the Younger Dryas period in Hokkaido made resource use difficult for human adaptations, regardless of whether human groups could use aquatic resources for subsistence or not. Therefore, this means that the migration from Honshu with the non-microblade tradition did not play a significant role in the disappearance of microblade technology.

Reliable evidence of human activity in eastern Hokkaido in the initial Holocene has been found through excavations at the Taisho-6 site [102]. This site includes lithic assemblages composed of stone arrowheads, burins, side-scrapers and polished stone axes associated with flat-bottom potteries, defined as the Tenneru-Akatsuki type [27,89,100–102]. Radiocarbon dating of the charred materials from potteries shows that human activities at the Taisho-6 site began around 11,000 years ago. Many pit dwellings have also been discovered in sites in eastern Hokkaido, where the Tenneru-Akatsuki-type pottery was found, so we can assume that a more sedentary way of life followed in the second half of the Preboreal. A more intensive use of aquatic resources and nuts has been shown at the stage where the Tenneru-Akatsuki-type pottery appears and spreads [27,101,103].

Despite a few claims to the contrary [27,89], many argue that the appearance of Tenneru-Akatsuki-type pottery in eastern Hokkaido reflects the migration of groups from Honshu and their adaptation to Hokkaido [101,104]. At present, it is unreasonable to expect that the various cultural features observed in potteries and stone tools in the initial Holocene sites of Hokkaido originated in the Younger Dryas period. Though, archaeological sites chronologically located during the Younger Dryas may be discovered in the future. However, even if such discoveries become possible, it would not change the conclusion that a sudden decrease in the number of sites occurred during the Younger Dryas compared to other periods. In Honshu, there are some fluctuations in the number of sites, but human activities continued across all periods of the terminal Pleistocene without interruption [94,105]. No definite break in the archaeological record can be seen, distinguishing it from the trends seen in Hokkaido.

6. Conclusions

Advancements in the investigation of archaeological sites across northeast Asia have revealed changes in human adaptations from the terminal Pleistocene to the initial Holocene. Radiocarbon dating has allowed a highly precise chronological framework to be constructed. Moreover, the analysis of nitrogen and carbon isotope ratios in charred materials from potteries has provided important information on the emergence of new subsistence activities. A systematic discussion of the disappearance and continuation of microblade technology and its various developments, as well as the background to these events following the LGM, is now becoming possible, based on archaeological studies of stone tools and potteries as well as the accumulation of evidence from bioarchaeology and palaeoenvironmental studies.

This paper has discussed the period during which microblade technology in Hokkaido disappeared, as well as the background to this. Microblade technology, seen until the Bølling–Allerød in the terminal Pleistocene, disappeared during the Younger Dryas. It may be that continued habitation in Hokkaido was made difficult due to the significant re-cooling of the climate, which caused a striking decline or extinction of the resources that humans had relied upon for food. Despite some drastic changes caused by the Younger Dryas, long-lasting developments in human behaviour such as high mobility and the hunting of large mammals in North China and Siberia, indicating open forest and grassland zones, seemed to have been unaffected by the impact of this event. In contrast, discontinuity rather than continuity in human behaviour and populations during the Younger Dryas has been observed in the archaeological records from Hokkaido. It is apparent that this process occurred during the terminal Pleistocene and was influenced by a wide range of factors, including changes in landscape, climate, faunal community, subsistence and human populations.

Funding: This research was funded by JSPS KAKENHI, grant number 17K03202.

Acknowledgments: I would like to thank Kazunobu Ikeya and Pei-Lin Yu for giving me an opportunity to contribute to the Special Issue and Hisao Mori for help in permitting the use of photographs. I also thank the anonymous reviewers for improving the manuscript.

Conflicts of Interest: The author declares no conflict of interest.

References

1. Nakazawa, Y.; Izuho, M.; Takakura, J.; Yamada, S. Toward an understanding of technological variability in microblade assemblages in Hokkaido, Japan. *Asian Perspect.* **2005**, *44*, 276–292. [CrossRef]
2. Takakura, J. Emergence and development of the pressure microblade production: A view from the Upper Paleolithic of Northern Japan. In *The Emergence of Pressure Blade Making: From Origin to Modern Experimentation*; Desrosiers, P.M., Ed.; Springer: New York, NY, USA, 2012; pp. 285–306.
3. Nakazawa, Y.; Yamada, S. On the processes of diversification in microblade technocomplexes in Late Glacial Hokkaido. In *Emergence and Diversity of Modern Human Behavior in Paleolithic Asia*; Kaifu, Y., Izuho, M., Goebel, T., Sato, H., Ono, A., Eds.; Texas A&M University Press: College Station, TX, USA, 2015; pp. 418–433.
4. Clark, P.U.; Dyke, A.S.; Shakun, J.D.; Carlson, A.E.; Clark, J.; Wohlfarth, B.; Mitrovica, J.X.; Hostetler, S.W.; McCabe, A.M. The last glacial maximum. *Science* **2009**, *325*, 710–714. [CrossRef] [PubMed]
5. Kato, S.; Tsurumaru, T. *Fundamentals of Lithic Analysis*; Kashiwa-shobo: Tokyo, Japan, 1980. (In Japanese)
6. Chen, C. The microlithic of China. *J. Anthropol. Archaeol.* **1984**, *3*, 79–115. [CrossRef]
7. Gai, P. Microlithic industries in China. In *Palaeoanthropology and Palaeolithic Archaeology in the People's Republic of China*; Wu, R., Olsen, J.W., Eds.; Academic Press: Orlando, FL, USA, 1985; pp. 225–241.
8. Lu, T.L.D. The microblade tradition in China: Regional chronologies and significance in the transition to Neolithic. *Asian Perspect.* **1998**, *37*, 85–112.

9. Seong, C. Microblade technology in Korea and adjacent Northeast Asia. *Asian Perspect.* **1998**, *37*, 245–278.
10. Tsurumaru, T. Microlithic culture in Hokkaido district. *Sundai Hist. Rev.* **1979**, *47*, 23–50. (In Japanese)
11. Takakura, J. Refitted material and consideration of lithic reduction sequence among the microblade assemblages: A view from the Okushirataki-1 site, Northern Japan. *Asian Perspect.* **2010**, *49*, 332–347. [CrossRef]
12. Elston, R.G.; Brantingham, J.P. Microbalde technology in Northern Asia: A risk-minimizing strategy of the Late Paleolithic and Early Holocene. In *Thinking Small: Global Perspectives on Microlithization*; Elston, R.G., Kuhn, S.L., Eds.; Archaeological Papers of the American Anthropological Association 12; American Anthropological Association: Washington, DC, USA, 2002; pp. 103–116.
13. Yi, M.I.; Gao, X.; Li, F.; Chen, F.Y. Rethinking the origin of microblade technology: A chronological and ecological perspective. *Quat. Int.* **2015**, *400*, 130–139. [CrossRef]
14. Chen, C.; Wang, X. Upper Paleolithic microblade industries in North China and their relationships with Northeast Asia and North America. *Arct. Anthropol.* **1989**, *26*, 127–156.
15. Keats, S. Microblade technology in Siberia and neighboring regions. In *Origin and Spread of Microblade Technology in Northern Asia and Northe America*; Kuzmin, Y.V., Keats, S.G., Shen, C., Eds.; Archaeology Press: Burnaby, BC, Canada, 2007; pp. 125–146.
16. Kuzmin, Y.V. Geoarchaeological aspects of the origin and spread of microblade technology in Northern and Central Asia. In *Origin and Spread of Microblade Technology in Northern Asia and Northe America*; Kuzmin, Y.V., Keats, S.G., Shen, C., Eds.; Archaeology Press: Burnaby, BC, Canada, 2007; pp. 115–124.
17. Seong, C. Late Pleistocene microblade assemblages in Korea. In *Origin and Spread of Microblade Technology in Northern Asia and Northe America*; Kuzmin, Y.V., Keats, S.G., Shen, C., Eds.; Archaeology Press: Burnaby, BC, Canada, 2007; pp. 103–114.
18. Bae, K. Origin and patterns of the Upper Paleolithic industries in the Korean Peninsula and movement of modern humans in East Asia. *Quat. Int.* **2010**, *211*, 103–112. [CrossRef]
19. Tabarev, A.V. Blades and microblades, percussion and pressure: Toward the evolution of lithic technologies of the Stone Age period, Russian Far East. In *The Emergence of Pressure Blade Making: From Origin to Modern Experimentation*; Desrosiers, P.M., Ed.; Springer: New York, NY, USA, 2012; pp. 329–346.
20. Qu, T.; Bar-Yosef, O.; Wang, Y.; Wu, X. The Chinese Upper Paleolithic: Geography, chronology, and techno-typology. *J. Archaeol. Res.* **2013**, *21*, 1–73.
21. Chen, S.; Yu, P. Variations in the Upper Paleolithic adaptations of North China: A review of the evidence and implications for the onset of food production. *Archaeol. Res. Asia* **2017**, *9*, 1–12. [CrossRef]
22. Gómez Coutouly, Y.A. The emergence of pressure knapping microblade technology in Northeast Asia. *Radiocarbon* **2018**, *60*, 821–855. [CrossRef]
23. Goebel, T. The "microblade adaptation" and recolonization of Siberia during the Late Upper Pleistocene. In *Thinking Small: Global Perspectives on Microlithization*; Elston, R.G., Kuhn, S.L., Eds.; Archaeological Papers of the American Anthropological Association 12; American Anthropological Association: Washington, DC, USA, 2002; pp. 117–131.
24. Struiver, M.; Grootes, P.M.; Braziunas, T.F. The GISP $\delta^{18}O$ climate record of the past 16,500 years and the role the sun, ocean, and volcanoes. *Quat. Res.* **1995**, *44*, 341–354. [CrossRef]
25. Straus, L.; Goebel, T. Humans and Younger Dryas: Dead end, short detour, or open road to the Holocene? *Quat. Int.* **2011**, *242*, 259–261. [CrossRef]
26. Nakazawa, Y.; Iwase, I.; Akai, F.; Izuho, M. Human responses to the Younger Dryas in Japan. *Quat. Int.* **2011**, *242*, 416–433. [CrossRef]
27. Yamahara, T. The terminal Pleistocene in Hokkaido and Early Holocene in Eastern Hokkaido. In *Transformation of Structure during the Emergence of Jomon*; Sato, H., Ed.; Rokuichi Shobou: Tokyo, Japan, 2008; pp. 35–52. (In Japanese)
28. Kunikita, D.; Shevkomud, I.; Yoshida, K.; Onuki, S.; Yamahara, T.; Matsuzaki, H. Dating charred remains on pottery and analyzing food habits in the Early Neolithic period in Northeast Asia. *Radiocarbon* **2013**, *55*, 1334–1340. [CrossRef]
29. Natsuki, D. Incipient Jomon culture in Hokkaido. *Ronshu Oshorokko* **2018**, *5*, 59–78, (In Japanese with English abstract).
30. Imamura, K. *Prehistoric Japan: New Perspectives on Insular East Asia*; UCL Press: London, UK, 1997.

31. Hayashi, K. *Jomon Period (Part II)*; Yuzankaku: Tokyo, Japan, 2004. (In Japanese)
32. Ono, Y. The northern land bridge of Japan. *Quat. Res.* **1990**, *29*, 183–192, (In Japanese with English abstract). [CrossRef]
33. Takahashi, K. The formative history of the terrestrial mammalian fauna of the Japanese islands during the Plio-Pleistocene. *Palaeolithic Res.* **2007**, *3*, 5–14, (In Japanese with English abstract).
34. Kawamura, A. Mammal faunas, paleogeography, and paleoenvironment in Japan since the Middle Pleistocene reconstructed or inferred from fossil records. *Palaeolithic Res.* **2019**, *15*, 13–30, (In Japanese with English abstract).
35. Sato, H. *Palaeolithic: Beginning of the Japanese Culture*; Keibunsha: Tokyo, Japan, 2019. (In Japanese)
36. Fukusawa, H. Varved lacustrine sediments in Japan: Recent progress. *Quat. Res.* **1999**, *38*, 237–243. [CrossRef]
37. Nakagawa, T.; Kiragawa, H.; Yasuda, Y.; Tarasov, P.E.; Gotanda, K.; Sawai, Y. Pollen event stratigraphy of the varved sediments of Lake Suigetsu, central Japan from 15,701 to 10,217 SG vyr BP (Suigetsu varve years before present): Description, interpretation, and correlation with other regions. *Quat. Sci. Rev.* **2005**, *24*, 1691–1701. [CrossRef]
38. Smith, V.C.; Staff, R.A.; Blockley, S.P.E.; Ramsey, C.B.; Nakagawa, T.; Mark, D.F.; Takemura, K.; Danhara, T. Identification and correlation of visible tephras in the Lake Suigetsu SG06 sedimentary archive, Japan: Chronostratigraphic markers for synchronizing of east Asian/west Pacific paleoclimatic records across the last 150 ka. *Quat. Sci. Rev.* **2013**, *67*, 121–137. [CrossRef]
39. Yoshida, A. Paleoenvironmental studies during the Last Glacial period in Japanese archipelago: Recent trends and problems focused on palynological study. *Palaeolithic Res.* **2015**, *11*, 12–21, (In Japanese with English abstract).
40. Yoshikawa, M. Vegetation history in northeastern Japan from the end of the Pleistocene epoch to the early Holocene epoch. *Palaeolithic Res.* **2016**, *12*, 1–12, (In Japanese with English abstract).
41. Igarashi, Y.; Igarashi, T.; Daimaru, H.; Yamada, O.; Miyagi, T. Vegetation history of Kenbuchi Basin and Furano Basin in Hokkaido, North Japan, since 32,000 yrs BP. *Quat. Res.* **1993**, *32*, 89–105, (In Japanese with English abstract). [CrossRef]
42. Igarashi, Y. A Late Glacial climatic reversion in Hokkaido, northeast Asia, inferred from the Larix pollen record. *Quat. Sci. Rev.* **1996**, *15*, 989–995. [CrossRef]
43. Igarashi, Y. Climate and vegetation changes since 40,000 years BP in Hokkaido and Sakhalin. In *Human Ecosystem Changes in the Northern Circum Japan Sea Area (NCJSA) in Late Pleistocene*; Sato, H., Ed.; Research Institute for Humanity and Nature: Kyoto, Japan, 2008; pp. 27–41.
44. Kawamura, Y.; Nakagawa, R. Terrestrial mammal faunas in the Japanese islands during OIS 3 and 2. In *Environmental Changes and Human Occupation in East Asia during OIS 3 and OIS 2*; Ono, A., Izuho, M., Eds.; British Archaeological Report Internationa Series 2352; Archaeopress: Oxford, UK, 2012; pp. 33–54.
45. Iwase, A.; Hashizume, J.; Izuho, M.; Takahashi, K.; Sato, H. Timing of megafaunal extinction in the late Late Pleistocene on the Japanese islands. *Quat. Int.* **2012**, *255*, 114–124. [CrossRef]
46. Iwase, A.; Takahashi, K.; Izuho, M. Further study on the Late Pleistocene megafaunal extinction in the Japanese archipelago. In *Emergence and Diversity of Modern Human Behavior in Paleolithic Asia*; Kaifu, Y., Izuho, M., Goebel, T., Sato, H., Ono, A., Eds.; Texas A&M University Press: College Station, TX, USA, 2015; pp. 325–344.
47. Yamada, S. *A Study of the Microblade Industries in Hokkaido*; Rokuichi Shobou: Tokyo, Japan, 2006. (In Japanese)
48. Niimi, F. Diachronic change in birds and mammalian species. In *The Archaeology of Jomon Period 4: The Relationships between Humans and Animals*; Kosugi, Y., Taniguchi, Y., Nishida, Y., Mizunoe, W., Yano, K., Eds.; Douseisha: Tokyo, Japan, 2010; pp. 131–148. (In Japanese)
49. Wang, X.; Xie, F.; Mei, H.; Gao, X. Intensive exploitation of animal resources during Deglacial time in North China: A case study from the Yujiagou site. *Archaeol. Anthropol. Sci.* **2019**, *11*, 4983–5000. [CrossRef]
50. Terasaki, Y. Regional chronology of Hokkaido. In *A Study of Regional Chronology in the Palaeolithic*; Anzai, M., Sato, H., Eds.; Douseisha: Tokyo, Japan, 2006; pp. 277–314. (In Japanese)
51. Terasaki, Y.; Yamahara, T. Hokkaido region. *Paleolit. Archaeol.* **1999**, *58*, 3–10. (In Japanese)
52. Kimura, H. *Reexamination of the Yubetsu Technique and Study of the Horokazawa Toma Lithic Culture*; Archaeological Museum of Sapporo University: Sapporo, Japan, 1992; (In Japanese and English).

53. Yoshizaki, M. The Shirataki site and the preceramic culture in Hokkaido. *Jpn. J. Ethnol.* **1961**, *26*, 13–23. (In Japanese)
54. Naoe, Y. Chronology and radiocarbon dates from the Early Upper Paleolithic to the Incipient Jomon in Hokkaido. *Palaeolithic Res.* **2014**, *10*, 23–40, (In Japanese with English abstract).
55. Aiba, K. (Ed.) *Ankaritou-7 and Ankaritou-9 Sites*; Hokkaido Archaeological Operations Center: Ebetsu, Japan, 2010. (In Japanese)
56. Naganuma, T.; Sato, T. (Eds.) *Nittou Site*; Hokkaido Archaeological Operations Center: Ebetsu, Japan, 2000. (In Japanese)
57. Takakura, J. Reconsideration of the microblade assemblage at Yoshiizawa B site, Kitami City, Hokkaido. *J. Hokkaido Palaeolithic Res.* **2000**, *5*, 1–34. (In Japanese)
58. Ohya, Y. (Ed.) *Oribe-17 Site*; Kamishihoro Town Board of Education: Kamishihoro, Japan, 2001. (In Japanese)
59. Natsuki, D. Settlement of the Late Glacial humans in Hokkaido: A case from the Yoshiizawa site. In *Humans in the Late Glacial: Adaptive Behaviors and Settlement Patterns of Prehistoric Northern Hunter-Gatherers*; Sato, H., Ed.; Rokuichi-Shobou: Tokyo, Japan, 2016; pp. 43–64. (In Japanese)
60. Yamada, S. Aspects of the exploitation of lithic raw materials resources and the production and transportation of lithic artefacts in the Late Glacial: Viewed from the materials from the Yoshiizawa site. In *Humans in the Late Glacial: Adaptive Behaviors and Settlement Patterns of Prehistoric Northern Hunter-Gatherers*; Sato, H., Ed.; Rokuichi-Shobou: Tokyo, Japan, 2016; pp. 65–84. (In Japanese)
61. Kitazawa, M. (Ed.) *Ozora Site*; Obihiro City Board of Education: Obihiro, Japan, 1993. (In Japanese)
62. Imai, S. (Ed.) *Nishimachi-1 Site*; Shimokawa Town Board of Education: Shimokawa, Japan, 1999. (In Japanese)
63. Kitazawa, M. (Ed.) *Kawanishi-C Site*; Obihiro City Board of Education: Obihiro, Japan, 2000; Volume 2. (In Japanese)
64. Oda, N. Technological efficiency of the boat-shaped tools among the microblade industries in Hokkaido, Japan. *Palaeolithic Res.* **2017**, *13*, 17–34, (In Japanese with English abstract).
65. Iwase, A.; Sato, H.; Yamada, S.; Natsuki, D. A use-wear analysis of the Late Glacial microblade assemblage from Hokkaido, Northern Japan: A case study based on the Yoshiizawa site. *Jpn. J. Archaeol.* **2016**, *4*, 3–28.
66. Takakura, J.; Izuho, M. Identification of flaking techniques: From the analysis of fracture wings. *Quat. Res.* **2004**, *43*, 37–48, (In Japanese with English abstract). [CrossRef]
67. Takakura, J. Identification of blade and microblade flaking techniques in the lithic assemblage of the Okushirataki-1 site, Hokkaido, Japan. *Cult. Antiq.* **2007**, *58*, 98–109, (In Japanese with English abstract).
68. Takakura, J. Identification of the flaking techniques in the Paleolithic assemblage of the Kamihoronai-Moi site, Hokkaido (Japan). *Ronshu Oshorokko* **2008**, *2*, 41–48, (In Japanese with English abstract).
69. Takakura, J. New insights into the reduction process of the Hirosato type microblade core: An identification of flaking techniques for the microblade production and the core rejuvenations. *Ronshu Oshorokko* **2015**, *4*, 103–118, (In Japanese with English abstract).
70. Takakura, J. An identification of flaking techniques for the microblade production in the Oshorokko type microblade cores: A case study from the Ozora and Shouwa site in Hokkaido, northern Japan. *Ronshu Oshorokko* **2018**, *5*, 79–90, (In Japanese with English abstract).
71. Takakura, J. Rethinking the Tougeshita type microblade cores: Identification of microblade flaking techniques and their implication. *J. Jpn. Archaeol. Assoc.* **2020**, *50*, 1–26, (In Japanese with English abstract).
72. Kanomata, Y. Similarities in tool use activities in microblade industries between Hokkaido and Northeastern Honshu: A functional analysis of lithic tools from the Akatsuki site loc.1. *J. Archaeol. Assoc.* **2013**, *35*, 27–48, (In Japanese with English abstract).
73. Kanomata, Y. Hafting and function of microblades in Japan: Based on the analysis of materials from Araya site and Point C at Tachikarushunai-V site. *J. Archaeol. Soc. Nippon* **2004**, *88*, 1–27. (In Japanese)
74. Kanomata, Y. Functional change of burin after disappearing of microblade: A comparative study of the Late Upper Palaeolithic sites in Obihiro City. *Palaeolithic Archaeol.* **2015**, *80*, 51–65, (In Japanese with English abstract).
75. Iwase, A.; Nakazawa, Y. Lithic use-wear analysis on the LGM blade assemblage in Hokkaido, Northern Japan: A case study based on the Kashiwadai C site. *Palaeolithic Res.* **2017**, *13*, 35–56, (In Japanese with English abstract).

76. Iwase, A. Afunctional analysis of the LGM microblade assemblage in Hokkaido, northern Japan: A case study of Kashiwadai 1. *Quat. Int.* **2016**, *425*, 140–157. [CrossRef]
77. Kanomata, Y. A functional study of early microblade industry in Hokkaido: Use-wear analysis of lithic artifacts at the Kashiwadai 1 site in Chitose city. *Palaeolithic Res.* **2013**, *9*, 27–42, (In Japanese with English abstract).
78. Takakura, J. New evidence of endscraper reduction in Upper Paleolithic Japan. *Curr. Res. Pleistocene* **2007**, *24*, 40–43.
79. Takakura, J. Lithic refitting and its implication for the integrity and duration of site occupation: The case of the Late Upper Paleolithic site of the Kiusu-5 in Hokkaido, Northern Japan. *Quat. Int.* **2018**, *474*, 156–167. [CrossRef]
80. Kato, S. Historical and regional characteristics in preceramic age. In *Study of Local History and Archaeology*; Wakamori, T., Ed.; Asakura Shoten: Tokyo, Japan, 1970; pp. 58–92. (In Japanese)
81. Izuho, M.; Hirose, W. A review of archaeological obsidian studies on Hokkaido Island. In *Crossing the Straits: Prehistoric Obsidian Source Exploitation in the North Pacific Rim*; Kuzmin, Y.V., Glasscock, M.D., Eds.; BAR International Series 2152; Archaeopress: Oxford, UK, 2010; pp. 9–26.
82. Kimura, H. Obsidian, humans, technology. In *Paleoekologiya Pleistotseha I Kultury Kamennogo Veka Severnoi Azii I Sopredelnyikh Territorii. Tom 2*; Derevianko, A.P., Ed.; IAE RAS: Novosibirsk, Russia, 1998; pp. 302–314.
83. Hall, M.; Kimura, H. Quantitative EDXRF studies of obsidian sources in Northern Hokkaido. *J. Archaeol. Sci.* **2002**, *29*, 259–266. [CrossRef]
84. Kuzmin, Y.V.; Glasscock, M.D.; Sato, H. Sources of archaeological obsidian on Sakhalin Island (Russian Far East). *J. Archaeol. Sci.* **2002**, *29*, 741–749. [CrossRef]
85. Kuzmin, Y.V. Crossing mountains, rivers, and straits: A review of the current evidence for prehistoric obsidian exchange in Northeast Asia. In *Crossing the Straits: Prehistoric Obsidian Source Exploitation in the North Pacific Rim*; Kuzmin, Y.V., Glasscock, M.D., Eds.; BAR International Series 2152; Archaeopress: Oxford, UK, 2010; pp. 137–157.
86. Yakushige, M.; Sato, H. Shirataki obsidian exploitation and circulation in prehistoric northern Japan. *J. Lithic Stud.* **2014**, *1*, 319–342. [CrossRef]
87. Sato, H.; Yakushige, M. Obsidian source exploitation and its circulation in the Upper Paleolithic Hokkaido. *Palaeolithic Res.* **2013**, *9*, 1–26, (In Japanese with English abstract).
88. Naoe, Y. Procurement of obsidian in Shirataki region and its distribution. *Palaeolithic Res.* **2009**, *5*, 11–22, (In Japanese with English abstract).
89. Yamahara, T. Lithic cultures during the transition between the Pleistocene and the Holocene in eastern Hokkaido. In *Symposium on Emergence of Jomon Culture: From Incipient Jomon to Initial Jomon*; Sato, H., Ed.; Department of Archaeology, the University of Tokyo: Tokyo, Japan, 2007; pp. 8–26. (In Japanese)
90. Kitazawa, M.; Yamahara, T. (Eds.) *Taisho Site Group*; Obihiro City Board of Education: Obihiro, Japan, 2006; Volume 2. (In Japanese)
91. Craig, O.; Saul, H.; Lucquin, A.; Nishida, Y.; Tache, K.; Clarke, L.; Thompson, A.; Altoft, D.T.; Uchiyama, J.; Ajimoto, M.; et al. Earliest evidence for the use of pottery. *Nature* **2013**, *496*, 351–354. [CrossRef]
92. Natsuki, D. A preliminary report of excavation at locality M-I of Tachikarushunai site. *J. Hokkaido Archaeol. Soc.* **2020**, *56*, 21–33. (In Japanese)
93. Taniguchi, Y. *Reconstruction of the Origin of Jomon Culture*; Douseisha: Tokyo, Japan, 2011. (In Japanese)
94. Nagai, K. *Archaeology of Lithic Manufacturing: Experimental Archaeology and the Emergence of Jomon*; Douseisha: Tokyo, Japan, 2009. (In Japanese)
95. Naoe, Y. Conclusion. In *The Shirataki Site Group 9*; Naoe, Y., Ed.; Hokkaido Archaeological Operations Center: Ebetsu, Japan, 2008; pp. 241–272. (In Japanese)
96. Suzuki, H. Procurement of Shirataki obsidian and its transition during MIS 2 and 3 in the Paleo-Hokkaido Peninsula. *Palaeolithic Res.* **2016**, *12*, 23–46, (In Japanese with English abstract).
97. Yamada, S. Change and behavioral interpretation of microblade assemblages and technologies in Hokkaido (Japan). In *International Symposium on Human Ecosystem Changes in the Northern Circum Japan Sea Area (NCJSA) in Late Pleistocene*; Sato, H., Ed.; Research Institute for Humanity and Nature: Kyoto, Japan, 2008; pp. 115–138, (In Japanese with English abstract).

98. Morisaki, K.; Natsuki, D. Human behavioral change and the distributional dynamics of early Japanese pottery. *Quat. Int.* **2017**, *441*, 91–101. [CrossRef]
99. Buvid, I.; Terry, K. The twilight of Paleolithic Siberia: Humans and their environments east of Lake Baikal at the Late-glacial/Holocene transition. *Quat. Int.* **2011**, *242*, 379–400. [CrossRef]
100. Nishi, Y. The potteries of the early stage of the Initial Jomon in eastern Hokkaido. In *Archaeology of Production*; Douseisha: Tokyo, Japan, 1997; pp. 21–33. (In Japanese)
101. Sawa, S. *Prehistory of Kushiro*; Kushiro City: Kushiro, Japan, 1987. (In Japanese)
102. Kitazawa, M.; Yamahara, T. (Eds.) *Taisho Site Group*; Obihiro City Board of Education: Obihiro, Japan, 2005; Volume 1. (In Japanese)
103. Yamada, G. On the botanical remains, mainly of nuts, from prehistorical sites in Hokkaido. *Cult. Antiq.* **1993**, *45*, 13–22, (In Japanese with English abstract).
104. Yokoyama, E. Formation of Jomon culture in Hokkaido. In *Archaeology of Northern Area*; The Executive Committee for Publishing of Takashi Nomura's 60th Anniversary Commemorative Essays: Sapporo, Japan, 1998; pp. 29–78. (In Japanese)
105. Kaner, S. Long-term innovation: Appearance and spread of pottery in the Japanese archipelago. In *Ceramics Before Farming: The Dispersal of Pottery among Prehistoric Eurasian Hinter-Gatherers*; Jordan, P., Zvelebil, M., Eds.; Left Coast Press: Walnut Creek, CA, USA, 2010; pp. 93–120.

quaternary

Commentary

New Discoveries and Theoretical Implications for the Last Foraging and First Farming in East Asia

Robert L. Kelly

Department of Anthropology, University of Wyoming, Laramie, WY 82070, USA; rlkelly@uwyo.edu

Abstract: A brief summation of the issue's articles is presented. This leads to a discussion of thematic issues of concepts, methods, and theory that crosscut the articles. These include use of the EnvCalc2.1 program, some issues of terminology, the theoretical approaches of niche construction as opposed to human behavioral ecology (HBE), and the linkage between technology and subsistence change, notably the difference between biface and microblade production.

Keywords: East Asia; origins of agriculture; paleolithic to Neolithic transition

Citation: Kelly, R.L. New Discoveries and Theoretical Implications for the Last Foraging and First Farming in East Asia. *Quaternary* **2021**, *4*, 40. https://doi.org/10.3390/quat4040040

Academic Editors: Pei-Lin Yu, Ikeya Kazunobu and Meng Zhang

Received: 3 July 2020
Accepted: 6 August 2020
Published: 22 November 2021

1. Introduction

Let me first make clear that I am an expert on neither Asian archaeology nor the origins of agriculture. I cannot comment on the specifics of particular Asian archaeological sites or on the botanical histories of the various plants involved in the transition from foraging to agriculture in Asia. I do know something about hunter-gatherers, and about how archaeologists have been thinking about the origins of agriculture in North America, where, as in these Asian cases, we have both domesticated indigenous plants (in the eastern US, e.g., Chenopodium) and already-domesticated imports (maize, from southern Mexico/Central America). My comments here, therefore, result from taking a step back from the specifics of the papers to look more at methods, data, and theoretical constructs. However, first I make some observations on the individual contributions.

2. The Case in Asia

The precise pathways taken toward agriculture in any given prehistoric case—the woodlands of the eastern US, Taiwan, Japan, or mainland China—depend on numerous factors of geography, climate, plant genetics and habitat, other subsistence options, human population growth, and others. However, what I see in these papers is a story that has occurred repeatedly in various places in the world over the past 15,000 years. Human population grew to the point where diet had to be supplemented, first, with lower return-rate-resources, which in many cases included plants that produce small seeds (or below-ground storage vessels, such as tubers). These resources responded to their human use and manipulation by becoming more productive in density and/or seed size; eventually, for better or for worse (a steady diet of maize or rice, e.g., is not good for long-term health), they became the focus of diet—and the plants that support the modern world's 7.8 billion people.

This collection of papers, however, does show that this general, global process has variations produced by local environmental, historical, botanical, and prior cultural conditions. Investigation of this variation helps us understand variety in the specific pathways to agriculture, and consequently which variables are important under what conditions.

There are different theoretical tools at our disposal to understand this variation. Agriculture was imported from Southeast China to Taiwan about 6000 years BP, and, using the combined perspectives of human behavioral ecology (HBE) and niche variation theory (NVT), Yu [1] argues those Neolithic agriculturalists met a population divided into two basic adaptations: one focused on coastal fishing and the other on mountain hunting. The arrival of agriculturalists almost certainly increased the island's population density. The result was most likely increased competition, and niche contraction. HBE would predict increased diet breadth under these conditions, and that appears to be what happened, as previous hunter-gatherers incorporated low-ranked/low-cost cultigens into diet. I will add that the ideal free distribution (IFD) model might be a useful next step, as it would predict the order in which different regions should witness the transition to agriculture. Right now, that appears to be western coastal areas, the mountains, and then the east coast.

Fujio [2] also looks at an instance where agriculture (and, later, irrigation technology) was imported, in this case, to Japan from the Korean peninsula before about 3000 years ago. This seems to have been driven by population packing and competition for land on the Korean peninsula. It is often said that hunter-gatherers solve problems by moving; agriculturalists do, too, but not as a matter of course (hunter-gatherers expect to move, agriculturalists do not), and not after exhausting other possibilities. A hunter-gatherer can live in many places; suitable places for agriculture might be few and far between. In Japan, the Jomon population appears to have grown in size, especially after 5500 years ago, when large pithouse villages formed, and then declined, with movement to dispersed groups about 4900 years ago. This process may reflect a "rigidity trap" one based on more focused use of labor-intensive resources such as horse chestnuts [3]. If so, it might signal that the Jomon population ~3000 years ago was in need of a new food source, and rice agriculture from Korea provided it; the same might have happened in the eastern US woodlands, with the appearance of full-time maize horticultural systems ~1100 years ago (maize appears ~2000 years ago but was not consequential for a long time), after which eastern US societies made rapid changes in social and political organization (the "Mississippian culture"). Fujio [2] is able to examine the timing of settlement of different regions of Japan in some detail, and, again, this offers the potential to test whether this process can be described by the IFD. For both Japan and Taiwan, if the IFD fails, then it is probably pointing to some other factor—such as the presence of foragers—that affected agriculturalists' choice of where to settle.

Likewise, in Japan, Ikeya [4] is concerned with the process whereby Yayoi agriculturalists replaced or subsumed the existing Jomon hunter-gatherers. Across the globe, agriculturalists gradually replaced hunter-gatherers. In fact, this process is not complete, and is still on-going, with very few populations left who live even partially by hunting and gathering (virtually no population today lives entirely by foraging). Again, the IFD model could help explain the spread of agriculture (keeping in mind the "starting" point, where Korean migrants landed may not have been the "best" place to start on the islands of Japan). Nevertheless, Ikeya [4] is more interested in what governs whether the indigenous (Jomon) people of Japan, simply became Yayoi agriculturalists or remained hunter-gatherers to one extent or another. That is a complicated question, and no doubt has to do with the returns from rice agriculture versus the returns from foraging in different parts of Japan. Determining those returns will require further ethnographic and/or experimental work with foraged foods versus cultigens.

Takakura [5] looks at the archaeology of Hokkaido, Japan's northern island. Here he finds, as others have, that microblade technology disappears during the Younger Dryas (YD) cooling, perhaps, quite possibly, because the human population itself disappears at this time (although the ages of the boat-shaped tool assemblages are not well dated and may fall within the YD). When people returned to the island after the YD (or when their numbers rebounded), they used bifacial technology, without microblades, and ceramics. The Taisho-3 site assemblage shows that the shift away from microblade technology (and accompanied by ceramics) had happened even before the YD, but after the demise of large Pleistocene fauna on the island. At Tasiho-3 there is evidence for the use of aquatic resources, a subsistence shift that in other places (according to the accompanying papers) was a precursor to an agricultural economy.

In North China, Paleolithic foragers shifted to a more Neolithic-like adaptation, including sedentism in large aggregated settlements, large-scale food storage, and so on, before full agriculture (millet) appears. In this regard, North China is similar to the Near

East; and also portions of the eastern US, where the "Hopewell culture" (much later in time) had a predominantly foraging adaptation, with some low-level use of indigenous cultigens. Hopewell people lived in large, sedentary villages that show evidence of social and political "complexity" in the form of elaborate log-tomb burials beneath mounds, and large, sophisticated earthen enclosures, many in geometric forms. Zhao [6] shows that the transition to a fully Neolithic lifeway with millet agriculture (rice in some portions of North China) appears after the 8.2 k climatic cooling event (that increased aridity in northern latitudes). I suspect the adaptive system was moving in the direction of a fully-agricultural lifeway due simply to population pressure and resulting competition. Population pressure, of course, results from an increase in population and/or a decline in resource abundance. I suspect Zhao [6] is right: the 8.2 k event was a severe "trigger" that sped the Neolithization process up by decreasing food availability in North China. Obviously, what is needed are tighter chronological studies to sort assemblages into pre-8.2 k, 8.2 k, and post-8.2 k intervals, e.g., [3].

Liu and her colleagues [7] show that the Paleolithic-to-Neolithic transition was not a simple phenomenon, especially in those zones that are somewhat marginal for agriculture (marginal without considerable labor investment). They see a geographic mosaic of at least three adaptations in the Middle Yangtze region: rice agriculture, hunting and gathering (a continuation of Paleolithic lifeways), and low-level horticulture coupled with hunting, gathering, and fishing, the last indicated by the appearance of a simple, but consistent and apparently efficient flake tool technology incorporating "ridged-hammer bipolar flaking." It is regions of such adaptive mosaics that, in the future, can provide some of the best tests of hypotheses, given that they allow us to see the effect of different combinations of environment, geography, and population density.

Zhang [8] employs Binford's [9] macroecological approach to help reconstruct changes in diet, mobility and group size in northern China at the last glacial maximum (LGM), YD, and afterwards. He finds that the evidence matches Binford's now >40-year-old hypothesis that agriculture should appear in the areas adjacent to those with the greatest density of the wild ancestors of domesticated plants. Binford was thinking specifically of the Near East, and the domestication of wheat and barley. However, the best evidence now suggests that domestication in the Near East began in those areas with high densities of the wild ancestors of domesticated plants, not in the adjacent areas. The North China case, therefore, might be another place to test Binford's hypothesis rigorously.

Yu [1], Liu and colleagues [7], and Zhang [8] all see an interplay between aquatic resources (riverine in one case, marine in the others) and the incorporation of agriculture into the diet. Cross-cultural studies do suggest that aquatic resources are used in lieu of storable gathered foods, but also in lieu of hunted foods [9,10]. In both cases, an increasing population density would restrict access to food, probably first to large game and then to storable plant foods (but this could depend on specific resource distributions). In any case, Asia offers the potential to examine the trade-offs between the plant, animal, and aquatic components of diet in understanding why some people added agriculture to the mix when they did.

3. EnvCalc2.1 Program

Two papers (Yu [1] and Zhang [8]) make use of the EnvCalc2.1 program [11,12], based on Binford's [9] exhaustive set of ethnographic and environmental data. With this program one can hypothesize the nature of prehistoric foraging societies by using modern climate data at specific locations to reconstruct past climates, such as the LGM, and then, based on correlations between environmental variables and ethnographic data on foraging societies, predict a number of aspects of prehistoric hunting and gathering societies. It is potentially remarkable, but I have three concerns with the use of this approach.

First, and briefly, the program is not designed to serve in place of reconstructions of the past based on archaeological data. Yu [1] and Zhang [8] are both aware of this, and they instead offer the approach as a way to frame hypotheses about how we expect prehistoric hunter-gatherers to have behaved, hypotheses that can then be tested against archaeological data. I make this comment only to reiterate and hence avoid any misunderstanding of Binford's goal: to create a way to use the prehistoric record to improve our understanding of why hunter-gatherers acted the way they did and consequently to understand why prehistory took the particular path it did in a particular region. Yu [1], for example, uses Binford's approach combined with HBE and NVT to provide some testable ideas of what happened in Taiwan's prehistory; Zhang [8] uses the approach to suggest when LGM hunter-gatherers might have increased exploitation of storable food resources, such as seeds, again a testable idea.

Second, a shortcoming of Binford's "macroecological" approach is that no explicit theory stands behind it. Instead, it is empirical generalizations, correlations between environmental and human behavioral variables. Make no mistake: these are enormously useful and, to use one of Binford's favorite terms, provocative. The idea that stands behind much of Binford's approach is cultural ecology, or, its updated version, HBE. Although, inexplicably, in my opinion, Binford always rejected HBE, and thus he rejected the most obvious theory to account for the many patterns he demonstrated.

Third, I have concerns about the ethnographic data that stand behind EnvCalc's predictive capacity. I have only checked one of Binford's datasets [9], the mobility data presented in his Table 8.04 (I am indebted to Brigid Grund for her dogged assistance in this task). There are 303 cases listed in this table. Of these, 26 (8.5%) are presented as "LRB estimates," and another 13 (4.3%) are based on personal communications from a relevant ethnographer; the remainder list published references.

I have found no one who knows what "LRB estimate" refers to; they are obviously not based on Binford's own fieldwork or the field notes of others. In some of those cases it is clear that the group had ceased being nomadic long before any ethnographer could have been on the scene. I have also searched Binford's archives at Truman State University, but I could find no personal letters from ethnographers in the folders. Although, to be precise, all of Binford's materials were not yet archived when I made my visit, so they may yet be discovered.

Of the 264 cases with references listed we could verify the exact data in only 3 (1.1%) cases. We found mobility data in 85 (32%) cases but came up with (sometimes only slightly) different estimates; and in some cases, Binford presents data as miles, but they are actually kilometers. However, some of the cases are ones where I am certain there are no mobility data. For example, the Toedökadö (Cattail-Eater) Paiute in western Nevada, is listed as having 5 residential moves/yr and a total annual movement of 90 miles/yr, and the reference is Fowler's [13] compilation of Willard Park's 1930s fieldnotes. There are no mobility data in that compilation; indeed, there cannot be, for the Paiute no longer lived a nomadic lifeway in the 1930s. Perhaps Binford made a mistake and listed the wrong reference. That is possible, and I cannot cast the first stone here as there is at least one such mistake in my own mobility data compilation [10] (for the Netsilik). However, I did my dissertation fieldwork in Toedökadö territory, and consequently I am very familiar with the literature, and there are no mobility data in any of the region's ethnographic or ethnohistoric references. I do not know where Binford's Toedökadö data came from.

Binford's data were gathered over many years, and most likely from the efforts of many students. No doubt some errors have crept in as happens with all large compilations of data. Whether they are enough to negate EnvCalc is something that ought to be investigated. I sincerely hope the answer is no.

4. Terminology: Intensification, Complexity, Climate Amelioration/Deterioration

Archaeology uses terms whose meanings are often thought to be clear, but, upon reflection, are not. Morgan [14] makes clear that one such term is intensification. Part of the problem is that "intensification" is not a substantive thing, but a concept. Concepts are good to think with, but not if they mean substantially different things to different researchers. Intensification can refer to increased productivity, increased efficiency, or increased productivity with increased efficiency. It almost always implies a process or trend over time, as in "increasing intensification."

Since the *concept* of intensification is not a *definition*, it is not always clear what evidence will prove or disprove an argument for intensification. I have long preferred the language of HBE, whose hypotheses entail expectations such as an increase or decrease in diet breadth, which can be measured by relative changes in the range of food resources evidenced in archaeological assemblages; whether it entails a decline in efficiency (it usually does) and/or an increase in productivity (it can, but usually at an increase in time input). In such cases an increase in diet breadth can be defined as "intensification" but what matters is that the hypothesis permits its testing. Agriculture is a special case because it begins with an increase in diet breadth (including, e.g., small, initially low-return-rate seeds in the diet) that implies a decline in efficiency and a consequent increase in workload [10] (pp. 50,51)). However, these small seeds respond to human use and ingenuity by an increase in productivity (through selection of larger seeds to generate next season's crop, or productivity-enhancing behaviors such as irrigation or terracing—all of which require increased labor input). Being part of a package, a plant with an enhanced return rate due to genetics, and to human behavior aimed at exploiting those genetics can leapfrog other resources, especially when imported in their fully domesticated form to a new region (e.g., rice to Japan, or maize to the US eastern woodlands). What is needed, then, is knowledge of how plants of different potential (genetic) productivities coupled with different human behaviors aimed at exploiting those potentials (e.g., simple floodplain farming, irrigation, fertilizing, etc.) affect the return-rates of different horticultural systems. Barlow [15] does this for indigenous maize farming systems in the American southwest.

Other terms have similar difficulties: climate amelioration/deterioration should be made more specific in terms of precipitation and temperature and, when discussing subsistence, their linkages to food availability (or seasonality, etc.) be made clear. Likewise, "complexity" has long bothered me. On the one hand, it implies that the behavior of nomadic hunter-gatherers was simple, which simply is not true (even an hour's reading on indigenous Australians' social organization or Dreamtime theology will leave the reader's mind spinning); even what appears to be "simple" technology and subsistence is predicated on detailed knowledge of natural history and physics. More to the point, "complexity" masks the actual behaviors at work in producing the artifacts that lead archaeologists to label a case of prehistoric hunter-gatherers (or anyone) as "complex" for these are usually artifacts pointing to the importance of formalized social relations between individuals or groups, or to social, political, and/or economic inequality. A concern with terms is not trivial because imprecise terms lead to imprecise explanations.

5. Niche Construction versus Human Behavioral Ecology

Long before maize appeared in the eastern US, inhabitants in the mid-continent cultivated indigenous plants, including squash (*Cucurbita pepo*), sunflower (*Helianthus annuus*), marshelder (*Iva annua*), and chenopod (*Chenopodium berlandieri*). Other possible cultigens (of lesser importance if they were cultivated) are giant ragweed (*Ambrosia trifida*), little barley (*Hordeum pusillum*), maygrass (*Phalaris caroliniana*), and knotweed (*Polygonum erectum*). The process of cultivation was underway sometime between 5800 and 3200 cal BP, but was not significant until well after 3200 cal BP. By ~1100 cal BP, maize supplanted these cultivated indigenous plants and long-time horticulturalists/foragers in the eastern US became agriculturalists.

In recent years, some researchers have used niche construction theory (NCT) to explain the origin of agriculture in the eastern US [16]. This approach is not used heavily in these papers (see Yu [1]), but I am certain my Asian colleagues are reading this literature, especially those interested in the origins of agriculture.

Briefly, NCT focuses on how behaviors emerge from the effects that interactions among organisms have on one another. Exploitation of a plant leads to a change in the plant's productivity, which leads to a change in human use of the plant, which further affects the plant's productivity, which further changes how humans use it. Such a process may entail both intentional and unintentional behaviors. Many of the early cultigens of the eastern US grew in the floodplains of rivers, and especially in disturbed habitats. Human occupation of those floodplains, and the removal of trees, e.g., for house construction, or burning to enhance forage for wild game, would have increased the abundance/density of these weeds, possibly increasing their return rate, making them a more attractive food choice. Human foragers might realize that they could enhance this process, and horticulture/domestication would be underway.

An alternative to NCT is HBE. HBE sees subsistence change as driven largely by changes in the availability of high-return-rate resources, changes produced by climate—linked environmental change and/or human population growth that results in a food's over—exploitation and consequent decline in abundance, a situation described as "resource depression." Changes in the abundance of low-return-rate resources (such as small seeds), changes that do not increase a resource's return rate, are not anticipated to have an effect on diet composition.

Although proponents present NCT as an alternative to HBE, it is not. In fact, the predictions of NCT fall quite comfortably within the models used by HBE [17,18], although HBE has admittedly focused on resource depression and not on how human use of resources might have the opposite effect—increasing their return rate by increasing their abundance and/or by technological innovations. (Note, however, that increasing abundance does not automatically increase a food's return rate. However, under some conditions, an increase in density can increase a resource's return rate by making the resource more amenable to mass collection methods that raise its return rate by raising the efficiency of collection. A new technology is usually involved: Imagine the return rate from scattered seed-bearing plants, harvested plant by plant as opposed to that of a dense field of the same plant, harvested with a wicker paddle and basket. This process can be enhanced by humans through selective breeding, e.g., for larger seed size.)

The difference between NCT and HBE matters because NCT proponents argue that human population growth was not responsible for the origins of agriculture. Zeanah [17] argues there is good reason to expect that small seeds such as chenopod, even with a density enhanced by human activity, would not be incorporated into the diet until high- return-rate foods were diminished in abundance. He shows that the archaeologically- documented patterns of use in the eastern woodlands—heavy use of hickory, with the later addition of walnut over small seeds, and then diminished use of walnut in favor of the enhanced returns from cultivated small seed plants—fits with an HBE model that predicts cultivation as a response to declines in the abundance of higher-return-rate resources. Miller [19] shows that intensive use of lower-return-rate seeds and cultivation of indigenous plants appears in the eastern US after evidence appears of both population growth and the reduced use of high-return-rate resources.

In other words, the origins of agriculture in the eastern US seems to be completely consistent with a model of changing subsistence driven by population growth, reduction of high-return-rate food resources, coupled with the effects of human activity on the abundance of weedy, small-seed-producing indigenous plants. NCT is an important addition to our understanding of the origins of agriculture, in both primary and secondary settings, but its legitimate concern with the effects of human behavior on the environment and their subsequent effects on subsistence choices falls comfortably within HBE, which provides a sounder foundation in evolutionary theory.

6. Population

Several papers rely implicitly or explicitly on the relationship agriculture had to population growth. It has long been logically assumed that agriculture leads to population growth, what is known as the Neolithic Demographic Transition, and recent studies demonstrate this fact from the data of prehistory [20–22]. At the same time, studies show that the long-term rate of growth of hunter-gatherer and agriculturalists is the same,

~0.04% [23], but with a difference: agricultural populations can grow at high rates over short periods of time but are more susceptible to periodic population crashes [23,24]. Future research should give specific attention to measures of population growth/density to test demographic hypotheses about the relationships among agriculture, social and political organization, warfare, and agriculturalists' foraging neighbors. Asia offers an excellent laboratory for this project.

7. Technology

The stone component of the technology that immediately precedes agriculture in eastern Asia entails microblades. To an archaeologist of the continental US, microblades are quite foreign and I am not qualified to discuss their various methods of production or use. However, let me make two comments that might be useful in understanding why microblade technology disappears. First, Takakura [5] and Zhang [8] associate the decline in microblade technology with the demise of the large Pleistocene fauna of Hokkaido—and that association indeed appears to be true. Although in the US, large fauna, including mammoths and mastodons were hunted with Clovis spearpoints, large bifacial points, a technology that, invented by a population that migrated to North America from Asia about 15,000 years ago, almost certainly came out of a microblade tradition. Microblade technology is certainly not *necessary* for the hunting of large game. In fact, I have long thought that microblades and bifaces are two different ways to accomplish the same goal of ensuring maintainability, the former relying on modularization-with-replacement and the latter on resharpening.

Second, my sense, and it is only a sense, is that microblade technology is more difficult and time-consuming to learn than bifacial technology. The best candidate in my opinion for the precursor to Clovis technology is the Yubetsu microblade method, which first requires production of a biface, one edge of which is then removed (with one or more "ski spalls") to create a platform for the removal of standard-sized microblades. That is, the Yubetsu method requires biface technology *and* creation of a core to produce pressure-flaked, standard-sized microblades. The simple fact that there is a second production stage tells us the microblade method will be more difficult to master. (Clovis retained some of this difficulty with the production of flutes, and, in some places, macroblades from cores, but the fluting technique was dropped, after it was used most prominently in the slightly later-in-time Folsom spearpoints. Macroblade production also disappears.)

What these two facts suggest is that microblade-using populations replaced one technology, microblades, with another, bifaces and flake tools, that (probably) turned out to work just as well for hunting and yet allowed time for children to be enculturated in other tasks that had become necessary, and thus more important than microblade "education," due to environmentally-induced changes in food resources and/or an increased population, e.g., ceramics, plant gathering, and the hunting of a wider range of animals with all their various behaviors (I will note that the lack of blade standardization at Donghulin may point to less enculturation in proper production methods). Thus, the key to the decline in microblades may lie in a shift in the importance of tasks that children had to learn to be successful adults in a post-Pleistocene world.

8. Conclusions

From an outsider's perspective, the Near East has dominated our understanding of the origins and development of agricultural and of early agricultural society. Clearly Asia, as an independent center of agricultural origins, and one that contains plants of quite different needs (millet and wet-farming of rice) offers the possibility of testing hypotheses generated by research in the Near East and elsewhere. To do so will require: (1) a better understanding of how climate affects the distribution and abundance of the local, non-domesticated food resources over the past 18,000 years; (2) the collection of data (such as radiocarbon dates, but also measures of site frequency and artifact densities over time) that will permit testing hypotheses relating agriculture's inception and development to population growth; (3)

better understanding, generated by ethnographic, ethnohistoric, and experimental work on the productivity, constraints and possibilities of different agricultural plants and systems; and (4) better understanding of technologies, their constraints and possibilities. Although NCT certainly points to the importance of understanding how one generation's behavior alters the environment to which the next generation must adapt, I think that HBE is a productive theoretical approach within which to understand the material possibilities and consequences of such a process. I, for one, look forward to this research in Asia with enthusiasm.

Funding: This article was not supported by external funding.

Acknowledgments: The author thanks the co-editors for the opportunity to offer his thoughts on a subject that he is now better educated on.

Conflicts of Interest: The author declares no conflict of interest.

References

1. Yu, P.-L. Modeling Incipient Use of Neolithic Cultigens by Taiwanese Foragers: Perspectives from Niche Variation Theory, the Prey Choice Model, and the Ideal Free Distribution. *Quaternary* **2020**, *3*, 26. [CrossRef]
2. Fujio, S. Early Grain Cultivation and Starting Processes in the Japanese Archipelago. *Quaternary* **2021**, *4*, 3. [CrossRef]
3. Crema, E.R.; Kobayashi, K. A multi-proxy inference of Jomon population dynamics using Bayesian phase models, residential data, and summed probability distribution of ^{14}C dates. *J. Archaeol. Sci.* **2020**, *117*, 105136. [CrossRef]
4. Ikeya, K. Ethnoarchaeology of Introducing Agriculture and Social Continuity among Sedentarised Hunter–Gatherers: The Transition from the Jomon to the Yayoi Period. *Quaternary* **2021**, *4*, 28. [CrossRef]
5. Takakura, J. Rethinking the disappearance of microblade technology in the Terminal Pleistocene of Hokkaido, northern Japan: Looking at archaeological and palaeoenvironmental evidence. *Quaternary* **2020**, *3*, 21. [CrossRef]
6. Zhao, C. The climate fluctuation of the 8.2 ka BP cooling event and the transition into Neolithic lifeways in North China. *Quaternary* **2020**, *3*, 23. [CrossRef]
7. Liu, R.; Liu, H.; Chen, S. Alternative Adaptation Strategy during the Paleolithic–Neolithic Transition: Potential Use of Aquatic Resources in the Western Middle Yangtze Valley, China. *Quaternary* **2020**, *3*, 28. [CrossRef]
8. Zhang, M. Microblade–Based Societies in North China at the End of the Ice Age. *Quaternary* **2020**, *3*, 20. [CrossRef]
9. Binford, L.R. *Constructing Frames of Reference: An Analytical Method for Archaeological Theory Building Using Ethnographic and Environmental Data Sets*; University of California Press: Berkeley, CA, USA, 2001.
10. Kelly, R.L. *The Lifeways of Hunter-Gatherers: The Foraging Spectrum*; Cambridge University Press: Cambridge, UK, 2013.
11. Binford, L.R.; Johnson, A. Program for Calculating Environmental and Hunter-Gatherers Frames of Reference (ENVCALC2); Java Version, 2006. Available online: http://ajohnson.sites.truman.edu/data-and-program/ (accessed on 10 November 2021).
12. Johnson, A. Exploring adaptive variation among hunter-gatherers with Binford's frames of reference. *J. Archaeol. Res.* **2014**, *22*, 1–42. [CrossRef]
13. Fowler, C. *In the Shadow of Fox Peak: An Ethnography of the Cattail-Eater Northern Paiute People of the Stillwater Marsh*; US Department of the Interior, Fish and Wildlife Service: Reno, NV, USA, 1992.
14. Morgan, C. Is it intensification yet? Current archaeological perspectives on the evolution of hunter-gatherer economies. *J. Archaeol. Res.* **2015**, *23*, 163–213. [CrossRef]
15. Barlow, K.R. Predicting maize agriculture among the Fremont: An economic comparison of farming and foraging in the American southwest. *Am. Antiq.* **2002**, *67*, 65–88. [CrossRef]
16. Smith, B. A comparison of niche construction theory and diet breadth models as explanatory frameworks for the initial domestication of plants and animals. *J. Archaeol. Res.* **2015**, *23*, 215–262. [CrossRef]
17. Zeanah, D. Foraging models, niche construction, and the eastern agricultural complex. *Am. Antiq.* **2017**, *82*, 3–24. [CrossRef]
18. Broughton, J.M.; Cannon, M.D.; Bartelink, E.J. Evolution ecology, resource depression, and niche construction theory: Applications to central California hunter-gatherers and Mimbres-Mogollon agriculturalists. *J. Archaeol. Method Theory* **2010**, *17*, 371–421. [CrossRef]
19. Miller, D.S. *From Colonization to Domestication: Population, Environment, and the Origins of Agriculture in Eastern North America*; University of Utah Press: Salt Lake City, UT, USA, 2018.
20. Bocquet-Appel, J.-P.; Bar-Yosef, O. *The Neolithic Demographic Transition and Its Consequences*; Springer: Dordrecht, The Netherlands, 2008.
21. Downey, S.S.; Boceaege, E.; Kerig, T.; Edinborough, K.; Shennan, S. The neolithic demographic transition in Europe: Correlation with juvenile index supports interpretation of the summed calibration radiocarbon date probability distribution (SCDPD) as a valid demographic proxy. *PLoS ONE* **2014**, *9*, e105730. [CrossRef] [PubMed]

22. Kohler, T.A.; Reese, K.M. Long and spatially variable Neolithic demographic transition in the North American southwest. *Proc. Natl. Acad. Sci. USA* **2014**, *111*, 10101–10106. [CrossRef] [PubMed]
23. Zahid, H.J.; Robinson, E.; Kelly, R.L. Agriculture, Population Growth and Statistical Analysis of the Radiocarbon Record. *Proc. Natl. Acad. Sci. USA* **2016**, *113*, 931–935. [CrossRef] [PubMed]
24. Bird, D.; Freeman, J.; Robinson, E.; Maughan, G.; Finley, J.B.; Lambert, P.M.; Kelly, R.L. A first empirical analysis of population stability in North America using radiocarbon records. *Holocene* **2020**, *30*, 1345–1359. [CrossRef]

25. Morrison, K.E.; Rodgers, A.B.; Morgan, C.P.; Bale, T.L. Epigenetic Mechanisms in Pubertal Brain Maturation. *Neuroscience* **2014**, *264*, 17–24. [CrossRef] [PubMed]
26. Selemon, L.D. A Role for Synaptic Plasticity in the Adolescent Development of Executive Function. *Transl. Psychiatry* **2013**, *3*, e238. [CrossRef] [PubMed]
27. Kim, J.Y.; Kang, S.W. Relationships between Dietary Intake and Cognitive Function in Healthy Korean Children and Adolescents. *J. Lifestyle Med.* **2017**, *7*, 10–17. [CrossRef] [PubMed]
28. Julvez, J.; Gignac, F.; Fernández-Barrés, S.; Romaguera, D.; Sala-Vila, A.; Ranzani, O.T.; Persavento, C.; Delgado, A.; Carol, A.; Torrent, J.; et al. Walnuts, Long-Chain Polyunsaturated Fatty Acids, and Adolescent Brain Development: Protocol for the Walnuts Smart Snack Dietary Intervention Trial. *Front. Pediatr.* **2021**, *9*, 593847. [CrossRef]
29. Julvez, J.; Davey Smith, G.; Ring, S.; Grandjean, P. A Birth Cohort Study on the Genetic Modification of the Association of Prenatal Methylmercury With Child Cognitive Development. *Am. J. Epidemiol.* **2019**, *188*, 1784–1793. [CrossRef]
30. Tahaei, H.; Gignac, F.; Pinar, A.; Fernandez-Barrés, S.; Romaguera, D.; Vioque, J.; Santa-Marina, L.; Subiza-Pérez, M.; Llop, S.; Soler-Blasco, R.; et al. Omega-3 Fatty Acid Intake during Pregnancy and Child Neuropsychological Development: A Multi-Centre Population-Based Birth Cohort Study in Spain. *Nutrients* **2022**, *14*, 518. [CrossRef]
31. Sala-Vila, A.; Fleming, J.; Kris-Etherton, P.; Ros, E. Impact of α-Linolenic Acid, the Vegetable ω-3 Fatty Acid, on Cardiovascular Disease and Cognition. *Adv. Nutr.* **2022**, *13*, 1584–1602. [CrossRef]
32. Pinar-Martí, A.; Fernández-Barrés, S.; Gignac, F.; Persavento, C.; Delgado, A.; Romaguera, D.; Lázaro, I.; Ros, E.; López-Vicente, M.; Salas-Salvadó, J.; et al. Red Blood Cell Omega-3 Fatty Acids and Attention Scores in Healthy Adolescents. *Eur. Child Adolesc. Psychiatry*, 2022; *Epub ahead of print*. [CrossRef]
33. Pribis, P.; Bailey, R.N.; Russell, A.A.; Kilsby, M.A.; Hernandez, M.; Craig, W.J.; Grajales, T.; Shavlik, D.J.; Sabatè, J. Effects of Walnut Consumption on Cognitive Performance in Young Adults. *Br. J. Nutr.* **2012**, *107*, 1393–1401. [CrossRef]
34. Feigin, V.L.; Nichols, E.; Alam, T.; Bannick, M.S.; Beghi, E.; Blake, N.; Culpepper, W.J.; Dorsey, E.R.; Elbaz, A.; Ellenbogen, R.G.; et al. Global, Regional, and National Burden of Neurological Disorders, 1990–2016: A Systematic Analysis for the Global Burden of Disease Study 2016. *Lancet Neurol.* **2019**, *18*, 459–480. [CrossRef]
35. Bettio, L.E.B.; Rajendran, L.; Gil-Mohapel, J. The Effects of Aging in the Hippocampus and Cognitive Decline. *Neurosci. Biobehav. Rev.* **2017**, *79*, 66–86. [CrossRef] [PubMed]
36. Folstein, M.F.; Folstein, S.E.; McHugh, P.R. Mini-Mental State. *J. Psychiatr. Res.* **1975**, *12*, 189–198. [CrossRef] [PubMed]
37. Langa, K.M.; Levine, D.A. The Diagnosis and Management of Mild Cognitive Impairment: A Clinical Review. *JAMA J. Am. Med. Assoc.* **2014**, *312*, 2551–2561. [CrossRef] [PubMed]
38. Nurk, E.; Refsum, H.; Drevon, C.A.; Tell, G.S.; Nygaard, H.A.; Engedal, K.; Smith, A.D. Cognitive Performance among the Elderly in Relation to the Intake of Plant Foods. the Hordaland Health Study. *Br. J. Nutr.* **2010**, *104*, 1190–1201. [CrossRef] [PubMed]
39. Nooyens, A.C.J.; Bueno-De-Mesquita, H.B.; van Boxtel, M.P.J.; van Gelder, B.M.; Verhagen, H.; Verschuren, W.M.M. Fruit and Vegetable Intake and Cognitive Decline in Middle-Aged Men and Women: The Doetinchem Cohort Study. *Br. J. Nutr.* **2011**, *106*, 752–761. [CrossRef] [PubMed]
40. Valls-Pedret, C.; Lamuela-Raventós, R.M.; Medina-Remón, A.; Quintana, M.; Corella, D.; Pintó, X.; Martínez-González, M.Á.; Estruch, R.; Ros, E. Polyphenol-Rich Foods in the Mediterranean Diet Are Associated with Better Cognitive Function in Elderly Subjects at High Cardiovascular Risk. *J. Alzheimer's Dis.* **2012**, *29*, 773–782. [CrossRef]
41. Samieri, C.; Grodstein, F.; Rosner, B.A.; Kang, J.H.; Cook, N.R.; Manson, J.E.; Buring, J.E.; Willett, W.C.; Okereke, O.I. Mediterranean Diet and Cognitive Function in Older Age. *Epidemiology* **2013**, *24*, 490–499. [CrossRef]
42. O'Brien, J.; Okereke, O.; Devore, E.; Rosner, B.; Breteler, M.; Grodstein, F. Long-Term Intake of Nuts in Relation to Cognitive Function in Older Women. *J. Nutr. Health Aging* **2014**, *18*, 496–502. [CrossRef]
43. Arab, L.; Ang, A. A Cross Sectional Study of the Association between Walnut Consumption and Cognitive Function among Adult Us Populations Represented in NHANES. *J. Nutr. Health Aging* **2015**, *19*, 284–290. [CrossRef]
44. Dong, L.; Xiao, R.; Cai, C.; Xu, Z.; Wang, S.; Pan, L.; Yuan, L. Diet, Lifestyle and Cognitive Function in Old Chinese Adults. *Arch. Gerontol. Geriatr.* **2016**, *63*, 36–42. [CrossRef]
45. De Amicis, R.; Leone, A.; Foppiani, A.; Osio, D.; Lewandowski, L.; Giustizieri, V.; Cornelio, P.; Cornelio, F.; Fusari Imperatori, S.; Cappa, S.F.; et al. Mediterranean Diet and Cognitive Status in Free-Living Elderly: A Cross-Sectional Study in Northern Italy. *J. Am. Coll. Nutr.* **2018**, *37*, 494–500. [CrossRef] [PubMed]
46. Li, M.; Shi, Z. A Prospective Association of Nut Consumption with Cognitive Function in Chinese Adults Aged 55+ _ China Health and Nutrition Survey. *J. Nutr. Health Aging* **2019**, *23*, 211–216. [CrossRef] [PubMed]
47. Rabassa, M.; Zamora-Ros, R.; Palau-Rodriguez, M.; Tulipani, S.; Miñarro, A.; Bandinelli, S.; Ferrucci, L.; Cherubini, A.; Andres-Lacueva, C. Habitual Nut Exposure, Assessed by Dietary and Multiple Urinary Metabolomic Markers, and Cognitive Decline in Older Adults: The InCHIANTI Study. *Mol. Nutr. Food Res.* **2020**, *64*, e1900532. [CrossRef] [PubMed]
48. Tan, S.Y.; Georgousopoulou, E.N.; Cardoso, B.R.; Daly, R.M.; George, E.S. Associations between Nut Intake, Cognitive Function and Non-alcoholic Fatty Liver Disease (NAFLD) in Older Adults in the United States: NHANES 2011-14. *BMC Geriatr.* **2021**, *21*, 313. [CrossRef]
49. Jiang, Y.W.; Sheng, L.T.; Feng, L.; Pan, A.; Koh, W.P. Consumption of Dietary Nuts in Midlife and Risk of Cognitive Impairment in Late-Life: The Singapore Chinese Health Study. *Age Ageing* **2021**, *50*, 1215–1221. [CrossRef]

50. Bishop, N.J.; Zuniga, K.E. Investigating Walnut Consumption and Cognitive Trajectories in a Representative Sample of Older US Adults. *Public Health Nutr.* **2021**, *24*, 1741–1752. [CrossRef]
51. Chen, X.; Liu, Z.; Sachdev, P.S.; Kochan, N.A.; O'Leary, F.; Brodaty, H. Dietary Patterns and Cognitive Health in Older Adults: Findings from the Sydney Memory and Ageing Study. *J. Nutr. Health Aging* **2021**, *25*, 255–262. [CrossRef]
52. Li, F.; Jiang, W.; Wang, J.; Zhang, T.; Gu, X.; Zhai, Y.; Wu, M.; Xu, L.; Lin, J. Beneficial Effects of Nut Consumption on Cognitive Function Among Elderly: Findings From a 6-Year Cohort Study. *Front. Aging Neurosci.* **2022**, *14*, 816443. [CrossRef]
53. Estruch, R.; Ros, E.; Salas-Salvadó, J.; Covas, M.-I.; Corella, D.; Arós, F.; Gómez-Gracia, E.; Ruiz-Gutiérrez, V.; Fiol, M.; Lapetra, J.; et al. Primary Prevention of Cardiovascular Disease with a Mediterranean Diet Supplemented with Extra-Virgin Olive Oil or Nuts. *N. Engl. J. Med.* **2018**, *378*, e34. [CrossRef]
54. Willett, W. *Nutritional Epidemiology*, 3rd ed.; Chapter 11; Oxford University: New York, NY, USA, 2012; p. 306. ISBN 9780199754038.
55. Martínez-Lapiscina, E.H.; Clavero, P.; Toledo, E.; Estruch, R.; Salas-Salvadó, J.; San Julián, B.; Sanchez-Tainta, A.; Ros, E.; Valls-Pedret, C.; Martinez-Gonzalez, M.Á. Mediterranean Diet Improves Cognition: The PREDIMED-NAVARRA Randomised Trial. *J. Neurol. Neurosurg. Psychiatry* **2013**, *84*, 1318–1325. [CrossRef]
56. Valls-Pedret, C.; Sala-Vila, A.; Serra-Mir, M.; Corella, D.; de La Torre, R.; Martínez-González, M.Á.; Martínez-Lapiscina, E.H.; Fitó, M.; Pérez-Heras, A.; Salas-Salvadó, J.; et al. Mediterranean Diet and Age-Related Cognitive Decline: A Randomized Clinical Trial. *JAMA Intern. Med.* **2015**, *175*, 1094–1103. [CrossRef] [PubMed]
57. Barbour, J.A.; Howe, P.R.C.; Buckley, J.D.; Bryan, J.; Coates, A.M. Cerebrovascular and Cognitive Benefits of High-Oleic Peanut Consumption in Healthy Overweight Middle-Aged Adults. *Nutr. Neurosci.* **2017**, *20*, 555–562. [CrossRef] [PubMed]
58. Dhillon, J.; Tan, S.Y.; Mattes, R.D. Effects of Almond Consumption on the Post-Lunch Dip and Long-Term Cognitive Function in Energy-Restricted Overweight and Obese Adults. *Br. J. Nutr.* **2017**, *117*, 395–402. [CrossRef] [PubMed]
59. Sala-Vila, A.; Valls-Pedret, C.; Rajaram, S.; Coll-Padrós, N.; Cofán, M.; Serra-Mir, M.; Pérez-Heras, A.M.; Roth, I.; Freitas-Simoes, T.M.; Doménech, M.; et al. Effect of a 2-Year Diet Intervention with Walnuts on Cognitive Decline. The Walnuts and Healthy Aging (WAHA) Study: A Randomized Controlled Trial. *Am. J. Clin. Nutr.* **2020**, *111*, 590–600. [CrossRef]
60. Mustra Rakic, J.; Tanprasertsuk, J.; Scott, T.M.; Rasmussen, H.M.; Mohn, E.S.; Chen, C.Y.O.; Johnson, E.J. Effects of Daily Almond Consumption for Six Months on Cognitive Measures in Healthy Middle-Aged to Older Adults: A Randomized Control Trial. *Nutr. Neurosci.* **2022**, *25*, 1466–1476. [CrossRef]
61. Rost, N.S.; Brodtmann, A.; Pase, M.P.; van Veluw, S.J.; Biffi, A.; Duering, M.; Hinman, J.D.; Dichgans, M. Post-Stroke Cognitive Impairment and Dementia. *Circ. Res.* **2022**, *130*, 1252–1271. [CrossRef]
62. Rosenberg, A.; Mangialasche, F.; Ngandu, T.; Solomon, A.; Kivipelto, M. Multidomain Interventions to Prevent Cognitive Impairment, Alzheimer's Disease, and Dementia: From FINGER to World-Wide FINGERS. *J. Prev. Alzheimer's Dis.* **2020**, *7*, 29–36. [CrossRef]
63. Blinkouskaya, Y.; Caçoilo, A.; Gollamudi, T.; Jalalian, S.; Weickenmeier, J. Brain Aging Mechanisms with Mechanical Manifestations. *Mech. Ageing Dev.* **2021**, *200*, 111575. [CrossRef]
64. Harada, C.N.; Natelson Love, M.C.; Triebel, K.L. Normal Cognitive Aging. *Clin. Geriatr. Med.* **2013**, *29*, 737–752. [CrossRef]
65. Hou, Y.; Dan, X.; Babbar, M.; Wei, Y.; Hasselbalch, S.G.; Croteau, D.L.; Bohr, V.A. Ageing as a Risk Factor for Neurodegenerative Disease. *Nat. Rev. Neurol.* **2019**, *15*, 565–581. [CrossRef]
66. NIH National Institute on Aging (NIA). Alzheimer´s Disease and Related Dementias: 11 Myths About Alzheimer's Disease. Available online: https://www.nia.nih.gov/health/11-myths-about-alzheimers-disease (accessed on 11 January 2023).
67. Iqbal, K.; Grundke-Iqbal, I. Alzheimer's Disease, a Multifactorial Disorder Seeking Multitherapies. *Alzheimer's Dement.* **2010**, *6*, 420–424. [CrossRef] [PubMed]
68. Jack, C.R.; Bennett, D.A.; Blennow, K.; Carrillo, M.C.; Dunn, B.; Haeberlein, S.B.; Holtzman, D.M.; Jagust, W.; Jessen, F.; Karlawish, J.; et al. NIA-AA Research Framework: Toward a Biological Definition of Alzheimer's Disease. *Alzheimer's Dement.* **2018**, *14*, 535–562. [CrossRef] [PubMed]
69. World Health Organization (WHO). Implementation Roadmap 2023–2030 for the Global Action Plan for the Prevention and Control of NCDs 2013–2030. Available online: https://www.who.int/teams/noncommunicable-diseases/governance/roadmap (accessed on 14 November 2022).
70. Alzheimer´s Association. Alzheimer´s Association: FY23-FY25 Strategic Plan. Available online: https://www.alz.org/media/Documents/strategic-plan-fy2023-fy2025.pdf (accessed on 14 November 2022).
71. Stefaniak, O.; Dobrzyńska, M.; Drzymała-Czyż, S.; Przysławski, J. Diet in the Prevention of Alzheimer's Disease: Current Knowledge and Future Research Requirements. *Nutrients* **2022**, *14*, 4564. [CrossRef] [PubMed]
72. Gorji, N.; Moeini, R.; Memariani, Z. Almond, Hazelnut and Walnut, Three Nuts for Neuroprotection in Alzheimer's Disease: A Neuropharmacological Review of Their Bioactive Constituents. *Pharmacol. Res.* **2018**, *129*, 115–127. [CrossRef]
73. Yassine, H.N.; Samieri, C.; Livingston, G.; Glass, K.; Wagner, M.; Tangney, C.; Plassman, B.L.; Ikram, M.A.; Voigt, R.M.; Gu, Y.; et al. Nutrition State of Science and Dementia Prevention: Recommendations of the Nutrition for Dementia Prevention Working Group. *Lancet Healthy Longev.* **2022**, *3*, e501–e512. [CrossRef]
74. Hardy, D.; Higgins, G. Alzheimer's Disease: The Amyloid Cascade Hypothesis. *Science* **1992**, *256*, 184–185. [CrossRef] [PubMed]
75. Murphy, M.P.; Levine, H. Alzheimer's Disease and the Amyloid-β Peptide. *J. Alzheimer's Dis.* **2010**, *19*, 311–323. [CrossRef]

76. Shankar, G.M.; Li, S.; Mehta, T.H.; Garcia-Munoz, A.; Shepardson, N.E.; Smith, I.; Brett, F.M.; Farrell, M.A.; Rowan, M.J.; Lemere, C.A.; et al. Amyloid-β Protein Dimers Isolated Directly from Alzheimer's Brains Impair Synaptic Plasticity and Memory. *Nat. Med.* **2008**, *14*, 837–842. [CrossRef]
77. Grothe, M.J.; Barthel, H.; Sepulcre, J.; Dyrba, M.; Sabri, O.; Teipel, S.J. In Vivo Staging of Regional Amyloid Deposition. *Neurology* **2017**, *89*, 2031–2038. [CrossRef]
78. Thal, D.R.; Rüb, U.; Orantes, M.; Braak, H. Phases of Aβ-Deposition in the Human Brain and Its Relevance for the Development of AD. *Neurology* **2002**, *58*, 1791–1800. [CrossRef]
79. Díaz, G.; Lengele, L.; Sourdet, S.; Soriano, G.; de Souto Barreto, P. Nutrients and Amyloid β Status in the Brain: A Narrative Review. *Ageing Res. Rev.* **2022**, *81*, 101728. [CrossRef] [PubMed]
80. Esselun, C.; Dieter, F.; Sus, N.; Frank, J.; Eckert, G.P. Walnut Oil Reduces Aβ Levels and Increases Neurite Length in a Cellular Model of Early Alzheimer Disease. *Nutrients* **2022**, *14*, 1694. [CrossRef]
81. Althobaiti, N.A.; Menaa, F.; Dalzell, J.J.; Albalawi, A.E.; Ismail, H.; Alghuthaymi, M.A.; Aldawsari, R.D.; Iqbal, H.; McAlinney, C.; Green, B.D. Ethnomedicinal Plants with Protective Effects against Beta-Amyloid Peptide (Aβ)1-42 Indicate Therapeutic Potential in a New In Vivo Model of Alzheimer's Disease. *Antioxidants* **2022**, *11*, 1865. [CrossRef]
82. Chauhan, N.; Wang, K.; Wegiel, J.; Malik, M. Walnut Extract Inhibits the Fibrillization of Amyloid Beta-Protein, and Also Defibrillizes Its Preformed Fibrils. *Curr. Alzheimer Res.* **2004**, *1*, 183–188. [CrossRef] [PubMed]
83. Amtul, Z.; Westaway, D.; Cechetto, D.F.; Rozmahel, R.F. Oleic Acid Ameliorates Amyloidosis in Cellular and Mouse Models of Alzheimer's Disease. *Brain Pathol.* **2011**, *21*, 321–329. [CrossRef] [PubMed]
84. Youn, K.; Yun, E.Y.; Lee, J.; Kim, J.Y.; Hwang, J.S.; Jeong, W.S.; Jun, M. Oleic Acid and Linoleic Acid from Tenebrio Molitor Larvae Inhibit BACE1 Activity in Vitro: Molecular Docking Studies. *J. Med. Food* **2014**, *17*, 284–289. [CrossRef]
85. Ali, W.; Ikram, M.; Park, H.Y.; Jo, M.G.; Ullah, R.; Ahmad, S.; bin Abid, N.; Kim, M.O. Oral Administration of Alpha Linoleic Acid Rescues Aβ-Induced Glia-Mediated Neuroinflammation and Cognitive Dysfunction in C57BL/6N Mice. *Cells* **2020**, *9*, 667. [CrossRef]
86. Wang, J.; Wu, F.; Shi, C. Substitution of Membrane Cholesterol with β-Sitosterol Promotes Nonamyloidogenic Cleavage of Endogenous Amyloid Precursor Protein. *Neuroscience* **2013**, *247*, 227–233. [CrossRef]
87. Kim, E.J.; Yang, S.J. Nicotinamide Reduces Amyloid Precursor Protein and Presenilin 1 in Brain Tissues of Amyloid Beta-Tail Vein Injected Mice. *Clin. Nutr. Res.* **2017**, *6*, 130. [CrossRef]
88. Feng, Y.; Yang, S.G.; Du, X.T.; Zhang, X.; Sun, X.X.; Zhao, M.; Sun, G.Y.; Liu, R.T. Ellagic Acid Promotes Aβ42 Fibrillization and Inhibits Aβ42-Induced Neurotoxicity. *Biochem. Biophys. Res. Commun.* **2009**, *390*, 1250–1254. [CrossRef]
89. Zhong, L.; Liu, H.; Zhang, W.; Liu, X.; Jiang, B.; Fei, H.; Sun, Z. Ellagic Acid Ameliorates Learning and Memory Impairment in APP/PS1 Transgenic Mice via Inhibition of β-Amyloid Production and Tau Hyperphosphorylation. *Exp. Ther. Med.* **2018**, *16*, 4951–4958. [CrossRef] [PubMed]
90. Gupta, S.; Dasmahapatra, A.K. Destabilization Potential of Phenolics on Aβ Fibrils: Mechanistic Insights from Molecular Dynamics Simulation. *Phys. Chem. Chem. Phys.* **2020**, *22*, 19643–19658. [CrossRef] [PubMed]
91. Zhang, Z.X.; Li, Y.B.; Zhao, R.P. Epigallocatechin Gallate Attenuates β-Amyloid Generation and Oxidative Stress Involvement of PPARγ in N2a/APP695 Cells. *Neurochem. Res.* **2017**, *42*, 468–480. [CrossRef] [PubMed]
92. Shimmyo, Y.; Kihara, T.; Akaike, A.; Niidome, T.; Sugimoto, H. Multifunction of Myricetin on Aβ: Neuroprotection via a Conformational Change of Aβ and Reduction of Aβ via the Interference of Secretases. *J. Neurosci. Res.* **2008**, *86*, 368–377. [CrossRef]
93. Ono, K.; Yoshiike, Y.; Takashima, A.; Hasegawa, K.; Naiki, H.; Yamada, M. Potent Anti-Amyloidogenic and Fibril-Destabilizing Effects of Polyphenols in Vitro: Implications for the Prevention and Therapeutics of Alzheimer's Disease. *J. Neurochem.* **2003**, *87*, 172–181. [CrossRef]
94. Chang, W.; Huang, D.; Lo, Y.M.; Tee, Q.; Kuo, P.; Wu, J.S.; Huang, W.; Shen, S. Protective Effect of Caffeic Acid against Alzheimer's Disease Pathogenesis via Modulating Cerebral Insulin Signaling, β-Amyloid Accumulation, and Synaptic Plasticity in Hyperinsulinemic Rats. *J. Agric. Food Chem.* **2019**, *67*, 7684–7693. [CrossRef]
95. Milà-Alomà, M.; Suárez-Calvet, M.; Molinuevo, J.L. Latest Advances in Cerebrospinal Fluid and Blood Biomarkers of Alzheimer's Disease. *Ther. Adv. Neurol. Disord.* **2019**, *12*, 1756286419888819. [CrossRef]
96. Bateman, R.J.; Xiong, C.; Benzinger, T.L.S.; Fagan, A.M.; Goate, A.; Fox, N.C.; Marcus, D.S.; Cairns, N.J.; Xie, X.; Blazey, T.M.; et al. Clinical and Biomarker Changes in Dominantly Inherited Alzheimer's Disease. *N. Engl. J. Med.* **2012**, *367*, 795–804. [CrossRef]
97. Blennow, K.; Zetterberg, H. Fluid Biomarker-Based Molecular Phenotyping of Alzheimer's Disease Patients in Research and Clinical Settings. In *Progress in Molecular Biology and Translational Science*; Elsevier: Amsterdam, The Netherlands, 2019; Volume 168, pp. 3–23. ISBN 9780128178744.
98. Lopes Alves, I.; Collij, L.E.; Altomare, D.; Frisoni, G.B.; Saint-Aubert, L.; Payoux, P.; Kivipelto, M.; Jessen, F.; Drzezga, A.; Leeuwis, A.; et al. Quantitative Amyloid PET in Alzheimer's Disease: The AMYPAD Prognostic and Natural History Study. *Alzheimer's Dement.* **2020**, *16*, 750–758. [CrossRef]
99. Barbier, P.; Zejneli, O.; Martinho, M.; Lasorsa, A.; Belle, V.; Smet-Nocca, C.; Tsvetkov, P.O.; Devred, F.; Landrieu, I. Role of Tau as a Microtubule-Associated Protein: Structural and Functional Aspects. *Front. Aging Neurosci.* **2019**, *11*, 204. [CrossRef]

100. Neddens, J.; Temmel, M.; Flunkert, S.; Kerschbaumer, B.; Hoeller, C.; Loeffler, T.; Niederkofler, V.; Daum, G.; Attems, J.; Hutter-Paier, B. Phosphorylation of Different Tau Sites during Progression of Alzheimer's Disease. *Acta Neuropathol. Commun.* **2018**, *6*, 52. [CrossRef] [PubMed]
101. Wang, Y.; Mandelkow, E. Tau in Physiology and Pathology. *Nat. Rev. Neurosci.* **2016**, *17*, 5–21. [CrossRef] [PubMed]
102. Berron, D.; Vogel, J.W.; Insel, P.S.; Pereira, J.B.; Xie, L.; Wisse, L.E.M.; Yushkevich, P.A.; Palmqvist, S.; Mattsson-Carlgren, N.; Stomrud, E.; et al. Early Stages of Tau Pathology and Its Associations with Functional Connectivity, Atrophy and Memory. *Brain* **2021**, *144*, 2771–2783. [CrossRef]
103. Desale, S.E.; Dubey, T.; Chinnathambi, S. α-Linolenic Acid Inhibits Tau Aggregation and Modulates Tau Conformation. *Int. J. Biol/ Macromol.* **2021**, *166*, 687–693. [CrossRef]
104. Desale, S.E.; Chinnathambi, S. α-Linolenic Acid Induces Clearance of Tau Seeds via Actin-Remodeling in Microglia. *Mol. Biomed.* **2021**, *2*, 4. [CrossRef]
105. Desale, S.E.; Chinnathambi, S. α–Linolenic Acid Modulates Phagocytosis and Endosomal Pathways of Extracellular Tau in Microglia. *Cell Adhes. Migr.* **2021**, *15*, 84–100. [CrossRef] [PubMed]
106. Sul, D.; Kim, H.S.; Lee, D.; Joo, S.S.; Hwang, K.W.; Park, S.Y. Protective Effect of Caffeic Acid against Beta-Amyloid-Induced Neurotoxicity by the Inhibition of Calcium Influx and Tau Phosphorylation. *Life Sci.* **2009**, *84*, 257–262. [CrossRef]
107. Giraldo, E.; Lloret, A.; Fuchsberger, T.; Viña, J. Aβ and Tau Toxicities in Alzheimer's Are Linked via Oxidative Stress-Induced P38 Activation: Protective Role of Vitamin E. *Redox Biol.* **2014**, *2*, 873–877. [CrossRef]
108. Ossenkoppele, R.; van der Kant, R.; Hansson, O. Tau Biomarkers in Alzheimer's Disease: Towards Implementation in Clinical Practice and Trials. *Lancet Neurol.* **2022**, *21*, 726–734. [CrossRef]
109. Theodore, L.E.; Kellow, N.J.; Mcneil, E.A.; Close, E.O.; Coad, E.G.; Cardoso, B.R. Nut Consumption for Cognitive Performance: A Systematic Review. *Adv. Nutr.* **2021**, *12*, 777–792. [CrossRef]
110. Food and Drug Administration (FDA). *Qualified Health Claims: Letter of Enforcement Discretion–Nuts and Coronary Heart Disease*; Docket No 02P-0505; FDA: Washington, DC, USA, 2003. Available online: Http://Wayback.Archive-It.Org/7993/20171114183724/Https://Www.Fda.Gov/Food/IngredientsPackagingLabeling/LabelingNutrition/Ucm072926.Htm (accessed on 11 February 2023).
111. Food and Drug Administration (FDA). *Qualified Health Claims: Letter of Enforcement Discretion Walnuts and Coronary Heart Disease*; (Docket No. 02P-0292); FDA: Washington, DC, USA, 2004. Available online: http://Wayback.Archive-It.Org/7993/20171114183725/Https://Www.Fda.Gov/Food/IngredientsPackagingLabeling/LabelingNutrition/Ucm072910.Htm (accessed on 11 February 2023).
112. Food and Drug Administration (FDA). *Qualified Health Claims: Letter of Enforcement Discretion–Macadamia Nuts and Reduced Risk of Coronary Heart Disease*; FDA: Washington, DC, USA, 2017. Available online: https://Www.Fda.Gov/Media/106201/Download (accessed on 11 February 2023).

nutrients

MDPI

Review

Dried Fruits, Nuts, and Cancer Risk and Survival: A Review of the Evidence and Future Research Directions

Bradley W. Bolling [1,*], Dagfinn Aune [2,3,4], Hwayoung Noh [5,6], Kristina S. Petersen [7] and Heinz Freisling [6,*]

1 Department of Food Science, University of Wisconsin-Madison, 1605 Linden Dr., Madison, WI 53706, USA
2 Department of Epidemiology and Biostatistics, School of Public Health, Imperial College London, London W2 1PG, UK
3 Department of Nutrition, Oslo New University College, Lovisenberggata 13, 0456 Oslo, Norway
4 Department of Endocrinology, Morbid Obesity and Preventive Medicine, Oslo University Hospital Ullevål, 0424 Oslo, Norway
5 Department of Cancer Prevention and Environment, INSERM U1296, Léon Bérard Cancer Center, 28 Rue Laennec, 69008 Lyon, France
6 Nutrition and Metabolism Branch, International Agency for Research on Cancer (IARC-WHO), 25 Avenue Tony Garnier, CS 90627, CEDEX 07, 69366 Lyon, France
7 Department of Nutritional Sciences, Texas Tech University, 508 Human Sciences Building, Lubbock, TX 79409, USA
* Correspondence: bwbolling@wisc.edu (B.W.B.); freislingh@iarc.who.int (H.F.); Tel.: +1-608-890-0212 (B.W.B.); +33-698557239 (H.F.)

Abstract: Dried fruits and nuts contain high amounts of nutrients and phytochemicals—all of which may have anticarcinogenic, anti-inflammatory, and antioxidant properties. This narrative review summarizes the evidence for dried fruits and nuts and cancer incidence, mortality, and survival and their potential anticancer properties. The evidence for dried fruits in cancer outcomes is limited, but existing studies have suggested an inverse relationship between total dried fruit consumption and cancer risk. A higher consumption of nuts has been associated with a reduced risk of several site-specific cancers in prospective cohort studies, including cancers of the colon, lung, and pancreas, with relative risks per 5 g/day increment equal to 0.75 (95% CI 0.60, 0.94), 0.97 (95% CI 0.95, 0.98), and 0.94 (95% CI 0.89, 0.99), respectively. A daily intake of total nuts of 28 g/day has also been associated with a 21% reduction in the rate of cancer mortality. There is also some evidence that frequent nut consumption is associated with improved survival outcomes among patients with colorectal, breast, and prostate cancer; however, further studies are needed. Future research directions include the investigation of additional cancer types, including rare types of cancer. For cancer prognosis, additional studies with pre- and postdiagnosis dietary assessment are warranted.

Keywords: dried fruits; tree nuts; peanuts; cancer; cancer survivors; mortality; review

Citation: Bolling, B.W.; Aune, D.; Noh, H.; Petersen, K.S.; Freisling, H. Dried Fruits, Nuts, and Cancer Risk and Survival: A Review of the Evidence and Future Research Directions. *Nutrients* **2023**, *15*, 1443. https://doi.org/10.3390/nu15061443

Academic Editor: Peter Pribis

Received: 31 January 2023
Revised: 9 March 2023
Accepted: 11 March 2023
Published: 16 March 2023

1. Introduction

Globally, an estimated 19.3 million new cancer cases and almost 10.0 million cancer deaths occurred in 2020 [1]. In the next two decades, the cancer burden is projected to increase by 47% to 28.4 million new cancer cases in 2040 [1]. This estimate is solely based on the growth and aging of the population and may be further exacerbated by an increasing prevalence of risk factors in many parts of the world [1].

Factors such as consuming a healthful diet, being physically active, avoiding tobacco use, and maintaining a healthful weight can have a strong influence on cancer prevention [2]. It has been estimated that approximately 50% of cancer cases can be prevented [3]. However, international cancer statistics continue to show that up to 80% of the cancer burden in high-income countries could be preventable in principle [4]. Therefore, there is considerable interest in studying the impact of lifestyle changes, in particular, the impact of changes

in diet, on cancer development and progression [3]. There is compelling evidence that nutrition has substantial effects on the incidence and progression of cancer [5].

This narrative review summarizes the evidence for dried fruits and nuts and cancer incidence, mortality, and survival and their potential anticancer properties. In addition, the gaps that exist in the literature and recommendations for future research are discussed. The steps of study selection are described in Supplementary Figure S1.

2. Dietary Strategies to Prevent Cancer

Few countries have optimal diets for cancer prevention [6]. For example, analysis of diets by the Alternate Healthy Eating Index reveals suboptimal global levels due to either a relatively high consumption of red meat, added sugars, and transfat or a relatively low consumption of fruits, vegetables, nuts, and whole grains [7]. Suboptimal diets are estimated to cause 1.6 million preventable cancer deaths annually [7].

Nearly every authoritative health body recommends increased fruit intake for preventing cancer and chronic disease risk. The International Agency for Research on Cancer/World Health Organization (IARC/WHO) Report estimated that ~5% of cancer deaths in the U.S. are due to low fruit and vegetable consumption [6]. The U.S. Dietary Guidelines for Americans recommends 2 cups of fruit equivalents/day per 2000 kcal where $\frac{1}{2}$ cup of dried fruit is 1 equivalent [8]. The U.S. Food and Drug Administration (FDA) also allows for a fruit and cancer claim provided the fruit meets the nutrient content requirements for a "good source" of at least vitamin A, vitamin C, or dietary fiber [9]. The model wording of the claim reads, "*Development of cancer depends on many factors. Eating a diet low in fat and high in fruits and vegetables, foods that are low in fat and may contain vitamin A, vitamin C, and dietary fiber, may reduce your risk of some cancers. Oranges, a food low in fat, are a good source of fiber and vitamin C*" [9]. The IARC/WHO European Code Against Cancer recommends the consumption of whole grains, pulses, vegetables, and fruits for cancer prevention [10].

Nuts, encompassing tree nuts and compositionally related peanuts, are also widely recommended for consumption as nutrient-dense foods [6,8,11]. The World Cancer Report associates dietary patterns containing high intakes of fruits and nuts with the reduction of colorectal cancer risk [6]. Several FDA-qualified claims about diet and cancer risk are relevant to the components found in dried fruit and nuts. Furthermore, other agencies have highlighted the importance of dietary bioactives in cancer prevention. The American Cancer Society guidance for consumers states, "Fruit and non-starchy vegetables contain a large number of potential anti-tumorigenic agents, such as dietary fiber, carotenoids" [11,12]. The World Cancer Research Fund International has published, "Fruit and non-starchy vegetables contain a large number of potential anti-tumorigenic agents", among dietary guidance for cancer prevention [13].

3. Dried Fruits and Cancer

Many types of dried fruits are consumed worldwide. Globally, the most common dried fruits are raisins, dates, prunes, apricots, and figs [14]. However, other specialty dried fruits are produced, including sweetened-dried cranberry and high-value, freeze-dried fruits and powders. Globally, the Middle East and Europe account for half of dried fruit consumption [14]. In the U.S., dried fruit contributes <4% of total fruit intake, whereas juice and fresh or other nondried forms are more commonly consumed [15]. For this review, we mainly focus on cancer prevention through the most common dried fruits.

3.1. Preclinical Studies Relating Dried Fruits and Cancer

Cancer develops through many mechanisms. Preclinical studies have revealed that dried fruit may prevent cancer at the stage of initiation through the induction of detoxification enzymes and reducing the impact of carcinogens and environmental stress; during promotion by inhibiting oxidative stress and inflammation; and at the stage of progression by inducing apoptosis [16]. Mechanistic data from in vitro studies are available for raisins [17,18], apricots [19,20], figs [21,22], prunes [23,24], and dates [25,26] and in vivo for

figs [27], dates [28], prunes [29], and apricots [30]. As reviewed elsewhere, these studies cover gastric, colon, breast, liver, bone, prostate, renal, and testicular cancers [16]. Furthermore, fruit bioactives have a direct impact on cancer-related factors, including antioxidant and anti-inflammatory activity and improved gut immune function in preclinical studies [16,31–33]. These functionalities may reduce cancer by inhibiting the risk of cancer initiation and progression and by improving survival.

Dried fruits contain fiber, micronutrients, and bioactives that may contribute to cancer chemoprevention, as summarized in a companion article [34]. Dried fruits contain a diverse number of bioactives, including phenolics, carotenoids, and terpenoids [16]. The drying process itself impacts the profile, abundance, and possibly the bioaccessibility of bioactives in dried fruits. For example, freeze-drying may retain bioactives more so than forced-air or sun-drying processes [35]. Heat can increase the formation of melanoidins, especially during the production of raisins [36]. Fruit melanoidins are poorly described but function as antioxidants and accumulate polyphenols into complex polymers. Thermal processing of dried fruits (and nuts) also increases the content of advanced glycation endproducts (AGEs) as compared to unprocessed fruits and nuts [37]. AGEs may play a role in carcinogenesis [38]. However, a large, multinational, prospective cohort study across 20 anatomical cancer sites reported that a higher intake of dietary AGEs was not associated with an increased risk of overall cancer and most cancer types studied. A nonlinear, weak positive association was observed between higher AGE intake and the risk of prostate cancer [38]. The addition of sulfites during processing inhibits melanoidin formation [39] and therefore may alter the profile of bioactives when used in the fruit drying process.

Drying may also impact the structure and accessibility of soluble fibers and bioactives from fruit. In vitro digestion of different dried fruits may increase or decrease the release of antioxidants [40]. In mice, the consumption of nonextractable phenolics from dried berries increases colonic polyphenol content more so than the consumption of freely extractable polyphenols [41]. Therefore, drying may also impact the accessibility of fiber-associated phenolics and other bioactives that could subsequently affect the gut microbiota and intestinal immune system. Further studies are needed to link dried fruit processing parameters to cancer prevention activities.

3.2. Human Intervention Studies of Dried Fruits and Cancer

The evidence from human intervention studies for cancer prevention by dried fruit consumption is very limited. Thus far, human intervention studies with dried fruits have focused on cardiometabolic health or other chronic diseases [42–44]. Some evidence is available for the function of high-value, freeze-dried berry products in healthy individuals and those with colorectal, oral, and prostate cancers [45–50]. However, these forms of dried fruit are not commonly consumed and have a profile of bioactives similar to fresh fruits, which differs from more commonly available and consumed dried fruits produced via processes that utilize heat during drying.

3.3. Epidemiological Studies of Dried Fruits and Cancer

3.3.1. Cancer Incidence (or Risk)

Few epidemiological studies have directly assessed the relationship between dried fruits and cancer risk. A systematic review of observational studies published through 2018 found insufficient studies to perform a meta-analysis [51]. The review identified 16 studies with 12,732 cases from 437,298 participants that assessed cancer risk and the intake of total dried fruits or specifically raisins, prunes, dates, or figs [51]. Among the prospective studies, there was an inverse relation between total dried fruit consumption and cancer incidence in seven studies with a significant, dose-response trend identified in three studies on pancreatic cancer, prostate cancer, and colorectal polyps [51]. Among the case–control studies, there were inverse associations of cancer incidence with total dried fruits, raisins, or dates, with five of seven studies reporting significant associations [51]. Among these studies, dried fruit intake was associated with reduced stomach cancer,

pancreatic cancer, and colorectal cancer incidence (for dates but not other dried fruits), nasopharyngeal cancer incidence (raisins but not dried figs, which increased risk), and bladder cancer incidence [51]. Where dried fruit intakes were compared with fresh fruit intakes, dried fruit was more protective than fresh fruit in five of seven prospective studies and three of four case–control studies [51].

Prospective cohort studies conducted in the Netherlands have shown no association between total dried fruit intake and urothelial [52], stomach [53], and prostate [54] cancer incidence. In contrast, consumption of prunes was associated with increased colorectal cancer risk in the Nurses' Health Study and Health Professionals' Follow-up Study (RR (95% CI) for women: 1.46 (0.93, 2.31) and for men: 1.73 (1.20, 2.50)) [55]. Several prospective analyses conducted in cohorts of Californian Seventh Day Adventists have shown that the frequency of dried fruit intake is associated with a lower risk of cancer [56–58]. In the Adventists Health Study 1, intake of raisins, dates, and other dried fruits ≥3 times/week was associated with a 65% reduction (relative risk (RR) (95% CI): 0.35 (0.17, 0.73)) in the relative risk of fatal pancreatic cancer compared to intake less than once per month [56]. In an analysis of the Adventists Health Study 1 and 2, dried fruit intake ≥3 times/week was associated with 24% lower odds of rectal/colon polyps (odds ratio (OR) (95% CI): 0.76 (0.58, 0.99)) compared to intake less than once per week [57]. Finally, intake of raisins, dates, or other dried fruits ≥5 times/week was associated with a lower relative risk of prostate cancer after adjustment for age ((RR (95% CI): 0.51 (0.31, 0.85)) compared to intake <1 time/week; however, the relationship was attenuated after further adjustment for education and other dietary factors (RR (95% CI): 0.62 (0.36, 1.06)) [58]. In an analysis of Adventist Health Study 1, no association was observed between dried fruit intake and lung cancer [59].

Recent epidemiological studies in other populations have investigated dried fruit intake and cancer risk (Table 1).

Table 1. Epidemiological studies assessing the association between dried fruit intake and cancer risk published since 2019.

Study Type	Participants	Cancer Type	Outcome (95% CI)	Reference
Systematic review	*n* = 437,298 from 16 studies	Pancreatic, prostate, colorectal polyps	Dose-response trend from prospective studies	Mossine et al., 2020 [51]
		Stomach, pancreatic, colorectal, nasopharyngeal, bladder	Total dried fruit, raisins, or dates reduced incidence from case–control studies	
Cohort	UK Women's Cohort Study (*n* = 35,372 women aged 35–69 in England, Wales, and Scotland)	Breast	HR 1.04 (0.98,1.13)	Dunneram et al., 2019 [60]
		Endometrial	HR 0.60 (0.37, 0.97)	
		Ovarian	HR 1.06 (0.89, 1.26)	
Prospective cohort	National Institutes of Health-American Association of Retired Persons Diet and Health Study (*n* = 485,403 men and women aged 50–71 at baseline in the United States)	Liver	HR (Q5 vs. Q1) 0.73 (0.60, 0.89)	Zhao et al., 2022 [61]

Table 1. *Cont.*

Study Type	Participants	Cancer Type	Outcome (95% CI)	Reference
Mendelian randomization	UK Biobank (*n* ~500,000 men and women aged 49–69 in the United Kingdom)	Oral cavity/pharyngeal	IVW OR 0.17 (0.04, 0.69)	Jin et al., 2022 [62]
		Lung	IVW OR 0.33 (0.17, 0.64)	
		Squamous cell lung	IVW OR 0.23 (0.09, 0.60)	
		Breast	IVW OR 0.47 (0.32, 0.68)	
		Pancreatic	IVW OR 0.03 (0.001, 0.68)	
		Cervical	IVW OR 0.99 (0.9897, 0.9998)	
		Lung adenocarcinoma, endometrial, thyroid, prostate, bladder, brain	IVW OR not significant	

Abbreviations: HR, Hazard ratio; IVW, Inverse variance weighted; OR, Odds ratio.

The risk of breast, endometrial, and ovarian cancer was assessed in the prospective UK Women's Cohort study of $n = 35{,}372$ women [60]. After 18 y of follow-up in participants aged 35–69, total dried fruit intake was associated with reduced risks of endometrial cancer (hazards ratio (HR) (95% CI): 0.60 (0.37, 0.97)) and postmenopausal endometrial cancer (HR (95% CI) 0.55: (0.31, 0.98)). Total dried fruit intake was not related to breast or ovarian cancer risks; total fruit intake was not associated with breast, endometrial, or ovarian cancer risks.

Dried fruit intake and cancer risk were evaluated in the National Institutes of Health-American Association of Retired Persons Diet and Health study, which included 485,403 participants who were 50–71 years old [61]. In this study, dried fruit intake was associated with lower RR (95% CI) of both liver cancer (0.73 (0.60, 0.89)) and chronic liver disease mortality (0.59 (0.48, 0.73)) [61].

A two-sample Mendelian randomization study utilized ~500,000 samples from the UK Biobank database to investigate dried fruit intake and cancer risk [62]. In this data set, for one standard deviation increase in genetically predicted dried fruit intake, a reduced risk was observed for oral cavity/pharyngeal cancer, lung cancer, squamous cell lung cancer, breast cancer, pancreatic cancer, and cervical cancer [62].

Case–control analyses conducted in Australia, Jordan, Spain, and Turkey showed relationships between dried fruit intake and cancer incidence. An Australian case–control analysis reported that individuals with pancreatic cancer (cases) had a significantly lower intake of raisins than controls [63]. In a case–control study conducted in Jordan, a daily intake of dates (OR (95% CI): 0.52 (0.27, 0.98)) was associated with lower odds of colorectal cancer; total dried fruit intake was not associated with colorectal cancer risk [64]. In a case–control analysis conducted in Spain, a higher intake of dried fruits was associated with lower odds of gastric cancer (OR (95% CI): 0.40 (0.20, 0.80)) [65]. Likewise, in a Turkish case–control study, a higher intake of dried fruit was associated with a lower risk of gastric cancer [66].

3.3.2. Cancer Mortality and Survival

Increased fruit intake is associated with reduced cancer mortality. A meta-analysis of 26 cohort studies reported that an intake of five servings of fruits and vegetables relative to the reference level of two servings had a hazard ratio (HR) (95% CI) of 0.90 (0.86, 0.95) for cancer mortality [67]. Few studies have directly analyzed dried fruits and cancer mortality.

A meta-analysis published in 2017 identified only two studies on dried fruit and total cancer mortality risk and found a nonsignificant association (RR (95% CI): 0.89 (0.61, 1.30)) [68].

3.4. Research Gaps, Needs, and Priorities Related to Dried Fruits and Cancer

Although fruits are currently recommended for cancer prevention, more research is needed to understand the specific contributions and mechanisms of dried fruit intake and cancer prevention. Additional epidemiological studies are needed to assess dried fruit intake and additional cancer types. Dried fruit is often not treated as a separate subcategory of total fruit intake when assessing dietary intake, which hinders the ability to assess cancer risk and dried fruit intake in epidemiological studies [51]. Sufficient dried fruit intake in a study population should be considered. Although the intake on a g/day basis may be low, higher frequency intake of 3+ to 5+ times/week, such as in the AHS [57], may be considered reasonable intake frequencies to detect inverse associations between dried fruit consumption and cancer risk. At lower frequencies of intake, e.g., less than weekly, associations may be more difficult to detect. Furthermore, dried fruit intake is also difficult to describe since the composition and methods used to process dried fruit vary. Therefore, new assessment tools or biomarkers are needed to accurately assess intake.

Additional preclinical and mechanistic studies are needed. However, greater attention to the design of nutritionally relevant studies is needed considering that many fruit bioactives are metabolized before reaching tissues and organs [69]. Identifying specific mechanisms by which dried fruit may impact cancer risk can inform the design of mechanistic intervention studies. Relating specific bioactives in dried fruits to preventive mechanisms is desirable. For example, the World Cancer Report links the intake of fruit carotenoids with a reduced risk of estrogen receptor-negative breast tumors [6,70]. Additional human studies are needed to understand how other bioactives in dried fruits affect cancer risk and survival. Assessing dried fruit bioactives and their metabolites in tissues and plasma in human studies can lead to new mechanistic insights and dietary recommendations. Lastly, future human intervention studies are needed to clarify the role of dried fruit in primary prevention, secondary prevention, or improving the quality of life in cancer patients.

4. Tree Nuts, Peanuts, and Cancer

A holistic view of the evidence shows that most diets that are protective against cancer are rich in foods of plant origin [2]. Relatively unprocessed foods of plant origin are rich in nutrients and dietary fiber. Higher consumption of these foods instead of processed foods and sugars could protect against weight gain, overweightness, and obesity [71] and therefore protect against obesity-related cancers, such as postmenopausal breast, colorectum, liver, thyroid, pancreas, endometrium, and kidney cancer [2]. A higher consumption of nuts has been associated with less weight gain in adults as summarized in a systematic review and meta-analysis of prospective cohorts and randomized controlled trials [72]. Nuts, including tree nuts (almonds, hazelnuts, and walnuts) and peanuts, contain high amounts of nutrients such as unsaturated fats, protein, vitamins (e.g., α-tocopherol, folate, and niacin), nonsodium minerals (e.g., magnesium, calcium, and potassium), and phytochemicals—all of which may have anticarcinogenic, anti-inflammatory, and antioxidant properties. Nuts, and walnuts, in particular, have also been shown to modulate the microbiota by increasing gut microbial diversity [73], and new mechanistic hypotheses on diet and cancer relationships include interactions between host and environmental factors in selecting the microbiota that in turn influence carcinogenesis [74].

4.1. Preclinical Studies Related to Nuts and Cancer

In vitro and in vivo studies to determine whether nuts can help combat cancer are instrumental to understanding potential mechanisms and whether results are coherent with studies in humans.

Preclinical studies for breast cancer have demonstrated reduced growth and multiplicity of breast cancer tumors in relation to walnuts or the main compounds characteristic of

walnuts, such as melatonin [75,76]. Potential mechanisms for cancer prevention include the suppression of proliferation, alterations in cell signaling pathways involved in cell differentiation and apoptosis, and selective inhibition of some cyclooxygenase and lipoxygenase activities [75–77].

Preclinical studies for colon cancer suggest that consumption of mixed nuts, walnuts, and almonds inhibits DNA damage and tumor growth through the suppression of angiogenesis, proliferation, and inflammation, as well as increased apoptosis and favorable alterations to the gut bacteria and enterotype-like clusters [78–83]. In line with the latter, a cross-sectional study among 222 Koreans showed that a healthy dietary pattern characterized by higher intakes of nuts/seeds was related to higher α-diversity reflecting gut microbial health [84].

In vivo and in vitro models for prostate cancer suggest that diets containing walnuts and almonds may reduce the risk of prostate cancer through declines in plasma levels of IGF-1, resistin, LDL-cholesterol, oxidative stress, and inflammation, and increased expression of tumor suppressors [85–87].

Bioactives in nuts with anticancerogenic potential that have been studied in preclinical studies include lipid-associated components [82], such as ellagic acid, which is a dietary flavonoid polyphenol abundant in walnuts and pecans [80], and melatonin found in walnuts along with polyunsaturated fatty acids, which are abundant in nuts in general [76]. Another bioactive nutrient in nuts is selenium, for which Brazil nuts are one of the richest known food sources [88]. In a 2023 preclinical study in mice, selenium-rich Brazil nuts and selenomethionine dietary supplementation reduced mammary tumor growth [89]. Another aspect to highlight is that differences in the composition of bioactives across various types of nuts likely translate into different potential anticancer properties.

These potential anticancer properties of nuts based on preclinical studies are summarized in Table 2.

Table 2. Potential anticancer properties of nuts based on preclinical studies.

Author Year	Cancer Model	Putative Mechanism of Nuts Dietary Factor	Dietary Factor
Breast Cancer-Related Studies			
Hardman and Ion 2008 [90]	Human breast cancer tumors in nude mice	Suppression of cell proliferation or suppression of metastasis	18% of dietary calories from walnuts
Hardman et al., 2011 [75]	C(3)1 TAg transgenic mice, breast cancer	Alterations in cell signaling related to proliferation, differentiation, and apoptosis	Walnuts in the diet
Garcia et al., 2015 [76]	Implanted mammary gland adenocarcinoma in BALB/c mouse model	Inhibition of cyclooxygenase and lipoxygenase	6% walnut oil or 6% walnut flour containing phytomelatonin
Chen et al., 2015 [77]	Breast cancer cells	Growth inhibition of breast cancer cells through cell cycle arrest and inhibition of proliferation	Ellagic acid that is abundant in walnuts
Colorectal Cancer-Related Studies			
Hong et al., 2022 [78]	Colonic cell proliferation, apoptosis, and gene expression in rat model	Reduced DNA damage possibly via downregulation of RelA inflammation gene expression without changes to colonic cell proliferation and apoptosis	Mixed nuts in the diet

4.3.2. Cancer Mortality and Survival

In an umbrella review of epidemiological evidence that included 49,161 cancer deaths, a daily intake of 28 g/d nuts was inversely associated with cancer mortality with an RR of 0.89 (95% CI: 0.83–0.94) [104]. The proportion of variability due to between-study heterogeneity was low (I^2: 23%), and the strength of the evidence was rated as moderate (AMSTAR-2) [104]. The authors reported slightly stronger inverse associations with cancer mortality for tree nuts than for peanuts [104]. These findings are in line with previous systematic reviews and meta-analyses of prospective studies [100,105–108]. For example, Chen et al. [105] estimated a summary RRs for high (5+ servings/week) compared with low nut (never/almost never) consumption of 0.87 (95% CI: 0.80–0.93) for cancer mortality (11 studies 21,353 deaths), while Aune et al. reported a summary estimate of 0.83 (95% CI: 0.75–0.92) per 28 g/d for cancer mortality [106]. When updating the latter analysis by adding results from 6 additional cohort studies that have since been published [109–114] (Supplementary Table S1), the summary estimate was 0.87 (95% CI: 0.81–0.94, I^2 = 60%, n = 13) per 28 g/d (Figure 1A) based on data from 13 cohort studies (12 publications), and there was evidence of nonlinearity ($P_{nonlinearity}$ = 0.005) with most of the reductions in risk observed with an intake of 15–20 g/day (Figure 1B).

There is also evidence that among patients with cancer, frequent consumption of nuts may be linked to lower mortality from all causes. In the Health Professionals Follow-up Study, frequent nut intake (5 or more times per week; serving size: 28 g) among patients with prostate cancer was associated with a lower risk of dying from prostate cancer and from all causes by more than 30 percent compared to the men who ate nuts once or less a month [115].

In a prospective study among 3449 long-term breast cancer survivors with 374 deaths, including 252 breast cancer deaths, total nut intake $\geq$ 17 g/week compared to nonconsumption was inversely associated with overall survival (OS) (HR (95% CI): 0.74 (0.52, 1.05)) and disease-free survival (DFS) (0.48 (0.31, 0.73)), and these associations did not vary by nut type [116].

Among 826 patients with stage III colon cancer in a prospective, observational study, compared to nonconsumers, patients who consumed $\geq$ 2 servings of nuts (1 oz per serving) per week had an HR (95% CI) of 0.43 (0.25, 0.74) for OS, of 0.58 (0.37, 0.92) for DFS, and 0.70 (0.42, 1.16) for recurrence-free survival (RFS) compared to nonconsumers [117]. When cumulative averages of nut consumption before and after diagnosis were used for the statistical analysis, the corresponding HRs (95% CIs) were 0.43, (0.30–0.61), 0.45 (0.33–0.62), and 0.46 (0.32–0.64), respectively, suggesting the importance of repeated dietary assessments [110]. Subgroup analyses showed that the beneficial effects of nut intake were particularly attributable to tree nut intake [117]. Similarly, in a prospective, observational study among 1404 long-term colorectal cancer survivors, compared to the lowest nut intake (1st quintile), the highest nut intake (5th quintile) postdiagnosis was inversely associated with OS (HR (95% CI): 0.48 (0.31, 0.75) [118].

4.4. Research Gaps, Needs, and Priorities Related to the Study of Nuts and Cancer

Considering the current evidence of the relationship of nut consumption with cancer risk and mortality where nut consumption is often only assessed at baseline, and potential changes in consumption patterns over time are missed, more research with improved exposure assessment is warranted. This should include repeated assessment of the intake of specific types of nuts and improved quantification. Better biomarkers of nut consumption would also help to further investigate the promising observations of the putative chemopreventive effect of nuts in cancer development, including secondary and primary cancer outcomes. Furthermore, there is a need to investigate additional cancer types, including thyroid, kidney, or head and neck, and outcome-wide analyses across cancer sites could provide answers relatively quickly. The multifaceted nature of cancer risk and the variety of cancer chemopreventive mechanisms suggest that nuts may have different effects among

cancer subtypes. To address the heterogeneity across studies and investigate rare types of cancer, a pooling of prospective cohorts could be a way forward.

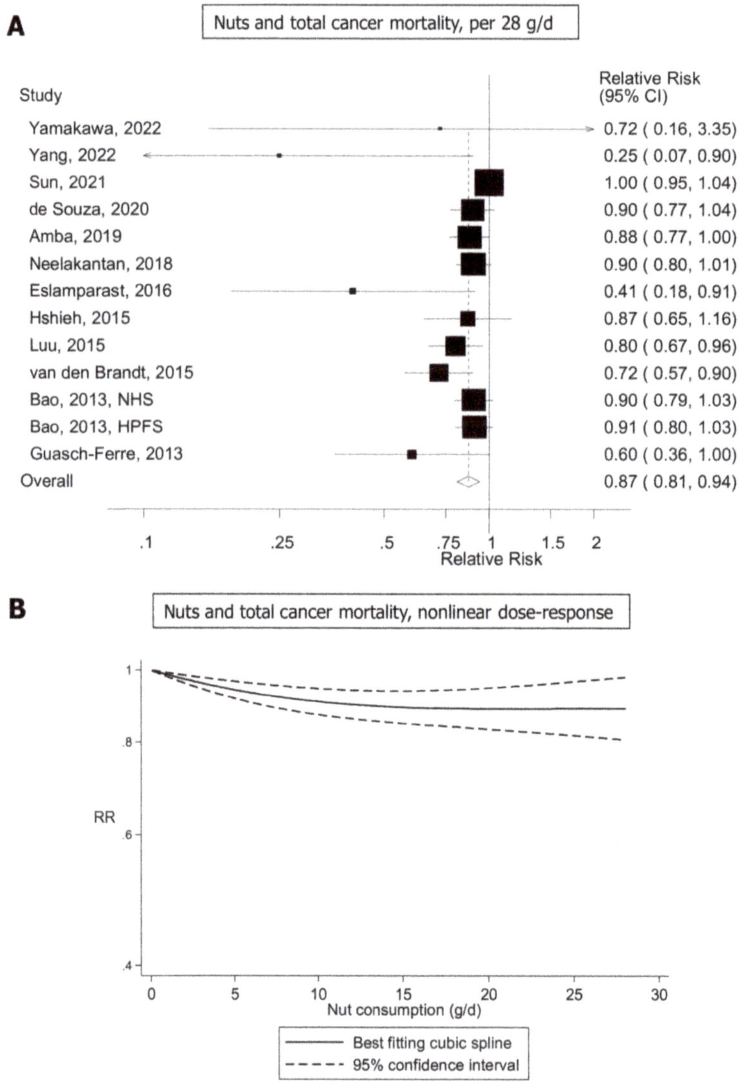

Figure 1. Nuts and total cancer mortality, linear (**A**), and nonlinear (**B**) dose-response analysis (updated analysis based on Aune et al.) [106,110–114,119–124].

For cancer prognosis, additional studies with pre- and postdiagnosis dietary assessment are warranted. It is likely that distinguishing the timing of nut consumption could provide a greater understanding of how nuts may modify risks during different stages of cancer development.

Few studies have addressed the impact of nuts on the co-occurrence of cancer and other long-term chronic diseases (e.g., diabetes) in an individual. For example, in a multinational cohort study, we showed that greater adherence to the Mediterranean diet was inversely associated with cardiometabolic multimorbidity after a first primary cancer [125].

5. Summary and Recommendations

The degree of evidence for dried fruit consumption and cancer prevention is more limited than what has been established for nuts. Preliminary studies have been promising and have begun to establish mechanistic links between bioactives and chemopreventive pathways. Commonly consumed dried fruits have different profiles of bioactives than their fresh counterparts. Several studies suggest that dried fruit intake is associated with a greater reduction of risk than that of fresh fruit. As with nuts, improved methods to track intake and document consumption across multiple types of dried fruit are needed to improve epidemiological studies for diet and cancer prevention.

Evidence from multiple lines of research encompassing cell line studies, animal models, observational studies, interventional studies, and meta-analyses is suggestive that a higher consumption of nuts is inversely associated with the risk of certain cancers and of dying from cancer. Among the 12 cancer sites investigated in the literature (i.e., esophagus, stomach, colorectum, liver, pancreas, lung, breast, ovary, endometrium, leukemia, prostate, and lymphomas), inverse associations were most consistent across studies for the incidence of colorectal cancer and, more specifically, with colon cancer. Studies differentiating between tree nuts and peanuts tend to report stronger inverse associations between tree nuts and cancer incidence and cancer mortality than for peanuts. There is limited, inconclusive evidence for an inverse association with the incidence of cancers of the pancreas, stomach, and lungs, while the most recent studies on breast cancer are null. Evidence is largely missing for other types of cancer; however, there is consistent evidence that nut consumption is associated with reduced total cancer mortality. Potential mechanisms include suppression of angiogenesis, proliferation, and inflammation, as well as increased apoptosis and favorable induced changes in gut bacteria. Indirectly, a higher consumption of nuts may be linked to a reduced risk of certain cancers or dying from cancer through reduced weight gain during adult life. Nuts in the diet may also have a role in tertiary prevention in cancer survivors where a higher nut intake was consistently related to better survival of cancers of the colorectum, breast, and prostate. Considering the observation that nut consumption appears to be more consistently associated with reduced total cancer mortality than total cancer incidence, it is also possible that an association with cancer is driven more so by improvements in survival after a cancer diagnosis than reductions in cancer incidence, but further studies are needed to clarify this, as the number of studies on total cancer incidence is limited.

Supplementary Materials: The following supporting information can be downloaded at: https://www.mdpi.com/article/10.3390/nu15061443/s1, Figure S1: PRISMA Flow Diagram of Study Selection; Table S1: Prospective cohort studies on total nut consumption and total cancer mortality. References [126,127] are cited in the supplementary materials.

Author Contributions: Writing—original draft preparation, B.W.B. and H.F.; writing—review and editing, B.W.B., D.A., H.N., K.S.P. and H.F. All authors have read and agreed to the published version of the manuscript.

Funding: The NUTS 2022 conference was planned by Rovira i Virgili University with all the expenses of the organization covered by Barcelo Congresos. Barcelo Congresos SA has received funds from the International Nut and Dried Fruit Council (INC). The researchers did not receive any additional financial support aside from travel and accommodations related to their participation in the NUTS 2022 conference. The sponsors did not have any role in selecting or inviting the speakers nor in the writing or reviewing the proceedings of the conference.

Institutional Review Board Statement: Not applicable.

Informed Consent Statement: Not applicable.

Data Availability Statement: Not applicable.

Acknowledgments: This review highlights the main discussion points from the scientific session "Nuts and Dried Fruits and Cancer" by speakers Heinz Freisling and Bradley W. Bolling at the NUTS2022 Conference held in Reus, Spain on 20–21 October 2022.

Disclaimer: Where authors are identified as personnel of the International Agency for Research on Cancer/World Health Organization, the authors alone are responsible for the views expressed in this article, and they do not necessarily represent the decisions, policy, or views of the International Agency for Research on Cancer/World Health Organization.

Conflicts of Interest: H.F., D.A. and H.N. have no conflict of interest to disclose. B.W.B. has received sponsored research funding from the Almond Board of California, the Peanut Institute, the National Dairy Council, Kikkoman USA R&D Laboratory, the UW Dairy Innovation Hub, and the Wisconsin Alumni Research Foundation. K.S.P. has received sponsored research funding from the Peanut Institute, the California Walnut Commission, the American Pecan Council, and the American Pistachio Growers.

References

1. Sung, H.; Ferlay, J.; Siegel, R.L.; Laversanne, M.; Soerjomataram, I.; Jemal, A.; Bray, F. Global Cancer Statistics 2020: GLOBOCAN Estimates of Incidence and Mortality Worldwide for 36 Cancers in 185 Countries. *CA Cancer J. Clin.* **2021**, *71*, 209–249. [CrossRef] [PubMed]
2. World Cancer Research Fund. Continuous Update Project Expert Report 2018. Recommendations and Public Health and Policy. 2018. Available online: https://www.wcrf.org/diet-activity-and-cancer/global-cancer-update-programme/about-the-third-expert-report/ (accessed on 26 January 2023).
3. Colditz, G.A.; Wolin, K.Y.; Gehlert, S. Applying What We Know to Accelerate Cancer Prevention. *Sci. Transl. Med.* **2012**, *4*, 127rv124. [CrossRef]
4. Brennan, P.; Davey-Smith, G. Identifying Novel Causes of Cancers to Enhance Cancer Prevention: New Strategies Are Needed. *J. Nat. Cancer Inst.* **2022**, *114*, 353–360. [CrossRef]
5. Zitvogel, L.; Pietrocola, F.; Kroemer, G. Nutrition, Inflammation and Cancer. *Nat. Immunol.* **2017**, *18*, 843–850. [CrossRef]
6. Wild, W.E.; Stewart, B.W. (Eds.) *World Cancer Report: Cancer Research for Cancer Prevention*; International Agency for Research on Cancer: Lyon, France, 2020. Available online: https://publications.iarc.fr/Non-Series-Publications/World-Cancer-Reports/World-Cancer-Report-Cancer-Research-For-Cancer-Prevention-2020 (accessed on 26 January 2023).
7. Wang, D.D.; Li, Y.; Afshin, A.; Springmann, M.; Mozaffarian, D.; Stampfer, M.J.; Hu, F.B.; Murray, C.J.L.; Willett, W.C. Global Improvement in Dietary Quality Could Lead to Substantial Reduction in Premature Death. *J. Nutr.* **2019**, *149*, 1065–1074. [CrossRef] [PubMed]
8. U.S. Department of Agriculture. *Dietary Guidelines for Americans, 2020–2025*, 9th ed.; U.S. Department of Health and Human Services: Washington, DC, USA, 2020. Available online: https://www.dietaryguidelines.gov/ (accessed on 26 January 2023).
9. United States Code of Federal Regulations. Health Claims: Fiber-Containing Grain Products, Fruits, and Vegetables and Cancer. 21 CFR 101.76. 1993. Available online: https://www.ecfr.gov/current/title-21/chapter-I/subchapter-B/part-101/subpart-E/section-101.76 (accessed on 26 January 2023).
10. International Agency for Research on Cancer; World Health Organization. European Code against Cancer. Available online: https://cancer-code-europe.iarc.fr/index.php/en/ (accessed on 26 January 2023).
11. Rock, C.L.; Thomson, C.; Gansler, T.; Gapstur, S.M.; McCullough, M.L.; Patel, A.V.; Andrews, K.S.; Bandera, E.V.; Spees, C.K.; Robien, K.; et al. American Cancer Society guideline for diet and physical activity for cancer prevention. *CA Cancer J. Clin.* **2020**, *70*, 245–271. [CrossRef]
12. The American Cancer Society Medical and Editorial Content Team. American Cancer Society Guideline for Diet and Physical Activity. Available online: https://www.cancer.org/healthy/eat-healthy-get-active/acs-guidelines-nutrition-physical-activity-cancer-prevention/guidelines.html# (accessed on 28 January 2023).
13. World Cancer Research Fund; American Institute for Cancer Research. Continuous Update Project Expert Report, Wholegrains, Vegetables and Fruit and the Risk of Cancer. 2018. Available online: https://www.wcrf.org/wp-content/uploads/2020/12/Wholegrains-veg-and-fruit.pdf (accessed on 26 January 2023).
14. International Nut & Dried Fruit Council. *Nuts & Dried Fruits Statistical Yearbook*; INC International Nut & Dried Fruit: Reus, Spain, 2022. Available online: https://inc.nutfruit.org/technical-projects/ (accessed on 26 January 2023).
15. Sullivan, V.K.; Na, M.; Proctor, D.N.; Kris-Etherton, P.M.; Petersen, K.S. Consumption of Dried Fruits Is Associated with Greater Intakes of Underconsumed Nutrients, Higher Total Energy Intakes, and Better Diet Quality in US Adults: A Cross-Sectional Analysis of the National Health and Nutrition Examination Survey, 2007–2016. *J. Acad. Nutr. Diet.* **2021**, *121*, 1258–1272. [CrossRef] [PubMed]
16. Bolling, B.; Liu, X.; Liu, J. Dried Fruit Consumption and Cancer. In *Health Benefits of Nuts and Dried Fruits*; Alasalvar, C., Salas-Salvado, J., Ros, E., Sabate, J., Eds.; CRC Press: Boca Raton, FL, USA, 2020; pp. 1–20, Chapter 19. [CrossRef]

17. Di Lorenzo, C.; Sangiovanni, E.; Fumagalli, M.; Colombo, E.; Frigerio, G.; Colombo, F.; Peres de Sousa, L.; Altindisli, A.; Restani, P.; Dell'Agli, M. Evaluation of the Anti-Inflammatory Activity of Raisins (*Vitis vinifera* L.) in Human Gastric Epithelial Cells: A Comparative Study. *Int. J. Mol. Sci.* **2016**, *17*, 1156. [CrossRef]
18. Kountouri, A.M.; Gioxari, A.; Karvela, E.; Kaliora, A.C.; Karvelas, M.; Karathanos, V.T. Chemopreventive Properties of Raisins Originating from Greece in Colon Cancer Cells. *Food Funct.* **2013**, *4*, 366–372. [CrossRef]
19. Mori, S.; Sawada, T.; Okada, T.; Ohsawa, T.; Adachi, M.; Keiichi, K. New Anti-Proliferative Agent, MK615, from Japanese Apricot "Prunus mume" Induces Striking Autophagy in Colon Cancer Cells In Vitro. *World J. Gastroenterol.* **2007**, *13*, 6512–6517. [CrossRef]
20. Nakagawa, A.; Sawada, T.; Okada, T.; Ohsawa, T.; Adachi, M.; Kubota, K. New Antineoplastic Agent, MK615, from UME (a Variety of) Japanese Apricot Inhibits growth of Breast Cancer Cells In Vitro. *Breast J.* **2007**, *13*, 44–49. [CrossRef] [PubMed]
21. Jasmine, R.; Manikandan, K.; Karthikeyan, K. Evaluating the Antioxidant and Cnticancer Property of *Ficus carica* Fruits. *Afr. J. Biotechnol.* **2015**, *14*, 634–641. [CrossRef]
22. Uz, R.I.; Bakar, N.H.A.; Swethadri, G.K.; Baig, A.; Idris, M.A.; Maryam, I.U. Non-Toxic Antiproliferative Effect of *Ficus carica* Fruit Extracts on Estrogen Receptor Positive Breast Cancer Cell (MCF-7). *J. Chem. Pharm. Res.* **2015**, *7*, 815–821.
23. Bu, S.Y.; Lerner, M.; Stoecker, B.J.; Boldrin, E.; Brackett, D.J.; Lucas, E.A.; Smith, B.J. Dried Plum Polyphenols Inhibit Osteoclastogenesis By Downregulating NFATc1 and Inflammatory Mediators. *Calcif. Tissue Int.* **2008**, *82*, 475–488. [CrossRef]
24. Fujii, T.; Ikami, T.; Xu, J.W.; Ikeda, K. Prune Extract (*Prunus domestica* L.) Suppresses the Proliferation and Induces the Apoptosis of Human Colon Carcinoma Caco-2. *J. Nutr. Sci. Vitaminol.* **2006**, *52*, 389–391. [CrossRef]
25. Khan, F.; Ahmed, F.; Pushparaj, P.N.; Abuzenadah, A.; Kumosani, T.; Barbour, E.; AlQahtani, M.; Gauthaman, K. Ajwa Date (*Phoenix dactylifera* L.) Extractinhibits Human Breast Adenocarcinoma (MCF7) cells *In Vitro* By Inducing Apoptosis and Cell Cycle Arrest. *PLoS ONE* **2016**, *11*, e0158963. [CrossRef]
26. Mirza, M.B.; Elkady, A.I.; Al-Attar, A.M.; Syed, F.Q.; Mohammed, F.A.; Hakeem, K.R. Induction of Apoptosis and Cell Cycle Arrest by Ethyl Acetate Fraction of *Phoenix dactylifera* L. (Ajwa dates) in Prostate Cancer Cells. *J. Ethnopharmacol.* **2018**, *218*, 35–44. [CrossRef] [PubMed]
27. Turan, A.; Celik, I. Antioxidant and Hepatoprotective Properties of Dried Fig Against Oxidative Stress and Hepatotoxicity in Rats. *Int. J. Biol. Macromol.* **2016**, *91*, 554–559. [CrossRef]
28. Khan, F.; Khan, T.J.; Kalamegam, G.; Pushparaj, P.N.; Chaudhary, A.; Abuzenadah, A.; Kumosani, T.; Barbour, E.; Al-Qahtani, M. Anti-cancer Effects of Ajwa Dates (*Phoenix dactylifera* L.) in Diethylnitrosamine Induced Hepatocellular Carcinoma in Wistar Rats. *BMC Complement. Altern. Med.* **2017**, *17*, 418. [CrossRef]
29. Yang, Y.; Gallaher, D.D. Effect of Dried Plums on Colon Cancer Risk Factors in Rats. *Nutr. Cancer* **2005**, *53*, 117–125. [CrossRef]
30. Ugras, M.Y.; Kurus, M.; Ates, B.; Soylemez, H.; Otlu, A.; Yilmaz, I. *Prunus armeniaca* L (Apricot) Protects Rat Testes from Detrimental Effects of Low-dose X-rays. *Nutr. Res.* **2010**, *30*, 200–208. [CrossRef] [PubMed]
31. Martin, D.A.; Smyth, J.A.; Liu, Z.; Bolling, B.W. Aronia Berry (*Aronia mitschurinii* 'Viking') Inhibits Colitis in Mice and Inhibits T Cell Tumour Necrosis Factor-α Secretion. *J. Funct. Foods* **2018**, *44*, 48–57. [CrossRef]
32. Martin, D.A.; Taheri, R.; Brand, M.H.; Draghi, A.; Sylvester, F.A.; Bolling, B.W. Anti-Inflammatory Activity of Aronia Berry Extracts in Murine Splenocytes. *J. Funct. Foods* **2014**, *8*, 68–75. [CrossRef]
33. Pei, R.; Liu, X.; Bolling, B. Flavonoids and Gut Health. *Curr. Opin. Biotechnol.* **2020**, *61*, 153–159. [CrossRef] [PubMed]
34. Alasalvar, C.; Chang, S.K.; Kris-Etherton, P.M.; Sullivan, V.K.; Petersen, K.S.; Guasch-Ferré, M.; Jenkins, D. Dried Fruits and Health. *Nutrients* **2023**, in press.
35. Raveendran, D.; Bhagwat, M.; Chidanand, D.V.; Anandakumar, S.; Sunil, C.K. Highlight on Drying Fruit Slices with Better Retention of Bioactive Compounds. *J. Food Proc. Eng.* **2022**, *45*, e14048. [CrossRef]
36. Serratosa, M.P.; Lopez-Toledano, A.; Merida, J.; Medina, M. Changes in Color and Phenolic Compounds during the Raisining of Grape Cv. Pedro Ximenez. *J. Agric. Food Chem.* **2008**, *56*, 2810–2816. [CrossRef]
37. Catak, J.; Yaman, M.; Ugur, H.; Servi, E.Y.; Mizrak, O.F. Investigation of the Advanced Glycation End Products Precursors in Dried Fruits and Nuts by HPLC using Pre-column Derivatization. *J. Food Nutr. Res.* **2022**, *61*, 81–88.
38. Córdova, R.; Mayén, A.-L.; Knaze, V.; Aglago, E.K.; Schalkwijk, C.; Wagner, K.-H.; Overvad, K.; Tjønneland, A.; Kyrø, C.; Katzke, V.A.; et al. Dietary Intake of Advanced Glycation Endproducts (AGEs) and Cancer Risk Across More Than 20 Anatomical Sites: A Multinational Cohort Study. *Cancer Commun.* **2022**, *42*, 1041–1045. [CrossRef]
39. McWeeny, D.J.; Biltcliffe, D.O.; Powell, R.C.T.; Spark, A.A. The Maillard Reaction and Its Inhibition by Sulfite. *J. Food Sci.* **1969**, *34*, 641–643. [CrossRef]
40. Scrob, T.; Covaci, E.; Hosu, A.; Tanaselia, C.; Casoni, D.; Torok, A.I.; Frentiu, T.; Cimpoiu, C. Effect of In Vitro Simulated Gastrointestinal Digestion on Some Nutritional Characteristics of Several Dried Fruits. *Food Chem.* **2022**, *385*, 132713. [CrossRef] [PubMed]
41. Liu, X.; Martin, D.A.; Valdez, J.C.; Sudakaran, S.; Rey, F.; Bolling, B.W. Aronia Berry Polyphenols Have Matrix-dependent Effects on the Gut Microbiota. *Food Chem.* **2021**, *359*, 129831. [CrossRef] [PubMed]
42. Sullivan, V.K.; Petersen, K.S.; Kris-Etherton, P.M. Dried Fruit Consumption and Cardiometabolic Health: A Randomised Crossover Trial. *Br. J. Nutr.* **2020**, *124*, 912–921. [CrossRef]
43. Viguiliouk, E.; Jenkins, A.L.; Blanco Mejia, S.; Sievenpiper, J.L.; Kendall, C.W.C. Effect of Dried Fruit on Postprandial Glycemia: A Randomized Acute-feeding Trial. *Nutr. Diabetes* **2018**, *8*, 59. [CrossRef] [PubMed]

44. George, K.S.; Munoz, J.; Ormsbee, L.T.; Akhavan, N.S.; Foley, E.M.; Siebert, S.C.; Kim, J.S.; Hickner, R.C.; Arjmandi, B.H. The Short-Term Effect of Prunes in Improving Bone in Men. *Nutrients* **2022**, *14*, 276. [CrossRef]
45. Mentor-Marcel, R.A.; Bobe, G.; Sardo, C.; Wang, L.S.; Kuo, C.T.; Stoner, G.; Colburn, N.H. Plasma cytokines as Potential Response Indicators to Dietary Freeze-dried Black Raspberries in Colorectal Cancer Patients. *Nutr. Cancer* **2012**, *64*, 820–825. [CrossRef]
46. Mallery, S.R.; Stoner, G.D.; Larsen, P.E.; Fields, H.W.; Rodrigo, K.A.; Schwartz, S.J.; Tian, Q.; Dai, J.; Mumper, R.J. Formulation and in-vitro and in-vivo Evaluation of a Mucoadhesive Gel Containing Freeze Dried Black Raspberries: Implications for Oral Cancer Chemoprevention. *Pharm. Res.* **2007**, *24*, 728–737. [CrossRef]
47. Wang, L.S.; Arnold, M.; Huang, Y.W.; Sardo, C.; Seguin, C.; Martin, E.; Huang, T.H.; Riedl, K.; Schwartz, S.; Frankel, W.; et al. Modulation of Genetic and Epigenetic Biomarkers of Colorectal Cancer in Humans by Black Raspberries: A Phase I Pilot Study. *Clin. Cancer Res.* **2011**, *17*, 598–610. [CrossRef]
48. Wang, L.S.; Burke, C.A.; Hasson, H.; Kuo, C.T.; Molmenti, C.L.; Seguin, C.; Liu, P.; Huang, T.H.; Frankel, W.L.; Stoner, G.D. A Phase Ib Study of the Effects of Black Raspberries on Rectal Polyps in Patients with Familial Adenomatous Polyposis. *Cancer Prev. Res.* **2014**, *7*, 666–674. [CrossRef]
49. McAnulty, L.S.; Collier, S.R.; Landram, M.J.; Whittaker, D.S.; Isaacs, S.E.; Klemka, J.M.; Cheek, S.L.; Arms, J.C.; McAnulty, S.R. Six Weeks Daily Ingestion of Whole Blueberry Powder Increases Natural Killer Cell Counts and Reduces Arterial Stiffness in Sedentary Males and Females. *Nutr. Res.* **2014**, *34*, 577–584. [CrossRef]
50. Student, V.; Vidlar, A.; Bouchal, J.; Vrbkova, J.; Kolar, Z.; Kral, M.; Kosina, P.; Vostalova, J. Cranberry Intervention in Patients with Prostate Cancer Prior to Radical Prostatectomy. Clinical, Pathological and Laboratory Findings. *Biomed. Pap. Med. Fac. Univ. Palacky Olomouc. Czech Repub.* **2016**, *160*, 559–565. [CrossRef] [PubMed]
51. Mossine, V.V.; Mawhinney, T.P.; Giovannucci, E.L. Dried Fruit Intake and Cancer: A Systematic Review of Observational Studies. *Adv. Nutr.* **2020**, *11*, 237–250. [CrossRef] [PubMed]
52. Zeegers, M.P.; Goldbohm, R.A.; van den Brandt, P.A. Consumption of Vegetables and Fruits and Urothelial Cancer Incidence: A Prospective Study. *Cancer Epidemiol. Biomark. Prev.* **2001**, *10*, 1121–1128.
53. Botterweck, A.A.; van den Brandt, P.A.; Goldbohm, R.A. A Prospective Cohort Study on Vegetable and Fruit Consumption and Stomach Cancer Risk in The Netherlands. *Am. J. Epidemiol.* **1998**, *148*, 842–853. [CrossRef] [PubMed]
54. Schuurman, A.G.; Goldbohm, R.A.; Dorant, E.; van den Brandt, P.A. Vegetable and Fruit Consumption and Prostate Cancer Risk: A Cohort Study in The Netherlands. *Cancer Epidemiol. Biomark. Prev.* **1998**, *7*, 673–680.
55. Michels, K.B.; Edward, G.; Joshipura, K.J.; Rosner, B.A.; Stampfer, M.J.; Fuchs, C.S.; Colditz, G.A.; Speizer, F.E.; Willett, W.C. Prospective Study of Fruit and Vegetable Consumption and Incidence of Colon and Rectal Cancers. *J. Natl. Cancer Inst.* **2000**, *92*, 1740–1752. [CrossRef]
56. Mills, P.K.; Beeson, W.L.; Abbey, D.E.; Fraser, G.E.; Phillips, R.L. Dietary Habits and Past Medical History as Related to Fatal Pancreas Cancer Risk among Adventists. *Cancer* **1988**, *61*, 2578–2585. [CrossRef]
57. Tantamango, Y.M.; Knutsen, S.F.; Beeson, W.L.; Fraser, G.; Sabate, J. Foods and Food Groups Associated with the Incidence of Colorectal Polyps: The Adventist Health Study. *Nutr. Cancer* **2011**, *63*, 565–572. [CrossRef]
58. Mills, P.K.; Beeson, W.L.; Phillips, R.L.; Fraser, G.E. Cohort Study of Diet, Lifestyle, and Prostate Cancer in Adventist Men. *Cancer* **1989**, *64*, 598–604. [CrossRef]
59. Fraser, G.E.; Beeson, W.L.; Phillips, R.L. Diet and Lung Cancer in California Seventh-day Adventists. *Am. J. Epidemiol.* **1991**, *133*, 683–693. [CrossRef]
60. Dunneram, Y.; Greenwood, D.C.; Cade, J.E. Diet and Risk of Breast, Endometrial and Ovarian Cancer: UK Women's Cohort Study. *Br. J. Nutr.* **2019**, *122*, 564–574. [CrossRef] [PubMed]
61. Zhao, L.; Jin, L.; Petrick, J.L.; Zeng, H.; Wang, F.; Tang, L.; Smith-Warner, S.A.; Eliassen, A.H.; Zhang, F.F.; Campbell, P.T.; et al. Specific Botanical Groups of Fruit and Vegetable Consumption and Liver Cancer and Chronic Liver Disease Mortality: A Prospective Cohort Study. *Am. J. Clin. Nutr.* **2022**, *17*, 278–285. [CrossRef] [PubMed]
62. Jin, C.; Li, R.; Deng, T.; Lin, Z.; Li, H.; Yang, Y.; Su, Q.; Wang, J.; Yang, Y.; Wang, J.; et al. Association between Dried Fruit Intake and Pan-cancers Incidence Risk: A Two-sample Mendelian Randomization Study. *Front. Nutr.* **2022**, *9*, 899137. [CrossRef] [PubMed]
63. Baghurst, P.A.; McMichael, A.J.; Slavotinek, A.H.; Baghurst, K.I.; Boyle, P.; Walker, A.M. A Case-Control Study of Diet and Cancer of the Pancreas. *Am. J. Epidemiol.* **1991**, *134*, 167–179. [CrossRef]
64. Tayyem, R.F.; Shehadah, I.; Abu-Mweis, S.S.; Bawadi, H.A.; Bani-Hani, K.E.; Al-Jaberi, T.; Al-Nusairr, M.; Heath, D.D. Fruit and Vegetable Intake among Jordanians: Results from a Case-control Study of Colorectal Cancer. *Cancer Control* **2014**, *21*, 350–360. [CrossRef]
65. González, C.A.; Sanz, J.M.; Marcos, G.; Pita, S.; Brullet, E.; Saigi, E.; Badia, A.; Riboli, E. Dietary Factors and Stomach Cancer in Spain: A Multi-centre Case-control Study. *Int. J. Cancer* **1991**, *49*, 513–519. [CrossRef] [PubMed]
66. Yassibaş, E.; Arslan, P.; Yalçin, S. Evaluation of Dietary and Life-style Habits of Patients with Gastric Cancer: A Case-control Study in Turkey. *Asian Pac. J. Cancer Prev.* **2012**, *13*, 2291–2297. [CrossRef]
67. Wang, D.D.; Li, Y.; Bhupathiraju, S.N.; Rosner, B.A.; Sun, Q.; Giovannucci, E.L.; Rimm, E.B.; Manson, J.E.; Willett, W.C.; Stampfer, M.J.; et al. Fruit and Vegetable Intake and Mortality: Results from 2 Prospective Cohort Studies of US Men and Women and a Meta-Analysis of 26 Cohort Studies. *Circulation* **2021**, *143*, 1642–1654. [CrossRef]

68. Aune, D.; Giovannucci, E.; Boffetta, P.; Fadnes, L.T.; Keum, N.; Norat, T.; Greenwood, D.C.; Riboli, E.; Vatten, L.J.; Tonstad, S. Fruit and Vegetable Intake and the Risk of Cardiovascular Disease, Total Cancer and All-cause Mortality-a Systematic Review and Dose-response Meta-analysis of Prospective Studies. *Int. J. Epidemiol.* **2017**, *46*, 1029–1056. [CrossRef]
69. Hollman, P.C. Unravelling of the Health Effects of Polyphenols is a Complex Puzzle Complicated by Metabolism. *Arch. Biochem. Biophys.* **2014**, *559*, 100–105. [CrossRef]
70. Bakker, M.F.; Peeters, P.H.; Klaasen, V.M.; Bueno-de-Mesquita, H.B.; Jansen, E.H.; Ros, M.M.; Travier, N.; Olsen, A.; Tjønneland, A.; Overvad, K.; et al. Plasma Carotenoids, Vitamin C, Tocopherols, and Retinol and the Risk of Breast Cancer in the European Prospective Investigation into Cancer and Nutrition Cohort. *Am. J. Clin. Nutr.* **2016**, *103*, 454–464. [CrossRef] [PubMed]
71. Cordova, R.; Kliemann, N.; Huybrechts, I.; Rauber, F.; Vamos, E.P.; Levy, R.B.; Wagner, K.H.; Viallon, V.; Casagrande, C.; Nicolas, G.; et al. Consumption of Ultra-Processed Foods Associated with Weight Gain and Obesity in Adults: A Multi-national Cohort Study. *Clin. Nutr.* **2021**, *40*, 5079–5088. [CrossRef] [PubMed]
72. Nishi, S.K.; Viguiliouk, E.; Blanco Mejia, S.; Kendall, C.W.C.; Bazinet, R.P.; Hanley, A.J.; Comelli, E.M.; Salas Salvadó, J.; Jenkins, D.J.A.; Sievenpiper, J.L. Are Fatty Nuts a Weighty Concern? A Systematic Review and Meta-Analysis and Dose–Response Meta-Regression of Prospective Cohorts and Randomized Controlled Trials. *Obes. Rev.* **2021**, *22*, e13330. [CrossRef]
73. Fitzgerald, E.; Lambert, K.; Stanford, J.; Neale, E.P. The Effect of Nut Consumption (Tree Nuts and Peanuts) on the Gut Microbiota of Humans: A Systematic Review. *Br. J. Nutr.* **2021**, *125*, 508–520. [CrossRef]
74. Schulz, M.D.; Atay, C.; Heringer, J.; Romrig, F.K.; Schwitalla, S.; Aydin, B.; Ziegler, P.K. High-Fat-Diet-Mediated Dysbiosis Promotes Intestinal Carcinogenesis Independently of Obesity. *Nature* **2014**, *514*, 508–512. [CrossRef]
75. Hardman, W.E.; Ion, G.; Akinsete, J.A.; Witte, T.R. Dietary Walnut Suppressed Mammary Gland Tumorigenesis in the C(3)1 TAg Mouse. *Nutr. Cancer Int. J.* **2011**, *63*, 960–970. [CrossRef] [PubMed]
76. Garcia, C.P.; Lamarque, A.L.; Comba, A.; Berra, M.A.; Silva, R.A.; Labuckas, D.O.; Das, U.N. Synergistic Anti-tumor Effects of Melatonin and PUFAs from Walnuts in a Murine Mammary Adenocarcinoma Model. *Nutrition* **2015**, *31*, 570–577. [CrossRef] [PubMed]
77. Chen, H.S.; Bai, M.H.; Zhang, T.; Li, G.D.; Liu, M. Ellagic acid Induces Cell Cycle Arrest and Apoptosis through TGF-beta/Smad3 Signaling Pathway in Human Breast Cancer MCF-7 Cells. *Int. J. Oncol.* **2015**, *46*, 1730–1738. [CrossRef]
78. Hong, M.Y.; Moore, J.; Nakagawa, A.; Nungaray, V. Effects of Mixed Nuts on Colonic Cell Proliferation and Ptgs2 and Rela Gene Expression. *Anticancer Res.* **2022**, *42*, 4285–4292. [CrossRef]
79. Chen, Y.; Nakanishi, M.; Bautista, E.J.; Qendro, V.; Sodergren, E.; Rosenberg, D.W.; Weinstock, G.M. Colon Cancer Prevention with Walnuts: A Longitudinal Study in Mice from the Perspective of a Gut Enterotype-like Cluster. *Cancer Prev. Res.* **2020**, *13*, 15–24. [CrossRef]
80. Nagel, J.M.; Brinkoetter, M.; Magkos, F.; Liu, X.; Chamberland, J.P.; Shah, S.; Zhou, J. Dietary Walnuts Inhibit Colorectal Cancer Growth in Mice by Suppressing Angiogenesis. *Nutrition* **2012**, *28*, 67–75. [CrossRef] [PubMed]
81. Nakanishi, M.; Chen, Y.F.; Qendro, V.; Miyamoto, S.; Weinstock, E.; Weinstock, G.M.; Rosenberg, D.W. Effects of Walnut Consumption on Colon Carcinogenesis and Microbial Community Structure. *Cancer Prev. Res.* **2016**, *9*, 692–703. [CrossRef] [PubMed]
82. Davis, P.A.; Iwahashi, C.K. Whole Almonds and Almond Fractions Reduce Aberrant Crypt Foci in a Rat Model of Colon Carcinogenesis. *Cancer Lett.* **2001**, *165*, 27–33. [CrossRef] [PubMed]
83. Tsoukas, M.A.; Ko, B.J.; Witte, T.R.; Dincer, F.; Hardman, W.E.; Mantzoros, C.S. Dietary Walnut Suppression of Colorectal Cancer in Mice: Mediation by miRNA Patterns and Fatty Acid Incorporation. *J. Nutr. Biochem.* **2015**, *26*, 776–783. [CrossRef] [PubMed]
84. Noh, H.; Jang, H.-H.; Kim, G.; Zouiouich, S.; Cho, S.-Y.; Kim, H.-J.; Kim, J.; Choe, J.-S.; Gunter, M.J.; Ferrari, P.; et al. Taxonomic Composition and Diversity of the Gut Microbiota in Relation to Habitual Dietary Intake in Korean Adults. *Nutrients* **2021**, *13*, 366. [CrossRef]
85. Davis, P.A.; Vasu, V.T.; Gohil, K.; Kim, H.; Khan, I.H.; Cross, C.E.; Yokoyama, W. A High-fat Diet Containing Whole Walnuts (*Juglans regia*) Reduces Tumour Size and Growth Along with Plasma Insulin-like Growth Factor 1 in the Transgenic Adenocarcinoma of the Mouse Prostate Model. *Br. J. Nutr.* **2012**, *108*, 1764–1772. [CrossRef] [PubMed]
86. Kim, H.; Yokoyama, W.; Davis, P.A. TRAMP Prostate Tumor Growth Is Slowed by Walnut Diets through Altered IGF-1 Levels, Energy Pathways, and Cholesterol Metabolism. *J. Med. Food* **2014**, *17*, 1281–1286. [CrossRef]
87. Reiter, R.J.; Tan, D.X.; Manchester, L.C.; Korkmaz, A.; Fuentes-Broto, L.; Hardman, W.E.; Rosales-Corral, S.A. A Walnut-Enriched Diet Reduces the Growth of LNCaP Human Prostate Cancer Xenografts in Nude Mice. *Cancer Investig.* **2013**, *31*, 365–373. [CrossRef]
88. Hu, Y.; McIntosh, G.H.; Le Leu, R.K.; Somashekar, R.; Meng, X.Q.; Gopalsamy, G.; Bambaca, L. Supplementation with Brazil Nuts and Green Tea Extract Regulates Targeted Biomarkers Related to Colorectal Cancer Risk in Humans. *Br. J. Nutr.* **2016**, *116*, 1901–1911. [CrossRef]
89. Pereira, M.A.N.; da Silva Junior, E.C.; Dayse da Silva, I.L.; de Carvalho, B.A.; Ferreira, E.; Andrade, E.F.; Guimarães Guilherme, L.R.; Pereira, L.J. Antitumor Effect of Selenium-Rich Brazil Nuts and Selenomethionine Dietary Supplementation on Pre-existing 4T1 Mammary Tumor Growth in Mice. *PLoS ONE* **2023**, *18*, e0278088. [CrossRef]
90. Hardman, W.E.; Ion, G. Suppression of Implanted MDA-MB 231 Human Breast Cancer Growth in Nude Mice by Dietary Walnut. *Nutr. Cancer Int. J.* **2008**, *60*, 666–674. [CrossRef] [PubMed]

91. Toledo, E.; Salas-Salvado, J.; Donat-Vargas, C.; Buil-Cosiales, P.; Estruch, R.; Ros, E.; Corella, D. Mediterranean Diet and Invasive Breast Cancer Risk among Women at High Cardiovascular Risk in the PREDIMED Trial: A Randomized Clinical Trial. *JAMA Intern. Med.* **2015**, *175*, 1752–1760. [CrossRef] [PubMed]
92. Hughes, D.J.; Fedirko, V.; Jenab, M.; Schomburg, L.; Méplan, C.; Freisling, H.; Bueno-de-Mesquita, H.B.; Hybsier, S.; Becker, N.-P.; Czuban, M.; et al. Selenium Status is Associated with Colorectal Cancer Risk in the European Prospective Investigation of Cancer and Nutrition Cohort. *Int. J. Cancer* **2015**, *136*, 1149–1161. [CrossRef] [PubMed]
93. Lippman, S.M.; Klein, E.A.; Goodman, P.J.; Lucia, M.S.; Thompson, I.M.; Ford, L.G.; Parnes, H.L.; Minasian, L.M.; Gaziano, J.M.; Hartline, J.A.; et al. Effect of Selenium and Vitamin E on Risk of Prostate Cancer and Other Cancers: The Selenium and Vitamin E Cancer Prevention Trial (SELECT). *JAMA* **2009**, *301*, 39–51. [CrossRef] [PubMed]
94. Simon, J.A.; Tanzman, J.S.; Sabate, J. Lack of Effect of Walnuts on Serum Levels of Prostate Specific Antigen: A Brief Report. *J. Am. Coll. Nutr.* **2007**, *26*, 317–320. [CrossRef]
95. Spaccarotella, K.J.; Kris-Etherton, P.M.; Stone, W.L.; Bagshaw, D.M.; Fishell, V.K.; West, S.G.; Lawrence, F.R. The Effect of Walnut Intake on Factors Related to Prostate and Vascular Health in Older Men. *Nutr. J.* **2008**, *7*, 13. [CrossRef]
96. Jia, X.D.; Li, N.; Zhang, W.Z.; Zhang, X.P.; Lapsley, K.; Huang, G.W.; Blumberg, J. A Pilot Study on the Effects of Almond Consumption on DNA Damage and Oxidative Stress in Smokers. *Nutr. Cancer* **2006**, *54*, 179–183. [CrossRef]
97. Bamberger, C.; Rossmeier, A.; Lechner, K.; Wu, L.; Waldmann, E.; Fischer, S.; Stark, R.G.; Altenhofer, J.; Henze, K.; Parhofer, K.G. A Walnut-Enriched Diet Affects Gut Microbiome in Healthy Caucasian Subjects: A Randomized, Controlled Trial. *Nutrients* **2018**, *10*, 244. [CrossRef]
98. Naghshi, S.; Sadeghian, M.; Nasiri, M.; Mobarak, S.; Asadi, M.; Sadeghi, O. Association of Total Nut, Tree Nut, Peanut, and Peanut Butter Consumption with Cancer Incidence and Mortality: A Comprehensive Systematic Review and Dose-Response Meta-Analysis of Observational Studies. *Adv. Nutr.* **2021**, *12*, 793–808. [CrossRef]
99. Long, J.; Ji, Z.; Yuan, P.; Long, T.; Liu, K.; Li, J.; Cheng, L. Nut Consumption and Risk of Cancer: A Meta-analysis of Prospective Studies. *Cancer Epidemiol. Biomark. Prev.* **2020**, *29*, 565–573. [CrossRef]
100. Cao, C.; Gan, X.; He, Y.; Nong, S.; Su, Y.; Liu, Z.; Zhang, Y.; Hu, X.; Peng, X. Association Between Nut Consumption and Cancer Risk: A Meta-Analysis. *Nutr. Cancer* **2023**, *75*, 82–94. [CrossRef] [PubMed]
101. Fang, Z.; Wu, Y.; Li, Y.; Zhang, X.; Willett, W.C.; Eliassen, A.H.; Rosner, B.; Song, M.; Mucci, L.A.; Giovannucci, E.L. Association of Nut Consumption with Risk of Total Cancer and 5 Specific Cancers: Evidence from 3 Large Prospective Cohort Studies. *Am. J. Clin. Nutr.* **2021**, *114*, 1925–1935. [CrossRef] [PubMed]
102. Nieuwenhuis, L.; van den Brandt, P.A. Nut and Peanut Butter Consumption and the Risk of Total Cancer: A Prospective Cohort Study. *Cancer Epidemiol. Biomark. Prev.* **2020**, *29*, 2100–2104. [CrossRef] [PubMed]
103. Von Ruesten, A.; Feller, S.; Bergmann, M.M.; Boeing, H. Diet and Risk of Chronic Diseases: Results from the First 8 Years of Follow-up in the EPIC-Potsdam Study. *Eur. J. Clin. Nutr.* **2013**, *67*, 412–419. [CrossRef] [PubMed]
104. Balakrishna, R.; Bjørnerud, T.; Bemanian, M.; Aune, D.; Fadnes, L.T. Consumption of Nuts and Seeds and Health Outcomes Including Cardiovascular Disease, Diabetes and Metabolic Disease, Cancer, and Mortality: An Umbrella Review. *Adv. Nutr.* **2022**, *13*, nmac077. [CrossRef] [PubMed]
105. Chen, G.C.; Zhang, R.; Martinez-Gonzalez, M.A.; Zhang, Z.L.; Bonaccio, M.; Dam, R.M.; Qin, L.Q. Nut Consumption in Relation to All-Cause and Cause-Specific Mortality: A Meta-analysis 18 Prospective Studies. *Food Funct.* **2017**, *8*, 3893–3905. [CrossRef]
106. Aune, D.; Keum, N.; Giovannucci, E.; Fadnes, L.T.; Boffetta, P.; Greenwood, D.C.; Tonstad, S.; Vatten, L.J.; Riboli, E.; Norat, T. Nut Consumption and Risk of Cardiovascular Disease, Total Cancer, All-Cause and Cause-Specific Mortality: A Systematic Review and Dose-Response Meta-Analysis of Prospective Studies. *BMC Med.* **2016**, *14*, 207. [CrossRef]
107. Grosso, G.; Yang, J.; Marventano, S.; Micek, A.; Galvano, F.; Kales, S.N. Nut Consumption on All-Cause, Cardiovascular, and Cancer Mortality Risk: A Systematic Review and Meta-Analysis of Epidemiologic Studies. *Am. J. Clin. Nutr.* **2015**, *101*, 783–793. [CrossRef]
108. Zhang, D.; Dai, C.; Zhou, L.; Li, Y.; Liu, K.; Deng, Y.J.; Li, N.; Zheng, Y.; Hao, Q.; Yang, S.; et al. Meta-Analysis of the Association between Nut Consumption and the Risks of Cancer Incidence and Cancer-Specific Mortality. *Aging* **2020**, *12*, 10772–10794. [CrossRef]
109. Amba, V.; Murphy, G.; Etemadi, A.; Wang, S.; Abnet, C.C.; Hashemian, M. Nut and Peanut Butter Consumption and Mortality in the National Institutes of Health-AARP Diet and Health Study. *Nutrients* **2019**, *11*, 1508. [CrossRef]
110. De Souza, R.J.; Dehghan, M.; Mente, A.; Bangdiwala, S.I.; Ahmed, S.H.; Alhabib, K.F.; Altuntas, Y.; Basiak-Rasała, A.; Dagenais, G.R.; Diaz, R.; et al. Association of Nut Intake with Risk Factors, Cardiovascular Disease, and Mortality in 16 Countries from 5 Continents: Analysis from the Prospective Urban and Rural Epidemiology (PURE) Study. *Am. J. Clin. Nutr.* **2020**, *112*, 208–219. [CrossRef] [PubMed]
111. Sun, Y.; Liu, B.; Snetselaar, L.G.; Wallace, R.B.; Shadyab, A.H.; Kroenke, C.H.; Haring, B.; Howard, B.V.; Shikany, J.M.; Valdiviezo, C.; et al. Association of Major Dietary Protein Sources with All-Cause and Cause-Specific Mortality: Prospective Cohort Study. *J. Am. Heart Assoc.* **2021**, *10*, e015553. [CrossRef] [PubMed]
112. Yang, J.; Yang, A.; Yeung, S.; Woo, J.; Lo, K. Joint Associations of Food Groups with All-Cause and Cause-Specific Mortality in the Mr. OS and Ms. OS Study: A Prospective Cohort. *Nutrients* **2022**, *14*, 3915. [CrossRef] [PubMed]

113. Yamakawa, M.; Wada, K.; Koda, S.; Uji, T.; Nakashima, Y.; Onuma, S.; Oba, S.; Nagata, C. Associations of Total Nut and Peanut Intakes with All-Cause and Cause-Specific Mortality in a Japanese Community: The Takayama Study. *Br. J. Nutr.* **2022**, *127*, 1378–1385. [CrossRef] [PubMed]
114. Neelakantan, N.; Koh, W.P.; Yuan, J.M.; van Dam, R.M. Diet-Quality Indexes Are Associated with a Lower Risk of Cardiovascular, Respiratory, and All-Cause Mortality among Chinese Adults. *J. Nutr.* **2018**, *148*, 1323–1332. [CrossRef] [PubMed]
115. Wang, W.; Yang, M.; Kenfield, S.A.; Hu, F.B.; Stampfer, M.J.; Willett, W.C.; Fuchs, C.S.; Giovannucci, E.L.; Bao, Y. Nut Consumption and Prostate Cancer Risk and Mortality. *Br. J. Cancer* **2016**, *115*, 371–374. [CrossRef] [PubMed]
116. Wang, C.; Gu, K.; Wang, F.; Cai, H.; Zheng, W.; Bao, P.; Shu, X.-O. Nut Consumption in Association with Overall Mortality and Recurrence/Disease-Specific Mortality among Long-Term Breast Cancer Survivors. *Int. J. Cancer* **2022**, *150*, 572–579. [CrossRef] [PubMed]
117. Fadelu, T.; Zhang, S.; Niedzwiecki, D.; Ye, X.; Saltz, L.B.; Mayer, R.J.; Mowat, R.B.; Whittom, R.; Hantel, A.; Benson, A.B.; et al. Nut Consumption and Survival in Patients with Stage III Colon Cancer: Results from CALGB 89803 (Alliance). *J. Clin. Oncol.* **2018**, *36*, 1112–1120. [CrossRef] [PubMed]
118. Ratjen, I.; Enderle, J.; Burmeister, G.; Koch, M.; Nöthlings, U.; Hampe, J.; Lieb, W. Post-Diagnostic Reliance on Plant-Compared with Animal-based Foods and All-Cause Mortality in Omnivorous Long-Term Colorectal Cancer Survivors. *Am. J. Clin. Nutr.* **2021**, *114*, 441–449. [CrossRef] [PubMed]
119. Eslamparast, T.; Sharafkhah, M.; Poustchi, H.; Hashemian, M.; Dawsey, S.M.; Freedman, N.D.; Boffetta, P.; Abnet, C.C.; Etemadi, A.; Pourshams, A.; et al. Nut consumption and total and cause-specific mortality: Results from the Golestan Cohort Study. *Int. J. Epidemiol.* **2017**, *46*, 75–85. [CrossRef] [PubMed]
120. Hshieh, T.T.; Petrone, A.B.; Gaziano, J.M.; Djoussé, L. Nut consumption and risk of mortality in the Physicians' Health Study. *Am. J. Clin. Nutr.* **2015**, *101*, 407–412. [CrossRef] [PubMed]
121. Luu, H.N.; Blot, W.J.; Xiang, Y.B.; Cai, H.; Hargreaves, M.K.; Li, H.; Yang, G.; Signorello, L.; Gao, Y.T.; Zheng, W.; et al. Prospective evaluation of the association of nut/peanut consumption with total and cause-specific mortality. *JAMA Intern. Med.* **2015**, *175*, 755–766. [CrossRef] [PubMed]
122. Van den Brandt, P.A.; Schouten, L.J. Relationship of tree nut, peanut and peanut butter intake with total and cause-specific mortality: A cohort study and meta-analysis. *Int. J. Epidemiol.* **2015**, *44*, 1038–1049. [CrossRef]
123. Bao, Y.; Han, J.; Hu, F.B.; Giovannucci, E.L.; Stampfer, M.J.; Willett, W.C.; Fuchs, C.S. Association of nut consumption with total and cause-specific mortality. *N. Engl. J. Med.* **2013**, *369*, 2001–2011. [CrossRef] [PubMed]
124. Guasch-Ferré, M.; Bulló, M.; Martínez-González, M.Á.; Ros, E.; Corella, D.; Estruch, R.; Fitó, M.; Arós, F.; Wärnberg, J.; Fiol, M.; et al. Frequency of nut consumption and mortality risk in the PREDIMED nutrition intervention trial. *BMC Med.* **2013**, *11*, 164. [CrossRef]
125. Freisling, H.; Viallon, V.; Lennon, H.; Bagnardi, V.; Ricci, C.; Butterworth, A.S.; Sweeting, M.; Muller, D.; Romieu, I.; Bazelle, P.; et al. Lifestyle Factors and Risk of Multimorbidity of Cancer and Cardiometabolic Diseases: A Multinational Cohort Study. *BMC Med.* **2020**, *18*, 1–11. [CrossRef]
126. Freisling, H.; Noh, H. Nut Consumption and Cancer. In *Health Benefits of Nuts and Dried Fruits*; Alasalvar, C., Salas-Salvado, J., Ros, E., Sabate, J., Eds.; CRC Press: Boca Raton, FL, USA, 2020; pp. 1–24, Chapter 9.
127. Moher, D.; Liberati, A.; Tetzlaff, J.; Altman, D.G.; PRISMA Group. Preferred Reporting Items for Systematic Reviews and Meta-Analyses: The PRISMA Statement. *PLoS Med.* **2009**, *6*, e1000097. [CrossRef]

nutrients

MDPI

Review

Dried Fruits: Bioactives, Effects on Gut Microbiota, and Possible Health Benefits—An Update

Cesarettin Alasalvar [1,*], Sui Kiat Chang [2], Penny M. Kris-Etherton [3], Valerie K. Sullivan [4], Kristina S. Petersen [5], Marta Guasch-Ferré [6,7] and David J. A. Jenkins [8,9,10,11]

1 Life Sciences, TÜBİTAK Marmara Research Center, Gebze 41470, Türkiye
2 Department of Allied Health Sciences, Faculty of Science, Universiti Tunku Abdul Rahman, Kampar 31900, Malaysia
3 Department of Nutritional Sciences, Penn State University, State College, PA 16802, USA
4 Department of Epidemiology, Johns Hopkins Bloomberg School of Public Health, Baltimore, MD 21205, USA
5 Department of Nutritional Sciences, Texas Tech University, Lubbock, TX 79409, USA
6 Department of Public Health and Novo Nordisk Foundation Center for Basic Metabolic Research, Faculty of Health and Medical Sciences, University of Copenhagen, 1356 Copenhagen, Denmark
7 Department of Nutrition, Harvard T.H. Chan School of Public Health, Boston, MA 02115, USA
8 Departments of Nutritional Sciences and Medicine, Temerty Faculty of Medicine, University of Toronto, Toronto, ON M5S 1A8, Canada
9 Clinical Nutrition Risk Factor Modification Centre, St. Michael's Hospital, Toronto, ON M5C 2T2, Canada
10 Li Ka Shing Knowledge Institute, St. Michael's Hospital, Toronto, ON M5B 1W8, Canada
11 Division of Endocrinology and Metabolism, St. Michael's Hospital, Toronto, ON M5C 2T2, Canada
* Correspondence: cesarettin.alasalvar@tubitak.gov.tr; Tel.: +90-262-677-3200

Abstract: Dried fruits contain many bioactive compounds broadly classified as phytochemicals including phenolics, flavonoids, carotenoids, proanthocyanidins, stilbenes, chalcones/dihydrochalcones, and phytoestrogens. These compounds have antioxidant effects that may benefit health. Dried fruits are also a diverse group of foods with varying fibre contents. The evaluation of the biological activity of these bioactive compounds, including their bioaccessibility and bioavailability, may contribute to the understanding of the health effects of dried fruits. Limited evidence suggests that dried fruits (raisins, cranberries, dates, and prunes) affect human gut microbiota composition in a potentially beneficial manner (in terms of effects on *Bifidobacteria*, *Faecalibacterium prausnitzii*, *Lactobacillus*, *Ruminococcaceae*, *Klebsiella* spp., and *Prevotella* spp.). There is little epidemiological evidence about the association of dried fruit consumption with cardiovascular disease incidence and mortality, as well as the risk of type 2 diabetes or obesity. Clinical trial evidence for the effects of dried fruit consumption on cardiovascular risk factors, including glycaemic control, is mixed. Clinical trial evidence suggests prunes might preserve bone mineral density in postmenopausal women. Consumption of dried fruits is associated with higher-quality diets. Studies are needed to increase our understanding of the health effects of dried fruits and the underlying biological mechanisms.

Keywords: dried fruits; gut health and microbiome; cardiometabolic diseases; bone health; dietary guidance

Citation: Alasalvar, C.; Chang, S.K.; Kris-Etherton, P.M.; Sullivan, V.K.; Petersen, K.S.; Guasch-Ferré, M.; Jenkins, D.J.A. Dried Fruits: Bioactives, Effects on Gut Microbiota, and Possible Health Benefits—An Update. *Nutrients* **2023**, *15*, 1611. https://doi.org/10.3390/nu15071611

Academic Editor: Peter Pribis

Received: 9 March 2023
Revised: 20 March 2023
Accepted: 22 March 2023
Published: 26 March 2023

1. Introduction

Dried fruits are enjoyed by populations worldwide as a shelf-stable, convenient alternative to fresh fruit. Epidemiological evidence suggests dried fruit consumption is associated with lower risk of cardiovascular disease (CVD), type 2 diabetes (T2D), as well as obesity, various cancers, and other chronic diseases, although the evidence is limited and sometimes contradictory. Nonetheless, dried fruits are nutrient dense and a good source of bioactives/phytochemicals [1].

The biological action of bioactives/phytochemicals in dried fruits is dependent on the food matrix release (e.g., bioaccessibility), bioavailability, and metabolism by colonic microbiota [2]. In 2014, the European Food Safety Authority (EFSA) authorized a health claim for dried plums/prunes and gastrointestinal health [3]. This claim states that "*Dried plums/prunes can contribute to normal bowel function*". To obtain the claimed effect, about 100 g/day (~8–12 prunes, depending on their size) of prunes should be consumed. More recently, the effects of dried fruits and the constituent phytochemicals on microbiota composition and functionality have been active areas of investigation. It is now recognized that microbiota contributes to metabolic health and, when aberrant, the development of cardiometabolic diseases. Thus, identifying dietary strategies to promote metabolic health through microbial modulation is a priority.

Evidence from epidemiological and clinical studies suggests that dried fruit intake may improve glucose metabolism and other cardiovascular risk factors, as well as a lower risk for osteoporosis [4,5]. The intake of dried fruits has also been proposed as a strategy to meet fruit recommendations, improve diet quality, and address nutrient deficiencies [6,7].

This review summarizes evidence on the relationship between dried fruit intake and gastrointestinal (GI) health. Bioactive/phytochemical composition and bioaccessibility and bioavailability of dried fruits are also highlighted. We also discuss the association between dried fruit intake and cardiometabolic diseases, bone health, and diet quality as well as the potential mechanisms involved. This is an emerging area of science, and current evidence suggests that the effects of dried fruits on the microbiome, cardiometabolic disease risk, bone health, and diet quality warrant further investigation.

2. Methodologies

To write this narrative review, a detailed literature review was conducted (via sources such as Web of Science, PubMed, SCOPUS, MEDLINE, and Google Scholar). Articles related to the following topics were included: (1) bioactives/phytochemicals present in most commonly consumed dried fruits, (2) bioaccessibility and bioavailability of compounds in dried fruits, (3) GI effects (gut health and microbiota) of dried fruits in animals (in vivo) or human clinical trials, (4) epidemiological evidence about the association of dried fruit consumption with CVD, T2D, and adiposity, (5) cardiometabolic and bone health effects of dried fruit consumption, (6) dietary guidance for dried fruits and benefits on diet quality, and finally, (7) potential mechanisms involved in the observed biological effects. To ensure that current and recent research was presented in this review, only articles published from 2000 onward were included (with a few exceptions due to the relevance of the work), with preference given to articles published between 2015 and 2022 in order to improve contemporary relevance.

Selected articles were examined in detail, and then the bioactives/phytochemicals, bioaccessibility and bioavailability of compounds, gut health and microbiota, epidemiological evidence, cardiometabolic diseases, bone health, and potential mechanisms involved for health benefits as well as diet quality and dietary recommendations for dried fruits were compiled and evaluated. This review is not a systematic review. The most innovative aspect of this review is the update of GI health and cardiometabolic effects of commonly consumed dried fruits (in vivo and in vitro studies). In addition, the update on bioactives/phytochemicals, potential mechanisms involved in the observed biological effects, recommendations for dried fruit consumption, and benefits on diet quality contribute to the novelty of this review.

3. Bioactives/Phytochemicals, Dietary Fibre, and Antioxidant Activity in Dried Fruits

Dried fruits contain a variety of bioactive compounds/phytochemicals such as flavonoids (anthocyanins, flavan-3-ols, flavonols, and flavones), proanthocyanidins (dimer, trimer, 4–6 m, and 7–10 m), phenolic acids (hydroxycinnamic acids and hydroxybenzoic acids), carotenoids (α-carotene, β-carotene, β-cryptoxanthin, lutein, and zeaxanthin), and stilbenes as well as phytoestrogens (isoflavones, lignans, and coumestan) and chalcones/

dihydrochalcones [1,8,9]. Among these bioactive phytochemicals, phenolic compounds are the major group (Table 1). Alasalvar et al. [9] reported various phenolic compounds (anthocyanins, flavan-3-ols, flavonols, flavones, phenolic acids, proanthocyanidins, chalcones/dihydrochalcones, and stilbenes) in nine dried fruits (apples, apricots, cranberries, dates, figs, peaches, pears, prunes, and raisins). Some dried fruits (such as apricots, cranberries, dates, figs, prunes, and raisins) have the most diverse phenolic profiles. Little information is available about the exact phenolic profiles of dried apples, peaches, and pears. With regard to carotenoids, which are plant pigments responsible for yellow, orange, and bright red hues in many fruits and vegetables, α-carotene, β-carotene, β-cryptoxanthin, and lutein + zeaxanthin are present in dried fruits except raisins (seedless), albeit in varying quantities. Of these, β-carotene is the most abundant in apricots (2163 μg/100 g), peaches (1074 μg/100 g), and prunes (394 μg/100 g) [10]. Phytoestrogens consist of isoflavones, lignans, and coumestans. Apricots, dates, prunes, and raisins have been reported to contain phytoestrogens. Total phytoestrogen content ranged from 30.3 μg/100 g in raisins (seedless) to 445 μg/100 g in apricots. No phytoestrogens have been reported in dried apples, cranberries, figs, and peaches (Table 1) [11]. Detailed quantitative analysis on different classes of phenolic compounds, carotenoids, and phytoestrogens in different forms and varieties of dried fruits are needed.

Dried fruits are a good source of dietary fibre (3.7–9.8 g/100 g) (Table 1). Consumption of dried fruits (around 20–30 g per/day recommended by many countries) provides 10–16% of the recommended daily intake of fibre (14 g/day), depending on the fruit [10,12,13].

The oxygen radical absorbance capacity (ORAC), a measure of antioxidant activity, of dried fruits is relatively high, although it varies by dried fruit type as well as by cultivar/variety (Table 1). For example, raisins (seedless) have the lowest ORAC values (3037 μmol trolox equivalents (TE)/100 g), whereas raisins (Golden seedless) have the highest ORAC value ((10,450 μmol TE)/100 g). Similarly, ORAC values vary appreciably for dates cultivars of Deglet noor and Medjool [14].

Table 1. Reported bioactives/phytochemicals, dietary fibre, and antioxidant activity in selected dried fruits.

	Phenolics	**Carotenoids (μg/100 g)**	**Phytoestrogens (μg/100 g)**	**Dietary Fibre (g/100 g)**	**Antioxidant Activity (μmol of TE/100 g) [a]**
Apples	Flavan-3-ols Flavonols Phenolic acids Chalcones/ dihydrochalcones	Lutein + zeaxanthin (18)	-	8.7	6681
Apricots	Flavan-3-ols Flavonols Flavones Phenolic acids Chalcones/ dihydrochalcones	β-Carotene (2163)	Isoflavones (39.8) Lignans (401) Coumestan (4.2)	7.3	3234
Cranberries	Anthocyanins Flavan-3-ols Flavonols Phenolic acids Proanthocyanidins	β-Carotene (27) Lutein + zeaxanthin (138)	-	5.3	-
Dates	Anthocyanins Flavonols Phenolic acids Proanthocyanidins	β-Carotene (6) Lutein + zeaxanthin (75)	Isoflavones (5.1) Lignans (324) Coumestan (0.8)	8.0	2387–3895 [b]

Table 1. *Cont.*

	Phenolics	Carotenoids (μg/100 g)	Phytoestrogens (μg/100 g)	Dietary Fibre (g/100 g)	Antioxidant Activity (μmol of TE/100 g) [a]
Figs	Anthocyanins Flavan-3-ols Flavonols Flavones Phenolic acids Proanthocyanidins	β-Carotene (6) Lutein + zeaxanthin (32)	-	9.8	3383
Peaches	Anthocyanins Flavan-3-ols Flavonols Phenolic acids	α-Carotene (3) β-Carotene (1074) β-Cryptoxanthin (444) Lutein + zeaxanthin (559)	-	8.2	4222
Pears	Flavan-3-ols Phenolic acids Chalcones/ dihydrochalcones	β-Carotene (2) Lutein + zeaxanthin (50)	-	7.5	9496
Prunes	Flavan-3-ols Flavonols Phenolic acids	α-Carotene (57) β-Carotene (394) β-Cryptoxanthin (93) Lutein + zeaxanthin (148)	Isoflavones (4.2) Lignans (178) Coumestan (1.8)	7.1	8578
Raisins	Anthocyanins Flavan-3-ols Flavonols Flavones Phenolic acids Stilbenes	-	Isoflavones (8.1) Lignans (22) Coumestan (0.2)	3.7	3037–10,450 [c]
References	[1,9]	[10]	[11]	[10]	[14]

[a] Based on oxygen radical absorbance capacity (ORAC). [b] Between Deglet noor and Medjool cultivars. [c] Among white, seedless, and Golden seedless raisins.

Several studies have reported the bioactive compounds and antioxidant activities of dried fruits are higher than those of their corresponding fresh counterparts [15–18]. This is due to bioactive compounds and antioxidants becoming concentrated after the drying process. However, losses (e.g., carotenoids and anthocyanins) or changes in some compounds occur during drying and storage. Therefore, drying types and duration, as well as storage and packaging are of great importance in terms of functional/nutritional quality and flavour (taste and aroma) of the final product for consumption.

4. Bioaccessibility and Bioavailability of Compounds in Dried Fruits

The bioaccessibility and bioavailability of compounds in dried fruits have been investigated using in vitro models. These models mimic human in vitro GI digestion (e.g., mouth (oral or salivary digestion), stomach (gastric digestion), small intestine (intestinal digestion), and colon or large intestine (colonic digestion)) [19–22]. Bioaccessibility refers to the level of a compound released from the food matrix during GI digestion that becomes available for absorption (bioavailability) in the intestine [23]. To exert health effects, ingested compounds, including phytochemicals and micronutrients (vitamins and minerals) contained in food, must be released from the food matrix in the GI tract and become bioavailable [21].

Evidence suggests that the phenolics contained within dried fruits are bioaccessible. Recently, Scrob et al. [24] investigated the bioaccessibility of constituents in six dried fruits (dates, raisins, coconuts, cranberries, prunes, and bananas) and demonstrated the highest bioaccessibility of phenolics was observed in prunes and the lowest in cranberries and dates. Total sugars content increased after in vitro digestion of coconuts, dates, and raisins, but it decreased for bananas, cranberries, and prunes. In vitro digestion led to an increase in

the antioxidant activity for most dried fruits. This study showed prunes, coconuts, bananas, and raisins are sources of high bioaccessible phenolics. However, the contribution of dried fruit consumption to the recommended dietary allowances (%) was less considering the bioaccessible fraction compared to the total content.

Polar phenol bioaccessibility of dates using a static model of in vitro digestion was also investigated by Panagopoulou et al. [22]. Simulated GI digestion revealed date polar phenols were found to be bioaccessible to an extent depending on the polar phenol class, the nature of the polar phenols, and the specific date matrix. A 37–70% release was observed post-oral digestion, in terms of total phenolic content, which further increased post-gastric digestion (>100%).

Ma et al. [20] investigated the biological activities of kiwifruits and kiwifruit products including dried slices under simulated GI in vitro digestion. Dried slices showed the lowest biological activity compared to those of other kiwifruit products (such as raw fruit, juice, vinegar, wine, yogurt, and jelly). However, dried slices and jam had the highest quantity of minerals (per unit weight). Thus, consuming dried slices and jam could supply more mineral elements than other forms of the fruit [20].

The impact of GI digestion on the total phenolic content (TPC) and antioxidant activities of dried apricots, figs, and raisins was evaluated by Kamiloglu et al. [25]. There was an increase in TPC (0.4–4.5-fold) for all samples after the gastric digestion. The antioxidant activities of dried apricots and figs were increased as determined by various antioxidant activity assays.

In conclusion, in vitro GI digestion studies have some advantages including being fast and inexpensive, without human ethics concerns. However, these digestion systems (static and dynamic) might not completely mimic human physiology. In vitro models need to be compared with in vivo models (particularly, human intervention studies) to better understand the biological effects. These comparative data are essential for demonstrating the biological relevance of bioactive compounds in the context of nutrition and human health [1,25].

5. Dried Fruits, Gut Health, and Microbiota

Diet is an important modulator of the gut microbiota and its metabolite production. The multiple interactions between food components and gut microbiota as well as the modification of the gut microbiota composition and activities by food components contribute to human health [26,27]. To the best of our knowledge, few studies have investigated the effects of dried fruit intake on gut microbiota. Recent findings from animal and human studies are reviewed.

5.1. In Vivo Animal Studies

A recent chapter by Muñoz and Lamuela-Raventós [28] reviewed the effects of different dried fruits on gut health and microbiota composition using in vitro and in vivo studies. These in vivo studies were conducted using rat, fish, or broiler chick models. The effects of dried fruits on the modulation of gut microbiota from preclinical studies published following the chapter by Muñoz and Lamuela-Raventós [28] are discussed in this section. Among dried fruits, gut health and the microbiota data are available for goji berries, prunes, and dried cranberries.

In a recent study conducted by Cremonesi et al. [29], New Zealand white rabbits fed chow with 3% goji berries had enrichment of *Ruminococcaceae*, *Lachnospiraceae*, *Lactobacillaceae*, and the genus *Lactobacillus*, all of which are considered to be beneficial bacteria, compared to a control group fed regular chow. In addition, the supplementation of goji berries enhanced lactic acid fermentation that contributes to the caecal fermentation [29]. Similarly, in a 10-week study involving mice, Tian et al. [30] demonstrated that supplementation of goji berries at 1.5 or 3% modulated the gut microbiota composition by enhancing the growth of beneficial bacteria such as *Verrucomicrobia*, *Bacteroidetes*, *Bacteroidales S24-7* group, *Anaerotruncus*, *Coprococcus 1*, *Ruminococcaceae UCG-014*, and *Akkermansia*, while

suppressing the growth of harmful bacteria such as *Firmicutes*, *Helicobacter*, *Bacteroides*, and *Mucispirillum*. Meanwhile, administration of goji berries promoted the growth of short-chain fatty acid (SCFA)-producing bacteria, increasing the production of SCFAs [30]. Finally, Kang et al. [31] studied interleukin (IL)-10-deficient mice and showed feeding with goji berries (1% of dry feed weight) for 14 days enhanced the abundance of *Bifidobacteria* and butyrate-producing bacteria, *Clostridium leptum*, and its dominant constituent *Fecalibacterium prazusnitzii,* compared to control chow fed mice. This resulted in an increase in faecal butyrate content [31].

In another study, the effects of freeze-dried cranberries in dextran sodium sulphate-treated (DSST) male CD-1 mice (to induce colitis) were evaluated [32]. This study showed supplementation with 1.5% (*w*/*w*) freeze-dried cranberries (equivalent to 7.5 g of whole cranberry powder) alleviated colitis in DSST mice by reducing the levels of numerous pro-inflammatory cytokines. In addition, treatment with freeze-dried cranberries alleviated the reduced α-diversity of the gut microbiota induced by DSST [32]. Specifically, treatment with freeze-dried cranberries enhanced the abundance of beneficial bacteria, such as *Bifidobacterium* and *Lactobacillus* while reducing the abundance of harmful bacteria, such as *Suterella* and *Bilophila* [32].

5.2. Human Clinical Trials

Wijayabahu et al. [33] conducted a human clinical trial evaluating the effect of three servings (28.3 g per serving) of sun-dried raisins daily for 14 days on gut microbiota composition in healthy adults. Overall gut microbiota composition was not different after raisin consumption, but specific operational taxonomic units (OTUs) were affected. For example, OTUs matching *Faecalibacterium prausnitzii* and *Ruminococcaceae* were significantly enhanced, while OTUs matching *Klebsiella* spp. and *Prevotella* spp. were reduced significantly. These taxa, *Faecalibacterium prausnitzii* and *Ruminococcaceae*, are important for the breakdown of complex carbohydrates in the gut microbiota [33]. Meanwhile, the reduction in OTUs matching *Klebsiella sp.* and *Prevotella sp.* indicated a reduced risk for urinary tract infections and chronic inflammation, respectively [34].

In another randomized, double-blind, cross-over, controlled trial, healthy adults consumed 30 g/day of freeze-dried whole cranberry powder or a placebo for 5 days [35]. Cranberry powder consumption decreased the abundance of *Firmicutes*, while increasing the abundance of *Bacteroidetes*. In addition, the consumption of freeze-dried cranberry powder reduced the production of secondary bile acids and prevented the reduction in SCFAs, relative to the control [35]. Bekiares et al. [36] demonstrated that intake of 42 g/d of sweetened dried cranberries (SDC) for 14 days increased the *Firmicutes*:*Bacteroidetes* ratio and the relative abundance of *Akkermansia*. The authors recommended that further studies be conducted using well-controlled study designs and larger sample sizes to better understand the effect of cranberries on the relative abundance of *Akkermansia* [36].

The effect of prunes on bowel function has also been investigated [37]. Healthy adults (n = 120) consumed either 80 g or 120 g of prunes daily for 4 weeks, with stool weight and frequency as the primary study outcomes. Participants who consumed both 80 g and 120 g of prunes daily had higher stool weight and frequency than the control group. Supplementation with prunes significantly enhanced the relative abundance of *Bifidobacteria* compared to the control group. However, supplementation of prunes did not affect the levels of SCFA or stool pH in the subjects studied. The authors postulated that the effect of prunes on the gut microbiota could be mediated by fibre content, sorbitol, or phytochemicals in prunes [37]. Hence, more research should be carried out to confirm this result. A recent randomized, open-label, controlled trial evaluated the effects of prune consumption in adult women after undergoing benign gynaecologic surgery [38]. Participants (n = 77) consumed 12 prunes with 100 g docusate sodium (widely used as medicine as laxative and as stool softener) twice daily vs. docusate alone for 3 days. Participants who consumed 12 prunes twice daily had an increased likelihood of a bowel movement and earlier hospital discharge than the control group [38].

Dates have also been tested in a randomized, controlled, cross-over, clinical trial for gut microbiota, and GI function [39]. Healthy adult participants consumed 50 g of dates per day or maltodextrin-dextrose as a control for 21 days, with a 14-day washout period. Adults who consumed prunes daily had higher stool weight and bowel movement frequency than the control group. Supplementation with dates did not cause any significant alterations in the SCFA levels or in the growth of selected bacteria [39].

Most studies conducted to date have examined the effects of dried fruits on microbiota composition, whereas studies on metabolite production and functionality are scarce. Figure 1 summarises the potential mechanisms by which intake of dried fruits may modulate gut microbiota to influence health. Phytochemicals from dried fruits undergo significant biotransformation by gut microbiota, and the resulting metabolites may influence health [40]. Future studies, including in vitro, animal, human, and mechanistic studies are needed to address this research gap.

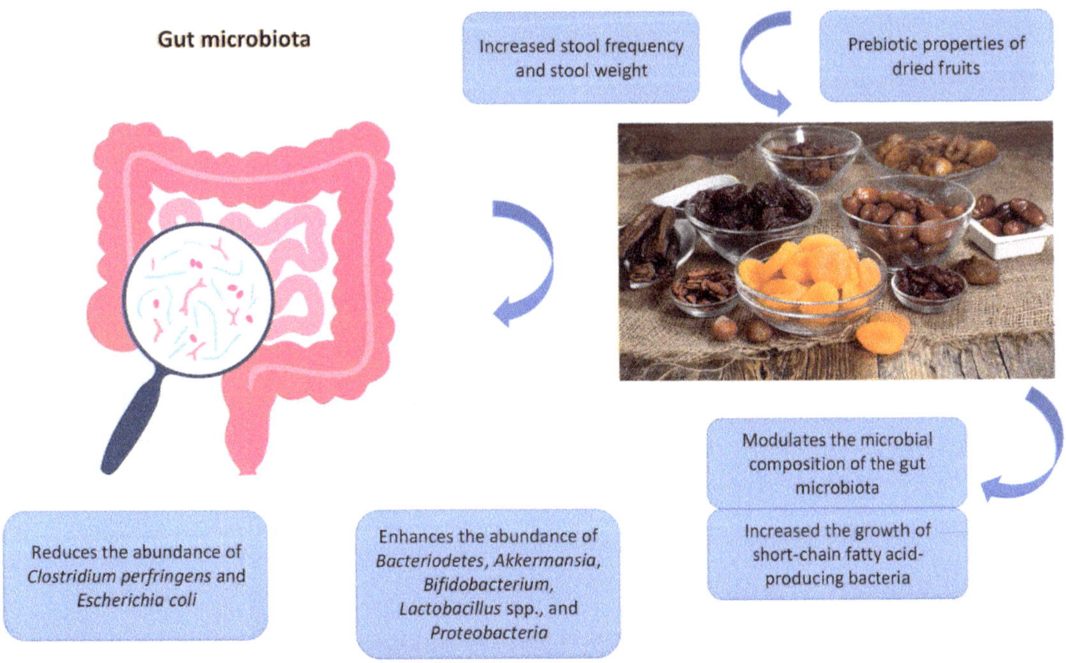

Figure 1. Potential mechanisms of action of dried fruit-related gut microbiota modulation.

6. Epidemiological Evidence for Health Benefits of Dried Fruits

6.1. CVD

CVDs are the leading cause of death worldwide [41]. A suboptimal diet is a major contributor to cardiovascular mortality, with low fruit intake ranked among the top three global dietary risk factors for cardiovascular deaths [42]. Individuals consuming dried fruits within the context of healthy dietary patterns generally have a healthier cardiometabolic risk profile, with lower lipid concentrations, blood glucose, and blood pressure [43–45]. However, there is limited evidence regarding the impact of dried fruit consumption on cardiovascular risk factors, CVD incidence, and mortality. Intake of grapes and raisins (queried together) $\geq$4 servings/week was associated with an 8% lower risk of hypertension in the Nurses' Health Study and Health Professionals Follow-Up Study cohorts, which included 187,453 individuals. Dried plum intake was not associated with incident hypertension after adjusting for other cardiovascular risk factors (including body mass index—BMI) and lifestyle and dietary factors [46]. Dried fruit consumption

(≥1 vs. <1 serving/day) was not associated with cardiovascular mortality in the Massachusetts Health Care Panel Study, though few people (~5%) consumed more than 1 serving of dried fruit daily [47]. In the UK Women's Cohort Study, combined fresh and dried fruit intake was associated with a lower risk of CVD mortality (8% lower risk per 80 g/day) [48]. However, just dried fruit intake was not significantly associated with cardiovascular mortality.

6.2. T2D

In the Nurses' Health Study and Health Professionals Follow-Up Study cohorts after adjusting for demographic, lifestyle, dietary factors, and diabetes-related risk factors including BMI, every three servings/week of grapes and raisins was associated with a 12% lower risk of T2D [49]. Greater dried plum intake was not associated with T2D incidence. Substituting equivalent portions (three servings/day) of dried plums or grapes and raisins for fruit juice was associated with an 18–19% lower diabetes risk [49]; however, the association may be attributable to fibre intake rather than dried fruit intake per se.

6.3. Body Weight

Observational evidence suggests that dried fruit intake is associated with a lower risk of excess adiposity. Based on the most recent cross-sectional analysis of data from the National Health and Nutrition Examination Survey (NHANES; from 2007 to 2016), dried fruit consumers had a lower mean BMI (−0.8, 99% CI −1.4 to −0.2; $p = 0.002$) and waist circumference (−2.6 cm, 99% CI −4.2 to −0.9 cm; $p < 0.001$) than non-consumers [50]. Mean dried fruit intake in US adults was 0.04 ± 0.001 cup-equivalents/day, which represented 3.7% of total daily fruit intake. In an earlier NHANES analysis (from 1999 to 2004), dried fruit consumers (≥1/8 cup-equivalent per day) had lower body weight (78.2 ± 0.6 vs. 80.7 ± 0.3 kg; $p < 0.01$) and BMI (27.1 ± 0.2 vs. 28.1 ± 0.2 kg/m^2; $p < 0.01$) than non-consumers [51]. In another NHANES analysis (from 2001 to 2012), raisin consumption (defined as having any amount during the first 24 h dietary recall) was associated with a lower body weight (−4.2%), BMI (−5.2%), and waist circumference (−3.8%) [52]. Raisin consumers were 39% less likely to have overweight or obesity.

In summary, few epidemiological studies report favourable associations between dried fruit and CVD, T2D, and body weight, but health benefits are not consistently shown. Observed associations between dried fruit intake and CVD, T2D, and body weight may be confounded by overall diet quality as well as other health-promoting behaviours. While many studies adjust for some foods and nutrients, the specific dietary components included in models vary widely. More comprehensive adjustment for dietary and lifestyle factors may strengthen future epidemiological studies investigating dried fruit consumption. In addition, the ability to detect associations between dried fruit intake and health may be limited by low observed consumption in these study populations. Investigating health associations in populations that routinely consume greater amounts of dried fruits could yield stronger evidence.

7. Clinical Trial Evidence for Dried Fruits and Health

7.1. Cardiometabolic Diseases

In several clinical studies, dried fruit intake has improved cardiovascular risk factors, including cholesterol, blood pressure, and glycaemic control (Table 2). However, the effects are inconsistent, which may be attributable to differences in the bioactive phytochemical and nutrient profiles of dried fruits, as well as differences in trial designs.

Table 2. Clinical trials reporting cardiometabolic effects of routine (≥4 weeks) dried fruit consumption.

References	Study Design	Duration (Week)	Participants (*n*)	Fruit (Dose)	Comparator	Findings
Sullivan et al. [45]	Crossover	4	Men and women with BMI 25–36 kg/m^2 and ≥1 additional cardiometabolic risk factor, *n* = 55	Equal parts (~28 g each) dried plums, Mission figs, Deglet Noor dates, and raisins totalling 3/4 cups/day	Energy-matched processed snacks (animal crackers and fruit snack gummies)	Dried fruits increased LDL-C (0.10 mmol/L) and non-HDL-C (0.12 mmol/L) and reduced HDL-C (−0.05 mmol/L) compared to baseline. Dried fruits increased fasting glucose compared to control (0.08 mmol/L). No between-group or within-group differences in total cholesterol, TAG, blood pressure, or insulin.
Tinker et al. [53]	Crossover	4	Men with elevated total cholesterol (5.2–7.5 mmol/L), *n* = 41	Dried plums, ~100 g/day (12 plums)	360 mL grape juice	Dried plums reduced LDL-C compared to grape juice (−0.17 mmol/L). No difference in total cholesterol, HDL-C, or TAG.
Clayton et al. [54]	Parallel	8	Men and women with BMI ≥ 25 kg/m^2, *n* = 45	Dried plums, ~84 g/day	Energy-matched portion (200 kcal) of low-fat muffins	Dried plums reduced LDL-C compared to low-fat muffins (−24.5 mg/dL). Dried plums increased C-peptide compared to baseline (+1.56 ng/mL). No between-group or within-group differences in total cholesterol, HDL-C, blood pressure, TAG, insulin, or glucose.
Alalwan et al. [55]	Parallel	16	Men and women with T2D, *n* = 96	Dates (Khudary cultivar, tamar stage), 3 dates/day	Usual diet	Dates reduced total cholesterol compared to baseline (−0.209 mmol/L). No between-group or within-group differences in HbA1c, TAG, HDL-C, or LDL-C.
Shishehbor et al. [56]	Parallel	5	Men and women with elevated total cholesterol (>200 mg/dL) or TAG (>200 mg/dL), *n* = 38	Raisins, 90 g/day	Usual diet	Raisins reduced DBP compared to control group (−1.56 mm Hg). Raisins reduced LDL-C (−0.68 mmol/L) and total cholesterol (−0.72 mmol/L) compared to baseline. No between-group or within-group differences in SBP, HDL-C, or TAG.
Kanellos et al. [57]	Parallel	24	Men and post-menopausal women with T2D, *n* = 48	Corinthian raisins, 36 g/day	Usual diet	Raisins reduced DBP compared to the control group (−6 mm Hg). No between-group or within-group differences in SBP, total cholesterol, LDL-C, HDL-C, TAG, fasting glucose, or HbA1c.

Table 2. *Cont.*

References	Study Design	Duration (Week)	Participants (*n*)	Fruit (Dose)	Comparator	Findings
Anderson et al. [58]	Parallel	12	Men and women with BMI 25–34.9 kg/m^2, blood pressure >120/80 mm Hg, and fasting glucose 90–150 mg/dL, *n* = 46	Raisins, 3 ounces/day	Energy-matched pre-packaged processed snacks (three 100 kcal packages)	Raisins reduced SBP (−5.4 mmHg vs. baseline; −6.3 mmHg vs. snacks), DBP (−5.5 mmHg vs. baseline; −3.6 mmHg vs. snacks), HDL-C (−3.6 mg/dL vs. baseline), and HbA1c (−0.12% vs. baseline; −0.08% vs. snacks). No between-group differences in total cholesterol, LDL-C, TAG, or fasting glucose.
Bays et al. [59]	Parallel	12	Men and women with T2D and BMI 25–50 kg/m^2, *n* = 46	Raisins, 3 ounces/day	Energy-matched pre-packaged processed snacks (three 100 kcal packages)	Raisins reduced SBP compared to snacks (−8.7 mm Hg). No between-group differences in fasting glucose, HbA1c, DBP, total cholesterol, LDL-C, HDL-C, or TAG.
Peterson et al. [60]	Crossover	5 (per arm)	Men and women with LDL-C 100–189 mg/dL and BMI 18.5–35 kg/m^2, *n* = 102	Dried California Mission figs (~120 g/day, 12–15 figs)	Usual diet	Figs increased total cholesterol compared to control (6 mg/dL). No difference in LDL-C, HDL-C, or TAG.

Abbreviations: BMI, body mass index; DBP, diastolic blood pressure; HbA1c, hemoglobin A1c; HDL-C, high-density lipoprotein cholesterol; LDL-C, low-density lipoprotein cholesterol; SBP, systolic blood pressure; T2D, type 2 diabetes; TAG, triacylglycerols.

Two studies reported improved low-density lipoprotein cholesterol (LDL-C) after prune consumption compared to energy-matched control foods. In a randomized crossover trial, LDL-C was 0.17 mmol/L (~6.6 mg/dL) lower after 41 hypercholesterolemic men consumed 12 (~100 g) prunes vs. a 360 mL portion of grape juice daily for 4 weeks [53]. In a parallel design trial, adults with overweight and obesity randomized to consume two daily 100-calorie prune snacks (~84 g/day) for 8 weeks had lower LDL-C (−24.5 mg/dL) compared to the control arm randomized to consume low-fat muffins [54]. In two other parallel design studies, cholesterol reductions were observed in participants who consumed dried fruits, but the changes did not differ relative to the control group. Consumption of three dates/day for 16 weeks reduced total cholesterol (−0.209 mmol/L) in adults with T2D, while changes in LDL-C and high-density lipoprotein cholesterol (HDL-C) were not significant [55]. Among adults with hyperlipidaemia, consuming 90 g/day of raisins for 5 weeks reduced total cholesterol (−0.72 mmol/L) and LDL-C (−0.68 mmol/L) [56]. In all four of these studies, diet records confirmed that participants maintained constant energy intakes across the study periods that were consistent with baseline intakes, and three of the four studies confirmed that weight remained stable throughout the study duration [54–56].

In contrast, several studies have shown either no change or increased cholesterol after dried fruit consumption. Three studies reported no effect of raisins on total or LDL-C compared to the usual diet [57] or energy-matched processed snacks [58,59]. In a randomized crossover trial in adults with above optimal or high LDL-C (100–189 mg/dL; *n* = 102), daily consumption of 120 g of dried Mission figs for 5 weeks increased total cholesterol compared to usual diet, though neither LDL-C nor HDL-C significantly differed between conditions [60]. Energy intake was approximately 200 kcal greater on the fig condition, resulting in a small statistically non-significant 0.4 kg weight gain. Among adults

with overweight and obesity, average total and LDL-C did not differ after 4 weeks of daily consumption of 3/4 cup of mixed dried fruits (comprising equal parts raisins, dates, prunes, and dried figs) compared to calorie- and carbohydrate-matched processed snacks, though LDL-C increased 0.10 mmol/L (~4 mg/dL) from baseline after dried fruit consumption [45]. While no diet records were collected, small (0.3–0.4 kg) weight gains were observed after both conditions, suggesting that study foods were not completely substituted for other dietary energy sources. Differences in energy balance may be an important explanatory factor distinguishing trials that demonstrate the cholesterol-lowering effects of dried fruits vs. those that do not.

Several studies have demonstrated blood pressure-lowering effects of raisins, but not other dried fruits. Both systolic and diastolic blood pressure (SBP and DBP) were reduced in adults with overweight or obesity who consumed three 1-ounce portions of raisins (84 g) daily for 12 weeks, compared to energy-matched processed snack foods [58]. SBP decreased with raisin consumption in a similarly designed study among adults with T2D [59], while a 36 g portion of Corinthian raisins consumed daily for 24 weeks improved DBP in adults with T2D, compared to the control arm consuming their usual diets [57]. Daily consumption of 90 g of raisins for 5 weeks also reduced DBP in adults with hyperlipidaemia, compared to the usual diet control arm [56]. In contrast, 3/4 cup of mixed dried fruits did not improve resting brachial, 24 h ambulatory, or central blood pressure compared to processed snacks in adults with overweight or obesity [45]. Prune consumption (84 g/day for 8 weeks) also did not reduce blood pressure compared to processed snacks [54].

The effect of dried fruit intake on glycaemic control is important given that they are high in natural sugars. Acutely, dried fruits have a low-to-moderate glycaemic index and can attenuate glycaemic response when substituted for refined carbohydrates [61,62], likely due to partial displacement of glucose with fructose. A lower glycaemic index diet has been associated with a lower risk of CVD and mortality [63].

Several studies have shown that routine dried fruit consumption does not adversely affect glycaemic control. In adults with overweight and obesity, a daily intake of 3 ounces of raisins for 12 weeks reduced haemoglobin A1c (HbA1c) (−0.08%) compared to energy-matched processed snacks, while fasting glucose was unchanged [58]. Among adults with T2D, daily consumption of 3 ounces/day of raisins for 12 weeks did not alter HbA1c or fasting glucose compared to calorie-matched snacks [59]. Similarly, compared to the usual diet, daily consumption of dates (3 g/day for 16 weeks) [55] or Corinthian raisins (36 g/day for 24 weeks) [57] did not affect HbA1c in adults with T2D.

In contrast, two studies showed adverse effects of dried fruits on glycaemic control. Daily prune intake (84 g) for 8 weeks increased C-peptide, which is released during insulin production [54]. However, fasting glucose and insulin concentrations were not altered, and the increase in C-peptide did not significantly differ from the control arm. In adults with overweight and obesity, a small increase in fasting glucose (0.08 mmol/L, ~1.4 mg/dL) was observed after 4 weeks of consuming 3/4 cup/day of mixed dried fruits compared to calorie- and carbohydrate-matched processed snacks [45]. Since glycaemic measures were not the primary focus of the trial, these findings should be interpreted with caution and require replication.

Overall, the evidence is mixed regarding the effect of dried fruit consumption on cardiovascular risk factors. While several clinical studies show reductions in cholesterol and blood pressure, without harm to glycaemic control, benefits are not consistently observed. Additional well-designed randomized controlled trials that account for the potential confounding effect of changes in energy intake and body weight are needed to confirm the cardiovascular benefits of dried fruit consumption.

7.2. Bone Health

Preclinical studies conducted in rodent models of osteopenia or osteoporosis show prune supplementation prevents and reverses bone loss by modulating oxidative and inflammatory pathways [64–66]. Inflammation and oxidative stress enhance bone resorption

by increasing osteoclast function and suppress bone formation through reducing osteoblast function [64]. These preclinical findings are generally supported by evidence from clinical trials. Other dried fruits have not been linked improved bone health.

Five clinical trials have been conducted in postmenopausal women and provide suggestive evidence that intake of prunes (50 or 100 g/day) for 3 to 12 months may have osteoprotective effects [67–71]. Four trials conducted by one research group show potential antioxidant and anti-inflammatory effects [67–69], as well as improvements in markers of bone formation [70] and resorption [67,68]. Hooshmand et al. [67] showed 100 g/day of prunes increased bone mineral density (BMD) of the ulna and spine compared to 75 g of dried apple after 12 months; no change in neck of femur, total hip, or total body BMD was observed. In a subsequent study, this group demonstrated 50 and 100 g/day of prunes attenuated loss of total BMD compared to the control group after 6 months; no effects were observed for total hip, L_1–L_4 lumbar vertebra, or ulna BMD [68]. In a recent, single-centre, parallel-arm 12-month randomized controlled trial, 50 g/day of prunes preserved total hip BMD (–0.3 ± 0.2%) compared to the control group (–1.1 ± 0.2%) in postmenopausal women. An intake of 100 g/day of prunes did not affect BMD; however, the dropout rate was 41% for this group, suggesting limited feasibility of this dose [71]. This large, well-conducted randomized controlled trial is generally confirmatory of previous trials, and the totality of the evidence suggests intake of 50 g/day of prunes might be an efficacious, non-pharmacological intervention to preserve BMD in postmenopausal women.

Recently, two small studies examining the effect of prune intake on markers of bone metabolism in older men have been conducted [72,73]. In a 3-month randomized controlled trial of men 55–80 years with mild bone loss, 50 g/day and 100 g/day of prunes had limited and inconsistent effects on markers of bone turnover compared to the control group [72]. In a subsequent 12-month study by the same group, 100 g/day of prunes did not affect total body, spine (L1–L4), hip, and ulna BMD [73]. The findings of these studies should be cautiously interpreted given the small sample sizes examined limiting statistical power.

8. Dried Fruits and Diet Quality

Individuals who consume dried fruits tend to have higher-quality diets, overall. Based on an analysis of NHANES data from 2007–2016, adult consumers of dried fruits had higher Healthy Eating Index (HEI)-2015 scores, representing better adherence to the 2015–2020 Dietary Guidelines for Americans [50]. Specifically, they had higher intakes of fruits, vegetables, whole grains, legumes, seafood, and plant proteins, and they had lower intakes of (and thus higher HEI-2015 component scores for) sodium, refined grains, and saturated fats (Figure 2) [50]. Consumed as snacks or incorporated into meals, dried fruits can contribute to a healthy dietary pattern. In US adults, dried fruit consumption adds to total fruit intake, rather than displacing other forms of fruit, and contributes to greater intakes of dietary fibre and potassium [50]. Thus, increasing dried fruit intake could be an effective strategy to increase intakes of fruit, fibre, and potassium.

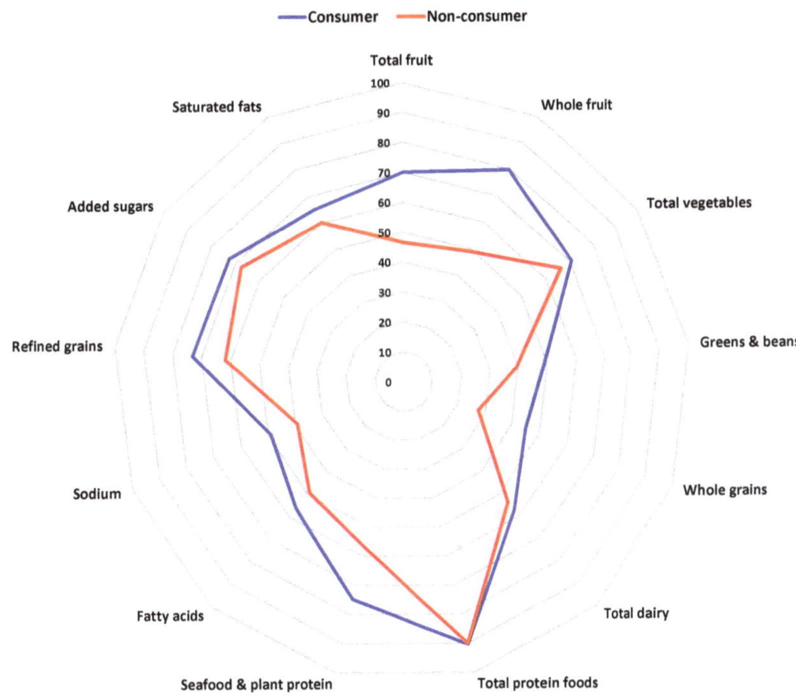

Figure 2. HEI-2015 component scores for dried fruit consumers and non-consumers, NHANES 2007–2016. Dried fruit consumers reported ≥1/4 cup-equivalent dried fruit intake on at least one of two 24 h diet recalls. Component scores are represented as percentages of maximum score. Data from [50]. Copyright Elsevier (2021).

9. Dietary Recommendations for Dried Fruit Consumption

A suboptimal diet is a leading cause of morbidity and mortality globally, and a suboptimal fruit intake is a major contributor to CVD, diabetes, and neoplasms [42]. A healthy dietary pattern that includes fruits is the basis for current dietary recommendations made by many organizations globally. In 2020, the World Health Organization recommended a healthy diet that includes the following: at least 400 g (e.g., five portions) of fruit and vegetables per day, excluding potatoes, sweet potatoes, cassava, and other starchy roots [74]. The 2020–2025 US Dietary Guidelines for Americans recommends two cup equivalents of fruits per day (per 2000 calories), which is equal to four servings per day [6]. With respect to dried fruits, 1/4 cup is equal to a 1/2 cup serving of fruit. The European Commission (the European Union as well as Iceland, Norway, Switzerland, and the UK) recommends two to three servings per day of fruit [7]. According to the Global Burden of Disease Study 2017, fruit consumption (94 g per day) falls short of current (two to three servings per day) dietary recommendations for fruit [42]. Similarly to the US, some European countries include dried fruit in the fruit recommendations, whereas others have specific recommendations for dried fruit in the range of 20–30 g per day [7]. According to 2020–2021 data from International Nuts Council, annual global per capita dried fruit consumption is about 1.2 g per day [75].

10. Potential Mechanisms Involved for Health Benefits of Dried Fruits

Among dried fruits, prunes are an excellent source of vitamin K, providing about 28 μg of vitamin K per serving of five prunes (47.5 g), which is 23% of the recommended dietary allowances for men and 31% for women [2,10]. Although no bioavailability studies have been conducted for vitamin K in prunes, vitamin K absorption is significantly increased in

the presence of some dietary fat. Thus, consuming prunes in the absence of fat, for example, alone as a snack, may result in relatively low absorption of their vitamin K [2]. In general, vitamin K plays a role in blood clotting, bone metabolism, and regulating blood calcium levels. Dried fruits are a good source of potassium [10]. It is likely that potassium is well absorbed from dried fruits and therefore would be a significant dietary source of the mineral. It has been reported that a high sodium:potassium ratio is associated with a significant increase in the risk of CVD and all-cause mortality in a US population [76]. Increased consumption of dried fruits would provide a means to reduce the sodium-potassium ratio, therefore potentially reducing CVD risk.

Dried fruits are rich in bioactives/phytochemicals, such as phenolic compounds, phytoestrogens, and carotenoids with potent antioxidant capacities [1,8,9]. These compounds scavenge free radicals and hence alleviate the oxidative stress that causes tissue damage, aging, and other chronic diseases [9]. Studies using simulated in vitro GI track digestion models indicate that there is little absorption of bioactives/phytochemicals from dried fruits after intestinal digestion. It has been hypothesized that unabsorbed bioactives/phytochemicals might be active in the digestive tract, rather than systemically. Because the digestive tract is a major organ involved in the immune response, effects within it may still contribute significantly to its overall health indirectly. In the colon, bioactives/phytochemicals are metabolized by the gut microbiota to form a wide range of metabolites, some of which would be responsible for health benefits attributed to the parent compounds. Therefore, measuring nutrient bioavailability from dried fruits is an open area of investigation [2]. Furthermore, the consumption of dried fruits modulates the diversity of gut microbiota by enhancing the relative abundance of beneficial microbes while reducing the relative abundance of harmful microbes [31,32]. Modulation of the gut microbiota by dried fruit consumption alleviated chronic inflammation, which in turn reduces the severity of metabolic disorders, such as CVD, T2D, and obesity [34].

In vitro and in vivo studies have been conducted to elucidate the mechanisms by which dried fruits may promote improvement in glycaemic control and insulin sensitivity. In human studies, dried fruits have a low-to-moderate glycaemic index. Mechanisms that may help to explain the benefits of dried fruits may relate to their relatively lower glycaemic index and insulin index potential, high mineral content of potassium and magnesium, and increased fibre content, as well as high levels of antioxidant and bioactives/phytochemicals [77].

Obesity is a disease characterized by chronic accumulation of excessive fat in adipose tissues, which leads to the production of pro-inflammatory cytokines. Chronic inflammation causes endothelial dysfunction, accompanied by insulin resistance. In addition, obesity, high blood pressure, high glucose levels, abnormal lipids/lipoproteins [total cholesterol, LDL-C, and triacylglycerols (TAG)], and chronic inflammation are major risk factors for atherosclerosis. Inhibition of inflammatory pathways by numerous bioactives/phytochemicals contained in dried fruits may beneficially affect inflammation-related diseases (e.g., metabolic disorders, such as CVD and T2D) [45,54].

Evidence suggests that prunes both prevent and reverse bone loss in postmenopausal women and potentially in men. Dried fruits, in general, contain several bioactives/ phytochemicals (including but not limited to resveratrol, kaemferol, proanthocyanidins, quercetin, cholorogenic acid, and catechin) with potential osteoprotective effects; however, the mechanisms by which these effects occur remain unclear [5].

In short, the potential of dried fruit to be a therapeutic strategy to prevent the severity of numerous chronic metabolic diseases warrants further investigation.

The relationships between bioactive compounds/phytochemicals present in dried fruits and health outcomes are summarized in Figure 3.

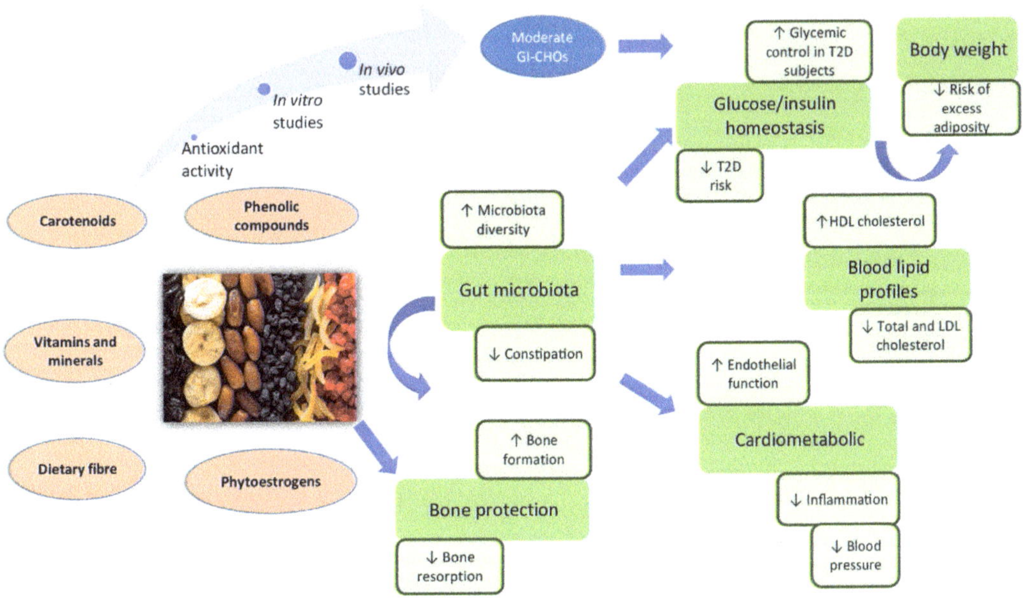

Figure 3. Summary of the health benefits ascribed to dried fruits. Frequent consumption of dried fruits benefits cardiovascular, gut microbiota, and bone health due to their unique composition of nutrients, bioactives/phytochemicals, and fibre. Abbreviations: CHOs, carbohydrates; HDL, high-density lipoprotein; GI, glycaemic index; LDL, low-density lipoprotein; T2D, type 2 diabetes.

11. Limitation of Studies in Dried Fruits and Future Recommendations

The bioactives/phytochemicals, gut microbiota, and bioavailability as well as health benefits of dried fruits have been less explored compared to their fresh counterparts. Although the bioactive/phytochemical profiles of some dried fruits (such as apricots, cranberries, dates, figs, prunes, and raisins) are well known, limited evidence is available for dried berries and tropical/non-tropical dried fruits. In addition, information about the bioavailability of minerals, vitamins, and bioactives/phytochemicals from dried fruits is scarce [1,8,9]. Measuring bioavailability of nutrients and bioactives/phytochemicals is an active area of research. More research needs to be conducted to determine circulating metabolite profiles after ingestion of dried fruits compared to fresh fruit counterparts [2]. Few studies have investigated the effect of dried fruits on gut microbiota, and further research is needed to understand the health implications of dried fruit related gut microbiota modulation [28]. Further research is also needed to clarify the extent to which bioactives/phytochemicals are altered by processing and whether this affects their bioactivity. Elucidating the mechanisms and bioactives/phytochemicals responsible can also help to identify processing techniques (such as sun-drying vs. heat-drying vs. freeze-drying) or particular fruits that best promote cardiometabolic and bone health [78]. Finally, evidence suggests that prunes have beneficial effects on bone health and may prevent osteoporosis. Further research is needed to examine the effectiveness of this potential non-pharmacological intervention given the side effects of pharmacological therapy for osteopenia [5].

12. Conclusions

Research about the health benefits (e.g., specifically related to the microbiota, cardiometabolic diseases, and bone health) of dried fruits is in its early stages. The phytochemical profiles of different dried fruits have been investigated; however, our understanding of their bioaccessibility and bioavailability is not well understood. Furthermore, a better

understanding of the biological effects of dried fruits and their bioactive compounds on cardiometabolic diseases, their risk factors, bone health, and the microbiome is needed. Despite this, we have an understanding that is evolving about the health benefits of some of the bioactive compounds in dried fruits and also the many health benefits of fresh fruits and juices. The encouraging results from the studies with both dried fruits, as well as fresh fruits and juices, justify further research. Additional scientific investigations will provide a better understanding of the biological effects of dried fruits on major chronic diseases and their biological mechanisms of action will be useful for future dietary guidance for dried fruits.

Author Contributions: C.A., S.K.C., P.M.K.-E., V.K.S. and K.S.P. designed the review, performed the literature research, wrote the first draft of the manuscript, and contributed to the editing of the manuscript. M.G.-F. and D.J.A.J. reviewed the manuscript. All authors have read and agreed to the published version of the manuscript.

Funding: The NUTS 2022 conference was planned by the Rovira i Virgili University, with all the expenses of the organization covered by Barcelo Congresos. Barcelo Congresos SA has received funds from the International Nut & Dried Fruit Council (INC). The researchers did not receive any additional financial support, aside from travel and accommodation, related to their participation in the NUTS 2022 conference. The sponsors did not have any role in selecting or inviting the speakers, nor in the writing or reviewing the proceedings of the conference.

Institutional Review Board Statement: Not applicable.

Informed Consent Statement: Not applicable

Data Availability Statement: No new data were created in this study. Data sharing is not applicable to this article.

Conflicts of Interest: P.M.K.-E reports current and recent research funding from the American Pistachio Growers, the American Pecan Council, the Peanut Institute, and the California Walnut Commission. K.S.P. has received sponsored research funding from The Peanut Institute, the California Walnut Commission, the American Pecan Council, and the American Pistachio Growers. C.A., S.K.C., V.K.S. and M.G.-F. have no conflicts of interest to disclose on dried fruits. D.J.A.J (his other COIs have been published in *Diabetes Care* **2022**, Dec 1, *45* (12), 2862–2870) has no conflicts of interest to disclose on dried fruits.

References

1. Chang, S.K.; Alasalvar, C.; Shahidi, F. Review of Dried Fruits: Phytochemicals, Antioxidant Efficacies, and Health Benefits. *J. Funct. Foods* **2016**, *21*, 113–132. [CrossRef]
2. Carughi, A.; Gallaher, D.; Mandalari, G. Bioavailability of Nutrients and Phytochemicals from Dried Fruits. In *Health Benefits of Nuts and Dried Fruits*; Alasalvar, C., Salas-Salvadó, J., Ros, E., Sabaté, J., Eds.; CRC Press, Taylor & Francis Group: Boca Raton, FL, USA, 2020; pp. 369–396.
3. European Food Safety Authority (EFSA). Scientific Opinion on the Substantiation of a Health Claim Related to Prunes and Contribution to Normal Bowel Function Pursuant to Article 14 of Regulation (EC) No 1924/2006. *EFSA J.* **2014**, *12*, 3892. [CrossRef]
4. Alasalvar, C.; Salas-Salvadó, J.; Ros, E.; Sabaté, J. Health Benefits of Nuts and Dried Fruits: An Overview. In *Health Benefits of Nuts and Dried Fruits*; Alasalvar, C., Salas-Salvadó, J., Ros, E., Sabaté, J., Eds.; CRC Press, Taylor & Francis Group: Boca Raton, FL, USA, 2020; pp. 1–9.
5. Arjmandi, B.H.; George, K.S. Bone Health and Osteoprotection. In *Health Benefits of Nuts and Dried Fruits*; Alasalvar, C., Salas-Salvadó, J., Ros, E., Sabaté, J., Eds.; CRC Press, Taylor & Francis Group: Boca Raton, FL, USA, 2020; pp. 469–486.
6. U.S. Department of Agriculture; U.S. Department of Health and Human Services. Dietary Guidelines for Americans, 2020–2025. 9th Edition. Available online: https://www.dietaryguidelines.gov (accessed on 23 January 2023).
7. Health Promotion Knowledge Gateway. Food-Based Dietary Guidelines in Europe. Available online: https://knowledge4policy.ec.europa.eu/health-promotion-knowledge-gateway/topic/food-based-dietary-guidelines-europe_en (accessed on 18 January 2023).
8. Alasalvar, C.; Salas-Salvadó, J.; Ros, E. Bioactives and Health Benefits of Nuts and Dried Fruits. *Food Chem.* **2020**, *314*, 126192. [CrossRef]

9. Alasalvar, C.; Chang, S.K.; Shahidi, F. Dried Fruits: Nutrients, Natural Antioxidants, and Phytochemicals. In *Health Benefits of Nuts and Dried Fruits*; Alasalvar, C., Salas-Salvadó, J., Ros, E., Sabaté, J., Eds.; CRC Press, Taylor & Francis Group: Boca Raton, FL, USA, 2020; pp. 335–368.
10. U.S. Department of Agriculture (USDA). National Nutrient Database for Standard Reference Legacy Release. 2018. Available online: https://ndb.nal.usda.gov/ndb/search/list (accessed on 1 March 2023).
11. Thompson, L.U.; Boucher, B.A.; Liu, Z.; Cotterchio, M.; Kreiger, N. Phytoestrogen Content of Foods Consumption in Canada, Including Isoflavones, Lignans, and Coumestan. *Nutr. Cancer* **2006**, *54*, 184–201. [CrossRef]
12. U.S. Department of Agriculture (USDA). *Dietary Reference Intakes for Energy, Carbohydrate, Fiber, Fat, Fatty Acids, Cholesterol, Protein, and Amino Acids*; The National Academies Press: Washington, DC, USA, 2005.
13. Silva Caldas, A.P.; Bressan, J. Dried Fruits as Components of Health Dietary Patters. In *Health Benefits of Nuts and Dried Fruits*; Alasalvar, C., Salas-Salvadó, J., Ros, E., Sabaté, J., Eds.; CRC Press, Taylor & Francis Group: Boca Raton, FL, USA, 2020; pp. 513–526.
14. U.S. Department of Agriculture (USDA). *Database for the Oxygen Radical Absorbance Capacity (ORAC) of Selected Foods, Release 2.0*; U.S. Department of Agriculture: Beltsville, MD, USA, 2010.
15. Ishiwata, K.; Yamaguchi, T.; Takamura, H.; Matoba, T. DPPH Radical-Scavenging Activity and Polyphenol Content in Dried Fruits. *Food Sci. Technol. Res.* **2004**, *10*, 152–156. [CrossRef]
16. Vinson, J.A.; Zubik, L.; Bose, P.; Samman, N.; Proch, J. Dried Fruits: Excellent *In Vitro* and *In Vivo* antioxidants. *J. Am. Coll. Nutr.* **2005**, *24*, 44–50. [CrossRef]
17. Rababah, T.M.; Ereifej, K.; Howard, L. Effect of Ascorbic Acid and Dehydration on Concentrations of Total Phenolics, Antioxidant Capacity, Anthocyanins, and Color in Fruits. *J. Agric. Food Chem.* **2005**, *53*, 4444–4447. [CrossRef]
18. Threlfall, R.; Morris, J.; Meullenet, J.F. Product Development and Nutraceutical Analysis to Enhance the Value of Dried Fruit. *J. Food Qual.* **2007**, *30*, 552–566. [CrossRef]
19. McClements, D.J.; Li, Y. Review of In Vitro Digestion Models for Rapid Screening of Emulsion-based Systems. *Food Funct.* **2010**, *1*, 32–59. [CrossRef] [PubMed]
20. Ma, T.; Lan, T.; Geng, T.; Ju, Y.; Cheng, G.; Que, Z.; Gao, G.; Fang, Y.; Sun, X. Nutritional Properties and Biological Activities of Kiwifruit (Actinidia) and Kiwifruit Products under Simulated Gastrointestinal In Vitro Digestion. *Food Nutr. Res.* **2019**, *63*, 1674. [CrossRef]
21. Scrob, T.; Hosu, A.; Cimpoiu, C. The Influence of *In Vitro* Gastrointestinal Digestion of Brassica Oleracea Florets on the Antioxidant Activity and Chlorophyll, Carotenoid and Phenolic Content. *Antioxidants* **2019**, *8*, 212. [CrossRef]
22. Panagopoulou, E.A.; Chiou, A.; Kasimatis, T.-D.; Bismpikis, M.; Mouraka, P.; Karathanos, V.T. Dried Dates: Polar Phenols and Their Fate during *in Vitro* Digestion. *J. Food Meas. Charact.* **2021**, *15*, 1899–1906. [CrossRef]
23. Schmite, B.d.F.P.; Bitobrovec, A.; Hacke, A.C.M.; Pereira, R.P.; Los Weinert, P.; Dos Anjos, V.E. In Vitro Bioaccessibility of Al, Cu, Cd, and Pb Following Simulated Gastro-Intestinal Digestion and Total Content of These Metals in Different Brazilian Brands of Yerba Mate Tea. *Food Chem.* **2019**, *281*, 285–293. [CrossRef]
24. Scrob, T.; Covaci, E.; Hosu, A.; Tanaselia, C.; Casoni, D.; Torok, A.I.; Frentiu, T.; Cimpoiu, C. Effect of In Vitro Simulated Gastrointestinal Digestion on Some Nutritional Characteristics of Several Dried Fruits. *Food Chem.* **2022**, *385*, 132713. [CrossRef]
25. Kamiloglu, S.; Pasli, A.A.; Ozcelik, B.; Capanoglu, E. Evaluating the *In Vitro* Bioaccessibility of Phenolics and Antioxidant Activity during Consumption of Dried Fruits with Nuts. *LWT-Food Sci. Technol.* **2014**, *56*, 284–289. [CrossRef]
26. Moles, L.; Otaegui, D. The Impact of Diet on Microbiota Evolution and Human Health. Is Diet an Adequate Tool for Microbiota Modulation? *Nutrients* **2020**, *12*, 1654. [CrossRef] [PubMed]
27. Zhang, N.; Ju, Z.; Zuo, T. Time for Food: The Impact of Diet on Gut Microbiota and Human Health. *Nutrition* **2018**, *51*, 80–85. [CrossRef] [PubMed]
28. Muñoz, M.M.; Lamuela-Raventós, R.M. Gut Health and Microbiota. In *Health Benefits of Nuts and Dried Fruits*; Alasalvar, C., Salas-Salvadó, J., Ros, E., Sabaté, J., Eds.; CRC Press, Taylor & Francis Group: Boca Raton, FL, USA, 2020; pp. 487–495.
29. Cremonesi, P.; Curone, G.; Biscarini, F.; Cotozzolo, E.; Menchetti, L.; Riva, F.; Marongiu, M.L.; Castiglioni, B.; Barbato, O.; Munga, A. Dietary Supplementation with Goji Berries (*Lycium barbarum*) Modulates the Microbiota of Digestive Tract and Caecal Metabolites in Rabbits. *Animals* **2022**, *12*, 121. [CrossRef]
30. Tian, B.; Zhang, Z.; Zhao, J.; Ma, Q.; Liu, H.; Nie, C.; Ma, Z.; An, W.; Li, J. Dietary Whole Goji Berry (*Lycium barbarum*) Intake Improves Colonic Barrier Function by Altering Gut Microbiota Composition in Mice. *Int. J. Food Sci. Technol.* **2021**, *56*, 103–114. [CrossRef]
31. Kang, Y.; Yang, G.; Zhang, S.; Ross, C.F.; Zhu, M. Goji Berry Modulates Gut Microbiota and Alleviates Colitis in IL-10-deficient Mice. *Mol. Nutr. Food Res.* **2018**, *62*, 1800535. [CrossRef]
32. Cai, X.; Han, Y.; Gu, M.; Song, M.; Wu, X.; Li, Z.; Li, F.; Goulette, T.; Xiao, H. Dietary Cranberry Suppressed Colonic Inflammation and Alleviated Gut Microbiota Dysbiosis in Dextran Sodium Sulfate-Treated Mice. *Food Funct.* **2019**, *10*, 6331–6341. [CrossRef]
33. Wijayabahu, A.T.; Waugh, S.G.; Ukhanova, M.; Mai, V. Dietary Raisin Intake Has Limited Effect on Gut Microbiota Composition in Adult Volunteers. *Nutr. J.* **2019**, *18*, 14. [CrossRef] [PubMed]
34. Clemente, J.C.; Manasson, J.; Scher, J.U. The Role of the Gut Microbiome in Systemic Inflammatory Disease. *BMJ* **2018**, *360*, j5145. [CrossRef] [PubMed]

35. Rodríguez-Morató, J.; Matthan, N.R.; Liu, J.; de la Torre, R.; Chen, C.-Y.O. Cranberries Attenuate Animal-Based Diet-Induced Changes in Microbiota Composition and Functionality: A Randomized Crossover Controlled Feeding Trial. *J. Nutr. Biochem.* **2018**, *62*, 76–86. [CrossRef] [PubMed]
36. Bekiares, N.; Krueger, C.G.; Meudt, J.J.; Shanmuganayagam, D.; Reed, J.D. Effect of Sweetened Dried Cranberry Consumption on Urinary Proteome and Fecal Microbiome in Healthy Human Subjects. *Omi. J. Integr. Biol.* **2017**, *22*, 145–153. [CrossRef]
37. Lever, E.; Scott, S.M.; Louis, P.; Emery, P.W.; Whelan, K. The Effect of Prunes on Stool Output, Gut Transit Time and Gastrointestinal Microbiota: A Randomised Controlled Trial. *Clin. Nutr.* **2019**, *38*, 165–173. [CrossRef] [PubMed]
38. Rasouli, M.A.; Dancz, C.E.; Dahl, M.; Volpe, K.A.; Horton, C.J.; Ozel, B.Z. Effect of Prunes on Gastrointestinal Function after Benign Gynecological Surgery: A Randomized Control Trial. *Langenbeck's Arch. Surg.* **2022**, *407*, 3803–3810. [CrossRef] [PubMed]
39. Eid, N.; Osmanova, H.; Natchez, C.; Walton, G.; Costabile, A.; Gibson, G.; Rowland, I.; Spencer, J.P.E. Impact of Palm Date Consumption on Microbiota Growth and Large Intestinal Health: A Randomised, Controlled, Cross-over, Human Intervention Study. *Br. J. Nutr.* **2015**, *114*, 1226–1236. [CrossRef] [PubMed]
40. Luca, S.V.; Macovei, I.; Bujor, A.; Miron, A.; Skalicka-Woźniak, K.; Aprotosoaie, A.C.; Trifan, A. Bioactivity of Dietary Polyphenols: The Role of Metabolites. *Crit. Rev. Food Sci. Nutr.* **2020**, *60*, 626–659. [CrossRef]
41. Roth, G.A.; Mensah, G.A.; Johnson, C.O.; Addolorato, G.; Ammirati, E.; Baddour, L.M.; Barengo, N.C.; Beaton, A.Z.; Benjamin, E.J.; Benziger, C.P. Global Burden of Cardiovascular Diseases and Risk Factors, 1990–2019: Update from the GBD 2019 Study. *J. Am. Coll. Cardiol.* **2020**, *76*, 2982–3021. [CrossRef]
42. Afshin, A.; Sur, P.J.; Fay, K.A.; Cornaby, L.; Ferrara, G.; Salama, J.S.; Mullany, E.C.; Abate, K.H.; Abbafati, C.; Abebe, Z. Health Effects of Dietary Risks in 195 Countries, 1990–2017: A Systematic Analysis for the Global Burden of Disease Study 2017. *Lancet* **2019**, *393*, 1958–1972. [CrossRef] [PubMed]
43. Centritto, F.; Iacoviello, L.; di Giuseppe, R.; De Curtis, A.; Costanzo, S.; Zito, F.; Grioni, S.; Sieri, S.; Donati, M.B.; de Gaetano, G. Dietary Patterns, Cardiovascular Risk Factors and C-Reactive Protein in a Healthy Italian Population. *Nutr. Metab. Cardiovasc. Dis.* **2009**, *19*, 697–706. [CrossRef] [PubMed]
44. Czekajlo, A.; Rozanska, D.; Zatonska, K.; Szuba, A.; Regulska-Ilow, B. Association between Dietary Patterns and Cardiovascular Risk Factors in a Selected Population of Lower Silesia (PURE Study Poland). *Ann. Agric. Environ. Med.* **2018**, *25*, 635–641. [CrossRef]
45. Sullivan, V.K.; Petersen, K.S.; Kris-Etherton, P.M. Dried Fruit Consumption and Cardiometabolic Health: A Randomised Crossover Trial. *Br. J. Nutr.* **2020**, *124*, 912–921. [CrossRef]
46. Borgi, L.; Muraki, I.; Satija, A.; Willett, W.C.; Rimm, E.B.; Forman, J.P. Fruit and Vegetable Consumption and the Incidence of Hypertension in Three Prospective Cohort Studies. *Hypertension* **2016**, *67*, 288–293. [CrossRef] [PubMed]
47. Gaziano, J.M.; Manson, J.E.; Branch, L.G.; Colditz, G.A.; Willett, W.C.; Buring, J.E. A Prospective Study of Consumption of Carotenoids in Fruits and Vegetables and Decreased Cardiovascular Mortality in the Elderly. *Ann. Epidemiol.* **1995**, *5*, 255–260. [CrossRef]
48. Lai, H.T.M.; Threapleton, D.E.; Day, A.J.; Williamson, G.; Cade, J.E.; Burley, V.J. Fruit Intake and Cardiovascular Disease Mortality in the UK Women's Cohort Study. *Eur. J. Epidemiol.* **2015**, *30*, 1035–1048. [CrossRef]
49. Muraki, I.; Imamura, F.; Manson, J.E.; Hu, F.B.; Willett, W.C.; van Dam, R.M.; Sun, Q. Fruit Consumption and Risk of Type 2 Diabetes: Results from Three Prospective Longitudinal Cohort Studies. *BMJ* **2013**, *347*, f5001. [CrossRef]
50. Sullivan, V.K.; Na, M.; Proctor, D.N.; Kris-Etherton, P.M.; Petersen, K.S. Consumption of Dried Fruits Is Associated with Greater Intakes of Underconsumed Nutrients, Higher Total Energy Intakes, and Better Diet Quality in US Adults: A Cross-Sectional Analysis of the National Health and Nutrition Examination Survey, 2007–2016. *J. Acad. Nutr. Diet.* **2021**, *121*, 1258–1272. [CrossRef]
51. Keast, D.R.; O'Neil, C.E.; Jones, J.M. Dried Fruit Consumption Is Associated with Improved Diet Quality and Reduced Obesity in US Adults: National Health and Nutrition Examination Survey, 1999–2004. *Nutr. Res.* **2011**, *31*, 460–467. [CrossRef]
52. Fulgoni, V.L., III; Painter, J.; Carughi, A. Association of Raisin Consumption with Nutrient Intake, Diet Quality, and Health Risk Factors in US Adults: National Health and Nutrition Examination Survey 2001–2012. *Food Nutr. Res.* **2017**, *61*, 1378567. [CrossRef]
53. Tinker, L.F.; Schneeman, B.O.; Davis, P.A.; Gallaher, D.D.; Waggoner, C.R. Consumption of Prunes as a Source of Dietary Fiber in Men with Mild Hypercholesterolemia. *Am. J. Clin. Nutr.* **1991**, *53*, 1259–1265. [CrossRef] [PubMed]
54. Clayton, Z.S.; Fusco, E.; Schreiber, L.; Carpenter, J.N.; Hooshmand, S.; Hong, M.Y.; Kern, M. Snack Selection Influences Glucose Metabolism, Antioxidant Capacity and Cholesterol in Healthy Overweight Adults: A Randomized Parallel Arm Trial. *Nutr. Res.* **2019**, *65*, 89–98. [CrossRef]
55. Alalwan, T.A.; Perna, S.; Mandeel, Q.A.; Abdulhadi, A.; Alsayyad, A.S.; D'Antona, G.; Negro, M.; Riva, A.; Petrangolini, G.; Allegrini, P. Effects of Daily Low-Dose Date Consumption on Glycemic Control, Lipid Profile, and Quality of Life in Adults with Pre-and Type 2 Diabetes: A Randomized Controlled Trial. *Nutrients* **2020**, *12*, 217. [CrossRef]
56. Shishehbor, F.; Joola, P.; Malehi, A.S.; Jalalifar, M.A. The Effect of Black Seed Raisin on Some Cardiovascular Risk Factors, Serum Malondialdehyde, and Total Antioxidant Capacity in Hyperlipidemic Patients: A Randomized Controlled Trials. *Ir. J. Med. Sci.* **2022**, *191*, 195–204. [CrossRef]
57. Kanellos, P.T.; Kaliora, A.C.; Tentolouris, N.K.; Argiana, V.; Perrea, D.; Kalogeropoulos, N.; Kountouri, A.M.; Karathanos, V.T. A Pilot, Randomized Controlled Trial to Examine the Health Outcomes of Raisin Consumption in Patients with Diabetes. *Nutrition* **2014**, *30*, 358–364. [CrossRef] [PubMed]

58. Anderson, J.W.; Weiter, K.M.; Christian, A.L.; Ritchey, M.B.; Bays, H.E. Raisins Compared with Other Snack Effects on Glycemia and Blood Pressure: A Randomized, Controlled Trial. *Postgr. Med.* **2014**, *126*, 37–43. [CrossRef]
59. Bays, H.; Weiter, K.; Anderson, J. A Randomized Study of Raisins versus Alternative Snacks on Glycemic Control and Other Cardiovascular Risk Factors in Patients with Type 2 Diabetes Mellitus. *Phys. Sport.* **2015**, *43*, 37–43. [CrossRef] [PubMed]
60. Peterson, J.M.; Montgomery, S.; Haddad, E.; Kearney, L.; Tonstad, S. Effect of Consumption of Dried California Mission Figs on Lipid Concentrations. *Ann. Nutr. Metab.* **2011**, *58*, 232–238. [CrossRef]
61. Esfahani, A.; Lam, J.; Kendall, C.W.C. Acute Effects of Raisin Consumption on Glucose and Insulin Reponses in Healthy Individuals. *J. Nutr. Sci.* **2014**, *3*, E1. [CrossRef] [PubMed]
62. Viguiliouk, E.; Jenkins, A.L.; Blanco Mejia, S.; Sievenpiper, J.L.; Kendall, C.W.C. Effect of Dried Fruit on Postprandial Glycemia: A Randomized Acute-Feeding Trial. *Nutr. Diabetes* **2018**, *8*, 59. [CrossRef]
63. Jenkins, D.J.A.; Dehghan, M.; Mente, A.; Bangdiwala, S.I.; Rangarajan, S.; Srichaikul, K.; Mohan, V.; Avezum, A.; Díaz, R.; Rosengren, A. Glycemic Index, Glycemic Load, and Cardiovascular Disease and Mortality. *N. Engl. J. Med.* **2021**, *384*, 1312–1322. [CrossRef]
64. Damani, J.J.; De Souza, M.J.; VanEvery, H.L.; Strock, N.C.A.; Rogers, C.J. The Role of Prunes in Modulating Inflammatory Pathways to Improve Bone Health in Postmenopausal Women. *Adv. Nutr.* **2022**, *13*, 1476–1492. [CrossRef] [PubMed]
65. Wallace, T.C. Dried Plums, Prunes and Bone Health: A Comprehensive Review. *Nutrients* **2017**, *9*, 401. [CrossRef]
66. Arjmandi, B.H.; Johnson, S.A.; Pourafshar, S.; Navaei, N.; George, K.S.; Hooshmand, S.; Chai, S.C.; Akhavan, N.S. Bone-Protective Effects of Dried Plum in Postmenopausal Women: Efficacy and Possible Mechanisms. *Nutrients* **2017**, *9*, 496. [CrossRef] [PubMed]
67. Hooshmand, S.; Chai, S.C.; Saadat, R.L.; Payton, M.E.; Brummel-Smith, K.; Arjmandi, B.H. Comparative Effects of Dried Plum and Dried Apple on Bone in Postmenopausal Women. *Br. J. Nutr.* **2011**, *106*, 923–930. [CrossRef]
68. Hooshmand, S.; Kern, M.; Metti, D.; Shamloufard, P.; Chai, S.C.; Johnson, S.A.; Payton, M.E.; Arjmandi, B.H. The Effect of Two Doses of Dried Plum on Bone Density and Bone Biomarkers in Osteopenic Postmenopausal Women: A Randomized, Controlled Trial. *Osteoporos. Int.* **2016**, *27*, 2271–2279. [CrossRef] [PubMed]
69. Hong, M.Y.; Kern, M.; Nakamichi-Lee, M.; Abbaspour, N.; Ahouraei Far, A.; Hooshmand, S. Dried Plum Consumption Improves Total Cholesterol and Antioxidant Capacity and Reduces Inflammation in Healthy Postmenopausal Women. *J. Med. Food* **2021**, *24*, 1161–1168. [CrossRef]
70. Arjmandi, B.H.; Khalil, D.A.; Lucas, E.A.; Georgis, A.; Stoecker, B.J.; Hardin, C.; Payton, M.E.; Wild, R.A. Dried Plums Improve Indices of Bone Formation in Postmenopausal Women. *J. Women's Health Gend. Based Med.* **2002**, *11*, 61–68. [CrossRef]
71. De Souza, M.J.; Strock, N.C.A.; Williams, N.I.; Lee, H.; Koltun, K.J.; Rogers, C.; Ferruzzi, M.G.; Nakatsu, C.H.; Weaver, C. Prunes Preserve Hip Bone Mineral Density in a 12-Month Randomized Controlled Trial in Postmenopausal Women: The Prune Study. *Am. J. Clin. Nutr.* **2022**, *116*, 897–910. [CrossRef]
72. George, K.S.; Munoz, J.; Ormsbee, L.T.; Akhavan, N.S.; Foley, E.M.; Siebert, S.C.; Kim, J.-S.; Hickner, R.C.; Arjmandi, B.H. The Short-Term Effect of Prunes in Improving Bone in Men. *Nutrients* **2022**, *14*, 276. [CrossRef] [PubMed]
73. Hooshmand, S.; Gaffen, D.; Eisner, A.; Fajardo, J.; Payton, M.; Kern, M. Effects of 12 Months Consumption of 100 g Dried Plum (Prunes) on Bone Biomarkers, Density, and Strength in Men. *J. Med. Food* **2022**, *25*, 40–47. [CrossRef] [PubMed]
74. World Health Organization. Healthy Diet. Available online: https://www.who.int/news-room/fact-sheets/detail/healthy-diet (accessed on 19 January 2023).
75. International Nut and Dried Fruit Council. INC Nuts & Dried Fruits Statistical Yearbook 2021/2022. Available online: https://inc.nutfruit.org/inc-releases-2021-2022-statistical-yearbook/ (accessed on 16 January 2023).
76. Yang, Q.; Liu, T.; Kuklina, E.V.; Flanders, W.D.; Hong, Y.; Gillespie, C.; Chang, M.-H.; Gwinn, M.; Dowling, N.; Khoury, M.J.; et al. Sodium and Potassium Intake and Mortality among US Adults: Prospective Data from the Third National Health and Nutrition Examination Survey. *Arch. Intern. Med.* **2011**, *171*, 1183–1191. [CrossRef] [PubMed]
77. Srichaikul, K.; Ong, M.; Prasla, Z.; Kohen, Y.; Mandalozano, I.; Paquette, M.; Sahye-Pudaruth, S.; Patel, D.; Kendal, C.W.; Sievenpiper, J.L.; et al. *Dried Fruits in the Prevention and Control of Diabetes (Insulin Resistance and Prediabetics)*; Alasalvar, C., Salas-Salvadó, J., Ros, E., Sabaté, J., Eds.; CRC Press, Taylor & Francis Group: Boca Raton, FL, USA, 2020; pp. 436–447.
78. Sullivan, V.; Petersen, K.; Kris-Etherton, P. Dried Fruits and Cardio-Metabolic Syndrome (Endothelial Function, Inflammation, and Blood Pressure). In *Health Benefits of Nuts and Dried Fruits*; Alasalvar, C., Salas-Salvadó, J., Ros, E., Sabaté, J., Eds.; CRC Press, Taylor & Francis Group: Boca Raton, FL, USA, 2020; pp. 413–436.

nutrients

MDPI

Review

Novel Lines of Research on the Environmental and Human Health Impacts of Nut Consumption

Linda Tapsell [1], Joan Sabaté [2], Raquel Martínez [2], Marc Llavanera [3,4], Elizabeth Neale [1] and Albert Salas-Huetos [5,6,7,*]

1 School of Medical Indigenous and Health Sciences, University of Wollongong, Wollongong, NSW 2522, Australia
2 Center for Nutrition, Lifestyle and Disease Prevention, School of Public Health, Loma Linda University, Loma Linda, CA 92350, USA
3 Unit of Cell Biology, Department of Biology, Faculty of Sciences, University of Girona, 17003 Girona, Spain
4 Biotechnology of Animal and Human Reproduction (TechnoSperm), Institute of Food and Agricultural Technology, University of Girona, 17003 Girona, Spain
5 Unit of Preventive Medicine and Public Health, Faculty of Medicine and Health Sciences, Universitat Rovira i Virgili, 43201 Reus, Spain
6 Department of Nutrition, Harvard T.H. Chan School of Public Health, Harvard University, Boston, MA 02115, USA
7 Consorcio CIBER, M.P., Fisiopatología de la Obesidad y Nutrición (CIBERobn), Instituto de Salud Carlos III (ISCIII), 28029 Madrid, Spain
* Correspondence: albert.salas@urv.cat; Tel.: +34-977-75-93-79

Abstract: Nuts have formed part of human diets throughout the ages. In recent decades, research has shown they are key foods in dietary patterns associated with lower chronic disease risk. The current state of climate change, however, has introduced an imperative to review the impact of dietary patterns on the environment with a shift to plant-based diets. Nuts emerge as a significant source of protein in plant-based diets and are a minimally processed and sustainable food. Research in this area is evolving to drive better production methods in varying climate conditions. Nevertheless, nut consumption remains an important contributor to human health. The mechanisms of action can be explained in terms of the nutrients they deliver. Studies of nut consumption have linked components such as monounsaturated fatty acids, plant omega-3 fatty acids, antioxidants, and plant sterols to improved lipoprotein profiles, lower blood pressure, and reduced cardiovascular disease risk. Preliminary research also indicates possible beneficial effects of nut consumption on reproductive health. In any case, the ultimate effects of foods on health are the results of multiple interactive factors, so where nuts fit within dietary patterns is a significant consideration for research translation. This has implications for research methodologies, including categorization within food groups and inclusion in Healthy Dietary Indices. The aim of this narrative review is to outline new focal points for investigation that examine the environmental and some novel human health impacts of nut consumption and discuss future directions for research.

Keywords: nuts; environment; sustainability; reproduction; sexual function; diet; dietary patterns

Citation: Tapsell, L.; Sabaté, J.; Martínez, R.; Llavanera, M.; Neale, E.; Salas-Huetos, A. Novel Lines of Research on the Environmental and Human Health Impacts of Nut Consumption. *Nutrients* **2023**, *15*, 955. https://doi.org/10.3390/nu15040955

Academic Editor: Licia Iacoviello

Received: 17 January 2023
Revised: 10 February 2023
Accepted: 13 February 2023
Published: 14 February 2023

1. Introduction

As naturally occurring edible and nutritious foods, nuts have been part of the human diet throughout the ages [1]. Modern nutrition science provides evidence of their health benefits, but foods have not always been the object of nutrition research, with a focus in past decades on nutrients contained in foods. At the same time, the industrialization of the food supply saw the emergence of chronic lifestyle-related diseases, such as obesity, cardiovascular disease (CVD), and type 2 diabetes. Research identified so-called 'negative' components of the diet including dietary fat and excess energy consumption [2], which

translated to dietary advice and created an ambiguous position for nuts as a high-energy, high-fat food.

Population-based dietary guidelines appeared in the 1980s aimed at providing adequate nutrition and preventing chronic disease. Early guidelines referred to staple food groups, with advice to avoid foods high in fat, sugar, and salt [2]. The position of nuts in food groups was variable, but the value of naturally occurring foods, captured in the concept of food synergy [3], and an appreciation of the relationship between nutrients, foods, and dietary patterns [4] led to today's guidelines having a greater focus on dietary patterns. Research on nuts followed this direction, expanding beyond their nutritional contributions to nuts as a significant food in healthy dietary patterns. Direct clinical evidence of health effects came from trials involving at-risk populations. Basic science research provided insights into the molecular pathways underlying health effects, and epidemiological studies confirmed that associations between nut consumption and health outcomes occurred in the broader population. Each of these types of research were important in building the body of evidence, despite challenges in providing timely and consistent studies to support nutrition policy and practice [5–7]. At the same time, clinical evidence review methodology has developed further to consider the quality as well as quantity of research. This development recognized that research practices and study designs require sufficient scrutiny to assure confidence in results and valid translation to practice [8].

Today, evidence supporting nut consumption is extensive. It goes beyond chronic disease prevention to other forms of human health—including reproductive health—and then to the planet's health. The environmental impact of healthy dietary patterns is part of the evidence analysis, as nuts are significant foods in plant-based diets. The global imperative to address climate change calls for additional research methodologies that address the environmental impact of foods [8]. Today, there are strong calls to combine imperatives for human health with that of the environment [9].

The aim of this narrative review is to outline new focal points for research that examine the environmental and novel human health impacts of the consumption of nuts and discuss future directions.

2. Nuts and Environmental Sustainability

Environmental sustainability is one of the most pressing new areas of research on food today. Food production and consumption face unprecedented scrutiny as their impact on the natural environment and human health becomes more evident. How foods are grown (including water use), processed, sold, prepared, cooked, consumed, and disposed of is crucial. In studying nuts and environmental sustainability, all these stages of production and consumption (the 'food system') must be considered.

The global food system produces enough calories for a growing world population. However, about half of the population is malnourished with almost 1 billion people not consuming enough food and experiencing hunger, and almost 2 billion overconsuming foods low in nutritional quality associated with micronutrient and phytonutrient deficiencies, obesity, and increased incidence of chronic disease. Globally, the non-optimal intake of foods in the diet is estimated to account for approximately 22% of all deaths among adults and 15% of disability-adjusted life-years [10]. At the same time, agriculture uses ~70% of the global freshwater [11]. The food supply chain is responsible for ~26% of global greenhouse gas emissions (GHGe), occupies ~43% of habitable land, causes ~78% of the ocean and freshwater eutrophication, and ~32% of terrestrial acidification worldwide [12].

Several strategies have been proposed to decrease the environmental pressures exerted by the food systems. These include (1) improving agricultural technologies to enhance productivity and reduce harmful emissions; (2) reducing food loss and waste to decrease food production requirements and waste emissions in landfills; and (3) shifting to the production and consumption of foods that support human and planetary health [13–16]. Thus, the type and amount of food produced and consumed are major determining factors in promoting human health within planetary boundaries [17,18]. Research can identify

where nuts fit within these parameters. Addressing the environmental impacts of the lifecycle of nuts (LCA, Lifecycle Analysis), from production to consumption, is one way to approach this.

According to the Food and Agriculture Organization (FAO), environmentally sustainable diets are "those diets with low environmental impacts which contribute to food and nutrition security and to healthy life for present and future generations (. . .) while optimizing natural and human resources [19]." We have previously identified four determinants for, or dimensions of, a sustainable diet from the consumer's perspective. These dimensions are based on the ratios of dietary characteristics. They are (1) the proportion of foods in the diet of animal versus plant origin, (2) the proportion of processed versus whole foods, (3) the proportion of seasonal/locally sourced foods versus out-of-season/context, and (4) the proportion of foods consumed versus wasted [18] (Figure 1).

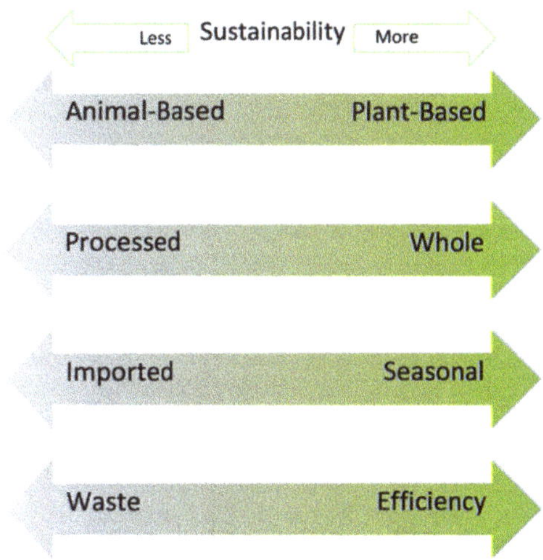

Figure 1. Characteristics of foods in a diet that determines its sustainability. The graphic illustrates that a sustainable diet has a higher proportion of foods that are plant-based, whole, in-season, and consumed with no or minimal waste [18].

Most diets (and most meals) have a mix of foods, each having different characteristics. The inclusion of nuts in the diet is variable. The larger the amount and proportion of foods in a meal or diet whose constituent foods are animal-sourced, processed, out of season or context (requiring transportation or refrigeration for storage), and wasted, the less sustainable the diet is. Reciprocally, the higher the proportion of foods of plant origin consumed, minimally processed, in season, and locally sourced, the more sustainable the diet is. Thus, nuts would appear to have a place in sustainable diets.

2.1. Nuts as Sustainable Foods

At the consumption level, nuts are plant foods consumed whole or very minimally processed; they have a "long season" (since they do not require refrigeration for storage, can be transported and stored with minimal energy use), and have little waste. Thus, nuts appear to be sustainable foods.

At the production level, however, concerns have been raised regarding the high water usage and chemical inputs in the production of nuts under intensive agricultural practices. Importantly, this is not the case when nuts are grown in extensive or traditional agricultural practices.

and fat composition of nuts, variations in food group allocations present challenges when comparing population intakes to recommendations.

Despite the current inclusion in dietary guidelines, population intakes do not appear to meet recommended levels for nut consumption. The 2017 Global Burden of Disease Study noted that global consumption of nuts was approximately 12% of the optimal intake of nuts and seeds (considered to be 21 g per day) [10]. Results from national surveys similarly highlight a common issue of low nut consumption. For instance, a secondary analysis of a subset of the 2005–2018 National Health and Nutrition Examination Survey (NHANES) from the United States of America found 12.9% of adult males and 9.1% of adult females met recommendations to consume 30 g or more of nuts per day [102]. Analysis of the 2011–2013 National Nutrition and Physical Activity Survey in Australia found 5.6% of individuals consuming nuts met the recommendation to eat 30 g of nuts per day [103]. Of note, under 40% of Australians reported consuming nuts during the survey, despite nut consumption including nuts in mixed dishes such as breakfast cereals or muesli bars. Similarly, the European Prospective Investigation into Cancer and Nutrition (EPIC) study found that on the day of the 24 h recall, less than 30% of respondents consumed nuts from any source, although it should be noted that nut intake did vary substantially among countries [104]. These results suggest that population intakes globally do not match current recommendations for nut consumption and highlight the need for increased nut consumption as part of a healthy dietary pattern.

4.2. Positioning Nuts in the Diet

Translating the evidence supporting nuts in healthy dietary patterns requires an understanding of nut consumption with other foods and within meals. The lack of congruence between recommendations and actual intakes suggests more work is required in that area. Nuts can serve as snack foods, and while research has exposed their impact on diet quality when substituted for poor-quality snacks [105], positioning them as snacks can align them with incidental, non-staple foods. In addition, theoretical positions of nuts in trials of healthy dietary patterns does not always translate to their inclusion as a staple food. For example, nuts were critical foods in the PREDIMED study, which demonstrated the CVD-preventive effects of the Mediterranean diets supplemented with extra-virgin olive oil or mixed nuts [62], but were less obvious in a study of staple foods in the context of food insecurity in the USA [106]. On the other hand, nuts are emerging as key foods in sustainable diets [107].

Behind this issue lies the question of collecting and managing research data on nuts. One suggestion is to create a separate food group of nuts, possibly with seeds as contemporaries [108]. This would clarify measurement in dietary surveys, albeit with a need to address the name of the group, serving size, and frequency of consumption. There are also implications for how nuts might be included in dietary indices that evaluate diet quality and how they might fit within a cuisine pattern. Even if this were the case, there have been major shifts away from naming food groups in terms of actual foods, with a greater focus on degree of processing.

The NOVA system of food categorization [109] places nuts favorably in the desired 'unprocessed or minimally processed' food category. This position is consistent with accompanying research that indicates a risk to cardiovascular health with ultra-processed foods, explainable through the loss of natural food synergy, displacement of healthy foods, and high content of saturated fat, sugar, and salt [110]. There is substantial debate on this methodological development, arguing the issue of misclassification with the NOVA system, the lack of associated conventional research, and the adequacy of current nutrient scoring systems [111]. Nevertheless, the minimally processed food category is consistent with staple foods recommended in dietary guidelines [112], which may assist in better compliance. The debate has brought into question not just food groupings, but food classification systems in general used in the review of dietary guidelines [113]. Whether these are nutrient-based rating or scoring systems, or food categories based on processing

(such as NOVA), or dominant nutrient contribution (as in dietary guidelines), there is wide variation in agreement in the way they present the health potential of individual foods.

Healthy dietary patterns tend to be investigated using forms of diet quality indices or scores (for example, Mediterranean diet scores) [114]. They address the whole of diet relationships with health outcomes, such as CVD risk, and take various approaches to the consumption of foods and nutrients and/or dietary patterns/cuisines. It is important to note, however, that they also serve various purposes: from health promotion activity to food labeling requirements and from measuring relationships in observational studies to effects seen in intervention trials. The positioning of nuts in instruments that address dietary patterns would need to take into consideration the purpose of the research activity.

4.3. Future Research on Nuts in Healthy Dietary Patterns

The ability to discern the impact of nuts in a dietary pattern, as with any food, will depend on how they are categorized and treated in the analysis [77]. Of the two main analytical approaches: a posteriori (looking for groups, identifying and naming patterns), or a priori (using pre-determined dietary patterns), those using an Index (a priori approach) are the commonest, followed by Factor or Principal Component Analysis (a posteriori approaches) [115]. This is also a developing area, with varying applications and reporting systems, as well as attention to foods and/or nutrients. Building an argument that deals with the interrelationship between foods and nutrients, however, strengthens the evidence of food effects. For example, using Principal Component Analysis in a study of baseline relationships between nutrients, foods, and dietary patterns with blood pressure, the dietary pattern categorized by nuts, seeds, fruit, and fish was significantly associated with lower blood pressure and with lower sodium:potassium ratio (a nutrient ratio related to blood pressure) [116]. In this case, the analyses based on healthy foods and intervening nutrients produced congruent outcomes.

Building the evidence base for nuts in healthy dietary patterns continues to rely on epidemiology to provide evidence of associations, clinical trials to demonstrate effects, and experimental laboratory studies to expose explanatory mechanisms [4]. Systematic reviews, meta-analyses, and quality assessments add rigor to the process, but this is a dynamic system requiring regular updates, oversight, and dedicated funding [7]. The purpose of building the evidence base will also inform the nature of the research, and this includes a consideration of transitional issues such as consumer communications. For example, research has shown the presence of health claims influences consumer purchasing, but there is more to do around how consumers understand and act on health claims in relation to meal contexts and overall dietary patterns [117].

5. Conclusions

Nuts are healthy foods: they are a source of important micronutrients, unsaturated fatty acids, protein, fiber, and plant sterols, and they form part of recognized healthy dietary patterns. Today, however, there is an imperative to review the impact of dietary patterns on the environment. This has led to a shift to plant-based diets, where nuts emerge as a significant source of protein. Health perspectives see nuts as a minimally processed and sustainable food, but research at the production level is evolving. Given their high nutritional value, environmental research is likely to drive better nut production methods in varying climate conditions. Nuts remain an important contributor to human health, with the mechanisms of action explained in terms of the nutrients they deliver. Studies have linked nut consumption to better blood lipoprotein profiles and lower CVD risk, but early research is now indicating possible beneficial effects of nut consumption at the other end of the life spectrum, namely reproductive health. This is a novel and interesting area of new research with many questions open for further investigation. Whether we consider the production of nuts or their consumption, the position of nuts in the dietary pattern remains an issue. The ultimate effects of food on health are the results of multiple interactive factors, so where nuts fit within dietary patterns is a significant consideration for research

translation. There are implications for research methodologies, including categorization within food groups and inclusion in Healthy Dietary Indices.

One of the most significant issues for research translation is that the consumption of nuts in many jurisdictions across the globe does not meet evidence-based recommendations. New areas of research, such as reproductive health discussed here, may help to increase the recognition of nuts as important foods in the diet. Likewise, their role in plant-based diets aimed at addressing environmental as well as health concerns may be casual. While continuing to build the evidence base on the health benefits of nuts, a focus should remain on methodology affecting the positioning of nuts in dietary assessment instruments, which may, in turn, influence forms of communication to consumers.

In dietary surveys, a separate category of nuts (and possibly seeds) may address the problem, as could the inclusion of nuts in healthy diet indices. Translational targets also require clarity of purpose in research. The nutrition science community recognizes the diversity of research methods to 'advance discovery, interpretation and application of knowledge' [118]. This includes an appreciation of how different layers of knowledge create the evidence base that enables appropriate (dietary) recommendations. In research on the health benefits of nuts, expanding the scope of interest to health throughout the lifecycle, especially in the area of reproductive health, and integrating research on environmental issues and sustainable diets represent very positive ways forward.

Author Contributions: Conceptualization, J.S., L.T. and A.S.-H.; Writing—Original Draft Preparation, J.S., L.T., M.L., E.N., R.M. and A.S.-H.; Writing—Review and Editing, J.S., L.T. and A.S.-H.; Supervision, J.S., L.T. and A.S.-H. All authors have read and agreed to the published version of the manuscript.

Funding: The NUTS 2022 conference was planned by the Rovira i Virgili University, with all the expenses of the organization covered by Barcelo Congresos, including the fees related to publishing the Proceedings in Nutrients. Barcelo Congresos SA has received funds from the International Nut & Dried Fruit Council (INC). Neither the researchers attending the NUTS 2022 Conference, nor the Rovira i Virgili University have received funds from the INC for this manuscript. Researchers' travel expenses and accommodation have been covered by Barcelo Congresos. Researchers did not receive any additional financial support related to their participation in the NUTS 2022 conference. Sponsors did not have any role in organizing the conference, selecting, or inviting the speakers, nor in the writing or reviewing the proceedings of the conference.

Institutional Review Board Statement: Not applicable.

Informed Consent Statement: Not applicable.

Data Availability Statement: Not applicable.

Acknowledgments: We are grateful to the organizers and expert researchers participating in the NUTS 2022: Where are we and where are we going? conference held in Reus, Spain.

Conflicts of Interest: L.T., J.S., E.N. and A.S.-H. have received research funding through their institutions from several nut commodity boards. L.T. and J.S. were non-paid members of California Walnut Commission Scientific (CWC) Advisory Council. J.S. has received honoraria from the CWC for presentations. All other authors declare no competing interest.

References

1. Salas-Salvadó, J.; Casas-Agustench, P.; Salas-Huetos, A. Cultural and Historical Aspects of Mediterranean Nuts with Emphasis on Their Attributed Healthy and Nutritional Properties. *Nutr. Metab. Cardiovasc. Dis.* **2011**, *21*, 1–6. [CrossRef]
2. Mozaffarian, D.; Rosenberg, I.; Uauy, R. History of Modern Nutrition Science-Implications for Current Research, Dietary Guidelines, and Food Policy. *BMJ* **2018**, *361*, k2392. [CrossRef] [PubMed]
3. Jacobs, D.R.; Tapsell, L.C. Food, Not Nutrients, Is the Fundamental Unit in Nutrition. *Nutr. Rev.* **2007**, *65*, 439–450. [PubMed]
4. Tapsell, L.C.; Neale, E.P.; Satija, A.; Hu, F.B. Foods, Nutrients, and Dietary Patterns: Interconnections and Implications for Dietary Guidelines. *Adv. Nutr.* **2016**, *7*, 445–454. [CrossRef] [PubMed]
5. Beyerbach, J.; Stadelmaier, J.; Hoffmann, G.; Balduzzi, S.; Bröckelmann, N.; Schwingshackl, L. Evaluating Concordance of Bodies of Evidence from Randomized Controlled Trials, Dietary Intake, and Biomarkers of Intake in Cohort Studies: A Meta-Epidemiological Study. *Adv. Nutr.* **2022**, *13*, 48–65.

6. Williams, C.M.; Ashwell, M.; Prentice, A.; Hickson, M.; Stanner, S. Nature of the Evidence Base and Frameworks Underpinning Dietary Recommendations for Prevention of Non-Communicable Diseases: A Position Paper from the Academy of Nutrition Sciences. *Br. J. Nutr.* **2021**, *126*, 1076–1090. [CrossRef]
7. Neale, E.P.; Tapsell, L.C. Perspective: The Evidence-Based Framework in Nutrition and Dietetics: Implementation, Challenges, and Future Directions. *Adv. Nutr.* **2019**, *10*, 1–8. [CrossRef]
8. Obbagy, J.; Raghavan, R.; English, L.K.; Spill, M.K.; Bahnfleth, C.L.; Bates, M.; Callahan, E.; Cole, N.C.; Güngör, D.; Kim, J.H.; et al. Strengthening Research That Answers Nutrition Questions of Public Health Importance: Leveraging the Experience of the USDA Nutrition Evidence Systematic Review Team. *J. Nutr.* **2022**, *152*, 1823–1830. [CrossRef]
9. Willett, W.; Rockström, J.; Loken, B.; Springmann, M.; Lang, T.; Vermeulen, S.; Garnett, T.; Tilman, D.; DeClerck, F.; Wood, A.; et al. Food in the Anthropocene: The EAT–Lancet Commission on Healthy Diets from Sustainable Food Systems. *Lancet* **2019**, *393*, 447–492. [CrossRef]
10. Afshin, A.; Sur, P.J.; Fay, K.A.; Cornaby, L.; Ferrara, G.; Salama, J.S.; Mullany, E.C.; Abate, K.H.; Abbafati, C.; Abebe, Z.; et al. Health Effects of Dietary Risks in 195 Countries, 1990–2017: A Systematic Analysis for the Global Burden of Disease Study 2017. *Lancet* **2019**, *393*, 1958–1972. [CrossRef]
11. The World Bank. Water in Agriculture. Available online: https://www.worldbank.org/en/topic/water-in-agriculture# (accessed on 9 April 2022).
12. Poore, J.; Nemecek, T. Reducing Food's Environmental Impacts through Producers and Consumers. *Science* **2018**, *360*, 987–992. [PubMed]
13. Foley, J.A.; Ramankutty, N.; Brauman, K.A.; Cassidy, E.S.; Gerber, J.S.; Johnston, M.; Mueller, N.D.; O'Connell, C.; Ray, D.K.; West, P.C.; et al. Solutions for a Cultivated Planet. *Nature* **2011**, *478*, 337–342. [PubMed]
14. Charles, H.; Godfray, J.; Beddington, J.R.; Crute, I.R.; Haddad, L.; Lawrence, D.; Muir, J.F.; Pretty, J.; Robinson, S.; Thomas, S.M.; et al. Food Security: The Challenge of Feeding 9 Billion People. *Science* **2010**, *327*, 812–818.
15. Sabaté, J.; Soret, S. Sustainability of Plant-Based Diets: Back to the Future. *Am. J. Clin. Nutr.* **2014**, *100*, 476S–482S. [CrossRef] [PubMed]
16. Springmann, M.; Clark, M.; Mason-D'Croz, D.; Wiebe, K.; Bodirsky, B.L.; Lassaletta, L.; de Vries, W.; Vermeulen, S.J.; Herrero, M.; Carlson, K.M.; et al. Options for Keeping the Food System within Environmental Limits. *Nature* **2018**, *562*, 519–525. [CrossRef]
17. Rosi, A.; Mena, P.; Pellegrini, N.; Turroni, S.; Neviani, E.; Ferrocino, I.; di Cagno, R.; Ruini, L.; Ciati, R.; Angelino, D.; et al. Environmental Impact of Omnivorous, Ovo-Lacto-Vegetarian, and Vegan Diet. *Sci. Rep.* **2017**, *7*, 6105. [CrossRef]
18. Sabaté, J.; Jehi, T. Determinants of Sustainable Diets. In *Environmental Nutrition: Connecting Health and Nutrition with Environmentally Sustainable Diets*; Sabaté, J., Ed.; Academic Press: Cambridge, MA, USA, 2019; Chapter 10; p. 181. ISBN 978-0-12-811660-9.
19. Burlingame, B.; Dernini, S. *Sustainable Diets and Biodiversity: Directions and Solutions for Policy, Research and Action*; FAO: Rome, Italy, 2012; ISBN 9789251072882.
20. Haddad, L.; Hawkes, C.; Waage, J.; Webb, P.; Godfray, C.; Toulmin, C. *Food Systems and Diets: Facing the Challenges of the 21st Century*; Global Panel on Agriculture and Food Systems for Nutrition: London, UK, 2016; ISBN 9780995622807.
21. Clune, S.; Crossin, E.; Verghese, K. Systematic Review of Greenhouse Gas Emissions for Different Fresh Food Categories. *J. Clean. Prod.* **2017**, *140*, 766–783.
22. Tilman, D.; Clark, M. Global Diets Link Environmental Sustainability and Human Health. *Nature* **2014**, *515*, 518–522.
23. Sabaté, J.; Sranacharoenpong, K.; Harwatt, H.; Wien, M.; Soret, S. The Environmental Cost of Protein Food Choices. *Public Health Nutr.* **2015**, *18*, 2067–2073.
24. Clark, M.A.; Springmann, M.; Hill, J.; Tilman, D. Multiple Health and Environmental Impacts of Foods. *Proc. Natl. Acad. Sci. USA* **2019**, *116*, 23357–23362. [CrossRef]
25. Watford, M.; Wu, G. Protein. *Adv. Nutr.* **2018**, *9*, 651–653. [CrossRef] [PubMed]
26. Mekonnen, M.M.; Hoekstra, A.Y. The Green, Blue and Grey Water Footprint of Crops and Derived Crop Products. *Hydrol. Earth Syst. Sci.* **2011**, *15*, 1577–1600. [CrossRef]
27. Naylor, T. Ditch the Almond Milk: Why Everything You Know about Sustainable Eating Is Probably Wrong. *Guardian*. 2018. Available online: https://www.theguardian.com/food/2018/sep/05/ditch-the-almond-milk-why-everything-you-know-about-sustainable-eating-is-probably-wrong (accessed on 10 February 2023).
28. Saner, E. Almond Milk: Quite Good for You—Very Bad for the Planet. *Guardian*. 2015. Available online: https://www.theguardian.com/lifeandstyle/shortcuts/2015/oct/21/almond-milk-quite-good-for-you-very-bad-for-the-planet (accessed on 10 February 2023).
29. Clothier, B.; Green, S.; Deurer, M. Green, Blue and Grey Waters: Minimising the Footprint Using Soil Physics. *Plant Food Res.* **2010**, 81–84.
30. Mekonnen, M.M.; Hoekstra, A.Y. A Global Assessment of the Water Footprint of Farm Animal Products. *Ecosystems* **2012**, *15*, 401–415. [CrossRef]
31. Vanham, D.; Mekonnen, M.M.; Hoekstra, A.Y. Treenuts and Groundnuts in the EAT-Lancet Reference Diet: Concerns Regarding Sustainable Water Use. *Glob. Food Secur.* **2020**, *24*, 100357. [CrossRef]
32. Liu, J.; Yang, H.; Gosling, S.N.; Kummu, M.; Flörke, M.; Pfister, S.; Hanasaki, N.; Wada, Y.; Zhang, X.; Zheng, C.; et al. Water Scarcity Assessments in the Past, Present, and Future. *Earths Future* **2017**, *5*, 545–559. [CrossRef]

33. Barreca, D.; Nabavi, S.M.; Sureda, A.; Rasekhian, M.; Raciti, R.; Silva, A.S.; Annunziata, G.; Arnone, A.; Tenore, G.C.; Süntar, İ.; et al. Almonds (*Prunus dulcis* Mill. D. A. Webb): A Source of Nutrients and Health-Promoting Compounds. *Nutrients* **2020**, *12*, 672. [CrossRef]
34. Fulton, J.; Norton, M.; Shilling, F. Water-Indexed Benefits and Impacts of California Almonds. *Ecol. Indic.* **2019**, *96*, 711–717. [CrossRef]
35. Volpe, R.; Messineo, S.; Volpe, M.; Messineo, A. Carbon Footprint of Tree Nuts Based Consumer Products. *Sustainability* **2015**, *7*, 14917–14934. [CrossRef]
36. Magliocca, N.R.; Brown, D.G.; Ellis, E.C. Exploring Agricultural Livelihood Transitions with an Agent-Based Virtual Laboratory: Global Forces to Local Decision-Making. *PLoS ONE* **2013**, *8*, e73241. [CrossRef]
37. Kyriakopoulou, K.; Keppler, J.K.; van der Goot, A.J. Functionality of Ingredients and Additives in Plant-Based Meat Analogues. *Foods* **2021**, *10*, 600. [CrossRef] [PubMed]
38. López-Uriarte, P.; Bulló, M.; Casas-Agustench, P.; Babio, N.; Salas-Salvadó, J. Nuts and Oxidation: A Systematic Review. *Nutr. Rev.* **2009**, *67*, 497–508. [CrossRef] [PubMed]
39. Practice Committee of the American Society for Reproductive Medicine. Definitions of Infertility and Recurrent Pregnancy Loss. *Fertil. Steril.* **2008**, *90*, S60. [CrossRef] [PubMed]
40. Carlsen, E.; Giwercman, A.; Keiding, N.; Skakkebaek, N.E. Evidence for Decreasing Quality of Semen during Past 50 Years. *BMJ* **1992**, *305*, 609–613. [CrossRef]
41. Swan, S.H.; Elkin, E.P.; Fenster, L. The Question of Declining Sperm Density Revisited: An Analysis of 101 Studies Published 1934–1996. *Environ. Health Perspect.* **2000**, *108*, 961–966. [CrossRef]
42. Mishra, P.; Negi, M.P.S.; Srivastava, M.; Singh, K.; Rajender, S. Decline in Seminal Quality in Indian Men over the Last 37 Years. *Reprod. Biol. Endocrinol.* **2018**, *16*, 103. [CrossRef]
43. Nelson, C.M.K.; Bunge, R.G. Semen Analysis: Evidence for Changing Parameters of Male Fertility Potential. *Fertil. Steril.* **1974**, *25*, 503–507. [CrossRef]
44. Levine, H.; Jørgensen, N.; Martino-Andrade, A.; Mendiola, J.; Weksler-Derri, D.; Mindlis, I.; Pinotti, R.; Swan, S.H. Temporal Trends in Sperm Count: A Systematic Review and Meta-Regression Analysis. *Hum. Reprod. Update* **2017**, *23*, 646–659. [CrossRef]
45. Pillai, R.N.; McEleny, K. Management of Male Infertility. *Obstet. Gynaecol. Reprod. Med.* **2021**, *31*, 192–198. [CrossRef]
46. Ilacqua, A.; Izzo, G.; Emerenziani, G.P.; Baldari, C.; Aversa, A. Lifestyle and Fertility: The Influence of Stress and Quality of Life on Male Fertility. *Reprod. Biol. Endocrinol.* **2018**, *16*, 115. [CrossRef]
47. Sharma, R.; Biedenharn, K.R.; Fedor, J.M.; Agarwal, A. Lifestyle Factors and Reproductive Health: Taking Control of Your Fertility. *Reprod. Biol. Endocrinol.* **2013**, *11*, 66. [CrossRef] [PubMed]
48. Salas-Huetos, A.; Bulló, M.; Salas-Salvadó, J. Dietary Patterns, Foods and Nutrients in Male Fertility Parameters and Fecundability: A Systematic Review of Observational Studies. *Hum. Reprod. Update* **2017**, *23*, 371–389. [CrossRef] [PubMed]
49. Salas-Huetos, A.; Mínguez-Alarcón, L.; Mitsunami, M.; Arvizu, M.; Ford, J.B.; Souter, I.; Yeste, M.; Chavarro, J.E.; Team, E.S. Paternal Adherence to Healthy Dietary Patterns in Relation to Sperm Parameters and Outcomes of Assisted Reproductive Technologies. *Fertil. Steril.* **2022**, *117*, 298–312. [CrossRef] [PubMed]
50. Salas-Huetos, A.; Babio, N.; Carrell, D.T.; Bulló, M.; Salas-Salvadó, J. Adherence to the Mediterranean Diet Is Positively Associated with Sperm Motility: A Cross-Sectional Analysis. *Sci. Rep.* **2019**, *9*, 3389. [CrossRef] [PubMed]
51. Ricci, E.; Bravi, F.; Noli, S.; Somigliana, E.; Cipriani, S.; Castiglioni, M.; Chiaffarino, F.; Vignali, M.; Gallotti, B.; Parazzini, F. Mediterranean Diet and Outcomes of Assisted Reproduction: An Italian Cohort Study. *Am. J. Obstet. Gynecol.* **2019**, *221*, e1–e627. [CrossRef]
52. Ricci, E.; Bravi, F.; Noli, S.; Ferrari, S.; De Cosmi, V.; La Vecchia, I.; Cavadini, M.; La Vecchia, C.; Parazzini, F. Mediterranean Diet and the Risk of Poor Semen Quality: Cross-Sectional Analysis of Men Referring to an Italian Fertility Clinic. *Andrology* **2019**, *7*, 156–162. [CrossRef]
53. Efrat, M.; Stein, A.; Pinkas, H.; Unger, R.; Birk, R. Dietary Patterns Are Positively Associated with Semen Quality. *Fertil. Steril.* **2018**, *109*, 809–816. [CrossRef]
54. Karayiannis, D.; Kontogianni, M.D.; Mendorou, C.; Douka, L.; Mastrominas, M.; Yiannakouris, N. Association between Adherence to the Mediterranean Diet and Semen Quality Parameters in Male Partners of Couples Attempting Fertility. *Hum. Reprod.* **2017**, *32*, 215–222. [CrossRef]
55. Cutillas-Tolín, A.; Adoamnei, E.; Navarrete-Muñoz, E.M.; Vioque, J.; Moñino-García, M.; Jørgensen, N.; Chavarro, J.E.; Mendiola, J.; Torres-Cantero, A.M. Adherence to Diet Quality Indices in Relation to Semen Quality and Reproductive Hormones in Young Men. *Hum. Reprod.* **2019**, *34*, 1866–1875. [CrossRef]
56. Danielewicz, A.; Morze, J.; Przybyłowicz, M.; Przybyłowicz, K. Association of the Dietary Approaches to Stop Hypertension, Physical Activity, and Their Combination with Semen Quality: A Cross-Sectional Study. *Nutrients* **2020**, *12*, 39. [CrossRef]
57. Montano, L.; Ceretti, E.; Donato, F.; Bergamo, P.; Zani, C.; Viola, G.C.V.; Notari, T.; Pappalardo, S.; Zani, D.; Ubaldi, S.; et al. Effects of a Lifestyle Change Intervention on Semen Quality in Healthy Young Men Living in Highly Polluted Areas in Italy: The FASt Randomized Controlled Trial. *Eur. Urol. Focus* **2022**, *8*, 351–359. [CrossRef] [PubMed]
58. Braga, D.P.D.A.F.; Halpern, G.; Figueira, R.D.C.S.; Setti, A.S.; Iaconelli, A.; Borges, E. Food Intake and Social Habits in Male Patients and Its Relationship to Intracytoplasmic Sperm Injection Outcomes. *Fertil. Steril.* **2012**, *97*, 53–59. [CrossRef] [PubMed]

59. Eslamian, G.; Amirjannati, N.; Rashidkhani, B.; Sadeghi, M.-R.; Hekmatdoost, A. Intake of Food Groups and Idiopathic Asthenozoospermia: A Case-Control Study. *Hum. Reprod.* **2012**, *27*, 3328–3336. [CrossRef] [PubMed]
60. Salas-Huetos, A.; Rosique-Esteban, N.; Becerra-Tomás, N.; Vizmanos, B.; Bulló, M.; Salas-Salvadó, J. The Effect of Nutrients and Dietary Supplements on Sperm Quality Parameters: A Systematic Review and Meta-Analysis of Randomized Clinical Trials. *Adv. Nutr.* **2018**, *9*, 833–848. [CrossRef]
61. Salas-Huetos, A.; James, E.R.; Aston, K.I.; Jenkins, T.G.; Carrell, D.T. Diet and Sperm Quality: Nutrients, Foods and Dietary Patterns. *Reprod. Biol.* **2019**, *19*, 219–224. [CrossRef]
62. Estruch, R.; Ros, E.; Salas-Salvadó, J.; Covas, M.-I.; Corella, D.; Arós, F.; Gómez-Gracia, E.; Ruiz-Gutiérrez, V.; Fiol, M.; Lapetra, J.; et al. Primary Prevention of Cardiovascular Disease with a Mediterranean Diet Supplemented with Extra-Virgin Olive Oil or Nuts. *N. Engl. J. Med.* **2018**, *378*, e34. [CrossRef]
63. Hernández-Alonso, P.; Camacho-Barcia, L.; Bulló, M.; Salas-Salvadó, J. Nuts and Dried Fruits: An Update of Their Beneficial Effects on Type 2 Diabetes. *Nutrients* **2017**, *9*, 673. [CrossRef]
64. Del Gobbo, L.C.; Falk, M.C.; Feldman, R.; Lewis, K.; Mozaffarian, D. Effects of Tree Nuts on Blood Lipids, Apolipoproteins, and Blood Pressure: Systematic Review, Meta-Analysis, and Dose-Response of 61 Controlled Intervention Trials. *Am. J. Clin. Nutr.* **2015**, *102*, 1347–1356. [CrossRef]
65. Kara, H.; Orem, A.; Yulug, E.; Yucesan, F.B.; Kerimoglu, G.; Yaman, S.O.; Bodur, A.; Turedi, S.; Alasalvar, C. Hazelnut Consumption Improves Testicular Antioxidant Function and Semen Quality in Young and Old Male Rats. *Food Chem.* **2019**, *294*, 1–8. [CrossRef]
66. Akomolafe, S.F.; Aina, B.; Bajulaye, J.; Ogundare, I.; Olulade, D.; Adeniji, R.; Fatuase, F.; Olojo-Kosoko, A.; Ganiyu, O.; Asogwa, N.T. Modulatory Effect of Cashew (*Anacardium occidentale* L.) Nut Supplemented Diet on Fertility Activity of Clomiphene Citrate in Male Rats. *Biomed. J.* **2021**, *44*, 190–200. [CrossRef]
67. Robbins, W.A.; Xun, L.; FitzGerald, L.Z.; Esguerra, S.; Henning, S.M.; Carpenter, C.L. Walnuts Improve Semen Quality in Men Consuming a Western-Style Diet: Randomized Control Dietary Intervention Trial. *Biol. Reprod.* **2012**, *87*, 1–8. [CrossRef]
68. Salas-Huetos, A.; Moraleda, R.; Giardina, S.; Anton, E.; Blanco, J.; Salas-Salvadó, J.; Bulló, M. Effect of Nut Consumption on Semen Quality and Functionality in Healthy Men Consuming a Western-Style Diet: A Randomized Controlled Trial. *Am. J. Clin. Nutr.* **2018**, *108*, 953–962. [CrossRef] [PubMed]
69. Garcia-Aloy, M.; Hulshof, P.J.M.; Estruel-Amades, S.; Osté, M.C.J.; Lankinen, M.; Geleijnse, J.M.; de Goede, J.; Ulaszewska, M.; Mattivi, F.; Bakker, S.J.L.; et al. Biomarkers of Food Intake for Nuts and Vegetable Oils: An Extensive Literature Search. *Genes Nutr.* **2019**, *14*, 1–21. [CrossRef] [PubMed]
70. Petrović-Oggiano, G.; Debeljak-Martačić, J.; Ranković, S.; Pokimica, B.; Mirić, A.; Glibetić, M.; Popović, T. The Effect of Walnut Consumption on N-3 Fatty Acid Profile of Healthy People Living in a Non-Mediterranean West Balkan Country, a Small Scale Randomized Study. *Nutrients* **2020**, *12*, 192. [CrossRef] [PubMed]
71. Salas-Huetos, A.; James, E.R.; Salas-Salvadó, J.; Bulló, M.M.; Aston, K.I.; Carrell, D.T.; Jenkins, T.G.; Salas-Huetos, A.; James, E.R.; Salas-Salvadó, J.; et al. Sperm DNA Methylation Changes after Short-term Nut Supplementation in Healthy Men Consuming a Western-style Diet. *Andrology* **2021**, *9*, 260–268. [CrossRef] [PubMed]
72. Esposito, K.; Ciotola, M.; Giugliano, F.; de Sio, M.; Giugliano, G.; D'Armiento, M.; Giugliano, D. Mediterranean Diet Improves Erectile Function in Subjects with the Metabolic Syndrome. *Int. J. Impot. Res.* **2006**, *18*, 405–410. [CrossRef] [PubMed]
73. Esposito, K.; Giugliano, F.; Maiorino, M.I.; Giugliano, D. Dietary Factors, Mediterranean Diet and Erectile Dysfunction. *J. Sex. Med.* **2010**, *7*, 2338–2345. [CrossRef]
74. Aldemir, M.; Okulu, E.; Neşelioğlu, S.; Erel, O.; Kayıgil, Ö. Pistachio Diet Improves Erectile Function Parameters and Serum Lipid Profiles in Patients with Erectile Dysfunction. *Int. J. Impot. Res.* **2011**, *23*, 32–38. [CrossRef]
75. Salas-Huetos, A.; Muralidharan, J.; Galiè, S.; Salas-Salvadó, J.; Bulló, M. Effect of Nut Consumption on Erectile and Sexual Function in Healthy Males: A Secondary Outcome Analysis of the FERTINUTS Randomized. *Nutrients* **2019**, *11*, 1372. [CrossRef]
76. Jacobs, D.R.; Gross, M.D.; Tapsell, L.C. Food Synergy: An Operational Concept for Understanding Nutrition. *Am. J. Clin. Nutr.* **2009**, *89*, 1543S–1548S. [CrossRef]
77. Tapsell, L.C.; Neale, E.P.; Probst, Y. Dietary Patterns and Cardiovascular Disease: Insights and Challenges for Considering Food Groups and Nutrient Sources. *Curr. Atheroscler. Rep.* **2019**, *21*, 1–8. [CrossRef]
78. Schwingshackl, L.; Hoffmann, G. Diet Quality as Assessed by the Healthy Eating Index, the Alternate Healthy Eating Index, the Dietary Approaches to Stop Hypertension Score, and Health Outcomes: A Systematic Review and Meta-Analysis of Cohort Studies. *J. Acad. Nutr. Diet.* **2015**, *115*, 780–800.e5. [CrossRef] [PubMed]
79. Galbete, C.; Schwingshackl, L.; Schwedhelm, C.; Boeing, H.; Schulze, M.B. Evaluating Mediterranean Diet and Risk of Chronic Disease in Cohort Studies: An Umbrella Review of Meta-Analyses. *Eur. J. Epidemiol.* **2018**, *33*, 909–931. [CrossRef] [PubMed]
80. Steffen, L.M.; Hootman, K.C. A Posteriori Data-Derived Dietary Patterns and Incident Coronary Heart Disease: Making Sense of Inconsistent Findings. *Curr. Nutr. Rep.* **2016**, *5*, 168–179. [CrossRef] [PubMed]
81. Schwingshackl, L.; Hoffmann, G.; Iqbal, K.; Schwedhelm, C.; Boeing, H. Food Groups and Intermediate Disease Markers: A Systematic Review and Network Meta-Analysis of Randomized Trials. *Am. J. Clin. Nutr.* **2018**, *108*, 576–586. [CrossRef] [PubMed]
82. Tapsell, L.C.; Lonergan, M.; Batterham, M.J.; Neale, E.P.; Martin, A.; Thorne, R.; Deane, F.; Peoples, G. Effect of Interdisciplinary Care on Weight Loss: A Randomised Controlled Trial. *BMJ Open* **2017**, *7*, e014533. [CrossRef]
83. Yong, L.C.; Gray, A.R.; Chisholm, A.; Leong, S.L.; Tey, S.L.; Brown, R.C. Barriers to and Facilitators and Perceptions of Nut Consumption among the General Population in New Zealand. *Public Health Nutr.* **2017**, *20*, 3166–3182. [CrossRef]

84. Pawlak, R.; London, H.A.; Colby, S.E.; Wall-Bassett, E.; Sira, N. Perception of Nut Intake among Individuals with or at Risk for Heart Disease and/or Diabetes. *J. Behav. Health* **2012**, *1*, 185–188. [CrossRef]
85. Nishi, S.K.; Viguiliouk, E.; Blanco Mejia, S.; Kendall, C.W.C.; Bazinet, R.P.; Hanley, A.J.; Comelli, E.M.; Salas Salvadó, J.; Jenkins, D.J.A.; Sievenpiper, J.L. Are Fatty Nuts a Weighty Concern? A Systematic Review and Meta-Analysis and Dose–Response Meta-Regression of Prospective Cohorts and Randomized Controlled Trials. *Obes. Rev.* **2021**, *22*. [CrossRef]
86. Guarneiri, L.L.; Cooper, J.A. Intake of Nuts or Nut Products Does Not Lead to Weight Gain, Independent of Dietary Substitution Instructions: A Systematic Review and Meta-Analysis of Randomized Trials. *Adv. Nutr.* **2021**, *12*, 384–401. [CrossRef]
87. Neale, E.; Tapsell, L. Nuts in Healthy Dietary Patterns and Dietary Guidelines. In *Health Benefits of Nuts and Dried Fruits*; CRC Press: Boca Raton, FL, USA, 2020; pp. 289–312. ISBN 9781315173337.
88. Kromhout, D.; Spaaij, C.J.K.; de Goede, J.; Weggemans, R.M.; Brug, J.; Geleijnse, J.M.; van Goudoever, J.B.; Hoes, A.W.; Hopman, M.T.E.; Iestra, J.A.; et al. The 2015 Dutch Food-Based Dietary Guidelines. *Eur. J. Clin. Nutr.* **2016**, *70*, 869–878. [CrossRef] [PubMed]
89. Nordic Nutrition Recommendations 2012. Available online: https://www.norden.org/en/publication/nordic-nutrition-recommendations-2012 (accessed on 17 January 2023).
90. Malaysian Dietary Guidelines 2020. Available online: https://nutrition.moh.gov.my/wp-content/uploads/2021/07/Web%20MDG.pdf (accessed on 17 January 2023).
91. The Official Dietary Guidelines-Good for Health and Climate. Available online: https://www.fao.org/nutrition/education-nutritionnelle/food-dietary-guidelines/regions/denmark/en/ (accessed on 17 January 2023).
92. Healthy Eating Dietary Guidelines for Maltese Adults. Available online: https://deputyprimeminister.gov.mt/en/health-promotion/documents/library/publications/healthy%20plate%20en.pdf (accessed on 17 January 2023).
93. From Plate to Guide: What, Why and How for the Eatwell Model. Available online: https://www.gov.uk/government/publications/the-eatwell-guide (accessed on 17 January 2023).
94. Lebanon Food-Based Dietary Guidelines. Available online: https://www.fao.org/nutrition/education/food-dietary-guidelines/regions/countries/lebanon/en/ (accessed on 17 January 2023).
95. Canada's Dietary Guidelines for Health Professionals and Policy Makers. Available online: https://food-guide.canada.ca/en/guidelines/ (accessed on 17 January 2023).
96. Dietary Guidelines for Indians—A Manual. Available online: https://www.nin.res.in/downloads/DietaryGuidelinesforNINwebsite.pdf (accessed on 17 January 2023).
97. Food Based Dietary Guidelines for Sri Lankans. Available online: https://www.fao.org/nutrition/education/food-dietary-guidelines/regions/sri-lanka/ru/ (accessed on 17 January 2023).
98. Sierra Leone Food-Based Dietary Guidelines for Healthy Eating. Available online: https://www.afro.who.int/publications/sierra-leone-food-based-dietary-guidelines-healthy-eating-2016 (accessed on 17 January 2023).
99. Dietary Guidelines for Americans Make Every Bite Count with the Dietary Guidelines. Available online: https://www.dietaryguidelines.gov/sites/default/files/2020-12/Dietary_Guidelines_for_Americans_2020-2025.pdf (accessed on 17 January 2023).
100. Eating and Activity Guidelines for New Zealand Adults. Available online: https://www.health.govt.nz/publication/eating-and-activity-guidelines-new-zealand-adults (accessed on 17 January 2023).
101. Eat for Health: Australian Dietary Guidelines. Available online: https://www.eatwellnutrition.com.au/general-nutrition/eat-for-health-new-australian-dietary-guidelines (accessed on 17 January 2023).
102. Cardoso, B.R.; Tan, S.Y.; Daly, R.M.; Via, J.D.; Georgousopoulou, E.N.; George, E.S. Intake of Nuts and Seeds Is Associated with a Lower Prevalence of Nonalcoholic Fatty Liver Disease in US Adults: Findings from 2005–2018 NHANES. *J. Nutr.* **2021**, *151*, 3507–3515. [CrossRef] [PubMed]
103. Nikodijevic, C.J.; Probst, Y.C.; Batterham, M.J.; Tapsell, L.C.; Neale, E.P. Nut Consumption in a Representative Survey of Australians: A Secondary Analysis of the 2011–2012 National Nutrition and Physical Activity Survey. *Public Health Nutr.* **2020**, *23*, 3368–3378. [CrossRef] [PubMed]
104. Jenab, M.; Sabaté, J.; Slimani, N.; Ferrari, P.; Mazuir, M.; Casagrande, C.; Deharveng, G.; Tjønneland, A.; Olsen, A.; Overvad, K.; et al. Consumption and Portion Sizes of Tree Nuts, Peanuts and Seeds in the European Prospective Investigation into Cancer and Nutrition (EPIC) Cohorts from 10 European Countries. *Br. J. Nutr.* **2006**, *96*, S12–S23. [CrossRef] [PubMed]
105. Rehm, C.D.; Drewnowski, A. Replacing American Snacks with Tree Nuts Increases Consumption of Key Nutrients among US Children and Adults: Results of an NHANES Modeling Study. *Nutr. J.* **2017**, *16*, 1–15. [CrossRef] [PubMed]
106. Laska, M.N.; Caspi, C.E.; Lenk, K.; Moe, S.G.; Pelletier, J.E.; Harnack, L.J.; Erickson, D.J. Evaluation of the First U.S. Staple Foods Ordinance: Impact on Nutritional Quality of Food Store Offerings, Customer Purchases and Home Food Environments. *Int. J. Behav. Nutr. Phys. Act.* **2019**, *16*, 1–20. [CrossRef]
107. Drewnowski, A.; Finley, J.; Hess, J.M.; Ingram, J.; Miller, G.; Peters, C. Toward Healthy Diets from Sustainable Food Systems. *Curr. Dev. Nutr.* **2020**, *4*, nzaa083. [CrossRef]
108. George, E.S.; Daly, R.M.; Tey, S.L.; Brown, R.; Wong, T.H.T.; Tan, S.Y. Perspective: Is It Time to Expand Research on "Nuts" to Include "Seeds"? Justifications and Key Considerations. *Adv. Nutr.* **2022**, *13*, 1016–1027. [CrossRef]
109. Monteiro, C.A.; Cannon, G.; Levy, R.B.; Moubarac, J.C.; Louzada, M.L.C.; Rauber, F.; Khandpur, N.; Cediel, G.; Neri, D.; Martinez-Steele, E.; et al. Ultra-Processed Foods: What They Are and How to Identify Them. *Public Health Nutr.* **2019**, *22*, 936–941. [CrossRef]

110. Lawrence, M. Ultraprocessed Foods and Cardiovascular Health: It's Not Just about the Nutrients. *Am. J. Clin. Nutr.* **2021**, *113*, 257–258. [CrossRef]
111. Astrup, A.; Monteiro, C.A. Does the Concept of "Ultra-Processed Foods" Help Inform Dietary Guidelines, beyond Conventional Classification Systems? NO. *Am. J. Clin. Nutr.* **2022**, *116*, 1482–1488. [CrossRef] [PubMed]
112. Monteiro, C.A.; Astrup, A. Does the Concept of "Ultra-Processed Foods" Help Inform Dietary Guidelines, beyond Conventional Classification Systems? YES. *Am. J. Clin. Nutr.* **2022**, *116*, 1476–1481. [CrossRef] [PubMed]
113. Dickie, S.; Woods, J.; Machado, P.; Lawrence, M. Nutrition Classification Schemes for Informing Nutrition Policy in Australia: Nutrient-Based, Food-Based, or Dietary-Based? *Curr. Dev. Nutr.* **2022**, *6*, nzac112. [CrossRef] [PubMed]
114. Zaragoza-Martí, A.; Cabañero-Martínez, M.J.; Hurtado-Sánchez, J.A.; Laguna-Pérez, A.; Ferrer-Cascales, R. Evaluation of Mediterranean Diet Adherence Scores: A Systematic Review. *BMJ Open* **2018**, *8*, e019033. [CrossRef] [PubMed]
115. Wingrove, K.; Lawrence, M.A.; McNaughton, S.A. A Systematic Review of the Methods Used to Assess and Report Dietary Patterns. *Front. Nutr.* **2022**, *9*, 892351. [CrossRef]
116. Ndanuko, R.N.; Tapsell, L.C.; Charlton, K.E.; Neale, E.P.; Batterham, M.J. Associations between Dietary Patterns and Blood Pressure in a Clinical Sample of Overweight Adults. *J. Acad. Nutr. Diet.* **2017**, *117*, 228–239. [CrossRef]
117. Neale, E.P.; Tapsell, L.C. Nutrition and Health Claims: Consumer Use and Evolving Regulation. *Curr. Nutr. Rep.* **2022**, *11*, 431–436. [CrossRef]
118. Mattes, R.D.; Rowe, S.B.; Ohlhorst, S.D.; Brown, A.W.; Hoffman, D.J.; Liska, D.A.J.; Feskens, E.J.M.; Dhillon, J.; Tucker, K.L.; Epstein, L.H.; et al. Valuing the Diversity of Research Methods to Advance Nutrition Science. *Adv. Nutr.* **2022**, *13*, 1324–1393. [CrossRef]

MDPI
St. Alban-Anlage 66
4052 Basel
Switzerland
Tel. +41 61 683 77 34
Fax +41 61 302 89 18
www.mdpi.com

Nutrients Editorial Office
E-mail: nutrients@mdpi.com
www.mdpi.com/journal/nutrients

www.ingramcontent.com/pod-product-compliance
Lightning Source LLC
Chambersburg PA
CBHW061522210926
51850CB00008B/2006
9783036578866

www.ingramcontent.com/pod-product-compliance
Lightning Source LLC
Chambersburg PA
CBHW061614210926
51850CB00009B/2323
9783036534077

MDPI
St. Alban-Anlage 66
4052 Basel
Switzerland
Tel. +41 61 683 77 34
Fax +41 61 302 89 18
www.mdpi.com

Quaternary Editorial Office
E-mail: quaternary@mdpi.com
www.mdpi.com/journal/quaternary

Nuts: Where We Are and Where We Are Going in Research. Proceedings from the NUTS 2022 International Conference

Nuts: Where We Are and Where We Are Going in Research. Proceedings from the NUTS 2022 International Conference

Editors

Jordi Salas-Salvadó
Emilio Ros
Joan Sabaté
Stephanie K. Nishi

MDPI • Basel • Beijing • Wuhan • Barcelona • Belgrade • Manchester • Tokyo • Cluj • Tianjin

Editors
Jordi Salas-Salvadó
Universitat Rovira i Virgili
Spain

Emilio Ros
Instituto de Salud Carlos III
Spain

Joan Sabaté
Loma Linda University
USA

Stephanie K. Nishi
Universitat Rovira i Virgili
Spain

Editorial Office
MDPI
St. Alban-Anlage 66
4052 Basel, Switzerland

This is a reprint of articles from the Special Issue published online in the open access journal *Nutrients* (ISSN 2072-6643) (available at: https://www.mdpi.com/journal/nutrients/special_issues/NUTS_2022).

For citation purposes, cite each article independently as indicated on the article page online and as indicated below:

LastName, A.A.; LastName, B.B.; LastName, C.C. Article Title. *Journal Name* **Year**, *Volume Number*, Page Range.

ISBN 978-3-0365-7886-6 (Hbk)
ISBN 978-3-0365-7887-3 (PDF)

Cover image courtesy of Prescott C. Ensign
The International Nut and Dried Fruit Council

Contents

About the Editors vii

Jordi Salas-Salvadó, Stephanie K. Nishi, Joan Sabaté and Emilio Ros
Where We Are and Where We Are Going in Nut Research
Reprinted from: *Nutrients* **2023**, *15*, 1691, doi:10.3390/nu15071691 1

David J. Baer, Michelle Dalton, John Blundell, Graham Finlayson and Frank B. Hu
Nuts, Energy Balance and Body Weight
Reprinted from: *Nutrients* **2023**, *15*, 1162, doi:10.3390/nu15051162 5

Stephanie K. Nishi, Effie Viguiliouk, Cyril W. C. Kendall, David J. A. Jenkins, Frank B. Hu, John L. Sievenpiper, et al.
Nuts in the Prevention and Management of Type 2 Diabetes
Reprinted from: *Nutrients* **2023**, *15*, 878, doi:10.3390/nu15040878 23

Marta Guasch-Ferré, Anne-Julie Tessier, Kristina S. Petersen, Philip A. Sapp, Linda C. Tapsell, Jordi Salas-Salvadó, et al.
Effects of Nut Consumption on Blood Lipids and Lipoproteins: A Comprehensive Literature Update
Reprinted from: *Nutrients* **2023**, *15*, 596, doi:10.3390/nu15030596 45

Giuseppina Mandalari, Teresa Gervasi, Daniel W. Rosenberg, Karen G. Lapsley and David J. Baer
Effect of Nuts on Gastrointestinal Health
Reprinted from: *Nutrients* **2023**, *15*, 1733, doi:10.3390/nu15071733 65

Sujatha Rajaram, Nagila Raquel Teixeira Damasceno, Ribanna Aparecida Marques Braga, Raquel Martinez, Penny Kris-Etherton and Aleix Sala-Vila
Effect of Nuts on Markers of Inflammation and Oxidative Stress: A Narrative Review
Reprinted from: *Nutrients* **2023**, *15*, 1099, doi:10.3390/nu15051099 81

Andrea J. Glenn, Dagfinn Aune, Heinz Freisling, Noushin Mohammadifard, Cyril W. C. Kendall, Jordi Salas-Salvadó, et al.
Nuts and Cardiovascular Disease Outcomes: A Review of the Evidence and Future Directions
Reprinted from: *Nutrients* **2023**, *15*, 911, doi:10.3390/nu15040911 95

Stephanie K. Nishi, Aleix Sala-Vila, Jordi Julvez, Joan Sabaté and Emilio Ros
Impact of Nut Consumption on Cognition across the Lifespan
Reprinted from: *Nutrients* **2023**, *15*, 1000, doi:10.3390/nu15041000 111

Bradley W. Bolling, Dagfinn Aune, Hwayoung Noh, Kristina S. Petersen and Heinz Freisling
Dried Fruits, Nuts, and Cancer Risk and Survival: A Review of the Evidence and Future Research Directions
Reprinted from: *Nutrients* **2023**, *15*, 1443, doi:10.3390/nu15061443 131

Cesarettin Alasalvar, Sui Kiat Chang, Penny M. Kris-Etherton, Valerie K. Sullivan, Kristina S. Petersen, Marta Guasch-Ferré and David J. A. Jenkins
Dried Fruits: Bioactives, Effects on Gut Microbiota, and Possible Health Benefits—An Update
Reprinted from: *Nutrients* **2023**, *15*, 1611, doi:10.3390/nu15071611 149

Linda Tapsell, Joan Sabaté, Raquel Martínez, Marc Llavanera, Elizabeth Neale and Albert Salas-Huetos
Novel Lines of Research on the Environmental and Human Health Impacts of Nut Consumption
Reprinted from: *Nutrients* **2023**, *15*, 955, doi:10.3390/nu15040955 . . . . **169**

About the Editors

Jordi Salas-Salvadó

Jordi Salas-Salvadó MD, Ph.D., is a Distinguished Professor of Nutrition and ICREA Academia Researcher at Rovira i Virgili University; Director of the Food, Nutrition and Mental Health Research Group recognized by the Government of Catalonia; Principal Investigator of the CIBER Network Physiopathology of Obesity and Nutrition (CIBERobn) of the Carlos III Health Institute; Director of the Catalan Nutrition Centre of the Institute of Catalan Studies; Chairman of the Professorship Tree Nut World Forum for Nutrition Research and Dissemination; member of the Group of Experts of the Public Health Agency of Catalonia. Prof. Salas-Salvadó is an expert in clinical trials evaluating the effect of diets and dietary compounds on cardiometabolic health. He was a lead researcher of the PREDIMED study, a landmark trial evaluating the effect of the Mediterranean Diet on the primary prevention of cardiovascular diseases. He is currently the Director and Chair of the Steering Committee of the PREDIMED-Plus trial. He has published >700 scientific articles (h-index 81; >35,000 citations; 2018–2021 Clarivate Analytics "Highly Cited Researcher") and has received multiple awards and recognitions.

Emilio Ros

Emilio Ros MD, Ph.D., is the founder and former head of the Lipid Clinic, Endocrinology Service, Hospital Clínic, Barcelona. He is an Emeritus Investigator, Institut d'Investigacions Biomèdiques August Pi Sunyer (DIBAPS) Barcelona, former Principal Investigator, and now associate member of the research group "Nutrition, Lipids and Cardiovascular Risk", CIBERobn, Instituto de Salud Carlos III, member and founder of the Spanish Arteriosclerosis Society (SEA), as well as a member of the European and International Atherosclerosis Societies and American College of Cardiology. Dr. Ros' research focus includes nutrition in the prevention of cardiovascular diseases and cognitive decline, with a particular interest in the following topics: Mediterranean diet and walnuts; plant sterols; blood membrane fatty acids; genetic dyslipidemias; cardiovascular risk assessment; vascular imaging techniques, especially carotid ultrasound. He led the nutritional intervention of the landmark PREDIMED trial of the Mediterranean diet for primary cardiovascular prevention and has published >600 papers (h-index 113; >37,000 citations; Clarivate Analytics 2018–2021 "Highly Cited Researcher").

Joan Sabaté

Joan Sabaté MD, Ph.D., is a Professor of Nutrition and Epidemiology at Loma Linda University School of Public Health and a board-certified physician in Internal Medicine. He was the principal investigator of a nutrition intervention trial that directly linked the consumption of walnuts to significant reductions in serum cholesterol, published in the New England Journal of Medicine in 1993. He is a co-investigator of Adventist Health Studies, the largest cohort of vegetarians relating dietary intake with health outcomes. For the past 25 years, he has been the principal investigator of many human nutrition intervention trials investigating the health effects of nuts, avocados, and other plant foods. Dr. Sabaté has authored >200 high-impact research articles (h-index of 65, >15,000 citations) and has contributed to public health, including being a member of the US 2020 Dietary Guidelines Advisory Committee. Dr. Sabaté directs the Environmental Nutrition research program at the Loma Linda University School of Public Health, which focuses on sustainable diets, explores the interrelationships between food choices' environmental and health impacts, and ultimately seeks to improve food systems' sustainability, health, and equity.

Stephanie K. Nishi

Stephanie K. Nishi, Ph.D., RD, is a Canadian Institutes of Health Research (CIHR)-funded postdoctoral fellow with the Unitat de Nutrició Humana, Departament de Bioquímica i Biotecnologia at the Universitat Rovira i Virgili, Centro de Investigación Biomédica en Red de la Fisiopatología de la Obesidad y Nutrición (CIBERobn), and Institut d´Investigació Sanitària Pere Virgili (IISPV) in Spain. She is a registered dietitian with a Ph.D. in Nutritional Sciences from the University of Toronto, Canada. Her research focuses on chronic disease prevention from a plant-based nutritional perspective using systematic reviews and meta-analyses, randomized controlled trials, and epidemiological methodologies. Her ongoing research assesses cognitive health within the framework of the multicenter PREDIMED-Plus trial. Dr. Nishi works to support evidence-based practice and knowledge dissemination in order to inform dietetic practice, public health policy, and nutrition guidelines via her research as an educator/dietetic preceptor and a medical-science-based podcast host for the not-for-profit organization Plant-Based Canada.

nutrients

MDPI

Editorial

Where We Are and Where We Are Going in Nut Research

Jordi Salas-Salvadó [1,2,3,*], Stephanie K. Nishi [1,2,3,4,5], Joan Sabaté [6] and Emilio Ros [3,7,8]

1 Universitat Rovira i Virgili, Departament de Bioquímica i Biotecnologia, Alimentació, Nutrició, Desenvolupament i Salut Mental (ANUT-DSM), 43201 Reus, Spain
2 Institut d´Investigación Sanitària Pere Virgili (IISPV), 43201 Reus, Spain
3 Centro de Investigació Biomédica en Red Fisiopatología de la Obesidad y la Nutrición (CIBEROBN), Institute of Health Carlos III, 28029 Madrid, Spain
4 Toronto 3D (Diet, Digestive Tract and Disease) Knowledge Synthesis and Clinical Trials Unit, Toronto, ON M5C 2T2, Canada
5 Clinical Nutrition and Risk Factor Modification Centre, St. Michael´s Hospital, Unity Health Toronto, Toronto, ON M5C 2T2, Canada
6 Center for Nutrition, Lifestyle and Disease Prevention, School of Public Health, Loma Linda University, Loma Linda, CA 92350, USA
7 Lipid Clinic, Endocrinology and Nutrition Service and Institute d´Investigacions Biomèdique August Pi Sunyer, Hospital Clínic, 08036 Barcelona, Spain
8 Institut Dínvestigació Sanitària Pere Virgili (IISPV), 43201 Reus, Spain
* Correspondence: jordi.salas@urv.cat

Nuts have been part of the human diet for thousands of years [1]. Traditionally, nuts have been incorporated as an ingredient in many dishes, and over the years, nuts have been consumed in various forms from raw or minimally processed to more processed forms and eaten as snacks as well as included within recipes for main dishes.

In the last decades of the 20th century, the prevailing belief that dietary fat was harmful was at the basis of nuts being discouraged due to their high fat content. However, this perspective started to change following the first scientific studies demonstrating the potential health benefits of nut consumption. In 1992–1993, seminal publications from Loma Linda University showed that walnut consumption significantly reduced serum cholesterol [2] and that the frequency of nut consumption was inversely associated with coronary heart disease incidence according to data from the Adventist Health Study cohort [3]. Since then, many randomized clinical trials, epidemiological studies, and in vitro/in vivo mechanistic studies have explored and described the role of the consumption of different types of nuts on reduced incidence of cardiovascular disease and all-cause mortality, management of lipid disorders, and glycaemic control, without undue effects on body weight or overall adiposity, among other cardiometabolic and health-related risk factors and conditions [4]. More recently, several studies have examined the potential beneficial effects of nuts on the gastrointestinal system, cognitive performance, fertility, and different types of cancer, as well as the potential mechanisms implicated in the observed benefits. Importantly, landmark studies, such as the Adventist Health Study, the Nurses' Health Study, the Health Professionals Follow-up Study, the Physicians' Health Study, and the PREDIMED trial, have consistently reported that frequent nut consumption was associated with a lower risk of different cardiovascular outcomes [4] and all-cause mortality.

Based on the available scientific evidence, specific health claims have been accepted for nuts. Particularly, the Food and Drug Administration (FDA) has authorized qualified health claims for nuts in general and for walnuts and macadamias in particular concerning heart disease prevention when daily consuming one and one-half oz (42 g). However, the European Food Safety Authority (EFSA) has only agreed on a specific health claim for walnuts regarding beneficial effects on endothelial function. At the same time, nuts have been recommended over the last two decades by several health organizations and agencies worldwide.

Citation: Salas-Salvadó, J.; Nishi, S.K.; Sabaté, J.; Ros, E. Where We Are and Where We Are Going in Nut Research. *Nutrients* **2023**, *15*, 1691. https://doi.org/10.3390/nu15071691

Received: 15 March 2023
Accepted: 27 March 2023
Published: 30 March 2023

Due to the accumulating evidence on nut consumption and health outcomes, we thought it would be important to recapitulate and examine in detail what is well known and established, and what avenues of knowledge are still lacking in nut research. It is for this reason that we organized the NUTS 2022 Conference with the slogan: *Where we are and where we are going in nut research.*

The NUTS 2022 Conference offered the unique opportunity to bring together experts in the field of nut research from around the world with the following aims: (a) to summarize all the evidence related to the beneficial effects of nuts on health; (b) to identify new topics, needs, and opportunities in nut research; (c) to share knowledge with food industry and set new primary objectives for the future; and (d) to develop these scientific proceedings summarizing the current knowledge and new opportunities of research in the nut–health axis.

We believed it would be important and extremely useful to summarize and discuss future lines of nut research in the context of a multidisciplinary group of investigators with expertise in different fields for the benefit of: (1) the investigators, since it allows us to interact, share new ideas, and establish collaborations in the future; (2) the food industry, because they need to know that we know relatively little and that knowledge needs to be invested in; and (3) health agencies, because they need the most up-to-date knowledge to establish appropriate public health recommendations.

The NUTS 2022 Conference took place on 20–21 October 2022, and was organized in Reus by the University Rovira i Virgili together with Institut d'Investigació Sanitària Pere i Virgili (IISPV) and the Ciber Fisiopatología de la Obesidad y Nutrición (CIBEROBN) of Instituto de Salud Carlos III of Spain.

Author Contributions: Writing—original draft preparation, J.S.-S. and S.K.N.; writing—review and editing, J.S.-S., S.K.N., J.S. and E.R. All authors have read and agreed to the published version of the manuscript.

Funding: J.S.-S. is partially supported by ICREA under the ICREA Academia program. S.K.N. is supported by a Canadian Institutes of Health Research (CIHR) Postdoctoral Fellowship. The NUTS 2022 conference was planned by the Rovira i Virgili University, with all the expenses of the organization covered by Barcelo Congresos. Barcelo Congresos SA has received funds from the International Nut & Dried Fruit Council (INC). Researchers did not receive any additional financial support, aside from travel and accommodation, related to their participation in the NUTS 2022 conference. Sponsors had no role in selecting or inviting the speakers, nor in writing or reviewing the proceedings of the conference.

Conflicts of Interest: J.S.-S. reports serving on the boards of the INC and receiving grant support from these entities through his institution. He has also received research funding (nuts for free to the PREDIMED participants) from the CWC; La Morella Nuts, Spain; and Borges SA, Spain. He has also received research funding (nuts for free to the PREDIMED-Plus participants) from the Almond Board of California, USA and Pistachio Growers of California, USA. He is a non-paid member of the Instituto Danone International and was a member of the executive committee of the Instituto Danone Spain. S.K.N. is a volunteer member of the not-for profit group Plant-Based Canada and has received a research grant from the INC (International Nut and Dried Fruit Council) through her institution. J.S. has received health research grant funding through his institution from several nut commodity boards. E.R. reports receiving grant support through his institution from the California Walnut Commission (Folsom, CA), in addition to personal funds for project supervision and advice, and serving as a non-paid member of its Scientific Advisory Committee; funds for travel and accommodation from the International Nut and Dried Fruit Council; and personal funds from Alexion for serving on the advisory committee.

References

1. Salas-Salvadó, J.; Casas-Agustench, P.; Salas-Huetos, A. Cultural and historical aspects of Mediterranean nuts with emphasis on their attributed healthy and nutritional properties. *Nutr. Metab. Cardiovasc. Dis.* **2011**, *21* (Suppl. S1), S1–S6. [CrossRef] [PubMed]
2. Sabaté, J.; Fraser, G.E.; Burke, K.; Knutsen, S.F.; Bennett, H.; Lindsted, K.D. Effects of walnuts on serum lipid levels and blood pressure in normal men. *N. Engl. J. Med.* **1993**, *328*, 603–607. [CrossRef] [PubMed]

3. Fraser, G.E.; Sabaté, J.; Beeson, W.L.; Strahan, T.M. A possible protective effect of nut consumption on risk of coronary heart disease. The Adventist Health Study. *Arch. Intern. Med.* **1992**, *152*, 1416–1424. [CrossRef] [PubMed]
4. Ros, E.; Singh, A.; O'Keefe, J.H. Nuts: Natural Pleiotropic Nutraceuticals. *Nutrients* **2021**, *13*, 3269. [CrossRef] [PubMed]

nutrients

MDPI

Review

Nuts, Energy Balance and Body Weight

David J. Baer [1,*], Michelle Dalton [2], John Blundell [3], Graham Finlayson [3] and Frank B. Hu [4]

1 US Department of Agriculture, Agricultural Research Service, Beltsville, MD 20705, USA
2 School of Psychology, Leeds Trinity University, Leeds LS18 5HD, UK
3 School of Psychology, Appetite Control and Energy Balance Research Group, University of Leeds, Leeds LS2 9JT, UK
4 Department of Nutrition, Harvard T.H. Chan School of Public Health; Boston, MA 02115, USA
* Correspondence: david.baer@usda.gov; Tel.: +1-(301)-504-8719

Abstract: Over several decades, the health benefits of consuming nuts have been investigated, resulting in a large body of evidence that nuts can reduce the risk of chronic diseases. The consumption of nuts, being a higher-fat plant food, is restricted by some in order to minimize weight gain. In this review, we discuss several factors related to energy intake from nuts, including food matrix and its impact on digestibility, and the role of nuts in regulating appetite. We review the data from randomized controlled trials and observational studies conducted to examine the relationship between nut intake and body weight or body mass index. Consistently, the evidence from RCTs and observational cohorts indicates that higher nut consumption does not cause greater weight gain; rather, nuts may be beneficial for weight control and prevention of long-term weight gain. Multiple mechanisms likely contribute to these findings, including aspects of nut composition which affect nutrient and energy availability as well as satiety signaling.

Keywords: energy; calories; mastication; appetite; food intake; body weight; obesity; nuts

Citation: Baer, D.J.; Dalton, M.; Blundell, J.; Finlayson, G.; Hu, F.B. Nuts, Energy Balance and Body Weight. *Nutrients* **2023**, *15*, 1162. https://doi.org/10.3390/nu15051162

Academic Editor: Peter Pribis

Received: 27 January 2023
Revised: 12 February 2023
Accepted: 13 February 2023
Published: 25 February 2023

1. Introduction

Achieving and maintaining a healthy body weight is a difficult goal for many individuals. Obesity is a global health issue. According to the WHO [1], in 2016, almost 40% of the world's population were overweight and over 10% had obesity (among adults aged 18 years and older). Obesity is largely preventable, and at the simplest level, it is a matter of appropriate energy balance. To lose weight, energy intake must be less than energy expenditure, and to maintain weight, energy intake and expenditure must be equal. Within the constructs of this simple energy balance problem, there are many interdependent and complex factors that make body weight maintenance difficult. These factors include factors related to food and macronutrient composition, as well as the food matrix and energy availability. Appetite regulation is even more complex with a multitude of organ systems involved in making decisions multiple times, every day, about what to eat, when to eat, when to stop eating and how much to eat. Ultimately, interactions amongst food and its consumption determines energy intake.

In the early 1990s, research was beginning to show beneficial health effects associated with nut consumption [2]. With the emerging evidence, the 1995 Dietary Guidelines for Americans mentions including nuts in the diet but cautioned that foods, including nuts, high in fat should be used sparingly [3]. In the subsequent decades, much additional research has been conducted to better understand the health benefits of nuts, including the role nuts play in body weight maintenance. Herein, we review the state of the science in regard to how the food matrix of nuts effects energy availability, how nuts effect ingestive behavior and the literature on the relationship between nut intake and body weight maintenance.

2. Energy Availability

The plant cell wall significantly effects the bioavailability of energy and nutrients from nuts. Plant cell walls are complex extracellular matrices containing cellulose, hemicellulose, pectin and some proteins (usually enzymes that play a role in cell wall integrity). Lignin, which are polymers made of phenylpropanoid units, are additionally found in secondary cell walls [4]. Together, the cell wall components provide structural integrity for the plant, encapsulate the cell membrane to protect the individual cell and play a role in water and nutrient transport. Nutritionally, the plant cell wall is the source of dietary fiber which is resistant to mammalian digestive enzymes. Microbial anerobic fermentation or physical breakage of the plant cell wall is needed in order to release the cellular contents and make those contents available for absorption.

Among the first studies to report on the bioaccessibility of fat from nuts was a study of peanuts, peanut butter and peanut oil [5]. In this small study, subjects consumed each treatment, feces were collected and daily fecal fat excretion was determined. Consumption of peanuts resulted in increased fecal fat excretion compared to peanut butter and oil, and consumption of peanut butter resulted in increased excretion of fecal fat compared to peanut oil [5]. As seen in other studies, the absorption of lipids from oils is generally quite high in humans [6,7]. However, when the same fat is consumed in the plant matrix, bioaccessibility is decreased, resulting with increased fecal fat excretion.

2.1. *Effect of Processing and Mastication on Almond Lipid Bioaccessibility*

Microscopic analyses of nuts demonstrate that several factors contribute to breakage of the plant cell wall. Prior to that breakage, these microscopic analyses show that the lipid vacuoles remain intact, encased by the plant cell wall. Mastication is one factor which influences lipid bioaccessibility. In one study, subjects masticated and expectorated natural (unroasted) or roasted almonds [8]. Post-mastication, there were differences in the particle size distribution between the natural and roasted almonds, with larger particles (1700 to >3350 μm) being more prevalent in the natural almond samples and smaller particles (<1700 μm) being more prevalent in the roasted almond samples [8]. Lipid was identified on the surface of ruptured cells. In smaller particles (approximately 250 μm), free lipid was identified in all areas of the particles, not just the surface. In this study, roasting effected lipid bioaccessibility, which was greater in the roasted almonds compared to the natural almonds, and the higher lipid bioaccessibility was related to the increased proportion of smaller particles observed in the roasted almonds [8].

In a study of different forms of almonds (natural, roasted, roasted diced and almond butter (made from roasted almonds)), following simulated oral digestion, particle size distribution was similar for natural, roasted and chopped almonds (most particles having a size $\geq$1000 μm), whereas the particle size distribution of the almond butter resulted in mostly smaller particles (<850 μm) [9].

2.1.1. Effect of Roasting

Roasting changes the physical properties of almonds, and these changes contribute to the degree of cell ruptures. Using three-point bending to determine fracture force (N) at load failure, roasted almonds required less force for load failure than natural almonds [10]. The hardiness of roasted almonds, quantified by maximum force (N) required for failure during uniaxial compression, was also lower than natural almonds [10]. Upon fracture, 8-bit, binary digitized photos were used to quantify particle area and total number of fragments. The median particle area was smaller for roasted almonds compared to natural almonds, whereas there was a greater number of particles from roasted compared to natural almonds. The physical changes associated with roasting impact lipid bioaccessibility by increasing the ratio of surface area to volume and making more cellular contents available for digestion and absorption.

2.1.2. Effect of Mastication

Mastication is one of the physical processes which plays an important role in bioaccessibility of fat and energy. In a study of controlled mastication, subjects were provided 5 g of natural almonds and instructed to chew them for 10, 25 or 40 times and then expectorated [11]. The number of particles recovered was measured post-mastication, with more particles recovered after 10 chews compared to 25 or 40 chews [11]. Moreover, the mean particle size of the recovered particles was larger after 10 chews than 25 or 40 chews [11]. In a separate study with these subjects, they were allowed to chew the almonds and swallow after 10, 25 or 40 chews. Fecal samples were collected and fecal energy and fat were measured. Fecal energy and fat extraction were higher after 10 chews compared to 25 or 40 chews [11]. In this study, chewing almonds 10 times resulted in differences from chewing almonds for 25 or 40 times, but additional chewing of almonds beyond 25 times did not significantly change the particle size distribution, fecal energy or fat excretion.

2.1.3. Observations with Walnuts and Pistachios

Much of the research on the effects of processing and mastication has been conducted with almonds. One study [12] used walnuts (unsalted pieces) and pistachios (roasted) in addition to almonds (roasted and salted), and focused on in vitro gastrointestinal digestion. In undigested samples, walnuts had thinner cell walls compared to almonds and pistachios, whereas pistachios had smaller oil bodies than walnuts and almonds. Transmission electron microscopy revealed that in walnuts and almonds, the lipid was stored in a single, dense agglomerate, whereas the lipid in pistachios was observed in smaller and dispersed droplets within the cell [12]. Following mastication and in vitro digestion, cell walls from all nuts showed fissures and free lipids in the extracellular space. Thus, the effect of mastication and digestion (in vitro) of walnuts and pistachios also results in the breakage of cell walls, the release of lipids and increased bioaccessibility.

2.2. *History of Determining the Energy Value of Foods*

For food labeling purposes, the metabolizable energy value of the food is typically used. Metabolizable energy is the gross energy of the food corrected for energy losses in feces and urine. Gross energy of food, feces and urine are measured by bomb calorimetry. For food labeling, direct measures of metabolizable energy are not performed, but rather, the metabolizable energy is estimated using energy density factors which represent the energy, adjusted for incomplete digestion. The energy density factors, commonly known as the Atwater general factors, were based on research conducted by Atwater and colleagues [13]. Based on these studies, Atwater proposed that the metabolizable energy value of protein, fat and carbohydrate could be estimated as 4, 9 and 4 kcal/g, respectively. These factors were further refined based on food groups, more targeted to improve digestibility estimates of macronutrients. These refined energy density factors are commonly known as the Atwater specific factors.

While there is no evidence that Atwater performed studies with nuts, Jaffa performed studies with walnuts, Brazil nuts, pecans and almonds [14]. These studies were conducted with two to three men, involved simple diets usually containing a few items and lasted for a few days. While the intention of Jaffa's and Atwater's research was to provide information on the energy and nutrient availability of mixed diets, their results have been applied to individual foods. After the seminal work of Atwater and Jaffa, few additional studies have been reported focusing on measuring the energy value of individual foods or simple diets. While the state-of-the-art nature of Atwater's work has been the foundation for nutrition labeling, the approach is not without limitations. Some of the limitations have been reviewed [15] and include the small sample size, short duration of collections and measurement errors.

2.3. Recent Measures of the Metabolizable Energy Value of Nuts

In order to better measure the metabolizable energy value of an individual food while it is being consumed as part of a mixed diet, Novotny developed an approach [16] that overcomes the limitations of the Atwater approach [15]. Briefly, this approach requires a pair of diets—one without the food of interest and the other an identical diet with the food of interest. Using this approach, the metabolizable energy value of pistachios [17], almonds [16], walnuts [18] and cashews [19] was investigated. Additionally, a study of different forms of almonds was conducted [10].

In all of these studies, the measured metabolizable energy value of the nuts was lower than the energy value calculated using Atwater general or specific factors. The difference between the measured and calculated metabolizable energy values were 6% for pistachios (whole, lighted roasted and lightly salted) [17], 19% for almonds (whole, unroasted, unsalted) [16], 21% for walnuts (pieces) [18] and 6% for cashews (whole, roasted, lightly salted) [19]. In all of these studies, the amount of nuts included in the diet was 42 g/day (1.5 oz/day), and this amount was selected to be consistent with the US FDA qualified health claim for nuts [20,21]. Furthermore, with the study of pistachios and almonds, a second amount of 84 g/day was used to investigate dose effects. There was no change in the measured metabolizable energy value between the two doses [16,17].

3. Appetite as a Complex System

The regulation of appetite is complex and influenced by various biological, nutritional, physical and social factors. Humans are omnivores, allowing them to make food choices from a wide range of available options, but this versatility can also be a challenge. Appetite can be broadly divided into tonic and episodic components, which are generally represented by the drive to eat and food choice behavior. The key determinants of the drive to eat are the body's lean mass and resting metabolic rate, but these are unrelated to food choice [22]. Food hedonics, or the experienced pleasure derived from eating food, has a major influence on food choice. The consumption of chosen foods inhibit the drive to eat through the processes of satiation and satiety, which form part of the Satiety Cascade [23]. Satiety is the post-prandial inhibitory component of appetite control and is mediated by complex physiological processes. There is huge individual variability in the way people experience satiety, and the strength of satiety is heavily influenced by the diet selected.

The effect of nuts on appetite can be assessed by scientifically investigating their effects on the processes of satiation and satiety in relation to an individual's pattern of satiety control. When people freely consume nuts as part of their diet, either within or between meals, it is important to enquire what effects this will have on their overall energy intake and their pattern of food consumption. To investigate this issue, it is necessary to understand some features of the appetite system and the mechanisms that mediate the effects of ingested foods.

3.1. Appetite and Satiety (and Satiation)

Appetite is not the opposite of satiety; rather, satiety is an important component of appetite control. Appetite encompasses various processes that influence food consumption, including the initiation of eating and the duration and termination of eating episodes. Such processes include the level of hunger, food availability and choice, food hedonics, psychological traits, situational factors and social factors. Satiety refers to the reduction in hunger and eating following a meal. Satiation, on the other hand, is the termination of a meal and affects meal size. The period after a meal involves a complex series of physiological events in the digestive tract, including gastric activity and hormone release, which control the digestion and absorption of nutrients. These physiological events depend on the type of the foods consumed in the diet. As they accompany the state of satiety, they are often referred to as satiety signals. Whether these satiety signals are biomarkers of satiety or the cause of it is an area of debate in the field. However, it can be concluded that there is no single unique satiety signal [24]. As satiety is an inhibitory process, it plays an important role in determining how much food is eaten and one's levels of hunger. Therefore, satiety may potentially influence body weight by either permitting or preventing overconsumption [25,26]. Weak satiety is seen as a major factor in obesity, while intensifying satiety through certain foods or drugs may support weight loss [27–29].

3.2. *Foods and the Satiety Cascade*

As omnivores, human beings have the capability to consume a vast variety of foods from around the world, leading to a range of unique dietary patterns. The foods chosen impact the levels of satiation and satiety felt. This phenomenon can be explained through the idea of the 'Satiety Cascade' (Figure 1) [23]. The satiety cascade provides a framework to understand the mechanisms involved in the short-term control of eating behavior.

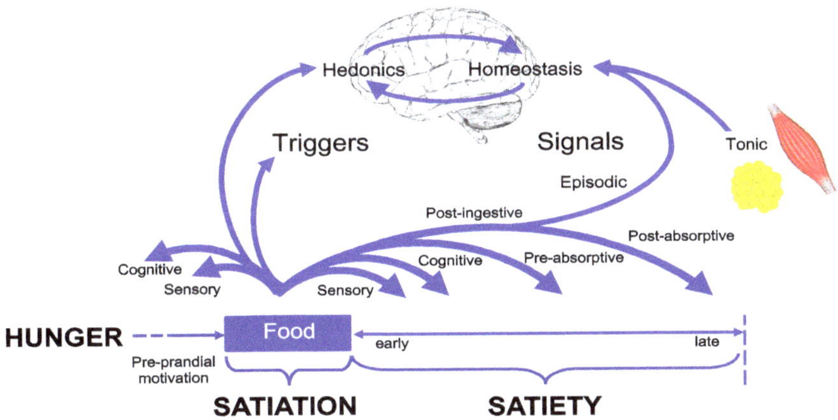

Figure 1. The Satiety Cascade depicts meal size and the time between meals is influenced by the processes of satiation and satiety. It also demonstrates the interaction between the homeostatic and hedonic influences on the processes of satiation and satiety. Adapted from Blundell and Finlayson [30].

The satiety cascade distinguishes between satiation and satiety and illustrates how a variety of signals, such as those arising from sensory, cognitive, post-ingestive and post-absorptive processes, affect the frequency and size of meals. The processes of satiation control meal size through their effect on the duration and termination of an eating episode. These processes, along with the nutritional content of the food consumed, determine the amount of energy consumed during the eating episode. Once the meal is finished, the desire to eat is temporarily suppressed by the physiological effects of the consumed food, especially in the stomach, and the hormones released by the gastrointestinal system during the digestion and absorption of food.

3.3. *The Nature of Satiety Signals*

After eating, the sensation of fullness (satiety) is produced by several features of the foods consumed, including volume, weight, sensory features (taste, texture, mouthfeel), enjoyment, appearance, nutritional composition, non-nutritional elements (such as fiber) and packaging/labeling. Therefore, satiety is a result of the combined effects of various components of the food consumed. Many studies have aimed to determine the specific characteristics of foods that have the most significant effect on satiety. Understanding these factors is crucial for the food industry in creating foods that can regulate hunger and enhance the feeling of fullness. There is evidence to suggest that high protein and fiber levels can increase satiety. However, energy density is a crucial factor, with low-energy-density diets producing stronger feelings of satiety [31].

Since the first investigations of satiety, it has been believed that the impact of food composition is influenced by post-meal physiological responses. These responses involve alterations in gastric distension, digestion and emptying, as well as the release of gastrointestinal peptides including cholecystokinin (CCK), glucagon-like peptide (GLP-1), peptide YY (PYY), insulin and others. For a long time, CCK was considered to be the sole satiety signal. It is important to note that all these peptides play important roles in the body's management of food through digestion and absorption processes, such as slowing down gastric emptying and releasing bile for fat emulsification. As a result, their impact on satiety may be secondary to their other functions. It remains a topic of discussion whether gut peptides are markers or the actual cause of satiety. The fact that different foods may have similar effects on satiety but produce distinct physiological profiles suggests that there is no uniform pattern behind satiety and that the same level of satiety may be linked to different post-prandial physiological changes [24]. In recent years, there has been considerable interest in the post-prandial physiological effects of raw foods compared with highly and ultra-processed foods [32].

3.4. A Note on (the Role of) Food Hedonics

Food is a reliable source of pleasure for most people, and the reward derived from food plays an important role in the initiation, maintenance and termination of an eating episode, in part through interaction with processes involved in hunger and satiety. Food hedonics is more than simply liking the taste of food or the experience of pleasure. Non-human animal research has demonstrated that the brain structures underpinning food hedonics comprise dissociable affective and motivational subcomponents, termed 'liking' and 'wanting', respectively. Liking refers to the sensory pleasure experienced while eating a food and is generated by the binding of opioids to specialized clusters of neurons in the reward pathway, particularly in the Nacc shell. Wanting refers to the process that assigns motivational value to finding and consuming a food and is mediated by the release of dopamine (DA) from the ventral tegmental area (VTA) to the nucleus accumbens (Nacc) and amygdala [33,34]. In human appetite research, the terms "liking" and "wanting" for food are often seen as explicit subjective states that correspond to their everyday meanings. Liking refers to the enjoyment of the sensory qualities of food that give it its hedonic impact, while wanting refers to a subjective state of desire or craving. People are generally good at estimating and reporting their liking for food, but are often inaccurate in their assessment of their implicit wanting for food, meaning why they are attracted to or craving a particular food over another [35,36].

After food is consumed, the sensory aspects of the food are registered by both cognitive and sensory processes before it is swallowed. Highly palatable food stimulates the reward pathways in the brain, causing the release of dopamine and endorphins. These reward pathways have connections to the hypothalamus, which triggers the release of hunger-inducing peptides such as NPY and orexins and suppresses the release of satiety-inducing peptides such as insulin, leptin and cholecystokinin. Thus, the consumption of highly palatable food can result in overeating, as the drive to eat is motivated by pleasure rather than actual hunger. The interplay between the hedonic and homeostatic systems of appetite regulation contributes to the overall pattern of eating behavior, and in an environment that promotes obesity, the hedonic drive to eat may have a stronger impact on food consumption compared to homeostatic mechanisms [29,37].

3.5. Individual Variability in Appetite Control and the Low Satiety Phenotype

The range of factors that contribute to a person's susceptibility to overconsume (and eventually weight gain and obesity) can include their genetics, physical and psychological characteristics, lifestyle habits and surrounding food and activity environment. Decades of research have pinpointed many aspects of the typical Western lifestyle and diet that interact with these factors, making it easier for people to overeat and gain weight. However, not everyone in a 'westernized' environment overconsumes food, and it is unlikely that one single factor can account for why some are more vulnerable than others; this has implications for appetite control and the prevention of weight gain.

One approach to characterize individual susceptibility is through the identification and characterization of phenotypes. One such phenotype may be characterized by a weakened satiety response to food, which has been proposed as a possible marker of susceptibility to overeating [38–40]. Research has shown that under controlled conditions, appetite sensations are a valid and reliable method for measuring the subjective motivation to eat [41]. However, not everyone reports a good relationship between their sensations of appetite (hunger and fullness) and their eating behavior. A weakened satiety response to the feeling of satiety may play a role in a lack of control over one's appetite. Individuals with such a reduced response to food are referred to as the "low satiety phenotype" [42]. The low satiety phenotype has largely been observed in people with obesity, but evidence suggests that a weakened satiety response to food may lead individuals to be vulnerable to future weight gain. Research examining the low satiety phenotype has demonstrated that it is characterized by increased Three Factor Eating Questionnaire disinhibition and hunger scores, lower levels of craving control, greater food wanting and increased energy intake under laboratory conditions [42–44]. In terms of weight management, studies have shown that individuals with a low satiety responsiveness tend to lose less weight, experience smaller decreases in abdominal fat, report lower control over cravings and face more challenges in sticking to a diet compared to those with a high satiety responsiveness [44–46].

3.6. Nuts and Appetite Control: A Case Study with Almonds

Snacking between meals is a common behavior [47], and snack foods make a significant contribution to total daily food intake [48]. Snack foods are often characterized as being low in nutritional quality, primarily comprising fats and carbohydrates [49] that contribute to overconsumption. However, research suggests that frequent snacking can promote feelings of satiety throughout the day, which results in overall lower daily energy intake [50]. This suggests that snacking behavior itself is not undesirable and may present an opportunity for the addition of healthy foods, such as nuts, into the diet [51]. A recent meta-analysis of randomized clinical trials found that regular consumption of nuts was associated with increased daily energy intake and lower hunger but had no effect on weight or feelings of fullness [52]. The increase in daily energy intake was lower than the amount of energy consumed from the nuts, which may be due to the lower amount of available energy from nuts following digestion [16,53].

Almonds are a natural food product that are high in protein and fiber as well as fat, but lower in metabolizable energy compared to the predicted levels (using Atwater factors) [16]. It is well established that proteins and fibers have prominent effects on appetite control [54,55], and since they act via different mechanisms, their effects may be additive. The unique structural properties and macronutrient composition of almonds may be beneficial for the control of hunger, strength of satiety and subsequent energy intake relative to other foods. The addition of almonds to a meal has been shown to increase satiety and decrease blood glucose concentrations in those with and without impaired glucose tolerance [56–58]. Furthermore, when consumed as a snack, almonds have been shown to reduce feelings of hunger and desire to eat [59,60]. The consumption of almonds as a snack does not seem to cause an increase in total daily energy intake [61] or result in significant weight change over time [62–64]. A recent study compared the effect of consuming almonds as a mid-morning snack compared to an energy- and weight-matched comparator snack (crackers) and a zero-energy, weight-matched control (water) on measures of subjective appetite, food intake and food hedonics. It was found that overall hunger was lower in the almonds condition, and almonds were more satiating than the crackers (Figure 2). There was also a reduction in implicit wanting for high-fat food following almond consumption suggesting a beneficial effect on hedonic hunger. Further to this, participants' perceptions of the almonds were favorable, with almonds being perceived as healthy, filling and good for weight management [61].

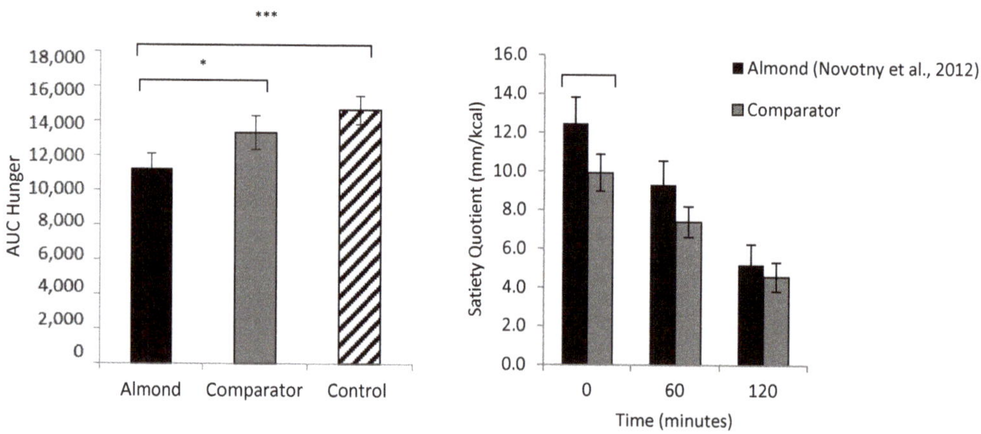

Figure 2. Left: Area under the curve hunger for the almonds condition, energy- and weight-matched comparator (crackers) and weight-matched comparator (water). Right: Satiating efficiency (measured by the Satiety Quotient) of the almonds compared to comparator for 120 min post-consumption [16]. Adapted from Hollingworth et al. [61]. Note: * $p < 0.05$; *** $p < 0.001$.

3.7. *Nuts and the Low Satiety Phenotype*

As outlined above, the low satiety phenotype is characterized by higher levels of hunger across the day, greater overall energy intake, increased liking and wanting for food, and poorer weight loss outcomes following structured weight management programs [42–46]. Foods that promote satiety have the potential to support individuals (in general and perhaps in particular those with a weakened satiety response to food) to control their appetite, eat fewer calories and manage their weight [65]. Research suggests that even when matched for calories, not all foods provide the same level of satiety [66], and a hierarchy of macronutrient satiating power has been established, with foods that are high in protein and fiber and low in energy density being more satiating [54,67–69]. The unique structural properties and macronutrient composition of nuts may be beneficial for the control of hunger, strength of satiety and subsequent energy intake relative to other foods; therefore, the consumption of nuts may support those with low satiety responsiveness in improving their appetite control.

Hollingworth [70] compared the effect of consuming almonds as a mid-morning snack compared to an energy- and weight-matched comparator snack food (crackers) on satiating efficiency, energy intake and feelings of hunger and fullness across the day in the low satiety phenotype compared to the high satiety phenotype. They found that almonds had a greater satiating efficiency, measured using the satiety quotient, in the low satiety phenotype compared to the comparator snack. In addition, when compared to the comparator food, almonds were perceived as being healthier, more filling and more favorable for weight management. Expectations about the satiating potential of food has been shown to play a role in expected satiety [71] and may present another mechanism by which almonds (and potentially other nuts) may support appetite control. Furthermore, while almonds and the comparator snack were rated as equally palatable, participants rated the almonds as more difficult to chew. The texture and chewiness of almonds may improve their satiating capacity, with evidence suggesting that oral processing plays an important role in food intake by affecting both satiation and satiety [72].

4. Overview of Nut Consumption and Body Weight

In order to understand the effects of almonds on appetite control, it is necessary to recognize the complex nature of human appetite as an emergent property of a complex system [73]. The act of food consumption in the real world is influenced by a diverse set of biological and environmental variables, with a greater complexity than can be achieved in laboratory investigations. Recognizing this complexity, it can be shown that changes to the whole diet (for example, by changing energy density) can exert effects on meal sizes, daily energy intake and profiles of hunger [31]. With this in mind, we can ask what is the likely strength of effect on appetite of manipulating a single food in the diet? One systematic review with a meta-analysis of laboratory and field trials has noted that nuts in general do not exert consistent effects on food intake or hunger [52]. However, seeking general effects in an unselected cohort or population will be too insensitive to discriminate effects on people with varying existing degrees of appetite control (satiety phenotypes). An enhancement of satiety is more likely to occur in individuals showing poor appetite control. As shown above, almonds can improve satiety in the low satiety phenotype. This is important since such people are the most likely to benefit from an improvement in control over their appetites (hunger drive and meal size). This action demonstrates how a single food in the diet can exert a meaningful effect. When consumed by people with normal or strong appetite control (high satiety phenotype), the most likely outcome is the maintenance of the habitual eating pattern and a prevention of overconsumption. In achieving these outcomes, almonds (and other nuts) benefit from a range of food factors that influence satiety, including taste and texture, postprandial physiology as well as expectations about satiety. Therefore, in weighing up how nuts can influence appetite control, it is important to manage expectations and not to anticipate the same effect in all types of eaters. Actions can be expected to vary according to the strength of a person's natural appetite control. Different types of benefit can be expected in people with different forms of satiety control. This approach could form the basis for future investigations of the effect of nuts on appetite.

4.1. Evidence from Prospective Cohort Studies

Several prospective cohort studies have examined the association between nut consumption and long-term weight change and obesity risk. Bes-Rastrollo et al. [74] examined the long-term association between nut consumption and weight change over 8 years among 51,188 women aged 20–45 years from the Nurses' Health Study (NHS) II. The analysis prospectively evaluated the dietary intake of nuts reported in 1989 and subsequent weight changes from 1991 to 1999. After adjusting for age, BMI, alcohol consumption, physical activity, smoking, postmenopausal hormone use, oral contraceptive use and dietary factors, this study found that women who reported eating nuts $\geq$2 times/week experienced a slightly lower mean ($\pm$SE) weight gain (5.04 $\pm$ 0.12 kg) than did women who rarely ate nuts (5.55 $\pm$ 0.04 kg) (p-trend < 0.001). The findings were similar when nut consumption was subdivided into peanuts and tree nuts as well as for participants who are normal-weight, overweight and have obesity.

In an analysis of three prospective cohorts that included 120,877 US women and men with follow-ups ranging from 1986 to 2006 [75], each four-year weight change was inversely associated with a one-serving increment in the intake of nuts (−0.26 kg), fruit (−0.22 kg), vegetables (−0.10 kg), whole grains (−0.17 kg) and yogurt (−0.37 kg), whereas weight gain was positively associated with the intake of potato chips (0.77 kg), potatoes or fries (0.58 kg), sugar-sweetened beverages (0.45 kg), unprocessed red meats (0.43 kg) and processed meats (0.42 kg). These data suggest that specific dietary factors including nuts and overall diet quality influence long-term weight gain.

In a Spanish cohort study consisting of 8865 adult men and women [76], regular nut consumption was significantly associated with a reduced risk of weight gain of $\geq$5 kg. After adjustment for age, sex, smoking status, physical activity and other covariates, participants who ate nuts $\geq$2 times/week had a significantly lower risk of weight gain (OR: 0.69; 95% CI: 0.53, 0.90; p-trend = 0.006) compared with those who never or almost never ate nuts.

In a prospective analysis of 3092 young adults enrolled in the Coronary Artery Risk Development in Young Adults (CARDIA) study that assessed consumption of walnuts and other nuts three times during the follow-up [77], higher walnut consumption was significantly associated with higher HEI-2015, lower BMI, waist circumference, blood pressure, and triglyceride concentration. Walnut consumers gained less weight since baseline than other nut consumers ($p \leq 0.05$).

Recently, Li and colleagues evaluated changes in total and different types of nut consumption and long-term weight change in three US cohorts [78]. These analyses included 27,521 men (Health Professionals Follow-up Study, 1986 to 2010), 61,680 women (Nurses' Health Study, 1986 to 2010) and 55 684 younger women (Nurses' Health Study II, 1991 to 2011) who were free of chronic disease at baseline in the analyses. The study found that increases in nut consumption, per 0.5 servings/day (14 g), was significantly associated with less weight gain per 4-year interval ($p < 0.01$ for all): −0.19 kg (95% CI −0.21 to −0.17) for total consumption of nuts, −0.37 kg (95% CI −0.45 to −0.30) for walnuts, −0.36 kg (95% CI −0.40 to −0.31) for other tree nuts and −0.15 kg (95% CI −0.19 to −0.11) for peanuts. In addition, increasing the intake of nuts, walnuts and other tree nuts was associated with a lower risk of obesity. In substitution analyses, substituting 0.5 servings/day of nuts for red meat, processed meat, French fries, desserts or potatoes and chips was associated with less weight gain ($p < 0.05$ for all) (Figure 3). This study provides further evidence that increasing daily consumption of total and different types of nuts is associated with less long-term weight gain and a lower risk of obesity in adults. More importantly, this study indicates that replacing "less healthful foods" with nuts may be an effective strategy to help prevent gradual long-term weight gain and obesity.

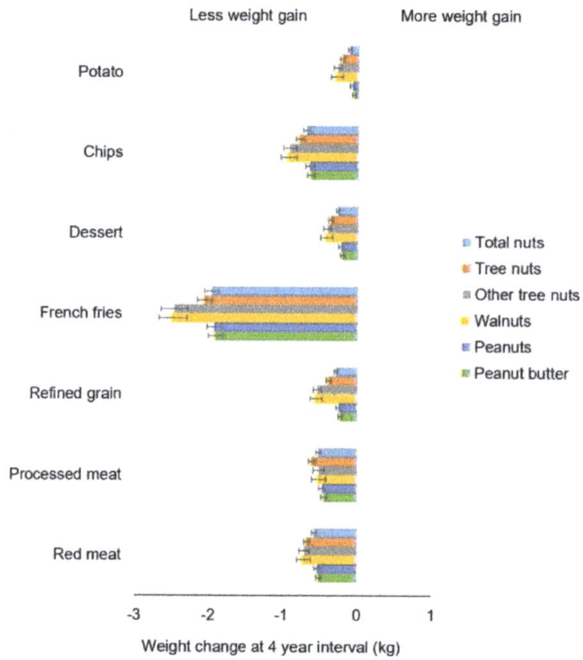

Figure 3. Association between weight change (kg) every 4 years and substitution of nuts and individual types of nuts, per 0.5 servings/day with equal serving of other food items among NHS, NHS II and HPFS. Weight changes are presented as solid bars; T bars represent 95% CI. Multivariate model was adjusted for age, menopausal status (pre- or postmenopausal) and hormone therapy use (never, past or current) in women; baseline BMI of every 4 years; hours of sleeping at baseline; changes in lifestyle factors: smoking status (never, former, current: 1 to 14, 15 to 24, or ≥25 cigarettes/day), physical activity (MET hours/week), hours of sitting (hours/week); and changes in dietary factors: fruits, vegetables, alcohol, snacks, dessert, French fries, red or processed meat, whole grain, refined grain products and sugar sweetened beverages.

Nishi et al. conducted a systematic review and meta-analysis of five prospective cohorts on nut consumption and weight gain and obesity among 520,331 participants [79]. It found that higher nut intake was associated with a decrease in overweight/obesity incidence (RR 0.93 [95% CI 0.88 to 0.98], $p < 0.01$; I2 = 90.0%, p-heterogeneity < 0.01). Similarly, higher nut consumption was associated with weight loss (MD 0.46 kg [95% CI 0.78 to 0.13 kg], $p < 0.01$; I2 = 95.9%, p-heterogeneity < 0.01) and reduced risk of weight gain ≥5 kg (RR 0.95 [95% CI, 0.94 to 0.96], $p < 0.01$; I2 = 46.7%, p-heterogeneity = 0.15). The certainty of evidence was rated moderate based on the GRADE criteria. In pooled analyses from models not adjusting for energy intake, higher nut consumption was associated with less weight gain (MD 0.64 kg [95% CI 1.12 to 0.15 kg]).

4.2. Evidence from RCTs

Few RCTs have specifically evaluated the role of nuts in weight loss and maintenance or obesity prevention. Wien et al. [80] evaluated the effect of an almond-enriched (84 g/day) or complex carbohydrate-enriched, formula-based, low-calorie diet (LCD) on anthropometric, body composition and metabolic parameters in a randomized 24-week trial among 65 adults with overweight and obesity (age: 27–79 y, BMI: 27–55). LCD supplementation with almonds, compared to complex carbohydrates, led to greater reductions in weight/BMI (−18 vs. −11%, $p < 0.0001$), waist circumference (WC) (−14 vs. −9%, $p < 0.05$), fat mass (−30 vs. −20%, $p < 0.05$), total body water (−8 vs. −1%, $p < 0.05$) and systolic blood pressure (−11 vs. 0%, $p < 0.02$). Ketone levels increased only in the almond-LCD group ($p < 0.02$). This study suggests that an almond-enriched LCD is beneficial for a sustained and greater weight reduction for the duration of the 24-week intervention.

Numerous small, short-term RCTs have examined the effects of nut-rich diets on a wide range cardiometabolic risk factors in which body weight or fatness were secondary outcomes. Fernández-Rodrígue et al. [81] conducted a systematic review and network meta-analysis on the relationship of tree nut and peanut consumption with adiposity measures including body weight (BW), BMI, waist circumference (WC) and body fat percentage (BF%). This study included a total of 105 RCTs with measures of BW (n = 6768 participants), BMI (n = 2918), WC (n = 5045) and BF% (n = 1226). Compared to a control diet, nut-enriched diets had no significant effects on the adiposity-related measures, except for a positive effect of hazelnut-enriched diets and an increase in WC. Moreover, almond-enriched diets significantly reduced WC compared to the control diet. In subgroup analyses with only RCTs designed to assess whether nut consumption affected weight loss, almond-rich diets significantly reduced BMI and walnut-rich diets significantly reduced %BF. This study provides evidence to supports that tree nut and peanut enriched diets do not increase adiposity. A similar conclusion was reached by a meta-analysis conducted by Nishi et al. [79], which found no adverse effect of nuts compared with control diets on body weight (105 trial comparisons involving 9655 participants, MD 0.09 kg, [95% CI 0.09 to 0.27 kg], $p = 0.34$; I2 = 63.2%, p-heterogeneity < 0.01).

In a systematic review and meta-analysis of 15 RCTs on almond consumption and cardiovascular risk factors [82], compared to control diets, almond-enriched diets significantly improved blood lipids and reduced inflammatory biomarkers. In the meantime, higher almond consumption of >42.5g/day significantly improved fasting blood glucose and reduced BMI.

In a systematic review and meta-analysis of 55 parallel-arm or crossover interventions of nuts (including mixed nuts, nut-based snack bar and individual nuts including almonds, cashews, hazelnuts, macadamia nut, peanut, pecan, pistachio and walnut) there was no change reported in body weight, BMI or waist circumference. The mean duration of these studies was 13.8 ± 21.5 weeks and the mean intake of nuts was 48.2 ± 20.8 g/d. The analysis included studies where no substitutions instructions were provided as well as studies which provided to the participants instruction on substitution. In the studies where substitution instructions were not provided, there was no change in body fat percentage.

In studies with dietary substitution instructions, there was a significant decrease in body fat percentage [83].

Only one long-term RCT examined the effects of a Mediterranean diet supplemented with nuts on body weight and waist circumference changes in the context of a Mediterranean dietary intervention and primary prevention of CVD [84]. The PREDIMED trial randomly assigned 7447 participants with high risk of CVD to one of three interventions: Mediterranean diet supplemented with extra-virgin olive oil (n = 2543); Mediterranean diet supplemented with mixed nuts including almonds, walnuts and hazelnuts (n = 2454); or a control diet (advice to reduce dietary fat; n = 2450). After a median 4.8 years of follow-up, participants in all three groups had marginally reduced bodyweight. After multivariable adjustment, the difference in 5-year changes in bodyweight in the olive oil group was −0.41 kg (95% CI −0.83 to 0.01; $p = 0{\cdot}06$) and −0.02 kg (−0.45 to 0.42; $p = 0.94$) in the nut group compared with the control group. The adjusted difference in 5-year changes in waist circumference was −0.47 cm (−1.11 to 0.18; $p = 0.15$) in the olive oil group and −0.92 cm (−1.60 to −0.24; $p = 0.008$) in the nut group compared with the control group. This study provides strong evidence that diets supplemented with either extra-virgin olive oil or nuts had no adverse effects on body weight or WC. In contrast, these diets may have beneficial effects on adiposity measures compared to a lower-fat diet.

4.3. Methodological Issues in Observational Studies and RCTs

Both observational studies and RCTs of diet and body weight fraught with methodologic problems (see Chapter 14 in [85]). RCTs should provide some of the most rigorous evaluations of dietary intake and body weight. However, long-term dietary intervention studies are seldom feasible because of the high cost and lack of compliance by study participants. In addition, lack of compliance and high dropout rates are common in dietary intervention trials. Most RCTs on body weight and other CVD risk factors are of short-duration, small sample sizes and use different control groups. In most RCTs, adiposity measures such as weight or waist circumference changes were considered as secondary outcomes.

Observational studies of nut consumption and body weight are also complicated by several methodologic issues. In particular, residual confounding by other dietary and lifestyle factors cannot be ruled out because regular nut consumers tend to follow a healthier diet and lifestyle than non-consumers. Dietary assessment tools such as the 24 h recalls, dietary records and FFQs that are widely used in epidemiologic studies are prone to both random and systematic measurement errors. Although carefully validated FFQs that are administered repeatedly during follow-up are best-suited to the assessment of long-term patterns in intake, few large-cohort studies assessed diets repeatedly. In addition, no study has specifically examined the influence of food processing methods on body weight outcomes (i.e., salted, raw, roasted). Finally, most studies have been conducted in white or European populations, and thus the results may not be generalizable to other racial and ethnic groups.

5. Potential Components of Nuts That Contribute to Weight Control

Several mechanisms have been proposed to explain the potential benefits of nut consumption on body weight [86] (Figure 4). Nuts are rich in (1) proteins and (2) dietary fiber, which are associated with increased satiety, and in (3) unsaturated fats, which may increase oxidation that potentially decreases body fat accumulation [87]. High amounts of protein and fiber in nuts may also increase thermogenesis and resting energy expenditure. Dietary fiber (especially viscous fiber) in nuts delays gastric emptying and subsequent absorption that potentially suppresses hunger and promote healthy gut microbiome that improves energy metabolism. In addition, incomplete mastication of nuts may lead to increased energy loss via feces, which contributes to energy availability of nuts and thus a lower energy intake. Furthermore, consuming nuts at expense of red meat and refined carbohydrates may also contribute to less weight gain and lower risk of chronic diseases.

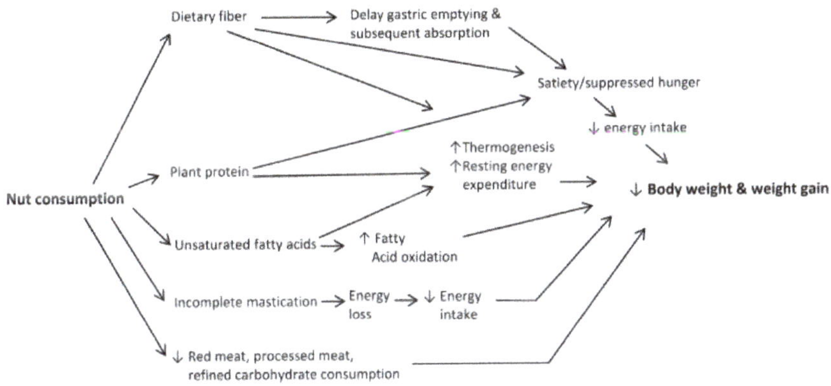

Figure 4. Conceptual framework of potential mechanisms linking nut consumption to decreased body weight and weight gain [86].

6. Clinical and Public Health Dietary Recommendations on Nuts and Weight Management

Cumulative evidence from long-term large cohort studies supports that an increased consumption of nuts, including total nuts and different types of nuts, is associated with less weight gain and lower risk of obesity, despite being calorically dense. The benefits to body weight are more pronounced when nuts are used to replace unhealthy foods such as red meat, processed meat, French fries, desserts or potato, chips. In addition, short-term RCTs suggest that nut-enriched diets had no adverse effects on body weight or other adiposity measures compared to control diets. There is some evidence that nuts may have beneficial effects on weight loss and maintenance, although more research is needed. Healthy dietary patterns rich in nuts, such as the Mediterranean diet, DASH diet and healthy plant-based diets, have been associated with age-related weight gain, although in these studies, the effects of nuts cannot be separated from other components of the dietary patterns [84].

7. Conclusions

To date, the plant cell wall factors that influence the energy available from nuts have mostly been investigated in almonds, with some research conducted in pistachios and walnuts. The effect of the plant cell wall, and its fermentation, on energy availability of other nuts has not been reported. Furthermore, the metabolizable energy value of nuts has been measured for almonds, walnuts, pistachios and cashews. Data from other nuts have not been reported. Additionally, the effect of processing on energy availability has only been investigated in almonds and peanuts. More information on dose response and individual variability may be useful to understand individual variability in energy intake, especially when trying to determine compensation of energy intake.

Evidence from RCTs and observational cohorts indicates higher nut consumption does not appear to cause greater weight gain; rather, nuts may be beneficial for weight control and prevention of long-term weight gain. Diet and lifestyle changes such as the replacement of less healthful food items (e.g., red or processed meats, refined grain products) with nuts and other healthy foods have the potential to reduce risk of obesity and obesity-related chronic diseases. In terms of future directions, more observational studies and RCTs are needed to examine the effects of nut consumption on different body depots, especially abdominal, visceral and liver fat. More studies are also needed to be conducted among individuals with type 2 diabetes, the metabolic syndrome and fatty liver disease and in diverse populations of different racial and ethnic groups and socio-economic status. Finally, research is needed to examine the role of nuts in healthy and sustainable eating patterns such as the Healthy Planetary Diet recommended by the Eat-Lancet Commission [88].

Author Contributions: D.J.B., M.D., J.B., G.F. and F.B.H. shared equally in the preparation of this manuscript. All authors have read and agreed to the published version of the manuscript.

Funding: The NUTS 2022 conference was organized by the Rovira i Virgili University, with all the expenses of the organization covered by Barcelo Congresos, including the fees related to publishing the Proceedings in Nutrients. Barcelo Congresos SA has received funds from the International Nut & Dried Fruit Council (INC). Neither the researchers attending the NUTS 2022 Conference nor the Rovira i Virgili University have received funds from the INC. FBH's travel expenses and accommodation have been covered by Barcelo Congresos. Researchers did not receive any additional financial support related to their participation in the NUTS 2022 conference. Sponsors did not have any role in organizing the conference, selecting or inviting the speakers nor in the writing or reviewing the proceedings of the conference.

Institutional Review Board Statement: Not applicable.

Informed Consent Statement: Not applicable.

Data Availability Statement: Not applicable.

Conflicts of Interest: D.J.B. has received research grants related to this manuscript from Paramount Farms, Almond Board of California and the California Walnut Commission. He has served (non-compensated) on the California Walnut Commission's Health and Research Advisory Group. J.B., G.F. and M.D. have received research grants from the Almond Board of California for work cited in this manuscript.

References

1. WHO. *Obesity and Overweight;* World Health Organization: Geneva, Switzerland, 2003.
2. Fraser, G.E.; Sabate, J.; Beeson, W.L.; Strahan, T.M. A possible protective effect of nut consumption on risk of coronary heart disease. The Adventist Health Study. *Arch. Intern. Med.* **1992**, *152*, 1416–1424. [CrossRef]
3. United States. Dietary Guidelines Advisory Committee.; United States. Department of Agriculture.; United States. Department of Health and Human Services.; United States. Office of Disease Prevention and Health Promotion. Dietary guidelines for Americans. In *USDA Publication Number: Home and Garden Bulletin No. 232;* U.S. Dept. of Health and Human Services: U.S. Dept. of Agriculture: Washington, DC, USA, 1995; p. 45.
4. Alberts, B. *Molecular Biology of the Cell*, 4th ed.; Garland Science: New York, NY, USA, 2002.
5. Levine, A.S.; Silvis, S.E. Absorption of whole peanuts, peanut oil, and peanut butter. *N. Engl. J. Med.* **1980**, *303*, 917–918. [CrossRef]
6. Baer, D.J.; Judd, J.T.; Kris-Etherton, P.M.; Zhao, G.; Emken, E.A. Stearic acid absorption and its metabolizable energy value are minimally lower than those of other fatty acids in healthy men fed mixed diets. *J. Nutr.* **2003**, *133*, 4129–4134. [CrossRef]
7. Baer, D.J.; Rumpler, W.V.; Miles, C.W.; Fahey, G.C., Jr. Dietary fiber decreases the metabolizable energy content and nutrient digestibility of mixed diets fed to humans. *J. Nutr.* **1997**, *127*, 579–586.
8. Grundy, M.M.; Grassby, T.; Mandalari, G.; Waldron, K.W.; Butterworth, P.J.; Berry, S.E.; Ellis, P.R. Effect of mastication on lipid bioaccessibility of almonds in a randomized human study and its implications for digestion kinetics, metabolizable energy, and postprandial lipemia. *Am. J. Clin. Nutr.* **2015**, *101*, 25–33. [CrossRef] [PubMed]
9. Mandalari, G.; Parker, M.L.; Grundy, M.M.; Grassby, T.; Smeriglio, A.; Bisignano, C.; Raciti, R.; Trombetta, D.; Baer, D.J.; Wilde, P.J. Understanding the effect of particle size and processing on almond lipid bioaccessibility through microstructural analysis: From mastication to faecal collection. *Nutrients* **2018**, *10*, 213. [CrossRef] [PubMed]
10. Gebauer, S.K.; Novotny, J.A.; Bornhorst, G.M.; Baer, D.J. Food processing and structure impact the metabolizable energy of almonds. *Food Funct.* **2016**, *7*, 4231–4238. [CrossRef]
11. Cassady, B.A.; Hollis, J.H.; Fulford, A.D.; Considine, R.V.; Mattes, R.D. Mastication of almonds: Effects of lipid bioaccessibility, appetite, and hormone response. *Am. J. Clin. Nutr.* **2009**, *89*, 794–800. [CrossRef]
12. McArthur, B.M.; Mattes, R.D. Energy extraction from nuts: Walnuts, almonds and pistachios. *Br. J. Nutr.* **2020**, *123*, 361–371. [CrossRef]
13. Merrill, A.L.; Watt, B.K. Energy Value of Foods: Basis and Derivation. Agriculture Handbook No. 74, ARS United States Department of Agriculture, Washington DC. 1973. Available online: https://www.ars.usda.gov/arsuserfiles/80400535/data/classics/usda%20handbook%2074.pdf (accessed on 15 November 2022).
14. Jaffa, M.E. *Further Investigations among Fruitarians at the California Agricultural Experiment Station, 1901–1902;* Government Printing Office: Washington, DC, USA, 1903.
15. Sanchez-Pena, M.J.; Marquez-Sandoval, F.; Ramirez-Anguiano, A.C.; Velasco-Ramirez, S.F.; Macedo-Ojeda, G.; Gonzalez-Ortiz, L.J. Calculating the metabolizable energy of macronutrients: A critical review of Atwater's results. *Nutr. Rev.* **2017**, *75*, 37–48. [CrossRef]

16. Novotny, J.A.; Gebauer, S.K.; Baer, D.J. Discrepancy between the Atwater factor predicted and empirically measured energy values of almonds in human diets. *Am. J. Clin. Nutr.* **2012**, *96*, 296–301. [CrossRef]
17. Baer, D.J.; Gebauer, S.K.; Novotny, J.A. Measured energy value of pistachios in the human diet. *Br. J. Nutr.* **2012**, *107*, 120–125. [CrossRef]
18. Baer, D.J.; Gebauer, S.K.; Novotny, J.A. Walnuts Consumed by Healthy Adults Provide Less Available Energy than Predicted by the Atwater Factors. *J. Nutr.* **2016**, *146*, 9–13. [CrossRef]
19. Baer, D.J.; Novotny, J.A. Metabolizable energy from cashew nuts is less than that predicted by Atwater factors. *Nutrients* **2018**, *11*, 33. [CrossRef]
20. US Food and Drug Adminstration. Qualified Health Claims: Letter of Enforcement Discretion—Nuts and Coronary Heart Disease (Docket No 02P-0505). Available online: http://wayback.archive-it.org/7993/20171114183724/https://www.fda.gov/Food/IngredientsPackagingLabeling/LabelingNutrition/ucm072926.htm (accessed on 15 November 2022).
21. US Food and Drug Adminstration. Qualified Health Claims: Letter of Enforcement Discretion—Walnuts and Coronary Heart Disease (Docket No 02P-0292). Available online: http://wayback.archive-it.org/7993/20171114183725/https://www.fda.gov/Food/IngredientsPackagingLabeling/LabelingNutrition/ucm072910.htm (accessed on 15 November 2022).
22. Blundell, J.E.; Gibbons, C.; Beaulieu, K.; Casanova, N.; Duarte, C.; Finlayson, G.; Stubbs, R.J.; Hopkins, M. The drive to eat in homo sapiens: Energy expenditure drives energy intake. *Physiol. Behav.* **2020**, *219*, 112846. [CrossRef]
23. Blundell, J. Pharmacological approaches to appetite suppression. *Trends Pharmacol. Sci.* **1991**, *12*, 147–157. [CrossRef] [PubMed]
24. Gibbons, C.; Hopkins, M.; Beaulieu, K.; Oustric, P.; Blundell, J.E. Issues in measuring and interpreting human appetite (satiety/satiation) and its contribution to obesity. *Curr. Obes. Rep.* **2019**, *8*, 77–87. [CrossRef] [PubMed]
25. Cecil, J.; Dalton, M.; Finlayson, G.; Blundell, J.; Hetherington, M.; Palmer, C. Obesity and eating behaviour in children and adolescents: Contribution of common gene polymorphisms. *Int. Rev. Psychiatry* **2012**, *24*, 200–210. [CrossRef]
26. Blundell, J.E.; Stubbs, R.J.; Golding, C.; Croden, F.; Alam, R.; Whybrow, S.; Le Noury, J.; Lawton, C.L. Resistance and susceptibility to weight gain: Individual variability in response to a high-fat diet. *Physiol. Behav.* **2005**, *86*, 614–622. [CrossRef]
27. Aaseth, J.; Ellefsen, S.; Alehagen, U.; Sundfor, T.M.; Alexander, J. Diets and drugs for weight loss and health in obesity—An update. *Biomed Pharm.* **2021**, *140*, 111789. [CrossRef] [PubMed]
28. Jeong, D.; Priefer, R. Anti-obesity weight loss medications: Short-term and long-term use. *Life Sci.* **2022**, *306*, 120825. [CrossRef]
29. Blundell, J.E.; Finlayson, G. Is susceptibility to weight gain characterized by homeostatic or hedonic risk factors for overconsumption? *Physiol. Behav.* **2004**, *82*, 21–25. [CrossRef] [PubMed]
30. Blundell, J.; Finlayson, G. Mechanisms and biomarkers of appetite control. *Agro Food Ind. Hi-Tech* **2008**, *19*, 18–20.
31. Buckland, N.J.; Camidge, D.; Croden, F.; Lavin, J.H.; Stubbs, R.J.; Hetherington, M.M.; Blundell, J.E.; Finlayson, G. A low energy–dense diet in the context of a weight-management program affects appetite control in overweight and obese women. *J. Nutr.* **2018**, *148*, 798–806. [CrossRef] [PubMed]
32. Hall, K.D.; Ayuketah, A.; Brychta, R.; Cai, H.; Cassimatis, T.; Chen, K.Y.; Chung, S.T.; Costa, E.; Courville, A.; Darcey, V. Ultra-processed diets cause excess calorie intake and weight gain: An inpatient randomized controlled trial of ad libitum food intake. *Cell Metab.* **2019**, *30*, 67–77.e63. [CrossRef] [PubMed]
33. Peciña, S.; Smith, K.S.; Berridge, K.C. Hedonic hot spots in the brain. *Neurosci* **2006**, *12*, 500–511. [CrossRef]
34. Berridge, K.C.; Kringelbach, M.L. Pleasure systems in the brain. *Neuron* **2015**, *86*, 646–664. [CrossRef]
35. Finlayson, G.; King, N.; Blundell, J.E. Liking vs. wanting food: Importance for human appetite control and weight regulation. *Neurosci. Biobehav. Rev.* **2007**, *31*, 987–1002. [CrossRef]
36. Oustric, P.; Thivel, D.; Dalton, M.; Beaulieu, K.; Gibbons, C.; Hopkins, M.; Blundell, J.; Finlayson, G. Measuring food preference and reward: Application and cross-cultural adaptation of the Leeds Food Preference Questionnaire in human experimental research. *Food Qual. Prefer.* **2020**, *80*, 103824. [CrossRef]
37. Berthoud, H.-R.; Münzberg, H.; Morrison, C.D. Blaming the brain for obesity: Integration of hedonic and homeostatic mechanisms. *Gastroenterology* **2017**, *152*, 1728–1738. [CrossRef]
38. Schachter, S. Obesity and eating: Internal and external cues differentially affect the eating behavior of obese and normal subjects. *Science* **1968**, *161*, 751–756. [CrossRef]
39. Barkeling, B.; King, N.; Näslund, E.; Blundell, J. Characterization of obese individuals who claim to detect no relationship between their eating pattern and sensations of hunger or fullness. *Int. J. Obes.* **2007**, *31*, 435–439. [CrossRef]
40. Drapeau, V.; Hetherington, M.; Tremblay, A. Impact of eating and lifestyle behaviors on body weight: Beyond energy value. In *Handbook of Behavior, Food and Nutrition*; Springer: Berlin/Heidelberg, Germany, 2011; pp. 693–706.
41. Stubbs, R.J.; Hughes, D.A.; Johnstone, A.M.; Rowley, E.; Reid, C.; Elia, M.; Stratton, R.; Delargy, H.; King, N.; Blundell, J. The use of visual analogue scales to assess motivation to eat in human subjects: A review of their reliability and validity with an evaluation of new hand-held computerized systems for temporal tracking of appetite ratings. *Br. J. Nutr.* **2000**, *84*, 405–415. [CrossRef] [PubMed]
42. Drapeau, V.; Blundell, J.; Gallant, A.; Arguin, H.; Després, J.-P.; Lamarche, B.; Tremblay, A. Behavioural and metabolic characterisation of the low satiety phenotype. *Appetite* **2013**, *70*, 67–72. [CrossRef] [PubMed]
43. Dalton, M.; Hollingworth, S.; Blundell, J.; Finlayson, G. Weak satiety responsiveness is a reliable trait associated with hedonic risk factors for overeating among women. *Nutrients* **2015**, *7*, 7421–7436. [CrossRef] [PubMed]

44. Buckland, N.J.; Camidge, D.; Croden, F.; Myers, A.; Lavin, J.H.; Stubbs, R.J.; Blundell, J.E.; Finlayson, G. Women with a low-satiety phenotype show impaired appetite control and greater resistance to weight loss. *Br. J. Nutr.* **2019**, *122*, 951–959. [CrossRef]
45. Arguin, H.; Tremblay, A.; Blundell, J.E.; Després, J.-P.; Richard, D.; Lamarche, B.; Drapeau, V. Impact of a non-restrictive satiating diet on anthropometrics, satiety responsiveness and eating behaviour traits in obese men displaying a high or a low satiety phenotype. *Br. J. Nutr.* **2017**, *118*, 750–760. [CrossRef]
46. Drapeau, V.; Jacob, R.; Panahi, S.; Tremblay, A. Effect of energy restriction on eating behavior traits and psychobehavioral factors in the low satiety phenotype. *Nutrients* **2019**, *11*, 245. [CrossRef]
47. Piernas, C.; Popkin, B.M. Snacking increased among US adults between 1977 and 2006. *J. Nutr.* **2010**, *140*, 325–332. [CrossRef]
48. Duffey, K.J.; Popkin, B.M. Energy density, portion size, and eating occasions: Contributions to increased energy intake in the United States, 1977–2006. *PLoS Med.* **2011**, *8*, e1001050. [CrossRef]
49. Zizza, C.A.; Xu, B. Snacking is associated with overall diet quality among adults. *J. Acad. Nutr. Diet.* **2012**, *112*, 291–296.
50. Leidy, H.J.; Campbell, W.W. The effect of eating frequency on appetite control and food intake: Brief synopsis of controlled feeding studies. *J. Nutr.* **2011**, *141*, 154–157. [CrossRef] [PubMed]
51. Hartmann, C.; Siegrist, M.; van der Horst, K. Snack frequency: Associations with healthy and unhealthy food choices. *Public Health Nutr.* **2013**, *16*, 1487–1496. [CrossRef]
52. Akhlaghi, M.; Ghobadi, S.; Zare, M.; Foshati, S. Effect of nuts on energy intake, hunger, and fullness, a systematic review and meta-analysis of randomized clinical trials. *Crit. Rev. Food Sci. Nutr.* **2020**, *60*, 84–93. [CrossRef] [PubMed]
53. Mandalari, G.; Grundy, M.M.-L.; Grassby, T.; Parker, M.L.; Cross, K.L.; Chessa, S.; Bisignano, C.; Barreca, D.; Bellocco, E.; Lagana, G. The effects of processing and mastication on almond lipid bioaccessibility using novel methods of in vitro digestion modelling and micro-structural analysis. *Br. J. Nutr.* **2014**, *112*, 1521–1529. [CrossRef] [PubMed]
54. Clark, M.J.; Slavin, J.L. The effect of fiber on satiety and food intake: A systematic review. *J. Am. Coll. Nutr.* **2013**, *32*, 200–211. [CrossRef]
55. Fromentin, G.; Darcel, N.; Chaumontet, C.; Marsset-Baglieri, A.; Nadkarni, N.; Tomé, D. Peripheral and central mechanisms involved in the control of food intake by dietary amino acids and proteins. *Nutr. Res. Rev.* **2012**, *25*, 29–39. [CrossRef]
56. Jenkins, D.J.; Kendall, C.W.; Josse, A.R.; Salvatore, S.; Brighenti, F.; Augustin, L.S.; Ellis, P.R.; Vidgen, E.; Rao, A.V. Almonds decrease postprandial glycemia, insulinemia, and oxidative damage in healthy individuals. *J. Nutr.* **2006**, *136*, 2987–2992. [CrossRef]
57. Josse, A.R.; Kendall, C.W.; Augustin, L.S.; Ellis, P.R.; Jenkins, D.J. Almonds and postprandial glycemia—A dose-response study. *Metabolism* **2007**, *56*, 400–404. [CrossRef]
58. Mori, A.M.; Considine, R.V.; Mattes, R.D. Acute and second-meal effects of almond form in impaired glucose tolerant adults: A randomized crossover trial. *Nutr. Metab.* **2011**, *8*, 6.
59. Tan, S.Y.; Mattes, R. Appetitive, dietary and health effects of almonds consumed with meals or as snacks: A randomized, controlled trial. *Eur. J. Clin. Nutr.* **2013**, *67*, 1205–1214. [CrossRef] [PubMed]
60. Hull, S.; Re, R.; Chambers, L.; Echaniz, A.; Wickham, M.S. A mid-morning snack of almonds generates satiety and appropriate adjustment of subsequent food intake in healthy women. *Eur. J. Nutr.* **2015**, *54*, 803–810. [CrossRef] [PubMed]
61. Hollingworth, S.; Dalton, M.; Blundell, J.E.; Finlayson, G. Evaluation of the influence of raw almonds on appetite control: Satiation, satiety, hedonics and consumer perceptions. *Nutrients* **2019**, *11*, 2030. [CrossRef] [PubMed]
62. Fraser, G.E.; Bennett, H.W.; Jaceldo, K.B.; Sabaté, J. Effect on body weight of a free 76 kilojoule (320 calorie) daily supplement of almonds for six months. *J. Am. Coll. Nutr.* **2002**, *21*, 275–283. [CrossRef]
63. Sabaté, J. Nut consumption and body weight. *Am. J. Clin. Nutr.* **2003**, *78*, 647S–650S. [CrossRef]
64. Hollis, J.; Mattes, R. Effect of chronic consumption of almonds on body weight in healthy humans. *Br. J. Nutr.* **2007**, *98*, 651–656. [CrossRef] [PubMed]
65. Halford, J.C.; Harrold, J.A. Satiety-enhancing products for appetite control: Science and regulation of functional foods for weight management. *Proc. Nutr. Soc.* **2012**, *71*, 350–362. [CrossRef]
66. Holt, S.H.; Brand Miller, J.C.; Petocz, P.; Farmakalidis, E. A satiety index of common foods. *Eur. J. Clin. Nutr.* **1995**, *49*, 675–690.
67. Blundell, J.E.; MacDiarmid, J.I. Fat as a risk factor for overconsumption: Satiation, satiety, and patterns of eating. *J. Am. Diet. Assoc.* **1997**, *97*, S63–S69. [CrossRef]
68. Rolls, B.J. Dietary energy density: Applying behavioural science to weight management. *Nutr. Bull.* **2017**, *42*, 246–253. [CrossRef]
69. Buckland, N.J.; Stubbs, R.J.; Finlayson, G. Towards a satiety map of common foods: Associations between perceived satiety value of 100 foods and their objective and subjective attributes. *Physiol. Behav.* **2015**, *152*, 340–346. [CrossRef] [PubMed]
70. Hollingworth, S.L. *Biopsychological Investigation of Satiety Responsiveness and Its Implications for Appetite Control*; University of Leeds: Leeds, UK, 2020.
71. Forde, C.G.; Almiron-Roig, E.; Brunstrom, J.M. Expected satiety: Application to weight management and understanding energy selection in humans. *Curr. Obes. Rep.* **2015**, *4*, 131–140. [CrossRef]
72. Hogenkamp, P.S.; Schiöth, H.B. Effect of oral processing behaviour on food intake and satiety. *Trends Food Sci. Technol.* **2013**, *34*, 67–75. [CrossRef]
73. Dalton, M.; Buckland, N.; Blundell, J. Psychobiology of Obesity: Eating Behavior and Appetite Control. In *Clinical Obesity in Adults and Children*; John Wiley & Sons Ltd.: Hoboken, NJ, USA, 2022; pp. 99–112.

74. Bes-Rastrollo, M.; Wedick, N.M.; Martinez-Gonzalez, M.A.; Li, T.Y.; Sampson, L.; Hu, F.B. Prospective study of nut consumption, long-term weight change, and obesity risk in women. *Am. J. Clin. Nutr.* **2009**, *89*, 1913–1919. [CrossRef]
75. Mozzaffarian, D.; Hao, T.; Rimm, E.B.; Willett, W.C.; Hu, F.B. Changes in diet and lifestyle and long-term weight gain in women and men. *N. Engl. J. Med.* **2011**, *364*, 2392–2404. [CrossRef]
76. Bes-Rastrollo, M.; Sabate, J.; Gomez-Gracia, E.; Alonso, A.; Martinez, J.A.; Martinez-Gonzalez, M.A. Nut consumption and weight gain in a Mediterranean cohort: The SUN study. *Obes. (Silver Spring Md.)* **2007**, *15*, 107–116. [CrossRef] [PubMed]
77. Yi, S.Y.; Steffen, L.M.; Zhou, X.; Shikany, J.M.; Jacobs, D.R., Jr. Association of nut consumption with CVD risk factors in young to middle-aged adults: The Coronary Artery Risk Development in Young Adults (CARDIA) study. *Nutr. Metab. Cardiovasc. Dis. NMCD* **2022**, *32*, 2321–2329. [CrossRef] [PubMed]
78. Liu, X.; Li, Y.; Guasch-Ferre, M.; Willett, W.C.; Drouin-Chartier, J.P.; Bhupathiraju, S.N.; Tobias, D.K. Changes in nut consumption influence long-term weight change in US men and women. *BMJ Nutr. Prev. Health* **2019**, *2*, 90–99. [CrossRef] [PubMed]
79. Nishi, S.K.; Viguiliouk, E.; Blanco Mejia, S.; Kendall, C.W.C.; Bazinet, R.P.; Hanley, A.J.; Comelli, E.M.; Salas Salvado, J.; Jenkins, D.J.A.; Sievenpiper, J.L. Are fatty nuts a weighty concern? A systematic review and meta-analysis and dose-response meta-regression of prospective cohorts and randomized controlled trials. *Obes. Rev.* **2021**, *22*, e13330. [CrossRef]
80. Wien, M.A.; Sabate, J.M.; Ikle, D.N.; Cole, S.E.; Kandeel, F.R. Almonds vs complex carbohydrates in a weight reduction program. *Int. J. Obes. Relat. Metab. Disord. J. Int. Assoc. Study Obes.* **2003**, *27*, 1365–1372. [CrossRef] [PubMed]
81. Fernandez-Rodriguez, R.; Mesas, A.E.; Garrido-Miguel, M.; Martinez-Ortega, I.A.; Jimenez-Lopez, E.; Martinez-Vizcaino, V. The Relationship of Tree Nuts and Peanuts with Adiposity Parameters: A Systematic Review and Network Meta-Analysis. *Nutrients* **2021**, *13*, 2251. [CrossRef] [PubMed]
82. Lee-Bravatti, M.A.; Wang, J.; Avendano, E.E.; King, L.; Johnson, E.J.; Raman, G. Almond Consumption and Risk Factors for Cardiovascular Disease: A Systematic Review and Meta-analysis of Randomized Controlled Trials. *Adv. Nutr.* **2019**, *10*, 1076–1088. [CrossRef]
83. Guarneiri, L.L.; Cooper, J.A. Intake of Nuts or Nut Products Does Not Lead to Weight Gain, Independent of Dietary Substitution Instructions: A Systematic Review and Meta-Analysis of Randomized Trials. *Adv. Nutr.* **2021**, *12*, 384–401. [CrossRef] [PubMed]
84. Estruch, R.; Martinez-Gonzalez, M.A.; Corella, D.; Salas-Salvado, J.; Fito, M.; Chiva-Blanch, G.; Fiol, M.; Gomez-Gracia, E.; Aros, F.; Lapetra, J.; et al. Effect of a high-fat Mediterranean diet on bodyweight and waist circumference: A prespecified secondary outcomes analysis of the PREDIMED randomised controlled trial. *Lancet. Diabetes Endocrinol.* **2019**, *7*, e6–e17. [CrossRef] [PubMed]
85. Hu, F.B. *Obesity Epidemiology*; Oxford University Press: Oxford, NY, USA, 2008; p. xiii.
86. Jackson, C.L.; Hu, F.B. Long-term associations of nut consumption with body weight and obesity. *Am. J. Clin. Nutr.* **2014**, *100* (Suppl. S1), 408S–411S. [CrossRef] [PubMed]
87. Alasalvar, C.; Salas-Salvadó, J.; Ros, E.; Sabaté, J. *Health Benefits of Nuts and Dried Fruits*; CRC Press: Boca Raton, FL, USA, 2020; p. xxi.
88. Willett, W.; Rockstrom, J.; Loken, B.; Springmann, M.; Lang, T.; Vermeulen, S.; Garnett, T.; Tilman, D.; DeClerck, F.; Wood, A.; et al. Food in the Anthropocene: The EAT-Lancet Commission on healthy diets from sustainable food systems. *Lancet* **2019**, *393*, 447–492. [CrossRef] [PubMed]

nutrients MDPI

Review

Nuts in the Prevention and Management of Type 2 Diabetes

Stephanie K. Nishi [1,2,3,4,5,*], Effie Viguiliouk [4], Cyril W. C. Kendall [4,5,6,7], David J. A. Jenkins [4,5,6,8,9,10], Frank B. Hu [11,12,13], John L. Sievenpiper [4,5,6,8,9,10], Alessandro Atzeni [1,2,3], Anoop Misra [14,15,16] and Jordi Salas-Salvadó [1,2,3,*]

1 Universitat Rovira i Virgili, Departament de Bioquímica i Biotecnologia, Unitat de Nutrició, 43201 Reus, Spain
2 Institut d'Investigació Sanitària Pere Virgili (IISPV), 43201 Reus, Spain
3 Centro de Investigación Biomédica en Red Fisiopatología de la Obesidad y la Nutrición (CIBEROBN), Institute of Health Carlos III, 28029 Madrid, Spain
4 Toronto 3D (Diet, Digestive Tract and Disease) Knowledge Synthesis and Clinical Trials Unit, Toronto, ON M5C 2T2, Canada
5 Clinical Nutrition and Risk Factor Modification Centre, St. Michael's Hospital, Unity Health Toronto, Toronto, ON M5C 2T2, Canada
6 Department of Nutritional Sciences, Temerty Faculty of Medicine, University of Toronto, Toronto, ON M5S 1A1, Canada
7 College of Pharmacy and Nutrition, University of Saskatchewan, Saskatoon, SK S7N 5E5, Canada
8 Department of Medicine, Temerty Faculty of Medicine, University of Toronto, Toronto, ON M5S 1A1, Canada
9 Division of Endocrinology and Metabolism, Department of Medicine, St. Michael's Hospital, Toronto, ON M5B 1W8, Canada
10 Li Ka Shing Knowledge Institute, St. Michael's Hospital, Toronto, ON M5B 1T8, Canada
11 Department of Nutrition, Harvard T.H. Chan School of Public Health, Boston, MA 02115, USA
12 Department of Epidemiology, Harvard T.H. Chan School of Public Health, Boston, MA 02115, USA
13 Channing Division of Network Medicine, Department of Medicine, Brigham and Women's Hospital and Harvard Medical School, Boston, MA 02115, USA
14 Fortis C-DOC Hospital for Diabetes & Allied Sciences, New Delhi 110048, India
15 National Diabetes, Obesity and Cholesterol Foundation, New Delhi 110016, India
16 Diabetes Foundation (India), New Delhi 110070, India
* Correspondence: stephanie.nishi@urv.cat (S.K.N.); jordi.salas@urv.cat (J.S.-S.)

Abstract: Diabetes is a continuously growing global concern affecting >10% of adults, which may be mitigated by modifiable lifestyle factors. Consumption of nuts and their inclusion in dietary patterns has been associated with a range of beneficial health outcomes. Diabetes guidelines recommend dietary patterns that incorporate nuts; however, specific recommendations related to nuts have been limited. This review considers the epidemiological and clinical evidence to date for the role of nut consumption as a dietary strategy for the prevention and management of type 2 diabetes (T2D) and related complications. Findings suggest nut consumption may have a potential role in the prevention and management of T2D, with mechanistic studies assessing nuts and individual nut-related nutritional constituents supporting this possibility. However, limited definitive evidence is available to date, and future studies are needed to elucidate better the impact of nuts on the prevention and management of T2D.

Keywords: nuts; diabetes; glycemic control; insulin resistance

Citation: Nishi, S.K.; Viguiliouk, E.; Kendall, C.W.C.; Jenkins, D.J.A.; Hu, F.B.; Sievenpiper, J.L.; Atzeni, A.; Misra, A.; Salas-Salvadó, J. Nuts in the Prevention and Management of Type 2 Diabetes. *Nutrients* **2023**, *15*, 878. https://doi.org/10.3390/nu15040878

Academic Editor: Peter Pribis

Received: 9 January 2023
Revised: 30 January 2023
Accepted: 7 February 2023
Published: 9 February 2023

1. Introduction

Type 2 diabetes is one of the most globally challenging and prevalent metabolic disorders affecting an estimated 1 in 10 adults (10.5% of adults worldwide) [1]. Within the past 2 years, type 2 diabetes prevalence has risen by 16%, indicating an alarming growth rate [1]. Complications of diabetes, such as cardiovascular disease, chronic kidney disease, neuropathy, and retinopathy, and its high medical and other economic expenditures are a serious cause of concern [2]. Excluding mortality risks associated with COVID-19, approximately 12.2% of global adult deaths from all-cause are estimated to have occurred due to diabetes or its complications in 2021 [2]. Further, 10.6% of adults worldwide have impaired glucose tolerance, placing them at high risk for developing type 2 diabetes (T2D) [2].

Lifestyle changes, such as those related to nutrition, underpin a general approach to diabetes risk minimization and management. Current diabetes guidelines recommend dietary patterns, such as Mediterranean and vegetarian patterns, which encourage the consumption of nuts [3–6]. Nuts, represented by tree nuts (almonds, Brazil nuts, cashews, hazelnuts, macadamias, pecans, pine nuts, pistachios, walnuts) and peanuts (technically a legume, but sharing a similar nutritional and culinary profile to tree nuts, hence, their inclusion in the "nuts" classification), are nutrient-dense foods with complex matrices providing unsaturated fatty acids, plant-protein, non-sodium minerals, phenolic and other bioactive compounds [7–9].

In this narrative review, we summarize the human evidence currently available ("Where we are") for the role of nuts in the prevention and management of T2D and discuss future directions ("Where we are going") in terms of what questions may still need to be addressed and how research may address and inform any knowledge gaps. For this narrative review, a comprehensive search of PubMed and Cochrane databases through November 2022 for English language articles of epidemiological, clinical studies, and the latest reviews and meta-analyses assessing nut consumption (tree nuts and peanuts) and their components on T2D and related risk factors was conducted. The present article is not a systematic review; thus, some studies may not have been identified; further, the possibility of publication bias should be acknowledged. Nonetheless, the authors independently conducted literature searches, and these findings were further shared and discussed among an assembly of experts in the field of nut and health research.

2. Effect of Nut Consumption on Measures of Glucose Metabolism

Few epidemiological studies have assessed the association between nut consumption and markers of glycemic control. Table 1 summarizes epidemiological and clinical findings related to nut consumption and measures of glucose metabolism. To our knowledge, there is a lack of prospective cohort studies that have analyzed markers of glucose metabolism in individuals with or without diabetes. One prospective population-based study, conducted within the framework of the Tehran Lipid and Glucose study (TLGS), presented fasting serum glucose measures following a median 6.2-year follow-up across tertiles of nut consumption. At 6.2 years, findings showed lower fasting serum glucose levels in the highest tertile of nut consumption (nut intake: median 8.7 g/week, IQR, 5.3 to 15.8 g/week; fasting glucose: 4.7 ± 0.1 mmol/L) compared to the lowest tertile (nut intake: median 1.6 g/week, IQR, 0.7 to 2.8 g/week; fasting glucose: 5.3 ± 0.1 mmol/L) ($p = 0.02$) [10]. However, there appears to be a shortage of evidence in relation to other glucose-metabolism-related biomarkers and in individuals with diabetes. Cross-sectional studies have shown an association between nut consumption and markers of glucose/insulin homeostasis. One cross-sectional study assessing data from 16,784 American adults (51.8% women, aged ≥18 years) participating in the National Health and Nutrition Examination Survey (NHANES, 2005–2010) evaluated the association between nut intake and markers of glycemic control [11]. The authors observed that higher nut intake was associated with significantly lower levels of all diabetes-related biomarkers, including fasting blood glucose, plasma insulin, homeostasis model assessment-insulin resistance (HOMA-IR), HOMA-β, glycated hemoglobin (HbA1c), and oral glucose tolerance test (OGTT) ($p < 0.001$). Another cross-sectional study analyzed the association between the frequency of nut consumption and insulin resistance, measured by HOMA-IR, in 379,310 Koreans [12]. In this study, nut consumption ≥5 servings/week (where 1 serving = 15 g) compared to <1 serving/month was associated with lower HOMA-IR (odds ratio [OR]: 0.90; 95% confidence interval [CI] 0.86 to 0.94). This association was observed to be more prominent in women, participants with normal glycaemia, and younger age (<40 years).

Table 1. Summary of findings related to nuts and diabetes related prevention and management.

Variables	Finding [1]	Level of Evidence [2]	Reference
Epidemiological Evidence			
Fasting blood glucose	↓	+	[11]
Plasma insulin	↓	+	
HOMA-IR	↓	+	[11,12]
HOMA-B	↓	+	[11]
HbA1c	↓	+	
OGTT	↓	+	
Diabetes incidence	↓/↔	+	[13,14]
Diabetes prevalence	↓/↔	+	[13,15–17]
CVD incidence in participants with T2D	↓	+	[18]
Diabetes mortality	↓	++	[19]
Clinical Trial Evidence			
Acute Trial Evidence			
In participants free of T2D:			
Postprandial glycemia	↓	++	[20–26]
Postprandial insulinemia	↓/↔	+	[21,27,28]
In participants with T2D:			
Postprandial glycemia	↓	+	[21,23,29,30]
Postprandial insulinemia	↓/↔	+	[21,29]
Glucose metabolic clearance rate	↑	+	[29]
Longer-term Trial Evidence			
In participants free of T2D at baseline:			
Diabetes incidence	↓/↔	+	[31–33]
In participants with T2D at baseline:			
Fasting glucose	↓	+	[30,34–39]
Fasting insulin	↔	+	
HbA1c	↓	+	
HOMA-IR	↔	+	
In participants with/without T2D at baseline:			
Fasting glucose	↔	+	[40]
Fasting insulin	↓	+	
HbA1c	↔	+	
HOMA-IR	↓	+	

Abbreviations: CVD, cardiovascular disease; HbA1c, glycated hemoglobin; HOMA, homoeostasis model assessment; IR, insulin resistance; OGTT, oral glucose tolerance test; T2D, type 2 diabetes mellitus. [1] Findings are based on the authors' review, and assessment of the noted literature, and hence could present some subjectivity. In general: ↓, majority of evidence indicated a decrease; ↑, majority of evidence indicated an increase; ↔, majority of evidence indicated no change observed; ↓/↔, majority of evidence was split between showing a decrease or no effect on the outcome. Where "majority of evidence" refers to the entirety of the evidence, if a relevant systematic review and meta-analysis was conducted these findings were used as the basis of this determination. [2] Level of Evidence is based on the authors' review and assessment and, hence, could present with some subjectivity. In general: +, limited and/or inconsistent evidence from few studies in the denoted type of study design; ++, consistent evidence in several studies in the denoted type of study design.

When considering evidence from clinical trials, consumption of nuts alone and when added to high glycemic index (GI) foods show a lowering in postprandial glycemia when compared to consumption of high GI foods alone. Several acute trials have assessed the effect of almond intake on postprandial glycemia. In healthy individuals, the consumption of almonds with white bread was shown to significantly lower the postprandial area under the insulin concentration vs. time curve when compared to a high GI meal (instant mashed potatoes) (n = 15) [41] and significantly lower the glucose peak height when compared with white bread (n = 9) [27]. In another acute randomized crossover trial conducted in healthy participants (n = 100 with available data, n = 106 randomized), consumption of at least 10% of energy from raw almonds resulted in the mean area under the blood glucose response curve being significantly lowered when compared to consumption of biscuits [28]. Similar findings were shown for individuals at higher risk of diabetes. In an acute randomized five-arm crossover trial conducted in individuals with impaired glucose tolerance (n = 14), participants were randomized to consume whole almonds, almond butter, defatted almond flour, almond oil, or no almonds that were incorporated into a 75 g available carbohydrate-matched breakfast meal. Whole almonds significantly diminished the second meal and daylong blood glucose incremental area under the curve and elicited a greater second-meal insulin response [20]. Another acute randomized crossover trial conducted in individuals with good health (n = 12) and individuals with T2D (n = 7) showed consumption of 28 g of almonds with a test meal (bagel, juice, and butter) significantly reduced postprandial glycemia in participants with diabetes but not in participants without diabetes when compared to the test meal without almonds [21]. For pistachio intake, an acute trial conducted in healthy individuals (n = 10) showed consumption of pistachios alone and, when added to white bread at different doses (28 g, 56 g, 84 g), significantly lowered glycemic responses in comparison to white bread [42]. The addition of pistachios to other commonly consumed carbohydrate-rich foods (parboiled rice, pasta, potatoes) also resulted in reduced glycemic responses [42]. Similarly, in an acute trial conducted on individuals with metabolic syndrome (n = 20), the consumption of pistachios with white bread significantly lowered the glycemic response and increased insulin secretagogue levels when compared to white bread alone [22]. For mixed nuts, an acute trial conducted in individuals with good health (n = 14) and in individuals with T2D (n = 10) showed mixed nuts at three different doses significantly reduced the glycemic response in comparison to white bread. The addition of mixed nuts to white bread progressively reduced the glycemic response of the meal; however, in individuals with T2D, the reduction in glycemic response was half that seen in healthy individuals [23]. In another acute trial, adults with overweight/obesity (n = 54) were randomized to consume either mixed nuts or pretzels and showed pretzel consumption increased glucose and insulin, whereas, with mixed nuts, no elevation was detected at 60 min post snack consumption [24]. For peanut intake, an acute trial conducted in men with overweight/obesity (n = 65) who consumed a test meal of a shake containing conventional peanuts, high-oleic peanuts, or a control biscuit showed a quicker return of insulin to basal concentrations after consumption of the shakes containing conventional peanuts and high-oleic peanuts [25].

Several systematic reviews and meta-analyses (SRMAs) of randomized controlled trials (RCTs) with a duration of at least 3 weeks have been conducted assessing the effect of a tree nut(s) on markers of glycemic control in people with different health statuses (the effect of tree nuts on markers of glycemic control in people with diabetes is discussed in Section 5.2). In 2014, SRMA of 49 RCTs (n = 2226) was conducted to assess the effect of tree nuts on metabolic syndrome criteria, including fasting glucose. Twenty-six trials were included (n = 1360) for fasting glucose, which showed tree nuts significantly lowered fasting glucose compared with the controls (mean difference [MD] = −0.08 mmol/L; 95% confidence interval [CI] −0.16 to −0.01 mmol/L) [43]. In 2018, a network meta-analysis of RCTs assessed the effect of different food groups on intermediate disease markers in adults, including fasting glucose, HbA1c, and HOMA-IR [44]. The results showed nuts were more effective at reducing fasting blood glucose when compared to red meat and fruits and vegetables, as well as HOMA-IR when compared to eggs and dairy. No significant effects were shown for HbA1c. In 2019, another SRMA of 40 RCTs (n = 2832) was conducted to assess the effect of tree nut or peanut intake in adults on glycemic control, including fasting glucose, fasting insulin, HbA1c, and HOMA-IR. Nut intake showed a significant lowering in fasting insulin (28 RCTs; weighted mean difference [WMD]: −0.40 μIU/mL; 95% CI: −0.73, −0.07 μIU/mL; I^2 = 49.4%) and HOMA-IR (19 RCTs; WMD: −0.23; 95% CI: −0.40, −0.06; I^2 = 51.7%), with no significant effect on fasting glucose or HbA1c [40].

Subgroup analysis by nut type showed a significant reduction in fasting blood glucose with pistachio consumption compared with the control (WMD: −5.18 mg/dL; 95% CI: −8.76, −1.60 mg/dL; I^2 = 67%). This was supported by another SRMA of RCTs published in 2020, assessing the effect of pistachio intake on glycemic control in individuals with different health statuses (type 2 diabetes, prediabetes, and metabolic syndrome), which showed a significant reduction in fasting glucose and HOMA-IR but not HbA1c or fasting insulin [34]. Tindall et al. also identified a small number of studies that measured outcomes related to insulin production and HOMA-β cell function (7 studies), glucose concentrations after a 75-g OGTT (5 studies), insulin concentrations after a 75-g OGTT (2 studies), insulin sensitivity (3 studies) and short-term glucose control (2 studies) [40]. Due to a limited number of trials that measured these endpoints and the heterogeneity in the measurements, a meta-analysis was not performed. These studies showed no impact of nut intake on outcomes related to insulin concentrations after a 75 g OGTT and insulin sensitivity, whereas there were mixed findings for outcomes related to insulin production and HOMA-β cell function, glucose concentrations after a 75 g OGTT, and short-term glucose control [40]. Several other SRMAs of RCTs have been conducted between 2020 and 2022, assessing the effect of a specific nut type and/or the effect of nuts in a specific group of people. These SRMAs assessed the effect of different types of nuts in healthy adults with overweight/obesity (10 RCTs) [45], walnuts in middle-aged and older adults (17 RCTs) [46] and individuals with different health statuses (16 RCTs) [47], cashews (6 RCTs) [48], peanuts (11 RCTs) [49], and 2 SRMAs investigating almonds in individuals with different health statuses (24 RCTs [50], 15 RCTs [51]), all of which showed no impact on markers of glycemic status.

Since the publication of the above-mentioned SRMAs, more recent RCTs in people without diabetes have been published. In a 6-month RCT, 107 individuals who were overweight and at moderate or high risk of T2D were randomized to either an energy-restricted diet, including 70 g/d of peanuts or an energy-restricted low-fat diet, which showed no significant differences between groups in regard to HbA1c, fasting glucose, fasting insulin, 2 h glucose, and HOMA-IR [52]. In an 8-week RCT, 40 women were randomized to an energy-restricted diet without nuts or to an energy-restricted diet containing 45 g/d of nuts (15 g of Brazil nuts + 30g of cashew nuts), which also showed no significant differences in markers of glycemic status [53].

3. Nuts and Diabetes Prevention

Table 1 summarizes epidemiological and clinical findings related to nut consumption and diabetes prevention.

3.1. Epidemiological Evidence

Epidemiological studies conducted to date have shown inconsistent and inconclusive evidence related to nut consumption and the incidence of T2D. A number of SRMAs involving cross-sectional or prospective cohort studies have been published investigating associations between the frequency of nut consumption and the prevalence and/or the incidence of T2D risk. Most have not reported a significant association when comparing the highest to the lowest categories of nut consumption, nor were dose-response relationships observed [54–58]. Only one of these meta-analyses of prospective cohort studies showed a significant inverse association with the risk of T2D [59]. However, a key limitation is that most of these SRMAs included studies combining nuts with other plant foods as the exposure (i.e., peas, seeds, or legumes) and, therefore, the associations cannot be extrapolated specifically to the possible role of nuts per se [55,58,59]. Additionally, in some of the observational studies, the associations were adjusted for body weight or BMI, a potential mediator of the associations [11] and, therefore, possibly attenuating an association.

In 2021, an updated SRMA of cross-sectional (n = 3) and prospective (n = 5) studies, including only those with nuts alone as an exposure, was published [13]. The included studies were conducted in the United States (5 studies), Europe (3 studies), and Asia

(1 study). Findings from the meta-analyses of the cross-sectional studies (n = 72,559; 7559 cases of T2D) showed no significant association with diabetes prevalence when the highest compared to the lowest categories of total nut consumption was assessed (OR: 0.91; 95% CI: 0.83 to 1.01). When the prospective cohort studies were analyzed, no associations with risk of T2D were observed with consumption of total nuts (relative risk [RR]: 1.04: 95% CI: 0.94 to 1.15), tree nuts (RR: 0.98; 95% CI 0.87 to 1.11), or peanuts (RR: 0.95; 95% CI: 0.87 to 1.04). When peanut butter consumption was specifically assessed, it was shown to be inversely associated with T2D incidence (RR: 0.87; 95% CI: 0.77 to 0.98). Furthermore, there was no evidence of a linear dose-response or nonlinear dose-response gradient for the total nut or peanut consumption in prospective cohort studies. Of note, these analyses were adjusted for baseline BMI. Across all nut exposures evaluated, the certainty of the evidence was considered to be very low. The reduction in the risk of T2D seen in sensitivity analyses of this meta-analysis suggested that weight loss or decreased weight might mediate the reduction in risk, although appropriate statistical mediation analyses using repeated assessments are needed to confirm this assumption. It is important to highlight that in relation to the type of tree nuts, only one cross-sectional (n = 27,563) [15] and one prospective cohort (n = 137,956) study [14] had analyzed the association between the frequency of walnut consumption and T2D risk, reporting in both cases an inverse association with the prevalence and incidence of T2D, respectively. Of note, the largest prospective cohort study involving American adults participating in the Nurses' Health Study (NHS; 58,063 women aged 52–77 [1998–2008]) and NHS II (79,893 women aged 35–52 years [1999–2009]), free of diabetes, cardiovascular disease, or cancer at baseline, observed that consumption of ≥2 servings/week (where 1 serving = 28 g) of walnuts had a 24% (95% CI: 6–38%) lower risk of developing T2D than those that never or almost never consumed walnuts after adjustment for baseline BMI [14].

Following the publication of the 2021 SRMA by Becerra-Tomás and colleagues, two additional cross-sectional studies from Italy and Spain involving community-dwelling adults have been published. Both studies reported no association between nut consumption and the prevalence of T2D [16,17].

In view of the studies published to date, new research is needed that prospectively assesses differences in existing cohorts. Moreover, dose-response analyses are warranted in the future to determine the total amount of nuts associated with possible diabetes-related health benefits to better inform guidelines and practice.

3.2. Clinical Trial Evidence

Unfortunately, to date, no clinical trials have been conducted with the primary aim of testing the ability of nut supplementation to reduce or prevent the incidence of diabetes, probably because such types of trials are very expensive and difficult to perform. However, data is available in relation to a secondary analysis conducted in the context of the PREDIMED (PREvención con DIeta MEDiterránea) study, a randomized controlled trial aiming to assess the effect of a Mediterranean diet supplemented with virgin olive oil or nuts in comparison to a low-fat diet on primary prevention of cardiovascular disease [31].

A sub-analysis of this RCT conducted in participants from one of the 23 recruiting study centers (located in Reus, Spain) reported a beneficial effect of the Mediterranean diet enriched with 30 g/day of tree nuts (walnuts, almonds, and hazelnuts) on T2D prevention [31]. Results from the PREDIMED trial as a whole showed a non-significant decrease in the incidence of T2D when compared to participants in the group receiving the low-fat dietary advice [32,33]. It is important to recognize that due to the study design, it is not possible to quantify the beneficial effects secondary to the Mediterranean diet intervention or the nuts that participants consumed throughout the trial.

4. Nuts and Diabetes Management

Table 1 also summarizes epidemiological and clinical findings related to nut consumption and diabetes management.

4.1. Epidemiological Evidence

There is a lack of epidemiological evidence for the role of nut consumption in individuals with T2D for glucose control and the management of complications.

Of the available evidence, one prospective analysis including 16,217 men and women, from the Health Professionals Follow-up Study (HPFS, 1986–2014) and NHS (1980–2014), respectively, with diabetes mellitus at baseline or diagnosed during follow-up, showed that higher total nut consumption was associated with a lower risk of cardiovascular disease (CVD) incidence and mortality [18]. Specifically, for participants who consumed ≥5 servings of total nuts per week (1 serving = 28 g), compared to those who consumed <1 serving per month, multivariate-adjusted hazard ratios (HR; 95% CIs), showed reductions in total CVD incidence (HR = 0.83; 95% CI: 0.71–0.98; p trend = 0.01), coronary heart disease incidence (HR = 0.80; 95% CI: 0.67–0.96; p trend = 0.005), CVD mortality (HR = 0.66; 95% CI: 0.52–0.84; p trend < 0.001), and all-cause mortality (HR = 0.69; 95% CI: 0.61–0.77; p trend < 0.001). For specific types of nuts, higher tree nut consumption was associated with a lower risk of total CVD, coronary heart disease incidence, and mortality because of CVD, cancer, and all causes, whereas peanut consumption was associated with lower all-cause mortality only (all p trend <0.001). This study showed that higher consumption of nuts, especially tree nuts, may be associated with lower CVD incidence and mortality among participants with T2D.

A SRMA of four prospective cohort studies (n = 202,751) assessed the relationship of nut consumption with diabetes-related mortality, indicating higher nut intake to be associated with reduced risk of mortality from diabetes compared to the lowest intake [19]. A similar response was observed in the dose-response analysis, with a 39% reduction in the relative risk of diabetes mortality being observed with a one-serving/day (1 serving = 28 g) increase in nut consumption. Based on the findings of this SRMA and the assumption that the associations observed between nut consumption and diabetes mortality are causal, the authors estimated that for the regions assessed (i.e., North and South America, Europe, Southeast Asia, and Western Pacific), 139,000 deaths due to diabetes may be attributed to a nut intake below 20 g/day.

4.2. Clinical Trial Evidence

In clinical trials, consumption of nuts alone and when added to high GI foods show a lowering in postprandial glycemia when compared to the high GI food alone in people with diabetes. In an acute trial conducted in healthy individuals (n = 14) and in individuals with T2D (n = 10), mixed nuts at three different doses significantly reduced the glycemic response in comparison to white bread [23]. As previously noted, this trial also showed that adding mixed nuts to white bread progressively reduced the glycemic response of the meal. However, in individuals with T2D, the reduction in glycemic response was half that seen in healthy individuals. In a randomized crossover trial, the acute effects of almond intake were assessed in men with T2D (n = 7) randomized to consume a control (white bread, butter, cheese) and a test (white bread, almonds) meal. The test meal was found to be associated with lower postprandial glycemia and insulinemia, and an increased estimated glucose metabolic clearance rate [29]. Almonds were also assessed in another acute randomized controlled trial involving participants with (n = 7) and without (n = 12) diabetes [21]. Findings showed consumption of 28 g of almonds with a test meal, composed of a bagel, juice, and butter, significantly reduced postprandial glycemia in participants with diabetes but not those without diabetes when compared to the test meal without almonds.

Several SRMAs of RCTs have been conducted assessing the effect of tree nut(s) on markers of glycemic control in people with diabetes. In 2014, an SRMA of 12 RCTs assessing the effect of tree nuts on glycemic control in people with diabetes (n = 240) showed a significant lowering in fasting glucose (8 comparisons, MD = −0.15 mmol/L; 95% CI: −0.27, −0.02 mmol/L; I^2 = 35%) and HbA1c (8 comparisons, MD = −0.07%: 95% CI: −0.10, −0.03%; I^2 = 37%), with no significant effect on fasting insulin or HOMA-IR [35].

A 2019 SRMA of 40 RCTs (n = 2832) assessing the effect of tree nut or peanut intake in people with and without diabetes showed no significant effect on fasting glucose or HbA1c, and subgroup analyses by diabetes status showed no deviation from the main findings for either outcome [40]. There were a few differences in the inclusion/exclusion criteria between the 2014 and 2019 SRMAs. The 2019 SRMA [40] included trials using nut oil or peanuts as a treatment arm, non-isocaloric comparison arms, and studies published only in English. The 2014 SRMA [35] included only studies using whole tree nuts as the treatment arm, isocaloric comparison arms, and included studies that were not published in English. These differences may explain the discrepancy in findings. A more recent SRMA of RCTs published in 2021 (15 RCTs) assessed the effect of tree nuts on markers of glycemic control in individuals with T2D and showed no significant impact on fasting glucose, HbA1c, or postprandial glucose levels; however, the analysis only included RCTs with a follow-up period of 3 months or less [30]. Between 2020 and 2022 (present day), several other SRMAs of RCTs have been conducted assessing the effect of a specific nut type on markers of glycemic control in people with T2D. Two SRMA's assessed the effect of almond intake; the first SRMA (8 RCTs) showed a significant lowering in HbA1c but no impact on fasting glucose, insulin, or HOMA-IR [36], whereas the second SRMA (9 RCTs) showed no impact on markers of glycemic control, including HbA1c, fasting glucose, and insulin [37]. Another SRMA assessed the effect of pistachio intake in individuals with T2D, prediabetes, and metabolic syndrome (6 RCTs), which showed a significant lowering in fasting glucose and HOMA-IR, but not HbA1c or fasting insulin [34].

Since the publication of the above-mentioned SRMAs, more recent RCTs assessing the effect of nut consumption on glycemic control in people with diabetes have been published. In a 3-month RCT, 45 people with T2D were randomized to either an almond-based, low-carbohydrate diet group or a low-fat diet group. After 3 months, individuals in the almond-based, low carbohydrate diet group showed a significant improvement in HbA1c [38]. In another 3-month RCT, 204 individuals with stable coronary artery disease (~32% of which had diabetes) were randomized to one of three groups: a healthy diet, a healthy diet plus 30 g/d of pecans or a healthy diet plus 30 mL/d of EVOO. After 12-weeks there were no significant differences between groups in regard to fasting glucose, HbA1c, fasting insulin or HOMA-IR [39].

5. Possible Mechanisms of Action of Nuts in Diabetes Prevention and Management

While the possible protective role of nuts in diabetes prevention and management remains to be established with greater certainty, there is potential for a beneficial impact given the unique nutritional composition of nuts and the direct and indirect evidence to date relating relevant dietary constituents with diabetes prevention and management.

There are several proposed and speculative modulatory effects of the constituents of nuts in the prevention and management of T2D that may act synergistically (summarized in Figure 1).

Macronutrients, micronutrients, and other bioactive compounds found in nuts have been suggested to play a role in the regulation of postprandial glycemic and insulinemic levels. Furthermore, body weight management, control of cellular membrane fluidity and lipogenic gene expression, anti-inflammatory and antioxidant properties, and protection of β-cells against glucose toxicity and subsequent impacts on gene expression, microRNAs, and microbiota/metabolomics leading to regulation of postprandial glucose clearance, improving pancreatic insulin secretion, and decreasing insulin resistance have also been implicated with nut consumption or related factors. The following will briefly summarize available and relevant direct and indirect evidence for possible mechanisms for the impact of nut consumption on T2D related to macronutrients, micronutrients, other bioactive compounds, and resulting cellular and molecular mechanisms.

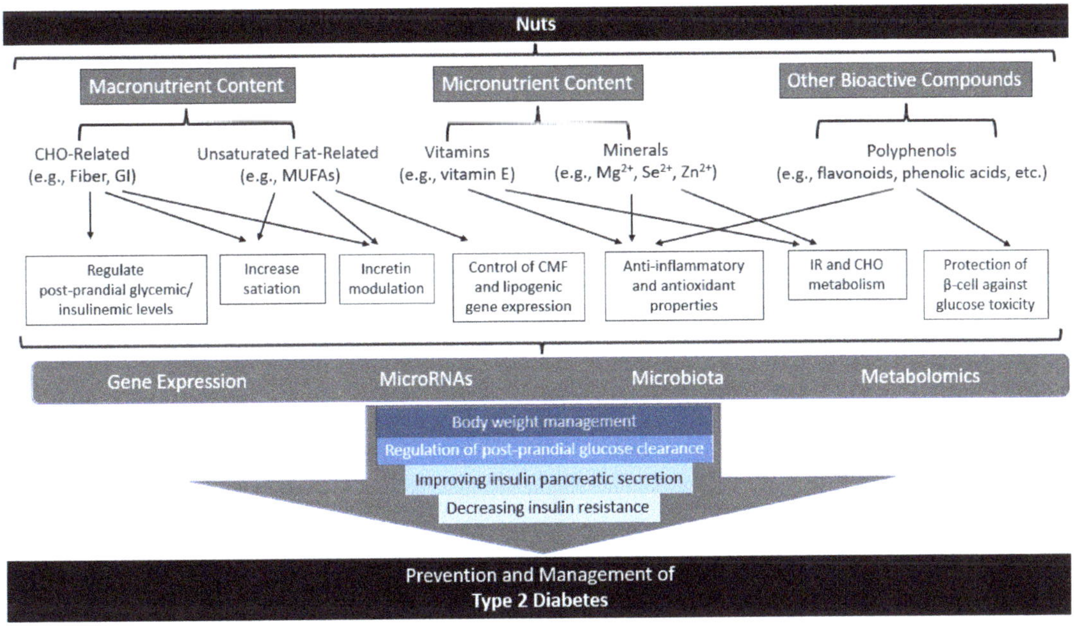

Figure 1. Summary of potential mechanisms of action for the role nuts may play in diabetes prevention and management. Adapted with permission from Ref. [60]. 2017, Hernández-Alonso et al. Abbreviations: CHO, carbohydrate; CMF, cellular membrane fluidity; GI, glycemic index; IR, insulin resistance; MUFAs, monounsaturated fatty acids; RNA, ribonucleic acid.

5.1. Related to Macronutrient Composition of Nuts

5.1.1. Low Glycemic Index and Fiber

Nuts contain low amounts of available carbohydrates, meaning they do not contribute significantly to postprandial glycemia [61,62]. However, when nuts are added to foods with a high available carbohydrate, they demonstrate a dose-dependent reduction in the glycemic index or relative glycemic response of the composite meal [27,42]. This is thought to be due to their fat and protein content, which are a source of additional energy when added to food with highly available carbohydrates [27,61]. Several studies conducted around 20 years ago demonstrated that an increase in energy density from high-fat, protein, and/or high-fiber containing foods decreases gastric emptying [27,42,63–65]. Therefore, as the dose of nuts is increased, the rate of gastric emptying decreases, which may increase feelings of satiety and would decrease the postprandial glycemic response [27].

Nuts are also a source of dietary fiber [8,9]. Soluble fiber has been shown to increase the viscosity of intestinal contents and slow down the absorption of nutrients in the gastrointestinal tract [66]. Consumption of meals/foods containing soluble fiber have been shown to lower postprandial glycemia [66].

Fiber has also been shown to be resistant to digestion by enzymes in the small intestine and, as a result, susceptible to fermentation by bacteria in the colon, which leads to the production of short-chain fatty acids (SCFA) [66]. SCFAs have been shown to reduce hepatic glucose output and stimulate the secretion of the incretin hormone glucagon-like peptide 1 (GLP-1) [67,68]. GLP-1, as well as other incretins such as gastric inhibitory polypeptide (GIP), promote the proliferation of beta-cells and their secretion of insulin, which favors the maintenance of blood glucose levels [69]. As such, the consumption of nuts may slow the absorption of carbohydrates and stimulate incretin secretion, which can positively impact glucose homeostasis.

Furthermore, repeated decreases in postprandial glucose peaks, such as that observed with nut consumption, have been hypothesized to contribute to decreased inflammation, oxidation processes, and mitochondrial toxicity, further contributing to reductions in the risk of diabetes [60].

5.1.2. Fatty Acids: Unsaturated vs. Saturated

Nuts have a high unsaturated fat content [8,9]. Substitution of carbohydrates or saturated fats (SFA) with unsaturated fats may be responsible for improvements in insulin sensitivity [40,70]. This was supported by an SRMA of 102 RCTs, which showed replacing carbohydrates or SFAs with monounsaturated or polyunsaturated fatty acids (MUFAs or PUFAs, respectively) improved markers of glycemic control, including HbA1c and HOMA-IR [70,71]. It should be noted that evidence involving the study of PUFAs tended to include a combination of these fatty acids, including eicosapentaenoic acid (EPA) and docosahexaenoic acid (DHA) along with alpha-linolenic acid (ALA); whereas, the PUFA found in nuts is ALA. Accordingly, additional investigation specific to ALA would better elucidate whether the PUFA content of nuts have the observed beneficial effects observed across overall PUFAs.

The quality of dietary fat can affect cell membrane composition and function, including membrane fluidity, insulin receptor binding/affinity, as well as facilitating the movement of the glucose receptor to the cell surface, which in turn can affect insulin sensitivity [72–75].

Dietary fat quality may also be involved in regulating gene expression and enzyme activity [72,76]. A diet high in unsaturated fatty acids, in particular, long-chain omega-6 and omega-3 polyunsaturated fatty acids, has been shown to lead to the suppression of lipogenic genes (genes of lipid synthesis) and induction of genes involved with fatty acid oxidation, which may reduce hepatic insulin resistance. Saturated fat and monounsaturated fat, on the other hand, do not appear to impact these same mechanisms. This was supported by a recent SRMA of 30 RCTs assessing the effect of omega-3 fatty acid supplementation on several cardiometabolic markers in people with T2D, which showed a significant lowering in HbA1c, fasting glucose and HOMA-IR [77].

Unsaturated fatty acids from nuts may also stimulate the secretion of GLP-1, which stimulates the secretion of insulin from beta-cells and promotes the proliferation of beta-cells and, therefore, improves beta-cell efficiency [70,78,79].

5.2. Related to Micronutrients and Other Bioactive Components of Nuts

A number of minerals, vitamins, and other bioactive components, which may be found within the nutritional composition of nuts, have been suggested to be protective against T2D and beneficial in the management of related complications.

5.2.1. Vitamins and Minerals

Nuts, depending on the type, contain relatively high amounts of vitamin E, magnesium, and selenium, among other nutrients [7–9]. While direct evidence does not appear to be currently available for the impact of the content of micronutrients from nut consumption specifically on the risk and management of T2D, there is evidence from epidemiological and oral micronutrient supplementation studies suggesting nutrients that are found in relatively high amounts in nuts may be beneficial.

Evidence from multiple SRMAs has supported the association of specific vitamins and minerals, analogous to those found in nuts, with markers of glycemic control and prevention of T2D [80–85]. When oral supplementation of antioxidant vitamins and minerals (such as those found in nuts: vitamin E, selenium, and zinc) were assessed in a SRMA of RCTs considering people with T2D, supplementation of zinc (30 to 660 mg/day) and vitamin E (200 to 800 IU/day) reduced HbA1c, and zinc reduced fasting blood sugar. None of the nut-associated micronutrient supplements were effective in the reduction of insulin, HOMA-IR, or HOMA-B, and all evidence was considered to be low certainty [85].

Magnesium has also been associated with beneficial effects on glycemic control. Imbalances in magnesium status, specifically hypomagnesemia, have been shown to inhibit glucose transporter type 4 translocation, increase insulin resistance, and affect lipid metabolism, oxidative stress, and the antioxidant system of endothelial cells [82]. A SRMA of 12 observational studies showed significantly lower circulating magnesium levels in people with prediabetes compared to individuals with good health [83]. Further, a 100 mg/day increase in oral magnesium intake has been associated with a 15% reduction in T2D risk (SRMA prospective cohort studies, n = 286,668). For perspective, a 100 mg amount of magnesium is approximately equivalent to the magnesium content in about $\frac{1}{4}$ cup of nuts, depending on nut type [8,9]. When considering individuals with T2D, oral magnesium supplementation, equivalent to just over $\frac{1}{2}$ cup of nuts, significantly improved glycemic control indicators, including HbA1c, IL, C-peptide, HOMA-IR, and HOMA-B and insignificantly decreased fasting blood glucose [86]. However, the impact of magnesium when consumed as a constituent of nuts may respond differently within the body compared to an oral magnesium supplement and further investigation may shed light on the possible role magnesium and/or other micronutrients from the consumption of nuts may have in the prevention and management of type 2 diabetes.

5.2.2. Phenolics and Other Bioactive Compounds

Nuts are composed of a matrix of other important bioactive compounds, including polyphenols of various types (e.g., flavonoids, phenolic acids, stilbenes, lignans, other polyphenols) and concentrations (e.g., 126 to 1576 mg total polyphenols per 100 g nuts) [87–89]. There have been a number of investigations into the polyphenol characteristics of nuts and, independently, a number of studies have assessed the role of polyphenols in diabetes progression and management. For example, polyphenols may improve HbA1c and insulin resistance, in addition to having anti-inflammatory and antioxidant properties such as superoxide dismutase (SOD)-like activity, 1,1-diphenyl-2-picrylhydrazyl (DPPH), and radical scavenging activity. In these ways, various polyphenols may lower the risk of developing diabetes and its complications. However, there is limited evidence confirming the role of polyphenols from nuts in glycemic control, insulin sensitivity, and ultimately in the prevention and management of diabetes [87,90,91]. Polyphenols in nuts may also be protective against diabetes by modifying the gut microbiota (discussed further in Section 5.4.1). Currently, nuts appear to provide only a small percentage of polyphenols in the diet based on cohort data and global average nut intake levels [88,89,92]. Yet, consumption of approximately 50 g/day of nuts could provide the polyphenol dose observed with reduced T2D incidence [88].

5.3. Related to Body Weight and Adiposity

Approximately 60% to 90% of T2D has been attributable to obesity or weight gain; moreover, elevated weight can increase the risk of complications and comorbidities in people with diabetes [93,94] While nuts appear to be relatively high in calculated total calories and fat, their consumption has not been associated with weight gain nor an increased risk of overweight or obesity [95]. Conversely, despite their high energy density, a SRMA of six prospective cohort studies (n = 569,910) and 86 RCTs (n = 5873) indicated nut intake was associated with lower incidence of overweight/obesity (RR 0.93; 95% CI: 0.88 to 0.98), had no effect on body weight, and meta-regression showed higher nut consumption to be related to reductions in body weight and body fat [95]. Furthermore, adiposity factors (i.e., body mass index (BMI) and waist circumference) have been shown to play a role in mediating the association between nut consumption and markers of glycemic control (i.e., fasting blood glucose, plasma insulin, HOMA-IR, HbA1c, and OGTT) suggesting a potential mechanism for the prevention of diabetes risk [11].

5.4. Related to Cellular and Molecular Mechanisms of Nuts

As briefly noted in Sections 3.1 and 3.2, the nutrient composition and bioactive compounds contained in nuts may play a role in preventing and managing diabetes through different cellular and molecular mechanisms, including the modulation of gut microbiota, modifying gene expression, or mediating gene expression through microRNAs (miRNAs). The following discusses evidence available explicitly related to nuts and these aspects.

5.4.1. Gut Microbiota

Within the complex nutrient matrix of nuts, some of the components, such as fiber and polyphenols, can reach the colon intact and interact with the gut microbial population changing its composition and function [66]. The microbial colonic fermentation of undigested fiber and other nutrients from nuts can lead to the production of metabolites, such as SCFA (e.g., butyrate and propionate), with well-demonstrated positive effects for gut microbial homeostasis and may serve as a prebiotic [96,97].

SCFAs can induce their beneficial effects on glucose homeostasis by reducing gut motility and appetite stimulating the expression of peptide YY via the G-protein-coupled receptors (Gpr41 and Gpr43) [98]. Additionally, SCFAs may activate Gpr41 and Gpr43 on L-cells subsequently triggering the secretion of GLP-1, which improves glucose homeostasis by increasing the secretion of insulin and decreasing the secretion of glucagon. The activation of Gpr43 inhibits insulin signalling in adipocytes and fat accumulation in adipose tissue. Butyrate and propionate promote intestinal gluconeogenesis, reducing the risk of T2D. Butyrate also suppresses the action of histone deacetylase (HDAC), which induces insulin resistance by acting in different molecular pathways [99].

Nuts are also rich in polyphenols [87–89], and undigested polyphenols are thought to exert a prebiotic effect by stimulating the growth and activity of some bacteria, such as Bifidobacteria, in the digestive tract [100]. Increased levels of fecal Bifidobacteria have been associated with improved glucose tolerance and diminished inflammatory markers such as the interleukins IL-6, IL-1α and IL-1β, tumor necrosis factor α, and monocyte chemoattractant protein-1 [101].

Considering specific nut types, the effects of almonds on gut microbiota, glycometabolism, and inflammatory parameters in individuals with T2D have been explored in a systematic review conducted by Ojo et al. [36]. The results suggest that an almond-based diet could promote the growth of SCFA-producing bacteria in the gut. Walnuts are also rich in polyphenols and ellagitannins, which are metabolized by intestinal bacteria into urolithins. It has been shown that walnut supplementation, even if short-term, can impact the metabolism of ellagitannins to urolithins via gut microbiota by increasing the production of SCFA (such as acetate, butyrate, and propionate) [102]. This may impact the risk of T2D, as it has been observed that propionate can reduce serum cholesterol and improve insulin resistance, as well as promote satiety [103].

5.4.2. Gene Expressions

Some nut components or their metabolites may act at the cellular level, modifying gene expression. Few studies have analyzed the impact of nut consumption on changes in gene expression in cells or tissues related to proteins that have important potential effects on carbohydrate metabolism, insulin resistance, or adiposity. The crossover EPIRDERM Study assessed the effect of pistachio intake (57 g/d for 4 months) versus a nut-free control diet on insulin resistance and T2D in participants with prediabetes (n = 54) showing changes in peripheral leukocyte gene expression and cellular glucose update [78]. Gene expression data showed that pistachio consumption, compared to control, significantly decreased the expression of interleukin-6 and resistin. Moreover, pistachio intake was shown to facilitate glucose transporter gene expression as assessed by SLC2A3 and SLC2A4 which showed different patterns. For instance, SLC2A4 appeared to be significantly increased in the control compared to pistachio phases. The percentage of change in cellular glucose transport activity also differed between the pistachio and control groups. Similarly, a significantly increased SLC2A4 protein expression on the surface of lymphocytes has been described in both individuals with diabetes and impaired glucose metabolism [104]. Consistent with this, attenuation in the expression of glucose transporters, with pistachio consumption leukocytes, was observed to be significantly expressed in T2D. Therefore, these results suggest a potential mechanism by which pistachios could lead to an improved systemic inflammatory profile increasing insulin sensitivity, as has been observed in the EPIRDERM study.

In a clinical trial conducted in 24 healthy participants, the consumption of hazelnuts (40g/d for 6 weeks) did not lead to weight gain, possibly due to the improvement of the body's antioxidant capacity by the upregulation of genes [codifying superoxide dismutase 1 (SOD1), catalase (CAT), macrophage migration inhibitory factor (MIF), peroxisome proliferator-activated receptor gamma (PPAR-γ), vitamin D receptor (VDR) and methylenetetrahydrofolate reductase (MTHFR)] implied in oxidant reactions and inflammation [105]. Some of these genes have also been related to insulin resistance or diabetes.

5.4.3. MicroRNAs

Nuts may modulate the expression of genes related to glucose metabolism through the mediatory effect of nutrients on microRNAs (miRNAs), defined as small non-coding RNAs with 20 to 25 nucleotides that post-transcriptionally and negatively regulate gene expression.

In the EPIRDERM Study, seven human circulating miRNAs were selected for analysis, that are considered widely related to glucose metabolism, insulin resistance status, prediabetes status, and biomarkers of T2D. The miRNA expression data showed that of the seven miRNAs studied, after pistachio intervention, circulating miRNA-192 and 375 expressions were significantly lower than in the control phase [78]. Furthermore, changes in the circulating miRNA-192 and miRNA-375, were positively associated with plasma glucose, insulin, and HOMA-IR, indicating that an increase in these miRNA levels mirror an increase in insulin resistance.

Similarly, in another trial involving 10 healthy women, 8 weeks of following a PUFA-enriched (achieved via daily intake of 30 g of almonds and walnuts) normocaloric diet resulted in significant modifications to several common miRNAs [106]. Specifically, the authors found that changes in circulating PUFAs were associated with changes of plasma miRNA-106a; changes in plasma miRNA-130b and miRNA-221 were associated with changes in plasma C-reactive protein, and changes in plasma miRNA-125a-5p was associated with changes in plasma fasting triglycerides and adiponectin.

5.4.4. Metabolomics Modulation

Nutritional metabolomics is an emergent approach to obtaining deeper insights into diet–disease association that holds great promise in improving our understanding of the biological effects of nutritional factors and may help to identify potential novel markers of dietary intake and disease risk [107]. To date, a few studies have evaluated the impact of nut consumption on plasma or urinary metabolites [108].

Metabolomics has been used as an agnostic machine learning approach to identify plasma metabolites associated with walnut consumption using data from the PREvención con DIeta MEDiterránea (PREDIMED) study [109]. A metabolite profile including 19 metabolites (including lipids, purines, acylcarnitines, and certain amino acids) was associated with walnut consumption and with a lower risk of T2D incident in a Mediterranean population at high cardiovascular risk. These findings provide new insights into potential biological mechanisms explaining the effect of nut consumption on diabetes risk.

6. Current Strengths and Limitations

There are several strengths and limitations of the available evidence. Strengths, in general, include the relatively long follow-up duration of multiple years observed in the epidemiological studies and data from various countries allowing for potential generalizability of the findings. However, the cross-sectional and prospective cohort studies are limited by the inability to determine causation. A cross-sectional study design is also limited by the lack of ability to assess a temporal relationship between the exposure and the outcome. Additionally, most studies included in this review tended to obtain intake data via food frequency questionnaires (FFQs). FFQs have inherent weaknesses as they are subject to possible measurement error and recall bias [110]. Further, data from these FFQs were limited by the questions asked, as they often assessed a combination of nuts rather than a specific nut type and did not provide data on their preparation, such as whether the nuts were salted, spiced, roasted, or raw. Tree nuts and peanuts were often grouped together in the FFQs and were sometimes also combined within a question, including seeds and/or legumes. The doses of nut intake studied were also relatively low, even in the highest quantile of the analyses (with estimated median nut intakes ranging from 0 to 213 g/week), or were not sufficiently described, being presented as times or servings per day without an equivalent gram amount noted. Moreover, the majority of prospective cohorts evaluated nut intake at baseline as the dietary exposure; however, dietary habits may have changed over the course of the study follow-up period. This could have potentially resulted in misclassification of the exposure to nuts, hence biasing results and possibly explaining null associations observed with T2D. Of note, only a few large cohort studies have collected repeated measures of nuts and other dietary factors; these include the Nurses' Health Study and Health Professionals' Follow-up Study [18], in which diet was assessed every 4 years over 3–4 decades of follow-up. These repeated measures not only represent long-term dietary habits, but also can reduce measurement errors. Finally, as aforementioned, some prospective studies assessing the association between the frequency of nut consumption and diabetes risk have adjusted the analyses for body weight, which is an important determinant of diabetes and, thus, may lead to an attenuation of potential associations. Future cohort studies should carefully evaluate the role of body weight in mediating the association between total and different types of nut consumption and the risk of T2D.

While evidence from both acute and chronic RCTs in individuals with diabetes suggests nut consumption may improve glycemic control via reductions in fasting glucose and HbA1c, the effect still needs to be confirmed by updated SRMAs studying individuals with and without diabetes separately and without mixing interventions of whole nuts with nut oils or other extracts from nuts. It would also be useful to better understand which foods should be replaced with nuts in the diet for the most beneficial impact, as the current trials vary in this regard—some prescribe proportional reductions to all foods, and some suggest replacing for carbohydrate- or saturated fat-rich foods. In contrast, others provide no specific instructions on food replacement. Glycemic control assessment methods are also limited, and there is a lack of direct evidence from clamp studies or from Bergman´s minimal model, which may provide a greater understanding of metabolic regulation [111]. Moreover, there is limited trial evidence for the effect of pecans, pine nuts, Brazil nuts, macadamias, or peanuts in this area. However, since most nut types have similar nutrient profiles, the findings and associated recommendations are likely to be able to be extended to include all types of nuts.

In addition to the limitations to the currently available evidence, there are also a few potential barriers to nut consumption, such as nut-related allergies, cost, dental or swallowing issues, especially in older adults, and lack of knowledge of health benefits by health professionals [112]. This, in conjunction with the limited research evidence to support knowledge and potential recommendations, could potentially explain the relatively low intake levels of nuts by individuals worldwide [92].

7. Future Directions

Future research is needed to better elucidate the impact of nuts on the prevention and management of T2D. Given the current research limitations and limited epidemiological and clinical trial evidence available, there are several lines of research that could provide greater insight and better inform diabetes dietary guidelines.

Further investigation via prospective cohorts assessing the impact of nuts on diabetes incidence and pooled cohort analyses needs to be undertaken. Pooling data from currently conducted prospective cohorts may provide a relatively cost-effective and informative real-world way to explore the possible role of nuts in diabetes prevention and complications. Additionally, more studies in individuals with T2D are needed to demonstrate the impact of nuts on glycemic control (e.g., HbA1c, etc.). To determine a possible causal effect, conducting larger and longer RCTs (such as a multicentre RCT) evaluating markers of glycemic control as primary endpoints is needed in order to expand current knowledge to assess the effect of nuts on diabetes prevention in high-risk participants. Acutely, insulin sensitivity analysis testing the effect of nut consumption, such as replacing carbohydrates, using the Bergman Minimal Model of glucose regulation and clamps would aid in increasing the strength of available evidence.

Further explorations related to metabolomic and metagenomic signatures of nut consumption in clinical trials and assessing the association in long-term cohort studies of diabetes incidence and complications would additionally provide greater insights into a potential diet (nut)–disease (diabetes) association. Then a range of mechanistic molecular biological studies may be justified when a clear phenomenon, such as reduced insulin resistance and improved diabetes control, has been established.

8. Conclusions

Of the limited evidence currently available, overall findings suggest higher nut consumption may have beneficial effects on diabetes prevention and management. In particular, some but not all large cohort studies have found that higher consumption of total nuts, walnuts, and peanuts was significantly associated with a lower risk of T2D. Moreover, inclusion of nuts in the diets of individuals may have a beneficial effect on glycemic control and lower the risk of cardiovascular disease and mortality in those with T2D. In individuals with T2D, specifically, acute studies have demonstrated reductions in postprandial glucose levels, and long-term trials have indicated modest positive effects on blood glucose control, as shown by reductions in HbA1c and fasting blood glucose. Mechanistic pathways provide further promise for the potential role nut consumption may have in diabetes prevention and management. Despite all the potential diabetes-related health benefits nuts may pose, current evidence is not definitive, and there remains much opportunity for future research to address present weaknesses and limited data in this field to provide more conclusive evidence on the role of nuts in the prevention and management of diabetes.

Author Contributions: Writing—original draft preparation, S.K.N. and E.V.; writing—review and editing, S.K.N., E.V., C.W.C.K., D.J.A.J., F.B.H., J.L.S., A.M., A.A. and J.S.-S. All authors have read and agreed to the published version of the manuscript.

Funding: The NUTS 2022 conference was planned by the Rovira i Virgili University, with all the expenses of the organization covered by Barcelo Congresos. Barcelo Congresos SA has received funds from the International Nut & Dried Fruit Council (INC). Researchers did not receive any additional financial support, aside from travel and accommodation, related to their participation in the NUTS 2022 conference. Sponsors did not have any role in selecting or inviting the speakers, nor in the writing or reviewing the proceedings of the conference. S.K.N. is supported by a postdoctoral fellowship from the Canadian Institutes of Health Research (CIHR, MFE-171207).

Institutional Review Board Statement: Not applicable.

Informed Consent Statement: Not applicable.

Data Availability Statement: Not applicable.

Acknowledgments: We are grateful to the organizers and expert researchers participating in the NUTS 2022: Where are we and where are we going? conference held in Reus, Spain.

Conflicts of Interest: S.K.N. is a volunteer member of the not-for profit group Plant-Based Canada and has received a research grant from the INC International Nut and Dried Fruit Council. E.V. is an employee of Roche and has previously received consulting fees from New Era Nutrition. C.W.C.K. has received grants or research support from the Advanced Food Materials Network, Almond Board of California, Barilla, Loblaw Brands Ltd., the Peanut Institute, Pulse Canada, and Unilever. He has received in-kind research support from the Almond Board of California, Barilla, California Walnut Commission, Kellogg Canada, Loblaw Companies, Nutrartis, Quaker (PepsiCo), the Peanut Institute, Primo, Unico, Unilever, and WhiteWave Foods/Danone. He has received travel support and/or honoraria from the Barilla, California Walnut Commission, Canola Council of Canada, General Mills, International Nut and Dried Fruit Council, International Pasta Organization, Lantmannen, Loblaw Brands Ltd., Nutrition Foundation of Italy, Oldways Preservation Trust, Paramount Farms, the Peanut Institute, Pulse Canada, Sun-Maid, Tate & Lyle, Unilever, and White Wave Foods/Danone. He has served on the scientific advisory board for the International Tree Nut Council, International Pasta Organization, McCormick Science Institute, and Oldways Preservation Trust. He is a founding member of the International Carbohydrate Quality Consortium (ICQC), Chair of the Diabetes and Nutrition Study Group (DNSG) of the European Association for the Study of Diabetes (EASD), is on the Clinical Practice Guidelines Expert Committee for Nutrition Therapy of the EASD, and is a Director of Glycemia Consulting and of the Toronto 3D Knowledge Synthesis and Clinical Trials foundation. D.J.A.J. has received research grants from Saskatchewan & Alberta Pulse Growers Associations, the Agricultural Bioproducts Innovation Program through the Pulse Research Network, the Advanced Foods and Material Network, Loblaw Companies Ltd., Unilever Canada and Netherlands, Barilla, the Almond Board of California, Pulse Canada, Kellogg's Company, Canada, Quaker Oats, Canada, Procter & Gamble Technical Centre Ltd., Bayer Consumer Care, Springfield, NJ, Pepsi/Quaker, Soy Foods Association of North America, the Coca-Cola Company (investigator initiated, unrestricted grant), Solae, Haine Celestial, the Sanitarium Company, Orafti, and the Peanut Institute. He has been on the speaker's panel, served on the scientific advisory board, and/or received travel support and/or honoraria from Nutritional Fundamentals for Health (NFH)-Nutramedica, Saint Barnabas Medical Center, The University of Chicago, 2020 China Glycemic Index (GI) International Conference, Atlantic Pain Conference, Academy of Life Long Learning, the Almond Board of California, Canadian Agriculture Policy Institute, Loblaw Companies Ltd., the Griffin Hospital (for the development of the NuVal scoring system), the Coca-Cola Company, Epicure, Danone, Diet Quality Photo Navigation (DQPN), Better Therapeutics (FareWell), Verywell, True Health Initiative (THI), Heali AI Corp, Institute of Food Technologists (IFT), Soy Nutrition Institute (SNI), Herbalife Nutrition Institute (HNI), Saskatchewan & Alberta Pulse Growers Associations, Sanitarium Company, Orafti, the International Tree Nut Council Nutrition Research and Education Foundation, the Peanut Institute, Herbalife International, Pacific Health Laboratories, Barilla, Metagenics, Bayer Consumer Care, Unilever Canada and Netherlands, Solae, Kellogg, Quaker Oats, Procter & Gamble, Abbott Laboratories, Dean Foods, the California Strawberry Commission, Haine Celestial, PepsiCo, the Alpro Foundation, Pioneer Hi-Bred International, DuPont Nutrition and Health, Spherix Consulting and WhiteWave Foods, the Advanced Foods and Material Network, the Canola and Flax Councils of Canada, Agri-Culture and Agri-Food Canada, the Canadian Agri-Food Policy Institute, Pulse Canada, the Soy Foods Association of North America, the Nutrition Foundation of Italy (NFI), Nutra-Source Diagnostics, the McDougall Program, the Toronto Knowledge Translation Group (St. Michael's Hospital), the Canadian College of Naturopathic Medicine, The Hospital for Sick Children, the Canadian Nutrition Society (CNS), the American Society of Nutrition (ASN), Arizona State University, Paolo Sorbini Foundation, and the Institute of Nutrition, Metabolism and Diabetes. He is a member of the International Carbohydrate Quality Consortium (ICQC). His wife, Alexandra L Jenkins, is a director and partner of INQUIS Clinical Research for the food industry, his two daughters, Wendy Jenkins and Amy Jenkins, have published a vegetarian book that promotes the use of the foods described here, The Portfolio Diet for Cardiovascular Risk Reduction (Academic Press/Elsevier 2020 ISBN:978-0-12- 810510-8), and his sister, Caroline Brydson, received funding through a grant from the St. Michael's Hospital Foundation to develop a cookbook for one of his studies. J.L.S. has received research support from the Canadian Foundation for Innovation, Ontario Research Fund, Province of Ontario Ministry of Research and Innovation and Science, Canadian Institutes of health Research (CIHR), Diabetes Canada, American Society for Nutrition (ASN), International Nut and Dried Fruit Council (INC) Foundation, National Honey Board (U.S. Department of Agriculture [USDA] honey "Checkoff" program), Institute for the Advancement of Food and Nutrition Sciences (IAFNS; formerly ILSI North America), Pulse Canada, Quaker Oats Center of Excellence, The United Soybean Board (USDA soy "Checkoff" program), The

Tate and Lyle Nutritional Research Fund at the University of Toronto, The Glycemic Control and Cardiovascular Disease in Type 2 Diabetes Fund at the University of Toronto (a fund established by the Alberta Pulse Growers), The Plant Protein Fund at the University of Toronto (a fund which has received contributions from IFF), and The Nutrition Trialists Network Fund at the University of Toronto (a fund established by an inaugural donation from the Calorie Control Council). He has received food donations to support randomized controlled trials from the Almond Board of California, California Walnut Commission, Peanut Institute, Barilla, Unilever/Upfield, Unico/Primo, Loblaw Companies, Quaker, Kellogg Canada, WhiteWave Foods/Danone, Nutrartis, and Dairy Farmers of Canada. He has received travel support, speaker fees and/or honoraria from ASN, Danone, Dairy Farmers of Canada, FoodMinds LLC, Nestlé, Abbott, General Mills, Nutrition Communications, International Food Information Council (IFIC), Calorie Control Council, International Sweeteners Association, and International Glutamate Technical Committee. He has or has had ad hoc consulting arrangements with Perkins Coie LLP, Tate & Lyle, Phynova, and Inquis Clinical Research. He is a former member of the European Fruit Juice Association Scientific Expert Panel and former member of the Soy Nutrition Institute (SNI) Scientific Advisory Committee. He is on the Clinical Practice Guidelines Expert Committees of Diabetes Canada, European Association for the study of Diabetes (EASD), Canadian Cardiovascular Society (CCS), and Obesity Canada/Canadian Association of Bariatric Physicians and Surgeons. He serves or has served as an unpaid member of the Board of Trustees and an unpaid scientific advisor for the Carbohydrates Committee of IAFNS. He is a member of the International Carbohydrate Quality Consortium (ICQC), Executive Board Member of the Diabetes and Nutrition Study Group (DNSG) of the EASD, and Director of the Toronto 3D Knowledge Synthesis and Clinical Trials foundation. His spouse is an employee of AB InBev. A.M. has received funding from the Almond Board of California and Danon (India). J.S.-S. reported receiving research support from the Instituto de Salud Carlos III, Ministerio de Educación y Ciencia, the European Commission, the USA National Institutes of Health; receiving consulting fees or travel expenses from Instituto Danone and Abbott Laboratories, receiving nonfinancial support from Patrimonio Comunal Olivarero, the Almond Board of California, Pistachio Growers and Borges S.A; serving on the board of and receiving grant support through his institution from the International Nut and Dried Foundation; and personal fees from Instituto Danone; Serving in the Board of Danone Institute International. The rest of the authors have declared that no competing interests exist. The funders had no role in the design of the study; in the collection, analyses, or interpretation of data; in the writing of the manuscript, or in the decision to publish the results. All other authors declare no conflict of interest. The funders had no role in the design of the study; in the collection, analyses, or interpretation of data; in the writing of the manuscript; or in the decision to publish the results.

References

1. International Diabetes Federation. Diabetes Is "a Pandemic of Unprecedented Magnitude" Now Affecting One in 10 Adults Worldwide. *Diabetes Res. Clin. Pr.* **2021**, *181*, 109133. [CrossRef]
2. International Diabetes Federation IDF Diabetes Atlas 10th Edition. Available online: www.diabetesatlas.org (accessed on 20 November 2022).
3. Mann, J.I.; de Leeuw, I.; Hermansen, K.; Karamanos, B.; Karlström, B.; Katsilambros, N.; Riccardi, G.; Rivellese, A.A.; Rizkalla, S.; Slama, G.; et al. Evidence-Based Nutritional Approaches to the Treatment and Prevention of Diabetes Mellitus. *Nutr. Metab. Cardiovasc. Dis.* **2004**, *14*, 373–394. [CrossRef] [PubMed]
4. Lifestyle Management: Standards of Medical Care in Diabetes. *Diabetes Care* **2019**, *42*, S46–S60. [CrossRef] [PubMed]
5. Sievenpiper, J.L.; Chan, C.B.; Dworatzek, P.D.; Freeze, C.; Williams, S.L. Nutrition Therapy. *Can. J. Diabetes* **2018**, *42*, S64–S79. [CrossRef] [PubMed]
6. Dyson, P.A.; Twenefour, D.; Breen, C.; Duncan, A.; Elvin, E.; Goff, L.; Hill, A.; Kalsi, P.; Marsland, N.; McArdle, P.; et al. Diabetes UK Evidence-Based Nutrition Guidelines for the Prevention and Management of Diabetes. *Diabet. Med.* **2018**, *35*, 541–547. [CrossRef]
7. Venkatachalan, M.; Sathe, S.K. Chemical Composition of Selected Edible Nut Seeds. *J. Agric. Food Chem.* **2006**, *54*, 4705–4714. [CrossRef] [PubMed]
8. Government of Canada Canadian Nutrient File (CNF). Available online: https://food-nutrition.canada.ca/cnf-fce/index-eng.jsp (accessed on 20 November 2022).
9. U.S. Department of Agriculture (USDA), A.R.S. FoodData Central. Available online: https://fdc.nal.usda.gov/ (accessed on 20 November 2022).
10. Hosseinpour-Niazi, S.; Hosseini, S.; Mirmiran, P.; Azizi, F. Prospective Study of Nut Consumption and Incidence of Metabolic Syndrome: Tehran Lipid and Glucose Study. *Nutrients* **2017**, *9*, 1056. [CrossRef] [PubMed]

11. Mazidi, M.; Vatanparast, H.; Katsiki, N.; Banach, M. The Impact of Nuts Consumption on Glucose/Insulin Homeostasis and Inflammation Markers Mediated by Adiposity Factors among American Adults. *Oncotarget* **2018**, *9*, 31173–31186. [CrossRef] [PubMed]
12. Park, S.K.; Oh, C.M.; Jung, J.Y. The Association between Insulin Resistance and the Consumption of Nut Including Peanut, Pine Nut and Almonds in Working-Aged Korean Population. *Public Health Nutr.* **2022**, *25*, 1904–1911. [CrossRef]
13. Becerra-Tomás, N.; Paz-Graniel, I.; Hernández-Alonso, P.; Jenkins, D.J.A.; Kendall, C.W.C.; Sievenpiper, J.L.; Salas-Salvadó, J. Nut Consumption and Type 2 Diabetes Risk: A Systematic Review and Meta-Analysis of Observational Studies. *Am. J. Clin. Nutr.* **2021**, *113*, 960–971. [CrossRef]
14. Pan, A.; Sun, Q.; Manson, J.A.E.; Willett, W.C.; Hu, F.B. Walnut Consumption Is Associated with Lower Risk of Type 2 Diabetes in Women. *J. Nutr.* **2013**, *143*, 512–518. [CrossRef]
15. Arab, L.; Dhaliwal, S.K.; Martin, C.J.; Larios, A.D.; Jackson, N.J.; Elashoff, D. Association between Walnut Consumption and Diabetes Risk in NHANES. *Diabetes Metab. Res. Rev.* **2018**, *34*, e3031. [CrossRef]
16. Cubas-Basterrechea, G.; Elío, I.; Sumalla-Cano, S.; Aparicio-Obregón, S.; González-Antón, C.T.; Muñoz-Cacho, P. The Regular Consumption of Nuts Is Associated with a Lower Prevalence of Abdominal Obesity and Metabolic Syndrome in Older People from the North of Spain. *Int. J. Environ. Res. Public Health* **2022**, *19*, 1256. [CrossRef]
17. Micek, A.; Godos, J.; Cernigliaro, A.; Cincione, R.I.; Buscemi, S.; Libra, M.; Galvano, F.; Grosso, G. Total Nut, Tree Nut, and Peanut Consumption and Metabolic Status in Southern Italian Adults. *Int. J. Environ. Res. Public Health* **2021**, *18*, 1847. [CrossRef] [PubMed]
18. Liu, G.; Guasch-Ferré, M.; Hu, Y.; Li, Y.; Hu, F.B.; Rimm, E.B.; Manson, J.A.E.; Rexrode, K.M.; Sun, Q. Nut Consumption in Relation to Cardiovascular Disease Incidence and Mortality among Patients with Diabetes Mellitus. *Circ. Res.* **2019**, *124*, 920–929. [CrossRef] [PubMed]
19. Aune, D.; Keum, N.N.; Giovannucci, E.; Fadnes, L.T.; Boffetta, P.; Greenwood, D.C.; Tonstad, S.; Vatten, L.J.; Riboli, E.; Norat, T. Nut Consumption and Risk of Cardiovascular Disease, Total Cancer, All-Cause and Cause-Specific Mortality: A Systematic Review and Dose-Response Meta-Analysis of Prospective Studies. *BMC. Med.* **2016**, *14*, 207. [CrossRef] [PubMed]
20. Mori, A.M.; Considine, R.V.; Mattes, R.D. Acute and Second-Meal Effects of Almond Form in Impaired Glucose Tolerant Adults: A Randomized Crossover Trial. *Nutr. Metab.* **2011**, *8*, 6. [CrossRef]
21. Cohen, A.E.; Johnston, C.S. Almond Ingestion at Mealtime Reduces Postprandial Glycemia and Chronic Ingestion Reduces Hemoglobin A1c in Individuals with Well-Controlled Type 2 Diabetes Mellitus. *Metabolism* **2011**, *60*, 1312–1317. [CrossRef]
22. Kendall, C.W.C.; West, S.G.; Augustin, L.S.; Esfahani, A.; Vidgen, E.; Bashyam, B.; Sauder, K.A.; Campbell, J.; Chiavaroli, L.; Jenkins, A.L.; et al. Acute Effects of Pistachio Consumption on Glucose and Insulin, Satiety Hormones and Endothelial Function in the Metabolic Syndrome. *Eur. J. Clin. Nutr.* **2014**, *68*, 370–375. [CrossRef]
23. Kendall, C.W.C.; Esfahani, A.; Josse, A.R.; Augustin, L.S.A.; Vidgen, E.; Jenkins, D.J.A. The Glycemic Effect of Nut-Enriched Meals in Healthy and Diabetic Subjects. *Nutr. Metab. Cardiovasc. Dis.* **2011**, *21*, S34–S39. [CrossRef]
24. Godwin, N.; Roberts, T.; Hooshmand, S.; Kern, M.; Hong, M.Y. Mixed Nuts May Promote Satiety While Maintaining Stable Blood Glucose and Insulin in Healthy, Obese, and Overweight Adults in a Two-Arm Randomized Controlled Trial. *J. Med. Food* **2019**, *22*, 427–432. [CrossRef] [PubMed]
25. Moreira, A.P.B.; Teixeira, T.F.S.; Alves, R.D.M.; Peluzio, M.C.G.; Costa, N.M.B.; Bressan, J.; Mattes, R.; Alfenas, R.C.G. Effect of a High-Fat Meal Containing Conventional or High-Oleic Peanuts on Post-Prandial Lipopolysaccharide Concentrations in Overweight/Obese Men. *J. Hum. Nutr. Diet.* **2016**, *29*, 95–104. [CrossRef]
26. Jenkins, D.J.A.; Kendall, C.W.C.; Marchie, A.; Josse, A.R.; Nguyen, T.H.; Faulkner, D.A.; Lapsley, K.G.; Blumberg, J.; Mayer, J. Almonds reduce biomarkers of lipid peroxidation in older hyperlipidemic subjects. *J. Nutr.* **2008**, *138*, 908–913. [CrossRef] [PubMed]
27. Josse, A.R.; Kendall, C.W.C.; Augustin, L.S.A.; Ellis, P.R.; Jenkins, D.J.A. Almonds and Postprandial Glycemia-a Dose-Response Study. *Metabolism* **2007**, *56*, 400–404. [CrossRef]
28. Brown, R.; Ware, L.; Gray, A.R.; Chisholm, A.; Tey, S.L. Snacking on Almonds Lowers Glycaemia and Energy Intake Compared to a Popular High-carbohydrate Snack Food: An Acute Randomised Crossover Study. *Int. J. Environ. Res. Public Health* **2021**, *18*, 10989. [CrossRef]
29. Bodnaruc, A.M.; Prud'homme, D.; Giroux, I. Acute Effects of an Isocaloric Macronutrient-Matched Breakfast Meal Containing Almonds on Glycemic, Hormonal and Appetite Responses in Men with Type 2 Diabetes: A Randomized Cross-over Study. *Appl. Physiol. Nutr. Metab.* **2020**, *45*, 520–529. [CrossRef]
30. Muley, A.; Fernandez, R.; Ellwood, L.; Muley, P.; Shah, M. Effect of Tree Nuts on Glycemic Outcomes in Adults with Type 2 Diabetes Mellitus: A Systematic Review. *JBI. Evid. Synth.* **2021**, *19*, 966–1002. [CrossRef]
31. Salas-Salvadó, J.; Bulló, M.; Babio, N.; Martínez-González, M.Á.; Ibarrola-Jurado, N.; Basora, J.; Estruch, R.; Covas, M.I.; Corella, D.; Arós, F.; et al. Reduction in the Incidence of Type 2 Diabetes with the Mediterranean Diet: Results of the PREDIMED-Reus Nutrition Intervention Randomized Trial. *Diabetes Care* **2011**, *34*, 14–19, Erratum in *Diabetes Care* **2018**, *41*, 2259–2260. [CrossRef] [PubMed]
32. Salas-Salvadó, J.; Bulló, M.; Estruch, R.; Ros, E.; Covas, M.-I.; ria Ibarrola-Jurado, N.; Corella, D.; Aró, F.; Gó mez-Gracia, E.; Ruiz-Gutiérrez, V.; et al. Prevention of Diabetes with Mediterranean Diets. A Subgroup Analysis of a Randomized Trial. *Ann. Intern. Med.* **2014**, *160*, 1–10. [CrossRef] [PubMed]

33. Estruch, R.; Ros, E.; Salas-Salvadó, J.; Covas, M.-I.; Corella, D.; Arós, F.; Gómez-Gracia, E.; Ruiz-Gutiérrez, V.; Fiol, M.; Lapetra, J.; et al. Primary Prevention of Cardiovascular Disease with a Mediterranean Diet Supplemented with Extra-Virgin Olive Oil or Nuts. *N. Engl. J. Med.* **2018**, *378*, e34. [CrossRef]
34. Nowrouzi-Sohrabi, P.; Hassanipour, S.; Sisakht, M.; Daryabeygi-Khotbehsara, R.; Savardashtaki, A.; Fathalipour, M. The Effectiveness of Pistachio on Glycemic Control and Insulin Sensitivity in Patients with Type 2 Diabetes, Prediabetes and Metabolic Syndrome: A Systematic Review and Meta-Analysis. *Diabetes Metab. Syndr. Clin. Res. Rev.* **2020**, *14*, 1589–1595. [CrossRef] [PubMed]
35. Viguiliouk, E.; Kendall, C.W.C.; Mejia, S.B.; Cozma, A.I.; Ha, V.; Mirrahimi, A.; Jayalath, V.H.; Augustin, L.S.A.; Chiavaroli, L.; Leiter, L.A.; et al. Effect of Tree Nuts on Glycemic Control in Diabetes: A Systematic Review and Meta-Analysis of Randomized Controlled Dietary Trials. *PLoS ONE* **2014**, *9*, e103376, Corrected in *PLoS ONE* **2014**, *9*, e109224. [CrossRef] [PubMed]
36. Ojo, O.; Wang, X.H.; Ojo, O.O.; Adegboye, A.R.A. The Effects of Almonds on Gut Microbiota, Glycometabolism, and Inflammatory Markers in Patients with Type 2 Diabetes: A Systematic Review and Meta-Analysis of Randomised Controlled Trials. *Nutrients* **2021**, *13*, 3377. [CrossRef] [PubMed]
37. Moosavian, S.P.; Rahimlou, M.; Rezaei Kelishadi, M.; Moradi, S.; Jalili, C. Effects of Almond on Cardiometabolic Outcomes in Patients with Type 2 Diabetes: A Systematic Review and Meta-Analysis of Randomized Controlled Trials. *Phytother. Res.* **2022**, *36*, 1839–1853. [CrossRef] [PubMed]
38. Ren, M.; Zhang, H.; Qi, J.; Hu, A.; Jiang, Q.; Hou, Y.; Feng, Q.; Ojo, O.; Wang, X. An Almond-Based Low Carbohydrate Diet Improves Depression and Glycometabolism in Patients with Type 2 Diabetes through Modulating Gut Microbiota and Glp-1: A Randomized Controlled Trial. *Nutrients* **2020**, *12*, 3036. [CrossRef]
39. dos Santos, J.L.; Portal, V.L.; Markoski, M.M.; de Quadros, A.S.; Bersch-Ferreira, Â.; Marcadenti, A. Effect of Pecan Nuts and Extra-Virgin Olive Oil on Glycemic Profile and Nontraditional Anthropometric Indexes in Patients with Coronary Artery Disease: A Randomized Clinical Trial. *Eur. J. Clin. Nutr.* **2022**, *76*, 827–834. [CrossRef]
40. Tindall, A.M.; Johnston, E.A.; Kris-Etherton, P.M.; Petersen, K.S. The Effect of Nuts on Markers of Glycemic Control: A Systematic Review and Meta-Analysis of Randomized Controlled Trials. *Am. J. Clin. Nutr.* **2019**, *109*, 297–314. [CrossRef]
41. Jenkins, D.J.A.; Kendall, C.W.C.; Josse, A.R.; Salvatore, S.; Brighenti, F.; Augustin, L.S.A.; Ellis, P.R.; Vidgen, E.; Rao, A.V. Almonds Decrease Postprandial Glycemia, Insulinemia, and Oxidative Damage in Healthy Individuals. *J. Nutr.* **2006**, *136*, 2987–2992. [CrossRef]
42. Kendall, C.W.C.; Josse, A.R.; Esfahani, A.; Jenkins, D.J.A. The Impact of Pistachio Intake Alone or in Combination with High-Carbohydrate Foods on Post-Prandial Glycemia. *Eur. J. Clin. Nutr.* **2011**, *65*, 696–702. [CrossRef]
43. Blanco Mejia, S.; Kendall, C.W.C.; Viguiliouk, E.; Augustin, L.S.; Ha, V.; Cozma, A.I.; Mirrahimi, A.; Maroleanu, A.; Chiavaroli, L.; Leiter, L.A.; et al. Effect of Tree Nuts on Metabolic Syndrome Criteria: A Systematic Review and Meta-Analysis of Randomised Controlled Trials. *BMJ. Open* **2014**, *4*, 4660. [CrossRef]
44. Schwingshackl, L.; Hoffmann, G.; Iqbal, K.; Schwedhelm, C.; Boeing, H. Food Groups and Intermediate Disease Markers: A Systematic Review and Network Meta-Analysis of Randomized Trials. *Am. J. Clin. Nutr.* **2018**, *108*, 576–586. [CrossRef] [PubMed]
45. Eslami, O.; Khorramrouz, F.; Sohouli, M.; Bagheri, N.; Shidfar, F.; Fernandez, M.L. Effect of Nuts on Components of Metabolic Syndrome in Healthy Adults with Overweight/Obesity: A Systematic Review and Meta-Analysis. *Nutr. Metab. Cardiovasc. Dis.* **2022**, *32*, 2459–2469. [CrossRef]
46. Mateș, L.; Popa, D.S.; Rusu, M.E.; Fizeșan, I.; Leucuța, D. Walnut Intake Interventions Targeting Biomarkers of Metabolic Syndrome and Inflammation in Middle-Aged and Older Adults: A Systematic Review and Meta-Analysis of Randomized Controlled Trials. *Antioxidants* **2022**, *11*, 1412. [CrossRef] [PubMed]
47. Neale, E.P.; Guan, V.; Tapsell, L.C.; Probst, Y.C. Effect of Walnut Consumption on Markers of Blood Glucose Control: A Systematic Review and Meta-Analysis. *Br. J. Nutr.* **2020**, *124*, 641–653. [CrossRef]
48. Jamshidi, S.; Moradi, Y.; Nameni, G.; Mohsenpour, M.A.; Vafa, M. Effects of Cashew Nut Consumption on Body Composition and Glycemic Indices: A Meta-Analysis and Systematic Review of Randomized Controlled Trials. *Diabetes Metab. Syndr. Clin. Res. Rev.* **2021**, *15*, 605–613. [CrossRef]
49. Parilli-Moser, I.; Hurtado-Barroso, S.; Guasch-Ferré, M.; Lamuela-Raventós, R.M. Effect of Peanut Consumption on Cardiovascular Risk Factors: A Randomized Clinical Trial and Meta-Analysis. *Front. Nutr.* **2022**, *9*. [CrossRef] [PubMed]
50. Asbaghi, O.; Moodi, V.; Neisi, A.; Shirinbakhshmasoleh, M.; Abedi, S.; Oskouie, F.H.; Eslampour, E.; Ghaedi, E.; Miraghajani, M. The Effect of Almond Intake on Glycemic Control: A Systematic Review and Dose–Response Meta-Analysis of Randomized Controlled Trials. *Phytother. Res.* **2022**, *36*, 395–414. [CrossRef]
51. Lee-Bravatti, M.A.; Wang, J.; Avendano, E.E.; King, L.; Johnson, E.J.; Raman, G. Almond Consumption and Risk Factors for Cardiovascular Disease: A Systematic Review and Meta-Analysis of Randomized Controlled Trials. *Adv. Nutr.* **2019**, *10*, 1076–1088. [CrossRef] [PubMed]
52. Petersen, K.S.; Murphy, J.; Whitbread, J.; Clifton, P.M.; Keogh, J.B. The Effect of a Peanut-Enriched Weight Loss Diet Compared to a Low-Fat Weight Loss Diet on Body Weight, Blood Pressure, and Glycemic Control: A Randomized Controlled Trial. *Nutrients* **2022**, *14*, 2986. [CrossRef]
53. Caldas, A.P.S.; Rocha, D.M.U.P.; Dionísio, A.P.; Hermsdorff, H.H.M.; Bressan, J. Brazil and Cashew Nuts Intake Improve Body Composition and Endothelial Health in Women at Cardiometabolic Risk (Brazilian Nuts Study): A Randomised Controlled Trial. *Br. J. Nutr.* **2022**, *128*, 1747–1757. [CrossRef]

54. Luo, C.; Zhang, Y.; Ding, Y.; Shan, Z.; Chen, S.; Yu, M.; Hu, F.B.; Liu, L. Nut Consumption and Risk of Type 2 Diabetes, Cardiovascular Disease, and All-Cause Mortality: A Systematic Review and Meta-Analysis. *Am. J. Clin. Nutr.* **2014**, *100*, 256–269. [CrossRef]
55. Guo, K.; Zhou, Z.; Jiang, Y.; Li, W.; Li, Y. Meta-Analysis of Prospective Studies on the Effects of Nut Consumption on Hypertension and Type 2 Diabetes Mellitus. *J. Diabetes* **2015**, *7*, 202–212. [CrossRef]
56. Wu, L.; Wang, Z.; Zhu, J.; Murad, A.L.; Prokop, L.J.; Murad, M.H. Nut Consumption and Risk of Cancer and Type 2 Diabetes: A Systematic Review and Meta-Analysis. *Nutr. Rev.* **2015**, *73*, 409–425. [CrossRef]
57. Zhou, D.; Yu, H.; He, F.; Reilly, K.H.; Zhang, J.; Li, S.; Zhang, T.; Wang, B.; Ding, Y.; Xi, B. Nut Consumption in Relation to Cardiovascular Disease Risk and Type 2 Diabetes: A Systematic Review and Meta-Analysis of Prospective Studies. *Am. J. Clin. Nutr.* **2014**, *100*, 270–277. [CrossRef]
58. Schwingshackl, L.; Hoffmann, G.; Lampousi, A.M.; Knüppel, S.; Iqbal, K.; Schwedhelm, C.; Bechthold, A.; Schlesinger, S.; Boeing, H. Food Groups and Risk of Type 2 Diabetes Mellitus: A Systematic Review and Meta-Analysis of Prospective Studies. *Eur. J. Epidemiol.* **2017**, *32*, 363–375. [CrossRef]
59. Afshin, A.; Micha, R.; Khatibzadeh, S.; Mozaffarian, D. Consumption of Nuts and Legumes and Risk of Incident Ischemic Heart Disease, Stroke, and Diabetes: A Systematic Review and Meta-Analysis. *Am. J. Clin. Nutr.* **2014**, *100*, 278–288. [CrossRef]
60. Hernández-Alonso, P.; Camacho-Barcia, L.; Bulló, M.; Salas-Salvadó, J. Nuts and Dried Fruits: An Update of Their Beneficial Effects on Type 2 Diabetes. *Nutrients* **2017**, *9*, 673. [CrossRef]
61. Jenkins, D.J.A.; Hu, F.B.; Tapsell, L.C.; Josse, A.R.; Kendall, C.W.C. Possible Benefit of Nuts in Type 2 Diabetes. *J. Nutr.* **2008**, *138*, 1752S–1756S. [CrossRef]
62. Kendall, C.W.C.; Josse, A.R.; Esfahani, A.; Jenkins, D.J.A. Nuts, Metabolic Syndrome and Diabetes. *Br. J. Nutr.* **2010**, *104*, 465–473. [CrossRef]
63. Calbet, J.A.; MacLean, D.A. Role of Caloric Content on Gastric Emptying in Humans. *J. Physiol.* **1997**, *498*, 553–559. [CrossRef]
64. Hunt, J.N.; Stubbs, D.F. The Volume and Energy Content of Meals as Determinants of Gastric Emptying. *J. Physiol.* **1975**, *245*, 209–225. [CrossRef]
65. Peracchi, M.; Gebbia, C.; Ogliari, C.; Fraquelli, M.; Vigano, R.; Baldassarri, A.; Bianchi, P.A.; Conte, D. Influence of Caloric Intake on Gastric Emptying of Solids Assessed by 13C-Octanoic Acid Breath Test. *Scand. J. Gastroenterol.* **2000**, *35*, 814–818. [CrossRef]
66. Dikeman, C.L.; Fahey, G.C. Viscosity as Related to Dietary Fiber: A Review. *Crit. Rev. Food Sci. Nutr.* **2006**, *46*, 649–663. [CrossRef]
67. Tolhurst, G.; Heffron, H.; Lam, Y.S.; Parker, H.E.; Habib, A.M.; Diakogiannaki, E.; Cameron, J.; Grosse, J.; Reimann, F.; Gribble, F.M. Short-Chain Fatty Acids Stimulate Glucagon-like Peptide-1 Secretion via the G-Protein-Coupled Receptor FFAR2. *Diabetes* **2012**, *61*, 364–371. [CrossRef]
68. Russell, W.R.; Baka, A.; Björck, I.; Delzenne, N.; Gao, D.; Griffiths, H.R.; Hadjilucas, E.; Juvonen, K.; Lahtinen, S.; Lansink, M.; et al. Impact of Diet Composition on Blood Glucose Regulation. *Crit. Rev. Food Sci. Nutr.* **2016**, *56*, 541–590. [CrossRef]
69. Heppner, K.M.; Perez-Tilve, D. GLP-1 Based Therapeutics: Simultaneously Combating T2DM and Obesity. *Front. Neurosci.* **2015**, *9*, 92. [CrossRef]
70. Kim, Y.; Keogh, J.B.; Clifton, P.M. Benefits of Nut Consumption on Insulin Resistance and Cardiovascular Risk Factors: Multiple Potential Mechanisms of Actions. *Nutrients* **2017**, *9*, 1271. [CrossRef]
71. Imamura, F.; Micha, R.; Wu, J.H.Y.; de Oliveira Otto, M.C.; Otite, F.O.; Abioye, A.I.; Mozaffarian, D. Effects of Saturated Fat, Polyunsaturated Fat, Monounsaturated Fat, and Carbohydrate on Glucose-Insulin Homeostasis: A Systematic Review and Meta-Analysis of Randomised Controlled Feeding Trials. *PLoS Med.* **2016**, *13*, e1002087. [CrossRef]
72. Risérus, U.; Willett, W.C.; Hu, F.B. Dietary Fats and Prevention of Type 2 Diabetes. *Prog. Lipid Res.* **2009**, *48*, 44–51. [CrossRef]
73. Storlien A', L.H.; Pan, D.A.; Kriketos, A.D.; O'connor, J.; Caterson, I.D.; Cooney, G.J.; Jenkins, A.B.; Baur, L.A. Skeletal Muscle Membrane Lipids and Insulin Resistance. *Lipids* **1996**, *31*, S261–S265. [CrossRef]
74. Ginsberg, B.H.; Brown, T.J.; Simon, I.; Spector, A.A. Effect of the Membrane Lipid Environment on the Properties of Insulin Receptors. *Diabetes* **1981**, *30*, 773–780. [CrossRef]
75. Kien, C.L. Dietary Interventions for Metabolic Syndrome: Role of Modifying Dietary Fats. *Curr. Diab. Rep.* **2009**, *9*, 43–50. [CrossRef]
76. Clarke, S.D.; Williams, L. The Multi-Dimensional Regulation of Gene Expression by Fatty Acids: Polyunsaturated Fats as Nutrient Sensors. *Curr. Opin. Lipidol.* **2004**, *15*, 13–18. [CrossRef]
77. Khalili, L.; Valdes-Ramos, R.; Harbige, L.S. Effect of N-3 (Omega-3) Polyunsaturated Fatty Acid Supplementation on Metabolic and Inflammatory Biomarkers and Body Weight in Patients with Type 2 Diabetes Mellitus: A Systematic Review and Meta-Analysis of RCTs. *Metabolites* **2021**, *11*, 742. [CrossRef]
78. Hernández-Alonso, P.; Giardina, S.; Salas-Salvadó, J.; Arcelin, P.; Bulló, M. Chronic Pistachio Intake Modulates Circulating MicroRNAs Related to Glucose Metabolism and Insulin Resistance in Prediabetic Subjects. *Eur. J. Nutr.* **2017**, *56*, 2181–2191. [CrossRef]
79. Rajaram, S.; Sabaté, J. Nuts, Body Weight and Insulin Resistance. *Br. J. Nutr.* **2006**, *96*, S79–S86. [CrossRef]
80. Vinceti, M.; Filippini, T.; Rothman, K.J. Selenium Exposure and the Risk of Type 2 Diabetes: A Systematic Review and Meta-Analysis. *Eur. J. Epidemiol.* **2018**, *33*, 789–810. [CrossRef]
81. Zhao, J.V.; Schooling, C.M.; Zhao, J.X. The Effects of Folate Supplementation on Glucose Metabolism and Risk of Type 2 Diabetes: A Systematic Review and Meta-Analysis of Randomized Controlled Trials. *Ann. Epidemiol.* **2018**, *28*, 249–257.e1. [CrossRef]

82. Feng, J.; Wang, H.; Jing, Z.; Wang, Y.; Cheng, Y.; Wang, W.; Sun, W. Role of Magnesium in Type 2 Diabetes Mellitus. *Biol. Trace. Elem. Res.* **2020**, *196*, 74–85. [CrossRef]
83. Ebrahimi Mousavi, S.; Ghoreishy, S.M.; Hemmati, A.; Mohammadi, H. Association between Magnesium Concentrations and Prediabetes: A Systematic Review and Meta-Analysis. *Sci. Rep.* **2021**, *11*, 24388. [CrossRef]
84. Larsson, S.C.; Wolk, A. Magnesium Intake and Risk of Type 2 Diabetes: A Meta-Analysis. *J. Intern. Med.* **2007**, *262*, 208–214. [CrossRef]
85. Kazemi, A.; Ryul Shim, S.; Jamali, N.; Hassanzadeh-Rostami, Z.; Soltani, S.; Sasani, N.; Mohsenpour, M.A.; Firoozi, D.; Basirat, R.; Hosseini, R.; et al. Comparison of Nutritional Supplements for Glycemic Control in Type 2 Diabetes: A Systematic Review and Network Meta-Analysis of Randomized Trials. *Diabetes Res. Clin. Pr.* **2022**, *191*, 110037. [CrossRef]
86. Elderawi, W.A.; Naser, I.A.; Taleb, M.H.; Abutair, A.S. The Effects of Oral Magnesium Supplementation on Glycemic Response among Type 2 Diabetes Patients. *Nutrients* **2019**, *11*, 44. [CrossRef]
87. Vitale, M.; Masulli, M.; Rivellese, A.A.; Bonora, E.; Cappellini, F.; Nicolucci, A.; Squatrito, S.; Antenucci, D.; Barrea, A.; Bianchi, C.; et al. Dietary Intake and Major Food Sources of Polyphenols in People with Type 2 Diabetes: The TOSCA.IT Study. *Eur. J. Nutr.* **2018**, *57*, 679–688. [CrossRef]
88. Godos, J.; Marventano, S.; Mistretta, A.; Galvano, F.; Grosso, G. Dietary Sources of Polyphenols in the Mediterranean Healthy Eating, Aging and Lifestyle (MEAL) Study Cohort. *Int. J. Food. Sci. Nutr.* **2017**, *68*, 750–756. [CrossRef]
89. Tresserra-Rimbau, A.; Medina-Remón, A.; Pérez-Jiménez, J.; Martínez-González, M.A.; Covas, M.I.; Corella, D.; Salas-Salvadó, J.; Gómez-Gracia, E.; Lapetra, J.; Arós, F.; et al. Dietary Intake and Major Food Sources of Polyphenols in a Spanish Population at High Cardiovascular Risk: The PREDIMED Study. *Nutr. Metab. Cardiovasc. Dis.* **2013**, *23*, 953–959. [CrossRef]
90. Xiao, J.B.; Hogger, P. Dietary Polyphenols and Type 2 Diabetes: Current Insights and Future Perspectives. *Curr. Med. Chem.* **2014**, *22*, 23–38. [CrossRef]
91. Rienks, J.; Barbaresko, J.; Oluwagbemigun, K.; Schmid, M.; Nöthlings, U. Polyphenol Exposure and Risk of Type 2 Diabetes: Dose-Response Meta-Analyses and Systematic Review of Prospective Cohort Studies. *Am. J. Clin. Nutr.* **2018**, *108*, 49–61. [CrossRef]
92. INC International Nut & Dried Fruit Council. *Nuts & Dried Fruits Statistical Yearbook 2019/2020*; Reus, Spain, 2021. Available online: https://inc.nutfruit.org/technical-projects/ (accessed on 20 November 2022).
93. Wharton, S.; Pedersen, S.D.; Lau, D.C.W.; Sharma, A.M. Weight Management in Diabetes. *Can. J. Diabetes* **2018**, *42*, S124–S129. [CrossRef]
94. Anderson, J.W.; Kendall, C.W.C.; Jenkins, D.J.A. Importance of Weight Management in Type 2 Diabetes: Review with Meta-Analysis of Clinical Studies. *J. Am. Coll. Nutr.* **2003**, *22*, 331–339. [CrossRef]
95. Nishi, S.K.; Viguiliouk, E.; Blanco Mejia, S.; Kendall, C.W.C.; Bazinet, R.P.; Hanley, A.J.; Comelli, E.M.; Salas Salvadó, J.; Jenkins, D.J.A.; Sievenpiper, J.L. Are Fatty Nuts a Weighty Concern? A Systematic Review and Meta-Analysis and Dose–Response Meta-Regression of Prospective Cohorts and Randomized Controlled Trials. *Obes. Rev.* **2021**, *22*, e13330. [CrossRef]
96. Muralidharan, J.; Galiè, S.; Hernández-Alonso, P.; Bulló, M.; Salas-Salvadó, J. Plant-Based Fat, Dietary Patterns Rich in Vegetable Fat and Gut Microbiota Modulation. *Front. Nutr.* **2019**, *6*, 157. [CrossRef] [PubMed]
97. Fitzgerald, E.; Lambert, K.; Stanford, J.; Neale, E.P. The Effect of Nut Consumption (Tree Nuts and Peanuts) on the Gut Microbiota of Humans: A Systematic Review. *Br. J. Nutr.* **2021**, *125*, 508–520. [CrossRef]
98. Portincasa, P.; Bonfrate, L.; Vacca, M.; de Angelis, M.; Farella, I.; Lanza, E.; Khalil, M.; Wang, D.Q.H.; Sperandio, M.; di Ciaula, A. Gut Microbiota and Short Chain Fatty Acids: Implications in Glucose Homeostasis. *Int. J. Mol. Sci.* **2022**, *23*, 1105. [CrossRef]
99. Khan, S.; Jena, G. The Role of Butyrate, a Histone Deacetylase Inhibitor in Diabetes Mellitus: Experimental Evidence for Therapeutic Intervention. *Epigenomics* **2015**, *7*, 669–680. [CrossRef]
100. Cao, H.; Ou, J.; Chen, L.; Zhang, Y.; Szkudelski, T.; Delmas, D.; Daglia, M.; Xiao, J. Dietary Polyphenols and Type 2 Diabetes: Human Study and Clinical Trial. *Crit. Rev. Food Sci. Nutr.* **2019**, *59*, 3371–3379. [CrossRef]
101. Kim, Y.A.; Keogh, J.B.; Clifton, P.M. Polyphenols and Glycemic Control. *Nutrients* **2016**, *8*, 17. [CrossRef]
102. García-Mantrana, I.; Calatayud, M.; Romo-Vaquero, M.; Espín, J.C.; Selma, M.V.; Collado, M.C. Urolithin Metabotypes Can Determine the Modulation of Gut Microbiota in Healthy Individuals by Tracking Walnuts Consumption over Three Days. *Nutrients* **2019**, *11*, 2483. [CrossRef]
103. Rivière, A.; Selak, M.; Lantin, D.; Leroy, F.; de Vuyst, L. Bifidobacteria and Butyrate-Producing Colon Bacteria: Importance and Strategies for Their Stimulation in the Human Gut. *Front. Microbiol.* **2016**, *7*, 979. [CrossRef]
104. Bernat-Karpińska, M.; Piątkiewicz, P.; Czech, A.; Wierzbicki, P. The Expression of Particular Glucose Transporters and Insulin Resistance Indicators in the Risk Groups of Type 2 Diabetes–a Two-Year Follow-Up. *Endokrynol. Pol.* **2012**, *63*, 212–219.
105. di Renzo, L.; Cioccoloni, G.; Bernardini, S.; Abenavoli, L.; Aiello, V.; Marchetti, M.; Cammarano, A.; Alipourfard, I.; Ceravolo, I.; Gratteri, S. A Hazelnut-Enriched Diet Modulates Oxidative Stress and Inflammation Gene Expression without Weight Gain. *Oxid. Med. Cell. Longev.* **2019**, *2019*, 4683723. [CrossRef]
106. Ortega, F.J.; Cardona-Alvarado, M.I.; Mercader, J.M.; Moreno-Navarrete, J.M.; Moreno, M.; Sabater, M.; Fuentes-Batllevell, N.; Ramírez-Chávez, E.; Ricart, W.; Molina-Torres, J.; et al. Circulating Profiling Reveals the Effect of a Polyunsaturated Fatty Acid-Enriched Diet on Common MicroRNAs. *J. Nutr. Biochem.* **2015**, *26*, 1095–1101. [CrossRef] [PubMed]
107. Guasch-Ferre, M.; Bhupathiraju, S.N.; Hu, F.B. Use of Metabolomics in Improving Assessment of Dietary Intake. *Clin. Chem.* **2018**, *64*, 82–98. [CrossRef]

108. Garcia-Aloy, M.; Hulshof, P.J.M.; Estruel-Amades, S.; Osté, M.C.J.; Lankinen, M.; Geleijnse, J.M.; de Goede, J.; Ulaszewska, M.; Mattivi, F.; Bakker, S.J.L.; et al. Biomarkers of Food Intake for Nuts and Vegetable Oils: An Extensive Literature Search. *Genes Nutr.* **2019**, *14*, 7. [CrossRef]
109. Guasch-Ferré, M.; Hernández-Alonso, P.; Drouin-Chartier, J.P.; Ruiz-Canela, M.; Razquin, C.; Toledo, E.; Li, J.; Dennis, C.; Wittenbecher, C.; Corella, D.; et al. Walnut Consumption, Plasma Metabolomics, and Risk of Type 2 Diabetes and Cardiovascular Disease. *J. Nutr.* **2021**, *151*, 303–311. [CrossRef]
110. Willett, W. *Nutritional. Epidemiology*, 3rd ed.; Oxford University Press: New York, NY, USA, 2012; ISBN 9780199754038.
111. Bergman, R.N. Origins and History of the Minimal Model of Glucose Regulation. *Front. Endocrinol.* **2021**, *11*, 583016. [CrossRef]
112. Neale, E.P.; Tran, G.; Brown, R.C. Barriers and Facilitators to Nut Consumption: A Narrative Review. *Int. J. Environ. Res. Public Health* **2020**, *17*, 9127. [CrossRef]

nutrients

MDPI

Review

Effects of Nut Consumption on Blood Lipids and Lipoproteins: A Comprehensive Literature Update

Marta Guasch-Ferré [1,2,*], Anne-Julie Tessier [2], Kristina S. Petersen [3], Philip A. Sapp [4], Linda C. Tapsell [5,6], Jordi Salas-Salvadó [7,8,9], Emilio Ros [9,10] and Penny M. Kris-Etherton [4,*]

1 Department of Public Health, Novo Nordisk Foundation Center for Basic Metabolic Research, Faculty of Health and Medical Sciences, University of Copenhagen, 1014 Copenhagen, Denmark
2 Department of Nutrition, Harvard T.H. Chan School of Public Health, Boston, MA 02115, USA
3 Department of Nutritional Sciences, Texas Tech University, Lubbock, TX 79409, USA
4 Department of Nutritional Sciences, Penn State University, State College, PA 16802, USA
5 Faculty of Science, Medicine and Health, University of Wollongong, Wollongong, NSW 2522, Australia
6 llawarra Health and Medical Research Institute, Wollongong, NSW 2522, Australia
7 Unitat de Nutrició Humana, Departament de Bioquímica i Biotecnologia, Universitat Rovira i Virgili, 43204 Reus, Spain
8 Institut d'Investigació Sanitària Pere Virgili (IISPV), 43204 Reus, Spain
9 Consorcio Centro de Investigación Biomédica en Red CIBER, Fisiopatología de la Obesidad y Nutrición (CIBERObn), Instituto de Salud Carlos III (ISCIII), 28029 Madrid, Spain
10 Lipid Clinic, Endocrinology and Nutrition Service and Institut d'Investigacions Biomèdiques August Pi Sunyer, Hospital Clínic, 08036 Barcelona, Spain
* Correspondence: marta.guasch@sund.ku.dk (M.G.-F.); pmk3@psu.edu (P.M.K.-E.); Tel.: +45-35-32-71-80 (M.G.-F.); +814-863-2923 (P.M.K.-E.)

Abstract: In the present review, we provide a comprehensive narrative overview of the current knowledge on the effects of total and specific types of nut consumption (excluding nut oil) on blood lipids and lipoproteins. We identified a total of 19 systematic reviews and meta-analyses of randomized controlled trials (RCTs) that were available in PubMed from the inception date to November 2022. A consistent beneficial effect of most nuts, namely total nuts and tree nuts, including walnuts, almonds, cashews, peanuts, and pistachios, has been reported across meta-analyses in decreasing total cholesterol (mean difference, MD, −0.09 to −0.28 mmol/L), LDL-cholesterol (MD, −0.09 to −0.26 mmol/L), and triglycerides (MD, −0.05 to −0.17 mmol/L). However, no effects on HDL-cholesterol have been uncovered. Preliminary evidence indicates that adding nuts into the regular diet reduces blood levels of apolipoprotein B and improves HDL function. There is also evidence that nuts dose-dependently improve lipids and lipoproteins. Sex, age, or nut processing are not effect modifiers, while a lower BMI and higher baseline lipid concentrations enhance blood lipid/lipoprotein responses. While research is still emerging, the evidence thus far indicates that nut-enriched diets are associated with a reduced number of total LDL particles and small, dense LDL particles. In conclusion, evidence from clinical trials has shown that the consumption of total and specific nuts improves blood lipid profiles by multiple mechanisms. Future directions in this field should include more lipoprotein particle, apolipoprotein B, and HDL function studies.

Keywords: nuts; cholesterol; lipids; apolipoproteins

Citation: Guasch-Ferré, M.; Tessier, A.-J.; Petersen, K.S.; Sapp, P.A.; Tapsell, L.C.; Salas-Salvadó, J.; Ros, E.; Kris-Etherton, P.M. Effects of Nut Consumption on Blood Lipids and Lipoproteins: A Comprehensive Literature Update. *Nutrients* **2023**, *15*, 596. https://doi.org/10.3390/nu15030596

Academic Editor: Arrigo Cicero

Received: 22 December 2022
Revised: 16 January 2023
Accepted: 17 January 2023
Published: 23 January 2023

1. Introduction

Cardiovascular diseases (CVD), specifically coronary heart disease (CHD) and stroke, are leading causes of death and disability-adjusted life years worldwide [1]. Dyslipidemia (elevated levels of low-density lipoprotein cholesterol (LDL-cholesterol) or triglycerides (TG) in blood, decreased levels of high-density lipoprotein cholesterol HDL-cholesterol, or other lipoprotein disturbances) is a well-documented risk factor for the development of atherosclerotic CVD [2]. CVD and its related risk factors are largely preventable. Therefore, effective approaches for the prevention of CVD, including changes in lifestyle and diet, are key to reducing the consequences of dyslipidemia and the associated disease burden to improve population health.

Dietary interventions to lower blood cholesterol concentrations and modify blood lipoprotein levels are the cornerstone of prevention and treatment for CHD and other atherosclerotic diseases. Indeed, suboptimal dietary intake was responsible for an estimated one in five premature deaths globally from 1990 to 2016 [3]. In the United States (U.S.), suboptimal diets were associated with more deaths than any other risk factor. In 2016, dietary risk factors were responsible for an estimated 529,300 deaths, of which 84% were due to CVD [4]. Among individual dietary components, the largest estimated mortality was associated with an excessive sodium intake (9.5%) followed by the suboptimal consumption of nuts/seeds, among others [5]. In addition, in 2017 a diet low in nuts and seeds was the fourth leading risk factor for all-cause mortality globally, after a diet low in whole grains, a diet high in sodium, and a diet low in fruits [5].

Nuts and seeds, along with other plant foods such as whole grains, vegetables, fruits, and legumes, are key components of recommended healthy diet patterns worldwide such as the Mediterranean diet. Nuts are a good source of unsaturated fatty acids and are rich in fiber, minerals (potassium, calcium, and magnesium), vitamins (folate and vitamin E), phytosterols, and polyphenols. The fatty acid composition varies widely among different types of nuts [6]. Almonds, hazelnuts, pistachios, cashews, and peanuts are rich in monounsaturated fatty acids (MUFAs), whereas walnuts are rich in polyunsaturated fatty acids (PUFAs) [6,7].

In recent decades, an extensive body of evidence has linked nut consumption to a wide range of health benefits including reduced risk and prevention of cardiometabolic diseases [8], making them a key dietary recommendation for health promotion and disease reduction. In large prospective cohort studies, frequent nut consumption has been inversely associated with the risk of type 2 diabetes, metabolic syndrome, CVD, and total and cause-specific mortality [8]. These findings are consistent with the results of the PREvención con DIEta MEDiterránea (PREDIMED) study, a primary prevention trial that found a 28% reduction in incident cardiovascular events among participants randomly assigned to a Mediterranean diet supplemented with nuts [9]. In addition, short-term trials have demonstrated the beneficial effects of nut consumption on intermediate markers of CVD risk, including LDL-C [10]. Importantly, more than 60 human dietary intervention studies have been conducted investigating the effects of nut consumption on blood lipid levels. These studies differ in the type and quantity of the nuts consumed, placebo/diet control, study design, subject selection criteria, and duration.

In the present narrative review, we provide a comprehensive overview of the current knowledge on the effects of total and specific types of nut consumption (excluding nut oil) on blood lipids and lipoproteins in clinical trials, the potential mechanisms of the lipid effects of nuts, and the future directions for research in this area.

2. Effects of Nuts on Blood Lipids

We conducted a literature search in PubMed (https://pubmed.ncbi.nlm.nih.gov/, accessed on 22 December 2022) for systematic reviews and meta-analyses of randomized controlled trials (RCTs) that examined the effect of nut consumption on blood lipid biomarkers from the inception date through 20 November 2022. The search strategy was as follows: (nuts[MeSH] OR "tree nuts" OR nut OR almonds[MeSH] OR almond OR walnuts[MeSH] OR walnut OR cashews[MeSH] OR cashew OR pistachios[MeSH] OR pistachio OR peanuts[MeSH] OR peanut OR "peanut butter") AND (meta-analysis OR "systematic review") AND (English[lang]). We restricted the search to adult human trials and to articles published in English. Selected articles were required to focus on whole nuts or nuts-enriched food interventions, and to report on at least one lipid variable (TC, LDL cholesterol, HDL cholesterol, TGs, or apolipoproteins). We excluded articles that included nut oils as an intervention given their different nutrient matrix.

Numerous systematic reviews and meta-analyses of RCTs that evaluated nut consumption on blood lipid biomarkers have been published [11–29]. Table 1 summarizes 19 selected studies. Three focused on interventions of both nuts and peanuts together [11,12,21], three were on tree nuts [16,18,22], five on almonds [13,15,19,25,29], four on walnuts [17,23,24,26], two on cashews [20,28], one on pistachios [27] and one on peanuts including peanut butter [14]. The doses tested varied from 5 to 200 g/d and the study durations were from 4 days to 2 years.

A significant reduction in LDL-C (mean differences, MD: −0.09 to −0.26 mmol/L; 10 of the 18 meta-analyses analyzing LDL-C) was most consistently reported across meta-analyses, followed by a reduction in triglycerides (TG; MD: −0.05 to −0.17 mmol/L; 9/19 meta-analyses) and total cholesterol (TC; MD: −0.09 to −0.28 mmol/L; 8/19). However, none reported an effect on HDL-C. Del Gobbo et al. (2018) further examined the effect of nuts, specifically tree nuts, on apolipoproteins (Apo) and found a significant reduction in Apolipoprotein B (ApoB) (−0.042 g/L (95% CI: −0.057, −0.026); 13 RCTs) [22]. In one meta-analysis that reported significance, the observed effect of nuts on LDL-C was comparable to up to $\frac{1}{4}$ of the effect of statin medication in populations including primary prevention, hemodialysis, CHD, diabetes, heart failure, and in those at low vascular risk [30,31].

Although the evidence supports a modest effect of nuts in lowering blood lipids/lipoproteins, it is unclear whether some types of nuts are more effective than others. Among tree nuts, walnuts are especially rich in linoleic acid (18:2n–6) and α-linolenic acid (18:3n–3) (ALA) [6]. The meta-analyses demonstrated the beneficial effects of walnuts on reducing TC, LDL-C, and TG [17,18,23,24]. This finding corroborates previous studies showing slightly reduced fasting serum TG [MD: −0.03 mmol/L (−0.11, −0.05)] with increasing ALA intake [32]. Pistachios are particularly rich in phytosterols and dietary fiber, and are high in MUFAs [33,34]. The only meta-analysis (of 12 RCTs) on pistachios found a significant effect of 32–126 g/d during 3–24 weeks in reducing TC (−0.19 mmol/L [95% CI: −0.33, −0.06]), LDL-C (−0.1 mmol/L [95% CI: −0.14, −0.06]), and TG (−0.13 mmol/L [95% CI: −0.16, −0.09]) [27]. Like pistachios, cashews have a high proportion of MUFAs but are lower in tocopherols, phytosterols, and dietary fiber. Few studies have examined the effect of cashew consumption on blood lipids. Two meta-analyses did not find any effect of 28–108 g cashews/d on lipid biomarkers in adult populations (3 RCTs; n = 384 to 392, duration: 4–12 weeks) [20,28]. These results were also confirmed in another meta-analysis [18]. The absence of an effect may be attributed to its differing food matrix or to limited available studies. Almonds are especially rich in alpha-tocopherol [34] and dietary fiber compared with other nuts. A consistent beneficial effect of almonds (10–168 g/d; 5–27 RCTs, n = 120–2,049 healthy or at risk of CVD individuals; 3–77 weeks duration) was reported in LDL-C (−0.15 to −0.18 mmol/L) [13,19,25], TC (−0.13 to −0.28 mmol/L) [19,25], and TG (−0.08 mmol/L) [25]. However, the evidence is less consistent in populations with type 2 diabetes [15,29]. Peanuts, although classified as a legume, have a comparable food matrix and fatty acid composition to those of tree nuts. The effect of 25–200 g/d peanuts or peanut butter consumption during 2–24 weeks on blood lipids was examined in a recent meta-analysis and demonstrated a significant reduction in TG [−0.13 mmol/L (95% CI: −0.2, −0.07)]; 9 RCTs, and 643 participants) [14].

To our knowledge, one RCT was published following the last meta-analysis on nuts and blood lipids that we summarize herein. The Brazilian Nut Study tested the effect of an energy-restricted diet with 45 g nuts (15 g Brazil nuts + 30 g cashews) and without nuts on various biomarkers including blood lipids in 40 women at risk of cardiometabolic disease [35]. After the 8-week intervention, the authors reported a decrease in TC and LDL-C in both groups, but no difference between groups; this is possibly due to the significant weight loss achieved in both the intervention (−3.5 +/−0.6 kg; $p < 0.001$) and the control (−1.8 ± 0.6; $p < 0.05$) groups at the end of the trial.

Table 1. Summary of meta-analyses on the effect of nuts and peanuts on lipid and lipoprotein biomarkers.

Publication	Search Dates	Population	Study Design	Sample Size	Duration of Intervention	Intervention	Control	Outcome Measures	Results Mean Change in mmol/L (95%CI)
Phung, 2009 [13]	through Jul 2008	Non-specified	RCTs with parallel or crossover design	5 RCTs 142 participants	4 weeks	Almonds 25–168 g/d	NCEP step II, usual diet, NCEP step I, high-fat diet, low-fat diet	Lipid profile: TC, LDL-C, HDL-C, TG, LDL/HDL; ApoA-I and apoB Lp(a)	↓ LDL-C −0.18 (−0.34, −0.02) (5 RCTs)
Banel, 2009 [24]	through May 2008	All patient populations and age groups	RCTs with parallel or crossover design	13 RCTs 365 participants	4–24 weeks	Walnuts 15–108 g/d	Controlled diet, Western diet, Med diet, modified low-fat diet, habitual diet, low-fat diet, cholesterol lowering meals	Lipid profile: TC, LDL-C, HDL-C, TG	↓ TC −0.27 (−0.38, −0.15) ↓ LDL-C −0.24 (−0.34, −0.14) (11 RCTs)
Sabaté, 2010 [12]	1992–2004	No recent exposure to lipid-lowering medications	Controlled trials; duration of intervention ≥ 3 weeks; no body weight change between diets at the end of intervention	25 trials 583 participants (pooled analysis with individual participant data)	3–8 weeks	Tree nuts and peanuts 34–100 g/d	Western diet, Med diet, low total, and saturated fat	Lipid profile: TC, LDL-C, HDL-C, TG	↓ TC −0.28 (−0.36, −0.2) ↓ LDL-C −0.26 (−0.34, −0.19) (25 trials)
Mejia, 2014 [16]	through Apr 2014	Non-specified	RCTs; duration of intervention ≥ 3 weeks	47 RCTs 2211 participants	3 weeks–18 months	Tree nuts (almonds, Brazil nuts, cashews, hazelnuts, macadamia nuts, pecans, pine nuts, pistachios, walnuts, and mixed nuts) 30–85.5 g/d	Habitual diet, diet for diabetes, Western diet, low-fat diet, muffin, NCEP step I diet, AHA step 1 diet, ad libitum diet, NCEP step II diet, NCEP step II diet + muffin, cheese	At least one criterion of MetS (waist circumference, TG, HDL-C, blood pressure, glycemic control)	↓ TG −0.06 (−0.09, −0.03) (43 RCTs)
Del Gobbo, 2015 [22]	through Mar 2013	Free of known CVD; Not receiving medication for diabetes, obesity, MetS, hypertension or hyperlipidemia; ≥18 yo	Randomized and nonrandomized controlled trials with parallel or crossover design	42 RCTs and 18 nonrandomized trials 2582 participants	3–26 weeks	Tree nuts 5–100 g/d	Habitual diet, healthy diet, low-fat diet, high-CHO diet, olive oil diet, habitual diet + red meat, low saturated fat diet with cereals and canola oil, AHA step 1 diet, American diet, isocaloric controlled diet, NCEP step 1 or 2 diet, salted pretzels, isocaloric high cholesterol diet, NCEP step 1, Med diet, AHA step 1 diet, ADA diet (with and without nuts)	Lipid profile: TC, LDL-C, HDL-C, TG; apolipoproteins	↓ TC −0.09 (−0.11, −0.07) ↓ LDL-C −0.11 (−0.13, −0.09) (38 RCTs) ↓ ApoB (g/L) −0.042 (−0.065, −0.026) (13 RCTs)

Table 1. *Cont.*

Publication	Search Dates	Population	Study Design	Sample Size	Duration of Intervention	Intervention	Control	Outcome Measures	Results Mean Change in mmol/L (95%CI)
Guasch-Ferré, 2018 [23]	through Jan 2018	Adults	RCTs with a parallel or crossover design; Duration of intervention ≥ 3 weeks	26 RCTs 1059 participants	4 weeks–1 year	Walnuts 15–108 g/d	ad libitum control diet, Med diet, ADA diet, low-fat diet, habitual diet, controlled diet (walnut-free)	At least one of the lipid markers: TC, LDL-C, HDL-C, TG apolipoproteins	↓ TC −0.18 (−0.24, −0.12) ↓ LDL-C −0.14 (−0.2, −0.09) ↓ TG −0.05 (−0.1, −0.01) (23 RCTs)
Lee-Bravatti, 2019 [19]	2015-June 2017 for lipid outcomes	Healthy or with CVD risk factors; ≥18 yo	RCTs; Duration of intervention ≥ 3 weeks	15 RCTs 534 participants	4–16 weeks	Almonds 37–113 g/d	NCEP step II diet, low-fat diet, high-fat diet custom diet, Med diet, NCEP step I diet, ADA diet	Lipid profile: TC, LDL-C, HDL-C, TG, TC/HDL, HDL/LDL; apolipoproteins, Lp(a)	↓ TC −0.28 (−0.43, −0.12) ↓ LDL-C −0.15 (−0.26, −0.05) (13 RCTs)
Morvaridzadeh, 2020 [28]	through June 2019	Non-specified	RCTs with a parallel or crossover design	3 RCTs 384 participants	4–12 weeks	Cashews 28–108 g	Isocaloric diet, baked potato chips	Lipid profile: TC, LDL-C, HDL-C, TG	No change
Liu, 2020 [18]	through June 2019	≥18 yo	RCTs; duration of intervention ≥ 3 weeks	34 RCTs 1677 participants	3–24 weeks	Tree nuts (walnuts, pistachios, hazelnuts, cashews, or almonds) 15–168 g/d	Control diet (nut-free)	Lipid profile: TC, LDL-C, HDL-C, TG	Walnut-enriched ↓ LDL-C −0.09 (−0.12, −0.07) ↓ TG −0.09 (−0.11, −0.07) (16 RCTs) Pistachio-enriched ↓ LDL-C −0.17 (−0.28, −0.06) Hazelnut-enriched No change Almond-enriched No change
Jalali, 2020 [20]	through Nov 2019	≥18 yo	RCTs	3 RCTs 392 participants	4–12 weeks	Cashews 30–42 g/d	Diet for diabetes, isocaloric controlled diet (nut-free)	Lipid profile: TC, LDL-C, HDL-C, TG	No change
Hadi, 2021 [27]	through June 2019	≥18 yo	RCTs; duration of intervention ≥ 3 weeks	12 RCTs 771 participants	3–24 weeks	Pistachios 32–126 g/d	Control diet	Lipid profile: TC, LDL-C, HDL-C, TG	↓ TC −0.19 (−0.33, −0.06) (10 RCTs) ↓ LDL-C −0.1 (−0.14, −0.06) (12 RCTs) ↓ TG −0.13 (−0.16, −0.09) (10 RCTs)

Table 1. *Cont.*

Publication	Search Dates	Population	Study Design	Sample Size	Duration of Intervention	Intervention	Control	Outcome Measures	Results Mean Change in mmol/L (95%CI)
Asbaghi, 2021 [25]	through Sept 2020	Healthy or otherwise; ≥18 yo	RCTs with a parallel or crossover design; Duration of intervention ≥ 3 weeks	27 RCTs 2049 participants	3–77 weeks	Almonds 10–168 g/d	No almond consumption or dietary substitutions containing no almond were used	At least one of the lipid markers: TC, LDL-C, HDL-C, TG	↓TC −0.13 (−0.2, −0.05) ↓TG −0.08 (−0.13, −0.02) (27 RCTs) ↓LDL-C −0.15 (−0.23, −0.07) (26 RCTs)
Wang, 2021 [29]	through Jan 2020	Adults with T2DB	RCTs with a parallel or crossover design; Duration of intervention ≥ 2 weeks	5 RCTs 120 participants	3–12 weeks	Almonds 30–60 g/d	Control diet, NCEP step II diet, peanuts, sunflower kernels	Lipid profile: TC, LDL-C, HDL-C, TG	No change
Xia, 2021 [11]	through June 2021	Patients with T2DB	RCTs	16 RCTs 1041 participants	6–52 weeks	Peanuts and tree nuts (walnuts, pistachios, macadamia nuts, pecans, cashews, almonds, hazelnuts, pine nuts, and Brazil nuts) 6–128 g/d	High-fat diet, low-fat diet, normal-fat diet, habitual diet, diet for diabetes, ADA meal plan (nut-free)	Lipid profile: TC, LDL-C, HDL-C, TG	↓TC −0.14 (−0.26, −0.02) (14 RCTs) ↓TG −0.1 (−0.17, −0.02) (12 RCTs)
Moosavian, 2022 [15]	through Mar 2021	Patients with T2DB; ≥18 yo	RCTs; Duration of intervention ≥ 3 weeks	9 RCTs 264 participants	4–12 weeks	Almonds 29–113g/d	NCEP step II diet, cheese, raw peanut with low carbohydrate diet, high-fat diet, low-fat diet, sunflower kernels with diabetic diet, custom diet (almond-free)	Lipid profile: TC, LDL-C, HDL-C, TG	↓LDL −0.14 (−0.26, −0.02) (8 RCTs)
Arabi, 2022 [26]	through Dec 2021	Diagnosed with MetS; ≥18 yo	RCTs with a parallel or crossover design	8 RCTs 506 participants	4–112 days	Walnuts (all forms, plain, or walnut-fortified food) 30 g–108 g/d	Standardized shakes, control diet, isocaloric white bread, ad libitum diet without walnuts, lifestyle counseling	Lipid profile: TC, LDL-C, HDL-C, TG	↓TG −0.17 (−0.32, −0.03) (5 RCTs)
Mates, 2022 [17]	through Nov 2021	Middle-aged and older adults ≥40 yo or mean age ≥50 yo	RCTs with a parallel or crossover design; Duration of intervention ≥ 3 weeks	17 RCTs 2466 participants	4 weeks-2 years	Walnuts (including plain or walnut-fortified food) 19.3–75 g/d	Med diet, modified low-fat diet, Western-type diet, habitual diet, CKD patients' diet (walnut-free)	Lipid profile: TC, LDL-C, HDL-C, TG	↓TC −0.13 (−0.2, −0.07) ↓LDL-C −0.15 (−0.2, −0.11) (12 RCTs) ↓TG −0.08 (−0.12, −0.04) (13 RCTs)

Table 1. *Cont.*

Publication	Search Dates	Population	Study Design	Sample Size	Duration of Intervention	Intervention	Control	Outcome Measures	Results Mean Change in mmol/L (95%CI)
Eslami, 2022 [21]	through Apr 2021	Overweight/obese (BMI: 25–40 kg/m^2); free of chronic diseases; ≥18 yo	RCTs with a parallel or crossover design; Duration of intervention ≥ one week	10 RCTS 711 participants	4–72 weeks	Peanuts and tree nuts (almonds, walnuts, hazelnuts, pistachios, cashews, macadamia nuts, Brazil nuts, pine nuts, pecans, mixed nuts) 20–60 g/d	Isocaloric nut-free diet	At least one of the following: Serum lipid profile: TC, LDL-C, HDL-C, TG	↓ TG −0.15 (−0.29, −0.01) (9 RCTs)
Parilli-Moser, 2022 [14]	through July 2021	Healthy or with MetS or at high risk of MetS	RCTs	9 RCTs 643 participants	2–24 weeks	Peanuts, peanut butter or high oleic acid peanuts 25–200 g/d	Hypocaloric diet, habitual diet, ADA meal plan, substitute snack (grain bar, white rice bar, candy, or almonds) (peanut-free)	Lipid profile: TC, LDL-C, HDL-C TG	↓ TG −0.13 (−0.2, −0.07) (9 RCTs)

RCT, randomized controlled trial; NCEP, National Cholesterol Education Program; TC, total cholesterol; LDL-C, low-density lipoprotein cholesterol; HDL-C, high-density lipoprotein cholesterol; TG, triglycerides; CHO, carbohydrates; ANDEAP, Academy of Nutrition and Dietetics Evidence Analysis Process; AHA, American Heart Association; ADA, American Diabetes Association; Med Diet, Mediterranean diet; BW, body weight; WC, waist circumference; BMI, body mass index; Lp (a), lipoprotein (a); MetS, metabolic syndrome; T2DB, type 2 diabetes. Units for blood lipids are presented as mmol/L, except for ApoB which is presented as g/L. Only lipid outcomes are reported.

2.1. Dose-Response Effects of Nuts on Blood Lipids

In eight of the 19 identified meta-analyses, dose-response analyses were conducted [12,16,22–25]. Some evidence suggests nuts dose-dependently improve TC, LDL-C, and TG. In a meta-analysis of 61 clinical trials, Del Gobbo et al. reported that tree nut intake lowered TC and LDL-C in a nonlinear manner such that stronger effects were observed in trials where >60 g/d of tree nuts were provided. For ApoB and TGs, linear dose-response effects were observed [22]. Similarly, Blanco Mejia et al. observed a borderline non-significant linear relationship between increasing tree nut doses and TG reductions; no dose-response relationship was observed for HDL-C [16]. In a pooled analysis of 25 RCTs, Sabaté et al. observed dose-dependent reductions in TC, LDL-C, and TG with higher consumption of tree nuts and peanuts [12].

In five meta-analyses, dose-response analyses were conducted to examine the relationship between the intake of a single nut type and lipid/lipoprotein responses [14,23–26]. In a meta-analysis of 24 clinical trials, higher walnut intake was dose-dependently associated with reductions in TC (−0.01 mmol/L per 1 g/d increase) [23]. A trend towards a dose-response relationship was observed for LDL-C (−0.01 mmol/L per 1 g/d increase), while no dose-response relationship was observed for TG or HDL-C. A recent meta-analysis of five studies conducted in cohorts with metabolic syndrome showed a non-linear association between walnut consumption and HDL-C whereby an intake of up to 50 g/d was associated with an increase in HDL-C [26]. Additionally, a trend toward a linear dose-response relationship was observed between the consumption of walnuts and TG reduction; no dose-response relationship was observed for TC or LDL-C. Less dose-response specific evidence is available for other nuts including almonds and peanuts. In a meta-analysis examining the effect of almond intake on lipids and lipoproteins, an inverse linear relationship was observed between almond dose and TG. TC, LDL-C, and HDL-C were not linearly related to almond intake dose [25]. A meta-analysis of nine RCTs showed no dose-response relationship between peanut intake and TC, LDL-C, HDL-C, or TG [14].

In 11 of the 19 included meta-analyses, subgroup analyses were conducted to examine the effect of higher vs. lower consumption of nuts on lipid/lipoprotein levels [11,13,15–17,19,21,23,25,26,29]. The highest consumption category was >63 g/d in one meta-analysis [26], ≥50 g/d in five meta-analyses [13,15,16,21,29], ≥45g/d in two meta-analyses [11,25], >42.5 g/d in one meta-analysis [19], ≥42 g/d in one meta-analysis [17], and ≥28 g/d in one meta-analysis [23]. In a meta-analysis of studies on healthy individuals, the consumption of ≥45 g/d of almonds lowered TC, LDL-C, and TG to a greater extent than <45 g/d [25]. Similarly, in a meta-analysis of five studies conducted in patients with type 2 diabetes, greater reductions in TC, LDL-C, HDL-C, and TG were observed with the consumption of ≥50 g/d of almonds [29]. In two meta-analyses including healthy participants and individuals at high risk for CVD, no effect modification by almond dose (≥50 g/d and >42 g/d) was observed [13,19].

Three meta-analyses examined the effect of walnut dose levels on changes in lipids and lipoproteins [17,23,26]. Guasch-Ferré et al. reported similar TC and LDL-C lowering with a walnut consumption of ≥28 g/d compared with <28 g/d [23]. However, greater reductions in TC and LDL-C were observed when walnut intake comprised 10–25% of total energy compared with 5–10% of total energy. No dose-related effect modification was observed for TG or HDL-C. In a meta-analysis of studies including middle-age and older adults, TC and TG reductions were only observed when walnut consumption was ≥42 g/d; LDL-C was lowered to a similar magnitude at both doses [17]. In a meta-analysis of studies including participants with metabolic syndrome, walnut dose (>63 g/d vs. ≤63 g/d) did not affect LDL-C differently [26].

Inconsistent findings were reported in three meta-analyses examining lipid/lipoprotein effect modification by doses of tree nuts and peanuts or tree nuts only [11,16,21]. A meta-analysis including studies involving patients with type 2 diabetes showed that higher tree nut and peanut consumption (≥45 g/d) lowered TC and LDL-C significantly, whereas lower consumption was not associated with TC and LDL-C lowering [11]. In contrast, in a meta-analysis of 11 studies including participants with overweight/obesity, no difference in lipid responses by peanut and tree nut dose (≥50 g/d vs. <50g/d) was observed [21]. In a meta-analysis of studies conducted in participants that were healthy or had dyslipidemia, metabolic syndrome, or type 2 diabetes, TGs were reduced with a higher intake of tree nuts (≥50g/d); HDL-C findings were not different by the dose consumed and LDL-C and TC were not assessed [16].

Collectively, there is evidence supporting that nuts dose-dependently improve lipids and lipoproteins. However, many of the meta-analyses reviewed included a relatively small number of studies, which limits the statistical power to examine dose-response relationships. In addition, across the meta-analyses reviewed, higher vs. lower consumption was inconsistently defined, and limited rationale was provided in most cases for the cut points used.

2.2. Subgroup Analyses: Effects of Nuts on Blood Lipids

Across the meta-analyses reviewed, several subgroup analyses were conducted to assess the potential for sex, age, BMI, baseline lipid/lipoprotein concentrations, and health status to influence the effect of nuts on lipids and lipoproteins. Broadly, sex and age do not appear to be effect modifiers; BMI and baseline lipid/lipoprotein concentrations may influence the lipids/lipoprotein lowering effects of nuts.

2.2.1. Sex

In two of the included meta-analyses, subgroup analyses evaluating effect modification by sex were reported [12,22]. In both meta-analyses, no differences in the effect of nuts on lipids/lipoproteins were observed by sex [12] or by the proportion of the study sample that was men (≥50% or <50%) [22].

2.2.2. Age

Seven of the included meta-analyses conducted subgroup analyses to assess the effect of nuts on lipids/lipoproteins in different age categories [11,12,17,22,23,25,26]. Limited evidence suggests that TG lowering in response to tree nut intake may be greater in those aged <50 years; however, most of the evidence evaluated suggests age is not a strong effect modifier. In four meta-analyses, including studies examining the effect of tree nuts and peanuts [12], tree nuts only [22], and walnuts [23,26] on lipids and lipoproteins, effects did not differ across age categories. In two meta-analyses, TG reductions were only observed in response to almond [25] and walnut [17] consumption in the <50 y age category; no intervention effect was observed in ≥50 y age category. In both meta-analyses, no differences in TC or LDL-C were observed by age category. Xia et al., however, observed only TC lowering in response to tree nut and peanut consumption in the ≥55 y age category; there was no difference in LDL-C by age category [11].

2.2.3. BMI

Across the five meta-analyses that conducted subgroup analyses to evaluate effect modification by BMI, the evidence suggests nuts may induce greater lipid/lipoprotein improvements when BMI is <30 kg/m^2 [12,15,21,23,25]. Sabaté et al. observed greater improvements in the LDL-C/HDL-C ratio and TC/HDL-C ratio when BMI was <25 kg/m^2 and 25–30 kg/m^2 compared with BMI >30 kg/m^2 [12]. Similar trends were observed for LDL-C, TC, and TG. Eslami et al. reported TG lowering with the consumption of tree nuts and peanuts only when BMI was <30 kg/m^2; in this meta-analysis effect modification was not observed for TC, LDL-C, and HDL-C [21]. Guasch-Ferré et al. did not observe any difference in the effect of walnuts on lipids/lipoproteins in studies where BMI was <25 kg/m^2 vs. ≥25 kg/m^2 [23]. Similarly, Moosvian et al. reported that the effect of almonds on lipids and lipoproteins in studies including patients with type 2 diabetes did not differ by BMI category (<30 vs. ≥30 kg/m^2) [15]. Conversely, in a meta-analysis including studies conducted in generally healthy populations, reductions in TC and LDL-C were only observed with almond consumption in individuals with overweight; no differences in TG or HDL-C were observed across BMI categories [25].

2.2.4. Baseline Lipids/Lipoprotein Concentrations

In nine of the included meta-analyses subgroup analyses were conducted to examine effect modification by baseline lipid/lipoprotein concentrations or hyperlipidemia/dyslipidemia status [12,13,15,16,22–25,27]. Evidence suggests greater improvements in lipids and lipoproteins in response to nut consumption when the baseline TC and/or LDL-C is higher [12,25,27] or in participants with hyperlipidemia or dyslipidemia [13]. However, in meta-analyses that examined effect modification by baseline TG concentrations, reductions in TGs were only observed when the baseline TG concentrations were lower (<1.69 mmol/L) [16,25,27]. In four meta-analyses, no effect modification was observed by the baseline lipid/lipoprotein level or the hyperlipidemia/dyslipidemia status [15,22–24]. Across the evaluated meta-analyses, inconsistent cut points were used to define higher vs. lower baseline lipid/lipoprotein concentrations and dyslipidemia/hyperlipidemia, which likely contributes to the variability observed.

2.2.5. Health Status

Across the six meta-analyses that conducted subgroup analyses to assess effect modification by health status (healthy vs. metabolic impairment), inconsistent findings were reported, with no clear pattern of effect modification by health status [14,17,19,22,25,27]. In the largest meta-analysis including 61 trials, no heterogeneity in the effect of tree nuts on TC, LDL-C, HDL-C, or TG was observed by disease status (healthy, type 2 diabetes, metabolic syndrome, high cholesterol, obesity) [22]. In this analysis, ApoB reductions were greater in those with type 2 diabetes (−0.115 g/L, 95% CI −0.162, −0.068) compared with healthy populations (−0.025 g/L; 95% CI −0.047, −0.003). Given the variation in the methodology used in these meta-analyses and the aggregate nature, subgroup analyses have limited power to identify true differences between subgroups. To further explore effect modification by health status, individual participant data meta-analyses are needed.

2.3. Effects of Nut Processing on Blood Lipid Profile

Few investigations of the effect of nut processing on lipids/lipoproteins have been conducted. The available evidence from RCTs suggests almond [36,37], hazelnut [38], and peanut [14,39,40] processing (i.e., roasting, or production of oil or butter) does not alter lipid and lipoprotein responses. In an RCT including adults with normolipidemia, the consumption of ~14% of energy from almond oil or whole almonds for 6 weeks improved TC, LDL-C, HDL-C, and TG with no difference in almond processing [36]. Similarly, in a study of individuals with hypercholesterolemia, the consumption of 100 g/d of roasted salted almonds, roasted almond butter, or raw almonds improved TC and LDL-C after 4 weeks and the magnitude of the effect was not impacted by almond processing [37]. Comparable findings were observed in an RCT whereby the intake of 30 g/d of either raw or dry roasted, lightly salted hazelnuts for 4 weeks did not differentially affect TC and LDL-C [38].

Findings from three RCTs suggest peanut processing does not influence lipid/lipoprotein changes [14,39,40]. In a five-arm randomized trial, the intake of 56 g of whole raw unsalted peanuts, whole roasted unsalted peanuts, whole roasted salted peanuts, whole honey roasted peanuts, or peanut butter for 4 weeks did not affect TC, LDL-C, HDL-C, or TG differently [40]. This is consistent with findings from crossover RCTs where the intake of a diet enriched with peanut butter/peanuts similarly improved TC, LDL-C, TG, and ApoB compared with an average American diet [39]. Similar findings were observed in a 6-month RCT examining the intake of 25 g/d of skin-roasted peanuts, two tablespoons (32 g)/day of peanut butter, or two tablespoons (32 g)/day of peanut oil [14]. In this trial, no differences in TC, LDL-C, HDL-C, or TG were observed per peanut form. Thus, from the limited evidence available, nut processing does not appear to alter lipid/lipoprotein responses.

3. Proposed Mechanisms of Action of Cholesterol-Lowering by Nuts

Nuts are a good source of MUFAs and PUFAs, and they also contain dietary fiber, phytosterols, and polyphenols. In isolation, all these nutrients and bioactive compounds may have a modest cholesterol-lowering effect; however, when these molecules combine in the matrix of nuts and synergize to potentiate cardiometabolic pathways, they have the capacity to reduce LDL-cholesterol beyond the effects predicted by equations based solely on fatty acid profiles [41]. Figure 1 summarizes the potential mechanisms for the beneficial effects of nuts consumption on lipid metabolism with the ensuing reduction of the atherogenic lipid/lipoprotein profile.

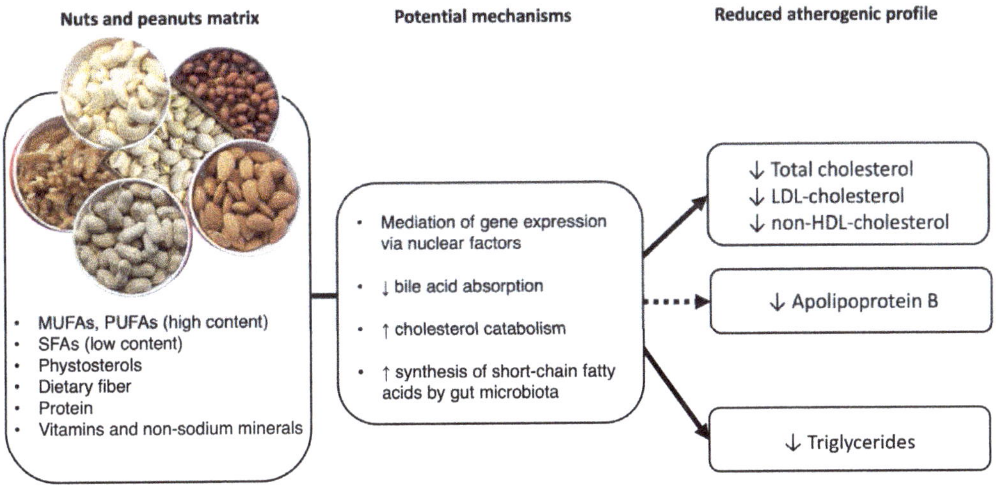

Figure 1. Potential mechanisms by which tree nuts and peanuts reduce atherogenic lipid/lipoprotein profile. A plain arrow indicates a strong level of evidence to support the effect, whereas a dashed arrow indicates a lower level of evidence. LDL, low density lipoprotein; HDL, high density lipoprotein.

Specifically, nuts have favorable effects on serum lipids primarily because of their high content of unsaturated fatty acids (both MUFAs and PUFAs), while they have a low content of saturated fatty acids (SFAs; 4 to 15%) [42]. The unique fatty acid profile of nuts facilitates a favorable shift in the dietary fatty acids when nuts are substituted for foods that are high in SFAs or carbohydrates. Dietary PUFAs have been shown to reduce ApoB while MUFAs increase ApoA1, which mediates the efflux of cholesterol associated with HDL particles [41]. Experimental and clinical studies have shown that the intake of unsaturated fatty acids enhances the hepatic receptor-dependent clearance of LDL and concomitantly reduces plasma LDL-C levels [43]. Unsaturated fats from nuts replacing SFA in lipid bilayers increase membrane fluidity, flexibility, and elasticity, while reducing membrane thickness [44]; these physical changes impact the interaction of membrane-bound receptors with their ligands, such as the affinity of LDL receptors for ApoB-100 in LDL particles, thus enhancing LDL-C uptake. Additionally, PUFA can mediate the expression of several genes involved in lipid metabolism via nuclear factors, including the peroxisomal proliferator-activated nuclear receptors gamma (PPARγ), liver X-receptor (LXR), hepatocyte nuclear factor-(HNF)-4α, nuclear factor kappa B (NFκB), and sterol-regulatory element binding proteins (SREBPs) [45]. Particularly, PUFAs downregulate the expression of SREBPs and enzymes for cholesterol synthesis, thus decreasing the body cholesterol pool [45].

Improvement in blood lipids is attributable mainly to the favorable fatty acid profile of nuts, but other nut components, namely dietary fiber, and plant sterols [42], may also play a significant role. Nuts contain ~7 g/100 g dietary fiber, of which ~25% is soluble fiber [46]. In a meta-analysis of 67 clinical trials to quantify the cholesterol-lowering effect of dietary fiber, 2–10 g/d of soluble fiber was associated with modest but significant reductions in total cholesterol and LDL-C [47]. According to a recent umbrella meta-analysis, total dietary fiber (independently of type) also has cholesterol-lowering properties [48]. Dietary fiber, particularly soluble fiber, exerts its hypocholesterolemic effect through several mechanisms: (1) increased intestinal viscosity, which reduces bile acid absorption and promotes cholesterol catabolism; (2) enhanced synthesis of short-chain fatty acids by gut microbiota, particularly butyrate and propionate, which reduce de novo cholesterol synthesis via 3-hydroxy-3-methylglutaryl-CoA (HMG-CoA) reductase inhibition; and (3) interference with micelle formation in the intestinal lumen and enhanced fecal excretion of fat, cholesterol, and bile acids [48].

Like all plant foods, nuts are cholesterol-free, but their fat fraction contains chemically related plant sterols or phytosterols [42]. The phytosterol content is variable ranging from approximately 72 to 272 mg/100 g with pistachios, almonds, and walnuts containing the most. These compounds play a structural role in their cell membranes just as cholesterol does in animal cell membranes [49]. Phytosterols interfere with cholesterol absorption in the intestinal lumen and thus help lower blood cholesterol. Interestingly, evidence has demonstrated that phytosterols contribute to the cholesterol-lowering effect of nut consumption [50]. In a systematic review and meta-analysis of 61 trials with 2582 participants, with nut intakes from 0.2 to 3.5 servings/d (equivalent to about 5 to 100 g/d) and phytosterol doses ranging from 4.8 to 279 mg/d, the total phytosterol dose from nuts was inversely correlated with a decrease in LDL-C ($r = -0.60$) [50]. Of note was that the predominant LDL-C lowering effect was due to the quantity of nuts consumed; hence representing a greater quantity of both phytosterols and unsaturated fat consumed. A review by Cofan and Ros [51] summarized the LDL-C lowering effects of phytosterols and reported that one meta-analysis [52] concluded that the cholesterol-lowering effects were greater when phytosterols were consumed with fat in the food matrix. Collectively, the evidence suggests that the LDL-C lowering effects of nuts are primarily due to their fatty acid profile but also due to the effects of dietary fiber and plant sterols, as well.

Finally, beyond LDL-C lowering, nuts contain highly bioactive polyphenols, representing one of the richest food sources [42,53]. Although data from clinical studies are few and inconclusive, the lipid effects of polyphenols appear to be limited to reducing LDL oxidation [54], which concurs with the well-known antioxidant and anti-inflammatory properties of these phytochemicals and likely contributes to their atheroprotective role. Oxidized LDL (oxLDL) is important because it is involved in several steps of atherogenesis (endothelial injury, leukocyte recruitment/retention, foam cell formation, etc.) [55,56]. However, the evidence for the benefits of nut consumption on oxLDL and other biomarkers of LDL oxidation is inconsistent, as shown in a recent review of 12 RCTs using a variety of nuts and showing results for changes in blood oxidation markers, usually as secondary outcomes [56].

Another important consideration of the mechanisms to explain the health benefits of nuts pertains to their effects on reverse cholesterol transport, an important mechanism whereby cholesterol is removed and transported from peripheral tissues by HDL to the liver for disposal. A recent review on the role of HDL in atherosclerotic disease notes the shift from measuring HDL-C concentrations to focusing on the more functional measures of HDL (e.g., HDL particle number and cholesterol efflux capacity), which is more predictive of future atherosclerotic CVD events [57]. There is evidence that nuts increase HDL function (i.e., increased cholesterol efflux capacity). This has been demonstrated for walnuts, pistachios, and mixed nuts (walnuts, almonds, and hazelnuts) [58–60]. Nut consumption may shift HDL distribution to improve reverse cholesterol transport. A shift, or increase, in large HDL particles, could indicate increased reverse cholesterol transport due to their affinity to sequester cholesterol that effluxes from macrophages via ATP Binding Cassette Subfamily G Member 1 (ABCG1) [61]. Although the evidence is limited and the underlying mechanisms are not completely understood, nut intake may increase the reverse cholesterol transport capacity of HDL leading to increased removal of cholesterol from peripheral tissues.

Lipoprotein(a) [Lp(a)] is an independent, causal, risk factor for atherosclerotic CVD [62]. Lp(a) concentrations are primarily genetically predetermined with minimal effects from dietary interventions [62]. Several RCTs have assessed the effects of diets enriched with nuts at doses of 1.5 servings (42.5 g)/d or more compared with control diets in various populations (healthy, type 2 diabetes, hyperlipidemia, at risk of CVD). The results have been inconsistent, with modest reductions of Lp(a) in three trials using walnuts [63], pecans [64], or almonds [65], and no discernible effect in four further trials, two with almonds [66,67] and two with walnuts [68,69].

3.1. Emerging Evidence of the Effects of Nuts on Lipoprotein Particle Size

Lipoproteins, assessed by nuclear magnetic resonance, are categorized by particle sizes and densities, or lipoprotein subclasses, and many of these subclasses are associated with CVD outcomes [63]. A review by Qiao et al. on the role of LDL-C and LDL particles in atherogenesis concluded that LDL particles/density (e.g., oxLDL and small dense LDL) may be superior to LDL-C for predicting atherosclerotic CVD risk [64]. In fact, all ApoB-containing lipoproteins are atherogenic, and the small dense LDL particles are even more proatherogenic than larger LDL particles. In the Women's Health Initiative, large very low-density lipoprotein (VLDL) particles increased CVD risk more so than small VLDL particles [65]. Interestingly, lipoprotein particle sizes are being evaluated in diet and nut studies. Observational studies and clinical trials have demonstrated a consistent relationship between improved diet quality and less atherogenic lipoprotein subclass profiles (lower large VLDL, small HDL, and small dense LDL) [66–68]. Moreover, the evidence to date indicates beneficial effects of nut consumption on lipoprotein profiles including particle sizes.

A cross-sectional and longitudinal analysis was conducted with 196 participants in the PREDIMED-Reus center to evaluate the associations of dietary intake (assessed by food frequency questionnaire [FFQ]) and plasma lipoprotein profiles at baseline and 1 year of follow-up [69]. Nut consumption for tertile 3 (highest nut consumption) was 26 g/d (total nuts); 14 g/d (walnuts); and 15 g/d (non-walnut nuts). The authors reported that the increased consumption of total nuts, walnuts, and non-walnut nuts was associated with decreased total and medium LDL particles, very large VLDL, and LDL-C; and decreased VLDL particle size, as well as increased HDL particles and HDL-C.

The Walnuts and Healthy Aging Study (WAHA), a multicenter study conducted in Spain and the U.S. with 628 participants (average age = 69 years) evaluated the effects of walnut consumption (15% of energy and 30 to 60 g/d) for two years on lipid and lipoproteins, including lipoprotein particle sizes [70]. The authors reported that the walnut diet decreased total LDL particles and small LDL particle numbers by 4.3% and 6.1%, respectively. In addition, the walnut diet significantly decreased total cholesterol, LDL-C, and intermediate-density lipoprotein cholesterol by 4.4%, 3.6%, and 16.8%, respectively. The take-home message from the WAHA Study is that the decrease in total LDL particles and small LDL particle number provides mechanistic insight into their cardiovascular benefit beyond changes in the conventional lipid/lipoprotein profile.

Several smaller clinical studies have shown similar benefits of tree nuts on lipoprotein particle size; however, likely because of the smaller sample sizes (and lower statistical power), significant diet effects were not consistently observed [59,71,72]. A study conducted with almonds (42.5 g/d) and dark chocolate (18 g cocoa power; 43 g dark chocolate/d) for four weeks reported that the almond diet significantly decreased LDL_{1+2}, and the dark chocolate plus almond diet significantly decreased LDL_{3+4} compared with the average American diet [71]. In another study conducted by Tindall et al. [72], a diet that provided 18% energy from walnuts (57–99 g/d) tended ($p < 0.1$) to decrease LDL subclasses LDL_{1+2} and LDL_4 compared with a fatty acid-matched diet and a diet where oleic acid was substituted for ALA in the comparator diets. Moreover, in a study conducted with pistachios [59] there was a significant decrease in small and dense LDL particles in response to a diet that provided 20% energy from pistachios (63–126 g/day) versus a diet with 10% of energy from pistachios as well as a lower-fat (25% of energy), low saturated fat (<8% of energy) control diet after 4 weeks. In addition, based on analysis of variance, there was a trend for an increase in α-1 and α-2 HDL (i.e., larger HDL particles) with the inclusion of pistachios. However, in a study conducted by Hernández-Alonso et al. [73] in participants with pre-diabetes, pistachios (57 g/d) increased small HDL particles and decreased medium and large HDL particles. The differences reported between the studies conducted by Holligan et al. [59] and Hernández-Alonso et al. [73] may be explained by differences in the study populations (i.e., healthy vs. pre-diabetes) and the known effects of elevated glucose levels on HDL function and HDL-C levels [74].

While research is still emerging about the effects of nuts on lipoprotein subparticle distribution and concentration, it is becoming clear that nuts favorably affect the conventional lipoprotein profile (i.e., reduced atherogenicity) with a consequent decreased risk of CVD. These findings are expanding our understanding of how tree nuts modulate lipoprotein metabolism and lower CVD risk.

3.2. Effects of Nut Consumption on Adiposity

Nuts are energy-dense foods containing high amounts of fat, a reason why there has been concern that their consumption may lead to weight gain and obesity. However, there is consistent evidence from large prospective studies, scientifically sound RCTs, and meta-analyses thereof that incorporation of substantial amounts of nuts into healthy diets do not lead to weight gain or increase the risk of abdominal obesity, and may even help promote weight loss and reduce waist circumference [75–78]. Several mechanisms explain the lack of the fattening effect of nuts, ranging from the effort required at mastication and chewing to increased satiety and the promotion of fullness due to delayed gastric emptying by the high fat and fiber content. Furthermore, the efficiency of energy absorption from nuts is reduced due to incomplete mastication and fat encasement within the unbroken cell walls in nut particles, thus limiting the bioaccessibility of fat from nuts in the intestine, with an ensuing increase in fecal fat losses [79].

4. Future Directions for Research on Nut Consumption and Blood Lipids

Non-communicable diseases such as CVD have multiple interacting dietary determinants, thus the effects of diet are likely to be dependent on the combination of foods rather than a single food [80]. Nevertheless, dyslipidemia remains a major risk factor contributing to CVD and the evidence supporting the effects of nut consumption on blood lipids and lipoproteins is compelling. Nut consumption improves lipid profiles by multiple mechanisms, and this understanding lays the groundwork for further research.

One of the challenges for this research is to integrate understanding at the level of key nutrients, foods, and dietary patterns. For example, a recent prospective study in the Coronary Artery Risk Development in Young Adults (CARDIA) study showed that individuals with higher walnut consumption also had higher diet quality (measured with Healthy Eating Index 2015), but also lower body mass index, waist circumference, blood pressure, and triglyceride concentration, and gained less weight since baseline than other nut consumers [81]. From a nutrient perspective, nuts make important contributions of unsaturated fatty acids, tocopherols, phytosterols, and dietary fiber. The relative composition varies by nut type, and this may explain inconsistencies in the research results. Separate studies may be required for mechanistic studies at the nutrient level, for example, on the role of PUFAs from nuts on gene expression related to lipid metabolism [44]. Further exploration of the effect of nut consumption on lipid particle number and size may need to focus on differences in fatty acid profiles and the varying doses of phytosterols and fiber provided by different nuts. This may be the case as the research progresses from the study of basic lipid profiles to sub-fractions, HDL function studies, and investigations around changes in ApoB.

Variations in results remain a problem for meta-analyses but this is often due to differences in study design, including dietary methodology. While age and sex do not appear to influence the effects of nut consumption on lipids, weight changes can confound results, so total diets are important. Likewise, studies may show that the processing of nuts does not appear to influence their relationship to blood lipids, but it may influence weight, due to the increased available energy from processing [82], so food form remains a consideration.

Other study design issues relate to the need for greater power in studies (larger sample sizes), further investigation of the linearity of effects (and determination of cut points), and study populations' health status. Given the multifaceted effects on blood lipids and the variations in disease profiles of study participants, more individual participant data meta-analyses may be required.

Further studies evaluating the association between nut consumption and the microbiome are needed [83,84], a new horizon for research with the potential to add to our knowledge of how nuts influence lipid profiles. Preliminary reports indicate little change, but modulatory effects are emerging. On the other hand, plasma metabolomics are providing a useful innovative path for research linking nut consumption with CVD risk [85] and providing insights into the underlying mechanisms. This provides added support for the growing evidence of the effects of nut consumption on lipids and lipid fractions.

5. Conclusions

In conclusion, evidence from clinical trials has shown that the consumption of total nuts and specific types of nuts improves blood lipid profiles by multiple mechanisms, as discussed herein. Specifically, nut-enriched diets are associated with lowering total cholesterol, LDL-C, and TG compared with control diets. Some RCTs have also shown benefits in reducing ApoB levels and improving the lipoprotein subparticle profile. The major determinant of cholesterol-lowering appears to be nut dose rather than nut type.

As summarized in Figure 1, many bioactive compounds of nuts might explain the beneficial effects of nut consumption on blood lipids and lipoproteins. Improvement in blood lipids is attributable mainly to the favorable fatty acid profile of nuts, but other nut components, namely dietary fiber, phytosterols, and bioactive polyphenols play a role.

Although more research is needed to better understand the biological mechanisms of cardiometabolic protection by nuts, increasing their consumption as part of a healthy diet improves cardiovascular risk factors and helps to reduce the risk of CVD in the general population as well as in individuals at high CVD risk. It goes without saying that an integral step for increasing nut consumption is to effectively educate consumers about the health benefits of nuts and, importantly, communicate how to substitute them for unhealthy foods in the diet to achieve the greatest possible CVD benefits.

Author Contributions: M.G.-F., A.-J.T., E.R. and P.M.K.-E. designed the review; A.-J.T. performed the literature research; M.G.-F. revised the literature review; M.G.-F., A.-J.T., K.S.P., P.A.S., L.C.T., J.S.-S., E.R. and P.M.K.-E. wrote the first draft of the manuscript and contributed to the editing of the manuscript. All authors have read and agreed to the published version of the manuscript.

Funding: The NUTS 2022 conference was organized by the Rovira i Virgili University, with all the expenses of the organization covered by Barcelo Congresos, including the fees related to publishing the Proceedings in Nutrients. Barcelo Congresos SA has received funds from the International Nut & Dried Fruit Council (INC). Neither the researchers attending the NUTS 2022 Conference, nor the Rovira i Virgili University have received funds from the INC. Researchers' travel expenses and accommodation have been covered by Barcelo Congresos. Researchers did not receive any additional financial support related to their participation in the NUTS 2022 conference. Sponsors did not have any role in organizing the conference, selecting, or inviting the speakers, nor in the writing or reviewing the proceedings of the conference. J.S-S is partially supported by ICREA under the ICREA Academia program. MG-F is supported by Novo Nordisk Research grant NNF18CC0034900. AJT is supported by a Canadian Institutes of Health Research Postdoctoral Fellowship Award.

Institutional Review Board Statement: Not applicable.

Informed Consent Statement: Not applicable.

Data Availability Statement: No new data were created or analyzed in this study. Data sharing is not applicable to this article.

Conflicts of Interest: M.G.-F., A.-J.T. and P.A.S. have no conflicts of interest to declare. E.R. declares receiving grant support through his institution from the California Walnut Commission (Folsom, CA), in addition to personal funds for project supervision and advice and serving as non-paid member of its Scientific Advisory Committee; funds for travel and accommodation from the International Nut and Dried Fruit Council (INC); and personal funds from Alexion for serving in the advisory committee. K.S.P. has received research grants from the California Walnut Commission, The Peanut Institute, the American Pecan Council, and the American Pistachio Growers. P.M.K.-E. reports current and recent research funding from the American Pistachio Growers, the American Pecan Council, the Peanut Institute, and the California Walnut Commission. L.C.T. has received grant support and funds for travel and accommodation from the California Walnut Commission (CWC), The International Nut and Dried Fruit Council (INC) and Nuts for Life (Horticulture Australia), and served as a non-paid member of the Scientific Advisory Committee of the CWC. J.S.-S. reported receiving research support from the Instituto de Salud Carlos III, Ministerio de Educación y Ciencia, the European Commission, the USA National Institutes of Health; receiving consulting fees or travel expenses from Instituto Danone and Abbott Laboratories, receiving nonfinancial support from Patrimonio Comunal Olivarero, the Almond Board of California, Pistachio Growers and Borges S.A; serving on the board of and receiving grant support through his institution from the International Nut and Dried Foundation; and personal fees from Instituto Danone; Serving in the Board of Danone Institute International.

Abbreviations

ALA: α-linolenic acid; Apo, apolipoproteins; BMI, body mass index; CHD, coronary heart disease; CVD, cardiovascular diseases; FFQ, food frequency questionnaire; HDL, high-density lipoprotein cholesterol; HMG-CoA, 3-hydroxy-3-methylglutaryl-CoA; HNF-4α, hepatocyte nuclear factor; LDL, low-density lipoprotein cholesterol; Lp(a), Lipoprotein(a); LXR, liver-X-receptor; MD, mean difference; MUFA, monounsaturated fat; NFκB, nuclear factor kappa B; PPARγ, peroxisomal proliferator-activated nuclear receptors gamma; PREDIMED, PREvención con DIeta MEDiterránea; PUFA, polyunsaturated fat; RCT, randomized controlled trial; SFA, saturated fat; SRBPs, sterol-regulatory element binding proteins; TG, triglycerides; TC, total cholesterol; VLDL, very low-density lipoprotein; WAHA, Walnuts and Healthy Aging Study.

References

1. Roth, G.A.; Abate, D.; Abate, K.H.; Abay, S.M.; Abbafati, C.; Abbasi, N.; Abbastabar, H.; Abd-Allah, F.; Abdela, J.; Abdelalim, A.; et al. Global, Regional, and National Age-Sex-Specific Mortality for 282 Causes of Death in 195 Countries and Territories, 1980–2017: A Systematic Analysis for the Global Burden of Disease Study 2017. *Lancet* **2018**, *392*, 1736–1788. [CrossRef]
2. Grundy, S.M.; Stone, N.J.; Bailey, A.L.; Beam, C.; Birtcher, K.K.; Blumenthal, R.S.; Braun, L.T.; de Ferranti, S.; Faiella-Tommasino, J.; Forman, D.E.; et al. 2018 AHA/ACC/AACVPR/AAPA/ABC/ACPM/ADA/AGS/APhA/ASPC/NLA/PCNA Guideline on the Management of Blood Cholesterol: A Report of the American College of Cardiology/American Heart Association Task Force on Clinical Practice Guidelines. *Circulation* **2019**, *139*, E1082–E1143. [CrossRef] [PubMed]
3. Gakidou, E.; Afshin, A.; Abajobir, A.A.; Abate, K.H.; Abbafati, C.; Abbas, K.M.; Abd-Allah, F.; Abdulle, A.M.; Abera, S.F.; Aboyans, V.; et al. Global, Regional, and National Comparative Risk Assessment of 84 Behavioural, Environmental and Occupational, and Metabolic Risks or Clusters of Risks, 1990–2016: A Systematic Analysis for the Global Burden of Disease Study 2016. *Lancet* **2017**, *390*, 1345–1422. [CrossRef] [PubMed]
4. Murray, C.J.L.; Mokdad, A.H.; Ballestros, K.; Echko, M.; Glenn, S.; Olsen, H.E.; Mullany, E.; Lee, A.; Khan, A.R.; Ahmadi, A.; et al. The State of US Health, 1990–2016: Burden of Diseases, Injuries, and Risk Factors among US States. *JAMA J. Am. Med. Assoc.* **2018**, *319*, 1444–1472. [CrossRef]
5. Micha, R.; Peñalvo, J.L.; Cudhea, F.; Imamura, F.; Rehm, C.D.; Mozaffarian, D. Association between Dietary Factors and Mortality from Heart Disease, Stroke, and Type 2 Diabetes in the United States. *JAMA J. Am. Med. Assoc.* **2017**, *317*, 912–924. [CrossRef]
6. Ros, E.; Tapsell, L.C.; Sabate, J. Nuts and Berries for Heart Health. *Curr. Atheroscler. Rep.* **2010**, *12*, 397–406. [CrossRef]
7. Li, L.; Tsao, R.; Yang, R.; Kramer, J.K.G.; Hernandez, M. Fatty Acid Profiles, Tocopherol Contents, and Antioxidant Activities of Heartnut (Juglans Ailanthifolia Var. Cordiformis) and Persian Walnut (*Juglans Regia* L.). *J. Agric. Food Chem.* **2007**, *55*, 1164–1169. [CrossRef]

8. Ros, E.; Hu, F.B. Consumption of Plant Seeds and Cardiovascular Health: Epidemiological and Clinical Trial Evidence. *Circulation* **2013**, *128*, 553–565. [CrossRef]
9. Estruch, R.; Ros, E.; Salas-Salvadó, J.; Covas, M.-I.; Corella, D.; Arós, F.; Gómez-Gracia, E.; Ruiz-Gutiérrez, V.; Fiol, M.; Lapetra, J.; et al. Primary Prevention of Cardiovascular Disease with a Mediterranean Diet Supplemented with Extra-Virgin Olive Oil or Nuts. *New Engl. J. Med.* **2018**, *378*, e34. [CrossRef]
10. Fito, M.; Guxens, M.; Corella, D.; Saez, G.; Estruch, R.; de la Torre, R.; Frances, F., Cabezas, C., Lopez-Sabater Mdel, C.; Marrugat, J.; et al. Effect of a Traditional Mediterranean Diet on Lipoprotein Oxidation: A Randomized Controlled Trial. *Arch. Intern. Med.* **2007**, *167*, 1195–1203. [CrossRef]
11. Xia, J.; Yu, J.; Xu, D.; Yang, C.; Xia, H.; Sun, G. The Effects of Peanuts and Tree Nuts on Lipid Profile in Type 2 Diabetic Patients: A Systematic Review and Meta-Analysis of Randomized, Controlled-Feeding Clinical Studies. *Front. Nutr.* **2021**, *8*, 765571. [CrossRef] [PubMed]
12. Sabaté, J. Nut Consumption and Blood Lipid Levels. *Arch. Intern. Med.* **2010**, *170*, 821. [CrossRef] [PubMed]
13. Phung, O.J.; Makanji, S.S.; White, C.M.; Coleman, C.I. Almonds Have a Neutral Effect on Serum Lipid Profiles: A Meta-Analysis of Randomized Trials. *J. Am. Diet Assoc.* **2009**, *109*, 865–873. [CrossRef] [PubMed]
14. Parilli-Moser, I.; Hurtado-Barroso, S.; Guasch-Ferré, M.; Lamuela-Raventós, R.M. Effect of Peanut Consumption on Cardiovascular Risk Factors: A Randomized Clinical Trial and Meta-Analysis. *Front. Nutr.* **2022**, *9*, 853378. [CrossRef] [PubMed]
15. Moosavian, S.P.; Rahimlou, M.; Rezaei Kelishadi, M.; Moradi, S.; Jalili, C. Effects of Almond on Cardiometabolic Outcomes in Patients with Type 2 Diabetes: A Systematic Review and Meta-Analysis of Randomized Controlled Trials. *Phytother. Res.* **2022**, *36*, 1839–1853. [CrossRef]
16. Blanco Mejia, S.; Kendall, C.W.C.; Viguiliouk, E.; Augustin, L.S.; Ha, V.; Cozma, A.I.; Mirrahimi, A.; Maroleanu, A.; Chiavaroli, L.; Leiter, L.A.; et al. Effect of Tree Nuts on Metabolic Syndrome Criteria: A Systematic Review and Meta-Analysis of Randomised Controlled Trials. *BMJ Open* **2014**, *4*, e004660. [CrossRef]
17. Mateș, L.; Popa, D.-S.; Rusu, M.E.; Fizeșan, I.; Leucuța, D. Walnut Intake Interventions Targeting Biomarkers of Metabolic Syndrome and Inflammation in Middle-Aged and Older Adults: A Systematic Review and Meta-Analysis of Randomized Controlled Trials. *Antioxidants* **2022**, *11*, 1412. [CrossRef]
18. Liu, K.; Hui, S.; Wang, B.; Kaliannan, K.; Guo, X.; Liang, L. Comparative Effects of Different Types of Tree Nut Consumption on Blood Lipids: A Network Meta-Analysis of Clinical Trials. *Am. J. Clin. Nutr.* **2020**, *111*, 219–227. [CrossRef]
19. Lee-Bravatti, M.A.; Wang, J.; Avendano, E.E.; King, L.; Johnson, E.J.; Raman, G. Almond Consumption and Risk Factors for Cardiovascular Disease: A Systematic Review and Meta-Analysis of Randomized Controlled Trials. *Adv. Nutr.* **2019**, *10*, 1076–1088. [CrossRef]
20. Jalali, M.; Karamizadeh, M.; Ferns, G.A.; Zare, M.; Moosavian, S.P.; Akbarzadeh, M. The Effects of Cashew Nut Intake on Lipid Profile and Blood Pressure: A Systematic Review and Meta-Analysis of Randomized Controlled Trials. *Complement Ther. Med.* **2020**, *50*, 102387. [CrossRef]
21. Eslami, O.; Khorramrouz, F.; Sohouli, M.; Bagheri, N.; Shidfar, F.; Fernandez, M.L. Effect of Nuts on Components of Metabolic Syndrome in Healthy Adults with Overweight/Obesity: A Systematic Review and Meta-Analysis. *Nutr. Metab. Cardiovasc. Dis.* **2022**, *32*, 2459–2469. [CrossRef]
22. del Gobbo, L.C.; Falk, M.C.; Feldman, R.; Lewis, K.; Mozaffarian, D. Effects of Tree Nuts on Blood Lipids, Apolipoproteins, and Blood Pressure: Systematic Review, Meta-Analysis, and Dose-Response of 61 Controlled Intervention Trials. *Am. J. Clin. Nutr.* **2015**, *102*, 1347–1356. [CrossRef]
23. Guasch-Ferré, M.; Li, J.; Hu, F.B.; Salas-Salvadó, J.; Tobias, D.K. Effects of Walnut Consumption on Blood Lipids and Other Cardiovascular Risk Factors: An Updated Meta-Analysis and Systematic Review of Controlled Trials. *Am. J. Clin. Nutr.* **2018**, *108*, 174–187. [CrossRef]
24. Banel, D.K.; Hu, F.B. Effects of Walnut Consumption on Blood Lipids and Other Cardiovascular Risk Factors: A Meta-Analysis and Systematic Review. *Am. J. Clin. Nutr.* **2009**, *90*, 56–63. [CrossRef]
25. Asbaghi, O.; Moodi, V.; Hadi, A.; Eslampour, E.; Shirinbakhshmasoleh, M.; Ghaedi, E.; Miraghajani, M. The Effect of Almond Intake on Lipid Profile: A Systematic Review and Meta-Analysis of Randomized Controlled Trials. *Food Funct.* **2021**, *12*, 1882–1896. [CrossRef]
26. Arabi, S.M.; Bahrami, L.S.; Milkarizi, N.; Nematy, M.; Kalmykov, V.; Sahebkar, A. Impact of Walnut Consumption on Cardio Metabolic and Anthropometric Parameters in Metabolic Syndrome Patients: GRADE-Assessed Systematic Review and Dose-Response Meta-Analysis of Data from Randomized Controlled Trials. *Pharmacol. Res.* **2022**, *178*, 106190. [CrossRef]
27. Hadi, A.; Asbaghi, O.; Kazemi, M.; Haghighian, H.K.; Pantovic, A.; Ghaedi, E.; Abolhasani Zadeh, F. Consumption of Pistachio Nuts Positively Affects Lipid Profiles: A Systematic Review and Meta-Analysis of Randomized Controlled Trials. *Crit. Rev. Food Sci. Nutr.* **2021**, *21*, 1–14. [CrossRef]
28. Morvaridzadeh, M.; Sepidarkish, M.; Farsi, F.; Akbari, A.; Mostafai, R.; Omidi, A.; Potter, E.; Heshmati, J. Effect of Cashew Nut on Lipid Profile: A Systematic Review and Meta-Analysis. *Complement Med. Res.* **2020**, *27*, 348–356. [CrossRef]
29. Wang, P.; Sheng, Y.; Samadi, M. Effects of Almond Consumption on Lipid Profile in Patients with Type 2 Diabetes: A Systematic Review and Meta-Analysis of Randomised Controlled Trials. *Arch. Physiol. Biochem.* **2021**, *126*, 1987477. [CrossRef]
30. The Effects of Lowering LDL Cholesterol with Statin Therapy in People at Low Risk of Vascular Disease: Meta-Analysis of Individual Data from 27 Randomised Trials. *Lancet* **2012**, *380*, 581–590. [CrossRef]

31. Efficacy and Safety of More Intensive Lowering of LDL Cholesterol: A Meta-Analysis of Data from 170000 Participants in 26 Randomised Trials. *Lancet* **2010**, *376*, 1670–1681. [CrossRef]
32. Abdelhamid, A.S.; Brown, T.J.; Brainard, J.S.; Biswas, P.; Thorpe, G.C.; Moore, H.J.; Deane, K.H.; Summerbell, C.D.; Worthington, H.V.; Song, F.; et al. Omega-3 Fatty Acids for the Primary and Secondary Prevention of Cardiovascular Disease. *Cochrane Database Syst. Rev.* **2020**, *3*, CD003177. [CrossRef]
33. Ryan, E.; Galvin, K.; O'Connor, T.P.; Maguire, A.R.; O'Brien, N.M. Fatty Acid Profile, Tocopherol, Squalene and Phytosterol Content of Brazil, Pecan, Pine, Pistachio and Cashew Nuts. *Int. J. Food Sci. Nutr.* **2006**, *57*, 219–228. [CrossRef]
34. Maguire, L.S.; O'Sullivan, S.M.; Galvin, K.; O'Connor, T.P.; O'Brien, N.M. Fatty Acid Profile, Tocopherol, Squalene and Phytosterol Content of Walnuts, Almonds, Peanuts, Hazelnuts and the Macadamia Nut. *Int. J. Food Sci. Nutr.* **2004**, *55*, 171–178. [CrossRef]
35. Caldas, A.P.S.; Rocha, D.M.U.P.; Dionísio, A.P.; Hermsdorff, H.H.M.; Bressan, J. Brazil and Cashew Nuts Intake Improve Body Composition and Endothelial Health in Women at Cardiometabolic Risk (Brazilian Nuts Study): A Randomized Controlled Trial. *Br. J. Nutr.* **2022**, *128*, 1747–1757. [CrossRef]
36. Hyson, D.A.; Schneeman, B.O.; Davis, P.A. Almonds and Almond Oil Have Similar Effects on Plasma Lipids and LDL Oxidation in Healthy Men and Women. *J. Nutr.* **2002**, *132*, 703–707. [CrossRef]
37. Spiller, G.A.; Miller, A.; Olivera, K.; Reynolds, J.; Miller, B.; Morse, S.J.; Dewell, A.; Farquhar, J.W. Effects of Plant-Based Diets High in Raw or Roasted Almonds, or Roasted Almond Butter on Serum Lipoproteins in Humans. *J. Am. Coll. Nutr.* **2003**, *22*, 195–200. [CrossRef]
38. Tey, S.L.; Robinson, T.; Gray, A.R.; Chisholm, A.W.; Brown, R.C. Do Dry Roasting, Lightly Salting Nuts Affect Their Cardioprotective Properties and Acceptability? *Eur. J. Nutr.* **2017**, *56*, 1025–1036. [CrossRef]
39. Kris-Etherton, P.M.; Pearson, T.A.; Wan, Y.; Hargrove, R.L.; Moriarty, K.; Fishell, V.; Etherton, T.D. High–Monounsaturated Fatty Acid Diets Lower Both Plasma Cholesterol and Triacylglycerol Concentrations. *Am. J. Clin. Nutr.* **1999**, *70*, 1009–1015. [CrossRef]
40. McKiernan, F.; Lokko, P.; Kuevi, A.; Sales, R.L.; Costa, N.M.B.; Bressan, J.; Alfenas, R.C.G.; Mattes, R.D. Effects of Peanut Processing on Body Weight and Fasting Plasma Lipids. *Br. J. Nutr.* **2010**, *104*, 418–426. [CrossRef]
41. Mensink, R.P.; Zock, P.L.; Kester, A.D.; Katan, M.B. Effects of Dietary Fatty Acids and Carbohydrates on the Ratio of Serum Total to HDL Cholesterol and on Serum Lipids and Apolipoproteins: A Meta-Analysis of 60 Controlled Trials. *Am. J. Clin. Nutr.* **2003**, *77*, 1146–1155. [CrossRef] [PubMed]
42. Ros, E.; Singh, A.; O'Keefe, J.H. Nuts: Natural Pleiotropic Nutraceuticals. *Nutrients* **2021**, *13*, 3269. [CrossRef] [PubMed]
43. Woollett, L.A.; Spady, D.K.; Dietschy, J.M. Saturated and Unsaturated Fatty Acids Independently Regulate Low Density Lipoprotein Receptor Activity and Production Rate. *J. Lipid Res.* **1992**, *33*, 77–88. [CrossRef] [PubMed]
44. Baccouch, R.; Shi, Y.; Vernay, E.; Mathelié-Guinlet, M.; Taib-Maamar, N.; Villette, S.; Feuillie, C.; Rascol, E.; Nuss, P.; Lecomte, S.; et al. The Impact of Lipid Polyunsaturation on the Physical and Mechanical Properties of Lipid Membranes. *Biochim. Biophys. Acta (BBA) Biomembr.* **2023**, *1865*, 184084. [CrossRef]
45. Sampath, H.; Ntambi, J.M. Polyunsaturated fatty acid regulation of genes of lipid metabolism. *Annu. Rev. Nutr.* **2005**, *25*, 317–340. [CrossRef] [PubMed]
46. Salas-Salvadó, J.; Bulló, M.; Pérez-Heras, A.; Ros, E. Dietary Fibre, Nuts and Cardiovascular Diseases. *Br. J. Nutr.* **2006**, *96*, S45–S51. [CrossRef]
47. Brown, L.; Rosner, B.; Willett, W.W.; Sacks, F.M. Cholesterol-Lowering Effects of Dietary Fiber: A Meta-Analysis. *Am. J. Clin. Nutr.* **1999**, *69*, 30–42. [CrossRef] [PubMed]
48. Fu, L.; Zhang, G.; Qian, S.; Zhang, Q.; Tan, M. Associations between Dietary Fiber Intake and Cardiovascular Risk Factors: An Umbrella Review of Meta-Analyses of Randomized Controlled Trials. *Front. Nutr.* **2022**, *9*, 972399. [CrossRef]
49. Moreau, R.A.; Nyström, L.; Whitaker, B.D.; Winkler-Moser, J.K.; Baer, D.J.; Gebauer, S.K.; Hicks, K.B. Phytosterols and Their Derivatives: Structural Diversity, Distribution, Metabolism, Analysis, and Health-Promoting Uses. *Prog. Lipid Res.* **2018**, *70*, 35–61. [CrossRef]
50. del Gobbo, L.C.; Falk, M.C.; Feldman, R.; Lewis, K.; Mozaffarian, D. Are Phytosterols Responsible for the Low-Density Lipoprotein–Lowering Effects of Tree Nuts? *J. Am. Coll. Cardiol.* **2015**, *65*, 2765–2767. [CrossRef]
51. Cofán, M.; Ros, E. Use of Plant Sterol and Stanol Fortified Foods in Clinical Practice. *Curr. Med. Chem.* **2019**, *26*, 6691–6703. [CrossRef]
52. Ferguson, J.J.A.; Stojanovski, E.; MacDonald-Wicks, L.; Garg, M.L. Fat Type in Phytosterol Products Influence Their Cholesterol-Lowering Potential: A Systematic Review and Meta-Analysis of RCTs. *Prog. Lipid Res.* **2016**, *64*, 16–29. [CrossRef] [PubMed]
53. Pérez-Jiménez, J.; Neveu, V.; Vos, F.; Scalbert, A. Identification of the 100 Richest Dietary Sources of Polyphenols: An Application of the Phenol-Explorer Database. *Eur. J. Clin. Nutr.* **2010**, *64*, S112–S120. [CrossRef] [PubMed]
54. Cicero, A.F.G.; Colletti, A. Polyphenols Effect on Circulating Lipids and Lipoproteins: From Biochemistry to Clinical Evidence. *Curr. Pharm. Des.* **2018**, *24*, 178–190. [CrossRef]
55. Meisinger, C.; Baumert, J.; Khuseyinova, N.; Loewel, H.; Koenig, W. Plasma Oxidized Low-Density Lipoprotein, a Strong Predictor for Acute Coronary Heart Disease Events in Apparently Healthy, Middle-Aged Men from the General Population. *Circulation* **2005**, *112*, 651–657. [CrossRef]
56. Matsuura, E.; Hughes, G.R.V.; Khamashta, M.A. Oxidation of LDL and Its Clinical Implication. *Autoimmun. Rev.* **2008**, *7*, 558–566. [CrossRef] [PubMed]

57. von Eckardstein, A.; Nordestgaard, B.G.; Remaley, A.T.; Catapano, A.L. High-Density Lipoprotein Revisited: Biological Functions and Clinical Relevance. *Eur. Heart J.* 2022; *ahead of print*. [CrossRef]
58. Hernáez, Á.; Castañer, O.; Elosua, R.; Pintó, X.; Estruch, R.; Salas-Salvadó, J.; Corella, D.; Arós, F.; Serra-Majem, L.; Fiol, M.; et al. Mediterranean Diet Improves High-Density Lipoprotein Function in High-Cardiovascular-Risk Individuals. *Circulation* **2017**, *135*, 633–643. [CrossRef]
59. Holligan, S.D.; West, S.G.; Gebauer, S.K.; Kay, C.D.; Kris-Etherton, P.M. A Moderate-Fat Diet Containing Pistachios Improves Emerging Markers of Cardiometabolic Syndrome in Healthy Adults with Elevated LDL Levels. *Br. J. Nutr.* **2014**, *112*, 744–752. [CrossRef]
60. Berryman, C.E.; Grieger, J.A.; West, S.G.; Chen, C.-Y.O.; Blumberg, J.B.; Rothblat, G.H.; Sankaranarayanan, S.; Kris-Etherton, P.M. Acute Consumption of Walnuts and Walnut Components Differentially Affect Postprandial Lipemia, Endothelial Function, Oxidative Stress, and Cholesterol Efflux in Humans with Mild Hypercholesterolemia. *J. Nutr.* **2013**, *143*, 788–794. [CrossRef]
61. Yvan-Charvet, L.; Wang, N.; Tall, A.R. Role of HDL, ABCA1, and ABCG1 Transporters in Cholesterol Efflux and Immune Responses. *Arterioscler. Thromb. Vasc. Biol.* **2010**, *30*, 139–143. [CrossRef]
62. Reyes-Soffer, G.; Ginsberg, H.N.; Berglund, L.; Duell, P.B.; Heffron, S.P.; Kamstrup, P.R.; Lloyd-Jones, D.M.; Marcovina, S.M.; Yeang, C.; Koschinsky, M.L.; et al. Lipoprotein(a): A Genetically Determined, Causal, and Prevalent Risk Factor for Atherosclerotic Cardiovascular Disease: A Scientific Statement From the American Heart Association. *Arterioscler. Thromb. Vasc. Biol.* **2022**, *42*, e48–e60. [CrossRef]
63. Krauss, R.M. Lipoprotein Subfractions and Cardiovascular Disease Risk. *Curr. Opin. Lipidol.* **2010**, *21*, 305–311. [CrossRef]
64. Qiao, Y.-N.; Zou, Y.-L.; Guo, S.-D. Low-Density Lipoprotein Particles in Atherosclerosis. *Front. Physiol.* **2022**, *13*, 931931. [CrossRef]
65. Mora, S.; Otvos, J.D.; Rifai, N.; Rosenson, R.S.; Buring, J.E.; Ridker, P.M. Lipoprotein Particle Profiles by Nuclear Magnetic Resonance Compared With Standard Lipids and Apolipoproteins in Predicting Incident Cardiovascular Disease in Women. *Circulation* **2009**, *119*, 931–939. [CrossRef]
66. Valkama, A.J.; Meinilä, J.M.; Koivusalo, S.B.; Lindström, J.; Rönö, K.; Stach-Lempinen, B.; Eriksson, J.G. Diet Quality as Assessed by the Healthy Food Intake Index and Relationship with Serum Lipoprotein Particles and Serum Fatty Acids in Pregnant Women at Increased Risk for Gestational Diabetes. *Br. J. Nutr.* **2018**, *120*, 914–924. [CrossRef]
67. Phillips, C.M.; Harrington, J.M.; Perry, I.J. Relationship between Dietary Quality, Determined by DASH Score, and Cardiometabolic Health Biomarkers: A Cross-Sectional Analysis in Adults. *Clin. Nutr.* **2019**, *38*, 1620–1628. [CrossRef]
68. Millar, S.R.; Navarro, P.; Harrington, J.M.; Shivappa, N.; Hébert, J.R.; Perry, I.J.; Phillips, C.M. Comparing Dietary Score Associations with Lipoprotein Particle Subclass Profiles: A Cross-Sectional Analysis of a Middle-to Older-Aged Population. *Clin. Nutr.* **2021**, *40*, 4720–4729. [CrossRef]
69. García-Gavilán, J.F.; Connelly, M.A.; Babio, N.; Mantzoros, C.S.; Ros, E.; Salas-Salvadó, J. Nut Consumption Is Associated with a Shift of the NMR Lipoprotein Subfraction Profile to a Less Atherogenic Pattern among Older Individuals at High CVD Risk. *Cardiovasc. Diabetol.* **2022**, *21*, 189. [CrossRef]
70. Rajaram, S.; Cofán, M.; Sala-Vila, A.; Haddad, E.; Serra-Mir, M.; Bitok, E.; Roth, I.; Freitas-Simoes, T.M.; Kaur, A.; Valls-Pedret, C.; et al. Effects of Walnut Consumption for 2 Years on Lipoprotein Subclasses Among Healthy Elders. *Circulation* **2021**, *144*, 1083–1085. [CrossRef]
71. Lee, Y.; Berryman, C.E.; West, S.G.; Chen, C.-Y.O.; Blumberg, J.B.; Lapsley, K.G.; Preston, A.G.; Fleming, J.A.; Kris-Etherton, P.M. Effects of Dark Chocolate and Almonds on Cardiovascular Risk Factors in Overweight and Obese Individuals: A Randomized Controlled-Feeding Trial. *J. Am. Heart Assoc.* **2017**, *6*, e005162. [CrossRef]
72. Tindall, A.M.; Kris-Etherton, P.M.; Petersen, K.S. Replacing Saturated Fats with Unsaturated Fats from Walnuts or Vegetable Oils Lowers Atherogenic Lipoprotein Classes Without Increasing Lipoprotein(a). *J. Nutr.* **2020**, *150*, 818–825. [CrossRef]
73. Hernández-Alonso, P.; Salas-Salvadó, J.; Baldrich-Mora, M.; Mallol, R.; Correig, X.; Bulló, M. Effect of Pistachio Consumption on Plasma Lipoprotein Subclasses in Pre-Diabetic Subjects. *Nutr. Metab. Cardiovasc. Dis.* **2015**, *25*, 396–402. [CrossRef]
74. Xepapadaki, E.; Nikdima, I.; Sagiadinou, E.C.; Zvintzou, E.; Kypreos, K.E. HDL and Type 2 Diabetes: The Chicken or the Egg? *Diabetologia* **2021**, *64*, 1917–1926. [CrossRef]
75. Fernández-Rodríguez, R.; Mesas, A.E.; Garrido-Miguel, M.; Martínez-Ortega, I.A.; Jiménez-López, E.; Martínez-Vizcaíno, V. The Relationship of Tree Nuts and Peanuts with Adiposity Parameters: A Systematic Review and Network Meta-Analysis. *Nutrients* **2021**, *13*, 2251. [CrossRef]
76. Guarneiri, L.L.; Cooper, J.A. Intake of Nuts or Nut Products Does Not Lead to Weight Gain, Independent of Dietary Substitution Instructions: A Systematic Review and Meta-Analysis of Randomized Trials. *Adv. Nutr.* **2021**, *12*, 384–401. [CrossRef]
77. Nishi, S.K.; Viguiliouk, E.; Blanco Mejia, S.; Kendall, C.W.C.; Bazinet, R.P.; Hanley, A.J.; Comelli, E.M.; Salas Salvadó, J.; Jenkins, D.J.A.; Sievenpiper, J.L. Are Fatty Nuts a Weighty Concern? A Systematic Review and Meta-Analysis and Dose-Response Meta-Regression of Prospective Cohorts and Randomized Controlled Trials. *Obes. Rev.* **2021**, *22*, 13330. [CrossRef]
78. Fernández-Rodríguez, R.; Martínez-Vizcaíno, V.; Garrido-Miguel, M.; Martínez-Ortega, I.A.; Álvarez-Bueno, C.; Eumann Mesas, A. Nut Consumption, Body Weight, and Adiposity in Patients with Type 2 Diabetes: A Systematic Review and Meta-Analysis of Randomized Controlled Trials. *Nutr. Rev.* **2022**, *80*, 645–655. [CrossRef]
79. McArthur, B.M.; Mattes, R.D. Energy Extraction from Nuts: Walnuts, Almonds and Pistachios. *Br. J. Nutr.* **2020**, *123*, 361–371. [CrossRef]

80. Tapsell, L.C.; Neale, E.P.; Satija, A.; Hu, F.B. Foods, Nutrients, and Dietary Patterns: Interconnections and Implications for Dietary Guidelines. *Adv. Nutr.* **2016**, *7*, 445–454. [CrossRef]
81. Yi, S.Y.; Steffen, L.M.; Zhou, X.; Shikany, J.M.; Jacobs, D.R. Association of Nut Consumption with CVD Risk Factors in Young to Middle-Aged Adults: The Coronary Artery Risk Development in Young Adults (CARDIA) Study. *Nutr. Metab. Cardiovasc. Dis.* **2022**, *32*, 2321–2329. [CrossRef]
82. Gebauer, S.K.; Novotny, J.A.; Bornhorst, G.M.; Baer, D.J. Food Processing and Structure Impact the Metabolizable Energy of Almonds. *Food Funct.* **2016**, *7*, 4231–4238. [CrossRef] [PubMed]
83. Creedon, A.C.; Hung, E.S.; Berry, S.E.; Whelan, K. Nuts and Their Effect on Gut Microbiota, Gut Function and Symptoms in Adults: A Systematic Review and Meta-Analysis of Randomised Controlled Trials. *Nutrients* **2020**, *12*, 2347. [CrossRef] [PubMed]
84. Fitzgerald, E.; Lambert, K.; Stanford, J.; Neale, E.P. The Effect of Nut Consumption (Tree Nuts and Peanuts) on the Gut Microbiota of Humans: A Systematic Review. *Br. J. Nutr.* **2021**, *125*, 508–520. [CrossRef] [PubMed]
85. Guasch-Ferré, M.; Hernández-Alonso, P.; Drouin-Chartier, J.-P.; Ruiz-Canela, M.; Razquin, C.; Toledo, E.; Li, J.; Dennis, C.; Wittenbecher, C.; Corella, D.; et al. Walnut Consumption, Plasma Metabolomics, and Risk of Type 2 Diabetes and Cardiovascular Disease. *J. Nutr.* **2021**, *151*, 303–311. [CrossRef]

nutrients

Article

Effect of Nuts on Gastrointestinal Health

Giuseppina Mandalari [1,*], Teresa Gervasi [2], Daniel W. Rosenberg [3], Karen G. Lapsley [4,†] and David J. Baer [5]

1 Department of Chemical, Biological, Pharmaceutical and Environmental Science, University of Messina, Viale Ferdinando Stagno d'Alcontres 31, 98166 Messina, Italy
2 Department of Biomedical and Dental Sciences and Morphofunctional Imaging, University of Messina, 98125 Messina, Italy
3 Centre for Molecular Oncology, University of Connecticut Health Center, Farmington, CT 06030-3101, USA
4 The Almond Board of California, Modesto, CA 94587, USA
5 USDA, Agricultural Research Service, Beltsville Human Nutrition Research Center, Beltsville, MD 20705, USA
* Correspondence: gmandalari@unime.it; Tel.: +39-(0)90-6766593
† Retired.

Abstract: Nuts are high nutrient-dense foods containing healthy lipids, dietary fiber, and bioactive phytochemicals, including vitamins and minerals. Although the beneficial effect of nut consumption on different chronic diseases has been well documented, especially in relation to their cardiometabolic benefits, less scientific evidence is available on their possible beneficial effects on gastrointestinal health. In this narrative review, we summarize the most important findings and new research perspectives in relation to the importance of nut consumption on gastrointestinal health. The integrity of the cell wall structure, cell size and particle size after mastication are known to play a crucial role in energy, nutrient and bioactive release from nuts during digestion, therefore affecting bioaccessibility. Other mechanisms, such as cell wall composition, thickness and porosity, as well as stability of the membranes surrounding the oil bodies within the cell, are also important for energy extraction. As the undigested nutrients and phytochemicals are delivered to the colon, effects on gut microbiota composition are predicted. Although the overall effect of nut consumption on microbial alpha- and beta-diversity has been inconsistent, some scientific evidence suggests an increase in fecal butyrate after almond consumption, and a beneficial role of walnuts on the prevention of ulcerative colitis and protection against the development of gastric mucosal lesions.

Keywords: nuts; gut health; microbiota; digestion

Citation: Mandalari, G.; Gervasi, T.; Rosenberg, D.W.; Lapsley, K.G.; Baer, D.J. Effect of Nuts on Gastrointestinal Health. *Nutrients* **2023**, *15*, 1733. https://doi.org/10.3390/nu15071733

Academic Editors: Emilio Ros, Jordi Salas-Salvadó, Joan Sabaté, Stephanie K Nishi and Young-Eun Cho

Received: 28 January 2023
Revised: 27 March 2023
Accepted: 29 March 2023
Published: 1 April 2023

1. Introduction

Nuts, including peanuts, are nutrient dense foods containing healthy lipids, beneficial phytonutrients and a range of essential vitamins and minerals [1,2]. After cereals, nuts are the plant food group highest in dietary fiber, which results in unique microstructure and physical properties. Since nuts resist digestion in the upper GI tract, their cellular structure retains intact lipids and polymerized polyphenols and plays a key role in how they are metabolized by gut microbiota in the colon to form bioactive molecules which could benefit human health [3]. The role for specific foods and dietary patterns in modifying gut microbiota and fecal metabolites and their impact on various aspects of human health is well known. Research to understand the composition and function of the microbiota has expanded dramatically in recent years with the development of increasingly sensitive analytical techniques. These tools have facilitated data mining to better understanding the relationship of the microbiome to physiology and health [4].

There are now four tree nuts (almond, cashews, pistachios, walnuts) for which human clinical trials have clearly shown that the measured (metabolizable) energy value is 5–25% lower that the calculated values used in food labelling [5]. Considering these nuts have varying cellular structures and macro, micro, and phytonutrient contents, the mechanisms for digestion and microbiota changes are not fully understood, although the evidence for lipid encapsulation is compelling [6]. Some studies have shown that an optimized diet rich in nuts may be an intervention that promotes a healthy microbial population and thereby improves overall physiology, but clinical trials to date are inconclusive.

In 2020, a systematic review [7] and meta-analysis [8] of a total of 10 randomized, controlled trials (RCTs) assessed the effects of various nuts on fecal microbiota for over 600 adults consuming western diets, with 40–100 g nuts daily, in the U.S.A., Germany, Italy and Spain. Fitzgerald et al. [7] concluded from nine RCTs (four almond, three walnut, one each hazelnut and pistachio) that the overall gut health benefits of nuts may be due, in part, to their unique composition and physical structure. However, the exact mechanisms by which nuts exert these modest modulatory effects on gut microflora remain unclear. Since specific microbial alterations were evident, but often inconsistent, the authors recommended future studies designed to address the baseline habitual dietary patterns and microbial composition to minimize inter-individual composition of the gut microflora. Creedon et al. [8] found the strength of evidence from their meta-analysis from nine RCTs (five almond, three walnut and one pistachio) to be generally inconclusive. Nut consumption affected gut microbiota composition at the genus level, but not at a phyla level nor on the diversity of the microbiome. However, nut type and, to some extent, their duration of consumption influenced the overall effects. They concluded that further parallel design RCTs, powered to detect changes in fecal microbiota and that incorporate functional and clinical outcomes, are still needed. Mead et al. [9] performed a systematic review of four studies that included children between the ages of 3 and 18 years (one almond, two hazelnut, one Brazil nut) who consumed between 15–30 g nuts daily for 8–16 weeks. Although they found nut consumption improved overall diet quality in this young population, there were inconsistent effects on gut health. They concluded that further studies were needed, with consideration given to higher doses and longer intervention periods.

Dietary pattern analysis has emerged as an alternative approach to study the relation between nutrition and disease. Nuts are typically included in different healthy food patterns, and, as part of the Mediterranean Diet (MedDiet), a dietary pattern widely recognized as a nutritional strategy that improves cardiometabolic health [10]. In Spain, Galie et al. [11] examined whether following a MedDiet modified gut microbiota composition and fecal metabolomics profiles, as well as cardiometabolic risk factors, compared with a non-MedDiet supplemented with 50 g nuts daily. They reported for the first time that the 50 participants with metabolic syndrome following the MedDiet, compared with a non-MedDiet diet supplemented with nuts, significantly changed specific microbial genera and fecal metabolites. However, it was concluded that further intervention studies were needed to understand the effects of different healthy dietary patterns on gut microbiota composition and functionality. In a separate study, Israeli researchers used a different approach to augment the MedDiet. In these studies, conducted in Spain and Italy, Rinott et al. [12] showed that within a 6 month controlled-feeding trial of 294 subjects, a green MedDiet, that included 28 g walnuts per day as well daily polyphenol-rich green tea and Mankai aquatic plant, led to more prominent compositional change in the gut microbiota.

They found both MedDiets induced substantial changes to the community structure of the gut microbiome, with the green MedDiet leading to more prominent compositional changes, largely driven by the low-abundant, "non-core", microorganisms [12]. They concluded that the diet microbiome–host interaction should be further explored in future studies that may guide the implementation of novel beneficial modifications of existing dietary patterns.

An overview on the digestibility of nut nutrients and phytochemicals and the impact of food matrices and processing on digestion in the upper gastrointestinal (GI) tract is provided. In addition, the effect of nuts on the composition and diversity of the gut microbiota and their impact on the production of microbially-derived short chain fatty acids and bile acids, as well as recent reports describing the prevention of gastrointestinal diseases associated with nut consumption, is described in this review.

2. Food Matrix and Digestion

2.1. Microstructure and Cell Properties

The great diversity of species, as well as of varieties within the same species, the cultivation methods, and the climatic characteristics where tree nuts are cultured, combine to exert profound effects on the chemical composition of nuts. The geometrical properties of nut shells and kernels, including length, width, thickness, density, surface area, volume, and specific gravity, influence the quality of nut products in the post-harvest process. Nut structure can influence lipid digestibility [13]. We have previously demonstrated that individual raw almond cells separated by the chelating agent CDTA are small (less than 50 μm in diameter) and the lipid is still within oleosomes (Figure 1), surrounding the protein bodies [14]. Roasting has an effect liberating the lipid from the oil bodies, which will then form large lipid droplets in unstained and Sudan IV-stained cells.

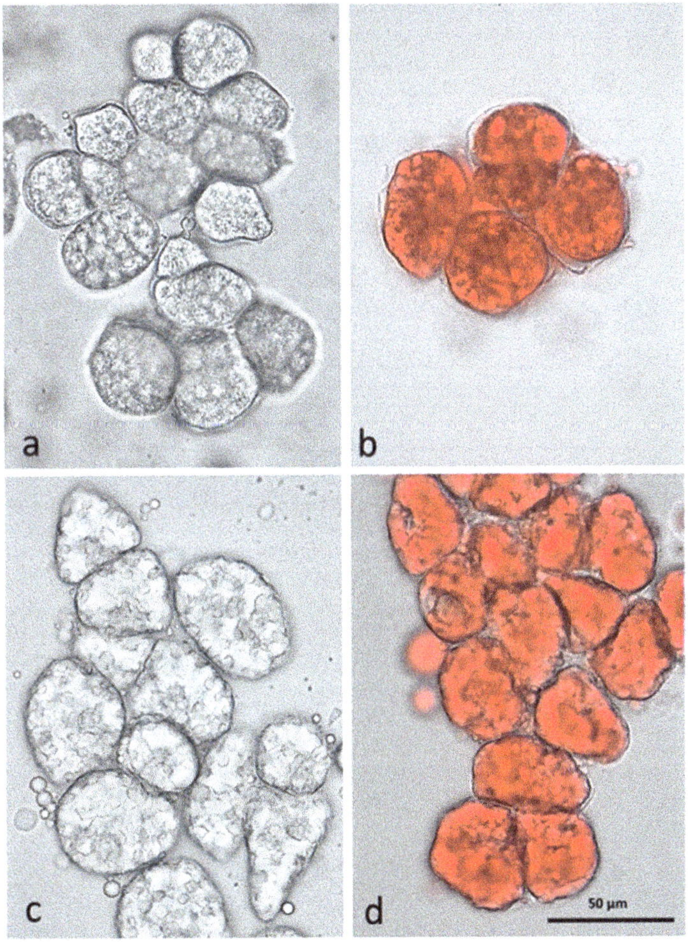

Figure 1. CDTA-separated cells of baseline natural raw almonds (**a**) unstained, (**b**) lipid stained with Sudan IV; and roasted almonds (**c**) unstained, (**d**) lipid stained with Sudan IV, showing lipid coalescence in the cells following roasting [14].

The physicochemical properties of the cell walls (e.g., dietary fiber), as well as their composition, mainly comprised of non-starch polysaccharides, are factors known to influence nutrient digestibility [15]. In addition, phospholipids and proteins can also limit the access of hydrolytic enzymes.

A theoretical model has been developed to predict lipid release from almonds in the gut: using simple geometry and data on cell dimensions and particle size, the number of ruptured cells in cut almond cubes was calculated [16]. The model has the potential to accurately predict nutrient bioaccessibility in a broad range of edible plants, based on their particle size and cell diameter. Grassby et al. [17] have also demonstrated that test meals containing almonds of different particle sizes behaved differently in the gut. Using a theoretical model, Creedon et al. [18] revealed a greater lipid bioaccessibility for ground almonds than whole almonds after mastication (10.4% ± 1.8% vs. 9.3% ± 2.0%, respectively; $p = 0.017$).

2.2. *Bioaccessibility of Nutrients and Phytochemicals in the Upper Gastrointestinal Tract*

During the multistage processing that occurs in the digestive system, mechanical and chemical mechanisms promote the breakdown of food molecules into smaller moieties, which can then be absorbed by the body. With the term "bioaccessibility", we refer to the proportion of nutrients and/or phytochemicals released from the upper GI tract, thereby becoming potentially available for absorption [19]. The physicochemical properties of nuts significantly affect the bioaccessibility of their constituent nutrients and phytochemicals [20,21].

2.2.1. Nutrient Bioaccessibility

Studies on almond digestion have shown that mechanical trituration or chewing breaks down a large fraction of the first outer layer of cells, while the majority of parenchyma cells, in which lipids and proteins are encapsulated, remains intact [19,22]. In a study with ileostomy volunteers, we showed that the lipids present in the intact cells located under the fractured layers appeared to 'leach' from the intact cells only after a protracted incubation in the upper GI tract [23]. This may be due to the increased porosity of the cells and to the degradation and solubilization of pectic compounds present in the cell wall and middle lamella [24]. The fractured surface may account for the lipid release that occurs after prolonged incubation in the GI tract. Although it is unclear to what extent lipolysis occurs inside almond cells and whether the lipids leave the cells as triacylglycerol molecules or hydrolyzed products, certainly mechanical processing (mainly grinding) or mastication is necessary for the cells to rupture and allow intracellular lipid and other nutrients (e.g., proteins) to be made available for digestion. Ellis et al. [19] observed the presence of almond tissues (cotyledon and testa) in fecal material after ingestion of almond kernels; some cells were still intact, whereas other cells had partially or totally lost their intracellular lipid. Recently, McArthur and Mattes [21] have subjected masticated samples of almonds, pistachios and walnuts obtained from healthy adults to a static model of gastric and intestinal digestion. While there was no significant difference in the total lipid release between the three nuts after intestinal digestion, walnuts produced a significantly larger particle size after chewing compared with almonds. Furthermore, the particle size after digestion was larger for walnuts compared with pistachios and almonds, indicating additional mechanisms, such as cell wall fissures and lipid storage properties, as relevant for energy extraction from nuts.

2.2.2. Phytochemicals Bioaccessibility

One of the main factors affecting the beneficial potential of polyphenols is their bioaccessibility and absorption in the upper GI tract, followed by their metabolism by the gut microbiota [25]. Polyphenols are a heterogeneous group of compounds characterized by complex structures and polymerization [26]. It is believed that only about 5–10% of the total polyphenol intake could be absorbed in the small intestine, mostly low molecular-weight polyphenols, starting with the removal of the sugar moiety from the glycoside [27]. The chemical structures and associated constituents largely influence their overall absorption, determining whether the polyphenols will be absorbed in the small intestine, or subsequently enter the colon where they could be metabolized by the colonic microbiota. Generally, hydrophobic forces and molecular hindering mechanisms are involved in the in vitro bioaccessibility of lipophilic phenolics, while hydrogen bonding and ionic forces are involved in the bioaccessibility of hydrophilic compounds [28].

We have demonstrated that polyphenols from pistachios are bioaccessible in the upper GI tract, with small differences between raw unsalted and roasted, salted pistachios [29]. It is believed that lutein and zeaxanthin bioavailability from pistachios are enhanced by the presence of fatty acids.

Clinical studies on the bioavailability of almond polyphenols are available. Urpi-Sarda et al. [30] analyzed the polyphenols and their metabolites in the plasma and urine of healthy human subjects after consumption of almond skin polyphenols. Products (O-methyl glucuronide, sulfate, glucuronide and O-methyl sulfate derivatives) of naringenin, (epi)catechin and isorhamnetin were identified in plasma and urine samples in the nanomolar range, together with the glucuronide and sulfate forms of 5-(dihydroxyphenyl)-γ-valerolactone and 5-(hydroxymethoxyphenyl)-γ-valerolactone. Bartolomé et al. [31] identified O-methyl glucuronide, O-methyl sulfate, sulfate and glucuronide derivatives of (epi)catechin, the glucuronide conjugates of isorhamnetin and naringenin, and sulfate conjugates of isorhamnetin, together with conjugates of hydroxyphenylvalerolactones and several products of microbial metabolization in plasma and urine samples. Garrido et al. [32] reported a maximum urinary excretion of naringenin and (epi)catechin conjugates between 2 and 6 h after consumption of almond skin polyphenols, while conjugated metabolites of isorhamnetin and hydroxyphenylvalerolactones reached their maximum levels between 10 and 24 h after consumption.

2.2.3. Effect of Processing and Food Matrix on Digestion

The type of nut and related processing methods greatly influences the damage incurred to the cell wall of the parenchyma, and, thus, the general bioaccessibility, the intracellular diffusion and lipase access to the oil bodies. Verghese et al. [33] have reviewed the effects of processing on the bioavailability of phytochemicals from a range of foods, including nuts, in relation to health benefits of bioactive compounds.

Amongst nut processing methods, dehydration through air or oil roasting can cause microstructural changes, such as lipid coalescence and chemical variations, which affects the integrity and structure of cell walls [34]. Roasting causes textural changes making nut mastication more efficient, which may be explained by the fact that tissues are more brittle when dehydrated. On the other hand, various types of roasting can influence the number of required chews before swallowing [24]. In a recent study, roasting of macadamia nuts changed the appearance of the cell walls and disrupted the oil body membrane, resulting in oil droplet coalescence [35].

Fewer studies have examined the influence of blanching on lipid digestibility, presumably due to the relatively mild process compared with roasting [13]. Oliveira et al. [36] reported that bioactive compounds and antioxidant activities increased with roasting and decreased with blanching. Both processing treatments positively affected the sensorial characteristics, increasing the content of polyunsaturated fatty acids, while saturated fatty acids, monounsaturated fatty acids and several health lipid indices decreased [36].

We have previously reported that incorporating natural and roasted salted pistachios in a food matrix (muffin) decreased the bioaccessibility of certain bioactive compounds, such as protocatechuic acid and luteolin, during in vitro gastric and duodenal digestion [29].

Different food matrices had a significant impact on bioaccessibility of polyphenols from almond skin using a dynamic gastric model. Use of full-fat milk lowered polyphenol recovery, influenced the free total phenols and associated antioxidant status, indicating that phenolics could bind protein within the matrix [37].

A pilot walnut supplementation study of urolithin bioavailability in healthy human volunteers demonstrated that ellagitannin (e.g., punicagalin) metabolism produced a highly variable profile of nine different urolithin metabolites in the urine [38]. Furthermore, the concentration of glucuronidated urolithins in blood and urine did not correlate with antioxidant capacity [39].

Overall, the available literature demonstrated that nutrient and phytochemical release from nuts during digestion is limited and influenced by several factors. Food matrix has an impact on bioaccessibility.

3. Effect of Nuts on Gastrointestinal Health

3.1. Microbiota Composition and Diversity

Understanding the importance of the gut microbiota (the collection of microorganisms present in a fecal sample) and the gut microbiome (genomes present in the fecal sample) is rapidly advancing. In healthy humans, gut microbiota and microbiome are usually assessed using fecal samples collected after dietary interventions. In these samples, microbial diversity and products of microbial metabolism are typically measured. Microbial end-products of metabolism can also be measured in other biospecimens such as blood or urine. There are many polyphenolic compounds (flavonoids and non-flavonoids) found in nuts. Although these compounds are generally poorly absorbed, they have a wide range of anti-bacterial, anti-inflammatory and anti-carcinogenic effects [40]. These anti-bacterial properties are of interest in how they may affect the host gut microbiota. For example, based on serving size, walnuts are the seventh largest source of total polyphenols among commonly consumed foods and beverages [41,42]. The phenolic profiles and antioxidant activities of free, esterified and bound phenolics in the walnut kernel reveal the presence of a remarkable array of phenolic compounds, including phenolic acids, flavonoids, tannins, phenolic lignans and stilbene-derivatives [43]. The main polyphenol found in walnuts is pedunculagin, an ellagitannin that has a wide range of antioxidant and anti-inflammatory properties [42]. After ingestion, ellagitannins are hydrolyzed to release ellagic acid, which is converted by the gut microflora into the urolithins [42]. With respect to nuts, analyses of microbial diversity, microbiota and microbial end-products have been performed in only a few studies, with more data becoming available as methodologies evolve and analytical costs decrease.

Alpha-diversity is the diversity within a defined microbial community. Typical measures of alpha-diversity are those that account for total species number (species richness) and the relative abundance of species (species evenness). One common measure of species richness is Chao-1, and measures of species richness and evenness include the Shannon index and Simpson index [44]. Chao-1 counts the number of different taxonomic groups (typically genus or species) in a sample, but does not take into account the abundance or relative distributions of the taxa. On the other hand, the Simpson index does consider relative abundance by weighing. Increased alpha-diversity is associated with improved health outcomes [45,46].

Studies of walnuts [47,48], almonds [49–51], and pistachios [51] have reported the effect on alpha-diversity of adding these nuts to the diet (intervention type, study design, sample size, dose and study duration are summarized in Table 1). In one study comparing the consumption of almonds or graham crackers as a snack, significant changes were reported in the Chao-1 index and Shannon index [52]. In this study, the snacks were provided for 8 weeks. The authors suggested that the 8 week provision of snacks was longer than many of the other studies, which typically last 3 weeks [47,49,51], and that perhaps these shorter interventions were not of sufficient length to affect alpha-diversity. On the other hand, an additional 8 week intervention of walnuts did not change the Simpson index [53]. Thus, it is unclear what length of feeding is important to affect alpha-diversity, and if tree nut dietary interventions have a substantial effect on alpha-diversity.

Table 1. Summary of studies evaluating nut intake and microbial changes.

Study Number (REF)	Intervention Nut	Study Design	Sample Size	Dose	Study Duration	Diversity Changes	Microbial Composition Change
[47] (Holscher et al., 2018)	Walnut	Crossover, controlled diet	18	42 g/d	3 wk	No effect on α diversity; β diversity, weighted principal coordinates analysis of UniFrac distances between samples based on their 97% OTU composition and abundances showed that bacterial communities were affected by walnut consumption.	Compared with after the control period, walnut consumption resulted in higher relative abundance of Faecalibacterium, Clostridium, Dialister, and Roseburia and lower relative abundances of Ruminococcus, Dorea, Oscillospira, and Bifidobacterium.
[48] (Tindall et al., 2020)	Walnut	Crossover, controlled diet	42	18% of energy	6 wk	No effect on α diversity; β diversity, weighted principal coordinates analysis of UniFrac distances between samples based on their 97% OTU composition and abundances showed that bacterial communities were affected by walnut consumption.	Compared with after the control period, walnut consumption resulted in higher relative abundance of Faecalibacterium, Clostridium, Dialister, and Roseburia and lower relative abundances of Ruminococcus, Dorea, Oscillospira, and Bifidobacterium.
[49] (Holscher et al., 2018)	Almond (whole, whole roasted, chopped roasted, butter)	Crossover, controlled diet	18	42 g/d	3 wk	No effect on α and β diversity.	Almond consumption increased the relative abundances of *Lachnospira*, *Roseburia*, and *Dialister*. Comparisons between control and the four almond treatments revealed that chopped almonds increased *Lachnospira*, *Roseburia*, and *Oscillospira* compared with the control; whole almonds increased *Dialister* compared with the control. There were no differences between almond butter and the control.
[50] (Burns et al., 2016)	Almond	Crossover, free-living	50	40 g/g	6 wk	No differences in overall microbiota diversity measures (Shannon diversity index and inverse Simpson diversity index).	Targeted qPCR analysis did not show almond intake-associated changes in the quantities of Bifidobacteria spp or lactic acid bacteria. When individual OTUs from 16S rRNA were combined at the phylum level, there were no significant differences in abundances correlating with almond intake. Some changes in the prevalence of various bacterial signatures at the genus and species levels were observed with the almond intervention at final vs. baseline.
[51] (Ukhanova et al., 2014)	Almond	Crossover, controlled diet	18	42 g/d and 84 g/d	18 d	α-diversity was not affected by the intake of almonds.	Numbers of bifidobacteria were not affected by the consumption of almonds.

Table 1. *Cont.*

Study Number (REF)	Intervention Nut	Study Design	Sample Size	Dose	Study Duration	Diversity Changes	Microbial Composition Change
[51] (Ukhanova et al., 2014)	Pistachio	Crossover, controlled diet	16	42 g/d and 84 g/d	18 d	α-diversity was not affected by the intake of pistachios.	Numbers of bifidobacteria were not affected by the consumption of pistachio. Pistachio consumption appeared to decrease the number of lactic acid bacteria.
[52] (Dhillon et al., 2022)	Almond	Parallel arm, free-living	73	57 g/d	8 wk		Microbial amino acid biosynthesis, and amino sugar and nucleotide sugar metabolism pathways were differentially enriched at the end of the intervention.
[53] (Bamberger et al., 2018)	Walnut	Crossover, free-living	142	43 g/d	4 wk	Supplementing walnuts in the diet did not significantly affect bacterial diversity measured by Shannons effective, and Simpsons effective counts. There was no significant difference in evenness as well as in richness for the walnut diet compared with the control diet. Beta-diversity increased with walnut consumption.	The abundance of Ruminococcaceae and Bifidobacteria increased significantly while Clostridium sp. cluster XIVa species (Blautia; Anaerostipes) decreased significantly during walnut consumption.
[54] (Dhillon et al., 2019)	Almond	Parallel arm, free-living	73	57 g/d	8 wk	Almond snacking resulted in 3% greater quantitative alpha-diversity (Shannon index) and 8% greater qualitative alpha-diversity (Chao1 index) than the cracker group.	Almond snacking decreased overall Bacteroides fragilis relative abundance by 48%.
[55] (Sapp et al., 2022)	Peanut	Crossover, controlled diet	50	28 g/d	6 wk	No between-condition differences in alpha- or beta- diversity were observed.	Following peanut intake, Ruminococcaceae were significantly more abundant compared with a lower-fat higher-carbohydrate snack. Metatranscriptomics showed increased expression of the K03518 (aerobic carbon-monoxide dehydrogenase small subunit) gene following peanut intake, and Roseburia intestinalis L1-82 was identified as a contributor to the increased expression.
[56] (Choo et al., 2021)	Almond	Parallel arm, free-living	69	56 g/d	8 wk	In the almond intervention group, there were significant increases in bacterial community richness, evenness and diversity.	Increases in both the relative and absolute abundance of operational taxonomic units in the Ruminococcaceae family, including Ruminiclostridium, Ruminococcaceae NK4A214, and Ruminococcaceae UCG-003 were the principal drivers of microbiota-level changes.

Beta-diversity is the diversity among different communities. For some approaches, such as UniFrac distances, qualitative plots are created to show beta-diversity. On the other hand, weighted UniFrac distances are quantitative. Both approaches have been used to measure beta-diversity [44,57]. Increased beta-diversity is associated with improvement in some health outcomes and reduction in BMI.

In four studies of almonds, two studies have reported no effect of almond consumption on beta-diversity using weighted and unweighted UniFrac distances [49,54], one study reported an increase in beta-diversity using unweighted UniFrac distances [51], and one study reported that beta-diversity was measured, but no data were presented [50].

In three studies of walnuts, two studies reported an increase in beta-diversity using weighted principal coordinates analysis of UniFrac distances [47,53]. Additionally, Bamberger et al. also measured beta-diversity with unweighted UniFrac distances, which was also significantly changed with walnut consumption [53]. One study [48] of walnuts used weighted UniFrac distances and did not report significant effects of the walnut diet compared with diets that were matched in fatty acid composition, but did not contain walnuts, or a diet replacing alpha-linolenic acid with oleic acids (also not containing walnuts) after 6 weeks of consuming each diet. The results observed in this study [48] may reflect the similarity of the composition of the diets, and that primarily fatty acid concentrations were manipulated. In one study of pistachios, beta-diversity was reportedly increased [53], and in one study of peanuts, changes in beta-diversity were unchanged between the peanut-containing diet and the control diet [55].

Overall, the effect of nut consumption on alpha- and beta-diversity was inconsistent. Reasons for these reported inconsistencies may be the variability in the length of intervention, the amount of nuts fed, dietary control of the intervention, comparator diets and sample size. The length of intervention for these various studies was 3 to 8 weeks. Since diversity was not a primary outcome, these studies may not have been designed or powered sufficiently to detect changes in diversity. The optimal length of feeding for these types of dietary interventions is unknown. Furthermore, a limited number of studies use provisioned diets which provide all the food consumed by the research volunteers, whereas other studies provide dietary guidance. The latter will likely result in more diet heterogeneity and variability which are two factors that may independently impact microbial diversity. Overall, the amount of nuts offered ranged from 42 to 99 g/d. The differences in these dose levels will likely affect substrate availability for fermentation in the large intestine, given the decreased digestibility of macronutrients in nuts and hence the associated increase in substrate reaching the large intestine [58–62]. Finally, the beta-diversity measured among diets will depend upon the differences in the composition of the diets, and perhaps some of the inconsistencies observed in beta-diversity is a reflection of the similarity of diet comparisons.

3.1.1. Changes in Relative Proportion at the Phyla Level

Changes in the relative abundance of bacteria can be determined at different phylogenetic levels. In a meta-analysis of nut studies, seven interventions investigated phlya-level changes in *Actinobacteria*, *Bacteroidetes*, *Firmicutes*, *Proteobacteria*, and *Verrucomicrobia* [47,49,50,54]. Additionally, two studies reported changes in relative abundance of the phyla *Tenericutes* [50,54]. Additional data were published but were not included in this meta-analysis [8]. Across these interventions, there were no significant effects of nut intake (five almond randomized control trials and one walnut RCT) on the relative abundance of these phlya. In fact, there was only one study in which any of these phyla were altered—with a significant change in the standard mean difference of *Proteobacteria* [50].

3.1.2. Changes in Relative Proportion at the Genus Level

At the genera-level, changes in the relative abundance of 19 phyla have been reported in several interventions [8]. Combining these data, the relative abundance of *Clostridium*, *Dialister*, *Lachnospira*, *Parabacteroides*, and *Roseburia* have been reported [8]. Nut consumption increased the relative abundance of *Clostridium*, *Dialister*, *Lachnospira*, and *Roseburia*. Further nut consumption decreased the relative abundance of *Parabacteroides*. These data represented five almond randomized control trials and one walnut RCT. When the data from the walnut trial were excluded from the meta-analysis, the effect of nut consumption on the relative abundance of *Clostridium* was no longer statistically significant. All of these studies were cross-over designed studies except one, the Dhillon study [54] which was a parallel arm intervention. When the parallel arm study (using almonds) was not included in the meta-analysis, the effect of nut consumption on the relative abundance of *Dialister*, *Lachnospira*, and *Parabacteroides* was no longer significant [8]. For many of these findings, there was heterogeneity, especially related to study duration (studies < 4 weeks vs. studies > 4 weeks), amount of nut consumed (<45 g/d vs. >45 g/d) and nut type (almond vs walnut). While the results of individual studies, and the results obtained from meta-analyses are intriguing, the total number of studies reported is limited, especially when it is likely that overall dose, nut type, and duration of intervention may all affect changes in relative abundance at the genera-level.

3.2. *Effect of Nuts on Microbial End Products*

3.2.1. Short Chain Fatty Acids

Short chain fatty acids (or volatile fatty acids) include acetate, butyrate and propionate. They are a microbial end-product of anerobic fiber fermentation. These short chain fatty acids can be used as an energy substrate by the microbes or host. Short chain fatty acids can block inflammatory processes via activation of G-protein-coupled receptors that are present within colonocytes [63]. These molecular alterations can subsequently activate intracellular signaling pathways that dampen NF-kB activation, modify downstream inflammatory mediators, and increase epithelial barrier function [64]. Species within the genera *Dialister*, *Lachnospira*, and *Roseburia* are known butyrate producers [65]. As mentioned above, the relative abundance of these genera has been shown to be increased with nut consumption.

Specific nut intervention studies that have measured the concentration of fecal short chain fatty acids are limited. In one such study, 87 subjects received 56 g/d of whole almonds or ground almonds (or no almonds as a control) for 4 weeks [18]. Compared with the baseline, there was no change in the fecal concentration of short chain fatty acids (acetate, butyrate, propionate, isobutyrate, valerate or isovalerate) in either the control diet or in subjects consuming either form of almond. However, when the data from the two forms of almonds were combined, there was a higher concentration in fecal butyrate compared with the controls. In a study of 69 subjects also receiving 56 g/d of whole almonds for 8 weeks, there was no observed effect of almond consumption on fecal concentrations of short chain fatty acids [66]. In a study of 63 subjects fed 25 g/d of peanuts, 32 g/d of peanut butter or 32 g/d of control butter made with peanut oil for 6 months, consumption of peanuts and peanut butter increased the fecal concentration of acetate, propionate and butyrate compared with baseline, with no changes in the control group fed butter [56]. In a crossover study of 50 subjects fed 28 g/d of dry roasted, unsalted peanuts or a lower-fat, higher-carbohydrate peanut-free snack for 6 weeks, short chain fatty acids were not measured; however, meta-transcriptomics analysis found that there was an increase in the expression of the bacterial K03518 gene that is directly involved in butyrate production [55]. In a recent short-term study of walnuts, a 3 day consumption in healthy individuals was found to modify the gut microbiome, while also increasing short chain fatty acid levels [67]. Importantly, these effects were dependent upon the composition of the individual microbiome [68]. Walnuts were found to modify the microbiome in an urolithin metabolite-dependent manner. Microbiota analysis further showed significant increases in two bacterial species, namely, *Coprococcus* and *Anaerostipes*, each established producers of butyrate [69]. In addition, *Phascolarbacterium*, a known producer of acetate and propionate, was also increased by walnut consumption [65]. Finally, this study identified significant variability in the metabolism of the polyphenols, differences that were present between the distinct urolithin metabotypes [68].

3.2.2. Bile Acids

The primary bile acids (cholic acid and chenocholic acid) are produced in the liver, while the secondary bile acids (lithocholic acid and deoxycholic acid) are produced in the large intestine by bacterial metabolism. Many bacteria are involved in the conversion of primary to secondary bile acids, including *Bifidobacterium*, *Lactobacillus*, *Clostridium*, *Enterococcus*, *Bacteroides*, *Eubacterium*, and *Escherichia* [70]. The microbially-produced secondary bile acids can bind to nuclear and membrane-bound receptors, activating a complex network of signaling cascades [71,72]. Through these cellular mechanisms, the secondary bile acids have been implicated in various disease etiologies, including several types of cancer, inflammatory bowel disease, cardiovascular disease and non-alcoholic fatty liver disease [72].

In a study of 18 subjects fed 42 g/d of walnuts or an identical control diet without walnuts for 3 weeks, fecal bile acids were measured at the end of each treatment [47]. There were no differences in the concentration of the primary bile acids measured between the two diets. However, after consumption of the diet containing walnuts, the concentration of the secondary bile acids was significantly lower [47]. These walnut-mediated changes in bile acid concentration raise the possibility that walnuts can affect multiple cell-signaling pathways, and possibly disease outcomes, through these microbially-derived end-products [47].

3.3. *Walnut Consumption and Gastrointestinal Disease*

Extensive research has been undertaken to determine whether walnuts may contribute to the mitigation of gastrointestinal disease, particularly with respect to ulcerative colitis and cancer. Walnut constituents contribute to decreased inflammation within the intestinal mucosa, related in part, to the microbial conversion of walnut-derived ellagitannins into a complex family of anti-inflammatory molecules, the urolithins [73]. Of course, walnuts also contain alpha-linolenic acid, a fatty acid that can be readily converted into eicosapentaenoic acid and docosahexaenoic acid, both associated with anti-inflammatory properties [74]. Studies in animal models and in several cell culture systems have uncovered a variety of health benefits that may be attributed to walnuts. A unifying mechanism is likely to involve at least some aspect of effects on immune-related and inflammatory cells. Defining the health benefits of dietary walnut consumption and the influence of its phytochemical composition may stimulate further research into underlying mechanisms that account for disease prevention.

In a preclinical animal model designed to recapitulate the pathology of ulcerative colitis (UC), ellagic acid was found to inhibit disease progression, while reducing associated intestinal inflammation in treated mice [75]. Furthermore, urolithin A, a microbial metabolite of ellagic acid, and its potent synthetic analogue, UAS03, were also found to mitigate DSS-induced intestinal inflammation, with reduced oxidative tissue damage and enhanced intestinal barrier function repair [76]. Both urolithin A and UAS03 provided significant protection against both acute and chronic colitis. This protection was caused by a number of distinct molecular mechanisms, including direct effects on inflammatory mediators, up-regulation of the ligand-activated transcription factor, AhR, and the remarkable ability of these compounds to enhance barrier function by eliciting an up-regulation of claudin 4, a critical tight junction protein [76]. These investigators also evaluated the effects of urolithin A on the direct activation of murine CD4-positive T cells and found a significant repression of their proliferative capacity that was associated with increased miR-10a-5p levels and down-regulation of Orai1/STIM1/STIM2 expression [76]. Koh et al. tested a walnut phenolic extract in both acute and chronic colitis models in mice [77]. This extract was found to inhibit NF-κB signaling, an effect directly associated with reduced expression of pro-inflammatory mediators [78]. Furthermore, Koh et al. also reported that their walnut phenolic extract inhibited colitis-associated colon cancer induced by treatment with the colon carcinogen, azoxymethane, followed by three cycles of 2% DSS for 5 days [77]. Overall, the therapeutic potential of walnuts to positively impact the severity of inflammatory diseases and possibly even inflammation-associated cancer has been established. Finally, Bartoszek et al. have tested the ability of walnut oil to stabilize tight junction proteins and to reduce the levels of pro-inflammatory cytokines commonly present within the inflamed mouse colon following treatment with the ulcerogenic agent, dextran sodium sulfate (DSS) [78]. Promising data from this research group have shown that walnut oil improves overall disease activity and restores normal ion transport and colonic wall permeability [78].

Nakanishi et al. used a similar mouse model to evaluate dietary supplementation with walnuts on colonic mucosal injury induced by DSS [74]. Mice were fed a "Total Western Diet" supplemented with walnuts (ranging from 0 to 14 g walnuts/100 g diet) for two weeks prior to DSS administration. After DSS administration, walnut supplementation significantly protected the colonic mucosa 10 days post-injury. Based on this observed protection against experimentally-induced colitis by walnuts, a follow-up study was conducted to determine the effect of walnuts on metabolites present in the colon [74]. Fecal and colonic samples were analyzed using discovery-based metabolite profiling two weeks post-walnut consumption. Nakanishi et al. found that walnuts caused a significant increase in fecal polyunsaturated fatty acids, including DHA and 9-oxo-10(E),12(E)-octadecadienoic acid (9-oxoODA), as well as kynurenic acid. In the colon, there was a significant increase in S-adenosylhomocysteine and betaine, two important mediators of fatty acid β-oxidation. Together, these findings suggest that metabolic changes caused by walnut consumption may contribute to protection against DSS-induced inflammatory tissue injury [74]. Additional studies are needed to confirm these findings and to better define the precise role of these metabolic changes on colonic inflammation.

Finally, walnut fractions have been found to protect against the development of gastric mucosal lesions, including gastritis, gastric ulcer, and gastric carcinoma [79]. Liu reported gastro-protective and cancer preventive effects of walnut constituents on alcohol-induced inflammation, with fewer gastric lesions and decreased gastric inflammation associated with decreased inflammatory cytokines [80]. Park tested the anti-inflammatory and anti-tumorigenic effects of walnuts in an *H. pylori* gastric cancer model [81]. Mice were infected with *H. pylori* and fed a high-salt diet to promote gastric cancer, and were supplemented with walnuts for nine months. Walnut supplementation caused a significant reduction in gastric cancer frequency with markedly reduced levels of PGE2 and COX-2, important pro-inflammatory mediators that play a key role in tumor promotion [80,81].

Overall, we have demonstrated that, although some nutrients and phytochemicals from nuts are absorbed in the upper GI tract and will reach the colon, clinical studies on their effect on the gut microbiota are still inconclusive. There is literature available on the beneficial effect of walnuts on the prevention of ulcerative colitis and gastric mucosal lesions.

4. Conclusions

In the present review, we have outlined the physiological processes that contribute to the digestion of tree nuts. Cell wall composition, thickness and porosity, as well as lipid encapsulation, may slow down or completely prevent some enzymes from entering the cell. It is clear that some fraction of nutrients and phytochemicals present in the nut are not digested in the upper GI tract and could reach the colon, where they may be fermented by the gut microbiota. Although some studies have demonstrated that nut consumption promotes a healthy microbiota, clinical trials are still inconclusive. Importantly, research focused on how nut consumption may affect microbial communities is at an early stage, is further confounded by the wide variability in overall quality of trial design, research methods used, age and health status of subjects, and the amount, type, and duration of nut intake.

5. Research Gaps and Future Directions

Future clinical trials must include key measures of microbial community structure, such as species diversity and composition, as well as changes to the microbiome that may be directly related to human health and disease risk. This information will be useful for comparing the beneficial effects of nut consumption across the population. While certain nuts have been investigated more than others for their impact on the GI transit, limited literature is available on the effects of regular consumption of mixed nuts. Research with additional types of nuts is needed to understand their broader effects.

To more accurately assess the health benefits and functionality associated with nut consumption, further studies are needed to better define the mechanisms responsible for their limited energy extraction during digestion, and how the physical structure of individual nuts may ultimately affect bioavailability. Clearly, epidemiological and clinical studies analyzing the potential beneficial effects of nut consumption on prevalent gastrointestinal diseases are warranted in the future.

Author Contributions: Conceptualization, G.M., T.G., D.W.R., K.G.L. and D.J.B.; writing—original draft preparation, G.M., T.G., D.W.R., K.G.L. and D.J.B.; writing—review and editing, G.M., T.G., D.W.R., K.G.L. and D.J.B. All authors have read and agreed to the published version of the manuscript.

Funding: The NUTS 2022 conference was organized by the Rovira i Virgili University, with all expenses of the organization covered by Barcelo Congresos, including the fees related to publishing the Proceedings in Nutrients. Barcelo Congresos SA has received funds from the International Nut & Dried Fruit Council (INC). Neither the researchers attending the NUTS 2022 Conference, nor the Rovira i Virgili University have received funds from the INC. Researchers' travel expenses and accommodation have been covered by Barcelo Congresos. Researchers did not receive any additional financial support related to their participation in the NUTS 2022 conference. Sponsors did not have any role in organizing the conference, selecting, or inviting the speakers, nor in the writing or reviewing the proceedings of the conference. DWR received funding from NIH grant R01 CA252045, The American Institute for Cancer Research, and generous financial support from the California Walnut Commission.

Institutional Review Board Statement: Not applicable.

Informed Consent Statement: Not applicable.

Data Availability Statement: Not applicable.

Conflicts of Interest: The authors declare no conflict of interest.

References

1. Alasalvar, C.; Bolling, B.W. Review of nut phytochemicals, fat-soluble bioactives, antioxidant components and health effects. *Br. J. Nutr.* **2015**, *113*, S68–S78. [CrossRef]
2. Alasalvar, C.; Salas-Salvado, J.; Ros, E. Bioactives and health benefits of nuts and dried fruits. *Food Chem.* **2020**, *314*, 126192. [CrossRef] [PubMed]
3. Dagbasi, A.; Lett, A.M.; Murphy, K.; Frost, G. Understanding the interplay between food structure, intestinal bacterial fermentation and appetite control. *Proc. Nutr. Soc.* **2020**, *79*, 514–530. [CrossRef] [PubMed]
4. ILSI Europe. Dietary probiotics, prebiotics and the gut microflora in human health. *ILSI Eur. Concise Monogr. Ser.* **2022**, 1–50.
5. McArthur, B.M.; Higgins, K.A.; Hunter, S.R.; Mattes, R.D. Energetics of Nut Consumption. In *Health Benefits of Nuts and Dried Fruits*; Alasalvar, C., Salas-Salvado, J., Ros, E., Sabate, S., Eds.; CRC Press: Boca Raton, FL, USA, 2020; pp. 126–156.
6. Lamuel-Raventos, R.M.; St. Onge, M.P. Prebiotic nut compounds and human microbiota. *Crit. Rev. Food Sci. Nutr.* **2017**, *57*, 3154–3163. [CrossRef] [PubMed]
7. Fitzgerald, E.; Lambert, K.; Stanford, J.; Neale, E.P. The effect of nut consumption (treenuts and peanuts) on the gut microbiota of humans: A systematic review. *Br. J. Nutr.* **2021**, *125*, 508–520. [CrossRef]
8. Creedon, A.C.; Hung, E.S.; Berry, S.E.; Whelan, K. Nuts and their Effect on Gut Microbiota, Gut Function and Symptoms in Adults: A Systematic Review and Meta-Analysis of Randomised Controlled Trials. *Nutrients* **2020**, *12*, 2347–2368. [CrossRef]
9. Mead, L.C.; Hill, A.M.; Carter, S.; Coates, A.M. The Effect of Nut Consumption on Diet Quality, Cardiometabolic and Gastrointestinal Health in Children: A Systematic Review of Randomized Controlled Trials. *Int. J. Environ. Res. Public Health* **2021**, *18*, 454–469. [CrossRef]
10. Neale, E.P.; Tapsell, L.C. Nuts in Healthy Dietary Patterns and Dietary Guidelines. In *Health Benefits of Nuts and Dried Fruits*; Alasalvar, C., Salas-Salvado, J., Ros, E., Sabate, S., Eds.; CRC Press: Boca Raton, FL, USA, 2020; pp. 290–314.
11. Galie, S.; Garcia-Gavilan, J.; Camacho-Barcia, L.; Atzenia, A.; Muralidharan, J.; Papandreou, C.; Arcelin, P.; Palau-Galindo, A.; Garcia, D.; Basora, J.; et al. Effects of the Mediterranean Diet or Nut Consumption on Gut Microbiota Composition and Fecal Metabolites and their Relationship with Cardiometabolic Risk Factors. *Mol. Nutr. Food Res.* **2021**, *65*, 2000982–2000991. [CrossRef]

12. Rinott, E.; Yaskolka, M.; Tsaban, G.; Zelicha, H.; Kaplan, A.; Knights, D.; Tuohy, K.; Scholz, M.U.; Koren, O.; Stampfer, M.J.; et al. The effects of the Green-Mediterranean diet on cardiometabolic health are linked to gut microbiome modifications: A randomized controlled trial. *Genome Med.* **2022**, *14*, 29–44. [CrossRef]
13. Li, C.H.; Shelp, G.; Wright, A.J. Influence of Nut Structure and Processing on Lipid Bioaccessibility and Absorption. *Curr. Opin. Food Sci.* **2022**, *49*, 100966. [CrossRef]
14. Mandalari, G.; Parker, M.L.; Grundy, M.M.; Grassby, T.; Smeriglio, A.; Bisignano, C.; Raciti, R.; Trombetta, D.; Baer, D.J.; Wilde, P.J. Understanding the Effect of Particle Size and Processing on Almond Lipid Bioaccessibility through Microstructural Analysis: From Mastication to Faecal Collection. *Nutrients* **2018**, *10*, 213. [CrossRef] [PubMed]
15. Holland, C.; Ryden, P.; Edwards, C.H.; Grundy, M.M. Plant Cell Walls: Impact on Nutrient Bioaccessibility and Digestibility. *Foods* **2020**, *9*, 201. [CrossRef] [PubMed]
16. Grassby, T.; Picout, D.R.; Mandalari, G.; Faulks, R.M.; Kendall, C.W.; Rich, G.T.; Wickham, M.S.J.; Lapsley, K.; Ellis, P.R. Modelling of nutrient bioaccessi-bility in almond seeds based on the fracture properties of their cell walls. *Food Funct.* **2014**, *5*, 3096–3106. [CrossRef] [PubMed]
17. Grassby, T.; Mandalari, G.; Grundy MM, L.; Edwards, C.H.; Bisignano, C.; Trombetta, D.; Smeriglio, A.; Chessa, S.; Ray, S.; Sanderson, J.; et al. In vitro and in vivo modeling of lipid bioaccessibility and digestion from almond muffins: The importance of the cell-wall barrier mechanism. *J. Funct. Foods* **2017**, *37*, 263–271. [CrossRef] [PubMed]
18. Creedon, A.C.; Dimidi, E.; Hung, E.S.; Rossi, M.; Probert, C.; Grassby, T.; Miguens-Blanco, J.; Marchesi, J.R.; Scott, S.M.; Berry, S.E.; et al. The impact of almonds and almond processing on gastrointestinal physiology, luminal microbiology and gastrointestinal symptoms: A randomized controlled trial and mastication study. *Am. J. Clin. Nutr.* **2022**, *116*, 1790–1804. [CrossRef] [PubMed]
19. Ellis, P.R.; Kendall, C.W.; Ren, Y.; Parker, C.; Pacy, J.F.; Waldron, K.W.; Jenkins, D.J. Role of cell walls in the bioaccessibility of lipids in almond seeds. *Am. J. Clin. Nutr.* **2004**, *80*, 604–613. [CrossRef]
20. Kumari, S.; Gray, A.R.; Webster, K.; Bailey, K.; Reid, M.; Kelvin, K.A.H.; Tey, S.L.; Chisholm, A.; Brown, R.C. Does 'activating' nuts affect nutrient bioavailability? *Food Chem.* **2020**, *319*, 126529. [CrossRef]
21. McArthur, B.M.; Mattes, R.D. Energy extraction from nuts: Walnuts, almonds and pistachios. *Br. J. Nutr.* **2020**, *123*, 361–371. [CrossRef]
22. Grundy, M.M.; Grassby, T.; Mandalari, G.; Waldron, K.W.; Butterworth, P.J.; Berry, S.E.; Ellis, P.R. Effect of mastication on lipid bioaccessibility of almonds in a randomized human study and its implications for digestion kinetics, metabolizable energy, and postprandial lipemia. *Am. J. Clin. Nutr.* **2015**, *101*, 25–33. [CrossRef]
23. Mandalari, G.; Faulks, R.M.; Rich, G.T.; Lo Turco, V.; Picout, D.R.; Lo Curto, R.B.; Bisignano, G.; Dugo, P.; Dugo, G.; Waldron, K.W.; et al. Release of protein, lipid, and vitamin E from almond seeds during digestion. *J. Agric. Food Chem.* **2008**, *56*, 3409–3416. [CrossRef] [PubMed]
24. Grundy, M.M.L.; Carrière, F.; Mackie, A.R.; Gray, D.A.; Butterworth, P.J.; Ellis, P.R. The role of plant cell wall encapsulation and porosity in regulating lipolysis during the digestion of almond seeds. *Food Funct.* **2016**, *7*, 69–78. [CrossRef] [PubMed]
25. Stevens, J.F.; Maier, C.S. The chemistry of gut microbial metabolism of polyphenols. *Phytochem. Rev.* **2016**, *15*, 425–444. [CrossRef]
26. Williamson, G.; Clifford, M.N. Role of the small intestine, colon and microbiota in determining the metabolic fate of polyphenols. *Biochem. Pharmacol.* **2017**, *139*, 24–39. [CrossRef] [PubMed]
27. Harborne, J.B. (Ed.) *The Flavonoids—Advances in Research since 1986*; Chapman & Hall: London, UK, 1993.
28. Stevens-Barrón, J.C.; de la Rosa, L.A.; Wall-Medrano, A.; Álvarez-Parrilla, E.; Rodríguez-Ramirez, R.; Robles-Zepeda, R.E.; Astiazaran-García, H. Chemical composition and in vitro bioaccessibility of antioxidant phytochemicals from selected edible nuts. *Nutrients* **2019**, *11*, 2303. [CrossRef]
29. Mandalari, G.; Bisignano, C.; Filocamo, A.; Chessa, S.; Sarò, M.; Torre, G.; Faulks, R.M.; Dugo, P. Bioaccessibility of pistachio polyphenols, xanthophylls, and tocopherols during simulated human digestion. *Nutrition* **2013**, *29*, 338–344. [CrossRef]
30. Urpi-Sarda, M.; Garrido, I.; Monagas, M.; Gómez-Cordovés, C.; Medina-Remón, A.; Andres-Lacueva, C.; Bartolomé, B. Profile of plasma and urine metabolites after the intake of almond [*Prunus dulcis* (Mill.) D.A. Webb] polyphenols in humans. *J. Agric. Food Chem.* **2009**, *57*, 10134–10142. [CrossRef]
31. Bartolomé, B.; Monagas, M.; Garrido, I.; Gómez-Cordovés, C.; Martín-Alvarez, P.J.; Lebrón-Aguilar, R.; Urpí-Sardà, M.; Llorach, R.; Andrés-Lacueva, C. Almond (*Prunus dulcis* (Mill.) D.A. Webb) polyphenols: From chemical characterization to targeted analysis of phenolic metabolites in humans. *Arch. Biochem. Biophys.* **2010**, *501*, 124–133. [CrossRef]
32. Garrido, M.; Urpi-Sarda, M.; Monagas, C.; Gomez-Cordoves, P.J.; Martin-Alvarez, R.; Llorach, B.; Bartolome, C.; Andres-Lacueva, C. Targeted analysis of conjugated and microbial-derived phenolic metabolites in human urine after consumption of an almond skin phenolic extract. *J. Nutr.* **2010**, *140*, 1799–1807. [CrossRef]
33. Verghese, M.; Willis, S.; Boateng, J.; Gomaa, A.; Kaur, R. Effect of food processing on antioxidant potential, availability and bioavailability. *Ann. Rev. Food Sci. Technol.* **2012**, *12*, 307–329. [CrossRef]
34. Grundy, M.M.-L.; Wilde, P.J.; Butterworth, P.J.; Gray, R.; Ellis, P.R. Impact of cell wall encapsulation of almonds on in vitro duodenal lipolysis. *Food Chem.* **2015**, *185*, 405–412. [CrossRef] [PubMed]
35. Tu, X.; Wu, B.; Xie, Y.; Xu, S.-L.; Wu, Z.-Y.; Lv, X.; Wei, F.; Du, L.-Q.; Chen, H. A comprehensive study of raw and roasted macadamia nuts: Lipid profile, physicochemical, nutritional, and sensory properties. *Food Sci. Nutr.* **2021**, *9*, 1688–1697. [CrossRef] [PubMed]

36. Oliveira, I.; Meyer, A.S.; Afonso, S.; Sequeira, A.; Vilela, A.; Goufo, P.; Trindade, H.; Gonçalves, B. Effects of Different Processing Treatments on Almond (*Prunus dulcis*) Bioactive Compounds, Antioxidant Activities, Fatty Acids, and Sensorial Characteristics. *Plants* **2020**, *9*, 1627. [CrossRef] [PubMed]
37. Mandalari, G.; Vardakou, M.; Faulks, R.; Bisignano, C.; Martorana, M.; Smeriglio, A.; Trombetta, D. Food Matrix Effects of Polyphenol Bioaccessibility from Almond Skin during Simulated Human Digestion. *Nutrients* **2016**, *8*, 568. [CrossRef]
38. Provatas, A.A.; Ayers, S.A.; Callas, A.A.; Birk, J.W.; Lacson, T.A.; Rosenberg, D.W. Quantitative determination of selected urolithin metabolites in human urine by simple sample preparation and UPLC-MS/MS analysis. *Curr. Top. Anal. Chem.* **2021**, *13*, 69–80.
39. Pfundstein, B.; Haubner, R.; Würtele, G.; Gehres, N.; Ulrich, C.M.; Owen, R.W. Pilot walnut intervention study of urolithin bioavailability in human volunteers. *J. Agric. Food Chem.* **2014**, *62*, 10264–10273. [CrossRef]
40. Bhosale, P.B.; Ha, S.E.; Vetrivel, P.; Kim, H.H.; Kim, S.M.; Kim, G.S. Functions of polyphenols and its anticancer properties in biomedical research: A narrative review. *Transl. Cancer Res.* **2020**, *9*, 7619–7631. [CrossRef]
41. Vinson, J.A.; Cai, Y. Nuts, especially walnuts, have both antioxidant quantity and efficacy and exhibit significant potential health benefits. *Food Funct.* **2012**, *3*, 134–140. [CrossRef]
42. Sanchez-Gonzalez, C.; Ciudad, C.J.; Noe, V.; Izquierdo-Pulido, M. Health benefits of walnut polyphenols: An exploration beyond their lipid profile. *Crit. Rev. Food Sci. Nutr.* **2017**, *57*, 3373–3383. [CrossRef]
43. Wu, S.; Shen, D.; Wang, R.; Li, Q.; Mo, R.; Zheng, Y.; Zhou, Y.; Liu, Y. Phenolic profiles and antioxidant activities of free, esterified and bound phenolic compounds in walnut kernel. *Food Chem.* **2021**, *350*, 129217. [CrossRef]
44. Lozupone, C.A.; Knight, R. Species divergence and the measurement of microbial diversity. *FEMS Microbiol. Rev.* **2008**, *32*, 557–578. [CrossRef] [PubMed]
45. Manor, O.; Dai, C.L.; Kornilov, S.A.; Smith, B.; Price, N.D.; Lovejoy, J.C.; Gibbons, S.M.; Magis, A.T. Health and disease markers correlate with gut microbiome composition across thousands of people. *Nat. Commun.* **2020**, *11*, 5206. [CrossRef] [PubMed]
46. Human Microbiome Project, Consortium. Structure, function and diversity of the healthy human microbiome. *Nature* **2012**, *486*, 207–214. [CrossRef] [PubMed]
47. Holscher, H.D.; Guetterman, H.M.; Swanson, K.S.; An, R.; Matthan, N.R.; Lichtenstein, A.H.; Novotny, J.A.; Baer, D.J. Walnut Consumption Alters the Gastrointestinal Microbiota, Microbially Derived Secondary Bile Acids, and Health Markers in Healthy Adults: A Randomized Controlled Trial. *J. Nutr.* **2018**, *148*, 861–867. [CrossRef]
48. Tindall, A.M.; McLimans, C.J.; Petersen, K.S.; Kris-Etherton, P.M.; Lamendella, R. Walnuts and Vegetable Oils Containing Oleic Acid Differentially Affect the Gut Microbiota and Associations with Cardiovascular Risk Factors: Follow-up of a Randomized, Controlled, Feeding Trial in Adults at Risk for Cardiovascular Disease. *J. Nutr.* **2020**, *150*, 806–817. [CrossRef]
49. Holscher, H.D.; Taylor, A.M.; Swanson, K.S.; Novotny, J.A.; Baer, D.J. Almond Consumption and Processing Affects the Composition of the Gastrointestinal Microbiota of Healthy Adult Men and Women: A Randomized Controlled Trial. *Nutrients* **2018**, *10*, 126. [CrossRef]
50. Burns, A.M.; Zitt, M.A.; Rowe, C.C.; Langkamp-Henken, B.; Mai, V.; Nieves, C., Jr.; Ukhanova, M.; Christman, M.C.; Dahl, W.J. Diet quality improves for parents and children when almonds are incorporated into their daily diet: A randomized, crossover study. *Nutr. Res.* **2016**, *36*, 80–89. [CrossRef]
51. Ukhanova, M.; Wang, X.; Baer, D.J.; Novotny, J.A.; Fredborg, M.; Mai, V. Effects of almond and pistachio consumption on gut microbiota composition in a randomised cross-over human feeding study. *Br. J. Nutr.* **2014**, *111*, 2146–2152. [CrossRef]
52. Dhillon, J.; Newman, J.W.; Fiehn, O.; Ortiz, R.M. Almond Consumption for 8 Weeks Altered Host and Microbial Metabolism in Comparison to a Control Snack in Young Adults. *J. Am. Nutr. Assoc.* **2022**, *42*, 242–254. [CrossRef]
53. Bamberger, C.; Rossmeier, A.; Lechner, K.; Wu, L.; Waldmann, E.; Fischer, S.; Stark, R.G.; Altenhofer, J.; Henze, K.; Parhofer, K.G. A Walnut-Enriched Diet Affects Gut Microbiome in Healthy Caucasian Subjects: A Randomized, Controlled Trial. *Nutrients* **2018**, *10*, 244. [CrossRef]
54. Dhillon, J.; Li, Z.; Ortiz, R.M. Almond Snacking for 8 wk Increases Alpha-Diversity of the Gastrointestinal Microbiome and Decreases Bacteroides fragilis Abundance Compared with an Isocaloric Snack in College Freshmen. *Curr. Dev. Nutr.* **2019**, *3*, nzz079. [CrossRef] [PubMed]
55. Sapp, P.A.; Kris-Etherton, P.M.; Arnesen, E.A.; Chen See, J.R.; Lamendella, R.; Petersen, K.S. Peanuts as a nighttime snack enrich butyrate-producing bacteria compared to an isocaloric lower-fat higher-carbohydrate snack in adults with elevated fasting glucose: A randomized crossover trial. *Clin. Nutr.* **2022**, *41*, 2169–2177. [CrossRef] [PubMed]
56. Parilli-Moser, I.; Dominguez-Lopez, I.; Trius-Soler, M.; Castellvi, M.; Bosch, B.; Castro-Barquero, S.; Estruch, R.; Hurtado-Barroso, S.; Lamuela-Raventos, R.M. Consumption of peanut products improves memory and stress response in healthy adults from the ARISTOTLE study: A 6-month randomized controlled trial. *Clin. Nutr.* **2021**, *40*, 5556–5567. [CrossRef] [PubMed]
57. Zouiouich, S.; Loftfield, E.; Huybrechts, I.; Viallon, V.; Louca, P.; Vogtmann, E.; Wells, P.M.; Steves, C.J.; Herzig, K.H.; Menni, C.; et al. Markers of metabolic health and gut microbiome diversity: Findings from two population-based cohort studies. *Diabetologia* **2021**, *64*, 1749–1759. [CrossRef] [PubMed]
58. Gebauer, S.K.; Novotny, J.A.; Bornhorst, G.M.; Baer, D.J. Food processing and structure impact the metabolizable energy of almonds. *Food Funct.* **2016**, *7*, 4231–4238. [CrossRef] [PubMed]
59. Baer, D.J.; Gebauer, S.K.; Novotny, J.A. Walnuts consumed by healthy adults provide less available energy than predicted by the Atwater factors. *J. Nutr.* **2016**, *146*, 9–13. [CrossRef]

60. Baer, D.J.; Gebauer, S.K.; Novotny, J.A. Measured energy value of pistachios in the human diet. *Br. J. Nutr.* **2012**, *107*, 120–125. [CrossRef]
61. Novotny, J.A.; Gebauer, S.K.; Baer, D.J. Discrepancy between the Atwater factor predicted and empirically measured energy values of almonds in human diets. *Am. J. Nutr.* **2012**, *96*, 296–301. [CrossRef]
62. Baer, D.J.; Novotny, J.A. Metabolizable energy from cashew nuts is less than that predicted by Atwater factors. *Nutrients* **2018**, *11*, 33. [CrossRef]
63. Husted, A.S.; Trauelsen, M.; Rudenko, O.; Hjorth, S.A.; Schwartz, T.W. GPCR-Mediated Signaling of Metabolites. *Cell. Metab.* **2017**, *25*, 777–796. [CrossRef]
64. Liu, P.; Wang, Y.; Yang, G.; Zhang, Q.; Meng, L.; Xin, Y.; Jiang, X. The role of short-chain fatty acids in intestinal barrier function, inflammation, oxidative stress, and colonic carcinogenesis. *Pharmacol. Res.* **2021**, *165*, 105420. [CrossRef] [PubMed]
65. Louis, P.; Flint, H.J. Formation of propionate and butyrate by the human colonic microbiota. *Environ. Microbiol.* **2017**, *19*, 29–41. [CrossRef] [PubMed]
66. Choo, J.M.; Tran, C.D.; Luscombe-Marsh, N.D.; Stonehouse, W.; Bowen, J.; Johnson, N.; Thompson, C.H.; Watson, E.J.; Brinkworth, G.D.; Rogers, G.B. Almond consumption affects fecal microbiota composition, stool pH, and stool moisture in overweight and obese adults with elevated fasting blood glucose: A randomized controlled trial. *Nutr. Res.* **2021**, *85*, 47–59. [CrossRef] [PubMed]
67. Reifen, R.; Karlinsky, A.; Stark, A.H.; Berkovich, Z.; Nyska, A. alpha-Linolenic acid (ALA) is an anti-inflammatory agent in inflammatory bowel disease. *J. Nutr. Biochem.* **2015**, *26*, 1632–1640. [CrossRef] [PubMed]
68. Garcia-Mantrana, I.; Selma-Royo, M.; Gonzalez, S.; Parra-Llorca, A.; Martinez-Costa, C.; Collado, M.C. Distinct maternal microbiota clusters are associated with diet during pregnancy: Impact on neonatal microbiota and infant growth during the first 18 months of life. *Gut Microbes* **2020**, *11*, 962–978. [CrossRef] [PubMed]
69. Martin-Gallausiaux, C.; Marinelli, L.; Blottiere, H.M.; Larraufie, P.; Lapaque, N. SCFA: Mechanisms and functional importance in the gut. *Proc. Nutr. Soc.* **2021**, *80*, 37–49. [CrossRef]
70. Thomas, J.P.; Modos, D.; Rushbrook, S.M.; Powell, N.; Korcsmaros, T. The Emerging Role of Bile Acids in the Pathogenesis of Inflammatory Bowel Disease. *Front. Immunol.* **2022**, *13*, 829525. [CrossRef]
71. Martinot, E.; Sedes, L.; Baptissart, M.; Lobaccaro, J.M.; Caira, F.; Beaudoin, C.; Volle, D.H. Bile acids and their receptors. *Mol. Asp. Med.* **2017**, *56*, 2–9. [CrossRef]
72. Rodriguez-Morato, J.; Matthan, N.R. Nutrition and Gastrointestinal Microbiota, Microbial-Derived Secondary Bile Acids, and Cardiovascular Disease. *Curr. Atheroscler. Rep.* **2020**, *22*, 47. [CrossRef]
73. Singh, R.; Chandrashekharappa, S.; Bodduluri, S.R.; Baby, B.V.; Hegde, B.; Kotla, N.G.; Hiwale, A.A.; Saiyed, T.; Patel, P.; Vijay-Kumar, M.; et al. Enhancement of the gut barrier integrity by a microbial metabolite through the Nrf2 pathway. *Nat. Commun.* **2019**, *10*, 89. [CrossRef]
74. Nakanishi, M.; Matz, A.; Klemashevich, C.; Rosenberg, D.W. Dietary Walnut Supplementation Alters Mucosal Metabolite Profiles During DSS-Induced Colonic Ulceration. *Nutrients* **2019**, *11*, 1118. [CrossRef] [PubMed]
75. Marin, M.; Maria Giner, R.; Rios, J.L.; Recio, M.C. Intestinal anti-inflammatory activity of ellagic acid in the acute and chronic dextrane sulfate sodium models of mice colitis. *J. Ethnopharmacol.* **2013**, *150*, 925–934. [CrossRef] [PubMed]
76. Zhang, S.; Al-Maghout, T.; Cao, H.; Pelzl, L.; Salker, M.S.; Veldhoen, M.; Cheng, A.; Lang, F.; Singh, Y. Gut Bacterial Metabolite Urolithin A (UA) Mitigates Ca(2+) Entry in T Cells by Regulating miR-10a-5p. *Front. Immunol.* **2019**, *10*, 1737. [CrossRef] [PubMed]
77. Koh, S.J.; Choi, Y.I.; Kim, Y.; Kim, Y.S.; Choi, S.W.; Kim, J.W.; Kim, B.G.; Lee, K.L. Walnut phenolic extract inhibits nuclear factor kappaB signaling in intestinal epithelial cells, and ameliorates experimental colitis and colitis-associated colon cancer in mice. *Eur. J. Nutr.* **2019**, *58*, 1603–1613. [CrossRef]
78. Bartoszek, A.; Makaro, A.; Bartoszek, A.; Kordek, R.; Fichna, J.; Salaga, M. Walnut Oil Alleviates Intestinal Inflammation and Restores Intestinal Barrier Function in Mice. *Nutrients* **2020**, *12*, 1302. [CrossRef]
79. Arab, H.H.; Salama, S.A.; Omar, H.A.; Arafa el, S.A.; Maghrabi, I.A. Diosmin protects against ethanol-induced gastric injury in rats: Novel anti-ulcer actions. *PLoS ONE* **2015**, *10*, e0122417. [CrossRef]
80. Liu, R.; Hao, Y.T.; Zhu, N.; Liu, X.R.; Kang, J.W.; Mao, R.X.; Hou, C.; Li, Y. The Gastroprotective Effect of Small Molecule Oligopeptides Isolated from Walnut (*Juglans regia* L.) against Ethanol-Induced Gastric Mucosal Injury in Rats. *Nutrients* **2020**, *12*, 1138. [CrossRef]
81. Park, J.M.; Han, Y.M.; Park, Y.J.; Hahm, K.B. Dietary intake of walnut prevented Helicobacter pylori-associated gastric cancer through rejuvenation of chronic atrophic gastritis. *J. Clin. Biochem. Nutr.* **2021**, *68*, 37–50. [CrossRef]

nutrients

Review

Effect of Nuts on Markers of Inflammation and Oxidative Stress: A Narrative Review

Sujatha Rajaram [1,*], Nagila Raquel Teixeira Damasceno [2], Ribanna Aparecida Marques Braga [2], Raquel Martinez [1], Penny Kris-Etherton [3] and Aleix Sala-Vila [4]

1 School of Public Health, Loma Linda University, Loma Linda, CA 92354, USA
2 Department of Nutrition, School of Public Health, University of São Paulo, São Paulo 05508-220, Brazil
3 Department of Nutritional Sciences, Pennsylvania State University, State College, PA 16801, USA
4 Cardiovascular Epidemiology and Genetics Group, IMIM Hospital del Mar Medical Research Institute, 08003 Barcelona, Spain
* Correspondence: srajaram@llu.edu; Tel.: +1-909-558-4500 (ext. 47228)

Abstract: Oxidative stress and inflammation are mediators in the pathophysiology of several non-communicable diseases (NCDs). Tree nuts and peanuts lower risk factors of cardiometabolic disease, including blood lipids, blood pressure and insulin resistance, among others. Given their strong antioxidant/anti-inflammatory potential, it is plausible that nuts may also exert a favorable effect on inflammation and oxidative stress. Evidence from systematic reviews and meta-analyses of cohort studies and randomized controlled trials (RCTs) suggest a modest protective effect of total nuts; however, the evidence is inconsistent for specific nut types. In this narrative review, the state of evidence to date is summarized for the effect of nut intake on biomarkers of inflammation and oxidative stress, and an attempt is made to define the gaps in research while providing a framework for future research. Overall, it appears that some nuts, such as almonds and walnuts, may favorably modify inflammation, and others, such as Brazil nuts, may favorably influence oxidative stress. There is a pressing need for large RCTs with an adequate sample size that consider different nut types, and the dose and duration of nut intervention, while evaluating a robust set of biomarkers for inflammation and oxidative stress. Building a stronger evidence base is important, especially since oxidative stress and inflammation are mediators of many NCDs and can benefit both personalized and public health nutrition.

Keywords: inflammation; non-communicable diseases; oxidative stress; peanuts; tree nuts

Citation: Rajaram, S.; Damasceno, N.R.T.; Braga, R.A.M.; Martinez, R.; Kris-Etherton, P.; Sala-Vila, A. Effect of Nuts on Markers of Inflammation and Oxidative Stress: A Narrative Review. *Nutrients* **2023**, *15*, 1099. https://doi.org/10.3390/nu15051099

Academic Editor: Manuel Sánchez Santos

Received: 6 February 2023
Revised: 18 February 2023
Accepted: 20 February 2023
Published: 22 February 2023

1. Introduction

Non-communicable diseases (NCDs) are related to lifestyle factors including smoking, sedentary habits, and an unhealthy dietary pattern [1]. Oxidative stress and low-grade chronic inflammation are common mediators in the pathophysiology of many age-related NCDs, including obesity, cardiovascular disease (CVD), type-2 diabetes (T2D), some cancers, and neurodegenerative diseases [2]. Diet and lifestyle modification strategies are cornerstones for preventing NCDs. The evidence supports the role of plant-based diet patterns and plant foods such as tree nuts and peanuts in preventing NCDs and comorbidities, thereby favorably modifying the incidence and mortality of these diseases [3–7].

Reactive oxygen species (ROS) and nitrogen oxidative species (NOS) are part of the normal cellular processes, but the oxidative stress induced by the imbalance of antioxidant systems triggers chronic low-grade inflammation (Figure 1) that contributes to the development of atherosclerosis, CVD, and insulin resistance-related diseases [2,8,9]. In the setting of oxidative stress there is the activation of the regulatory transcription factor, nuclear factor kappa beta (NFκβ), resulting in the release of pro-inflammatory cytokines and the inhibition of nuclear factor-erythroid factor 2-related factor 2 (Nrf2), all leading to more inflammation [10,11]. Plant foods like nuts are rich in phytochemicals and other antioxidant nutrients, with a potential to counteract oxidative stress and inflammation. Tree nuts and peanuts lower major risk factors of cardiometabolic disease, including blood lipids/lipoproteins, blood pressure, endothelial dysfunction and insulin resistance [12–15]. Given their strong antioxidant/anti-inflammatory potential, nuts may also exert a favorable effect on other risk factors of cardiometabolic disease, such as inflammation and oxidative stress.

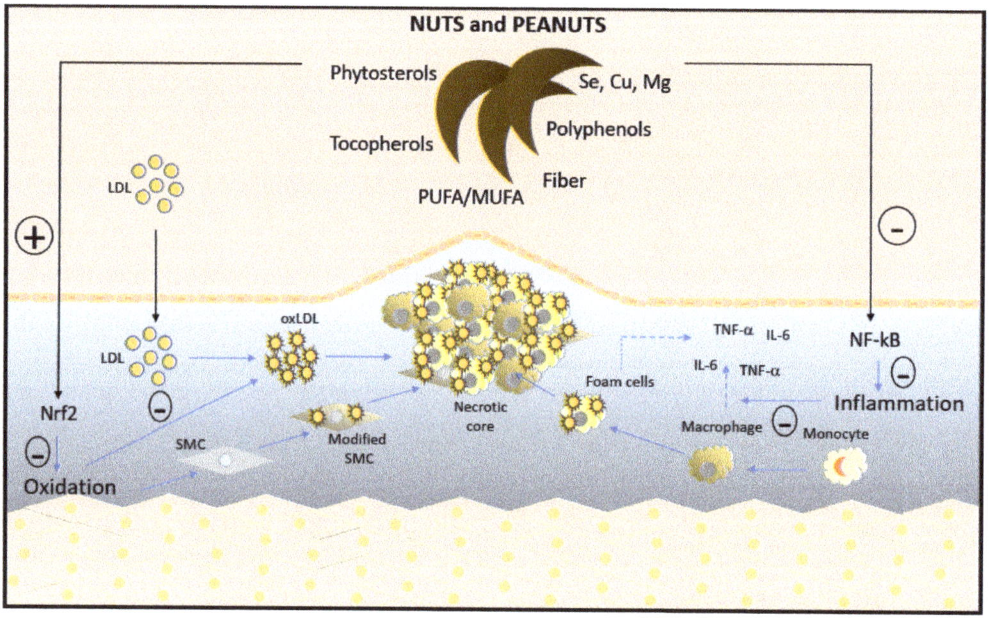

Figure 1. Inflammatory and oxidative processes modulated by nutrients and bioactive substances in tree nuts and peanuts. Fiber, phytosterols, PUFA/MUFA and mineral antioxidants such as Se, Cu, and Mg in tree nuts and peanuts can improve lipid metabolism through a cholesterol-lowering effect on low-density lipoprotein (LDL). Additionally, antioxidants (polyphenols and tocopherols) quench free radicals, avoiding the oxidative modification of LDL (oxLDL). Together, this process prevents monocyte migration and macrophage differentiation, and the consequent endocytosis of oxLDL by macrophages and the formation of foam cells. Increased levels of free radicals in the sub endothelial space stimulate changes in the smooth muscle cell (SMC) phenotype, favoring uptake of oxLDL. This complex process is essential to induce and maintain the low-grade inflammation characterized by the continuous synthesis of inflammatory cytokines (TNF-α, IL-6, and other markers) common in atherosclerosis and present in many other chronic diseases. Tree nuts and peanuts rich in bioactive substances can modulate multiple inflammation and oxidation pathways. Additionally, nut antioxidants can directly quench reactive oxygen species produced in the cell, reduce oxidative stress, and suppress NFκβ expression and downstream pro-inflammatory cytokine generation. Nut antioxidants are cofactors of several antioxidant enzymes. They can increase their activity, upregulate the gene expression of Nrf2, and increase the expression of antioxidant response element gene expression.

Tree nuts (almonds, Brazil nuts, cashew, hazelnuts, macadamia, pecans, pine nuts, pistachios, and walnuts) and peanuts (classified as a legume but considered a nut due to its similar nutrient profile and health benefits) are plant foods that offer a unique combination of macronutrients, micronutrients, and phytonutrients. They are rich in unsaturated fat, which includes polyunsaturated fat (PUFA), both linoleic and α-linolenic acid (ALA), and monounsaturated fat (MUFA). Walnuts and pine nuts have a significant amount of PUFAs, while all nuts are high in MUFAs. Nuts also have a plethora of powerful antioxidant and anti-inflammatory bioactives, including tocopherols, selenium, zinc, magnesium, fiber, phytosterols, and polyphenols [16–19]. Many of these nutrients and non-nutrients present in nuts exert antioxidant and anti-inflammatory effects independently [18,19], but their presence in a whole food matrix may promote synergistic effects.

Early nut intervention trials primarily observed a lipid-lowering effect of nuts; interestingly, the CVD risk reduction was greater than predicted based on low-density lipoprotein cholesterol (LDL-C) lowering alone [12–14,19,20]. To gain a deeper understanding of the other mechanisms beyond blood lipids that might contribute to the overall CVD risk reduction, the effect of nuts on biomarkers of inflammation and oxidative stress were evaluated mostly as secondary outcomes. This narrative review summarizes the current understanding of the role of tree nuts and peanuts on oxidative stress and inflammation biomarkers, identifies the gaps in this area, and provides a framework for future research. This review was presented at an international conference, "NUTS 2022, Where we are and where we are going in research", in the session on "nuts, inflammation and oxidation". While not exhaustive, this review has considered all relevant systematic reviews, cohort studies and randomized controlled trials (RCTs) published to date.

2. Nuts and Inflammation and Oxidative Stress

This section explores evidence from cohort studies and RCTs on the association or effect, respectively, of the various tree nuts and peanuts on markers of inflammation and oxidative stress. Serum markers of inflammation frequently assayed include C-reactive protein (CRP), tumor-necrosis factor-α (TNF-α), interleukin-6 (IL-6) and interleukin-1β (IL-1β), as well as adhesion molecules E-selectin, intercellular (ICAM-1), and vascular cell adhesion (VCAM-1) molecules. Enzymes play an integral part in managing antioxidant and oxidant homeostasis, from which superoxide dismutase (SOD) converts superoxide into hydrogen peroxide and dioxygen [21], while glutathione S-transferase (GST) participates in the phase II detoxification process [22]. The markers of oxidative stress frequently assessed in studies are oxidized LDL (oxLDL), lipid peroxides malondialdehyde (MDA), DNA damage marker, 8-hydroxy deoxy guanine (8-OHdG), antioxidant enzyme activity, and antioxidant capacity. The evidence to date from cohort studies and RCTs is presented below.

2.1. Evidence from Cohort Studies

Systematic reviews and meta-analyses of prospective cohort studies investigated the association between nut consumption and coronary heart disease (CHD) risk, stroke, hypertension, and T2D [5,6,23–25]. Irrespective of follow-up time (3.8 to 26 years), geographical regions, and variation in cohort size (175,000 to >300,000), higher nut intake was inversely associated with CHD risk (relative risk [RR] = 0.81; 95% CI = 0.72–0.91, heterogeneity [I^2] = 56.8%, $p = 0.018$) [6]. In cohort studies of hypertension cases versus control, a marginal association between higher nut intake and reduced risk of hypertension was noted (RR = 0.66; 95% CI = 0.44–1.00, $I^2 = 75.9\%$, $p = 0.006$) but no association was observed for T2D or stroke risk. With respect to nut consumption and colorectal cancer risk, the pooled relative risk for the highest versus lowest (never) categories of nut consumption was 0.91 (95% CI = 0.79–1.05, $I^2 = 49.1\%$, $p = 0.08$). In case–control studies, there was a significant reduction in colorectal cancer risk (RR = 0.84; 95%CI = 0.71–0.99) with nut consumption [6].

The association of nuts and peanuts on the incidence of total mortality, mortality related to CVD, cancer and all-cause among T2D subjects was evaluated in a cross-sectional sample from the Nurses' Health Study (NHS) (1980–2014) and Health Professionals Follow-Up Study (HPFS) (1986–2014) cohorts including 16,217 men and women with T2D at baseline or diagnosed during follow-up [24]. Nuts were associated with a significant reduction in the incidence of total CVD (hazard ratio [HR] = 0.80; 95% CI = 0.70–0.92), and CHD (HR = 0.77; 95% CI = 0.65–0.79), and CVD-specific mortality (HR = 0.61; 95% CI = 0.49–0.79), cancer mortality (HR = 0.73; 95% CI = 0.60–0.90) and all-cause mortality

(HR = 0.67; 95% CI = 0.60–0.74). However, peanuts were associated only with a decrease in all-cause mortality. Collectively, the cohort studies, which have considered many relevant covariates (age, BMI, physical activity, smoking, alcohol, energy intake) in their statistical models, confirmed the positive relationship of nuts and peanuts on mortality incidence; however, neither oxidative stress nor inflammation outcomes were considered.

The cohort studies that have determined the association of nuts and peanut intake on inflammatory markers and their role in risk, incidence, and mortality from NCDs are presented in Table 1. Li et al. [25] examined the association between nut consumption and incident CVD in a subsample from the NHS composed of 6309 women with T2D. From a semi-quantitative food frequency questionnaire (FFQ), habitual nut consumption was grouped into four categories based on the number of servings (serving size, 28 g (1 ounce) for nuts and 16 g for peanut butter): mostly never, 1–3 servings/month to 1 serving/week, 2–4 servings/week, or ≥5 servings/week. Although a higher consumption of nuts and peanuts was associated with lower CVD risk in women with T2D, it was not significantly associated with the inflammatory markers, including tumor necrosis factor receptor 2 (TNFR), ICAM-1, E-selectin, CRP, or fibrinogen. A cross-sectional study from the NHS cohort including 987 women with diabetes assessed the adherence level to the Mediterranean dietary (MedDiet) pattern on a 9-point scale. The higher (6–9 score) adherence level to the MedDiet was associated with higher adiponectin level in comparison to individuals in the lower category of adherence (0–3 score), independent of age, total energy intake, BMI, waist circumference, physical activity, and smoking status. To investigate which food groups in the MedDiet were able to explain the improvement in adiponectin level, using an age and energy-adjusted model, women with diabetes in the highest quintile of nut consumption had adiponectin levels significantly higher, by 23%, than those in the lowest quintile, and this remained significant even when adjusted for BMI, smoking status, activity level, and waist circumference [26].

Table 1. Cohort Studies: Associations between nuts and inflammatory biomarkers.

Reference	Study Country	Sample Size Gender	Follow-Up	Nut Type Amount	Outcome	Inflammatory Biomarkers
Li et al., 2009 [25]	NHS, USA	6.309 Female	1989–1990	Peanuts and mixed nuts Almost never or ≥5 servings/week (Portion size—28 g/day for nuts and 16 g/day for peanut butter)	Incident CVD	TNFR-2, ICAM-1, E-selectin, CRP, and fibrinogen (No changes)
Jiang et al., 2006 [27]	MESA, USA	6.080 Female, Male	Baseline	Mixed nuts, seeds, or peanuts/peanut butter Never/rare or ≥5 times/week (Portion size—no data)	Inflammation biomarkers	↓CRP, ↓IL-6 and ↓fibrinogen
Mantzoros et al., 2006 [26]	NHS, USA	987 Female	1989–1990	Mixed nuts Quintile of nuts intake (Portion size—no data)	Adipocytokine	↑Adiponectin
Bonaccio et al., 2015 [28]	Moli-sani Study, Italy	19,386 Female, Male	4.3 years	Walnuts, hazelnuts, almonds, and peanuts Never or ≥8 times/month (Portion size—no data)	Total and specific mortality	↓CRP, ↓platelet count and ↓neutrophil to lymphocyte ratio
Yu et al., 2016 [29]	NHS HPFS, USA	NHS (3654) HPFS (1359) Female, Male	NHS (1989–1990) HPFS (1993–1995)	Peanuts, mixed nuts, and peanut butter Almost never or ≥5 times/week (Portion size—28 g/day)	Inflammatory biomarkers	↓TNFR-2, ↓CRP and ↓IL-6

NHS—Nurses' Health Study; MESA—Multiethnic study of atherosclerosis; HPFS—Health professional follow-up study; CVD—cardiovascular disease; TNFR—Tumor necrosis factor receptor; CRP—C-reactive protein; IL-6—Interleukin 6. ↓—higher consumption of nuts associated with lower levels of the inflammatory markers; ↑—higher consumption of nuts associated with higher levels of the anti-inflammatory marker.

In the Multi-Ethnic Study of Atherosclerosis cohort, when compared to those with lower total tree nut consumption (never/rarely), a higher consumption (≥5 times/week; portion size not shown) was associated with lower levels of inflammatory biomarker CRP (1.98 vs. 1.69 mg/L). For peanuts and peanut butter, a higher level of consumption (≥5 times/week) compared to never/rarely consumed was associated with lower IL-6 (1.11

vs. 1.24 pg/mL) [27]. The same cohort also investigated the influence of anthropometric measures (BMI and waist-to-hip ratio (WHR)) and ethnicity on the relationship between total nuts and seed consumption and inflammatory markers. Lower levels of CRP and IL-6 and WHR < 0.94 cm were observed in groups with higher nut and seed consumption, while individuals with BMI $\geq$ 28 kg/m^2 and WHR $\geq$ 0.94 cm had lower fibrinogen values. Regarding ethnicity, lower CRP, IL-6, and fibrinogen levels were found in the Caucasian population with higher total nut and seed consumption.

Cross-sectional data from the NHS and HPFS demonstrated that higher nut consumption ($\geq$5 times/week; portion size—28 g/day) was associated with lower CRP (OR = 0.84; 95% CI = 0.74–0.95; *p* trend = 0.006) and IL-6 (OR = 0.88; 95% = 0.79–0.99; *p*-trend = 0.016) [27], but not TNFR-2 [29]. In addition, the Moli-Sani Study cohort [28] with a follow-up of 4.3 years showed that individuals with higher nut consumption ($\geq$8 times/month; portion size not shown) had lower levels of low-grade inflammation (CRP, platelet count, and neutrophil-to-lymphocyte ratio). Higher nut consumption was also associated with all-cause and cancer mortality when inflammatory markers were combined with other variables. Compared with those who reported never consuming nuts, participants with regular nut consumption had a 34% lower risk of all-cause mortality and a 36% reduction in cancer mortality.

In summary, although tree nuts and peanuts are rich in antioxidants and other bioactive components, cohort studies have not consistently considered the association between oxidative stress and inflammatory markers and the risk of fatal and non-fatal disease outcomes when diet evaluations occurred at baseline. Thus, we rely on evidence from RCTs (albeit inconclusive currently) to clarify the relationship between nut consumption and oxidative stress and inflammation.

2.2. Evidence from RCTs

A few systematic and narrative reviews and meta-analyses have summarized findings from RCTs on the role of nut consumption (almonds, Brazil nuts, hazel nuts, pistachios, walnuts) on inflammatory biomarkers [30–36]. There is some evidence, although inconsistent, that nuts may ameliorate inflammation.

When considering inflammatory markers, a meta-analysis on almond consumption [32] mainly noted a significant decrease in serum CRP. This effect was observed more consistently among adults free of obesity and generally healthy than those with metabolic disorders [37–39], and when the intervention was at least 12 weeks. Among other inflammatory cytokines assessed, almond consumption seemed to lower IL-6 significantly in some [40,41], but not all studies [38,42], with no significant changes in TNF-α or endothelial adhesion molecules [32]. The decrease in serum IL-6 attenuated when adjusted for weight loss [42]. Whether almond consumption affects adhesion molecules is unknown, since these markers have not been measured in studies conducted thus far. The evidence appears to primarily suggest a modest lowering of serum CRP with almond intake [32]. However, it is critical to highlight that CRP levels may be associated with numerous factors, including infections, trauma, and non-dietary lifestyle factors [43]. Consequently, researchers should assess multiple inflammatory parameters and adjust for potential confounding variables.

In contrast, a meta-analysis of nine RCTs [33] on walnut intake on CRP that ranged in duration from four weeks to twelve months, with comparators varying from protein foods (eggs, meat), dietary fat (olive oil) to a walnut-free habitual diet, revealed only a non-significant change. The effects of walnuts on other inflammatory cytokines such as IL-6, TNF-α, and IL-1β are inconsistent, mainly tending towards a null effect [33], except when the duration exceeded 24 weeks [44]. Regarding adhesion molecules, although the meta-analyses concluded no significant effect on walnuts (331), individual RCTs suggest a somewhat favorable effect on adhesion molecules [45–47]. The discrepancies in the findings may be partly attributed to the small sample size and short duration of most of these RCTs [33]. Meta-analyses performed with a small number of studies that vary in study design, duration, and subject characteristics, and are limited in sample size, may introduce sampling error that leads to bias in reporting [48]. A key consideration is the intensity and duration of the intervention, which was two years in the Walnuts and Healthy Aging (WAHA) study [44]. This significantly exceeded even the longest trial of 24 weeks cited in the meta-analysis on walnuts [33], highlighting the relevancy of the exposure period. In the WAHA study, the daily ingestion of walnuts over two years in healthy older adults significantly reduced several inflammatory biomarkers including E-Selectin, IL-6, TNF-α, granulocyte-macrophage colony stimulating factor, interferon-γ and, most strikingly, IL-1β (which are associated with CHD).

In a recent systematic review of hazelnut consumption on cardiometabolic risk factors [36], only a few studies assessed the outcomes of inflammation and/or oxidative stress markers and revealed null effects. The seemingly favorable effect on CRP and VCAM-1 was from a study that utilized a less rigorous sequential or single intervention design, and/or combined hazelnuts with cocoa as the intervention [49]. Similarly, for oxidative stress markers, oxLDL decreased and antioxidant enzyme gene expression increased, but only post-prandially [50] or was noted only in uncontrolled studies [36]. The impact of other tree nuts (Brazil nuts, cashews, macadamias, pecans, pine nuts, pistachios) and peanuts on inflammatory biomarkers has been explored less [30,31,34,35]. However, many of these nuts are rich in phytochemicals and other anti-inflammatory compounds similar to walnuts and almonds [16] and warrant further exploration.

Collectively, the evidence on the effect of nuts and inflammation indicates a significant gap in research. Overall, there is a trend towards modest reductions in CRP and IL-6 with almond intake, although this is attenuated when adjusted for weight loss for IL-6 [32]. Short-term studies ($\leq$24 weeks) with walnuts mostly suggest a significant reduction in endothelial adhesion molecules [33]. For other tree nuts and peanuts, conclusions cannot be drawn due to either insufficient data or inconclusive evidence. However, these inconsistencies and seeming null effects on inflammatory biomarkers may be explained by the study designs implemented, since favorable changes on multiple inflammatory biomarkers were noted when the exposure to a single nut was longer, as was seen with walnuts when consumed for two years [44].

A systematic review of 16 RCTs showed an overall favorable effect of nuts on oxidative stress markers [17,51,52]. However, the strength of the evidence is weak, due mostly to the variations in the markers assessed. A modest beneficial effect was specifically observed for almonds, Brazil nuts and mixed nuts. Almonds at doses ranging from 37 g/d to 74 g/d for at least four weeks consistently lowered oxLDL levels, especially in adults with hyperlipidemia and T2D [53], but other lipid peroxidation products such as MDA, or antioxidant capacity, were not responsive to almond intervention. In a recent meta-analysis on Brazil nuts [35], five RCTs that examined the effect on antioxidant enzyme activity noted a positive effect on the selenium-containing glutathione peroxidase activity (GPx) and the increased expression of Nrf2 [54,55]. Brazil nuts have high selenium content [16,35], which is reflected in high serum selenium levels following the intake of just one Brazil nut a day [54]. Another significant marker of oxidative stress, 8-OHdG, reflecting DNA damage, was reduced as a result of Brazil nut intervention in hemodialysis patients [56], indicating that individuals with this disease state may have amplified oxidative stress. For walnuts specifically, there seems to be a positive impact on oxidative stress markers such as oxLDL and antioxidant capacity, but only postprandially following the ingestion of a walnut meal [57]. In the short term ($\leq$12 weeks), walnuts had no effect on oxLDL [51] unless provided along with almonds and hazelnuts, in which case they lowered the urinary excretion of 8-OHdG following a 12-week intervention in those with metabolic syndrome [58]. For other tree nuts (cashew, pecan, pistachio) and peanuts there are mostly null findings concerning oxidative stress markers, largely because $\leq$ 2 RCTs have evaluated these markers for these nut types [51].

An increase in one or more of the antioxidant enzymes has been noted for some nuts [51], although mostly among adults with metabolic conditions accompanied by a heightened state of oxidative stress [53,56]. Dose and duration may be more important than the type of nuts, since studies of 8 weeks or longer, and at least 40–60 g/d (except for Brazil nuts, with just 1 unit/d) seem to have some clinically relevant impacts on antioxidant enzymes. A less common assay for antioxidant capacity is the ferric-reducing antioxidant property (FRAP), which has been used in a few studies and is increased with nut intervention [57]. Different forms of a nut, specifically walnuts, (whole, skin, oil, defatted) improve FRAP acutely [59]. However, the defatted walnuts have the least effect, suggesting that some of the polyphenol antioxidants may be removed during the fat removal, lowering the antioxidant potential. Inconsistencies across studies on the choice of biomarkers or analytical processes used to measure antioxidant capacity and oxidative stress preclude drawing any conclusions.

The evidence from the various meta-analyses and systematic reviews discussed in this section, and the authors' interpretations of the current knowledge and gaps in research, are summarized in Table 2. The discrepant findings for most nut types on oxidative stress markers illustrate several shortcomings in these studies, including small sample size, short duration of the intervention (mostly $\leq$ 8 weeks), varied nut dose (<10 g/d to >100 g/d), form of the nut (whole nut, nut butter, nut oil), participant characteristics (age, smoking status, health status e.g., healthy, presence of diabetes, metabolic disorders, high CVD risk), and the control diet choice (habitual nut free diet, specific high fat or high protein food, low-fat or MedDiet). While there are several limitations, the most significant current gap in knowledge is the lack of assessment of a more robust profile of oxidative stress biomarkers.

Table 2. Nuts and inflammation and oxidative stress biomarkers—current evidence from randomized controlled trials and research gaps.

Nut Type Reference [#]	CRP	IL-6	TNF-α	Adhesion Molecules	OxLDL	Antioxidant Enzymes	Oxidized Metabolites
Almonds [32,51]	↓	↓	↔	↔	↓	-	↔
Brazil nuts [35,51]	↔	-	-	-	↔	↑	↔
Hazelnuts [36,51,52]	↔	-	-	-	-	-	-
Pistachios [34,51]	↔	-	-	-	↔	-	-
Walnuts * [33,44,51,52]	↔	↓	↓	↓	↔	-	↔
Other ** [30,31,51]	-	-	-	-	-	-	-

Adhesion molecules (E-selectin, ICAM-1, VCAM-1); antioxidant enzymes include glutathione peroxidase or catalase, superoxide dismutase activity or gene expression; oxidized metabolites include 8-hydroxy deoxy guanine, urinary isoprostane, MDA. ↑ Significant increase; ↓ Significant decrease; ↔ Null impact or conflicting findings; - Not sufficient evidence (i.e., gaps in research) * Walnuts affect inflammatory biomarkers while mostly null according to a meta-analysis [33]; exposure to walnuts for two years produced a more robust anti-inflammatory effect [44], which is reflected in this table. ** Other tree nuts (cashews, macadamia, pecans, pine nuts) and peanuts.

3. Nuts: Antioxidant and Anti-Inflammatory Mechanisms

Inflammation and oxidative stress, two of the common mediators of NCDs, result from a dysregulation caused by an increased production of reactive oxygen species (due to an imbalance in oxidative and antioxidative pathways in the cell), producing pro-inflammatory cytokines [2,8]. These processes are regulated by two transcriptional factors, NFκβ and Nrf2, the former induced when oxidative stress is high and resulting in a pro-inflammatory state, and the latter responsible for mitigating the state of inflammation and oxidative stress. The plethora of bioactive compounds in nuts, such as unsaturated fatty acids, tocopherol, selenium, copper, fiber, phytosterols, polyphenols, and other phytonutrients, independently and synergistically modulate inflammation and oxidative stress by influencing one or more of the pathways involving these two nuclear factors (Figure 1).

With respect to oxidation, PUFAs act as a pro-oxidant whereas MUFAs are mostly neutral or do not contribute to oxidative stress. Nuts predominantly have MUFAs, except walnuts, which are high in PUFAs. Studies of nuts lowering oxLDL have mostly been observed with hazelnuts, almonds, and Brazil nuts, but not walnuts [51]. Importantly, no adverse effects have been reported, possibly due to the presence of other powerful antioxidants that might counteract the oxidant potential of PUFAs [16,17]. Some of the phytochemicals in nuts (phytosterols, and polyphenols), as well as selenium, can upregulate the Nrf2 pathway which stimulates the antioxidant response element (ARE) gene transcription and the various antioxidant enzymes it encodes, including GPx, superoxide dismutase and catalase [60]. When Nrf2 is activated, it directly counteracts the NFκβ pathway, and reduces the pro-inflammatory state [60,61]. Antioxidants such as γ-tocopherol and phytonutrients directly inhibit the NFκβ pathway and suppress the pro-inflammatory state [62]. Unsaturated fatty acids, particularly ALA found in walnuts, may exhibit anti-inflammatory activity, potentially via the modulation of cyclooxygenase and lipoxygenase pathways [63]. Vitamin E (tocopherols), polyphenols, phytosterols and selenium are powerful antioxidants that are mostly responsible for the antioxidant capacity of tree nuts and peanuts [16–18]. Selenium and copper, notedly high in some nuts (such as Brazil nuts), are cofactors of antioxidant enzymes, including GPx that suppresses oxidative stress [58]. Overall, the bioactive compounds in the nut matrix, both independently and synergistically, may be exerting antioxidant and anti-inflammatory effects.

4. Scope for the Future

Routine tree nut and peanut consumption improves health and reduces the risk of NCDs through multiple pathways, including a reduction in inflammation and oxidative stress [6,7,12–15]. Although the relationship between inflammation and oxidative stress is well recognized, cohort studies conducted to date have only evaluated the association between nut intake and inflammatory biomarkers, not oxidative markers, or their association with disease outcomes. Thus, future cohort studies could consider both oxidative stress and inflammatory markers as intermediate mediators between exposure and disease outcomes. Furthermore, cohort studies should include repeated measures of diet during the follow-up to reduce bias associated with changes that may be made in lifestyle behaviors over time.

The comprehensive evidence from RCTs on the role of nuts on inflammation and oxidative stress biomarkers remains unresolved, although there is evidence of beneficial effects for some nuts such as almonds and walnuts on select markers of inflammation, and for Brazil nuts on oxidative stress. Historically, RCTs have considered inflammation and oxidative stress biomarkers as secondary outcomes [30,31]. However, in nut intervention trials, the CVD risk reduction achieved was greater than predicted only by the LDL-C lowering [19]. This indicates that other mechanisms could contribute to the overall CVD risk reduction, which perhaps encouraged the evaluation of inflammation and oxidative biomarkers to explain this gap. These studies were likely underpowered to assess these biomarkers, as they were not a prespecified primary outcome. Well-designed RCTs with a large sample size, evaluating inflammation and oxidative stress biomarkers as primary outcomes, are vital to clarify some of the inconsistencies that exist at present. These RCTs should consider the duration (≥24 weeks), dose (40–60 g/d), how nuts are incorporated (added to the habitual diet, substitution for nutrient/food or displacement), the comparator diet (habitual nut-free diet or high carbohydrate snack, or other fat source), potential confounders (including age, sex, genotype and lifestyle factors), and the participants' health status (healthy, high risk for T2D or CVD, obesity).

One of the major confounders in studies assessing the association between nut intake and inflammation and oxidative stress biomarkers is body weight. While most studies adjust for weight or BMI, the exposure's effect may differ among those with and without obesity [30–36]. In addition, studies show that nuts, when part of a low-calorie diet

(LCD), may produce similar weight loss compared to a nut-free LCD [64]; however, nut-enriched LCD may have additional benefits, as in favorably modifying inflammatory biomarkers. This implies that nuts may influence inflammatory biomarkers independent of body weight, but future studies have to untangle the complex connections between body weight, inflammation and nut consumption.

The biggest gap in this area of nut research is the apparent lack of evaluation of multiple markers of inflammation and oxidative stress. No single biomarker is ideal, as each marker may be associated with different metabolic conditions (high CRP associated with CVD, for example) or originate from different processes (8-OHdG for DNA damage, F2 isoprostanes for lipid peroxidation). The current consensus summarized in Table 2 is that for some nuts there are few to no RCTs that have evaluated these outcomes. For others, comparing outcomes from different studies is challenging, since they have not always measured the same markers, or have used different study designs, sometimes of less rigor [30–36]. In the WAHA study, one of the strengths besides the long duration (two years) of exposure to walnuts was the inclusion of ten inflammatory biomarkers [44], where six showed a favorable modification with walnut consumption. There is also a concern that there may be large inter- and intra-individual variability for these biomarkers, that would make it difficult to detect any real change caused by diet modification [65]—this makes a case for including a large sample size. The validity of measuring these markers only in a fasting state, which is currently the practice [30–36], is considered insensitive, especially in healthy individuals. Instead, the use of inflammatory challenges has been proposed [66] and much work is needed in this area before considering for future studies on diet and inflammation/oxidative stress. In addition, metabolomic signatures that better represent low-grade inflammation could be identified and may be more useful than single markers.

Along with assessing multiple biomarkers, it would be useful to explore other mechanisms by which nuts influence inflammation and oxidative stress. Genetics, epigenetics, and omics would be relevant considerations for the future [67]. We have limited evidence that Brazil nuts increase the gene expression of antioxidant enzymes and suppress the NFκβ pathway [55], and that pistachios lower IL-6 gene expression [68]. This needs to be verified with more well-designed studies that include other nut types. Epigenetic changes can also modify inflammatory genes, as demonstrated in a cohort study of high CVD-risk patients following a MedDiet with mixed nuts [67]. While DNA methylation assays are expensive, they may be well worth the investment as plant food bioactives seem to favorably modify inflammation and oxidative stress through DNA methylation changes [69]. Finally, the connection between the gut microbiome, whole-body inflammation, and oxidative stress cannot be ignored. Given the beneficial role of nuts on the gut microbiome [70], future studies should include the inter-relationships between the gut axis and immune system in nut exposure studies.

In secondary prevention trials of CVD and type-2 diabetes, new frontiers to explore would include inflammation and oxidative stress. One would expect patients with NCDs to have elevated inflammation, and thus have a better response to an intervention. Although a small study in adults with CHD failed to show any change in inflammation with 30 g/d of pecans [71], another study showed that a mixed nuts or extra virgin olive oil enriched MedDiet was associated with increased atheroma plaque stability and reduced vascular inflammation compared to a low-fat MedDiet [72]. Thus, in healthy individuals with normal values of these biomarkers it may be challenging to see the intervention effect, and hence choosing individuals with elevated inflammation and oxidative stress as an inclusion criterion may be a better option.

Cardiometabolic disease is often associated with non-alcoholic fatty liver disease (NAFLD), with oxidative stress and inflammation considered mediators of this condition [2]. Preliminary evidence suggests that frequent nut consumption ($\geq$1 time/week) is inversely associated with NAFLD, at least in men [73]. Additionally, other polyphenol-rich plant foods (green tea, fruits, and spices) have been shown to protect against fatty liver [74]. With NAFLD becoming a significant public health problem, and given that nuts have abundant polyphenols and other anti-inflammatory nutrients, a critical next step would be to determine if nut intake could lower liver fat fraction in those with NAFLD. Alongside the nut type, duration and dose, such studies must carefully consider the background diet, include nuts in the context of LCDs, and use accurate techniques such as magnetic resonance imaging. Alongside NAFLD, future research can build on preliminary work highlighting a beneficial role for nuts in inflammatory/oxidative stress indices in women with polycystic ovary syndrome (PCOS) and bone health [75,76]. Another emerging trend since the COVID-19 pandemic has been research focused on identifying plant foods to boost the immune system. Oxidative stress and inflammation are potent modulators of the immune response [77] and, thus, evaluating the role of nuts in bolstering the immune response would be timely and critical.

Shifting our approach from utilizing biomarkers to assessing inflammation and oxidative stress, future studies could consider a dietary approach such as the dietary inflammatory index (DII). In the absence of a significant decrease in serum inflammatory markers, a lowering of the DII score may be a surrogate that implies an anti-inflammatory effect. In a six-month dietary intervention study (MedDiet versus low-fat control), a decrease in DII scores was observed with the MedDiet [78], pointing to the anti-inflammatory potential of the MedDiet. Studies exploring the DII of a diet with a single food (nut) intervention may thus be valuable to the overall understanding of the role of nuts in lowering inflammation. Moreover, higher DII scores seem to correlate with CVD and related adverse clinical events [79].

5. Conclusions

In conclusion, the current evidence from cohort studies and randomized clinical trials suggest that tree nuts and peanuts packed with potent bioactive nutrients (MUFAs, PUFAs, vitamin E, selenium and copper) and non-nutrients (fiber, polyphenols and phytosterols) have the potential to reduce inflammation and oxidative stress. However, the evidence is only modest for some, inconsistent for few, and has not been evaluated for many nut types. This creates excellent opportunities for future research to focus on well-designed RCTs that consider many of the limitations described in this narrative review, to further our understanding of the role of nuts in reducing inflammation and oxidative stress. A strong consensus is that including nuts in the habitual diet can help mitigate the risk of several chronic diseases. However, the evidence base for nuts in this area must be expanded to promote food (nut)-based strategies, to lower inflammation and oxidative stress for precision and public health nutrition.

Author Contributions: R.M. and R.A.M.B. conducted the literature review and summarized the evidence to date; S.R. and N.R.T.D. wrote the first draft and S.R. combined and finalized the manuscript; P.K.-E. and A.S.-V. reviewed and provided valuable feedback. All authors have read and agreed to the published version of the manuscript.

Funding: The NUTS 2022 conference was organized by the Rovira i Virgili University, with all the expenses of the organization covered by Barcelo Congresos, including the fees related to publishing the Proceedings in Nutrients. Barcelo Congresos SA has received funds from the International Nut and Dried Fruit Council (INC). Neither the researchers attending the NUTS 2022 Conference nor the Rovira i Virgili University have received funds from the INC. Researchers' travel expenses and accommodation have been covered by Barcelo Congresos. Researchers did not receive any additional financial support related to their participation in the NUTS 2022 conference. Sponsors did not have any role in organizing the conference, selecting, or inviting the speakers, nor in the writing or reviewing of the proceedings.

Conflicts of Interest: R.M., R.A.M.B. and N.R.T.D. have no conflict of interest to report. S.R. reports on current research funding from the Peanut Institute, California Walnut Commission, the Almond Board of California, and the International Nut and Dried Fruit Council. A.S.-V. has received research funding through his institution and support to attend professional meetings from the California Walnut Commission. P.K.-E. reports on current and recent research funding from the American Pistachio Growers, the American Pecan Council, the Peanut Institute, and the California Walnut Commission.

References

1. World Health Organization. Non Communicable Diseases. Available online: https://www.who.int/news-room/fact-sheets/detail/noncommunicable-diseases (accessed on 3 December 2022).
2. Seyedsadjadi, N.; Grant, R. The Potential Benefit of Monitoring Oxidative Stress and Inflammation in the Prevention of Non-Communicable Diseases (NCDs). *Antioxidants* **2020**, *10*, 15. [CrossRef] [PubMed]
3. Orlich, M.J.; Singh, P.N.; Sabaté, J.; Jaceldo-Siegl, K.; Fan, J.; Knutsen, S.; Beeson, W.L.; Fraser, G.E. Vegetarian Dietary Patterns and Mortality in Adventist Health Study. *JAMA Intern. Med.* **2013**, *173*, 1230–1238. [CrossRef] [PubMed]
4. Estruch, R.; Ros, E.; Salas-Salvadó, J.; Covas, M.-I.; Corella, D.; Arós, F.; Gómez-Gracia, E.; Ruiz-Gutiérrez, V.; Fiol, M.; Lapetra, J.; et al. Primary Prevention of Cardiovascular Disease with a Mediterranean Diet. *N. Engl. J. Med.* **2013**, *368*, 1279–1290. [CrossRef] [PubMed]
5. Qian, F.; Liu, G.; Hu, F.B.; Bhupathiraju, S.N.; Sun, Q. Associations between Plant-Based Dietary Patterns and Risk of Type 2 Diabetes: A Systematic Review and Meta-analysis. *JAMA Intern. Med.* **2019**, *179*, 1335–1344. [CrossRef]
6. Aune, D.; Keum, N.; Giovannucci, E.; Fadnes, L.T.; Boffetta, P.; Greenwood, D.C.; Tonstad, S.; Vatten, L.J.; Riboli, E.; Norat, T. Nut consumption and risk of cardiovascular disease, total cancer, all-cause and cause-specific mortality: A systematic review and dose-response meta-analysis of prospective studies. *BMC Med.* **2016**, *14*, 207. [CrossRef]
7. Kim, Y.; Keogh, J.; Clifton, P.M. Nuts and Cardio-Metabolic Disease: A Review of Meta-Analyses. *Nutrients* **2018**, *10*, 1935. [CrossRef]
8. Steven, S.; Frenis, K.; Oelze, M.; Kalinovic, S.; Kuntic, M.; Bayo Jimenez, M.T.; Vujacic-Mirski, K.; Helmstädter, J.; Kröller-Schön, S.; Münzel, T.; et al. Vascular Inflammation and Oxidative Stress: Major Triggers for Cardiovascular Disease. *Oxidative Med. Cell. Longev.* **2019**, *2019*, 7092151. [CrossRef]
9. Sharifi-Rad, M.; Kumar, N.V.A.; Zucca, P.; Varoni, E.M.; Dini, L.; Panzarini, E.; Rajkovic, J.; Tsouh Fokou, P.V.; Azzini, E.; Peluso, I.; et al. Lifestyle, Oxidative Stress, and Antioxidants: Back and Forth in the Pathophysiology of Chronic Diseases. *Front. Physiol.* **2020**, *11*, 1–21. [CrossRef]
10. Liu, T.; Zhang, L.; Joo, D.; Sun, S.-C. NF-κB signaling in inflammation. *Signal Transduct. Target. Ther.* **2017**, *2*, 17023. [CrossRef]
11. Juan, C.A.; de la Lastra, J.M.P.; Plou, F.J.; Pérez-Lebeña, E. The Chemistry of Reactive Oxygen Species (ROS) Revisited: Outlining Their Role in Biological Macromolecules (DNA, Lipids and Proteins) and Induced Pathologies. *Int. J. Mol. Sci.* **2021**, *22*, 4642. [CrossRef]
12. Guasch-Ferré, M.; Li, J.; Hu, F.B.; Salas-Salvadó, J.; Tobias, D.K. Effects of walnut consumption on blood lipids and other cardiovascular disease risk factors: An updated meta-analyses and systematic review of controlled trials. *Am. J. Clin. Nutr.* **2018**, *108*, 174–187. [CrossRef]
13. Xia, J.-Y.; Yu, J.-H.; Xu, D.-F.; Yang, C.; Xia, H.; Sun, G.-J. The Effects of Peanuts and Tree Nuts on Lipid Profile in Type 2 Diabetic Patients: A Systematic Review and Meta-Analysis of Randomized, Controlled-Feeding Clinical Studies. *Front. Nutr.* **2021**, *8*, 765571. [CrossRef]
14. Blanco Mejia, S.; Kendall, C.W.C.; Viguiliouk, E.; Augustin, L.S.; Ha, V.; Cozma, A.I.; Mirrahimi, A.; Maroleanu, A.; Chiavaroli, L.; Leiter, L.A.; et al. Effect of tree nuts on metabolic syndrome criteria: A systematic review and meta-analysis of randomised controlled trials. *BMJ Open* **2014**, *4*, e004660. [CrossRef]
15. Khalili, L.; A-Elgadir, T.M.E.; Mallick, A.K.; El Enshasy, H.A.; Sayyed, R.Z. Nuts as a Part of Dietary Strategy to Improve Metabolic Biomarkers: A Narrative Review. *Front. Nutr.* **2022**, *9*, 881843. [CrossRef]
16. Ros, E.; Singh, A.; O'Keefe, J.H. Nuts: Natural Pleiotropic Nutraceuticals. *Nutrients* **2021**, *13*, 3269. [CrossRef]
17. Lorenzon dos Santos, J.; Quadros, A.S.; Weschenfelder, C.; Garofallo, S.B.; Marcadenti, A. Oxidative Stress Biomarkers, Nut-Related Antioxidants, and Cardiovascular Disease. *Nutrients* **2020**, *12*, 682. [CrossRef]
18. Bolling, B.W.; Chen, C.-Y.O.; McKay, D.L.; Blumberg, J.B. Tree nut phytochemicals: Composition, antioxidant capacity, bioactivity, impact factors. A systematic review of almonds, Brazils, cashews, hazelnuts, macadamias, pecans, pine nuts, pistachios and walnuts. *Nutr. Res. Rev.* **2011**, *24*, 244–275. [CrossRef]
19. Sabaté, J.; Wien, M. Nuts, blood lipids and cardiovascular disease. *Asia Pac. J. Clin. Nutr.* **2010**, *19*, 131–136.
20. Rajaram, S.; Cofán, M.; Sala-Vila, A.; Haddad, E.; Serra-Mir, M.; Bitok, E.; Roth, I.; Freitas-Simoes, T.M.; Kaur, A.; Valls-Pedret, C.; et al. Effects of Walnut Consumption for 2 Years on Lipoprotein Subclasses among Healthy Elders: Findings from the WAHA randomized controlled trial. *Circulation* **2021**, *144*, 1083–1085. [CrossRef]
21. Kurutas, E.B. The importance of antioxidants which play the role in cellular response against oxidative/nitrosative stress: Current state. *Nutr. J.* **2016**, *15*, 71. [CrossRef]
22. Sheehan, D.; Meade, G.; Foley, V.M.; Dowd, C.A. Structure, function and evolution of glutathione transferases: Implications for classification of non-mammalian members of an ancient enzyme superfamily. *Biochem. J.* **2011**, *360*, 1–16. [CrossRef]
23. Lockyer, S.; de la Hunty, A.E.; Steenson, S.; Spiro, A.; Stanner, S.A. Walnut consumption and health outcomes with public health relevance—A systematic review of cohort studies and randomized controlled trials published from 2017 to present. *Nutr. Rev.* **2022**, *81*, 26–54. [CrossRef] [PubMed]
24. Liu, G.; Guasch-Ferré, M.; Hu, Y.; Li, Y.; Hu, F.B.; Rimm, E.B.; Manson, J.E.; Rexrode, K.M.; Sun, Q. Nut Consumption in Relation to Cardiovascular Disease Incidence and Mortality among Patients with Diabetes Mellitus. *Circ. Res.* **2019**, *124*, 920–929. [CrossRef] [PubMed]

25. Li, T.Y.; Brennan, A.M.; Wedick, N.M.; Mantzoros, C.; Rifai, N.; Hu, F.B. Regular Consumption of Nuts Is Associated with a Lower Risk of Cardiovascular Disease in Women with Type 2 Diabetes. *J. Nutr.* **2009**, *139*, 1333–1338. [CrossRef] [PubMed]
26. Mantzoros, C.S.; Williams, C.J.; Manson, J.E.; Meigs, J.B.; Hu, F.B. Adherence to the Mediterranean dietary pattern is positively associated with plasma adiponectin concentrations in diabetic women. *Am. J. Clin. Nutr.* **2006**, *84*, 328–335. [CrossRef]
27. Jiang, R.; Jacobs, D.R., Jr.; Mayer-Davis, E.; Szklo, M.; Herrington, D.; Jenny, N.S.; Kronmal, R.; Barr, R.G. Nut and Seed Consumption and Inflammatory Markers in the Multi-Ethnic Study of Atherosclerosis. *Am. J. Epidemiol.* **2006**, *163*, 222–231. [CrossRef]
28. Bonaccio, M.; Di Castelnuovo, A.; De Curtis, A.; Costanzo, S.; Bracone, F.; Persichillo, M.; Donati, M.B.; de Gaetano, G.; Iacoviello, L. Nut consumption is inversely associated with both cancer and total mortality in a Mediterranean population: Prospective results from the Moli-sani study. *Br. J. Nutr.* **2015**, *114*, 804–811. [CrossRef]
29. Yu, Z.; Malik, V.S.; Keum, N.; Hu, F.B.; Giovannucci, E.L.; Stampfer, M.J.; Willett, W.C.; Fuchs, C.S.; Bao, Y. Associations between nut consumption and inflammatory biomarkers. *Am. J. Clin. Nutr.* **2016**, *104*, 722–728. [CrossRef]
30. Neale, E.P.; Tapsell, L.C.; Guan, V.; Batterham, M.J. The effect of nut consumption on markers of inflammation and endothelial function: A systematic review and meta-analysis of randomised controlled trials. *BMJ Open* **2017**, *7*, e016863. [CrossRef]
31. Xiao, Y.; Xia, J.; Ke, Y.; Cheng, J.; Yuan, J.; Wu, S.; Lv, Z.; Huang, S.; Kim, J.H.; Wong, S.Y.-S.; et al. Effects of nut consumption on selected inflammatory markers: A systematic review and meta-analysis of randomized controlled trials. *Nutrition* **2018**, *54*, 129–143. [CrossRef]
32. Fatahi, S.; Daneshzad, E.; Lotfi, K.; Azadbakht, L. The effects of almond consumption on inflammatory biomarkers in adults: A systematic review and meta-analysis of randomized controlled trials. *Adv. Nutr.* **2021**, *13*, 1462–1475. [CrossRef]
33. Mateș, L.; Popa, D.-S.; Rusu, M.E.; Fizeșan, I.; Leucuța, D. Walnut Intake Interventions Targeting Biomarkers of Metabolic Syndrome and Inflammation in Middle-Aged and Older Adults: A Systematic Review and Meta-Analysis of Randomized Controlled Trials. *Antioxidants* **2022**, *11*, 1412. [CrossRef]
34. Asbaghi, O.; Hadi, A.; Campbell, M.S.; Venkatakrishnan, K.; Ghaedi, E. Effects of pistachios on anthropometric indices, inflammatory markers, endothelial function and blood pressure in adults: A systematic review and meta-analysis of randomised controlled trials. *Br. J. Nutr.* **2021**, *126*, 718–729. [CrossRef]
35. Godos, J.; Giampieri, F.; Micek, A.; Battino, M.; Forbes-Hernández, T.Y.; Quiles, J.L.; Paladino, N.; Falzone, L.; Grosso, G. Effect of Brazil nuts on selenium status, blood lipids, and biomarkers of oxidative stress and inflammation: A systematic review and meta-analysis of randomized controlled trials. *Antioxidants* **2022**, *11*, 403. [CrossRef]
36. Brown, R.; Ware, L.; Tey, S.L. Effects of Hazelnut Consumption on Cardiometabolic Risk Factors and Acceptance: A Systematic Review. *Int. J. Environ. Res. Public Health* **2022**, *19*, 2880. [CrossRef]
37. Berryman, C.E.; West, S.G.; Fleming, J.A.; Bordi, P.L.; Kris-Etherton, P.M. Effects of Daily Almond Consumption on Cardiometabolic Risk and Abdominal Adiposity in Healthy Adults with Elevated LDL-Cholesterol: A Randomized Controlled Trial. *J. Am. Heart Assoc.* **2015**, *4*, e000993. [CrossRef]
38. Rajaram, S.; Connell, K.M.; Sabaté, J. Effect of almond-enriched high-monounsaturated fat diet on selected markers of inflammation: A randomised, controlled, crossover study. *Br. J. Nutr.* **2010**, *103*, 907–912. [CrossRef]
39. Palacios, O.M.; Maki, K.C.; Xiao, D.; Wilcox, M.L.; Dicklin, M.R.; Kramer, M.; Trivedi, R.; Burton-Freeman, B.; Edirisinghe, I. Effects of Consuming Almonds on Insulin Sensitivity and Other Cardiometabolic Health Markers in Adults with Prediabetes. *J. Am. Coll. Nutr.* **2020**, *39*, 397–406. [CrossRef]
40. Madan, J.; Desai, S.; Moitra, P.; Salis, S.; Agashe, S.; Battalwar, R.; Mehta, A.; Kamble, R.; Kalita, S.; Phatak, A.G.; et al. Effect of Almond Consumption on Metabolic Risk Factors—Glucose Metabolism, Hyperinsulinemia, Selected Markers of Inflammation: A Randomized Controlled Trial in Adolescents and Young Adults. *Front. Nutr.* **2021**, *8*, 668622. [CrossRef]
41. Jung, H.; Chen, C.-Y.O.; Blumberg, J.B.; Kwak, H.-K. The effect of almonds on vitamin E status and cardiovascular risk factors in Korean adults: A randomized clinical trial. *Eur. J. Nutr.* **2018**, *57*, 2069–2079. [CrossRef]
42. Hou, Y.-Y.; Ojo, O.; Wang, L.-L.; Wang, Q.; Jiang, Q.; Shao, X.-Y.; Wang, X.-H. A Randomized Controlled Trial to Compare the Effect of Peanuts and Almonds on the Cardio-Metabolic and Inflammatory Parameters in Patients with Type 2 Diabetes Mellitus. *Nutrients* **2018**, *10*, 1565. [CrossRef] [PubMed]
43. Sproston, N.R.; Ashworth, J.J. Role of C-Reactive Protein at Sites of Inflammation and Infection. *Front. Immunol.* **2018**, *9*, 754. [CrossRef] [PubMed]
44. Cofán, M.; Rajaram, S.; Sala-Vila, A.; Valls-Pedret, C.; Serra-Mir, M.; Roth, I.; Freitas-Simoes, T.M.; Bitok, E.; Sabaté, J.; Ros, E. Effects of 2-Year Walnut-Supplemented Diet on Inflammatory Biomarkers. *J. Am. Coll. Cardiol.* **2020**, *76*, 2282–2284. [CrossRef] [PubMed]
45. Casas-Agustench, P.; Uriarte, P.J.L.; Bulló, M.; Ros, E.; Vila, J.J.C.; Salas-Salvadó, J. Effects of one serving of mixed nuts on serum lipids, insulin resistance and inflammatory markers in patients with the metabolic syndrome. *Nutr. Metab. Cardiovasc. Dis.* **2011**, *21*, 126–135. [CrossRef] [PubMed]
46. Chiang, Y.-L.; Haddad, E.; Rajaram, S.; Shavlik, D.; Sabaté, J. The effect of dietary walnuts compared to fatty fish on eicosanoids, cytokines, soluble endothelial adhesion molecules and lymphocyte subsets: A randomized, controlled crossover trial. *Prostaglandins Leukot. Essent. Fat. Acids* **2012**, *87*, 111–117. [CrossRef]

47. Canales, A.; Sánchez-Muniz, F.J.; Bastida, S.; Librelotto, J.; Nus, M.; Corella, D.; Guillen, M.; Benedi, J. Effect of walnut-enriched meat on the relationship between VACM, ICAM, and LTB4 levels and PON-1 activity in ApoA4 360 and PON-1 allele carriers at increased cardiovascular risk. *Eur. J. Clin. Nutr.* **2011**, *65*, 703–710. [CrossRef]
48. Lin, L. Bias caused by sampling error in meta-analysis with small sample sizes. *PLoS ONE* **2018**, *13*, e0204056. [CrossRef]
49. Orem, A.; Yucesan, F.B.; Orem, C.; Akcan, B.; Kural, B.V.; Alasalvar, C.; Shahidi, F. Hazelnut-Enriched diet improves cardiovascular risk biomarkers beyond a lipid-Lowering effect in hypercholesterolemic subjects. *J. Clin. Lipidol.* **2013**, *7*, 123–131. [CrossRef]
50. Di Renzo, L.; Merra, G.; Botta, R.; Gualtieri, P.; Manzo, A.; Perrone, M.A.; Mazza, M.; Cascapera, S.; De Lorenzo, A. Post-prandial effects of hazelnut-enriched high fat meal on LDL oxidatve status, oxidative and inflammatory gene expression in healthy subjects: A randomized trial. *Eur. Rev. Med. Pharm. Sci.* **2017**, *21*, 1610–1626.
51. Silveira, B.K.S.; da Silva, A.; Hermsdorff, H.H.M.; Bressan, J. Effect of chronic consumption of nuts on oxidative stress: A systematic review of clinical trials. *Crit. Rev. Food Sci. Nutr.* **2022**, *62*, 726–737. [CrossRef]
52. López-Uriarte, P.; Bulló, M.; Casas-Agustench, P.; Babio, N.; Salas-Salvadó, J. Nuts and oxidation: A systematic review. *Nutr. Rev.* **2009**, *67*, 497–508. [CrossRef]
53. Liu, J.-F.; Liu, Y.-H.; Chen, C.-M.; Chang, W.-H.; Chen, C.Y.O. The effect of almonds on inflammation and oxidative stress in Chinese patients with type 2 diabetes mellitus: A randomized crossover, controlled feeding trial. *Eur. J. Nutr.* **2013**, *52*, 927–935. [CrossRef]
54. Colpo, E.; Vilanova, C.D.D.A.; Reetz, L.G.B.; Duarte, M.M.M.F.; Farias, I.L.G.; Muller, E.I.; Muller, A.L.H.; Flores, E.M.M.; Wagner, R.; da Rocha, J.B.T. A Single Consumption of High Amounts of the Brazil Nuts Improves Lipid Profile of Healthy Volunteers. *J. Nutr. Metab.* **2013**, *2013*, 653185. [CrossRef]
55. Cardozo, L.F.M.F.; Stockler-Pinto, M.B.; Mafra, D. Brazil nut consumption modulates Nrf2 expression in hemodialysis patients: A pilot study. *Mol. Nutr. Food Res.* **2016**, *60*, 1719–1724. [CrossRef]
56. Stockler-Pinto, M.B.; Mafra, D.; Moraes, C.; Lobo, J.; Boaventura, G.T.; Farage, N.E.; Silva, W.S.; Cozzolino, S.F.; Malm, O. Brazil Nut (Bertholletia excelsa, H.B.K.) Improves Oxidative Stress and Inflammation Biomarkers in Hemodialysis Patients. *Biol. Trace Element Res.* **2014**, *158*, 105–112. [CrossRef]
57. Haddad, E.H.; Gaban-Chong, N.; Oda, K.; Sabaté, J. Effect of a walnut meal on postprandial oxidative stress and antioxidants in healthy individuals. *Nutr. J.* **2014**, *13*, 4. [CrossRef]
58. López-Uriarte, P.; Nogués, R.; Saez, G.; Bulló, M.; Romeu, M.; Masana, L.; Tormos, C.; Casas-Agustench, P.; Salas-Salvadó, J. Effect of nut consumption on oxidative stress and the endothelial function in metabolic syndrome. *Clin. Nutr.* **2010**, *29*, 373–380. [CrossRef]
59. Berryman, C.E.; Grieger, J.A.; West, S.G.; Chen, C.-Y.O.; Blumberg, J.B.; Rothblat, G.H.; Sankaranarayanan, S.; Kris-Etherton, P.M. Acute Consumption of Walnuts and Walnut Components Differentially Affect Postprandial Lipemia, Endothelial Function, Oxidative Stress, and Cholesterol Efflux in Humans with Mild Hypercholesterolemia. *J. Nutr.* **2013**, *143*, 788–794. [CrossRef]
60. Bayele, H.K.; Debnam, E.S.; Srai, K.S. Nrf2 transcriptional derepression from Keap1 by dietary polyphenols. *Biochem. Biophys. Res. Commun.* **2016**, *469*, 521–528. [CrossRef]
61. Wardyn, J.D.; Ponsford, A.H.; Sanderson, C.M. Dissecting molecular crosstalk between Nrf2 and NF-κB response pathways. *Biochem. Soc. Trans.* **2015**, *43*, 621–626. [CrossRef]
62. Ahn, K.S.; Sethi, G.; Krishnan, K.; Aggarwal, B.B. γ-Tocotrienol Inhibits Nuclear Factor-κB Signaling Pathway through Inhibition of Receptor-interacting Protein and TAK1 Leading to Suppression of Antiapoptotic Gene Products and Potentiation of Apoptosis. *J. Biol. Chem.* **2007**, *282*, 809–820. [CrossRef] [PubMed]
63. Anand, R.; Kaithwas, G. Anti-inflammatory Potential of Alpha-Linolenic Acid Mediated Through Selective COX Inhibition: Computational and Experimental Data. *Inflammation* **2014**, *37*, 1297–1306. [CrossRef] [PubMed]
64. Ghanavati, M.; Hosseinabadi, S.M.; Parsa, S.A.; Safi, M.; Emamat, H.; Nasrollahzadeh, J. Effect of a nut-enriched low-calorie diet on body weight and selected markers of inflammation in overweight and obese stable coronary artery disease patients: A randomized controlled study. *Eur. J. Clin. Nutr.* **2021**, *75*, 1099–1108. [CrossRef] [PubMed]
65. Mallard, A.R.; Hollekim-Strand, S.M.; Ingul, C.B.; Coombes, J.S. High day-to-day and diurnal variability of oxidative stress and inflammation biomarkers in people with type 2 diabetes mellitus and healthy individuals. *Redox Rep.* **2020**, *25*, 64–69. [CrossRef]
66. Calle, M.C.; Andersen, C.J. Assessment of Dietary Patterns Represents a Potential, Yet Variable, Measure of Inflammatory Status: A Review and Update. *Dis. Markers* **2019**, *2019*, 3102870. [CrossRef]
67. Ramos-Lopez, O.; Milagro, F.I.; Riezu-Boj, J.I.; Martinez, J.A. Epigenetic signatures underlying inflammation: An interplay of nutrition, physical activity, metabolic diseases, and environmental factors for personalized nutrition. *Inflamm. Res.* **2021**, *70*, 29–49. [CrossRef] [PubMed]
68. Hernández-Alonso, P.; Salas-Salvadó, J.; Baldrich-Mora, M.; Juanola-Falgarona, M.; Bulló, M. Beneficial Effect of Pistachio Consumption on Glucose Metabolism, Insulin Resistance, Inflammation, and Related Metabolic Risk Markers: A Randomized Clinical Trial. *Diabetes Care* **2014**, *37*, 3098–3105. [CrossRef]
69. Arpón, A.; Milagro, F.I.; Razquin, C.; Corella, D.; Estruch, R.; Fitó, M.; Marti, A.; Martínez-González, M.A.; Ros, E.; Salas-Salvadó, J.; et al. Impact of Consuming Extra-Virgin Olive Oil or Nuts within a Mediterranean Diet on DNA Methylation in Peripheral White Blood Cells within the PREDIMED-Navarra Randomized Controlled Trial: A Role for Dietary Lipids. *Nutrients* **2018**, *10*, 15. [CrossRef]

70. Creedon, A.C.; Hung, E.S.; Berry, S.E.; Whelan, K. Nuts and their Effect on Gut Microbiota, Gut Function and Symptoms in Adults: A Systematic Review and Meta-Analysis of Randomised Controlled Trials. *Nutrients* **2020**, *12*, 2347. [CrossRef]
71. Weschenfelder, C.; Gottschall, C.B.A.; Markoski, M.M.; Portal, V.L.; de Quadros, A.S.; Bersch-Ferreira, C.; Marcadenti, A. Effects of supplementing a healthy diet with pecan nuts or extra-virgin olive oil on inflammatory profile of patients with stable coronary artery disease: A randomised clinical trial. *Br. J. Nutr.* **2022**, *127*, 862–871. [CrossRef]
72. Casas, R.; Sacanella, E.; Urpí-Sardà, M.; Chiva-Blanch, G.; Ros, E.; Martinez-Gonzalez, M.A.; Covas, M.-I.; Salas-Salvadó, J.; Fiol, M.; Arós, F.; et al. The Effects of the Mediterranean Diet on Biomarkers of Vascular Wall Inflammation and Plaque Vulnerability in Subjects with High Risk for Cardiovascular Disease. A Randomized Trial. *PLoS ONE* **2014**, *9*, e100084. [CrossRef]
73. Semmler, G.; Bachmayer, S.; Wernly, S.; Wernly, B.; Niederseer, D.; Huber-Schönauer, U.; Stickel, F.; Aigner, E.; Datz, C. Nut consumption and the prevalence and severity of non-alcoholic fatty liver disease. *PLoS ONE* **2020**, *15*, e0244514. [CrossRef]
74. Li, H.-Y.; Gan, R.-Y.; Shang, A.; Mao, Q.-Q.; Sun, Q.-C.; Wu, D.-T.; Geng, F.; He, X.-Q.; Li, H.-B. Plant-Based Foods and Their Bioactive Compounds on Fatty Liver Disease: Effects, Mechanisms, and Clinical Application. *Oxidative Med. Cell. Longev.* **2021**, *2021*, 6621644. [CrossRef]
75. Kalgaonkar, S.; Almario, R.U.; Gurusinghe, D.; Garamendi, E.M.; Buchan, W.; Kim, K.; Karakas, S.E. Differential effects of walnuts vs. almonds on improving metabolic and endocrine parameters in PCOS. *Eur. J. Clin. Nutr.* **2010**, *65*, 386–393. [CrossRef]
76. Griel, A.E.; Kris-Etherton, P.M.; Hilpert, K.F.; Zhao, G.; West, S.G.; Corwin, R.L. An increase in dietary n-3 fatty acids decreases a marker of bone resorption in humans. *Nutr. J.* **2007**, *6*, 2. [CrossRef]
77. Iddir, M.; Brito, A.; Dingeo, G.; Fernandez Del Campo, S.S.; Samouda, H.; La Frano, M.R.; Bohn, T. Strengthening the Immune System and Reducing Inflammation and Oxidative Stress through Diet and Nutrition: Considerations during the COVID-19 Crisis. *Nutrients* **2020**, *12*, 1562. [CrossRef]
78. Mayr, H.L.; Itsiopoulos, C.; Tierney, A.C.; Ruiz-Canela, M.; Hebert, J.R.; Shivappa, N.; Thomas, C.J. Improvement in dietary inflammatory index score after 6-month dietary intervention is associated with reduction in interleukin-6 in patients with coronary heart disease: The AUSMED heart trial. *Nutr. Res.* **2018**, *55*, 108–121. [CrossRef]
79. Garcia-Arellano, A.; Ramallal, R.; Ruiz-Canela, M.; Salas-Salvadó, J.; Corella, D.; Shivappa, N.; Schröder, H.; Hébert, J.R.; Ros, E.; Gómez-Garcia, E.; et al. Dietary Inflammatory Index and Incidence of Cardiovascular Disease in the PREDIMED Study. *Nutrients* **2015**, *7*, 4124–4138. [CrossRef]

nutrients

Review

Nuts and Cardiovascular Disease Outcomes: A Review of the Evidence and Future Directions

Andrea J. Glenn [1,2,3], Dagfinn Aune [4,5,6], Heinz Freisling [7], Noushin Mohammadifard [8], Cyril W. C. Kendall [2,3,9], Jordi Salas-Salvadó [10,11], David J. A. Jenkins [2,3,12,13,14], Frank B. Hu [1,15,16] and John L. Sievenpiper [2,3,12,13,14,*]

1 Department of Nutrition, Harvard T.H. Chan School of Public Health, Boston, MA 02115, USA
2 Department of Nutritional Sciences, Temerty Faculty of Medicine, University of Toronto, Toronto, ON M5S 1A8, Canada
3 Toronto 3D Knowledge Synthesis and Clinical Trials Unit, Clinical Nutrition and Risk Factor Modification Centre, St. Michael's Hospital, Toronto, ON M5C 2T2, Canada
4 Department of Epidemiology and Biostatistics, School of Public Health, Imperial College London, London SW7 2AZ, UK
5 Department of Nutrition, Oslo New University College, 0372 Oslo, Norway
6 Department of Endocrinology, Morbid Obesity and Preventive Medicine, Oslo University Hospital, 0586 Oslo, Norway
7 Nutrition and Metabolism Branch, International Agency for Research on Cancer (IARC-WHO), 69366 Lyon, France
8 Isfahan Cardiovascular Research Center, Cardiovascular Research Institute, Isfahan University of Medical Sciences, Isfahan, Iran
9 College of Pharmacy and Nutrition, University of Saskatchewan, Saskatoon, SK S7N 5E5, Canada
10 Department of Biochemistry & Biotechnology, School of Medicine, Institut d'Investigacions Sanitàries Pere i Virgili, Rovira i Virgili University, 43204 Reus, Spain
11 CIBER Fisiopatología de la Obesidad y Nutrición (CIBEROBN), Instituto de Salud Carlos III, 28029 Madrid, Spain
12 Li Ka Shing Knowledge Institute, St. Michael's Hospital, Toronto, ON M5B 1T8, Canada
13 Division of Endocrinology and Metabolism, Department of Medicine, St. Michael's Hospital, Toronto, ON M5C 2T2, Canada
14 Department of Medicine, Temerty Faculty of Medicine, University of Toronto, Toronto, ON M5S 1A8, Canada
15 Channing Division of Network Medicine, Department of Medicine, Brigham and Women's Hospital and Harvard Medical School, Boston, MA 02115, USA
16 Department of Epidemiology, Harvard T.H. Chan School of Public Health, Boston, MA 02115, USA
* Correspondence: john.sievenpiper@utoronto.ca

Citation: Glenn, A.J.; Aune, D.; Freisling, H.; Mohammadifard, N.; Kendall, C.W.C.; Salas-Salvadó, J.; Jenkins, D.J.A.; Hu, F.B.; Sievenpiper, J.L. Nuts and Cardiovascular Disease Outcomes: A Review of the Evidence and Future Directions. *Nutrients* **2023**, *15*, 911. https://doi.org/10.3390/nu15040911

Academic Editor: Kalliopi Karatzi

Received: 20 January 2023
Revised: 1 February 2023
Accepted: 6 February 2023
Published: 11 February 2023

Abstract: Nuts are nutrient-rich foods that contain many bioactive compounds that are beneficial for cardiovascular health. Higher consumption of nuts has been associated with a reduced risk of several cardiovascular diseases (CVD) in prospective cohort studies, including a 19% and 25% lower risk of CVD incidence and mortality, respectively, and a 24% and 27% lower risk of coronary heart disease incidence and mortality, respectively. An 18% lower risk of stroke mortality, a 15% lower risk of atrial fibrillation, and a 19% lower risk of total mortality have also been observed. The role of nuts in stroke incidence, stroke subtypes, peripheral arterial disease and heart failure has been less consistent. This narrative review summarizes recommendations for nuts by clinical practice guidelines and governmental organizations, epidemiological evidence for nuts and CVD outcomes, nut-containing dietary patterns, potential mechanisms of nuts and CVD risk reduction, and future research directions, such as the use of biomarkers to help better assess nut intake. Although there are still some uncertainties around nuts and CVD prevention which require further research, as summarized in this review, there is a substantial amount of evidence that supports that consuming nuts will have a positive impact on primary and secondary prevention of CVD.

Keywords: tree nuts; peanuts; nutrition; cardiovascular diseases; review

1. Introduction

Cardiovascular diseases (CVD) are the leading cause of death worldwide and a major cause of premature mortality, causing an estimated 31% of all deaths globally [1]. Major modifiable risk factors for CVD include smoking, harmful alcohol use, physical inactivity, unhealthy diets, abdominal obesity, hypertension, dyslipidemia, high fasting glucose and diabetes, and kidney dysfunction [2]. Furthermore, there are many downstream complications of CVD that can significantly impact quality of life and cause disability and death, including dementia, peripheral arterial disease (PAD), heart failure (HF), kidney disease, frailty and poor aging, among others [3].

Despite a decline in CVD in several regions around the globe, it remains a major threat to public health as the absolute numbers continue to increase, with prevalent cases of total CVD nearly doubling from 1990 to 2019 [4]. These numbers are expected to increase even further in upcoming years due to population growth and aging. For example, a recent analysis in the United States projects large future increases in CVD risk factors and CVD prevalence (e.g., 31% increase in ischemic heart disease and 34% increase in stroke) by 2060 [5]. According to the Global Burden of Diseases (GBD) Study, unhealthy diets are the greatest contributor to premature morbidity and mortality worldwide, including CVD mortality [6]. The main dietary risk factors attributable to the global burden of diseases include diets low in whole grains, fruit, nuts/seeds, and vegetables and diets high in sodium and processed meat [7]. Tree nuts and peanuts, one of the top dietary risk factors noted by the GBD study, may be particularly beneficial for CVD prevention due to their bioactive components. Of note, peanuts are botanically defined as legumes; however, they have a similar nutrient composition and culinary use as tree nuts and are, therefore, usually included as nuts when estimating total nut intake [8]. The bioactive components of tree nuts and peanuts include their macronutrient, fat-soluble bioactive, fiber, vitamin, mineral and phenolic content [9]. Specifically, fat-soluble bioactives such as their fatty acid content (monounsaturated and polyunsaturated fatty acids), fiber, magnesium, tocopherols and tocotrienols, phytosterols, sphingolipids, carotenoids, chlorophylls and alkyl phenols, and phenolic compounds (including flavonoids, phenolic acids, stilbenes, lignans, among others) all likely contribute to their cardiovascular health-promoting effects [9].

In this narrative review, we describe the importance given to nuts in clinical practice guidelines, the evidence we have in relation to the beneficial effects of frequent consumption of nuts in the prevention of CVD, as well as the possible mechanisms involved. However, we also emphasize the gaps that exist in the literature and discuss the possible studies that we should develop in the future to increase the level of evidence and establish recommendations.

2. Nuts in Clinical Practice Guidelines for Cardiovascular Risk Reduction

Given their cardiovascular health-promoting properties, tree nuts and peanuts are recognized by several international health organizations for cardiovascular risk reduction for both primary and secondary prevention. Table 1 highlights the recommendations from cardiovascular clinical practice guidelines (CPGs), including the Canadian Cardiovascular Society, Joint British Societies for the Prevention of Cardiovascular Disease, the Australian Heart Foundation, the American Heart Association and the European Society of Cardiology [10–14]. Nuts are recommended as a healthy plant protein and fat source that should be frequently consumed to lower low-density lipoprotein cholesterol (LDL-C), improve the overall lipoprotein profile and decrease CVD risk.

Table 1. Examples of Recommendations for Nuts in CVD Clinical Practice Guidelines.

Guideline Association	Nuts Recommendation
American Heart Association 2021 [10]	"Choose healthy sources of protein, mostly protein from plants (legumes and nuts)"

Table 1. *Cont.*

Guideline Association	Nuts Recommendation
European Society of Cardiology and European Atherosclerosis Society 2019 [11]	"Food choices to lower low-density lipoprotein cholesterol and improve the overall lipoprotein profile are nuts and seeds"
Canadian Cardiovascular Society 2016 [12]	"We suggest that all individuals be encouraged to moderate energy (caloric) intake to achieve and maintain a healthy body weight (Conditional Recommendation; Moderate-Quality Evidence) and adopt a healthy dietary pattern to lower their CVD risk: Dietary patterns high in nuts (30 g/day) (Conditional Recommendation; Moderate-Quality Evidence)"
Joint British Society Consensus for prevention of Cardiovascular Disease 2014 [13]	"Consider regular consumption of whole grains and nuts"
Heart Foundation Australia 2019 [14]	"Eating patterns for heart health are based on: • A variety of healthy protein sources including fish, seafood, lean meat and poultry, legumes, nuts and seeds • Healthy fat choices with nuts, seeds, avocadoes, olives and their oils for cooking"

Abbreviations: CVD = cardiovascular disease.

3. Regulated Nut Health Claims Allowed for Cardiovascular Risk Reduction

To encourage the consumption of foods that may be beneficial for health, governmental agencies review the evidence and approve health claims that can provide consumers with reliable information about the relationship between the consumption of food and a specific health benefit. Several regulated disease risk reduction health claims have been approved for nuts and CVD by the U.S. Food and Drug Administration (FDA), the European Food Safety Authority (EFSA) and Food Standards Australia New Zealand (FSANZ), as shown in Table 2 [15,16]. Health Canada has not approved a disease risk reduction claim for nuts. The FDA and FSANZ both approved a health claim regarding nuts and CVD risk reduction, particularly for the LDL-C lowering effects of nuts. However, EFSA did not approve this type of claim and noted that, specifically for walnuts, the evidence provided did not establish that the consumption of walnuts had an effect on LDL-C beyond what could be expected from their fatty acid composition [17]. EFSA, however, established that a cause-and-effect relationship between the consumption of walnuts and the improvement of endothelium-dependent vasodilation exists and has approved a health claim for this finding (Table 2) [17].

Table 2. Examples of Regulated Health Claims for Nuts and CVD Risk Reduction.

FDA	EFSA	FSANZ
1.5 ounces per day of nuts, as part of a diet low in saturated fat and cholesterol and not resulting in increased intake of saturated fat or calories may reduce the risk of CHD [15]	30 g per day of walnuts may improve endothelium-dependent vasodilation [17]	General level health claim around heart health without causing weight gain allowed for tree nuts, peanuts, ground nuts/butters/pastes [16]

Abbreviations: EFSA = European Food Safety Authority; FDA = Food and Drug Administration; FSANZ = Food Standards Australia New Zealand.

4. Nuts and Cardiovascular Disease Outcomes

The scientific study of the role of nuts in preventing CVD started over 30 years ago, first with a discovery that individuals in the Adventist Health Study who consumed nuts more frequently (more than four times per week) had fewer coronary heart disease (CHD) events compared to those who consumed nuts less than once per week [18]. Investigators believed that this CHD risk reduction might be related to the favorable fatty acid profile of nuts. Researchers from the same group then assessed if consuming walnuts (20% of calories through replacing other fatty foods, meat and oils, margarine and butter) would affect blood lipids in healthy individuals [19]. They found that incorporating a moderate amount of walnuts in the diet decreased levels of total cholesterol (TC) and LDL-C [19]. The favorable modification of the lipid profile by frequent nut consumption has been confirmed in additional trials and systematic reviews and meta-analyses [20]. The role of nuts in preventing CVD outcomes in prospective cohort studies has also been extensively studied. Below we describe this evidence related to total CVD, CHD, stroke, HF, atrial fibrillation (AF), PAD and total mortality. The definition of nuts includes tree nuts, peanuts and seeds (sunflower, pumpkin, etc.) as a culinary definition for this review. An overview of the pooled summary data for each outcome is included in Figure 1.

Outcome	Studies	Cases	n	RR (95% CI)	Pooled RR (95% CI)
Cardiovascular Disease					
Incidence	11	18,655	376,228	0.81 (0.74, 0.89)	
Mortality	16	19,574	524,610	0.75 (0.71, 0.79)	
Coronary Heart Disease					
Incidence	11	12,331	315,397	0.76 (0.69, 0.84)	
Mortality	13	10,083	429,833	0.73 (0.67, 0.80)	
Stroke					
Incidence	7	12,646	302,888	1.00 (0.92, 1.09)	
Mortality	12	4396	449,293	0.82 (0.73, 0.92)	
Ischemic	7	8401	302,423	0.99 (0.89, 1.10)	
Hemorrhagic	2	5454	194,307	1.53 (0.97, 2.41)	
Heart Failure	2	4253	53,887	1.00 (0.86, 1.16)	
Atrial Fibrillation	2	10,867	53,965	0.85 (0.73, 0.99)	
Peripheral Arterial Disease*	3	6018	233,883	1.03 (0.96, 1.09)	
Total Mortality	15	85,870	819,448	0.81 (0.77, 0.85)	

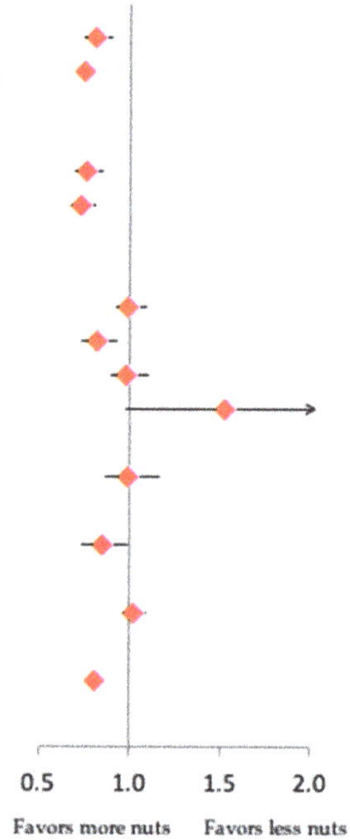

Figure 1. Summary of the pooled effect estimates of prospective cohort studies assessing the association between high and low consumption of nuts and risk of cardiovascular disease outcomes and total mortality. Figure adapted from Figure 2 in [21]. Pooled risk estimate is represented by the diamond. * To obtain summary estimates for peripheral arterial disease, we used generic inverse variance (fixed effects) to pool the natural log-transformed RRs of the extreme quantiles. Abbreviations: CI = confidence intervals; RR = relative risk.

4.1. Total Cardiovascular Disease

Total CVD is a composite outcome of CVD incidence (including only nonfatal or a combination of nonfatal and fatal outcomes of different CVD outcomes) or may include CVD mortality outcomes only, which is a composite of different fatal CVD endpoints. A recent systematic review and meta-analysis of prospective cohort studies that were commissioned to update the clinical practice guidelines for nutrition therapy for the European Association for the Study of Diabetes (EASD) found that in three cohort comparisons (including 210,839 participants and 14,136 events), high consumption of nuts was associated with a 15% lower risk of CVD incidence (relative risk [RR] = 0.85, 95% confidence intervals

[CI]: 0.80–0.91) compared to low consumption [22]. In 15 cohort comparisons (including 413,797 participants and 14,475 events), high consumption of nuts was associated with a 23% lower risk of CVD mortality (RR = 0.77, CI: 0.72–0.82) compared to low consumption [22]. The certainty of the evidence was assessed using the Grading Recommendations Assessment, Development and Evaluation (GRADE). The GRADE for CVD incidence was low quality due to a downgrade for indirectness (i.e., how applicable the evidence is to the general population) but an upgrade for a dose-response gradient. The GRADE for CVD mortality was moderate quality due to an upgrade for a dose-response gradient. The dose-response analysis showed that the reduction in risk of CVD incidence was observed up to 10 g/day with no further reduction in risk for higher consumption. For CVD mortality, there was a greater reduction in risk at 15–20 g/day, with no further reduction with higher consumption. When assessing different types of nuts, tree nuts, peanuts and walnuts were all associated with a 13–19% lower risk of CVD incidence, whereas the association for peanut butter was not significant (RR = 0.98, CI: 0.93–1.03). For CVD mortality, only peanuts have been analyzed, and higher consumption was associated with a lower risk of CVD mortality compared to low consumption [22]. The lack of association with peanut butter may be in part because many peanut butters on the market have added salt, fully hydrogenated oils or oils such as palm oil that can increase the saturated fat content, which could negatively impact their health benefits as compared with whole sources of peanuts; further research on natural peanut butter may provide additional insight into this hypothesis.

A more recent 2022 umbrella review assessing the role of nuts in preventing several chronic diseases found similar associations with CVD incidence and mortality [21]. For CVD incidence, 11 cohort comparisons were assessed (including 376,228 participants and 18,655 events), and a 19% lower risk was observed comparing high to low consumption (RR = 0.81, CI: 0.74–0.89) and 21% lower risk when assessing associations per serving (28 g/day) (RR = 0.79, CI: 0.70–0.89). For CVD mortality, 16 cohort comparisons were included (524,610 participants and 19,574 cases), and a 25% lower risk comparing high to low consumption (RR = 0.75, CI: 0.71, 0.79) was observed and a 6% lower risk when assessing nut intake by 28 g/day (RR = 0.94, CI: 0.93, 0.96) [21]. This umbrella review also assessed dose-response relationships with CVD mortality and found that the optimal intake levels of nuts are ~15–20 g/day and that there were limited further benefits of consuming up to one serving (28 g/day), similar to the earlier meta-analysis [22]. However, it should be mentioned that the high end of nut consumption across most cohorts is typically one serving (28 g)/day, and little is known about the dose-response relationship with higher intakes.

4.2. Coronary Heart Disease

Similar to total CVD, CHD incidence may include fatal and nonfatal events, whereas CHD mortality includes CHD mortality outcomes only. The systematic review and meta-analysis conducted to update the EASD nutrition therapy guidelines assessed 7 cohort comparisons (including 275,812 participants and 12,654 cases) and found that high nut consumption was associated with an 18% lower risk of CHD incidence (RR = 0.82, CI: 0.69–0.96) compared to low consumption [22]. For CHD mortality, 13 cohort comparisons were included (396,014 participants and 7877 cases), and high consumption of nuts was associated with a 24% lower risk (RR = 0.76, CI: 0.67–0.86) compared to low consumption. The certainty of evidence using GRADE was very low for CHD incidence owing to downgrades for inconsistency (i.e., unexplained heterogeneity), indirectness and imprecision (i.e., the minimally important difference for the clinical benefit [considered RR = 0.95 in this meta-analysis]), and an upgrade for a dose-response gradient. The certainty of the evidence was moderate for CHD mortality, owing to an upgrade for a dose-response gradient. When assessing different types of nuts, tree nuts, peanuts and walnuts were all associated with a 15–23% lower risk of CHD incidence, and similar to CVD incidence, no association was reported for peanut butter consumption (RR = 1.00, CI: 0.94–1.07). The findings were also similar between CVD and CHD mortality, where peanut consumption was inversely associated with the risk of CHD mortality. Comparable to CVD mortality, the CHD mortality dose-response analysis showed greater reductions in risk at around 15–20 g/day [13,22]. The 2022 umbrella review showed that for CHD incidence (including 12 cohort comparisons with 315,397 participants and 12,331 events), there was a 24% lower risk comparing high to low consumption (RR = 0.76, CI: 0.69–0.84) and 25% lower risk when assessing per servings associations of 28 g/day (RR = 0.75, CI: 0.64–0.88) [21]. For CHD mortality, 13 cohort comparisons were included (429,833 participants and 10,083 cases), and a 27% lower risk comparing low to high consumption (RR = 0.73, CI: 0.67, 0.80) was observed and a 6% lower risk when assessing nut intake by 28 g/day (RR = 0.94, CI: 0.93, 0.96) [21].

4.3. Stroke

Stroke outcomes include stroke incidence and mortality, and the main stroke subtypes, ischemic and hemorrhagic stroke. The association between nut consumption and stroke risk has been less consistent than that observed for total CVD and CHD. For stroke incidence, the EASD systematic review and meta-analysis included 7 cohort comparisons (of 302,888 participants and 12,646 events) and found no associations when comparing high to low consumption (RR = 1.00, CI: 0.92–1.09) [22]. In contrast, for stroke mortality, 12 cohort comparisons were analyzed (including 351,618 participants and 2332 cases) and comparing high vs. low categories of nut consumption was associated with a 17% lower risk (RR = 0.83, CI: 0.75–0.93). The certainty of evidence using GRADE was very low for stroke incidence owing to downgrades for indirectness and imprecision, and low for stroke mortality, owing to downgrades for imprecision but an upgrade for a dose-response gradient. Regarding specific types of nuts, peanut consumption was associated with a lower risk of stroke incidence and mortality, but other nut types and peanut butter were not significantly associated with either outcome. For stroke subtypes, no associations were seen with ischemic stroke (RR = 0.99, CI: 0.89–1.10 in 7 cohort comparisons including 302,423 participants and 8401 cases) or hemorrhagic stroke (RR = 1.02, CI: 0.77–1.34 in 5 cohort comparisons including 188,750 participants and 3088 cases) [22].

Another meta-analysis including 11 cohort studies (9272 stroke cases) and 396,768 participants also reported an inverse association in the high vs. low analysis (RR = 0.89, CI: 0.82–0.97), but not in the linear dose-response analysis (RR = 0.93, CI: 0.83–1.05); however, there was some indication of a non-linear J-shaped association with a reduction in risk up to approximately 10–15 g/day, but a slight positive association at 30 g/day [23]. There was no indication of an increased risk at high intakes when stroke incidence and stroke mortality were analyzed separately, suggesting that the direct association at high nut doses observation could be an artefact. When subtypes of nuts were examined, no association was observed for tree nuts in relation to the risk of stroke in the high vs. low and dose-response analyses, while slight inverse associations were observed for peanuts [23], which were both similar to those reported in the EASD meta-analysis [22]. It is unclear whether these differences in results between subtypes are real or simply because of chance variation due to the few studies available. Given largely overlapping confidence intervals between summary estimates, it is possible that chance variation is playing a role; therefore, further studies are needed. The 2022 umbrella review findings were also similar: no association was seen with stroke incidence (RR = 1.00, CI: 0.92–1.09) in 7 cohort comparisons, including 302,888 participants and 12,646 cases, with an inverse association seen with stroke mortality (RR = 0.82, CI: 0.73–0.92) in 12 cohort comparisons including 449,293 participants and 4398 events [21]. Although results regarding nut consumption and stroke risk have been more variable than for CHD, it seems there may be a modest inverse association between higher nut intake and stroke risk. Most of the individual studies may not have been sufficiently powered to detect an association. Nonetheless, a possible reason for the weaker association between nut consumption and stroke than for CHD could be the fact that many nuts are salted. Dietary salt consumption is one of the main determinants for elevated blood pressure, and it is possible that adding salt to nuts could dilute some of the benefits they have, particularly for stroke, similar to the possible reasons for no association seen with peanut butter and several CVD outcomes.

4.4. Heart Failure

Few prospective cohort studies have assessed the role of nut consumption in preventing one of the major complications of CVD, HF. The EASD systematic review and meta-analysis included two cohort studies with 53,877 participants and 4253 cases, and for high vs. low categories of nut consumption, the RR was 1.00 (CI: 0.86–1.16) [22]. The certainty of evidence by GRADE was very low for HF, owing to downgrades due to the risk of bias (i.e., study quality as assessed by the Newcastle Ottawa Scale), indirectness and imprecision. Similarly, the 2022 umbrella review included the same cohorts, and the same effect estimates were observed [21]. Interestingly, a large prospective Swedish Cohort study of 61,364 adults observed a non-linear inverse association with the risk of HF and a 12% reduction in risk (HR = 0.88, CI: 0.79–0.99) with consumption of nuts 1–2 times per week with 17 years of follow-up [24].

4.5. Atrial Fibrillation

Similar to HF, fewer prospective cohort studies have examined the role of nuts in preventing this important risk factor for stroke and HF. The EASD systematic review and meta-analysis assessed two prospective cohort studies, including 53,965 participants and 10,867 cases [22]. Comparing high vs. low consumption of nuts, there was a 15% lower risk of AF (RR = 0.85, CI: 0.73–0.99), similar to the 2022 umbrella review [21]. The certainty of the evidence was very low, owing to downgrades due to indirectness and imprecision [22]. Similar to HF, the Swedish Cohort study also found a non-linear inverse relationship with the risk of AF and consumption of nuts 3 or more times per week was associated with an 18% reduced risk of AF [24].

4.6. Peripheral Arterial Disease

Few studies have assessed the association between nut consumption and the risk of PAD [25]. In the Atherosclerosis Risk in Communities Study, 14,082 men and women were followed for 20 years, and 1569 incident cases of PAD were identified. There was no clear association between the frequency of nut consumption and the risk of PAD, and the HR was 1.04 (CI: 0.89–1.23) when comparing an intake of $\geq$2/week vs. almost never [26]. In an analysis from the Women's Health Initiative, including 138,506 postmenopausal women and 1036 PAD cases identified during 19 years of follow-up, there was no association between higher consumption of nuts and seeds and PAD (highest vs. lowest quartile of nuts and seeds consumption was 0.93 [CI: 0.78–1.10]) [27]. Similarly, in a combined analysis of 38,823 women in the Swedish Mammography Cohort and 45,472 men in the Cohort of Swedish Men (aged 45–83 years) with 22 years follow-up and 3413 PAD cases, there was no clear association between intake of nuts and PAD risk, and the HR was 1.05 (CI: 0.89–1.24) for the highest vs. lowest category of intake [28]. We pooled the data from the 3 cohorts, and the RR was 1.03 (CI: 0.96–1.09), also highlighting no clear association between nut consumption and the risk of PAD (Figure 1). However, we cannot exclude the possibility that there may be a U-shaped association between nut consumption and PAD.

4.7. Total Mortality

The 2022 umbrella review also assessed the association between nut consumption and overall total mortality. In 16 cohort comparisons (including 819,448 participants and 85,870 deaths), both high vs. low categories of nut consumption (RR = 0.81, CI: 0.77, 0.85) and per serving (28 g/day, RR = 0.78, CI: 0.72, 0.84) were associated with lower risk of total mortality [21]. Similar to total CVD, the dose-response analyses indicated optimal intake levels were approximately ~15–20 g/day, with limited further benefits up to 28 g/day. In addition, a meta-analysis of 15 cohort studies indicated that 4.4 million premature deaths in the Americas, Europe, Southeast Asia and Western Pacific would be attributable to a nut intake below 20 g/day [23].

4.8. Change in Nut Intake and Substitution Analyses

Changes in nut intake over time and the substitution of nuts for other dietary factors have also been associated with a lower risk of CVD. For instance, Liu et al. examined the association between 4-year changes in nut consumption and risk of CVD outcomes in the subsequent 4 years in the Health Professionals Follow-up Study (HPFS) and Nurses' Health Study (NHS) I and II [29]. They found that a per 1/2 serving/day increase in total nut consumption was associated with a lower risk of CVD (HR = 0.92, CI: 0.86–0.98), CHD (HR = 0.94, CI: 0.89–0.99) and stroke (HR = 0.89, CI: 0.83–0.95) and those that decreased their nut consumption over time had an increased risk of CVD, CHD and stroke. The 1/2 serving/day increase in consumption for a different type of nuts showed that most nut types were associated with a lower risk of CVD outcomes, with walnuts showing the strongest association and peanut butter showing no association. The researchers also examined the substitution effect per 1/2 serving/day of nuts and found that replacing red meat, processed meat, refined grains, French fries and dessert with nuts was associated with a lower risk of CVD, CHD and stroke [29]. In another analysis of HPFS and NHS, substituting both unprocessed and processed red meat for nuts was associated with the greatest reduction in total mortality [30].

5. Healthy Dietary Patterns That Contain Nuts

Several healthy dietary patterns that are recommended in CVD clinical practice guidelines contain nuts as a key food component. These include dietary patterns such as the Mediterranean, Nordic, Dietary Approaches to Stop Hypertension (DASH), vegetarian and Portfolio diets. Each of these dietary patterns has been shown to lower important CVD risk factors in RCTs and is associated with a lower risk of CVD in prospective cohort studies [31–38]. A healthy Mediterranean diet including nuts was also assessed with CVD endpoints in the landmark Prevención con Dieta Mediterránea (PREDIMED) trial. In this trial, over 7000 high-risk individuals for CVD were randomly assigned to an energy-unrestricted Mediterranean diet supplemented with either extra virgin olive oil (EVOO) or mixed nuts or the control diet (advice to curtail all types of fat) [39]. After approximately 5 years, both the Mediterranean diet groups supplemented with EVOO and nuts had a 30% reduction in CV events, mainly through reductions in stroke. Further analyses of the PREDIMED study also found a reduced risk of PAD both in the EVOO (HR = 0.36, CI: 0.21–0.65) and in the nuts group (HR = 0.54, CI: 0.32–0.92) when compared to the control group [40]. Due to the design of the PREDIMED study, it is not possible to separate the impact of nuts (or EVOO) from that of other dietary recommendations given to increase adherence to a Mediterranean diet, as well as with other healthy dietary patterns, and it is possible that other components of the diet could contribute in part to these findings.

6. Mechanisms Related to Nuts and Cardiovascular Risk Reduction

There are several mechanisms by which nuts can lower the risk of developing CVD, such as through positively impacting intermediate cardiovascular risk factors, including blood lipids, blood pressure, inflammation, and markers of glycemic control, among others. Nuts are rich in unsaturated fatty acids, plant protein, phytosterols, fiber, some minerals (including potassium, calcium and magnesium), vitamins (vitamin E and B6) and phenolic and bioactive compounds, all of which may contribute to their CV health-promoting benefits [9]. A systematic review and meta-analysis of 61 trials of tree nut consumption found that at a median dose of 56 g/day, TC, LDL-C, apolipoprotein B (ApoB) and triglycerides were significantly lowered, with no effect on high-density lipoprotein-cholesterol (HDL-C). A dose-dependent effect was also reported, with stronger effects seen for TC and LDL-C with nut intake over 60 g/day [20]. Importantly, the type of nut did not appear to be important for the cholesterol-lowering results observed [20]. The effects on blood pressure and inflammation have been less consistent than those observed for blood lipids, with the same systematic review and meta-analysis finding no significant effects of tree nut consumption on blood pressure and C-reactive protein [20]. Other meta-analyses have conversely shown that nut consumption, particularly pistachios, does have a modest blood pressure lowering effects in people without type 2 diabetes [41], with another meta-analysis finding that almond consumption lowered diastolic blood pressure [42]. This finding of stronger effects on blood lipids than on blood pressure is consistent with the studies showing that LDL-C and ApoB are causal in the development of CHD [43], while elevated blood pressure is a greater risk factor for stroke [44] and may explain the more consistent finding with a lower risk of CHD than with stroke seen in the prospective cohort studies described earlier. Furthermore, a systematic review and meta-analysis that included 12 trials found that tree nuts at a median dose of 56 g/day can improve markers of glycemic control in individuals with type 2 diabetes (including lowering HbA1c and fasting glucose) [45]. Another systematic review and meta-analysis of 40 RCTs at a median dose of 52 g/day, including diverse populations of adults (including healthy, those with type 2 diabetes or with CVD risk factors), found that tree nuts or peanuts improved markers of insulin sensitivity, however, the effect on fasting blood glucose and HbA1c was not significant [46]. Other potential mechanisms include their role in adiposity [47], possibly due to their satiating effect, increased efforts and/or time of mastication and hence incomplete digestion in the intestines, and alpha-linolenic acid content of nuts, especially walnuts, which might increase membrane fluidity of endothelial cells with the enhancement of nitric oxide synthesis and ensuing improvement of endothelial function [48,49]. Overall, there is good evidence that nut consumption consistently lowers atherogenic blood lipids and may improve insulin sensitivity and endothelial function, with less consistent effects on blood pressure, without adversely impacting adiposity. Regarding adiposity, a recent systematic review and meta-analysis highlighted that the median nut intake in the trials included in their analyses, as well as in the health claims noted in Table 2, that a dose of 42.5 g/day could be integrated into a daily dietary pattern without contributing to weight gain [47].

7. Future Directions

The evidence for nut consumption and total CVD and CHD is more consistent in prospective cohort studies compared to other CVD outcomes, with low to moderate certainty of evidence using the GRADE criteria. The GRADE criteria, however, may not be the best grading system to use when evaluating the certainty of evidence from observational studies, particularly in the field of nutritional epidemiology [50]. Future pooled analyses that assess the certainty of evidence should consider integrating ROBINS-I to assess the risk of bias [51] or consider also applying the NutriGrade system [52], both of which do not provide excessive downgrading of observational evidence. In addition, given the scarcity and sometimes conflicting results among studies, more studies are needed to clarify the role of nut consumption in stroke, particularly stroke subtypes, HF, AF and PAD. Further research on the type of nuts will eventually provide further insight into their role in CVD prevention, including studies of peanut and other nut butters (including natural) and salted vs. unsalted nuts. New analyses on nut intake should also report quantities of nut intake (i.e., grams/day) so that these data can be used in updated meta-analyses. The quantity of nuts is also more translatable for guiding dietary recommendations compared to high vs. low categories. Individual cohort pooled meta-analyses would also be useful to ensure consistency in analyses across cohort studies. Although RCTs of nut consumption and blood lipids support the results from observational cohort studies showing reduced CVD and CHD risk, there is some discrepancy between what doses of nuts have been shown to reduce blood lipids in RCTs and what doses lower CVD and CHD risk in observational cohort studies. For example, in a meta-analysis of RCTs, there was a steeper reduction in total and LDL-C between 50–100 g/d than at lower levels of intake [20], while in the observational cohort studies, maximum risk reductions have been observed around

15–20 g/d (approximately 4–5 servings/week) [21–23]. However, the highest nut intakes reported in cohort studies have typically been around one serving/day (28 g/d), and it is unknown whether CVD or CHD risk is reduced further with higher intakes. Considering that relatively few people consume more than one serving of nuts per day in most populations [23,53], pooled analyses may also be needed to explore, with sufficient statistical power, whether higher intakes are associated with further reductions in hard endpoints. Further studies are also needed to clarify if other mechanisms than reductions in lipids (e.g., antioxidant or anti-thrombotic effects) may contribute to the vascular benefits observed at the more modest nut consumption levels reported in the observational studies. Another important consideration related to the discrepancy of nut levels consumed in trials and cohort studies is the dietary assessment tool used, as food frequency questionnaires (FFQ) commonly administered in cohort studies may not be as accurate in quantifying absolute intake compared to trials, where diet records are typically used, and intervention groups are usually provided nuts to guarantee the desired consumption. As diet records are not feasible in large cohort studies, repeated measurements of FFQs will be important to represent long-term dietary habits and reduce measurement error, as well as allow assessment of change in nut consumption in relation to health outcomes. FFQs should also consider including more nut categories (walnuts, almonds, peanuts, seeds, etc.) to provide more detailed information on nuts and nut types.

Using objective biomarkers of nut consumption alongside dietary intake assessment methods will additionally be important in the future, as they are less prone to measurement error from FFQs or 24-h recalls [54]. For example, in the PREDIMED study, plasma alpha-linolenic acid (a polyunsaturated fatty acid that abounds in walnuts) levels were measured to confirm adherence in the group receiving mixed nuts alongside an FFQ [39]. Novel approaches, such as multi-omics, will likely play a larger role in the future for both assessing adherence to diet and precision nutrition [55]. Metabolomics, in particular, is a promising technique to help identify objective dietary biomarkers by providing a comprehensive representation of overall dietary intake by measuring the metabolites in biological samples (such as blood or urine). In prospective cohort studies, several metabolites, mainly lipid-related, have been found to be markers of nut intake in general [56] or of specific types of nuts, such as walnuts, and these metabolites have likewise been associated with a lower risk of CVD [57]. The metabolites associated with nut consumption may be helpful in identifying potential objective biomarkers of exposure to nuts in large prospective cohort studies, as well as in clarifying underlying mechanisms implicated in disease risk. Importantly though, these metabolomic profiles associated with nut consumption are not highly correlated with self-reported nut consumption and also reflect the metabolic response to consumption and are therefore not completely sensitive or specific markers. Many of these metabolites may also not be able to distinguish between different types of nuts. Thus, dietary intake assessment methods such as FFQs will still be important to determine more specific information on nut consumption.

Other future directions include undertaking large CV outcome trials of nut-containing dietary patterns, as was previously done with PREDIMED [39] and the current ongoing PREDIMED-Plus trials [58] (both primary CVD prevention trials) and the CORDIOPREV trial (a secondary CVD prevention trial) [59]. One limitation of these trials is that they cannot separate the effect of the Mediterranean diet from that of nuts on health outcomes. Thus, any further trials in this setting could benefit from having an additional intervention group on a Mediterranean diet only (without nuts). Large trials should also consider long-term measurements of renal function, as this area has been given insufficient attention and kidney dysfunction has been causally related to CHD risk [60], therefore highlighting the need for preventative approaches to also preserve renal function. Furthermore, determining metabolomic signatures that can reflect adherence and metabolic response to these nut-containing dietary patterns should be included within these types of trials. This method was previously assessed using the Mediterranean diet in the PREDIMED study, where a metabolic signature that robustly reflected adherence and metabolic response to the diet was determined [61]. The metabolic signature was then used to assess associations with CVD risk and showed stronger inverse associations with CVD risk compared to dietary intake alone in a Spanish and three US cohorts. Mendelian randomization analyses also showed that the genetically inferred metabolic signature was significantly associated with a lower risk of CHD and stroke [61]. These novel approaches hold promise for an objective and complete evaluation of both adherence and metabolic responses to diet, including nuts, and may allow more effective and individualized approaches to dietary interventions in the future; however, further research is also needed in this area. Overall, a combination of efforts, including well-conducted large prospective cohort studies, large RCTs of hard CV endpoints and incorporation of multi-omics approaches and genetics, will help us better understand the role of nuts in the primary and secondary prevention of CVD.

8. Summary and Conclusions

The overall findings and conclusions from this review are that nuts are a beneficial food for CVD risk reduction, with consistent findings for the benefit for total CVD and CHD in prospective cohort studies and likely benefits for stroke and AF, but additional research is needed for HF, AF, PAD and stroke subtypes. Considering all evidence from mechanistic studies, RCTs of intermediate risk factors and CV events, and prospective cohort studies, at least one serving per day of nuts should be considered for CV risk reduction, although further research on the optimal dose is needed. The type of nut may not be important, though more research is needed to confirm this finding. However, a general nut recommendation will provide more variety of options and be important for any availability and affordability concerns for consumers.

In conclusion, future research is urgently needed as outlined above, particularly for stroke subtypes, PAD and HF, and include individual cohort pooled analyses, large RCTs of nut-containing dietary patterns and using -omics methodologies to better capture adherence and metabolic responses to diet, including the consumption of nuts.

Author Contributions: Writing—original draft preparation, A.J.G. and D.A.; writing—review and editing, all authors. All authors have read and agreed to the published version of the manuscript.

Funding: AJG is supported by a Canadian Institutes of Health Research (CIHR) Postdoctoral Fellowship. DA has received funding from the South-Eastern Norway Regional Health Authority (grant no: 2017076). JSS is partially supported by ICREA under the ICREA Academia programme. The NUTS 2022 conference was organized by the Rovira i Virgili University, with all the expenses of the organization covered by Barcelo Congresos, including the fees related to publishing the Proceedings in Nutrients. Barcelo Congresos SA has received funds from the International Nut and Dried Fruit Council (INC). Neither the researchers attending the NUTS 2022 Conference, nor the Rovira i Virgili University have received funds from the INC. Researchers' travel expenses and accommodation have been covered by Barcelo Congresos. Researchers did not receive any additional financial support related to their participation in the NUTS 2022 conference. Sponsors did not have any role in organizing the conference, selecting, or inviting the speakers, nor in the writing or reviewing the proceedings of the conference.

Institutional Review Board Statement: Not applicable.

Informed Consent Statement: Not applicable.

Data Availability Statement: Not applicable.

Acknowledgments: This review highlights the main discussion points from the scientific session "Nuts and Cardiovascular Disease Outcomes" by speakers John Sievenpiper and Dagfinn Aune at the NUTS2022 Conference held in Reus, Spain on 20–21 October 2022.

Conflicts of Interest: **AJG** has received an honorarium from the Soy Nutrition Institute and the Academy of Nutrition and Dietetics. **JS-S** reports serving on the boards of the INC and receiving grant support from these entities through his institution. He has also received research funding (nuts for free to the PREDIMED participants) from the CWC; La Morella Nuts, Spain; and Borges SA, Spain. He has also received research funding (nuts for free to the PREDIMED-Plus participants) from the Almond Board of California, USA and Pistachio Growers of California, USA. He is a non-paid member of the Instituto Danone International and was a member of the executive committee of the Instituto Danone Spain. **DJAJ** has received research grants from Saskatchewan and Alberta Pulse Growers Associations, the Agricultural Bioproducts Innovation Program through the Pulse Research Network, the Advanced Foods and Material Network, Loblaw Companies Ltd., Unilever Canada and Netherlands, Barilla, the Almond Board of California, Agriculture and Agri-food Canada, Pulse Canada, Kellogg's Company, Canada, Quaker Oats, Canada, Procter and Gamble Technical Centre Ltd., Bayer Consumer Care, Springfield, NJ, Pepsi/Quaker, International Nut and Dried Fruit Council (INC), Soy Foods Association of North America, the Coca-Cola Company (investigator initiated, unrestricted grant), Solae, Haine Celestial, the Sanitarium Company, Orafti, the International Tree Nut Council Nutrition Research and Education Foundation, the Peanut Institute, Soy Nutrition Institute (SNI), the Canola and Flax Councils of Canada, the Calorie Control Council, the Canadian Institutes of Health Research (CIHR), the Canada Foundation for Innovation (CFI)and the Ontario Research Fund (ORF). He has received in-kind supplies for trials as a research support from the Almond board of California, Walnut Council of California, the Peanut Institute, Barilla, Unilever, Unico, Primo, Loblaw Companies, Quaker (Pepsico), Pristine Gourmet, Bunge Limited, Kellogg Canada, WhiteWave Foods. He has been on the speaker's panel, served on the scientific advisory board and/or received travel support and/or honoraria from Nutritional Fundamentals for Health (NFH)-Nutramedica, Saint Barnabas Medical Center, The University of Chicago, 2020 China Glycemic Index (GI) International Conference, Atlantic Pain Conference, Academy of Life Long Learning, the Almond Board of California, Canadian Agriculture Policy Institute, Loblaw Companies Ltd., the Griffin Hospital (for the development of the NuVal scoring system), the Coca-Cola Company, Epicure, Danone, Diet Quality Photo Navigation (DQPN), Better Therapeutics (FareWell), Verywell, True Health Initiative (THI), Heali AI Corp, Institute of Food Technologists (IFT), Soy Nutrition Institute (SNI), Herbalife Nutrition Institute (HNI), Saskatchewan and Alberta Pulse Growers Associations, Sanitarium Company, Orafti, the International Tree Nut Council Nutrition Research and Education Foundation, the Peanut Institute, Herbalife International, Pacific Health Laboratories, Barilla, Metagenics, Bayer Consumer Care, Unilever Canada and Netherlands, Solae, Kellogg, Quaker Oats, Procter and Gamble, Abbott Laboratories, Dean Foods, the California Strawberry Commission, Haine Celestial, PepsiCo, the Alpro Foundation, Pioneer Hi-Bred International, DuPont Nutrition and Health, Spherix Consulting and WhiteWave Foods, the Advanced Foods and Material Network, the Canola and Flax Councils of Canada, Agri-Culture and Agri-Food Canada, the Canadian Agri-Food Policy Institute, Pulse Canada, the Soy Foods Association of North America, the Nutrition Foundation of Italy (NFI), Nutra-Source Diagnostics, the McDougall Program, the Toronto Knowledge Translation Group (St. Michael's Hospital), the Canadian College of Naturopathic Medicine, The Hospital for Sick Children, the Canadian Nutrition Society (CNS), the American Society of Nutrition (ASN), Arizona State University, Paolo Sorbini Foundation and the Institute of Nutrition, Metabolism and Diabetes. He received an honorarium from the United States Department of Agriculture to present the 2013 W.O. Atwater Memorial Lecture. He received the 2013 Award for Excellence in Research from the International Nut and Dried Fruit Council. He received funding and travel support from the Canadian Society of Endocrinology and Metabolism to produce mini cases for the Canadian Diabetes Association (CDA). He is a member of the International Carbohydrate Quality Consortium (ICQC). His wife, Alexandra L Jenkins, is a director and partner of INQUIS Clinical Research for the Food Industry, his 2 daughters, Wendy Jenkins and Amy Jenkins, have published a vegetarian book that promotes the use of the foods described here, The Portfolio Diet for Cardiovascular Risk Reduction (Academic Press/Elsevier 2020 ISBN:978-0-12-810510-8)and his sister, Caroline Brydson, received funding through a grant from the St. Michael's Hospital Foundation to develop a cookbook for one of his studies. He is also a vegan. **CWCK** has received grants or research support from the Advanced Food Materials Network, Agriculture and Agri-Foods Canada (AAFC), Almond Board of California, Barilla, Canadian Institutes of Health Research (CIHR), Canola Council of Canada, International Nut and Dried Fruit Council, International Tree Nut Council Research and Education Foundation, Loblaw Brands Ltd., the Peanut Institute, Pulse Canada and Unilever. He has received in-kind research support from the Almond Board of California, Barilla, California Walnut Commission, Kellogg Canada, Loblaw Companies, Nutrartis, Quaker (PepsiCo), the Peanut Institute, Primo, Unico, Unilever, WhiteWave Foods/Danone. He has received travel support and/or honoraria from the Barilla, California Walnut Commission, Canola Council of Canada, General Mills, International Nut and Dried Fruit Council, International Pasta Organization, Lantmannen, Loblaw Brands Ltd., Nutrition Foundation of Italy, Oldways Preservation Trust, Paramount Farms, the Peanut Institute,

Pulse Canada, Sun-Maid, Tate and Lyle, Unilever and White Wave Foods/Danone. He has served on the scientific advisory board for the International Tree Nut Council, International Pasta Organization, McCormick Science Institute and Oldways Preservation Trust. He is a founding member of the International Carbohydrate Quality Consortium (ICQC), Chair of the Diabetes and Nutrition Study Group (DNSG) of the European Association for the Study of Diabetes (EASD), is on the Clinical Practice Guidelines Expert Committee for Nutrition Therapy of the EASD and is a Director of Glycemia Consulting and the Toronto 3D Knowledge Synthesis and Clinical Trials foundation. **JLS** has received research support from the Canadian Foundation for Innovation, Ontario Research Fund, Province of Ontario Ministry of Research and Innovation and Science, Canadian Institutes of health Research (CIHR), Diabetes Canada, PSI Foundation, Banting and Best Diabetes Centre (BBDC), American Society for Nutrition (ASN), INC International Nut and Dried Fruit Council Foundation, National Dried Fruit Trade Association, The Tate and Lyle Nutritional Research Fund at the University of Toronto, The Glycemic Control and Cardiovascular Disease in Type 2 Diabetes Fund at the University of Toronto (a fund established by the Alberta Pulse Growers), and the Nutrition Trialists Fund at the University of Toronto (a fund established by an inaugural donation from the Calorie Control Council). He has received in-kind food donations to support a randomized controlled trial from the Almond Board of California, California Walnut Commission, American Peanut Council, Barilla, Unilever, Unico/Primo, Loblaw Companies, Quaker, Kellogg Canada, and WhiteWave Foods. He has received travel support, speaker fees and/or honoraria from Diabetes Canada, Mott's LLP, Dairy Farmers of Canada, FoodMinds LLC, International Sweeteners Association, Nestlé, Pulse Canada, Canadian Society for Endocrinology and Metabolism (CSEM), GI Foundation, Abbott, Biofortis, ASN, Northern Ontario School of Medicine, INC Nutrition Research and Education Foundation, European Food Safety Authority (EFSA), Comité Européen des Fabricants de Sucre (CEFS), and Physicians Committee for Responsible Medicine. He has or has had ad hoc consulting arrangements with Perkins Coie LLP, Tate and Lyle, and Wirtschaftliche Vereinigung Zucker e.V. He is a member of the European Fruit Juice Association Scientific Expert Panel and Soy Nutrition Institute (SNI) Scientific Advisory Committee. He is on the Clinical Practice Guidelines Expert Committees of Diabetes Canada, European Association for the study of Diabetes (EASD), Canadian Cardiovascular Society (CCS), and Obesity Canada. He serves or has served as an unpaid scientific advisor for the Food, Nutrition, and Safety Program (FNSP) and the Technical Committee on Carbohydrates of the International Life Science Institute (ILSI) North America. He is a member of the International Carbohydrate Quality Consortium (ICQC), Executive Board Member of the Diabetes and Nutrition Study Group (DNSG) of the EASD, and Director of the Toronto 3D Knowledge Synthesis and Clinical Trials foundation. His wife is an employee of AB InBev. All other authors have no conflicts of interest to report.

Disclaimer: Where authors are identified as personnel of the International Agency for Research on Cancer/World Health Organization, the authors alone are responsible for the views expressed in this article and they do not necessarily represent the decisions, policy or views of the International Agency for Research on Cancer/World Health Organization.

References

1. World Health Organization. Cardiovascular Disease. Available online: http://www.who.int/cardiovascular_diseases/en/ (accessed on 9 May 2018).
2. Government of Canada. Heart Disease—Heart Health. Available online: https://www.canada.ca/en/public-health/services/diseases/heart-disease-heart-health.html (accessed on 21 November 2022).
3. Makover, M.E.; Shapiro, M.D.; Toth, P.P. There is urgent need to treat atherosclerotic cardiovascular disease risk earlier, more intensively, and with greater precision: A review of current practice and recommendations for improved effectiveness. *Am. J. Prev. Cardiol.* **2022**, *12*, 100371. [CrossRef]
4. Roth, G.A.; Mensah, G.A.; Johnson, C.O.; Addolorato, G.; Ammirati, E.; Baddour, L.M.; Barengo, N.C.; Beaton, A.Z.; Benjamin, E.J.; Benziger, C.P.; et al. Global Burden of Cardiovascular Diseases and Risk Factors, 1990-2019: Update From the GBD 2019 Study. *J. Am. Coll. Cardiol.* **2020**, *76*, 2982–3021. [CrossRef]
5. Mohebi, R.; Chen, C.; Ibrahim, N.E.; McCarthy, C.P.; Gaggin, H.K.; Singer, D.E.; Hyle, E.P.; Wasfy, J.H.; Januzzi, J.L., Jr. Cardiovascular Disease Projections in the United States Based on the 2020 Census Estimates. *J. Am. Coll. Cardiol.* **2022**, *80*, 565–578. [CrossRef]
6. GBD 2015 Risk Factors Collaborators. Global, regional, and national comparative risk assessment of 79 behavioural, environmental and occupational, and metabolic risks or clusters of risks, 1990-2015: A systematic analysis for the Global Burden of Disease Study 2015. *Lancet* **2016**, *388*, 1659–1724. [CrossRef]

7. GBD 2017 Diet Collaborators. Health effects of dietary risks in 195 countries, 1990-2017: A systematic analysis for the Global Burden of Disease Study 2017. *Lancet* **2019**, *393*, 1958–1972. [CrossRef]
8. Ros, E. Health benefits of nut consumption. *Nutrients* **2010**, *2*, 652–682. [CrossRef] [PubMed]
9. Alasalvar, C.; Salvadó, J.S.; Ros, E. Bioactives and health benefits of nuts and dried fruits. *Food Chem.* **2020**, *314*, 126192. [CrossRef]
10. Lichtenstein, A.H.; Appel, L.J.; Vadiveloo, M.; Hu, F.B.; Kris-Etherton, P.M.; Rebholz, C.M.; Sacks, F.M.; Thorndike, A.N.; Van Horn, L.; Wylie-Rosett, J. 2021 Dietary Guidance to Improve Cardiovascular Health: A Scientific Statement From the American Heart Association. *Circulation* **2021**, *144*, e472–e487. [CrossRef]
11. Mach, F.; Baigent, C.; Catapano, A.L.; Koskinas, K.C.; Casula, M.; Badimon, L.; Chapman, M.J.; De Backer, G.G.; Delgado, V.; Ference, B.A.; et al. 2019 ESC/EAS Guidelines for the management of dyslipidaemias: Lipid modification to reduce cardiovascular risk. *Eur. Heart J.* **2020**, *41*, 111–188. [CrossRef]
12. Anderson, T.J.; Gregoire, J.; Pearson, G.J.; Barry, A.R.; Couture, P.; Dawes, M.; Francis, G.A.; Genest, J., Jr.; Grover, S.; Gupta, M.; et al. 2016 Canadian Cardiovascular Society Guidelines for the Management of Dyslipidemia for the Prevention of Cardiovascular Disease in the Adult. *Can. J. Cardiol.* **2016**, *32*, 1263–1282. [CrossRef]
13. JBS Board. Joint British Societies' consensus recommendations for the prevention of cardiovascular disease (JBS3). *Heart* **2014**, *100* (Suppl. 2), ii1–ii67. [CrossRef] [PubMed]
14. Heart Foundation of Australia. Cardiac Rehabilitation Program Outline—Module 6. Healthy Eating and Weight Management. Available online: https://www.heartfoundation.org.au/getmedia/7d0cdb2c-6884-4912-9bf0-00cb207c71f1/CardRehabMod6_HealthEating_FINAL.pdf (accessed on 14 November 2021).
15. US Food and Drug Administration. Qualified Health Claims: Letter of Enforcement Discretion—Nuts and Coronary Heart Disease (Docket No 02P-0505). Available online: http://wayback.archive-it.org/7993/20171114183724/https:/www.fda.gov/Food/IngredientsPackagingLabeling/LabelingNutrition/ucm072926.htm (accessed on 21 November 2022).
16. Food Standards Australia and and New Zealand. Notified Food-Health Relationships to Make a Health Claim. Available online: https://www.foodstandards.gov.au/industry/labelling/fhr/Pages/default.aspx (accessed on 14 November 2021).
17. European Food Safety Authority. Scientific Opinion on the substantiation of health claims related to walnuts and maintenance of normal blood LDL-cholesterol concentrations (ID 1156, 1158) and improvement of endothelium-dependent vasodilation (ID 1155, 1157) pursuant to Article 13(1) of Regulation (EC) No 1924/2006. *EFSA J.* **2011**, *9*, 2074.
18. Fraser, G.E.; Sabaté, J.; Beeson, W.L.; Strahan, T.M. A possible protective effect of nut consumption on risk of coronary heart disease. The Adventist Health Study. *Arch. Intern. Med.* **1992**, *152*, 1416–1424. [CrossRef]
19. Sabate, J.; Fraser, G.E.; Burke, K.; Knutsen, S.F.; Bennett, H.; Lindsted, K.D. Effects of Walnuts on Serum Lipid Levels and Blood Pressure in Normal Men. *N. Engl. J. Med.* **1993**, *328*, 603–607. [CrossRef]
20. Del Gobbo, L.C.; Falk, M.C.; Feldman, R.; Lewis, K.; Mozaffarian, D. Effects of tree nuts on blood lipids, apolipoproteins, and blood pressure: Systematic review, meta-analysis, and dose-response of 61 controlled intervention trials. *Am. J. Clin. Nutr.* **2015**, *102*, 1347–1356. [CrossRef]
21. Balakrishna, R.; Bjørnerud, T.; Bemanian, M.; Aune, D.; Fadnes, L.T. Consumption of Nuts and Seeds and Health Outcomes Including Cardiovascular Disease, Diabetes and Metabolic Disease, Cancer, and Mortality: An Umbrella Review. *Adv. Nutr.* **2022**. [CrossRef] [PubMed]
22. Becerra-Tomás, N.; Paz-Graniel, I.; WC Kendall, C.; Kahleova, H.; Rahelić, D.; Sievenpiper, J.L.; Salas-Salvadó, J. Nut consumption and incidence of cardiovascular diseases and cardiovascular disease mortality: A meta-analysis of prospective cohort studies. *Nutr. Rev.* **2019**, *77*, 691–709. [CrossRef] [PubMed]
23. Aune, D.; Keum, N.; Giovannucci, E.; Fadnes, L.T.; Boffetta, P.; Greenwood, D.C.; Tonstad, S.; Vatten, L.J.; Riboli, E.; Norat, T. Nut consumption and risk of cardiovascular disease, total cancer, all-cause and cause-specific mortality: A systematic review and dose-response meta-analysis of prospective studies. *BMC. Med.* **2016**, *14*, 207. [CrossRef]
24. Larsson, S.C.; Drca, N.; Bjorck, M.; Back, M.; Wolk, A. Nut consumption and incidence of seven cardiovascular diseases. *Heart* **2018**, *104*, 1615–1620. [CrossRef]
25. Adegbola, A.; Behrendt, C.-A.; Zyriax, B.-C.; Windler, E.; Kreutzburg, T. The impact of nutrition on the development and progression of peripheral artery disease: A systematic review. *Clin. Nutr.* **2022**, *41*, 49–70. [CrossRef]
26. Ogilvie, R.P.; Lutsey, P.L.; Heiss, G.; Folsom, A.R.; Steffen, L.M. Dietary intake and peripheral arterial disease incidence in middle-aged adults: The Atherosclerosis Risk in Communities (ARIC) Study. *Am. J. Clin. Nutr.* **2017**, *105*, 651–659. [CrossRef] [PubMed]
27. Chen, G.-C.; Arthur, R.; Mossavar-Rahmani, Y.; Xue, X.; Haring, B.; Shadyab, A.H.; Allison, M.A.; Liu, S.; Tinker, L.F.; Saquib, N.; et al. Adherence to Recommended Eating Patterns Is Associated With Lower Risk of Peripheral Arterial Disease: Results From the Women's Health Initiative. *Hypertension* **2021**, *78*, 447–455. [CrossRef] [PubMed]
28. Yuan, S.; Bruzelius, M.; Damrauer, S.M.; Håkansson, N.; Wolk, A.; Åkesson, A.; Larsson, S.C. Anti-inflammatory diet and incident peripheral artery disease: Two prospective cohort studies. *Clin. Nutr.* **2022**, *41*, 1191–1196. [CrossRef] [PubMed]
29. Liu, X.; Guasch-Ferré, M.; Drouin-Chartier, J.P.; Tobias, D.K.; Bhupathiraju, S.N.; Rexrode, K.M.; Willett, W.C.; Sun, Q.; Li, Y. Changes in Nut Consumption and Subsequent Cardiovascular Disease Risk Among US Men and Women: 3 Large Prospective Cohort Studies. *J. Am. Heart Assoc.* **2020**, *9*, e013877. [CrossRef] [PubMed]
30. Pan, A.; Sun, Q.; Bernstein, A.M.; Schulze, M.B.; Manson, J.E.; Stampfer, M.J.; Willett, W.C.; Hu, F.B. Red meat consumption and mortality: Results from 2 prospective cohort studies. *Arch. Intern. Med.* **2012**, *172*, 555–563. [CrossRef]

31. Becerra-Tomás, N.; Blanco Mejía, S.; Viguiliouk, E.; Khan, T.; Kendall, C.W.C.; Kahleova, H.; Rahelić, D.; Sievenpiper, J.L.; Salas-Salvadó, J. Mediterranean diet, cardiovascular disease and mortality in diabetes: A systematic review and meta-analysis of prospective cohort studies and randomized clinical trials. *Crit. Rev. Food Sci. Nutr.* **2020**, *60*, 1207–1227. [CrossRef]
32. Massara, P.; Zurbau, A.; Glenn, A.J.; Chiavaroli, L.; Khan, T.A.; Viguiliouk, E.; Mejia, S.B.; Comelli, E.M.; Chen, V.; Schwab, U.; et al. Nordic dietary patterns and cardiometabolic outcomes: A systematic review and meta-analysis of prospective cohort studies and randomised controlled trials. *Diabetologia* **2022**, *65*, 2011–2031. [CrossRef]
33. Chiavaroli, L.; Viguiliouk, E.; Nishi, S.K.; Blanco Mejia, S.; Rahelić, D.; Kahleová, H.; Salas-Salvadó, J.; Kendall, C.W.; Sievenpiper, J.L. DASH Dietary Pattern and Cardiometabolic Outcomes: An Umbrella Review of Systematic Reviews and Meta-Analyses. *Nutrients* **2019**, *11*. [CrossRef]
34. Yokoyama, Y.; Nishimura, K.; Barnard, N.D.; Takegami, M.; Watanabe, M.; Sekikawa, A.; Okamura, T.; Miyamoto, Y. Vegetarian Diets and Blood Pressure: A Meta-analysis. *JAMA Intern. Med.* **2014**, *174*, 577–587. [CrossRef]
35. Viguiliouk, E.; Kendall, C.W.; Kahleová, H.; Rahelić, D.; Salas-Salvadó, J.; Choo, V.L.; Mejia, S.B.; Stewart, S.E.; Leiter, L.A.; Jenkins, D.J.; et al. Effect of vegetarian dietary patterns on cardiometabolic risk factors in diabetes: A systematic review and meta-analysis of randomized controlled trials. *Clin. Nutr.* **2019**, *38*, 1133–1145. [CrossRef]
36. Glenn, A.J.; Viguiliouk, E.; Seider, M.; Boucher, B.A.; Khan, T.A.; Blanco Mejia, S.; Jenkins, D.J.A.; Kahleová, H.; Rahelić, D.; Salas-Salvadó, J.; et al. Relation of Vegetarian Dietary Patterns With Major Cardiovascular Outcomes: A Systematic Review and Meta-Analysis of Prospective Cohort Studies. *Front. Nutr.* **2019**, *6*, 80. [CrossRef]
37. Chiavaroli, L.; Nishi, S.K.; Khan, T.A.; Braunstein, C.R.; Glenn, A.J.; Mejia, S.B.; Rahelić, D.; Kahleová, H.; Salas-Salvadó, J.; Jenkins, D.J.A.; et al. Portfolio Dietary Pattern and Cardiovascular Disease: A Systematic Review and Meta-analysis of Controlled Trials. *Prog. Cardiovasc. Dis.* **2018**, *61*, 43–53. [CrossRef]
38. Glenn, A.J.; Lo, K.; Jenkins, D.J.A.; Boucher, B.A.; Hanley, A.J.; Kendall, C.W.C.; Manson, J.E.; Vitolins, M.Z.; Snetselaar, L.G.; Liu, S.; et al. Relationship Between a Plant-Based Dietary Portfolio and Risk of Cardiovascular Disease: Findings From the Women's Health Initiative Prospective Cohort Study. *J. Am. Heart Assoc.* **2021**, *10*, e021515. [CrossRef]
39. Estruch, R.; Ros, E.; Salas-Salvadó, J.; Covas, M.-I.; Corella, D.; Arós, F.; Gómez-Gracia, E.; Ruiz-Gutiérrez, V.; Fiol, M.; Lapetra, J.; et al. Primary Prevention of Cardiovascular Disease with a Mediterranean Diet Supplemented with Extra-Virgin Olive Oil or Nuts. *N. Engl. J. Med.* **2018**, *378*, e34. [CrossRef]
40. Ruiz-Canela, M.; Martínez-González, M.A. Lifestyle and dietary risk factors for peripheral artery disease. *Circ. J.* **2014**, *78*, 553–559. [CrossRef]
41. Mohammadifard, N.; Salehi-Abargouei, A.; Salas-Salvadó, J.; Guasch-Ferré, M.; Humphries, K.; Sarrafzadegan, N. The effect of tree nut, peanut, and soy nut consumption on blood pressure: A systematic review and meta-analysis of randomized controlled clinical trials. *Am. J. Clin. Nutr.* **2015**, *101*, 966–982. [CrossRef]
42. Eslampour, E.; Asbaghi, O.; Hadi, A.; Abedi, S.; Ghaedi, E.; Lazaridi, A.V.; Miraghajani, M. The effect of almond intake on blood pressure: A systematic review and meta-analysis of randomized controlled trials. *Complement Ther. Med.* **2020**, *50*, 102399. [CrossRef]
43. Ference, B.A.; Ginsberg, H.N.; Graham, I.; Ray, K.K.; Packard, C.J.; Bruckert, E.; Hegele, R.A.; Krauss, R.M.; Raal, F.J.; Schunkert, H.; et al. Low-density lipoproteins cause atherosclerotic cardiovascular disease. 1. Evidence from genetic, epidemiologic, and clinical studies. A consensus statement from the European Atherosclerosis Society Consensus Panel. *Eur. Heart J.* **2017**, *38*, 2459–2472. [CrossRef]
44. Lewington, S.; Clarke, R.; Qizilbash, N.; Peto, R.; Collins, R. Age-specific relevance of usual blood pressure to vascular mortality: A meta-analysis of individual data for one million adults in 61 prospective studies. *Lancet* **2002**, *360*, 1903–1913. [CrossRef]
45. Viguiliouk, E.; Kendall, C.W.; Blanco Mejia, S.; Cozma, A.I.; Ha, V.; Mirrahimi, A.; Jayalath, V.H.; Augustin, L.S.; Chiavaroli, L.; Leiter, L.A.; et al. Effect of tree nuts on glycemic control in diabetes: A systematic review and meta-analysis of randomized controlled dietary trials. *PLoS ONE* **2014**, *9*, e103376. [CrossRef]
46. Tindall, A.M.; Johnston, E.A.; Kris-Etherton, P.M.; Petersen, K.S. The effect of nuts on markers of glycemic control: A systematic review and meta-analysis of randomized controlled trials. *Am. J. Clin. Nutr.* **2019**, *109*, 297–314. [CrossRef]
47. Nishi, S.K.; Viguiliouk, E.; Blanco Mejia, S.; Kendall, C.W.C.; Bazinet, R.P.; Hanley, A.J.; Comelli, E.M.; Salas Salvadó, J.; Jenkins, D.J.A.; Sievenpiper, J.L. Are fatty nuts a weighty concern? A systematic review and meta-analysis and dose–response meta-regression of prospective cohorts and randomized controlled trials. *Obes. Rev.* **2021**, *22*, e13330. [CrossRef]
48. Neale, E.P.; Tapsell, L.C.; Guan, V.; Batterham, M.J. The effect of nut consumption on markers of inflammation and endothelial function: A systematic review and meta-analysis of randomised controlled trials. *BMJ Open* **2017**, *7*, e016863. [CrossRef]
49. Smeets, E.; Mensink, R.P.; Joris, P.J. Effects of tree nut and groundnut consumption compared with those of l-arginine supplementation on fasting and postprandial flow-mediated vasodilation: Meta-analysis of human randomized controlled trials. *Clin. Nutr.* **2021**, *40*, 1699–1710. [CrossRef]
50. Tobias, D.K.; Wittenbecher, C.; Hu, F.B. Grading nutrition evidence: Where to go from here? *Am. J. Clin. Nutr.* **2021**, *113*, 1385–1387. [CrossRef]
51. Sterne, J.A.; Hernán, M.A.; Reeves, B.C.; Savović, J.; Berkman, N.D.; Viswanathan, M.; Henry, D.; Altman, D.G.; Ansari, M.T.; Boutron, I.; et al. ROBINS-I: A tool for assessing risk of bias in non-randomised studies of interventions. *BMJ* **2016**, *355*, i4919. [CrossRef]

52. Schwingshackl, L.; Knüppel, S.; Schwedhelm, C.; Hoffmann, G.; Missbach, B.; Stelmach-Mardas, M.; Dietrich, S.; Eichelmann, F.; Kontopantelis, E.; Iqbal, K.; et al. Perspective: NutriGrade: A Scoring System to Assess and Judge the Meta-Evidence of Randomized Controlled Trials and Cohort Studies in Nutrition Research. *Adv. Nutr.* **2016**, *7*, 994–1004. [CrossRef]
53. Micha, R.; Khatibzadeh, S.; Shi, P.; Andrews, K.G.; Engell, R.E.; Mozaffarian, D. Global, regional and national consumption of major food groups in 1990 and 2010: A systematic analysis including 266 country-specific nutrition surveys worldwide. *BMJ Open* **2015**, *5*, e008705. [CrossRef]
54. Satija, A.; Yu, E.; Willett, W.C.; Hu, F.B. Understanding nutritional epidemiology and its role in policy. *Adv. Nutr.* **2015**, *6*, 5–18. [CrossRef]
55. Guasch-Ferré, M.; Bhupathiraju, S.N.; Hu, F.B. Use of Metabolomics in Improving Assessment of Dietary Intake. *Clin. Chem.* **2018**, *64*, 82–98. [CrossRef]
56. Malik, V.S.; Guasch-Ferre, M.; Hu, F.B.; Townsend, M.K.; Zeleznik, O.A.; Eliassen, A.H.; Tworoger, S.S.; Karlson, E.W.; Costenbader, K.H.; Ascherio, A.; et al. Identification of Plasma Lipid Metabolites Associated with Nut Consumption in US Men and Women. *J. Nutr.* **2019**, *149*, 1215–1221. [CrossRef] [PubMed]
57. Guasch-Ferré, M.; Hernández-Alonso, P.; Drouin-Chartier, J.-P.; Ruiz-Canela, M.; Razquin, C.; Toledo, E.; Li, J.; Dennis, C.; Wittenbecher, C.; Corella, D.; et al. Walnut Consumption, Plasma Metabolomics, and Risk of Type 2 Diabetes and Cardiovascular Disease. *J. Nutr.* **2020**, *151*, 303–311. [CrossRef]
58. Sayón-Orea, C.; Razquin, C.; Bulló, M.; Corella, D.; Fitó, M.; Romaguera, D.; Vioque, J.; Alonso-Gómez, Á.M.; Wärnberg, J.; Martínez, J.A.; et al. Effect of a Nutritional and Behavioral Intervention on Energy-Reduced Mediterranean Diet Adherence Among Patients With Metabolic Syndrome: Interim Analysis of the PREDIMED-Plus Randomized Clinical Trial. *JAMA* **2019**, *322*, 1486–1499. [CrossRef] [PubMed]
59. Delgado-Lista, J.; Alcala-Diaz, J.F.; Torres-Pena, J.D.; Quintana-Navarro, G.M.; Fuentes, F.; Garcia-Rios, A.; Ortiz-Morales, A.M.; Gonzalez-Requero, A.I.; Perez-Caballero, A.I.; Yubero-Serrano, E.M.; et al. Long-term secondary prevention of cardiovascular disease with a Mediterranean diet and a low-fat diet (CORDIOPREV): A randomised controlled trial. *Lancet* **2022**, *399*, 1876–1885. [CrossRef]
60. Gaziano, L.; Sun, L.; Arnold, M.; Bell, S.; Cho, K.; Kaptoge, S.K.; Song, R.J.; Burgess, S.; Posner, D.C.; Mosconi, K.; et al. Mild-to-Moderate Kidney Dysfunction and Cardiovascular Disease: Observational and Mendelian Randomization Analyses. *Circulation* **2022**, *146*, 1507–1517. [CrossRef] [PubMed]
61. Li, J.; Guasch-Ferré, M.; Chung, W.; Ruiz-Canela, M.; Toledo, E.; Corella, D.; Bhupathiraju, S.N.; Tobias, D.K.; Tabung, F.K.; Hu, J.; et al. The Mediterranean diet, plasma metabolome, and cardiovascular disease risk. *Eur. Heart J.* **2020**, *41*, 2645–2656. [CrossRef] [PubMed]

nutrients

MDPI

Review

Impact of Nut Consumption on Cognition across the Lifespan

Stephanie K. Nishi [1,2,3,4,5], Aleix Sala-Vila [3,6,*], Jordi Julvez [7,8], Joan Sabaté [9] and Emilio Ros [3,10,*]

1 Universitat Rovira i Virgili, Departament de Bioquímica i Biotecnologia, Unitat de Nutrició, 43201 Reus, Spain
2 Institut d'investigació Sanitària Pere Virgili (IISPV), 43201 Reus, Spain
3 Centro de Investigación Biomédica en Red Fisiopatología de la Obesidad y la Nutrición (CIBEROBN), Institute of Health Carlos III, 28029 Madrid, Spain
4 Toronto 3D (Diet, Digestive Tract and Disease) Knowledge Synthesis and Clinical Trials Unit, Toronto, ON M5C 2T2, Canada
5 Clinical Nutrition and Risk Factor Modification Centre, St. Michael's Hospital, Unity Health Toronto, Toronto, ON M5C 2T2, Canada
6 Cardiovascular Risk and Nutrition, Hospital del Mar Medical Research Institute (IMIM), 08003 Barcelona, Spain
7 Clinical and Epidemiological Neuroscience Group (NeuroÈpia), Institut d' Investigació Sanitària Pere Virgili (IISPV), 43007 Reus, Spain
8 Barcelona Institute for Global Health (ISGlobal), 08036 Barcelona, Spain
9 Center for Nutrition, Lifestyle and Disease Prevention, School of Public Health, Loma Linda University, Loma Linda, CA 92350, USA
10 Lipid Clinic, Endocrinology and Nutrition Service and Institut d'Investigacions Biomèdiques August Pi Sunyer, Hospital Clínic, 08036 Barcelona, Spain
* Correspondence: asala3@imim.es (A.S.-V.); eros@clinic.cat (E.R.)

Abstract: Cognitive health is a life-long concern affected by modifiable risk factors, including lifestyle choices, such as dietary intake, with serious implications for quality of life, morbidity, and mortality worldwide. In addition, nuts are a nutrient-dense food that contain a number of potentially neuroprotective components, including monounsaturated and polyunsaturated fatty acids, fiber, B-vitamins, non-sodium minerals, and highly bioactive polyphenols. However, increased nut consumption relates to a lower cardiovascular risk and a lower burden of cardiovascular risk factors that are shared with neurodegenerative disorders, which is why nuts have been hypothesized to be beneficial for brain health. The present narrative review discusses up-to-date epidemiological, clinical trial, and mechanistic evidence of the effect of exposure to nuts on cognitive performance. While limited and inconclusive, available evidence suggests a possible role for nuts in the maintenance of cognitive health and prevention of cognitive decline in individuals across the lifespan, particularly in older adults and those at higher risk. Walnuts, as a rich source of the plant-based polyunsaturated omega-3 fatty acid alpha-linolenic acid, are the nut type most promising for cognitive health. Given the limited definitive evidence available to date, especially regarding cognitive health biomarkers and hard outcomes, future studies are needed to better elucidate the impact of nuts on the maintenance of cognitive health, as well as the prevention and management of cognitive decline and dementia, including Alzheimer disease.

Keywords: nuts; cognitive performance; brain health; dementia; lifespan; aging; epidemiological studies; clinical trials

Citation: Nishi, S.K.; Sala-Vila, A.; Julvez, J.; Sabaté, J.; Ros, E. Impact of Nut Consumption on Cognition across the Lifespan. *Nutrients* **2023**, *15*, 1000. https://doi.org/10.3390/nu15041000

Academic Editor: Hirofumi Tanaka

Received: 26 January 2023
Revised: 13 February 2023
Accepted: 14 February 2023
Published: 16 February 2023

1. Introduction

Cognitive health is a key component of healthy aging. Age-related cognitive decline and neurodegenerative disorders, such as dementia, which are a consequence of population aging and an increased lifespan, are a growing public health concern [1]. Dementia, with Alzheimer's Disease (AD) as the most common type, is currently one of the top ten leading causes of mortality among all diseases and one of the major causes of disability and dependency among older people worldwide [1]. As cognitive decline and dementias have physical, psychological, social, and economic impacts, not only for the people directly affected, but also for their caregivers, families, and society at large, it is a key public health concern to address [1].

While neurodegenerative disorders tend to occur at older ages, brain development and cognitive health are impacted across the lifespan, from the fetus stage in pregnancy through adulthood [2]. Increasing evidence indicates that a high cognitive reserve, a healthy lifestyle, and the control of modifiable cardiovascular risk factors may reduce the risk of developing cognitive decline and dementia, including AD [3]. Of the lifestyle components influencing brain health, nutrition holds much potential. Nutrition may directly affect the brain or indirectly influence risk factors shared by cardiovascular and neurodegenerative diseases, thereby possibly having a substantial influence on cognition and the risk of dementia [4,5]. For instance, oxidative stress and inflammation are thought to play a major role in the initiation and progression of AD and other neurodegenerative disorders [6]. Antioxidant-rich foods and dietary patterns are potential strategies to counteract cognitive decline and AD and promote healthy aging. As proof, evidence is accumulating from both prospective studies and randomized controlled trials (RCTs) that adherence to plant-based dietary patterns rich in antioxidant foods, such as vegetables, fruits, whole grains, legumes, and nuts, is associated with a delay in age-related cognitive decline among older adults from diverse populations, as summarized in recent systematic reviews [7,8].

The Mediterranean diet (MedDiet) has been by far the most investigated dietary pattern for associations with brain health outcomes, with findings suggesting a protective association with cognitive decline [9–11]. Neuroimaging studies have further supported the association between increasing adherence to the MedDiet and greater brain volumes, lesser changes of brain atrophy, and the preservation of structural connectivity in healthy older adults. Within these investigations, higher intake of specific nutrients such as unsaturated fatty acids, antioxidant vitamins, and polyphenols has been linked to larger brain volumes [12]. More limited research on other healthy plant-based diets, such as the dietary approach to stop hypertension (DASH) diet, the Mediterranean-DASH diet, the intervention for neurodegenerative delay (MIND) diet, and other anti-inflammatory diets, has shown beneficial associations with cognitive health in older adults as well [13,14].

Given the available evidence, the World Health Organization (WHO) guidelines for risk reduction of cognitive decline and dementia included nutrition-related recommendations relating to the MedDiet and healthy, balanced, plant-based dietary patterns, all of which included nuts [15]. In addition, the WHO guidelines state that unsaturated fats, such as those found in nuts, are preferred over saturated fats for brain health and that consumption of nuts has been associated with a reduced risk of cognitive impairment [15].

Nuts (i.e., tree nuts, including almonds, Brazil nuts, cashews, hazelnuts, macadamias, pecans, pine nuts, pistachios, and walnuts; and peanuts) are an integral part of plant-based diets and have an optimal nutrient profile, being particularly abundant in anti-inflammatory and antioxidant molecules, such as unsaturated fatty acids, non-sodium minerals, vitamins, and polyphenols; moreover, their frequent consumption is associated with a consistent reduction in the risk of cardiovascular disease (CVD) [16]. Due to the fact that cardiovascular risk factors and CVD have well-established links to neurodegeneration and unhealthy aging, nut consumption, already well known to benefit vascular function, has been hypothesized to also favor cognition and overall brain health [17].

The present narrative review aims to present up-to-date evidence regarding the association between nut consumption and cognitive health during different stages of life. Specifically, it reports on the proceedings from the "Nuts 2022, Where we are and where we are going in research" international conference session titled "Nuts, Ageing, and Cognition." In this session, epidemiological, clinical, and mechanistic evidence regarding nut consumption and cognition in different age groups was presented and discussed. This review is not a systematic review, and thus limitations should be acknowledged in that all studies may not have been identified. However, this review summarizes the available literature from independently conducted searches, and findings were further shared and discussed among the ensemble of experts in the field of nut and health research.

2. Nut Consumption and Neurodevelopment in Early Life (Gestation to Young Adulthood)

The early life years are critically important for cognitive development. Generally, brain development begins a few weeks after conception and is thought to be complete by early adulthood. The basic structure of the brain is believed to be shaped primarily during the prenatal period and early childhood, with the formation and refinement of neural networks and modification of functional abilities continuing over the long term [2]. Nutrition during each of the life stages of pregnancy, lactation, childhood, and adolescence can have a fundamental influence on development [18,19].

2.1. Nut Consumption and Prenatal Cognitive Development

The period of in-utero growth during gestation is considered to be particularly important for neurodevelopment since the brain undergoes several uniquely intense and complex processes [20,21]. Human brain development begins soon after conception with the inception of the formation of the neural tube and continues into early adulthood. The fetal brain begins to develop during the third week of gestation. By the end of the embryonic period (gestational week 10), the basics of the neural system are established. All the structures continue to develop throughout the fetal period and early childhood. By 6 years of age, the brain has reached 90% of its adult volume [22]. During this period of brain development, essential nutrients, such as the omega-3 fatty acid alpha-linolenic acid (ALA), which can be found in nuts, particularly walnuts, may alter the epigenetic control of neural processes, neuron formation, migration, axon and dendritic growth, synaptogenesis, and myelination [23]. In the long term, adequate nutrition that promotes neurodevelopment during the in-utero period may benefit a child's neuropsychological development, school performance, and future professional success [21]. Yet, the possible protective effects of nut consumption on cognitive health have hardly been explored in child neurodevelopment. At present, there is limited evidence from epidemiological and clinical studies assessing nut consumption and brain function during pregnancy, childhood, or adolescence.

In relation to the pregnancy period, to our knowledge, only one study has been published assessing the association between cognitive health and nut consumption during early development. This study involved the Spanish Childhood and Environment (Infancia y Medio Ambiente, INMA) Project, a large prospective, multicenter, population-based cohort of 2208 mother-child pairs, conducted in several regions of Spain [24]. Mothers were followed during pregnancy (first and third trimesters), and their children were enrolled at birth and followed until the age of 8 years. Twice during pregnancy and at the children's ages of 1.5, 5, and 8 years, dietary intake questionnaires and neuropsychological assessments were administered. The mean nut consumption among mothers in the first trimester of pregnancy was 41 g/week (standard deviation [SD], 74 g/week), and the median was 17 g/week (interquartile range [IQR]: 0 to 46 g/week), with a third of the total participants being non-consumers (n = 860, 33.5%). Overall, the authors found that higher maternal consumption of nuts in early pregnancy was associated with enhanced cognitive development in their children, compared to non-consumers, at 1.5, 5, and 8 years of age [24].

2.2. Nut Consumption and Early Life (Childhood and Adolescence) Cognitive Development

Nutrition during childhood is particularly important, as this is a period of relatively rapid brain development, and nutrients aid the brain in the creation of new synaptic connections during learning processes at school and in home environments [18]. While the structural components of the brain and the foundations of basic sensation and perception systems are fully developed by the time children reach kindergarten age, other systems such as those involved in memory, decision-making, and emotion continue to develop well into childhood. The foundations of many of these abilities, however, are constructed during the early years. Whereas the functional aspects of the brain can have varying developmental time frames and patterns, adolescence is an important period of brain development and remodeling to functionally develop for thinking and processing [2]. The brain reorganizes during this developmental stage with functional and structural changes resulting from the re-emergence of gonadotropin-releasing hormone, triggering a cascade of hormone-dependent processes. Other biological processes involve epigenetic factors, which are highly sensitive to the environment and may therefore make this period of growth more vulnerable to external insults [25]. Moreover, the prefrontal cortex, which carries out important functions such as internally guided behaviors (control of emotion), logical thinking, working memory, and organizing skills (executive function), is the last region of the brain to mature (in the early twenties). The synaptic plasticity of the prefrontal cortex is accentuated during adolescence, a process that involves loss of grey matter density and an increase in white matter volume, cerebral blood flow, and synaptic pruning. Adolescence is also a time of refinement of brain connectivity and complex behaviors [26]. It is widely recognized that the synaptic plasticity of the brain decreases with age [25,26], but this pattern does not seem to follow a linear trend, and adolescence is an important period during which brain development can be enhanced and protected from environmental hazards, from air pollution to unhealthy diets, with long-term consequences.

Considering the importance of brain structural and functional development during childhood and adolescence, very few studies have assessed the association between nut consumption and cognitive health during these life stages. One cross-sectional study conducted in 317 Korean children and adolescents (167 girls and 150 boys) with a mean age of 11.8 (range, 6 to 18) years and no prior diagnosis of neurologic or psychiatric disorders assessed the consumption of nuts, among other healthy foods estimated from diet questionnaires, in relation to cognitive performance [27]. Nut consumption was related to improved cognitive reaction time consistency and attention function as measured by the symbol-digit modality test (SDMT). However, no associations were observed with the other neuropsychological measures, specifically the verbal and visual memory tests, the shift attention test, the reasoning test, and the digit span forward and backward tasks, assessed as part of a computerized cognitive assessment battery. A limitation of this study is that the authors did not adjust the data for covariables known to influence cognitive performance in youth, such as parental social class, parity, type of delivery, breastfeeding, birthweight, maternal intellectual quotient (IQ), maternal mental health, maternal smoking and alcohol intake during pregnancy, clinical history during pregnancy, overall dietary pattern, and stress events [27].

While there is currently a lack of RCTs, there is promise for further evidence as a protocol for a RCT (the WALNUT study [WSS]) aiming to assess the effect of daily walnut consumption (30 g) over 6 months on cognitive function among nearly 700 healthy adolescents from several high schools in Barcelona, Spain, was recently published [28]. The results of this study will eventually fill a gap in our knowledge of the effect of nut consumption on cognitive health in adolescence.

Linking the previous findings relating the consequences of increased nut consumption by mothers during pregnancy to the later neuropsychological traits of their children [24], and further applying public health recommendations to an entire population, for example, recommending pregnant women to consume a daily serving of nuts throughout the prenatal period, might be hypothesized to increase the population mean IQ score by a few points. This is not clinically significant at the individual level; still, the impact on the IQ distribution at the population level might represent a significant reduction in the proportion of children with learning problems or low IQ scores [24,29]. Depending on the findings of future research, this could be a potentially impactful health promoting message for the population, possibly akin to the promotion of long-term breastfeeding and/or the recommendation to stop smoking during pregnancy.

2.3. Nut Consumption and Cognitive Health in Young Adulthood

As adolescence transitions into young adulthood, complex cognitive behaviors are refined. It is during this life stage that myelination of regions involved in higher cognitive abilities, such as the prefrontal cortex, is considered to be complete [2]. This process of myelination ultimately involves the axons of neurons being wrapped in fatty cells, which facilitates neuronal activity and communication for the transmission of electrical signals.

Nutrition, and particularly fatty acids, are critical for brain development. Essential fatty acids are long-chain polyunsaturated acids (LC-PUFAs) that the body cannot synthesize and must be obtained from the diet (mainly from oily fish, seeds, and nuts). LC-PUFAs are involved in the function and architecture of the central nervous system throughout the various life stages. It has been shown that the LC-PUFA docosahexaenoic acid (DHA) regulates neurotransmission systems such as serotonergic, dopaminergic, norepinephrinergic, and acetylcholinergic systems [23,30].

The omega-3 PUFA ALA, which is particularly abundant in walnuts, has been associated with cognitive function in older people [31]. However, it has been scarcely studied at younger ages. In a recent cross-sectional study of 332 healthy adolescents, the red blood cell proportions of ALA (an objective biomarker of walnut consumption) were inversely associated with impulsivity (a usually detrimental psychological trait and a key feature of many psychiatric disorders) [32]. The findings from this study support the hypothesis that nuts, particularly walnuts, could have a beneficial impact on cognition.

Additionally, to date, the only known RCT specifically investigating the effect of nut consumption on cognitive health in youth was that conducted by Pribis and colleagues in young adults aged 18 to 25 years [33]. In a crossover investigation, the consumption of 60 g of walnuts for 8 weeks by college students (n = 47) was associated with better critical thinking abilities as measured by the Watson-Glaser Critical Thinking Appraisal. However, no differences were observed for verbal reasoning, as measured by Raven's advanced progressive matrices, or memory, as assessed by the Wechsler Memory Scale, when compared with the placebo group. The authors acknowledged that these findings may be limited by the short duration of the intervention as well as by the fact that participants were college students, whose baseline cognitive functioning may be higher than that of the general population and hence may impact cognitive findings.

Early-life structural and functional development of the brain and the potential influence of dietary intake are important, yet little research has been conducted to assess the impact of nut consumption on cognitive health during childhood and adolescence. Although preliminary research is promising, further investigations are warranted to better determine the efficacy of consuming nuts on neurodevelopment, especially during life stages of growth and development.

3. Nut Consumption and Cognitive Performance in Adulthood

An essential component of healthy aging is normal cognitive function, which critically affects functional independence and health-related quality of life. Increased life expectancy and subsequent population aging entail a rising prevalence of age-associated cognitive impairment, a major public health concern given its frequent transition to mild cognitive impairment (MCI) and dementia, including AD [34]. Indeed, many older adults experience deteriorating cognitive function, usually with declining episodic memory and executive function that parallel volume losses in critical brain structures such as the hippocampus [35]. Cognitive domains that can be interrogated with specific neuropsychological tests include memory, executive function, attention, language, and visuospatial skills. The neuropsychological test most commonly used in epidemiologic studies is the Mini-Mental State Examination (MMSE), a brief test (it takes 7 to 10 min to complete) that is very useful to detect dementia when it is grossly abnormal but is limited in its ability to provide insight into subtler and much more frequent cognitive deficits [36]. A telephone-adaption of the MMSE (the TICS, or Telephone Interview for Cognitive Status) to assess overall cognitive performance has also been frequently used in the absence of more comprehensive tests.

While these screening instruments may offer an opportunity for cognitive comparisons, their accuracy in detecting cognitive impairment is a limitation in neurocognitive studies. MCI is said to be present when there is objective evidence at cognitive testing that one or more of these cognitive domains is impaired. As opposed to individuals with dementia, those with MCI maintain their independence in functional abilities and have no significant impairment in social or occupational functioning [37].

3.1. *Epidemiological Studies Examining the Association of Nut Consumption with Cognitive Performance*

In the review of epidemiological evidence evaluating nut consumption and cognitive performance, 15 studies were identified, including 7 cross-sectional and 8 prospective cohort studies. Table 1 lists these investigations by date of publication and summarizes their findings. Briefly, these observational studies involved men and women, with the majority aged ≥50 years, from Australia, China, Italy, the Netherlands, Norway, Spain, and the United States, with the prospective cohorts ranging from 3 to 20 years in duration and assessing quantiles of nut intake comparing none or low to higher intake dosages. The known factors influencing cognitive performance in adulthood, namely age, sex, educational level, body mass index, cardiovascular risk factors (smoking, hypertension, dyslipidemia), physical activity, overall dietary pattern, and depression [3], were treated as confounders and adjusted for in analyses of data from epidemiological studies.

Table 1. Epidemiological studies examining the association of nut consumption with cognitive performance.

Author, Year	Study Design (Source)	N	Age (Years)	Participant Characteristics	Neuro-Psychological Tests	Nut Dose/Day (Range)	FU (Years)	Outcome
Nurk, 2010 [38]	Cross-sectional (Hordaland Health Study)	2031	70–74	Men & women, general population	Complete battery	No consumption to high consumption	NA	No association
Nooyens, 2011 [39]	Prospective (Doetinchem cohort)	2613	43–70	Men & women, general population	Tests of memory, information processing, cognitive flexibility—sum of test scores (global cognition)	Quintiles of consumption	5	Higher nut consumption associated with cognitive flexibility and global cognition at baseline and trend to delayed cognitive decline at follow-up.
Valls-Pedret, 2012 [40]	Cross-sectional (PREDIMED study)	447	55–80	Men & women at high cardiovascular risk	Comprehensive battery	Total nuts (0–60) Walnuts 1 g (0–30)	NA	Walnuts, but not total nuts, associated with better working memory.
Samieri, 2013 [41]	Prospective (Women's Health Study)	6174	65+	Women	Comprehensive battery, including TICS	Quintiles of nut consumption within the Mediterranean diet	4	No association of nuts with cognitive changes
O'Brien, 2014 [42]	Prospective (Nurses' Health Study)	15,467	Mean 74	Women from a selected cohort of nurses	TICS	From never/<1/month to ≥5 servings/week	6	Higher long-term total nut intake associated with better average cognitive status for all cognitive outcomes.
Arab, 2015 [43]	Cross-sectional (NHANES)	5562 and 2975	2 groups: 20–59 ≥60	Men and women, general population	Various cognitive tests	Walnuts with high certainty/walnuts with other nuts	NA	Walnut consumption positively associated with cognitive function in the two groups.

Table 1. *Cont.*

Author, Year	Study Design (Source)	N	Age (Years)	Participant Characteristics	Neuro-Psychological Tests	Nut Dose/Day (Range)	FU (Years)	Outcome
Dong, 2016 [44]	Cross-sectional	894	50 to >80	Men and women from a population cohort	MoCa test	Tertiles of consumption	NA	Higher nut consumption associated with delayed memory. Cognitively healthy adults consumed more nuts than those with MCI.
De Amicis, 2018 [45]	Cross-sectional	279	>65	Men and women attending Nutrition center	MMSE	Highest vs. lowest nut consumption within the Mediterranean diet	NA	OR = 0.30; 95% CI, 0.13–0.69 of low MMSE
Li, 2019 [46]	Prospective	4822	>55	China Health and Nutrition survey	TICS	Consumers of nuts (mainly peanuts)	15	Nuts >10 g/d: OR 0.60, 95% CI 0.43–0.84) of poor cognition
Rabassa, 2020 [47]	Prospective	119	>65	InChianti population study	MMSE	Consumers vs. non-consumers	3	OR: 0.78; 95% CI: 0.61–0.99 of low MMSE
Tan, 2021 [48]	Cross-sectional (NHANES)	1848	60+	Men and women, general population	CERAD total, delayed recall, animal fluency and digit-symbol substitution test	4 groups, from no consumers to consumers meeting recom-mendations (>30 g/d)	NA	Cognitive scores higher from moderate intake (15.1–30.0 g/d), same in high intake
Jiang, 2021 [49]	Prospective	16,737	Mean 53,5	Singapore Chinese Health Study	MMSE	Nuts <1 serv/mo, 1–3 serv/mo, 1 serv/wk, and =>2 serv/wk	20	3 highest categories: 12% (CI 2–20%), 19% (CI 4–31%) and 21% (CI 2–36%) lower risk of cognitive impairment
Bishop, 2021 [50]	Prospective	3632	65+	Health and Retirement & Health Care and Nutrition studies	TICS	None, low or moderate intake of walnuts	4	Any walnut consumption had greater scores at baseline. No association with cognitive changes.
Chen, 2021 [51]	Cross-sectional	819	70–90	Sydney Memory and Ageing Study:	Comprehensive battery	Consumption of nuts and legumes	NA	Higher consumption related to global cognition (β = 0.117; CI: 0.052–0.181), visuospatial function (β = 0.105; CI: 0.047–0.163), and language (β = 0.113; CI: 0.038–0.189).
Li, 2022 [52]	Prospective	9028	Mean 69	Zhejiang Ageing and Health Cohort Study	MMSE (repeated)	None, <70 g/week, or =>70 g/week	6	Less cognitive impairment (RR = 0.83, 95% CI 0.75–0.91) for highest nut intake group

Abbreviations: CERAD, Consortium to Establish a Registry for Alzheimer's Disease; CI, confidence interval; FU, follow-up; MCI, mild cognitive impairment; MMSE, Mini-Mental State Examination; mo, month; MoCa, Montreal cognitive assessment (short-term memory recall ability, visuospatial abilities, executive functions, phonemic fluency ability, verbal abstraction ability, attention, concentration and working memory, language, and orientation); N, number of study participants; NA, not applicable; NHANES, National Health and Nutrition Examination Survey; OR, odds ratio; PREDIMED, PREvención con DIeta MEDiterránea; TICS, Telephone Interview for Cognitive Status, a telephone-adaption of the MMSE to assess overall cognitive status; serv, serving.

Out of a total of 15 observational studies, 13 showed a positive association between nut consumption and cognitive performance; however, beneficial relationships were not observed among all cognitive assessments conducted in each study. For instance, in a prospective Dutch study of middle-aged adults, cognitive performance was assessed at baseline and after a 5-year follow-up in relation to quintiles of consumption of plant foods. The highest nut intake was associated with better cognitive function (i.e., memory, speed, flexibility, and global cognitive function) at baseline but not lesser cognitive decline at follow-up when data were adjusted for cardiovascular risk factors [39]. In a cross-sectional study of Chinese adults, higher consumption of fruit, vegetables, and nuts was associated with delayed memory, while all other cognitive domains were unaffected; moreover, cognitively healthy participants consumed more nuts than those with MCI [44].

Further cross-sectional analyses have presented significant beneficial relationships between nut consumption and cognitive health. A cross-sectional study of nut consumption and cognitive performance was nested within a sub-cohort of the Prevención con Dieta Mediterránea (PREDIMED) study. This was a landmark 5-year RCT in which a MedDiet supplemented with either extra-virgin olive oil or mixed nuts (30 g/day: 15 g walnuts, 7.5 g almonds, and 7.5 g hazelnuts) versus a control diet (advice to follow a low-fat diet) resulted in nearly a 30% reduction in CVD events in older individuals at high cardiovascular risk [53]. The cross-sectional sub-study assessed the association of consumption of various foods with cognitive function. Of all the foods considered, only olive oil, coffee, wine, and walnuts—but not total nuts—were related to better cognitive function independently of known risk factors for cognitive decline, other food consumption, and energy intake. Of note, total urinary polyphenol excretion, an objective biomarker of intake of polyphenol-rich foods, was directly associated with working memory function [40]. In addition, within the context of the MedDiet, consumption of 1 serving of nuts (30 g)/week by older Italian adults was cross-sectionally associated with a reduced risk of low MMSE [45]. Cross-sectional analyses of the U.S. National Health and Nutrition Examination Survey (NHANES) indicated beneficial associations between nut consumption and cognition, determined based on 24-h dietary recalls [43,48]. Findings from the most recent relevant NHANES analyses of participants aged 65 and older who consumed nuts 15 to 30 g/d or met recommendations by consuming >30 g/day (either group accounting for 10% of the cohort) had better cognitive scores than non-consumers or low consumers [48]. A moderately sized cross-sectional study within the Sydney Memory and Ageing Study related consumption of different food groups to cognitive performance and found that higher consumption of nuts and legumes together related to higher global cognition, visuospatial function, and language [51].

Furthermore, prospective cohort analyses have demonstrated a potential beneficial relationship between nut consumption and cognitive function. In brief, a U.S. prospective study of a large sub-cohort of older women from the Nurses' Health Study (NHS) assessed total nut consumption in relation to cognitive function and found an association with better average status for all cognitive outcomes analyzed. The difference in the global composite score between women consuming at least 5 servings of nuts/week and non-consumers was equivalent to the mean difference observed in cognitive status between women 2 years apart in age [42]. Another sizable U.S. cohort framed within an observational study of an aging population found that participants with any walnut consumption had greater cognitive scores at baseline, but no association with 4-year cognitive change was observed. However, only 13% of the sample had moderate walnut consumption of around 1/2 oz (14 g) per day [50]. Along similar lines, three large Asian prospective cohort studies showed favorable associations between nut intake and cognitive function. A prospective large 15-year Chinese study of participants in a nutrition survey aged 55 years or more found that consumers of nuts (mostly peanuts), which made up only 17% of the cohort, had less cognitive decline, as measured with the telephone version of the MMSE, than those not consuming any nuts [46]. The large prospective Singapore Chinese Health Study, in which nut consumption was determined at baseline and cognitive function was measured by the MMSE after 20 years of follow-up, reported an inverse association between graded nut consumption and reduction of cognitive function [49]. However, adjustment for intake of unsaturated fatty acids attenuated the association to non-significance, suggesting mediation of the cognitive effect by this key nut component. Finally, a large Chinese cohort study assessed nut consumption at baseline and in relation to changes in the MMSE, administered one to three times during a 6-year follow-up; the results showed that higher nut consumption related to a lower risk of cognitive impairment [52]. Conversely, possibly due to the shorter duration, a small prospective Italian study reported that baseline nut consumers versus non-consumers had a borderline reduced rate of developing a low MMSE after follow-up for 3 years [47].

Regarding specific types of nuts in general, walnuts appeared to be the type most studied and reported to be associated with better cognitive function when compared to low or non-consumers [41,43,47]. However, most of the studies assessed the total nut consumption as a whole and did not or could not delineate the analyses by the different types of nuts.

Two of the 15 studies did not show statistically significant associations between nut consumption and cognitive performance. Specifically, in a cross-sectional investigation of an older Norwegian cohort, nuts were non-significantly associated with better cognitive performance, although only 16% of the participants were nut consumers [38]. Additionally, a large U.S. prospective cohort, the Women's Health Study, found no association for changes in cognitive performance in relation to quintiles of total nut consumption over a 4-year follow-up [41].

Overall, the epidemiological evidence suggests nut consumption may be positively associated with cognitive health. Still, the quality of the evidence from these epidemiological studies, which cannot determine causation, may be considered low for several reasons. First, six of the 13 studies with positive results were cross-sectional, and in four investigations reporting a beneficial association between increased exposure to nuts and cognitive performance, the outcome was assessed with the MMSE, exclusive of more precise neuropsychological tests. Second, most studies obtained exposure data from food frequency questionnaires (FFQs) or diet recalls, which have inherent weaknesses with regard to possible measurement error and recall bias [54]. Third, most prospective studies have an additional problem, i.e., nut exposure is assessed only once at baseline, thus missing the effect of any changes in consumption during follow-up. Fourth, tree nut and peanut data were also often reported in combination in the assessments, and information on nut preparation (salting, roasting, etc.) was lacking. Fifth, based on the available data, the prevalence of nut consumers was usually low, with some cohorts reporting only 13 to 30% of the study population consuming at least 1 serving (28 g)/week. Only 2 studies evaluated cohorts that met nut consumption recommendations of 30 or more g/d compared to non-consumers [44,45]. Finally, the amounts of nuts consumed by consumers tended to be relatively low (e.g., 2.9 g/d to ≥20 g/d) or were not sufficiently described to be able to provide comprehensive assessments and interpretations to help inform practice.

3.2. Randomized Controlled Trials of Nuts with Outcomes on Cognitive Performance

The results of scientifically sound RCTs are critical for formulating evidence-based dietary recommendations. However, few RCTs have examined the effects of nuts on cognitive outcomes in adults, and even fewer had sufficient statistical power or an intervention period lasting more than 6 months (Table 2). Hence, the level of evidence is still fragmentary.

Table 2. Randomized controlled trials of nuts with outcomes on cognitive performance in adults.

Author, Year	Study Design (Source)	N	Age (Years)	Participant Characteristics	Neuro-Psychological Tests	Nut Dose/Day (Range)	FU	Outcome
Martinez-Lapiscina, 2013 [55]	Parallel (Sub-sample of PREDIMED study)	522	55–80	Men & women at high cardiovascular risk	MMSE and Clock Drawing Test (Administered once at the end of the study)	Mixed nuts, 30 g with MedDiet	6.5 years	MedDiet + nuts: better global cognition compared to control diet.

Table 2. *Cont.*

Author, Year	Study Design (Source)	N	Age (Years)	Participant Characteristics	Neuro-Psychological Tests	Nut Dose/Day (Range)	FU	Outcome
Valls-Pedret, 2015 [56]	Parallel (Sub-sample of PREDIMED study)	334	55–80	Men & women at high cardiovascular risk	Tests of memory, executive function, global cognition (Administered at baseline and end of study)	Mixed nuts, 30 g with MedDiet	4.1 years	MedDiet + nuts: better memory and a tendency to improved executive function and global cognition compared to control diet.
Barbour, 2017 [57]	Crossover	61	Mean 65	Men and women with over-weight/obesity	Tests of memory, executive function, and processing speed	High-oleic acid peanuts, 56–84 g	12 weeks	Short-term memory and verbal fluency improved with the peanut diet compared to control diet.
Dhillon, 2017 [58]	Parallel	86	Mean 31	Men and women with over-weight/obesity	Tests of memory and attention	Almonds, dry-roasted at 15% energy	12 weeks	Cognition similarly improved with the almond and control diets.
Sala-Vila, 2020 [59]	Parallel (WAHA study)	708	63–79	Cognitively healthy	Complete test battery	Walnuts at 15% energy	2 years	No effect on cognitive scores in the whole cohort.
Mustra Rakic, 2022 [60]	Parallel	60	50–75	Healthy adults	CANTAB	Almonds/day: 1.5 oz, 3 oz or 3.5 oz snacks	6 months	No among-group changes in cognitive measures.

Abbreviations: CANTAB, Cambridge neuropsychological test automated battery: taps several cognitive domains, including memory, processing speed, and attention; FU, follow-up; MedDiet, Mediterranean diet; MMSE, Mini-Mental State Examination; N, number of study participants; PREDIMED, PREvención con DIeta MEDiterránea. WAHA, Walnuts And Healthy Aging trial.

Additionally, two of the largest RCTs with the longest follow-up were sub-studies of the PREDIMED trial. In a study conducted in the Navarra recruiting center, two neuropsychological tests assessing general cognition were administered [55]. However, these tests were only administered at the end of the study, following a median 6.5-year intervention period, thus changes over time were not evaluated. The results indicated that the two MedDiets (enriched with olive oil or mixed nuts) were associated with better cognitive outcomes compared to the control diet. In another PREDIMED sub-study carried out in the Barcelona center, a comprehensive cognitive battery was administered both at baseline and at the end of the trial after a median follow-up of 4.1 years [56]. The results showed that values for all cognitive domains declined in participants randomized to the control diet, while composites of memory performance, executive function, and global cognition improved above baseline with the two MedDiets. However, the improvement in executive function and global cognition observed with the nut diet did not reach statistical significance compared to the control diet. The findings demonstrated that a MedDiet supplemented with mixed nuts could delay the age-related decline of memory function.

Two small, short-term RCTs investigated the effect of peanuts [57] and almonds [58] on outcomes of cognitive performance in individuals with overweight or obesity. Surprisingly, given the intervention only lasted 12 weeks, in the study of Barbour et al. [57], the diet enriched with high-oleic acid peanuts resulted in improvements in short-term memory and verbal fluency compared to the control diet. In this trial, blood flow velocity in the middle cerebral artery was measured non-invasively with transcranial Doppler, and results showed that the peanut diet increased cerebrovascular reactivity (i.e., improved endothelial function of brain arteries). On the other hand, in the study by Dhillon et al. [58], almond consumption had no effect on cognitive performance compared to the control diet. In this trial, acute experiments examined whether a high-fat lunch rich in almonds would influence the well-known post-lunch dip in alertness, memory, and vigilance. The findings revealed that, compared with a high-carbohydrate meal, almond consumption at lunch counteracted in part the postprandial decline in memory, but not that of attention performance [58].

The large walnuts and healthy aging (WAHA) trial tested the 2-year effects of walnut consumption at 15% of daily energy on cognitive performance in healthy older adults from two sites, Barcelona (Spain) and Loma Linda (CA, USA) [59]. The WAHA study failed to find any differences in neurocognitive test scores for perception, language, memory, and frontal function domains or in a composite score of global cognition compared to the control diet. However, post hoc analyses by site revealed improved cognition in participants allocated to the walnut diet in Barcelona, who were more at risk of cognitive impairment than their California counterparts due to lower educational levels and more smoking. Functional brain magnetic resonance imaging (MRI) in a subset of the Barcelona cohort supported the benefit of walnuts on cognition [59]. Finally, a recent small RCT using different doses of almonds for 6 months in cognitively healthy middle-aged and older adults assessed with a complete neuropsychological test battery at baseline, 3 months, and 6 months found no differences in cognitive measures over time [60].

In summary, the findings of the two well powered, long-term, PREDIMED sub-studies examining the effects of MedDiets supplemented with nuts on cognitive performance indicated a beneficial effect in older individuals at high risk of CVD (thus, also at high risk of cognitive impairment and dementia); however, improved cognitive health might not be entirely attributable to nuts, as other components of the MedDiet changed in these studies [55,56]. Nevertheless, the MedDiet enriched with nuts reduced the relative risk of stroke by nearly 50% in the PREDIMED trial [39], which further supports the neuroprotective effect of nuts. Indeed, preventing stroke, post-stroke cognitive impairment, and dementia is critical for achieving optimal brain health [61]. The large, 2-year WAHA trial uncovered a salutary cognitive effect of walnuts only in participants at higher risk of cognitive impairment [59]. These findings concur with data collected in large multi-domain trials suggesting that individuals at high risk of cognitive impairment or who already have memory complaints, or MCI, are those who should be targeted for preventive interventions because they might obtain the largest benefit [62]. Clearly, larger and longer-term studies with nuts in populations at risk of dementia are warranted.

4. Potential Mechanisms of Action of Nuts in Cognitive Health

Normal aging involves many structural and functional brain changes. There are several hallmarks of cerebral atrophy (ventricular enlargement, cortical thinning, sulcal widening, and volume loss [63]), which are observed in parallel with declines in processing speed and certain memory, language, visuospatial, and executive function abilities [64]. Ageing is the primary risk factor for late-onset Alzheimer's disease (AD; occurring in people aged 65 and over) [65]. However, AD should not be considered part of normal aging [66]. What determines the transition from normal aging to AD remains to be elucidated, but there is a long-standing consensus supporting the view of AD as a multifactorial disease [67], with pathological brain changes taking place years (even decades) before symptomatology is present. However, knowledge of AD is rapidly evolving. For many years, AD was conceived as a clinical-pathological construct, defined by the presence of symptoms/signs. The advent of cost-effective biomarkers prompted the re-definition of AD as a clinical-biomarker construct, referring to an aggregate of neuropathologic changes that can be identified in vivo much before clinical symptoms appear [68]. This preclinical phase of AD represents a therapeutic window for preventive strategies, which are of utmost importance, as highlighted by several international organizations [69,70].

Evidence is accumulating on the many diet components that might have a significant impact on the progression and prevention of AD (reviewed in [71]). As discussed, nuts are rich in compounds with anti-inflammatory, antioxidant, and hypolipidemic effects, thereby reducing the risk of CVD [16]. Given that CVD and AD share many risk factors, particularly hypertension, obesity, type 2-diabetes, and smoking [3], it is reasonable to assume that regular nut consumption might also protect against AD. Abundant experimental research supports this hypothesis (reviewed in [72]). As summarized in the preceding section, epidemiologic evidence, albeit generally of low quality, also concurs with this view, while RCTs of nuts for cognitive outcomes are incipient and confined to changes in cognitive performance after short- or medium-term supplementations. In this section, we will summarize possible mechanisms underlying the putative effects of nut bioactives on the two major hallmarks of AD, namely the buildup in the brain of amyloid-beta (Aβ) plaques and neurofibrillary tangles. We will review experimental research involving either nuts, nut extracts, or nut bioactives given on their own. Additionally, and concurrently with the current needs for research on diet and dementia [73], we will also focus on how available biomarkers might help in better selecting participants to be included in RCTs, and/or as intermediate endpoints, allowing for the detection of subtle changes after short-term interventions.

4.1. Nuts and Extracellular Plaque Deposits of Aβ

4.1.1. Pathophysiologic Overview

According to the amyloid hypothesis [74], derangements in the production, accumulation, or disposal of Aβ are the main cause of AD. Aβ is a ~4 kDa peptide derived from the so-called β-amyloid precursor protein (APP), which is a transmembrane molecule. The enzymatic processes involving the metabolism of APP to Aβ involve a sequential cleavage by two membrane-bound endoproteases, β- and γ-secretase. In the first step, believed to be the rate limiting one, β-secretase (also known as BACE1) cleaves APP to release a large, secreted derivative (soluble peptide APPβ), while a 99-amino acid fragment (C99, also known as the carboxy-terminal fragment of beta [CTFβ]) remains bound to the fatty acid membrane. C99 undergoes a second cleavage by the action of γ-secretase, generating different species of Aβ, those ending at positions 40 (Aβ40) and 42 (Aβ42) being the most abundant ones (~80–90%, and ~5–10%, respectively) (reviewed in [75]). Aβ42 is more hydrophobic than Aβ40 and rapidly aggregates to form monomers and then mature fibrils and dense fibril meshes (senile plaques), the best-known hallmark of AD. However, soluble dimers, trimers, and small oligomeric Aβ42 aggregates other than monomers are increasingly believed to be more neurotoxic than Aβ42 mature fibrils [76]. Aβ deposition spreads from temporobasal and frontomedial areas to the remaining associative neocortex, primary sensory-motor areas, and the medial temporal lobe [77,78].

4.1.2. Nuts and Aβ

Nut bioactives could hamper Aβ plaque build-up by targeting BACE1, γ-secretase, and/or Aβ aggregation (Figure 1A). To the best of our knowledge, while there has been exploration of the association between adherence to the MedDiet and AD biomarkers [79], there has been no assessment of nut consumption and AD, and there are no available clinical nut supplementation studies on this specific topic. In a cellular model of early AD (human SH-SY5Y cells transfected with APP695), treatment with 10 μg/mL of a lipophilic walnut kernel extract for 24 h resulted in a significant reduction in Aβ40 levels when compared to control cells [80]. In line with this finding, in a study testing the in vitro inhibitory effects against BACE1 of 18 different hydroalcoholic ethnomedicinal plant extracts (1 g per 5 mL), leaves (nuts were not tested here) from *Juglans regia* (walnut tree) were found to display a BACE1 inhibitory activity in a concentration-dependent manner [81]. In another in vitro study, the methanolic extract of walnut kernels (4 g per 10 mL) was found to inhibit Aβ fibril formation and defibrillate preformed Aβ fibrils [82]. Finally, reduced Aβ burden was described in experimental studies testing isolated bioactives found in nuts, including oleic acid [83,84], linoleic acid [84], ALA [85], beta-sitosterol [86], nicotinamide [87], ellagic acid [88–90], epigallocatechin [91], myricetin [92,93], caffeic acid [94], and an array of other polyphenols [93].

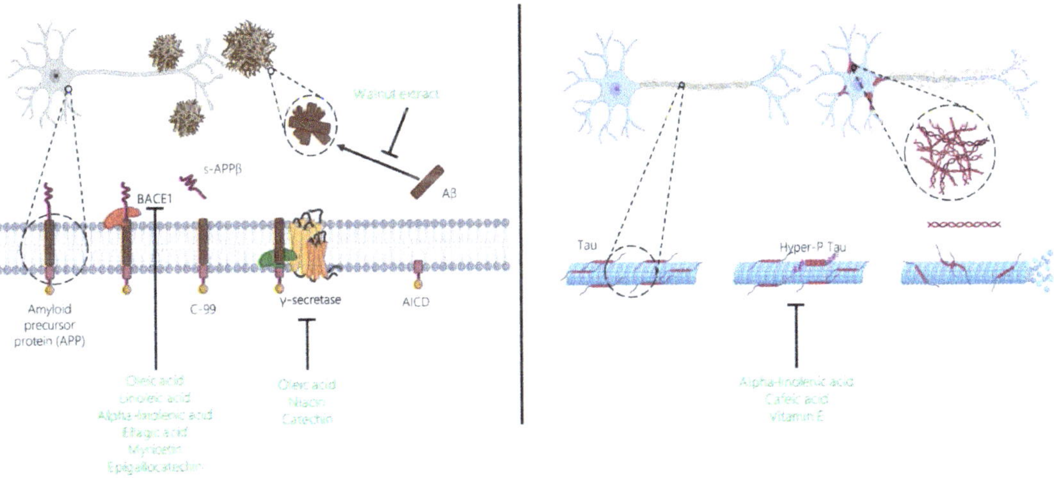

(A) (B)

Figure 1. Alzheimer´s Disease Hallmarks (**A**) amyloid-beta (Aβ) deposition and (**B**) neurofibrillary tangles of hyperphosphorylated tau. Effects of nut bioactives on critical pathways are shown.

4.1.3. Potential of Aβ Biomarkers in Future Research on the Neuroprotective Properties of Nuts

The accumulation of Aβ is considered the first detectable change of AD in the brain. It follows that an accurate determination of Aβ is relevant for the etiological diagnosis and for monitoring disease progression. Importantly, it might be a useful tool to identify changes in the rates of amyloid deposition without requiring long-term dietary interventions. Cerebrospinal fluid (CSF) analyses can indicate the presence of amyloid pathology in its earliest stages, although there is growing pressure to develop and validate less invasive blood-based biomarkers. A characteristic feature of AD is the reduction in CSF Aβ42 [95], which becomes evident about 15 years before clinical symptoms appear [96], although it plateaus relatively early in the AD continuum. The ratio Aβ42/Aβ40 in CSF has been repeatedly proven to improve prediction of clinical progression better than Aβ42 alone [97]. On the other hand, positron emission tomography (PET) imaging enables the non-invasive, in vivo assessment and quantification of continued build-up of amyloid burden beyond the CSF plateau, as well as providing information on the spatial distribution of the pathology in the brain [98]. Despite the potential interest of the Aβ42/Aβ40 ratio in CSF (or blood) and amyloid-PET imaging, none of them have been used as secondary outcomes in RCTs of nut supplementation with cognitive changes as primary outcomes.

4.2. Nuts and Neurofibrillary Tangles

4.2.1. Pathophysiologic Overview

The second histopathologic hallmark of AD is the presence of neurofibrillary tangles, which are entirely made up of hyperphosphorylated tau protein. Tau is a microtubule-associated protein that promotes the formation of axonal microtubules, stabilizing them [99]. Tau might undergo phosphorylation at over 70 potential sites [100]. In AD, there is an abnormal tau hyperphosphorylation, which decreases the capacity of the protein to bind microtubules, promoting the destabilization of axons. Hyperphosphorylation also contributes to the detachment of tau, which self-aggregates to form paired helical and straight filaments, leading to the formation of intracellular neurofibrillary tangles [101]. These tangles are initially found in the entorhinal region and subsequently progress to the limbic system and neocortical regions, greatly correlating with cognitive decline [102].

4.2.2. Nuts and Neurofibrillary Tangles

There is a current need to identify interventions capable of reducing tau aggregation by means of stabilizing microtubules, inhibiting tau phosphorylation, or inhibiting fibrilization. As in the case of Aβ, there is no clinical research on nut consumption and tau protein, while experimental research has tested only the bioactives that are present in nuts in isolation. Figure 1B presents the potential impact of specified nut bioactives on tau hyperphosphorylation. One of the most investigated is ALA, the vegetable omega-3 PUFA abundant in walnuts [16,31], which was found to inhibit tau aggregation in an in vitro study [103]. Furthermore, N9 (microglia) cells exposed to ALA increased their ability to better target [104], phagocyte, and degrade extracellular tau [105]. Another bioactive present in nuts tested in relation to tau is caffeic acid. In a study conducted in high-fat diet-induced hyperinsulinemic rats, the administration of caffeic acid (30 mg/kg body weight/day) for 30 weeks significantly reduced the expression of phosphorylated-tau protein in the hippocampus [94]. Similarly, pretreatment of P12 cells with caffeic acid prior to challenge with Aβ attenuated tau phosphorylation [106]. Finally, in a study including both in vitro (primary culture of cortical neurons) and in vitro (APP/PS1 double transgenic mice of AD), exposure to vitamin E reduced the formation of hyperphosphorylated tau through the inhibition of p38MAPK [107].

4.2.3. How Biomarkers Can Help in Future Research

Akin to the use of biomarkers to quantify amyloid burden, the tau landscape is rapidly evolving, with ultrasensitive immunoassays allowing the reliable measurement of tau biomarkers in blood, while second-generation tau-PET tracers are being developed [108]. Patients with AD show increased levels of total-tau and phosphorylated-tau in CSF and blood when compared to healthy controls. Phosphorylated-tau is a more specific AD biomarker than total-tau, which can be increased in neurodegenerative diseases other than AD. Phosphorylation at threonine 181 (so-called p-tau181) is the most widely used tau biomarker, although other tau species, including those phosphorylated in other mid-domain residues (threonine 217, threonine 231), are increasingly used [95]. However, to the best of our knowledge, no RCTs have assessed changes in these biomarkers using nut supplementation.

5. Summary and Future Directions

Epidemiological, clinical, and mechanistic evidence, while limited and inconclusive, suggests a possible role for nuts in the maintenance of cognitive health and prevention of cognitive decline in individuals across the lifespan, particularly in older adulthood. Given the potential beneficial impact of nuts on cognitive health, their consumption within a healthy dietary pattern may offer a simple public health strategy for the prevention of cognitive decline in most individuals.

Still, the limitations of the presently available evidence should be acknowledged and considered for future research. There is a dearth of research on nuts and cognition, especially in individuals under 60 years of age, and no study has examined whether nut containing diets influence hard clinical outcomes (e.g., dementia or AD) [109]. On the basis of the strength of the evidence on nut/walnut consumption and heart disease, the US Food and Drug Administration issued qualified health claims for nuts in 2003 [110] and for walnuts in 2004 [111], stating that "supportive but not conclusive research shows that eating 1.5 ounces per day of nuts/walnuts, as part of a low saturated fat and low cholesterol diet and not resulting in increased caloric intake, may reduce the risk of coronary heart disease". Research in the last two decades has indeed confirmed this beneficial effect for total nuts and walnuts [16], and more recently, a similar qualified health claim was issued for macadamia nuts [112]. Given the shared risk factors between common heart and brain diseases and the manifold salutary effects and safety of nut consumption, and pending additional evidence, this recommendation may also be applied for the prevention of cognitive decline and dementia.

Future large RCTs among healthy pregnant women, children or adolescents, and, particularly, older adults at risk of cognitive decline could provide a new and important public health dimension about simple nutritional recommendations for an entire population. There is a need to develop the area of research on nut consumption and neurodevelopment with a special focus on windows of vulnerability and opportunity, such as the pregnancy, childhood, and adolescence periods, as even if only a small positive cognitive effect is found, this may have significant implications at the population level from a public health perspective. Concerning the evidence currently available, a key methodological limitation is the low accuracy of the screening instruments commonly used to assess cognitive function, such as the MMSE, which compromises the ability to draw firm conclusions. Given that experimental research has tested only bioactives that are present in nuts in isolation, studying the build-up of Aβ plaques and neurofibrillary tangles may provide an avenue to address the limitations of neuropsychological tests. There is a current need to clinically identify if nut interventions may reduce Aβ plaque build-up and tau aggregation by stabilizing microtubules, inhibiting tau phosphorylation, or inhibiting fibrilization. Further research is warranted to elucidate the impact of nut consumption on brain health and better inform cognitive health-related practices and guidelines.

Author Contributions: Writing—original draft preparation, S.K.N., A.S.-V., J.J. and E.R.; writing—review and editing, S.K.N., A.S.-V., J.J., J.S. and E.R. All authors have read and agreed to the published version of the manuscript.

Funding: S.K.N. is supported by a Canadian Institutes of Health Research (CIHR) Postdoctoral Fellowship (MFE-171207). J.J. holds a Miguel Servet-II contract (grant CPII19/00015) awarded by the Instituto de Salud Carlos III (co-funded by the European Union, Social Fund "Investing in your future"). The NUTS 2022 conference was planned by the Rovira i Virgili University, with all the expenses of the organization covered by Barcelo Congresos. Barcelo Congresos SA has received funds from the International Nut & Dried Fruit Council (INC). Researchers did not receive any additional financial support, aside from travel and accommodation, related to their participation in the NUTS 2022 conference. Sponsors did not have any role in selecting or inviting the speakers, nor in writing or reviewing the proceedings of the conference.

Institutional Review Board Statement: Not applicable.

Informed Consent Statement: Not applicable.

Data Availability Statement: Not applicable.

Acknowledgments: We are grateful to the organizers and expert researchers participating in the NUTS 2022: Where are we and where are we going? conference held in Reus, Spain, on 20–21 October 2022.

Conflicts of Interest: S.K.N. is a volunteer member of the not-for profit group Plant-Based Canada and has received a research grant from the INC (International Nut and Dried Fruit Council) through her institution. A.S.-V. has received research funding through his institution and support to attend professional meetings from the California Walnut Commission (CWC, Folsom, CA, USA). J.J. has received research funding through his institutions, the Barcelona Institute for Global Health and Institut d'Investigació Sanitària Pere Virgili (IISPV), and support to attend professional meetings from the California Walnut Commission (CWC, Folsom, CA, USA). J.S. has received health research grant funding through his institution from several nut commodity boards. E.R. reports receiving grant support through his institution from the California Walnut Commission (Folsom, CA, USA), in addition to personal funds for project supervision and advice, and serving as a non-paid member of its Scientific Advisory Committee; funds for travel and accommodation from the International Nut and Dried Fruit Council; and personal funds from Alexion for serving on the advisory committee. The funders had no role in the design of the study; in the collection, analysis, or interpretation of data; in the writing of the manuscript, or in the decision to publish the results.

References

1. World Health Organization (WHO). Dementia. Available online: https://www.who.int/news-room/fact-sheets/detail/dementia#:~:text=Rates%20of%20dementia,and%20139%20million%20in%202050. (accessed on 11 January 2023).
2. Gabbianelli, R.; Damiani, E. Epigenetics and Neurodegeneration: Role of Early-Life Nutrition. *J. Nutr. Biochem.* **2018**, *57*, 1–13. [CrossRef] [PubMed]
3. Baumgart, M.; Snyder, H.M.; Carrillo, M.C.; Fazio, S.; Kim, H.; Johns, H. Summary of the Evidence on Modifiable Risk Factors for Cognitive Decline and Dementia: A Population-Based Perspective. *Alzheimer's Dement.* **2015**, *11*, 718–726. [CrossRef]
4. Solfrizzi, V.; Agosti, P.; Lozupone, M.; Custodero, C.; Schilardi, A.; Valiani, V.; Sardone, R.; Dibello, V.; di Lena, L.; Lamanna, A.; et al. Nutritional Intervention as a Preventive Approach for Cognitive-Related Outcomes in Cognitively Healthy Older Adults: A Systematic Review. *J. Alzheimer's Dis.* **2018**, *64*, S229–S254. [CrossRef] [PubMed]
5. Jennings, A.; Cunnane, S.C.; Minihane, A.M. Can Nutrition Support Healthy Cognitive Ageing and Reduce Dementia Risk? *BMJ* **2020**, *369*, m2269. [CrossRef] [PubMed]
6. Mecocci, P.; Boccardi, V.; Cecchetti, R.; Bastiani, P.; Scamosci, M.; Ruggiero, C.; Baroni, M. A Long Journey into Aging, Brain Aging, and Alzheimer's Disease Following the Oxidative Stress Tracks. *J. Alzheimer's Dis.* **2018**, *62*, 1319–1335. [CrossRef]
7. Rajaram, S.; Jones, J.; Lee, G.J. Plant-Based Dietary Patterns, Plant Foods, and Age-Related Cognitive Decline. *Adv. Nutr.* **2019**, *10*, 422–436. [CrossRef] [PubMed]
8. Gutierrez, L.; Folch, A.; Rojas, M.; Cantero, J.L.; Atienza, M.; Folch, J.; Camins, A.; Ruiz, A.; Papandreou, C.; Bulló, M. Effects of Nutrition on Cognitive Function in Adults with or without Cognitive Impairment: A Systematic Review of Randomized Controlled Clinical Trials. *Nutrients* **2021**, *13*, 3728. [CrossRef]
9. Charisis, S.; Ntanasi, E.; Yannakoulia, M.; Anastasiou, C.A.; Kosmidis, M.H.; Dardiotis, E.; Hadjigeorgiou, G.; Sakka, P.; Scarmeas, N. Mediterranean Diet and Risk for Dementia and Cognitive Decline in a Mediterranean Population. *J. Am. Geriatr. Soc.* **2021**, *69*, 1548–1559. [CrossRef]
10. Andreu-Reinón, M.E.; Chirlaque, M.D.; Gavrila, D.; Amiano, P.; Mar, J.; Tainta, M.; Ardanaz, E.; Larumbe, R.; Colorado-Yohar, S.M.; Navarro-Mateu, F.; et al. Mediterranean Diet and Risk of Dementia and Alzheimer's Disease in the Epic-Spain Dementia Cohort Study. *Nutrients* **2021**, *13*, 700. [CrossRef]
11. Radd-Vagenas, S.; Duffy, S.L.; Naismith, S.L.; Brew, B.J.; Flood, V.M.; Fiatarone Singh, M.A. Effect of the Mediterranean Diet on Cognition and Brain Morphology and Function: A Systematic Review of Randomized Controlled Trials. *Am. J. Clin. Nutr.* **2018**, *107*, 389–404. [CrossRef]
12. Drouka, A.; Mamalaki, E.; Karavasilis, E.; Scarmeas, N.; Yannakoulia, M. Dietary and Nutrient Patterns and Brain MRI Biomarkers in Dementia-Free Adults. *Nutrients* **2022**, *14*, 2345. [CrossRef] [PubMed]
13. Van den Brink, A.C.; Brouwer-Brolsma, E.M.; Berendsen, A.A.M.; van de Rest, O. The Mediterranean, Dietary Approaches to Stop Hypertension (DASH), and Mediterranean-DASH Intervention for Neurodegenerative Delay (MIND) Diets Are Associated with Less Cognitive Decline and a Lower Risk of Alzheimer's Disease-A Review. *Adv. Nutr.* **2019**, *10*, 1040–1065. [CrossRef] [PubMed]
14. Chen, X.; Maguire, B.; Brodaty, H.; O'Leary, F. Dietary Patterns and Cognitive Health in Older Adults: A Systematic Review. *J. Alzheimer's Dis.* **2019**, *67*, 583–619. [CrossRef]
15. WHO. *Risk Reduction of Cognitive Decline and Dementia: WHO Guidelines*; World Health Organization: Geneva, Switzerland, 2019; Licence: CC BY-NC-SA 3.0 IGO.
16. Ros, E.; Singh, A.; O'Keefe, J.H. Nuts: Natural Pleiotropic Nutraceuticals. *Nutrients* **2021**, *13*, 3269. [CrossRef] [PubMed]
17. Tan, S.-Y.; Tey, S.L.; Brown, R. Nuts and Older Adults' Health: A Narrative Review. *Int. J. Environ. Res. Public Health* **2021**, *18*, 1848. [CrossRef]
18. Alderman, H.; Behrman, J.R.; Glewwe, P.; Fernald, L.; Walker, S. Evidence of Impact of Interventions on Growth and Development during Early and Middle Childhood. In *Disease Control Priorities, Third Edition (Volume 8): Child and Adolescent Health and Development*; Bundy, D.A.P., de Silva, N., Horton, S., Jamison, D.T., Patton, G.C., Eds.; The World Bank: Washington, DC, USA, 2017; Volume 8, pp. 79–98.
19. Sizonenko, S.v.; Babiloni, C.; de Bruin, E.A.; Isaacs, E.B.; Jönsson, L.S.; Kennedy, D.O.; Latulippe, M.E.; Hasan Mohajeri, M.; Moreines, J.; Pietrini, P.; et al. Brain Imaging and Human Nutrition: Which Measures to Use in Intervention Studies? *Br. J. Nutr.* **2013**, *110*, S1–S30. [CrossRef]
20. Júlvez, J.; Paus, T.; Bellinger, D.; Eskenazi, B.; Tiemeier, H.; Pearce, N.; Ritz, B.; White, T.; Ramchandani, P.; Gispert, J.D.; et al. Environment and Brain Development: Challenges in the Global Context. *Neuroepidemiology* **2016**, *46*, 79–82. [CrossRef]
21. Moody, L.; Chen, H.; Pan, Y.X. Early-Life Nutritional Programming of Cognition-the Fundamental Role of Epigenetic Mechanisms in Mediating the Relation between Early-Life Environment and Learning and Memory Process. *Adv. Nutr.* **2017**, *8*, 337–350. [CrossRef]
22. Konkel, L. The Brain before Birth: Using FMRI to Explore the Secrets of Fetal Neurodevelopment. *Environ. Health Perspect.* **2018**, *126*, 112001. [CrossRef]
23. Karr, J.E.; Alexander, J.E.; Winningham, R.G. Omega-3 Polyunsaturated Fatty Acids and Cognition throughout the Lifespan: A Review. *Nutr. Neurosci.* **2011**, *14*, 216–225. [CrossRef]
24. Gignac, F.; Romaguera, D.; Fernández-Barrés, S.; Phillipat, C.; Garcia Esteban, R.; López-Vicente, M.; Vioque, J.; Fernández-Somoano, A.; Tardón, A.; Iñiguez, C.; et al. Maternal Nut Intake in Pregnancy and Child Neuropsychological Development up to 8 Years Old: A Population-Based Cohort Study in Spain. *Eur. J. Epidemiol.* **2019**, *34*, 661–673. [CrossRef]

applied sciences

MDPI

Editorial

Special Issue on "Human Health Engineering Volume II"

Jean-Marie Aerts

KU Leuven, Department of Biosystems, Division Animal and Human Health Engineering, Measure, Model & Manage Bioresponses Group, Kasteelpark Arenberg 30, 3000 Leuven, Belgium; jean-marie.aerts@kuleuven.be

Citation: Aerts, J.-M. Special Issue on "Human Health Engineering Volume II". *Appl. Sci.* **2021**, *11*, 7844. https://doi.org/10.3390/app11177844

Received: 3 August 2021
Accepted: 23 August 2021
Published: 26 August 2021

1. Referees for the Special Issue

A total of 23 manuscripts were received for our Special Issue (SI), of which 3 manuscripts were directly rejected without peer review. The remaining 20 articles were all strictly reviewed by no less than two reviewers in related fields. Finally, 12 of the manuscripts were recommended for acceptance and published in *Applied Sciences-Basel*. Referees from 16 different countries provided valuable suggestions for the manuscripts in our SI, the top five being the USA, Italy, UK, DE, and Spain. The names of these distinguished reviewers are listed in Table A1. We would like to thank all of these reviewers for their time and effort in reviewing the papers in our SI.

2. Main Content of the Special Issue

As explained by Aerts [1], we are standing on the shoulders of giants when looking to the development of the research field of Human Health Engineering, i.e., technology for monitoring the physical and/or mental health status of individuals in a variety of applications, as it evolved from seminal work of, among others, Walter Cannon [2], Norbert Wiener [3], and Ludwig von Bertalanffy [4]. For more detailed information on the background and history of the field of Human Health Engineering, the reader is referred to the editorial of Aerts [1] written for the first volume of the special issue on Human Health Engineering.

The 12 presented papers in this second volume are grouped according to their contributions to the main parts of the monitoring and control engineering scheme applied to human health applications, namely papers focusing on measuring/sensing of physiological variables [5–8], papers describing health monitoring applications [9–12], and finally, examples of control applications for human health [13–16]. As indicated by Aerts [1], it is envisioned that the field of human health engineering is complementary to the field of biomedical engineering as it not only contributes to developing technology for curing patients or supporting chronically ill people, but also covers applications on healthy humans (e.g., sports, sleep, and stress) and thus also focuses on disease prevention and optimizing human well-being more generally.

The first series of articles in this SI describes methods for (improved) measuring and/or sensing of health-related physiological signals. The work of Arciniega-Montiel et al. [5] contributes to developing methods for improving the measuring reliability of fetal monitors, used in the case of high-risk pregnancies, on the basis of probabilistic models. Al-Halhouli et al. [6] demonstrated the feasibility of using a wearable and stretchable inkjet-printed strain gauge sensor for estimating respiratory rate, which is a key vital sign variable in many medical and health applications. In their study, Alsayed et al. [7] developed an automated data acquisition system for measuring pinch and pulling forces in patients recovering from stroke. Nishio et al. [8] investigated a method for quantifying the functional decline in proprioceptors in patients with low back pain using vibrations with sweep frequencies covering the entire range of response frequencies of proprioceptors.

The second series of articles describes applications of human health monitoring. In their paper, Cheng et al. [9] described a method combining an electroencephalography image-based method with a convolutional neural network to predict drowsiness in drivers, which is a major cause of vehicle accidents. Yang et al. [10] developed a biometrics monitor allowing to verify individuals on the basis of an interval-based linear discriminant analysis algorithm applied to electrocardiogram signals. Visnovcova et al. [11] used linear and non-linear analyses of electrodermal activity signals for quantifying sympathetic dysregulation as a non-invasive potential biomarker for monitoring anorexia nervosa-linked cardiovascular risks. In the paper of Jin et al. [12], a system is described for monitoring gait in (elderly) people using toe-area activity as a predictor for stumbling risk.

Finally, the third series of articles in this SI discusses examples of control applications for human health management. Kaçar et al. [13] developed a prototype of a closed-loop controlled valveless piezoelectric insulin pump for keeping the blood glucose level of Type 1 diabetes mellitus patients within desired ranges. In their paper, Trappey et al. [14] developed a system for virtual reality-based exposure therapy for treating people suffering from driving phobia disorder using bioresponse measurements to optimise fear-based driving scenarios. Guttiérez et al. [15] reported about a serious game platform with haptic feedback and EMG monitoring developed for upper limb rehabilitation of spinal cord injury patients. Aprile et al. [16] demonstrated the feasibility of optimising the efficiency of a robot-assisted rehabilitation area for restoring upper limb function, allowing one physiotherapist to supervise four patients.

3. Conclusions

This SI presents recent contributions in the growing field of human health engineering. The contributions highlight research focusing on different aspects of the monitoring and control engineering scheme (sensors, sensing systems, monitoring approaches, and control algorithms) as applied to human health. Applications cover both healthy and (chronically) ill people and this is in relation to physical as well as mental processes.

The combination of ever growing possibilities in sensors and (wearable) sensing technology and powerful artificial intelligence tools is expected to further boost the field of human health engineering, which will become increasingly ubiquitous in our society and will increasingly assist people of all ages in living healthy, high quality, and productive lives.

Funding: This special issue frames within research funded by the European Union's Horizon 2020 Research and Innovation programme under the Marie Skłodowska-Curie grant number 645770.

Acknowledgments: We would like to sincerely thank our Section Managing Editor, Marin Ma (marin.ma@mdpi.com), for all the efforts she has made for this special issue in the past year.

Conflicts of Interest: The author declares no conflict of interest.

Appendix A

Table A1. Special Issue (SI) reviewer list.

Alessandro Nastro	Yun-Ju Lee	Wasim Ahmad
Joo Chuan Yeo	Mikail F. Lumentut	Oliver Pabst
Tareq Assaf	Mariya Aleksandrova	Juri Taborri
Shuang Song	Thanh Duc Nguyen	Chien-Chang Chen
Zhengguo Tan	Muhammad Nadeem	Fabio Stroppa
Jesús López Belmonte	Ghulam Hussain	Meytal Wilf
Simon Danner	Marco D'Alonzo	Astrid Gramstad
Lin Yao	Paul Farago	Andrea Viggiano
Isaac Macwan	Vahab Youssof Zadeh	Hyungju Kwon
Alfonso Javier Ibáñez-Vera	Yuvaraj Rajamanickam	Adam Gąska
		Zixuan Cang

References

1. Aerts, J.-M. Editorial—Special Issue on Human Health Engineering. *Appl. Sci.* **2020**, *10*, 564. [CrossRef]
2. Cannon, W.B. *The Wisdom of the Body*; Norton & Company Inc.: New York, NY, USA, 1932.
3. Wiener, N. *Cybernetics or Control and Communication in the Animal and the Machine*; MIT Press: Cambridge, MA, USA, 1948.
4. Von Bertalanffy, L. *General System Theory—Foundations, Development, Applications*; George Braziller Inc.: New York, NY, USA, 1968.
5. Arciniega-Montiel, S.; Ronquillo-Lomeli, G.; Salas-Zúñiga, R.; Salgado-Jiménez, T.; Barriga-Rodríguez, L. Compensation of measurement uncertainty in a remote fetal monitor. *Appl. Sci.* **2020**, *10*, 3274. [CrossRef]
6. Al-Halhouli, A.; Al-Ghussain, L.; El Bouri, S.; Habash, F.; Liu, H.; Zheng, D. Clinical evaluation of stretchable and wearable inkjet-printed strain gauge sensor for respiratory rate monitoring at different body positions. *Appl. Sci.* **2020**, *10*, 480. [CrossRef]
7. Alsayed, A.; Kamil, R.; Ramli, H.; As'arry, A. An automated data acquisition system for pinch grip assessment based on Fugl Meyer protocol: A feasibility study. *Appl. Sci.* **2020**, *10*, 3436. [CrossRef]
8. Nishio, R.; Ito, Y.; Morita, Y.; Ito, T.; Yamazaki, K.; Sakai, Y. Investigation of the functional decline in proprioceptors for low back pain using the sweep frequency method. *Appl. Sci.* **2020**, *9*, 4988. [CrossRef]
9. Cheng, E.J.; Young, K.-Y.; Lin, C.-T. Temporal EEG monitoring for drowsy driving prediction. *Appl. Sci.* **2020**, *9*, 5078. [CrossRef]
10. Yang, C.; Ku, G.W.; Lee, J.-G.; Lee, S.-H. Interval-based LDA algorithm for electrocardiograms for individual verification. *Appl. Sci.* **2020**, *10*, 6025. [CrossRef]
11. Visnovcova, Z.; Olexova, L.B.; Sekaninova, N.; Ondrejka, I.; Hrtanek, I.; Cesnekova, D.; Kelcikova, S.; Farsky, I.; Tonhajzerova, I. Spectral and nonlionear analysis of electrodermal activity in adolescent anorexia nervosa. *Appl. Sci.* **2020**, *10*, 4514. [CrossRef]
12. Jin, Y.; Shogenji, M.; Watanabe, T. Feature assessment of toe area activity during walking of elderly people with stumbling experiences through wearable clog-integrated plantar visualization system. *Appl. Sci.* **2020**, *10*, 1359. [CrossRef]
13. Kaçar, A.; Özer, M.B.; Taşcioğlu, Y. A novel artificial pancreas: Energy efficient valveless piezoelectric actuated closed-loop insulin pump for T1DM. *Appl. Sci.* **2020**, *10*, 5294. [CrossRef]
14. Trappey, A.; Trappey, C.V.; Chang, C.-M.; Kuo, R.R.T.; Lin, A.P.C.; Nieh, C.H. Virtual reality exposure therapy for driving phobia exposure: System design and development. *Appl. Sci.* **2020**, *10*, 4860. [CrossRef]
15. Guttiérez, Á.; Sepúlveda-Muñoz, D.; Gil-Agudo, Á.; de los Reyes Guzmán, A. Serious game platform with haptic feedback and EMG monitoring for upper limb rehabilitation and smoothness quantification on spinal cord injury patients. *Appl. Sci.* **2020**, *10*, 963. [CrossRef]
16. Aprile, I.; Pecchioli, C.; Loreti, S.; Cruciani, A.; Padua, L.; Germanotta, M. Improving the efficiency of robot-mediated rehabilitation by using a new organizational model: An observational feasibility study in an Italian rehabilitation center. *Appl. Sci.* **2020**, *9*, 5357. [CrossRef]

applied sciences

MDPI

Article

Compensation of Measurement Uncertainty in a Remote Fetal Monitor

Sadot Arciniega-Montiel [1], Guillermo Ronquillo-Lomeli [1,2,*], Roberto Salas-Zúñiga [3], Tomás Salgado-Jiménez [1] and Leonardo Barriga-Rodríguez [1]

1 Department of Energy, Center for Engineering and Industrial Development, Santiago de Queretaro 76125, Mexico; sadot@cidesi.edu.mx (S.A.-M.); tsalgado@cidesi.edu.mx (T.S.-J.); lbarriga@cidesi.edu.mx (L.B.-R.)
2 Faculty of Engineering, Autonomous University of Queretaro, Santiago de Queretaro 76010, Mexico
3 Engineering and Technology Center, Santiago de Queretaro 76079, Mexico; roberto.salas.correo@gmail.com
* Correspondence: gronquillo@cidesi.edu.mx; Tel.: +52-442-211-9800

Received: 10 April 2020; Accepted: 2 May 2020; Published: 8 May 2020

Abstract: The perinatal mortality rate is very high throughout the world. A fetal monitor may be used remotely, and this would tackle the problem of continuous monitoring of high-risk pregnancies. There is evidence that current technology is of low reliability, and, therefore, of low precision to identify fetal health. In medical technological implementation, a safe, efficient, and reliable operation must be guaranteed, and the main problem is that remote fetal monitor gathers just a few samples, so the hypothesis of classical theory is not met. We are proposing an approach that improves the data's lack of reliability that accompanies the use of a remote fetal monitor. The method refers to how, by using the existing technologies and the initial experimental data, it is possible to apply probabilistic models that are truly representative of each application. This leads to the characterization of properties of the statistics used to generate a representative probabilistic model without the need to consider the hard suppositions. Results show that, for every case study, it was possible to improve estimations of measurement uncertainty. The proposed method is a useful tool to increase the reliability of medical technology, especially for pieces of equipment where a health care professional is not available.

Keywords: remote fetal monitor; measurement uncertainty; standard deviation; Monte-Carlo method (MMC); efficient estimator

1. Introduction

The perinatal mortality rate is very high throughout the world. According to the World Health Organization (WHO), it is estimated that 800 women die every day due to pregnancy or birth-related causes. From all pregnancies, four out of ten women are at high risk. There is no hospital infrastructure for the timely diagnosis of fetal distress due to overpopulation.

Continuous remote monitoring technology would provide a solution to solve some problems in high-risk pregnancies. Patients would no longer need to be at the hospital, and, from the comfort of their homes, would directly carry out and forward the studies prescribed by their physicians. The monitoring equipment could be configured for providing the corresponding studies directly by the patient without the need of a specialist. Patients may feel calm and reassured, knowing that their studies, once transmitted by the monitor, will be checked and assessed by a specialist.

Fetal monitoring technology faces reliability issues. When compared against intermittent auscultation, the use of technology reduces perinatal mortality [1–3]. However, some other studies from the 1970s and 1980s did not show similar results [4], and the contribution of technology to reduce perinatal mortality, as observed in the last decades, is still under discussion. On the other hand, the increase in rates of births via Caesarian section has become a matter of public interest in many countries, and there is a general belief that continuous monitoring during childbirth (CTG) has contributed to this.

CTG is subject to a poor observer interpretation, particularly while assessing variability and de-acceleration, in addition to the classification of tracing [5–7]. Diagnosis depends on the criteria used for tracing analysis [8], so objective guidelines are required to provide a practical approach.

It is necessary to have an approach that improves the data reliability produced by the remote fetal monitor to increase the diagnosis precision [9,10]. Fetal monitoring systems should yield reliable data, regardless of the conditions of use, such as location, installation, number of samples, etc., for the revision of fetal health and the prompt response of healthcare professionals.

In medical technology implementation, a safe and efficient operation must be guaranteed [11]. Attention is focused on the practical implementation of technology within a sanitary environment. The simple fact of having the necessary technology is not a whole solution. The processes of planning, evaluation, selection, and implementation of new technologies, all in concept and context, must be considered. This may imply a direct solution to specific assessments of sanitary technology. When developing new technology, guaranteeing precision and reliability for the chosen environment contributes to achieving safer results and sound medical attention.

Under these conditions, the main problem is to reassure the reliability of the data obtained by the remote fetal monitor, since health-care professionals would not be present, and there would be only a few samples of variables (heart rate (HR), uterine activity) that lead to the detection of anomalies. Therefore, the analysis of measurement uncertainty becomes necessary.

For reliable estimation of measurement uncertainty, it is necessary to guarantee some properties (distribution independence, a large sample size, and a normal distribution). Health-care professionals are not familiar with the measurement process carried out by remote fetal monitoring, and, in general, only have a few samples. This means it is not possible to verify the hypotheses required so that the evaluator may have the properties required by a good estimator. By means of a statistical approach, several works related to uncertainty estimation have been conducted, for cases where the hypothesis of classical theory is not verified.

A fetal monitor is a stochastic process that has few samples available. Therefore, it is very difficult to model by using conventional methods. Different approaches have been used when the hypothesis of the classical theory is not verified, including the Monte-Carlo method (MCM) [12], parametric and non-parametric theory formulation [13], adaptive estimators [14], spectrum-based estimation [15], stochastic gradient [16], and mean square error [17] estimators with Cramer–Rao type restriction and reliability estimation under measures degradation [18].

The MCM is a widely used tool for modeling stochastic systems since with a large enough sample, the probability that an estimator will deviate from the expected value can be as small as required. The MCM is used to get information related to estimator properties. The parameters of softening and form of estimating function are adjusted to modify the performance of maximum probability estimators. The MCM can be used for modeling unknown distribution functions. The parametric and non-parametric models are wide probability function estimation methods; theory formulation has been developed considering local asymptotic length and restrictions in sample size. Adaptive estimators that guarantee the consistency in finite samples for any distribution function are helpful in the fetal monitor application. The term adaptive integrates the concept that such estimators adapt to the sample using data to estimate a non-parametric density function. Adaptive estimators have been applied using the least-squares method with finite samples, reaching substantial increases in efficiency, but with hypotheses that are difficult to verify for applications with a few samples. The spectrum-based method is useful when the relationship between observations and parameter model is noisy, potentially non-linear, and not invertible. The stochastic gradient estimation is a numeric method that is stable and robust for fitting parameters and predicting standard errors well. Even though only the typical case, with some modifications, was considered, the hypothesis that let the Cramer–Rao type limit such estimators have been changed. The mean square error can be improved by implementing estimators with parametric restrictions that do not satisfy certain conditions. With a new definition of bias in restricted surroundings, a Cramer-Rao restriction type was determined for the pondered mean square error estimators, for normal distribution only. Correlated perturbations were not considered. The reliability estimation for a system whose degradation measures are monotonous, can obtain the fault time distribution, and, therefore, system reliability.

The remote fetal monitor has a context application. The aim of this work is to obtain an integral maternal-fetal watch program, integrated by a technological platform that allows the timely detection of perturbations in baby heart rate, that enable us to remotely send information to an obstetric monitoring center for timely and reliable diagnosis of the patient.

When modeling a process, the most complicated component of the model is represented in probabilistic terms, and the hypotheses taken as valid in many situations are not met, and the results are not valid. This work presents a proposal of how, using the existing theory, with a deterministic model and the Monte-Carlo method [12], it is possible to apply probabilistic models that are genuinely representative of the application, and that characterize the statistics' properties used. In a specific application, the statistical properties of interest are generally mean, variance, extreme values, dependence on sample size, bias, etc. [13–17]. In general, only the mean and variance parameters are considered to obtain a probabilistic model because they characterize the uniform distributions that represent the worst-case scenario (total ignorance of the model) and the Gaussian distributions considered as typical or normal. However, to estimate these parameters, it is necessary to carry out a sampling to obtain more information about the system model. This additional information is not used in practice. This work proposes a new approach to include experimental data and to generate a representative probabilistic model without considering difficult suppositions. The suggested procedure is applied to a fetal monitor to correct the standard deviation estimation considering the noise type that occurs during the regular operation of the device.

2. Uncertainty Estimation Model

One of the main problems in metrology is the estimation of measurement uncertainty. According to [19], "estimated standard deviation associated to output, or measurement estimation is called combined standard deviation, and it is determined from the estimated standard deviation associated for each estimation of output, called standard uncertainty". In general, the process to be measured is ignored, and regularly, there are only just a few samples available. Measurements have defects that lead to an error in the result of such measurements. Although it is not possible to compensate for the random error in a measurement result, its effect can generally be reduced by increasing the number of observations until the expectations or expected value becomes zero.

Modeling of stochastic systems has been thoroughly studied, considering the hypothesis of the great numbers law. However, a measurement system is a stochastic system, with a limited or finite number of samples. The remote fetal monitor is a stochastic measurement system with a limited number of samples. Below an uncertainty estimation model for a measurement system with limited samples is presented.

A probability space $(\Omega, \mathcal{F}_0, P)$ for a random variable $X : \Omega \to \mathbb{R}^n$ where the conditional expectation $E\{(\cdot), \mathcal{F}_0\}$ concerning to sigma-algebra $\mathcal{F}_0 = \left\{\phi,\ \Omega\right\}$ [20] may be expressed as

$$E\{X|\mathcal{F}_0\} = E\{X\} \tag{1}$$

This means that the best estimation for the random variable, when there is no a priori knowledge of the process, is the mathematical expectation, and the associated standard uncertainty (u_c) may be obtained from the probability density function (PDF) of the random variable [21]. Therefore, the mean is proposed to estimate the following sample, and standard uncertainty is obtained from the standard deviation estimator.

Let A be the event where a device does not fail, the reliability is by definition $P(A)$ if the indicator function is considered χ_A,

$$\chi_A(\omega) = \begin{cases} 1, & \omega \in A \\ 0, & \omega \notin A \end{cases} \tag{2}$$

This means that the problem of calculating reliability is changed for the problem of estimating the expected value, for which the use of the Monte-Carlo method in this work is necessary.

In the estimation of the mathematical expectation, the equation regularly used is

$$\overline{X}_n = \frac{1}{n}\sum_{k=1}^{n} X_k, \tag{3}$$

being n the sample number.

In the normal case with independent samples, the mean estimator is optimal in the sense of the mean square error. In addition, the large number theorem is consistent for large samples, and its distribution is known by the central limit theorem [20], and is efficient [22].

In the general case, the estimator $\overline{X}_n$ is a random variable that depends on:

1. Sample size, n
2. Distribution, F_X
3. The degree of probabilistic dependence between samples, in addition to the fact that only linear or correlation dependence is considered $\rho_{X_iX_j}$.

To measure the quality of this, estimate the standard deviation is used (called uncertainty in the metrology area). This standard deviation is also estimated; this is done with S_n defined as

$$S_n := \sqrt{S_n^2}, \tag{4}$$

$$S_n^2 := \frac{1}{n-1}\sum_{k=1}^{n}(x_k - \overline{x}_n)^2, \tag{5}$$

for independent samples with the same distribution we get,

$$E\left\{S_{\overline{X}_n}^2\right\} = \frac{E\left\{S_X^2\right\}}{n} = \frac{\sigma_X^2}{n}. \tag{6}$$

So, the problem of determining the quality of the estimate focuses on determining σ_X. However, we must identify the restrictions in the estimation.

A widespread practice is to consider that if

$$\left\{S_n^2\right\} = \sigma^2, \tag{7}$$

then,

$$E\{S_n\} = \sigma. \tag{8}$$

Nonetheless, defining by z a random variable,

$$z(\omega) = 1,\ \forall\ \omega \in \Omega, \tag{9}$$

and applying Cauchy–Bunyakowskii–Schwartz inequality [20],

$$E\{S_n\} = E\{S_n z\} \le \sqrt{E\left\{S_n^2\right\}E\{z^2\}} = \sqrt{E\left\{S_n^2\right\}}, \tag{10}$$

that is

$$E\{S_n\} \le \sigma, \tag{11}$$

this inequality depends on all these three: sample size n, distribution function F_X, and correlation dependence $\rho_{X_iX_j}$.

For a sample, $X = (X_1, \cdots, X_N)$ corresponds a density function $f_X(X, \theta)$ for which the score $V(\theta)$ is defined

$$V(\theta) \equiv V_X(\theta) log f_X(X; \theta), \quad (12)$$

and the Fisher information matrix

$$I(\theta)E\left\{\left[\frac{\partial V(\theta)}{\partial \theta}\right]^2\right\}. \quad (13)$$

The Cramer–Rao theorem is fulfilled [20] where if $\hat{\theta}_n = \hat{\theta}(X_1, \ldots, X_n)$ is an unbiased estimator of θ^∞ then

$$E_\theta\left\{\| \hat{\theta}_n - \theta \|^2\right\} \geq \frac{1}{nI(\theta)} \quad (14)$$

and also depend on sample size n, distribution F_X, and correlation dependence $\rho_{X_i X_j}$.

This means that in the case of the fetal monitor where there are few samples, it is possible to determine the boundaries for the reliability estimation and its uncertainty using the Monte-Carlo method.

3. Uncertainty Estimation Approach

This section describes a proposed approach to obtain a reliability estimator for heart rate measurement.

Reliability R, the probability that the measurement is correct, is a function of the measurement device, the perturbations represented by its distribution function and its correlation, in addition to the number of data used in the estimation.

To estimate the reliability and the uncertainty in the measurement, the following procedure is proposed:

1. System model: device model and perturbation model.
2. Distribution function estimation of the perturbations and correlation function.
3. Estimators properties analysis using the Monte-Carlo method.
4. Correction factors calculation to compensate estimators.
5. Confidence intervals for the R estimation.

3.1. System Model

In the block diagram of Figure 1, a system of single input u_k and single output y_k (SISO) is shown in simplified form. Noise ξ_k presence is assumed a priori, and, in this case, white noise properties are attributed, which is presented as additive noise r_k in the system output.

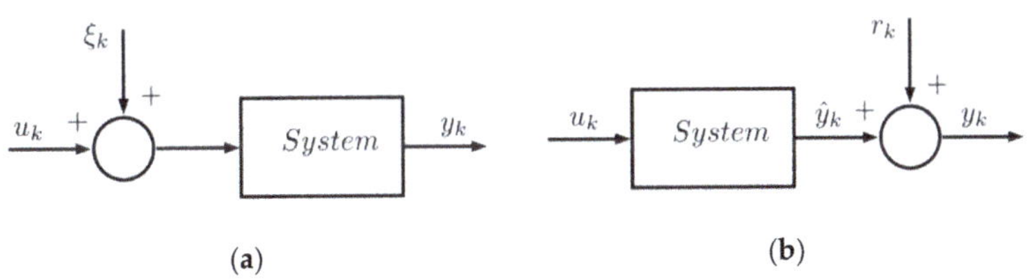

Figure 1. Single-input and single-output (SISO) system: (**a**) With white noise to the input; (**b**) With additive noise to the output.

It is considered that an input–output data set of the system is available $(y_k,\ u_k; k = 1,\ 2, \cdots, n)$ where n is the measurement number taken from the input and output for modeling.

3.1.1. Device Model

To describe the dynamic behavior from a set of input and output measurements, an auto-regressive and exogenous (ARX) [23] (p. 81) linear model structure is assumed. A scheme for system modeling is shown in Figure 2.

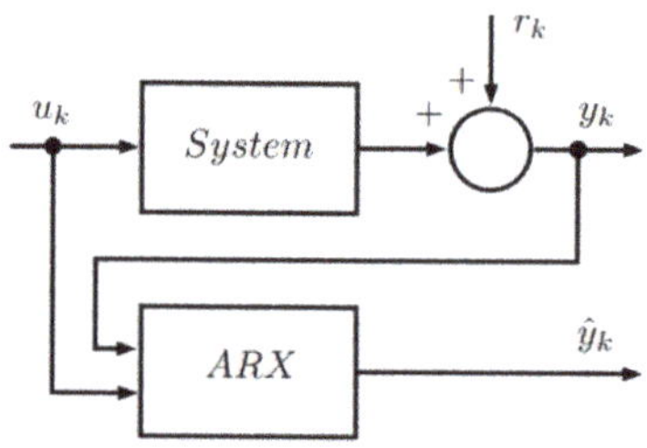

Figure 2. Scheme for linear system modeling based on auto-regressive and exogenous (ARX) structure.

The ARX structure model is

$$y_k = \phi_k^T \theta + \xi_k, \tag{15}$$

$\phi_k = [y_{k-1}, \cdots, y_{k-l}, u_k, \cdots, u_{k-l}]^T$ is the regression vector, $\theta = [a_1, \cdots, a_l, b_0, \cdots, b_l]^T$ is the parameter vector, while l is the system degree. Based on Equation (15), the relationship between the input and the output in a compact way can be written in the following vector form.

$$Y = \Phi\theta + \Xi, \tag{16}$$

where $Y = [y_1, y_2, \cdots, y_n]^T$, $\Phi = \left[\Phi_1^T, \Phi_2^T, \cdots, \Phi_n^T\right]^T$, and $\Xi = [\xi_1, \xi_2, \cdots, \xi_n]^T$.

With this information and applying the least-squares method [24] (pp. 17–27), a deterministic model is obtained which represents the fetal monitor dynamic, the solution is

$$\hat{\theta} = \left[\Phi^T \Phi\right]^{-1} \Phi^T Y. \tag{17}$$

Finally, the relationship between white noise ξ_k in the system input and additive noise r_k in the system output for the ARX model structure is

$$\mathrm{R}(z) = \frac{1}{1 - A(z)} \Xi(z), \tag{18}$$

where $A(z) = \hat{a}_1 z^{-1} + \cdots + \hat{a}_l z^{-l}$.

3.1.2. Perturbation Model

The probabilistic part is modeled by a stochastic process $\{\xi_k\}$, which must be strongly stationary, uncorrelated, with a cumulative distribution function (CDF) F_{Ξ_k}. These hypotheses are experimentally validated. The probabilistic model is used to determine the properties of the estimators used to determine the heart rate; considering that the noise ξ_k is naturally present and that it is transformed into r_k by the system during the route that the signal travels from the fetus heart through amniotic fluid, body fat, etc., to the fetal monitor output. Figure 3 shows a block diagram of the process for estimating the error distribution function F_{Ξ_k}.

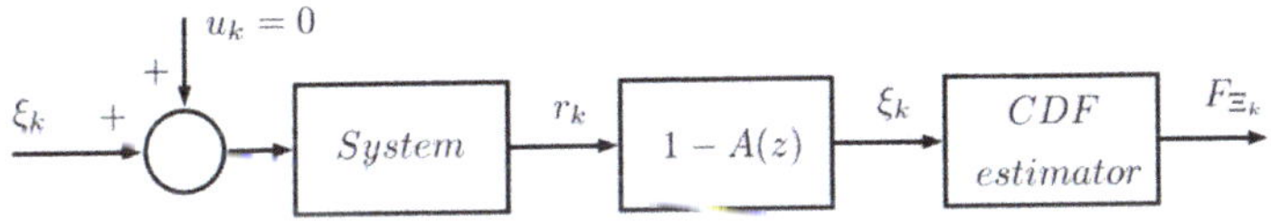

Figure 3. Estimation process of the error distribution function F_{Ξ_k}.

3.2. Distribution Function Estimation of the Perturbations and Correlation Function

For each patient, a stochastic model is determined, which describes the perturbation present in the fetal heart rate signal, which exists due to the individual anatomical characteristics of the patients, amniotic fluid, body fat, etc.

The stochastic model is obtained from data acquired by clinical studies using the patient's fetal heart rate monitor. Each clinical study (sample) is a set of measurements $(y_k,\ k = 1,\ 2, \cdots, n)$.

Using the data-normalized histogram, the probability density function (PDF) is estimated for n samples f_{Ξ_n}, numerically integrating and using some interpolator an estimator is obtained for the probability distribution function F_{Ξ_n} of the perturbations, which is used to build a ξ simulator required in the Monte-Carlo method. Each sample that is added improves the probability distribution function estimation, so that, the more samples available, the better the estimator.

To guarantee that the bias of the expectation estimate is zero, the correlation between the perturbation and the regression vector used in the least-squares model must be verified as zero. For this, a correlation degree estimate is made [20,22] that exists between the perturbations and the measured signal. The correlation is defined as follows

$$\rho_{XY}(k) := \frac{E\left\{X_{j+k}Y_j\right\}}{\sqrt{E\left\{X_{j+k}^2\right\}E\left\{Y_j^2\right\}}}, \tag{19}$$

where ρ_{XY} represents the correlation function between the random variables X and Y. This magnitude is estimated using

$$\hat{\rho}_{XY}(k) = \frac{\frac{1}{n}\sum_{j=1}^{n} X_{j+k}Y_j}{\sqrt{\frac{1}{n}\sum_{j=1}^{n} X_{j+k}^2 \frac{1}{n}\sum_{j=1}^{n} Y_j^2}}. \tag{20}$$

If there is a substantial correlation, a filter must be designed so that the perturbation behaves with white noise features.

3.3. Estimators Properties Analysis Using the Monte-Carlo Method

After obtaining the models of the system and the perturbations, it is necessary to estimate the value of the fetal heart rate. Simulated samples from the inverse distribution function $F_{\Xi_n}^{-1}$F_ (Ξ_n) were used for this estimation. To carry out this task, the sample mean $\overline{Y}_n$ and the sample deviation S_n were used. The mean is useful for determining the fetal heart rate value and the standard deviation for quantifying the quality of the estimate, known as uncertainty in the metrological area.

However, it should be noted that for these estimators to be used properly, it is necessary to determine their properties specifically for each patient. The most important property to determine is how the length of sample K impacts the distribution function.

On the other hand, the sample mean $\overline{Y}_n$ and the standard deviation S_n are in turn random variables that are characterized by their probability distribution function $F_{\overline{Y}_n}$ and F_{S_n}, respectively. These functions in classical theory can be approximated with large samples and hypotheses of probabilistic independence, which turned out to be difficult to verify in this application. To solve this mathematical problem, the Monte-Carlo method is proposed as a tool to determine these distribution functions without the use of these hypotheses. The procedure is presented in Figure 4, where N is the number of samples used to simulate Ξ_k and K the number of samples to determine $F_{\overline{Y}_n}$ and F_{S_n}.

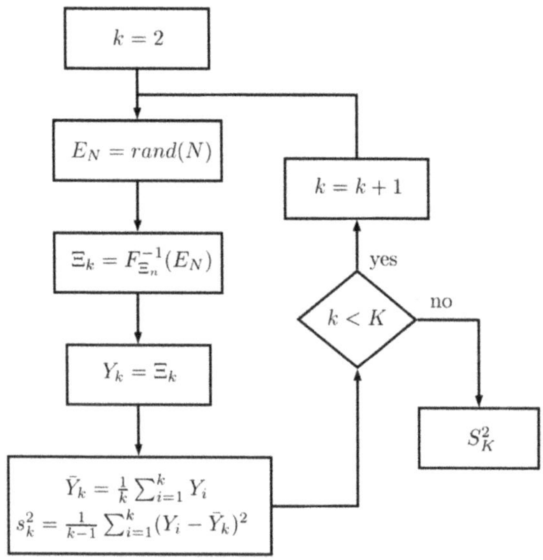

Figure 4. Simulation procedure with the Monte-Carlo method (MMC) to determine the variance s_k^2.

This procedure allows us to estimate the variance σ^2 because if K is large, the following is fulfilled

$$E\{Y\} = E\{\overline{Y}_K\}, \quad (21)$$

$$S_K^2 = \frac{1}{K-1}\sum_{i=1}^{K}[Y_i - E\{Y\}]^2, \quad (22)$$

$$\sigma^2 = E\{S_K^2\}. \quad (23)$$

3.4. Correction Factors Calculation to Compensate Estimators

This procedure removes the constraints that exist to achieve an adequate estimate of reliability and provides the factors that must be included to eliminate the estimators' bias. These factors are

1. Additive factor for $\overline{Y}_k$ $\left(\left[\Phi^T\Phi\right]^{-1}\Phi^T\Xi\right)$
2. Multiplicative factor for S_k (C_k)

In our case, the Φ and Ξ turned out to be uncorrelated, so this additive term is close to zero and was not considered in this study.

The multiplicative factor for S_k is obtained from the Cauchy–Bunyakowskii–Schwartz inequality described in Equation (10). In many cases where $E\{S_k\} < \sigma$, the question is how far is $E\{S_k\}$ from σ, so we will introduce a factor c_k that eliminates this bias, that is to say

$$E\{c_k S_k\} = \sigma \quad (24)$$

The new standard deviation estimator σ will be $c_k S_k$, this factor is estimated using the Monte-Carlo method and the relation

$$c_k = \frac{\sqrt{E\{(Y - E\{Y\})^2\}}}{\sqrt{\frac{1}{k-1}\sum_{i=1}^{k}\left(Y_i - \overline{Y}_k\right)^2}} = \frac{\sigma}{S_k}. \tag{25}$$

3.5. Confidence Intervals for the R Estimation

When clinical studies are performed on patients, new data are obtained, which must be analyzed to determine if they are correct. For carrying out this, confidence intervals are constructed on-line to ensure that the test is correct with a confidence level α.

The last sample's variations are analyzed. These variations are measured by standard deviation (s), which is calculated by means of the s_n estimator, considering the process defined by Y and the set of samples.

In a process defined by Y with N clinical studies set Y_n, $(n = 1, 2, \cdots, N)$, each clinical study has M samples $[Y_n]_m$, $(m = 1, 2, \cdots, M)$. The variations are measured by the standard deviation s_Y, which is estimated through the corrected s_n.

From a statistical point of view, a measurement is considered incorrect if it is an atypical value. The criteria to decide is built by considering the probability that the current measurement will fall outside a range defined by a standard deviation estimator, considering the necessary compensations for the used estimators.

Now, we will determine a criterion to eliminate measurements that, with a high probability, may be wrong, since values will be far from most of the measurements, known as atypical data.

To do so, the Chebyshev approximation [25] is used

$$P\left(\frac{|Y - \mu_Y|}{\sigma_Y} \geq k\right) \leq \frac{1}{k^2}, \tag{26}$$

$$P\left(\frac{|Y - \mu_Y|}{\sigma_Y} < k\right) \geq 1 - \frac{1}{k^2}, \tag{27}$$

where μ_Y and σ_Y are the mean and the standard deviation of the process, respectively.

If no further information is available but considering that new information is arriving every time a measurement is taken, the following statistics are considered

$$P\left(\frac{|Y - \mu_Y|}{c\, s_Y} < k\right) \geq 1 - \frac{1}{k^2}. \tag{28}$$

where c is a correction factor due to the finite number of samples, and to the law of standard deviation spread?

Since μ_Y, s_Y, and c are typically unknown, these parameters are replaced by their estimators $\overline{y}_n$, s_n, and c_n, respectively. Finally, the selection criterion is obtained

$$P\left(\frac{|Y_n - \overline{y}_n|}{c_n s_n} < k\right) \geq 1 - \frac{1}{k^2}, \tag{29}$$

which is equivalent to

$$\left(Y_n - \overline{y}_n\right)^2 < \beta s_n^2, \tag{30}$$

where $\beta = k^2 c_n^2$, and c_n is the correction factor that corresponds to the nth clinical study.

Figure 5 shows the flowchart of the reliability estimation algorithm based on Chebyshev's inequality.

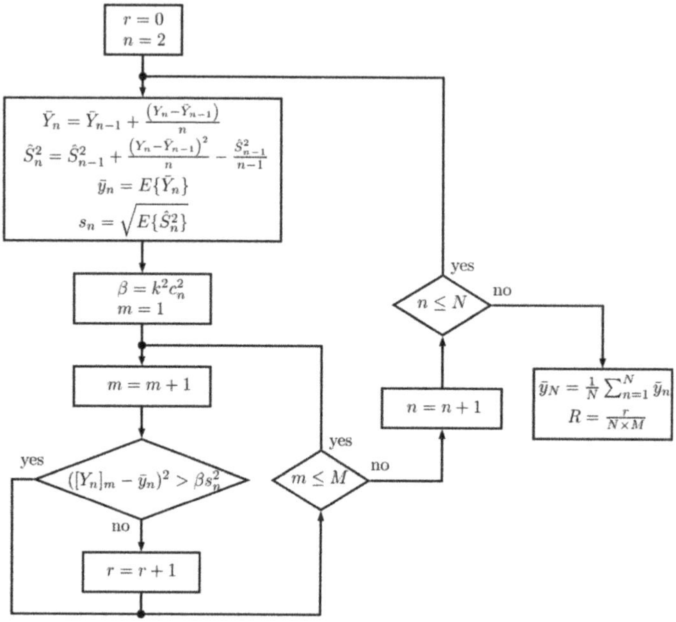

Figure 5. Reliability estimation algorithm.

3.6. Remote Fetal Monitoring Equipment

A fetal monitor was developed to provide a reliable remote monitoring of both mother and fetus during the pregnancy process. The fetal monitor has the capacity of recording fetal heart rate, uterine activity, and fetal movement. Its compact design allows the mother to use it herself. In addition, the fetal monitor allows the recording of studies, and features the capacity to send them directly to the obstetric monitoring center (OMC) to be evaluated by a specialist.

The purpose of this medical device is to provide monitoring of high-risk pregnancies, especially those in marginal areas, who cannot go to a clinical specialist periodically to reduce fetal distress, and, therefore, perinatal mortality in Mexico.

The fetal monitor is an embedded system, which includes a digital signal processor that allows for an adequate calculation to achieve higher precision in fetal heart rate reliability and signs of uterine activity. The algorithm developed included the proposed methodology of this work to guarantee the reliability of distance measurement. This instrument has enough hardware resources, such as remote connectivity by means of a mobile phone, through the general package radio service (GPRS), geo-referenced transmission, an alarm indicator for the patient, a speaker to listen to baby heartbeats, a flash memory to store studies and two Doppler ultrasound sensors, of 1.2 MHz in frequency, and 350 mW for output power, for the acquisition and processing of fetal monitoring variables. A block diagram with these functions is presented in Figure 6.

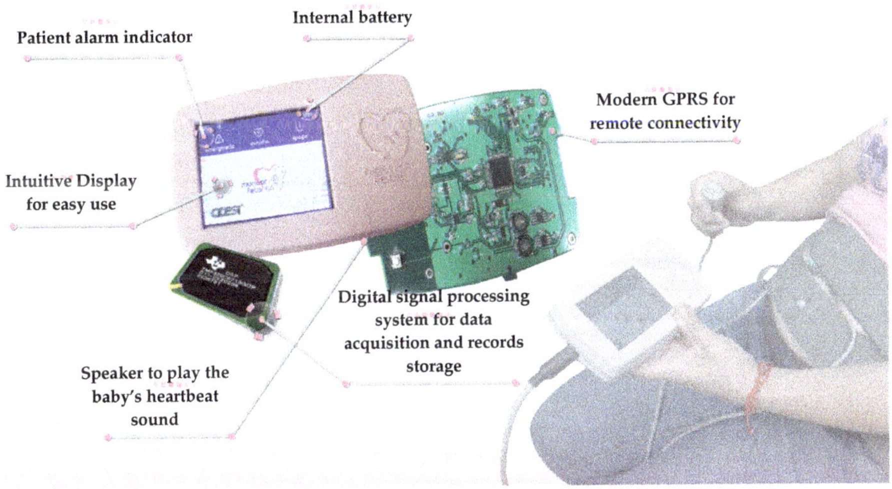

Figure 6. Scheme of the fetal monitor used for testing.

3.7. Experimental Clinical Studies

During experimental tests, several fetal monitoring instruments, in remote sites, were connected to patients and sent data signals to the OMC based at the Central Hospital. The patients were informed about their data usage and consented to publish their data for research purposes.

Data collected at OMC were primary signals without processing, that is, directly from the analog–digital converter of the instrument. For each test, a set of data with fetal heart rate and signs of uterine activity was stored. Table 1 shows the clinic studies in detail, to validate the proposed approach.

Table 1. Clinical Studies for approach validation.

Patient Name	Age (Year)	Studies Number	Place	Distance from OMC
Subject 1	37	42	Toliman	81.7 Km
Subject 2	41	45	Jalpan de Serra	187 Km
Subject 3	35	40	Querétaro	9.1 Km
Subject 4	26	43	Amealco de Bonfil	71.1 Km
Subject 5	43	32	Querétaro	9.4 Km
Subject 6	18	34	San Luis Potosí	215 Km
Subject 7	28	36	Querétaro	10.7 km
Subject 8	31	38	Querétaro	18.3 Km
Subject 9	19	43	Colima	576 Km
Subject 10	21	47	Querétaro	13.5 Km
Subject 11	39	51	Querétaro	30.4 Km
Subject 12	42	53	Querétaro	13.1 Km
Subject 13	29	54	El Marqués	7.1 Km
Subject 14	33	57	El Marqués	6.8 Km
Subject 15	36	58	Salamanca	89.7 Km
Subject 16	25	58	Puebla	329 Km
Subject 17	42	58	Guadalajara	378 Km
Subject 18	45	61	Querétaro	2.1 Km
Subject 19	38	75	Querétaro	17.4 Km

Experimental tests were conducted in the State of Querétaro. In the first stage, health care professionals of a Women and Child hospital in the city of Querétaro assessed patients. Patients were trained to handle the equipment at home, to step by step visualize data on the fetal monitor screen, and even to place two sensors on the abdominal region of the patient. One sensor measures the fetal heart rate, and the other measures the uterine activity of the patient. In the second stage, experimental tests for the fetal monitor were conducted with patients from remote sites. Data generated by each patient were sent to an OMC in the city of Querétaro. The locations of patients transmitting their records were in many different areas of the State of Queretaro, or in places nearby the State.

Distances from these sites to the OMC were 180 Km on average. Nineteen patients were evaluated, with around 40 studies for each patient. The number of measurements for each clinical study was 900, on average, to prove the approach proposed in this work, providing measurement reliability for fetal heart rate and uterine activity. Figure 7 shows some images of the sites where tests were conducted.

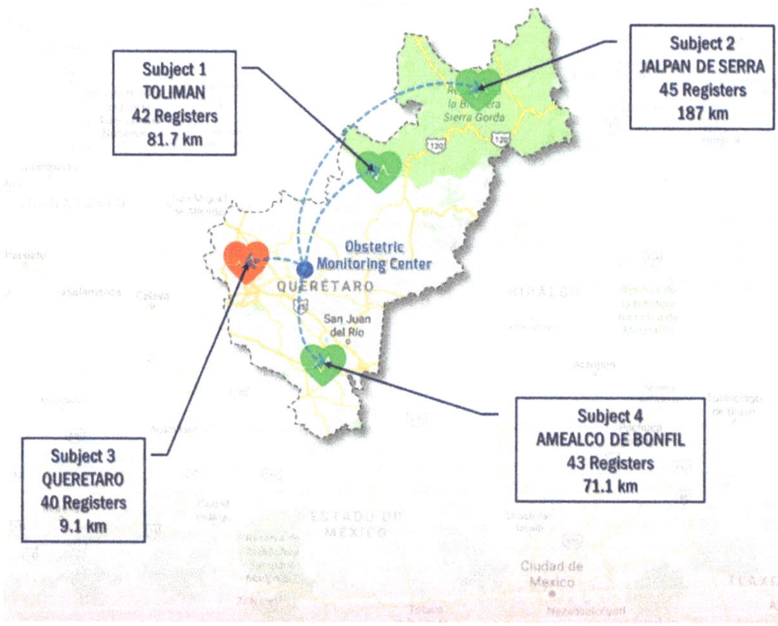

Figure 7. Map showing the location of the first four patients from Table 1.

Samples for each medical device were received and stored at OMC. The objective at this stage was to generate a database of unprocessed signals, containing bioelectrical perturbations, to implement and to validate the proposed methodology. Only fetal heart rate signals were processed.

4. Results and Discussion

Data from clinical studies carried out on 19 patients were used for system modeling and validating the proposed approach. The clinical study data was divided into two subsets, the first to model the system (data for modeling) and the second to validate the proposed approach (data for validation). The proposed approach was implemented in MATLAB® for building the model, calculating the compensated uncertainty, and evaluating the reliability.

4.1. Fetal Monitor Model

To obtain the dynamic model of the fetal monitor, a calibrated signal generator was used to generate an input signal u_k for obtaining an output y_k from the fetal monitor device used in this study. With these data and applying the least-squares method described in Section 3.1.1, a second-order linear deterministic model $l = 2$ was estimated, the ARX model parameters were $\theta = [1.1381, -0.2829, 0.0985, -0447, 0.0907]$. The dynamic model response of the fetal monitor is shown in Figure 8.

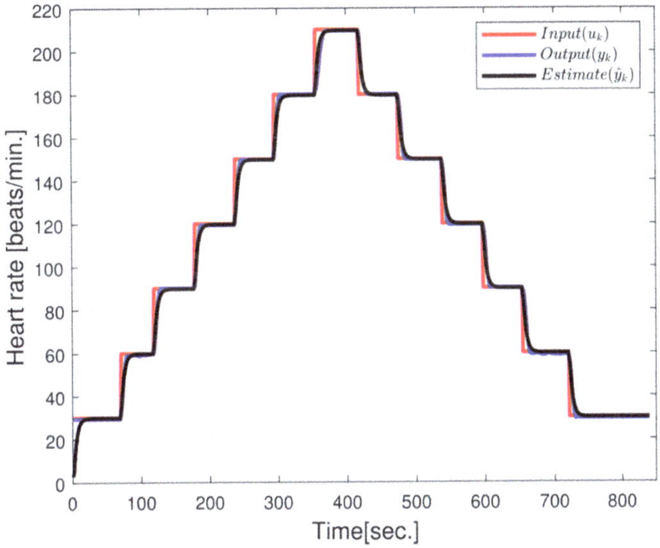

Figure 8. Fetal monitor model simulation results.

4.2. Stochastic Model

A probabilistic model for each patient was calculated from the experimental data set (data for modeling) acquired from clinical studies carried out on the 19 patients under study. From the fetal heart rate measurements, the perturbations ξ_k was obtained, using the dynamic model of the fetal monitor applying Equation (18), for each patient. Data for ten clinical studies per patient were used to obtain the probabilistic model. From perturbations histogram, estimations of probability density functions f_{Ξ_n} and distribution functions F_{Ξ_n} were obtained, finally a nonparametric model of the inverse function $F_{\Xi_n}^{-1}$ was available for each patient. The direct and inverse distribution functions were estimated by using a linear interpolation by parts [26]. Using the Monte-Carlo method with a large data number of the amount of $N = 10^5$ and $K = 10^3$, the probabilistic models $F_{\Xi_n}^{-1}$ were simulated applying the procedure described in Figure 4 and s_K^2 was estimated to calculate the variance using $\sigma^2 = E\{s_K^2\}$. The probability distribution function models of the patients used in the simulation are shown in Figure 9, and Table 2 shows the estimated standard deviations and data number n used in the F_{Ξ_n} modeling.

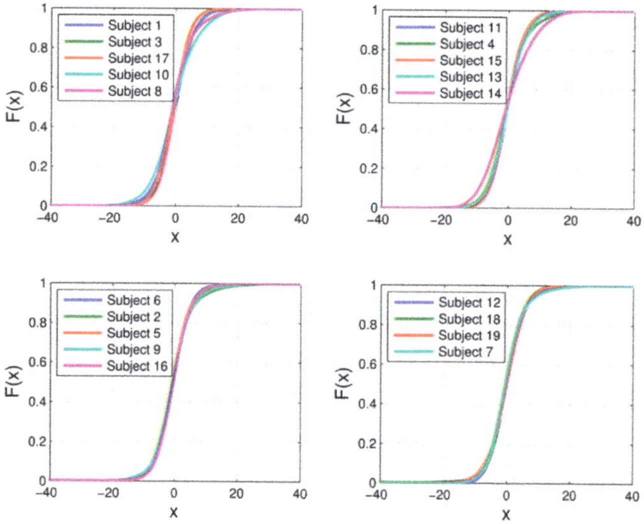

Figure 9. Modeled distribution functions of the patients.

Table 2. Estimated variance and data number used in stochastic modeling.

Patient Name	σ	n
Subject 1	7.75	12731
Subject 2	10.98	13821
Subject 3	12.77	9882
Subject 4	13.97	12062
Subject 5	9.19	13472
Subject 6	8.54	12778
Subject 7	10.52	13417
Subject 8	9.44	13125
Subject 9	11.13	12156
Subject 10	12.75	10243
Subject 11	8.64	13454
Subject 12	8.68	13357
Subject 13	9.61	13399
Subject 14	12.97	12410
Subject 15	6.88	13316
Subject 16	7.70	11677
Subject 17	7.49	13481
Subject 18	11.93	12010
Subject 19	8.07	14070

Finally, 100 variance correction coefficients were calculated for each patient, using the relationship defined in Equation (25).

4.3. Stochastic Model Validation

From the experimental data set (data for validation), the correction coefficients c_n(calculated during the modeling process) and the algorithm proposed in Figure 5, the corrected standard deviation and the expected fetal heart rate were estimated. For the validation process, $N = 5$ clinical studies with $M = 900$ samples per clinical study and β defined with $k = 1$ were considered to determine the minimum percentage of samples that fall within one standard deviation. Table 3 presents the results of the modeled standard deviation σ, estimated s_n, and corrected $c_n s_n$, and the expected value and $\overline{y}_n$ with samples from five clinical studies for all patients.

Table 3. Expected Value and Standard Deviation Estimation of Fetal Heart Rate.

Patient Name	σ	$\overline{y}_5$	s_2	c_2s_2	s_3	c_3s_3	s_4	c_4s_4	s_5	c_5s_5
Subject 1	7.75	148.6	3.70	7.29	4.51	7.65	4.85	7.54	5.15	7.52
Subject 2	10.98	142.9	5.10	10.23	6.44	10.98	7.23	11.27	7.75	11.37
Subject 3	12.77	157.9	6.23	12.45	7.75	13.24	8.67	13.49	9.14	13.36
Subject 4	13.97	140.7	6.77	13.37	8.31	14.12	9.41	14.61	9.89	14.48
Subject 5	9.19	139.4	4.36	8.68	5.48	9.32	6.02	9.38	6.35	9.32
Subject 6	8.54	140.6	3.99	7.95	4.99	8.51	5.50	8.57	6.05	8.86
Subject 7	10.52	131.6	5.08	10.17	6.31	10.77	7.02	10.93	7.38	10.82
Subject 8	9.44	138.3	4.48	8.92	5.57	9.49	6.31	9.83	6.68	9.80
Subject 9	11.13	139.5	5.25	10.37	6.65	11.27	7.29	11.26	7.89	11.43
Subject 10	12.75	132.9	6.26	12.67	7.78	13.32	8.68	13.56	9.30	13.63
Subject 11	8.64	123.7	4.05	8.03	5.17	8.77	5.68	8.84	6.17	9.02
Subject 12	8.68	135.7	3.96	7.92	5.12	8.70	5.70	8.87	6.07	8.87
Subject 13	9.61	148.0	4.51	9.03	5.67	9.65	6.40	9.93	6.82	9.95
Subject 14	12.97	141.3	6.39	12.72	8.10	13.83	8.96	13.95	9.50	13.92
Subject 15	6.88	129.7	3.14	6.22	3.92	6.68	4.36	6.80	4.58	6.70
Subject 16	7.70	134.2	3.52	7.07	4.42	7.55	4.99	7.76	5.38	7.86
Subject 17	7.49	141.5	3.40	6.82	4.27	7.30	4.72	7.34	5.03	7.35
Subject 18	11.93	132.8	6.04	11.84	7.37	12.49	8.11	12.58	8.59	12.51
Subject 19	8.07	150.6	3.66	7.29	4.67	7.96	5.29	8.26	5.62	8.25

The algorithm in Figure 5, additionally, allows the evaluation of the reliability improvement using the corrected and uncorrected standard deviation (uncertainty). To verify the reliability behavior, the proposed algorithm was executed using two bands level: $\beta = k^2 c_n$ to obtain the corrected reliability value and $\beta = k^2$ to obtain the reliability value without correction. Figure 10 shows the results of the corrected and uncorrected reliability for two clinical studies $n = \sqrt{2}$, Figure 10a, and for five clinical studies $n = 5$, Figure 10b. The reliability increases when considering the corrected standard deviation (uncertainty).

In the case of total ignorance of the system, the Chebyshev inequality indicates that at least $1 - \frac{1}{k^2}$ of the sample, data must be within k standard deviations of the mean, for $k = \sqrt{2}$ for at least 50% of the data, see Figure 10. However, having the distribution function models allow us, in addition to estimating the compensated uncertainty, to know the number of measurements that were within $\sqrt{2}$ standard deviations, which was close to 90%.

The parameter estimator behavior in traditional models [12–14], was performed considering some theoretical distributions [3], and their behavior was evaluated considering finite samples. The problems that arose were that the distributions used were not representative of the applications considered in this work, and the estimator properties within finite time windows depended on the distributions. The method proposed in this work allows defining distributions in practical applications with few samples; that solve the problems presented in classical models.

Reliability determination with few samples could be useful due to the visual analysis of fetal heart rate tracings being subjective and inconsistent [5]. There is disagreement in the clinical decision in fetal heart rate (FHR) classification [7]. A simpler and more objective set of guidelines could provide better reliability [10]. Given the inherent limitations of electronic fetal monitoring technology and of the ability of human beings to characterize fetal heart rate parameters and patterns, the low-reliability degree may represent the best-case scenario of the visual fetal heart rate interpretation reliability [6].

In the international federation of gynecology and obstetrics (FIGO), national institute for health and care (NICE), and American college of obstetrics and gynecology (ACOG) guidelines [8], there are two fundamental parameters to determine normal, suspicious, and pathological patterns. The first is the heart rate baseline, and the other is variability. The method proposed in this work can be applied directly to a correct estimate of the variability because it includes the specific distribution function of each patient. In the tests analyzed, differences of 10% were found in the estimate (n = 5). According to results, the baseline is not affected by the distribution.

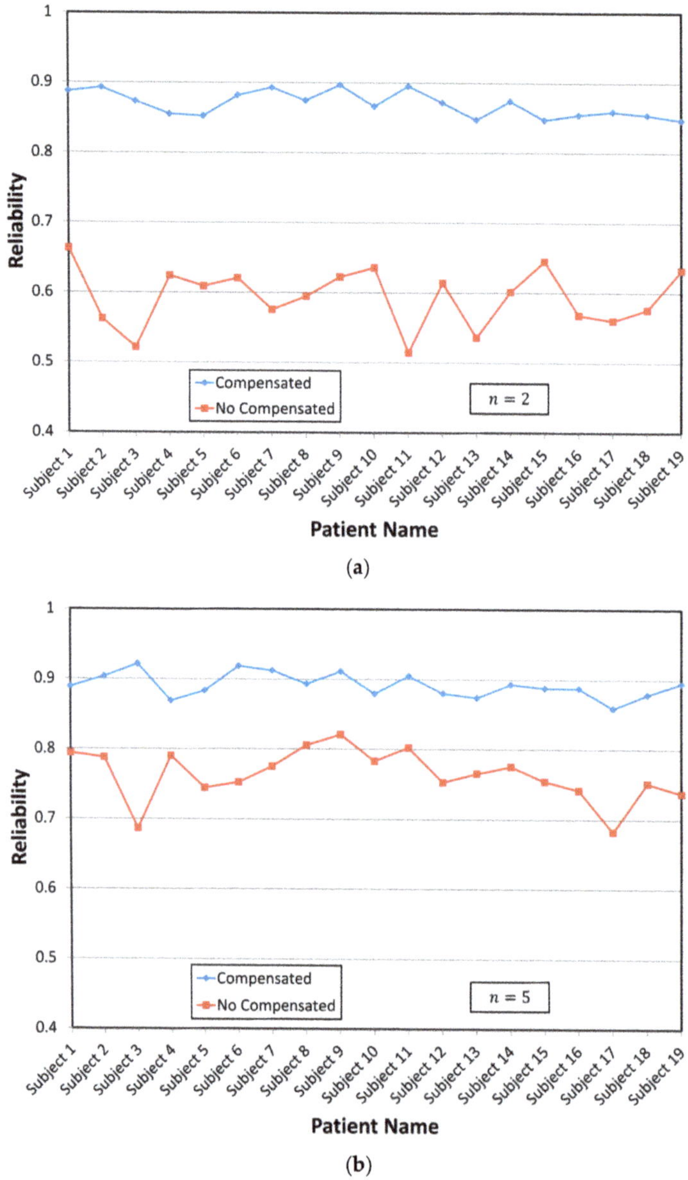

Figure 10. Corrected and uncorrected reliability (R) calculations; (**a**) with two clinical studies; (**b**) with five clinical studies.

5. Conclusions

Reliability has reached an important level for consideration at the beginning of product design, specifically for medical applications, which have been favored by including, as a quality criterion, the reliability level of the device.

Effect analysis of the distribution function and sample lengths in the typical estimators' calculation, mean, and standard deviation were presented. Specific distribution estimates of each patient were used; this procedure suggests a natural way to integrate the additional information obtained in each study. Several studies carried out on patients verified that the real distributions were far from the theoretical distributions that are usually used to determine the estimators' properties. Furthermore, these distributions were used to determine the factors necessary to obtain an unbiased estimator of uncertainty. This information was used to implement a simple algorithm that improves reliability estimation. With the available calculation tool, it is possible to obtain specific information that allows improving the estimators' behavior used in decision making without assuming hypotheses that are difficult to verify in daily applications. Applying classical theory without verifying that hypotheses are met may lead to incorrect results. However, it was demonstrated that these might be corrected by using compensation factors.

Considering how fast available electronic systems for the integration of medical devices change, the way reliability is assessed, restrictions imposed by classical estimation theory, and computer resources, the developed methodology attempts to be the spearhead of a series of works that would lead to a more accurate estimation of reliability.

Author Contributions: Conceptualization, S.A.-M. and R.S.-Z.; methodology, G.R.-L. and S.A.-M.; software, L.B.-R.; validation, G.R.-L., S.A.-M. and R.S.-Z.; formal analysis, R.S.-Z.; investigation, T.S.-J.; resources, S.A.-M.; data curation, R.S.-Z. and L.B.-R.; writing—original draft preparation, G.R.-L.; writing—review and editing, S.A.-M. and T.S.-J.; visualization, G.R.-L.; supervision, T.S.-J.; project administration, S.A.-M.; funding acquisition, S.A.-M. All authors have read and agreed to the published version of the manuscript.

Funding: This work was supported by the FOMIX, Queretaro State Mexico mixed foundation (Grant: QRO-2010-C02-149261).

Acknowledgments: Authors acknowledge Carlos Arturo Rebolledo Fernandez and Julio Cesar Ramirez Arguello for supporting the project logistic.

Conflicts of Interest: The authors declare no conflict of interest.

References

1. Parer, J. Fetal heart-rate monitoring. *Lancet* **1979**, *314*, 632–633. [CrossRef]
2. Yeh, S.-Y.; Diaz, F.; Paul, R.H. Ten-year experience of intrapartum fetal monitoring in Los Angeles County/University of Southern California Medical Center. *Am. J. Obstet. Gynecol.* **1982**, *143*, 496–500. [CrossRef]
3. Vintzileos, A.M.; Antsaklis, A.; Varvarigos, I.; Papas, C.; Sofatzis, I.; Montgomery, J.T. A randomized trial of intrapartum electronic fetal heart rate monitoring versus intermittent auscultation. *Obstet. Gynecol.* **1993**, *81*, 899–907. [PubMed]
4. Alfirevic, Z.; Devane, D.; Gyte, G.M.; Cuthbert, A. Continuous cardiotocography (CTG) as a form of electronic fetal monitoring (EFM) for fetal assessment during labour. *Cochrane Database Syst. Rev.* **2017**. [CrossRef] [PubMed]
5. Ayres-de-Campos, D.; Bernardes, J.; Costa-Pereira, A.; Pereira-Leite, L. Inconsistencies in classification by experts of cardiotocograms and subsequent clinical decision. *Bjog An. Int. J. Obs. Gynaecol.* **1999**, *106*, 1307–1310. [CrossRef] [PubMed]
6. Blackwell, S.C.; Grobman, W.A.; Antoniewicz, L.; Hutchinson, M.; Gyamfi Bannerman, C. Interobserver and intraobserver reliability of the NICHD 3-Tier Fetal Heart Rate Interpretation System. *Am. J. Obstet. Gynecol.* **2011**, *205*, 378. [CrossRef] [PubMed]
7. Chauhan, S.P.; Klauser, C.K.; Woodring, T.C.; Sanderson, M.; Magann, E.F.; Morrison, J.C. Intrapartum nonreassuring fetal heart rate tracing and prediction of adverse outcomes: Interobserver variability. *Am. J. Obs. Gynecol.* **2008**, *199*, 623.e1–623.e5. [CrossRef] [PubMed]

8. Santo, S.; Ayres-de-Campos, D.; Costa-Santos, C.; Schnettler, W.; Ugwumadu, A.; Da Graça, L.M. Agreement and accuracy using the FIGO, ACOG and NICE cardiotocography interpretation guidelines. *Acta Obs. Gynecol. Scand.* **2017**, *96*, 166–175. [CrossRef] [PubMed]
9. Dawes, G.S.; Rosevear, S.K.; Pello, L.C.; Moulden, M.; Redman, C.W.G. Computerized analysis of episodic changes in fetal heart rate variation in early labor. *Am. J. Obs. Gynecol.* **1991**, *165*, 618–624. [CrossRef]
10. Bernardes, J.; Moura, C.; Marques de Sa, J.P.; Pereira Leite, L. The Porto system for automated cardiotocographic signal analysis. *J. Perinat. Med.* **1991**, *19*, 61–65. [CrossRef] [PubMed]
11. Mytton, O.T.; Velazquez, A.; Banken, R.; Mathew, J.L.; Ikonen, T.S.; Taylor, K.; Painter, F.; Jean-Baptiste, R.; Poon, A.; Ruelas, E. Introducing new technology safely. *Qual. Saf. Heal. Care* **2010**, *19*, i9–i14. [CrossRef] [PubMed]
12. Hsieh, D.A.; Manski, C.F. Monte Carlo Evidence on Adaptive Maximum Likelihood Estimation of a Regression. *Ann. Stat.* **1987**, *15*, 541–551. [CrossRef]
13. Spokoiny, V. Parametric estimation. *Finite Sample Theory Ann. Stat.* **2012**, *40*, 2877–2909. [CrossRef]
14. Steigerwald, D.G. On the finite sample behavior of adaptive estimators. *J. Econ.* **1992**, *54*, 371–400. [CrossRef]
15. Yi, X.; Wang, Z.; Caramanis, C.; Liu, H. Optimal linear estimation under unknown nonlinear transform. In *Advances in Neural Information Processing Systems*; Cornell Univ: Ithaca, NY, USA, 2015.
16. Toulis, P.; Airoldi, E.M. Asymptotic and finite-sample properties of estimators based on stochastic gradients. *Ann. Stat.* **2017**, *45*, 1694–1727. [CrossRef]
17. Nitzan, E.; Routtenberg, T.; Tabrikian, J. Cramér-Rao bound for constrained parameter estimation using Lehmann-unbiasedness. *IEEE Trans. Signal Process* **2019**, *67*, 753–768. [CrossRef]
18. Siju, K.C.; Kumar, M. Bayesian estimation of reliability using time-to-failure distribution of parametric degradation models. *J. Stat. Comput. Simul.* **2018**, *88*, 1717–1748. [CrossRef]
19. *Evaluation of Measurement Data—Guide to the Expression of Uncertainty in Measurement, Joint Committee for Guides in Metrology 100*, 1st ed.; 2008; Available online: https://www.bipm.org/utils/common/documents/jcgm/JCGM_100_2008_E.pdf (accessed on 8 May 2020).
20. Shiryaev, A.N. *Probability*; Springer: New York, NY, USA, 1996; Volume 95. [CrossRef]
21. Lira, I.; Taylor, J.R. Evaluating the Measurement Uncertainty: Fundamentals and Practical Guidance. *Am. J. Phys.* **2003**, *71*, 93–94. [CrossRef]
22. Cramer, H. *Mathematical Methods of Statistics*; Princeton University Press: Princeton, NJ, USA, 1999; Volume 34. [CrossRef]
23. Ljung, L. *System Identification: Theory for the User*, 2nd ed.; Preniice Hall: Upper Saddle River, NJ, USA, 1999.
24. Ikonen, E.; Najim, K. *Advanced Process Identification and Control*; CRC Press: Hoboken, NJ, USA, 2001. [CrossRef]
25. Amidan, B.G.; Ferryman, T.A.; Cooley, S.K. Data outlier detection using the Chebyshev theorem. In Proceedings of the, Big Sky, MT, USA, 5–12 March 2005. [CrossRef]
26. Quarteroni, A.; Saleri, F.; Gervasio, P. *Scientific Computing with MATLAB and Octave*; Springer: Berlin/Heidelberg, Germany, 2014; Volume 2. [CrossRef]

applied sciences

MDPI

Article

Clinical Evaluation of Stretchable and Wearable Inkjet-Printed Strain Gauge Sensor for Respiratory Rate Monitoring at Different Body Postures

Ala'aldeen Al-Halhouli [1,2,3,*], Loiy Al-Ghussain [1,4], Saleem El Bouri [1], Fuad Habash [1], Haipeng Liu [5,6] and Dingchang Zheng [6]

1 Mechatronics Engineering Department/NanoLab, School of Applied Technical Sciences, German Jordanian University, P.O. Box 35247, Amman 11180, Jordan; loiy.al-ghussain@uky.edu (L.A.-G.); saleem.bouri@hotmail.com (S.E.B.); f.habash@gju.edu.jo (F.H.)
2 Institute of Microtechnology, Technische Universität Braunschweig, 38124 Braunschweig, Germany
3 Faculty of Engineering, Middle East University, Amman 11831, Jordan
4 Mechanical Engineering Department, University of Kentucky, Lexington, KY 40506, USA
5 Medical Device and Technology Research Laboratory, School of Allied Health, Faculty of Health, Education, Medicine and Social Care, Anglia Ruskin University, Chelmsford CM1 1SQ, UK; haipeng.liu@anglia.ac.uk
6 Research Centre of Intelligent Healthcare, Faculty of Health and Life Science, Coventry University, Coventry CV1 5FB, UK; dingchang.zheng@coventry.ac.uk
* Correspondence: alaaldeen.alhalhouli@gju.edu.jo

Received: 21 November 2019; Accepted: 11 December 2019; Published: 9 January 2020

Abstract: Respiratory rate (RR) is a vital sign with continuous, convenient, and accurate measurement which is difficult and still under investigation. The present study investigates and evaluates a stretchable and wearable inkjet-printed strain gauge sensor (IJP) to estimate the RR continuously by detecting the respiratory volume change in the chest area. As the volume change could cause different strain changes at different body postures, this study aims to investigate the accuracy of the IJP RR sensor at selected postures. The evaluation was performed twice on 15 healthy male subjects (mean ± SD of age: 24 ± 1.22 years). The RR was simultaneously measured in breaths per minute (BPM) by the IJP RR sensor and a reference RR sensor (e-Health nasal thermal sensor) at each of the five body postures namely standing, sitting at 90°, Flower's position at 45°, supine, and right lateral recumbent. There was no significant difference in measured RR between IJP and reference sensors, between two trials, or between different body postures (all $p > 0.05$). Body posture did not have any significant effect on the difference of RR measurements between IJP and the reference sensors (difference <0.01 BPM for each measurement in both trials). The IJP sensor could accurately measure the RR at different body postures, which makes it a promising, simple, and user-friendly option for clinical and daily uses.

Keywords: inkjet printing; respiratory rate; strain gauge; stretchable and wearable sensors; silver nanoparticles; clinical evaluation; body posture

1. Introduction

Clinically, respiratory rate (RR), defined as the times of breath per minute (BPM), is a vital sign whose abnormality could indicate various pathological conditions in cardiorespiratory system [1,2]. Compared with other vital signs such as heart rate and blood pressure, RR is more accurate in screening unstable patients [1,3,4]. Therefore, RR monitoring has been used in detecting sleep apnea, cardiac arrest, and sudden infant death syndrome [3,5]. Especially, compared with discrete RR measurement, continuous RR monitoring could detect RR abnormalities with more than 10 times of accuracy [2].

However, despite the clinical significance of RR, continuous, convenient, and accurate RR monitoring is still difficult. It was reported in [2] that the RR was the least documented vital sign even when the patients suffered from respiratory issues. Manual RR counting is widely used in clinical examination [3] but is not accurate [2]. Other traditional techniques of clinical RR monitoring include spirometry, pneumotachograph, and capnography [1,4,6]. These techniques could accurately measure RR, but are cumbersome, difficult to manipulate, or mask-based [4].

To achieve continuous and convenient RR monitoring, some new techniques have been developed [7–10]. Some researchers extract RR from other physiological signals such as photoplethysmogram and electrocardiogram [11]. Other studies derived respiration signals using non-contact techniques such as C-band sensing [12] and dual smartphone cameras [4]. Some researchers tried distant (less than 10 m) RR detecting based on Wi-Fi, radar, or thermal imaging [11]. Compared with these techniques, wearable and stretchable sensors could independently measure RR without limiting the daily activities of the users. Additionally, the stretchable and wearable sensors could achieve highly accurate RR measurement at a low cost [13].

For a wearable RR sensor, the high accuracy at different body postures could enable its application in various clinical and daily situations. Several studies [3,12,14] have investigated the accuracy of RR monitoring devices at different body postures (sitting, standing and supine and side). For instance, Huang et al. [3] embedded arrays of load sensors in an e-textile bed sheet to measure RR from the full body pressure distribution on the sheet. The sensor was tested on 14 human subjects in prone, side and supine postures during sleep. The results showed that RR measurement was more accurate in prone and supine postures compared with side posture. The authors deduced that this difference was due to the amplitude of respiratory movement which is more prominent in thoracic and abdominal areas than on both sides. Zhu et al. [14] developed a fiber bragg grating (FBG) sensor mat with three FBG arrays under the pillow, below upper chest, and below lower chest. The sensor was tested on 12 human subjects while lying in prone and side postures. For each of the three sensors, the mean error of RR measurement was lower than one BPM with standard deviation (STD) within ±0.25 BPM, compared with the reference RR measured by a polysomnography belt sensor. Guan et al. [12] developed a wireless system to measure RR based on C-band sensing. In prone, sitting, and standing postures, the error of RR detection was within 7%, 6.2% and 5.5% compared with a reference RR sensor. The author suggested that standing posture have the largest influence on the quality of C-band RR detection. Nam et al. [4] used dual cameras on a smartphone to measured RR by reflectance imaging on the chest and the abdomen, with average median error of 1.43% and 1.62%.

The majority of existing studies on the effect of body posture on RR detection accuracy were based on non-stretchable [15–17] or non-wearable RR sensors [3,12,14], with a lack of related study based on stretchable RR sensors. Compared with traditional RR sensors, stretchable RR sensors are in direct contact with body surface, and are therefore, more likely to be influenced by body posture. Considering the increasing amount of stretchable RR sensors and the possibility of application in different scenarios, there is an urgent need to investigate the effect of body posture on the accuracy of stretchable RR sensors.

The aim of this paper is to investigate the effect of different body postures (standing, sitting at 90°, Flower's position at 45°, supine and right lateral recumbent) on the accuracy of a new stretchable and wearable inkjet-printed RR sensor.

2. Methods

2.1. Physiological Measurements

2.1.1. Subjects

Fifteen healthy male subjects (mean ± SD of age: 24 ± 1.22 years) without any known cardiovascular or respiratory diseases participated in this study with written informed consent. The protocol adhered to the tenets of the Declaration of Helsinki. Before the experiment, the procedure was explained in detail to each subject.

2.1.2. Sensors

Inkjet printing (IJP) is a promising technology in the fabrication of stretchable and wearable sensors. Compared with other fabrication techniques, IJP has less sophisticated process and lower cost [18–21]. The IJP RR sensor was fabricated on polydimethylsiloxane (PDMS) substrate using conductive silver nanoparticle ink, as shown in Figure 1a which was presented in [22]. During respiration, the sensor acted as a variable resistance whose value increased during inhalation and decreased during exhalation, due to the volume change in the ribcage or the abdomen. The change in resistance was obtained by a Wheatstone bridge circuit with three fixed resistances, amplified by an instrumentational amplifier, and finally fed into an Arduino microcontroller to derive RR.

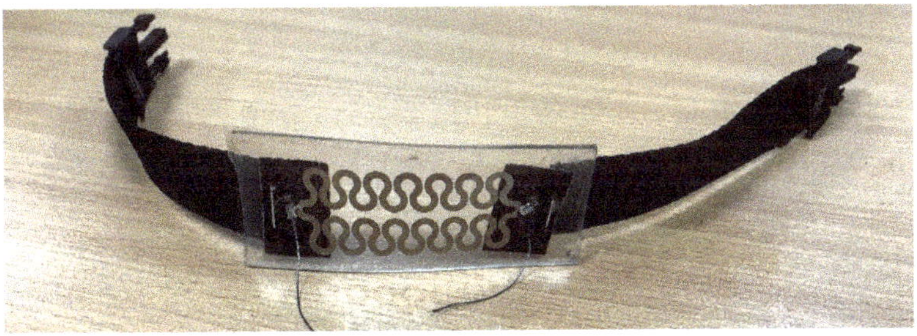

(a)

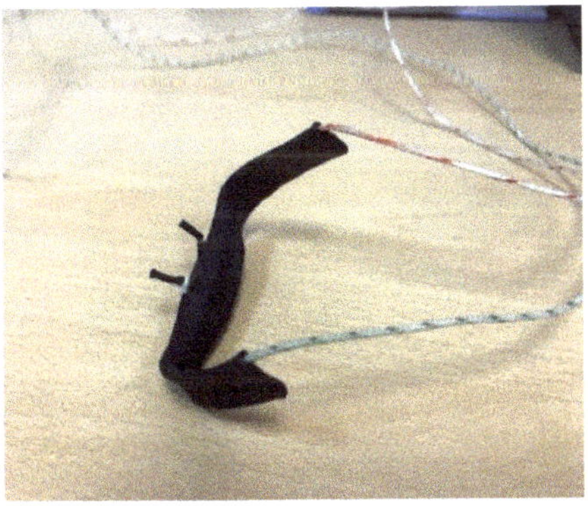

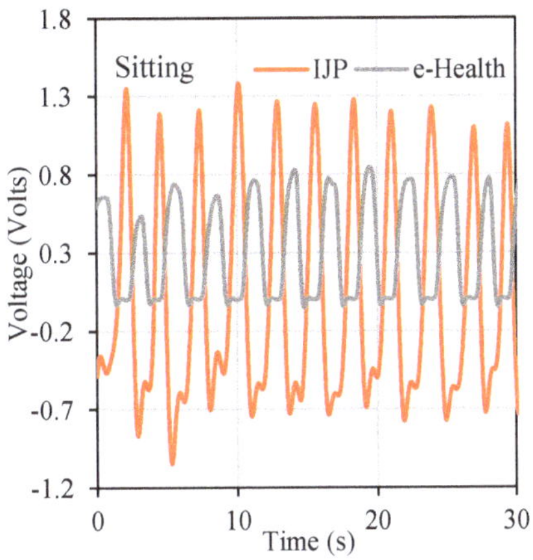

(b) (c)

Figure 1. (**a**) The Inkjet printing (IJP) respiration rate (RR) sensor. (**b**) The reference e-Health RR sensor. (**c**) The derived respiratory signals in 30s.

The reference RR value was measured by a thermal nasal e-Health sensor (e-Health AirFlow sensor, Cooking Hacks), shown in Figure 1b, which has been validated previously in the literature [23,24]. The e-Health sensor detects the thermal changes in airflow during inhalation and exhalation. Due to the difference in working principle between IJP and reference RR sensors, their respiratory signals fluctuated in opposite directions with a phase shift, as shown in Figure 1c.

2.1.3. Measurement Procedure

During the measurement, the IJP sensor was mounted at the xiphoid process using adjustable fabric belt with e-Health sensor mounted at the nostril as shown in Figure 2. The IJP sensor was tested at five different body postures in randomized order: sitting at 90°, 45° Fowler's position, standing, supine, and right lateral recumbent, as shown in Figure 3. In the lateral recumbent, Flower's position and supine, the test subjects had their heads resting on a pillow. Only the right lateral recumbent was included without the left one, because the bilateral respiratory movements have been proved to be similar [25]. At each posture, the test was repeated twice in order to investigate the repeatability. The RR recording lasted for one minute in each trial, with a 60 s break between the two trials. The test subjects were asked not to move during the test and to breathe normally.

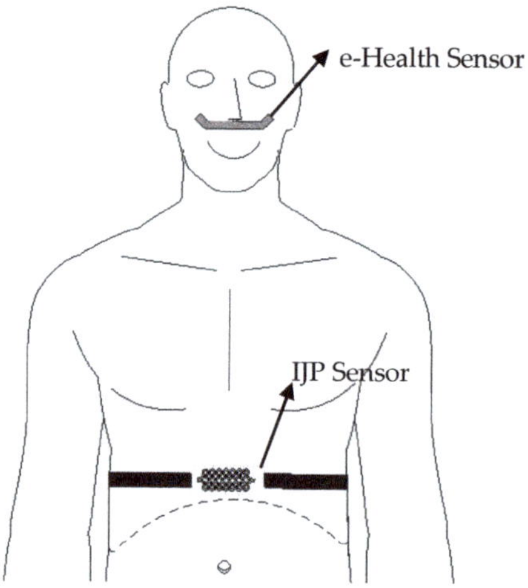

Figure 2. Mounting the IJP sensor at the xiphoid process using adjustable belt and the e-Health sensor at nostril.

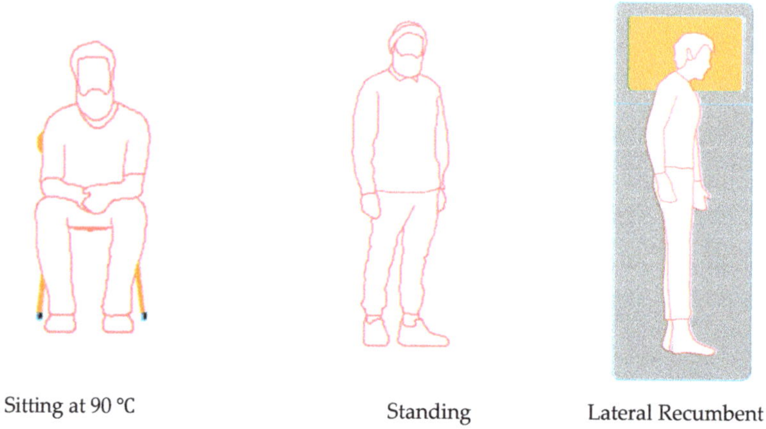

Figure 3. *Cont.*

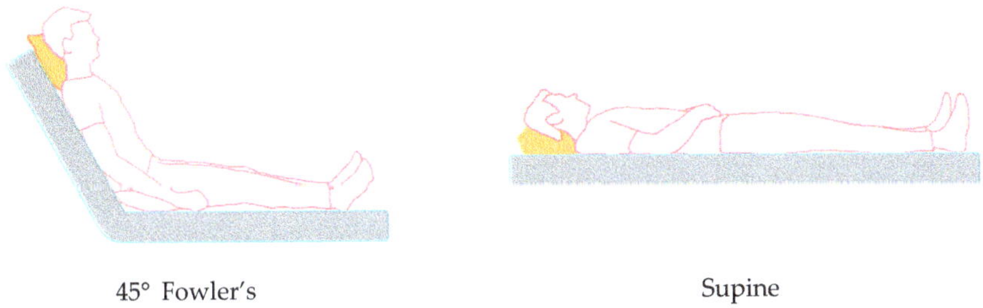

Figure 3. The RR measurement at different body postures.

2.2. Respiratory Rate Derivation

An algorithm was developed on MATLAB (R2018b, MathWorks Inc., Natick, MA, USA) to extract the RR values from the respiratory signals. Firstly, raw data were input into the bridge circuit at a sampling frequency of 100 Hz. The signal was then filtered to remove the direct current (DC) component and some extremely high and low frequencies using a band-pass filter with lower and higher cut-off frequencies of 0.05 Hz and 1.5 Hz. Finally, fast Fourier transform (FFT) was used to determine the respiration frequency, which is the frequency with the highest amplitude. RR was derived as: RR = 60 × respiratory frequency. Figure 4 shows the derivation procedure of the RR from the IJP sensor. Note in Figure 4 that the sharp peak is considered as noise since it exceeds the amplitudes of adjacent respiratory cycles with time period of less than 0.1 s, which is impossible for a respiratory cycle even for maximum inhalation or exhalation [11].

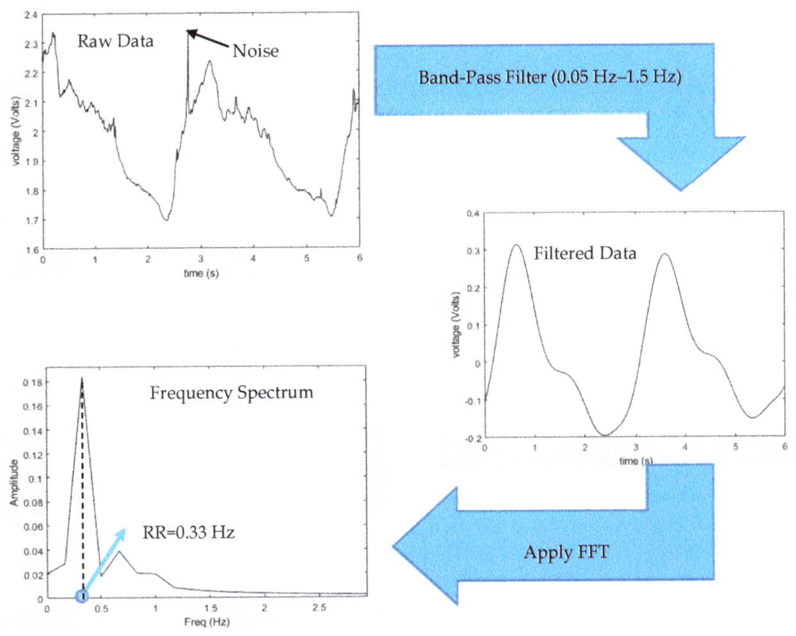

Figure 4. RR derivation flow chart from the IJP sensor. The raw data was input into band-pass filter to filter out the high frequencies and the noise. The RR was then extracted by Simulink (R2018b, MathWorks Inc., Natick, MA, USA). The arrow in the first subfigure shows the spike caused by noise.

2.3. Statistical Analysis

The statistical analysis was performed on SPSS (Version 20, SPSS Inc., Chicago, IL, USA). Firstly, to investigate if the RR measurement was significantly influenced by body postures, repeated trials, and different sensors, analysis of variance (ANOVA) was performed with Levene's test to validate the homogeneity of variance. Significance was defined as p-value less than 0.05. To further investigate the effect of body posture on the measurement accuracy, the difference between RR values derived by IJP sensor and reference sensor was calculated for each measurement in different postures. Moreover, post hoc test based on Tukey's test was used to evaluate the statistical significance of the RR between any two postures to figure out the effect of the posture on the RR.

3. Results

As shown in Figure 5, in each trial and posture, IJP and reference RR sensors derived nearly identical distributions of RR. The details are listed in Table A1, Appendix A. The Levene's test showed that the dataset satisfied the homogeneity of variance ($p > 0.05$). Hence, ANOVA was performed on RR values. The results showed no significant differences in measured RR between IJP and reference sensors, between two measurement trials, and among different body postures (all $p > 0.05$). Moreover, post hoc test showed no significant difference in the RR between any two posture (all $p > 0.05$).

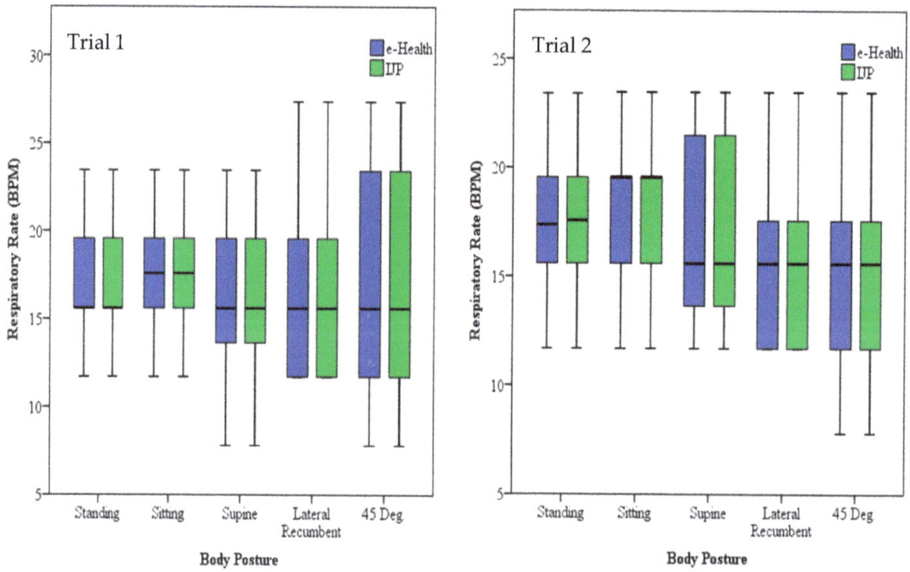

Figure 5. Box and whisker plots of the RR measured by the IJP sensor and the e-Health sensor for each trial at different body postures. The means are represented by the horizontal lines within the boxes while the first and third quartiles are represented by the boxes boundaries whereas the maximum and minimum values are represented by the whiskers.

4. Discussion

4.1. IJP Sensor Accuracy: Comparison with Other Sensors

The ANOVA results showed that IJP sensor had high accuracy, good repeatability, and stability at different body postures. With no observable deviation from the reference sensor at all body postures, the accuracy of the IJP sensor was better than many sensors reported in the literature where the RR measured by stretchable sensor deviated from the reference value from 1.7% to more than 6% in different postures [4,12,26–30].

4.2. RR Measurements at Different Body Postures

The body posture could significantly influence the lung volume [12,31–33] and the consequent RR detection. The lung has the largest volume capacity at the sitting posture [31]. In lateral recumbent posture, lung and chest wall compliances decrease, which limits the amplitude of thoracic respiration movement [25]. Small respiratory movement amplitude could result in miscounting of RR. In stretchable RR sensors, due to the difference in strain between different postures, the accuracy of RR could be different.

However, in this study, IJP sensor showed high accuracy at different postures. Firstly, the skin-like characteristics of the PDMS surface allow the IJP sensor to closely attach to the body surface and follow the topographical changes [34,35]. Secondly, the polymer-based strain gauge sensors are very sensitive (gauge factor > 100) compared with typical strain gauge (2 ≤ gauge factor ≤ 5). Thirdly, silver nanoparticles are highly sensitive which increased the accuracy of the IJP sensor [36]. Finally, during the measurement, test subjects did not move, talk or laugh, which largely reduced the movement artefact in respiratory signals.

4.3. Application of the IJP Sensor

In addition to its accuracy, the IJP sensor is a low-cost sensor with skin-like characteristics that enhances the wearability of the sensor without limiting the daily activities of the patients. All the aforementioned characteristics of the IJP sensor makes it suitable for remote and continuous RR monitoring in low-resources settings and refugee camps, decreasing the pressure on the healthcare facilities in these settings which suffer from limited healthcare resources [37]. Continuous RR measurement in different postures is clinically significant especially for the monitoring of hospitalized patients. Currently, the RR monitoring relies on inaccurate manual counting in a short period (often 30 s or 1 min) or is neglected [38]. The mask-based or contact RR sensors, such as the e-Health sensor in this study, could cause uncomfortableness which makes them inappropriate for long term continuous RR monitoring. The IJP sensor could accurately and continuously measure RR without any skin allergy or other clinical events, which makes it possible for clinical use. RR measurement at different body postures is important for patients with different physiological conditions. For example, some patients under physiotherapy or with plastic cast need to keep in mainly supine or lateral recumbent postures. Therefore, the IJP sensor provides the possibility of continuous RR monitoring in different patients.

The IJP RR sensor is also applicable for RR monitoring at home without disturbing the users' daily activities. Wearable sensors could provide RR monitoring during the night at the supine and lateral position to inspect the sleep quality and contribute in early diagnostic of several sleep-related issues such as sleep apnea and insomnia [39]. Moreover, such wearable sensors could be used to reduce sudden infant death syndrome by providing comfortable and continuous RR monitoring [40]. Additionally, wearable sensors can be integrated into safety belts in automobiles which eases the monitoring of the RR during driving [11,41]. The IJP sensor accurately measured the RR at all the inspected postures which paved the way for its further application in daily life.

4.4. Limitations and Future Work

The evaluation of the sensor was performed on 15 healthy male subjects only. Firstly, scale-up validation was needed. Secondly, to eliminate the error caused by movement artefact, the IJP sensor was firmly fixed on the xiphoid process. In different body positions, the sensitivity to posture change might be different, which needs further investigation. Thirdly, the evaluation of wider age groups and genders could be performed to investigate if the IJP sensor is suitable for people with different physiological conditions. Furthermore, the study was performed on healthy subjects. The performance of the sensor on subjects with pathological conditions needs further investigation. The IJP sensor is a low-cost and wearable sensor that provides the possibility of daily RR monitoring. However, motion artefact is a major concern regarding the accuracy of RR monitoring during daily activities [11]. For the further application especially in daily healthcare monitoring, full validation on different levels of motion artefact is needed. Finally, more comfortable mounting mechanism such as the use of biocompatible and skin-friendly adhesive should be investigated.

5. Conclusions

The IJP sensor was able to measure the RR accurately at the five postures (standing, sitting at 90°, Flower's position at 45°, supine and right lateral recumbent), with good repeatability which makes it user-friendly, and very convenient for clinical practices with minimal user efforts and training needs. This IJP sensor is promising for continuous RR monitoring in different clinical and daily situations.

Author Contributions: Conceptualization, A.A.-H., L.A.-G., S.E.B., H.L., and D.Z.; methodology, L.A.-G., H.L., and D.Z.; software, S.E.B..; validation, S.E.B., L.A.-G. and F.H.; formal analysis, L.A.-G., S.E.B., H.L.; investigation, L.A.-G.; resources, A.A.-H., D.Z.; data curation, S.E.B. and F.H.; writing—original draft preparation, L.A.-G., S.E.B.; writing—review and editing, L.A.-G., H.L., A.A.-H.; visualization, L.A.-G.; supervision L.A.-G., H.L., D.Z. and A.A.-H.; project administration, A.A.-H. and L.A.-G.; funding acquisition, A.A.-H. and D.Z. All authors have read and agreed to the published version of the manuscript.

Funding: This research was funded by the Royal Academy of Engineering under grant Ref. IAPP1R2\100204.

Acknowledgments: The authors would like to thank the Royal Academy of Engineering for Funding this project under grant Ref. IAPP1R2\100204. Moreover, the authors would like to thank the Industrial Research and Development Fund in the Higher Council for Science and Technology for their support of the project. In addition, the authors would like to thank Eng. Ali Al-Ghussein from the GJU Innovation and Entrepreneurship laboratory for providing the e-Health sensor.

Conflicts of Interest: The authors declare no conflict of interest.

Appendix A

Table A1. Average RR (breaths per minute, BPM) measured by the IJP and e-Health sensors at five different body postures.

Gender	Age	Standing		Sitting		45 Deg.		Supine		Lateral	
		Our	Ref	Our	Ref	Our	Ref	Our	Ref	Our	Ref
M	24	11.7	11.7	13.65	13.65	11.7	11.7	15.6	15.6	11.7	11.7
M	23	21.51	21.51	21.51	21.51	17.58	17.58	19.56	19.56	23.46	23.46
M	25	21.48	21.48	21.51	21.51	23.46	23.46	21.51	21.51	19.56	19.56
M	25	19.56	19.56	19.56	19.56	17.58	17.58	15.63	15.63	15.6	15.6
M	23	17.58	17.58	21.51	21.51	13.65	13.65	21.51	21.51	17.58	17.58
M	24	15.6	15.6	13.65	13.65	11.7	11.7	13.65	13.65	11.7	11.7
M	24	13.65	13.65	13.65	13.65	9.75	9.75	9.75	9.75	11.7	11.7
M	23	23.4	23.4	23.4	23.4	23.4	23.4	23.4	23.4	23.4	23.4
M	26	15.6	15.6	17.58	17.58	15.6	15.6	15.6	15.6	17.58	17.58
M	21	19.56	19.35	19.56	19.56	21.51	21.51	19.56	19.56	19.56	19.56
M	25	17.58	17.58	19.56	19.56	19.53	19.53	17.58	17.58	15.6	15.6
M	24	23.46	23.46	29.31	29.31	23.46	23.46	23.46	23.46	19.53	19.53
M	24	7.8	7.8	9.75	9.75	9.75	9.75	11.7	11.7	11.7	11.7
M	22	15.6	15.6	15.6	15.6	11.7	11.7	9.75	9.75	11.7	11.7
M	24	17.58	17.58	17.58	17.58	15.6	15.6	15.6	15.6	13.65	13.65

References

1. AL-Khalidi, F.Q.; Saatchi, R.; Burke, D.; Elphick, H.; Tan, S. Respiration rate monitoring methods: A review. *Pediatr. Pulmonol.* **2011**, *46*, 523–529. [CrossRef] [PubMed]
2. Lee, P.J. Clinical evaluation of a novel respiratory rate monitor. *J. Clin. Monit. Comput.* **2016**, *30*, 175–183. [CrossRef] [PubMed]

3. Huang, M.; Xu, W.; Liu, J.; Samy, L.; Vajid, A.; Alshurafa, N.; Sarrafzadeh, M. Inconspicuous on-bed respiratory rate monitoring. In Proceedings of the 6th International Conference on PErvasive Technologies Related to Assistive Environments, Rhodes, Greece, 29–31 May 2013; ACM Press: New York, NY, USA, 2013; pp. 1–8.
4. Nam, Y.; Kong, Y.; Reyes, B.; Reljin, N.; Chon, K.H. Monitoring of Heart and Breathing Rates Using Dual Cameras on a Smartphone. *PLoS ONE* **2016**, *11*, e0151013. [CrossRef]
5. Sharma, H.; Sharma, K.K. ECG-derived respiration using Hermite expansion. *Biomed. Signal Process. Control* **2018**, *39*, 312–326. [CrossRef]
6. Di Fiore, J.M. Neonatal cardiorespiratory monitoring techniques. *Semin. Neonatol.* **2004**, *9*, 195–203. [CrossRef] [PubMed]
7. Koch, E.; Dietzel, A. Stretchable sensor array for respiratory monitoring. In Proceedings of the 2017 19th International Conference on Solid-State Sensors, Actuators and Microsystems (TRANSDUCERS), Kaohsiung, Taiwan, 18–22 June 2017; pp. 2227–2230.
8. Jeong, J.W.; Jang, Y.W.; Lee, I.; Shin, S.; Kim, S. Wearable Respiratory Rate Monitoring using Piezo-resistive Fabric Sensor. In *World Congress on Medical Physics and Biomedical Engineering, 7–12 September 2009, Munich, Germany*; Dössel, O., Schlegel, W.C., Eds.; Springer: Berlin/Heidelberg, Germany, 2009; pp. 282–284. ISBN 978-3-642-03904-1.
9. Ciocchetti, M.; Massaroni, C.; Saccomandi, P.; Caponero, M.; Polimadei, A.; Formica, D.; Schena, E. Smart Textile Based on Fiber Bragg Grating Sensors for Respiratory Monitoring: Design and Preliminary Trials. *Biosensors* **2015**, *5*, 602–615. [CrossRef] [PubMed]
10. Chu, M.; Nguyen, T.; Pandey, V.; Zhou, Y.; Pham, H.N.; Bar-Yoseph, R.; Radom-Aizik, S.; Jain, R.; Cooper, D.M.; Khine, M. Respiration rate and volume measurements using wearable strain sensors. *NPJ Digit. Med.* **2019**, *2*, 8. [CrossRef]
11. Liu, H.; Allen, J.; Zheng, D.; Chen, F. Recent development of respiratory rate measurement technologies. *Physiol. Meas.* **2019**, *40*, 07TR01. [CrossRef]
12. Guan, H.; Yang, X.; Sun, W.; Ren, A.; Fan, D.; Zhao, N.; Guan, L.; Haider, D.; Abbasi, Q. Posture-Specific Breathing Detection. *Sensors* **2018**, *18*, 4443. [CrossRef]
13. Furtak, N.T.; Skrzetuska, E. Development of Screen-Printed Breathing Rate Sensors. *FIBRES Text. East. Eur.* **2013**, *6*, 84–88.
14. Zhu, Y.; Maniyeri, J.; Fook, V.F.S.; Zhang, H. Estimating respiratory rate from FBG optical sensors by using signal quality measurement. In Proceedings of the 2015 37th Annual International Conference of the IEEE Engineering in Medicine and Biology Society (EMBC), Milan, Italy, 25–29 August 2015; pp. 853–856.
15. Wu, D.; Wang, L.; Zhang, Y.; Huang, B.; Wang, B.; Lin, S.; Xu, X. A wearable respiration monitoring system based on digital respiratory inductive plethysmography. In Proceedings of the 2009 Annual International Conference of the IEEE Engineering in Medicine and Biology Society, Minneapolis, MN, USA, 3–6 September 2009; pp. 4844–4847.
16. Hesse, M.; Christ, P.; Hormann, T.; Ruckert, U. A respiration sensor for a chest-strap based wireless body sensor. In Proceedings of the IEEE SENSORS 2014 Proceedings, Valencia, Spain, 2–5 November 2014; pp. 490–493.
17. Sun, X.; Qiu, L.I.; Wu, Y.; Tang, Y.; Cao, G.; State, T.P. SleepMonitor: Monitoring Respiratory Rate and Body Position During Sleep Using Smartwatch. *PACM Interact. Mob. Wearable Ubiquitous Technol.* **2017**, *13*, 104. [CrossRef]
18. Al-Halhouli, A.; Qitouqa, H.; Alashqar, A.; Abu-Khalaf, J. Inkjet printing for the fabrication of flexible/stretchable wearable electronic devices and sensors. *Sens. Rev.* **2018**, *38*, 438–452. [CrossRef]
19. Abu-Khalaf, J.; Al-Ghussain, L.; Al-Halhouli, A. Fabrication of Stretchable Circuits on Polydimethylsiloxane (PDMS) Pre-Stretched Substrates by Inkjet Printing Silver Nanoparticles. *Materials* **2018**, *11*, 2377. [CrossRef] [PubMed]
20. Abu-khalaf, J.M.; Saraireh, R.; Eisa, S.M.; Al-halhouli, A. Experimental Characterization of Inkjet-Printed Stretchable Circuits for Wearable Sensor Applications. *Sensors* **2018**, *18*, 3476. [CrossRef] [PubMed]
21. Abu-khalaf, J.M.; Al-Ghussain, L.; Nadi, A.; Saraireh, R.; Rabayah, A.; Altarazi, S.; Al-Halhouli, A. Optimization of Geometry Parameters of Inkjet-Printed Silver Nanoparticle Traces on PDMS Substrates Using Response Surface Methodology. *Materials* **2019**, *12*, 3329. [CrossRef]

22. Al-Halhouli, A.; Al-Ghussain, L.; El Bouri, S.; Liu, H.; Zheng, D. Fabrication and Evaluation of a Novel Non-Invasive Stretchable and Wearable Respiratory Rate Sensor Based on Silver Nanoparticles Using Inkjet Printing Technology. *Polymers* **2019**, *11*, 1518. [CrossRef]
23. Bayo-Monton, J.L.; Martinez-Millana, A.; Han, W.; Fernandez-Llatas, C.; Sun, Y.; Traver, V. Wearable Sensors Integrated with Internet of Things for Advancing eHealth Care. *Sensors* **2018**, *18*, 1851. [CrossRef]
24. Rákay, R.; Višňovský, M.; Galajdová, A.; Šimšík, D. Testing Properties of E-health System Based on Arduino. *J. Autom. Control.* **2015**, *3*, 122–126.
25. Mezidi, M.; Guérin, C. Effects of patient positioning on respiratory mechanics in mechanically ventilated ICU patients. *Ann. Transl. Med.* **2018**, *6*, 384. [CrossRef]
26. Turnbull, H.; Kasereka, M.C.; Amirav, I.; Sahika, S.E.; Solomon, I.; Aldar, Y.; Hawkes, M.T. Development of a novel device for objective respiratory rate measurement in low-resource settings. *BMJ Innov.* **2018**, *4*, 185–191. [CrossRef]
27. Jarchi, D.; Salvi, D.; Tarassenko, L.; Clifton, D. Validation of Instantaneous Respiratory Rate Using Reflectance PPG from Different Body Positions. *Sensors* **2018**, *18*, 3705. [CrossRef] [PubMed]
28. Van Loon, K.; Breteler, M.J.M.; van Wolfwinkel, L.; Rheineck Leyssius, A.T.; Kossen, S.; Kalkman, C.J.; van Zaane, B.; Peelen, L.M. Wireless non-invasive continuous respiratory monitoring with FMCW radar: A clinical validation study. *J. Clin. Monit. Comput.* **2016**, *30*, 797–805. [CrossRef] [PubMed]
29. Guechi, Y.; Pichot, A.; Frasca, D.; Rayeh-Pelardy, F.; Lardeur, J.-Y.; Mimoz, O. Assessment of noninvasive acoustic respiration rate monitoring in patients admitted to an Emergency Department for drug or alcoholic poisoning. *J. Clin. Monit. Comput.* **2015**, *29*, 721–726. [CrossRef] [PubMed]
30. Shen, C.-L.; Huang, T.-H.; Hsu, P.-C.; Ko, Y.-C.; Chen, F.-L.; Wang, W.-C.; Kao, T.; Chan, C.-T. Respiratory Rate Estimation by Using ECG, Impedance, and Motion Sensing in Smart Clothing. *J. Med. Biol. Eng.* **2017**, *37*, 826–842. [CrossRef] [PubMed]
31. Moreno, F.; Lyons, H.A. Effect of body posture on lung volumes. *J. Appl. Physiol.* **2017**, *16*, 27–29. [CrossRef] [PubMed]
32. Leipälä, J.A.; Bhat, R.Y.; Rafferty, G.F.; Hannam, S.; Greenough, A. Effect of posture on respiratory function and drive in preterm infants prior to discharge. *Pediatr. Pulmonol.* **2003**, *36*, 295–300. [CrossRef] [PubMed]
33. Pal, A.K. Effect of Recumbent Body Positions on Dynamic Lung Function Parameters in Healthy Young Subjects. *J. Clin. Diagn. Res.* **2017**, *11*, CC08. [CrossRef]
34. Sun, J.; Jiang, J.; Bao, B.; Wang, S.; He, M.; Zhang, X.; Song, Y. Fabrication of Bendable Circuits on a Polydimethylsiloxane (PDMS) Surface by Inkjet Printing Semi-Wrapped Structures. *Materials* **2016**, *9*, 253. [CrossRef]
35. Nag, A.; Afasrimanesh, N.; Feng, S.; Mukhopadhyay, S.C. Strain induced graphite/PDMS sensors for biomedical applications. *Sensors Actuators A Phys.* **2018**, *271*, 257–269. [CrossRef]
36. Amjadi, M.; Kyung, K.-U.; Park, I.; Sitti, M. Stretchable, Skin-Mountable, and Wearable Strain Sensors and Their Potential Applications: A Review. *Adv. Funct. Mater.* **2016**, *26*, 1678–1698. [CrossRef]
37. Al-Rousan, T.; Schwabkey, Z.; Jirmanus, L.; Nelson, B.D. Health needs and priorities of syrian refugees in camps and urban settings in jordan: Perspectives of refugees and health care providers | Besoins et priorités sanitaires des réfugiés syriens dans les camps et en milieu urbain en jordanie: Perspectives de. *East. Mediterr. Health J.* **2018**, *24*, 243–253. [CrossRef] [PubMed]
38. Kelly, C. Respiratory rate 1: Why measurement and recording are crucial. *Nurs. Times* **2018**, *114*, 23–24.
39. Kwak, Y.H.; Kim, J.; Kim, K. Sleep monitoring sensor using flexible metal strain gauge. *Jpn. J. Appl. Phys.* **2018**, *57*, 05GD03. [CrossRef]
40. Mathew, J.; Semenova, Y.; Farrell, G. A miniature optical breathing sensor. *Biomed. Opt. Express* **2012**, *3*, 3325. [CrossRef] [PubMed]
41. Hamdani, S.; Fernando, A. The Application of a Piezo-Resistive Cardiorespiratory Sensor System in an Automobile Safety Belt. *Sensors* **2015**, *15*, 7742–7753. [CrossRef]

applied sciences

MDPI

Article

An Automated Data Acquisition System for Pinch Grip Assessment Based on Fugl Meyer Protocol: A Feasibility Study

Abdallah Alsayed [1,*], Raja Kamil [1,2,*], Hafiz Ramli [1] and Azizan As'arry [3]

1 Department of Electrical and Electronic Engineering, Faculty of Engineering, Universiti Putra Malaysia, Serdang 43400, Selangor, Malaysia; hrhr@upm.edu.my
2 Laboratory of Computational Statistics and Operations Research, Institute for Mathematical Research, Universiti Putra Malaysia, Serdang 43400, Selangor, Malaysia
3 Department of Mechanical and Manufacturing Engineering, Faculty of Engineering, Universiti Putra Malaysia, Serdang 43400, Selangor, Malaysia; zizan@upm.edu.my
* Correspondence: eng.abdallah.2013@gmail.com (A.A.); kamil@upm.edu.my (R.K.); Tel.: +6-03-9769-4360 (R.K.)

Received: 27 February 2020; Accepted: 28 March 2020; Published: 15 May 2020

Featured Application: This work was conducted to develop an automated data acquisition system to monitor the pinch grip in rehabilitation and assessment applications after a stroke.

Abstract: The Upper Extremity Fugl Meyer Assessment (UE-FMA) is the most comprehensive assessment for pinch impairment after stroke. The pinch test of UE-FMA is manually performed by pulling a pincer object away from the patient's fingers while providing a visual observation that results in a subjective assessment. In this study, an automated data acquisition system that consists of a linear electric actuator applying automatic pulling to the customized pincer object held by the volunteer was developed. The pinch force was measured such that a strain gauge was placed on the pincer object while pulling force was measured using pulling force load cell connected in between the linear electric actuator and customized pincer object. The pincer object's slip onset was detected using a displacement slip sensor. The mean pinch and pulling force values at the slip onset were 12.17 and 6.25 N for right hands, while mean pinch and pulling force values were 11.67 and 5.92 N for left hands of 50 healthy volunteers, respectively. Based on the paired *t*-test, there is no significant difference between right and left hands. The automated data acquisition system can objectively apply a pulling force, detect the slip onset, and measure the pinch and pulling forces.

Keywords: automated assessment; UE-FMA; pinch force; pulling force; slip onset; strain gauge; stroke

1. Introduction

Globally, stroke is the first leading cause of long-term disability [1]. Many daily activities, for example, pinching an object, require dynamic manipulation of the thumb and index finger. Pinch impairment is defined as the inability of the thumb and index finger to produce strength with sufficient magnitude and directional control, leading to object slipping [2]. Several studies showed that the recovery of pinch grip takes a longer time than handgrip, and it requires long-term rehabilitation after stroke [3]. Besides, a recent study conducted by Katab et al. [4] showed that pinch strength can be used as an alternative assessment to hand strength for assessing muscle strength in patients. Thus, an effective evaluation of pinch grip is needed to determine the pinch deficits accurately and objectively [5]. Accurate and objective pinch assessment can be achieved via automation of traditional standard assessment systems based on available technology. Current technology involves pinch gauge and dynamometers to measure the maximum static pinch force which is not related to object manipulation involved in daily activities [6]. Therapists are still using the standard manual assessments to evaluate the pinch grip, which is subjective due to the different human interpretations of the procedures as well as the use of qualitative ordinal scoring. There is no single gold standard pinch assessment used in clinical and research practices [7]. Pinch evaluation is usually listed as part of upper limb standard assessments [8–11]. The UE-FMA is arguably the most comprehensive clinical tool for measuring pinch impairment after stroke [12,13]. The UE-FMA provides step-by-step procedures and protocols, which are performed by a UE-FMA therapist. The therapist tests the patient's ability to pinch a pincer object (in which a pencil or pen is customarily used) and exert enough force to stabilize the pincer object against gentle pull generated manually by the therapist, as in Figure 1 [14]. A score of 0, 1, or 2 is assigned, which represents severe impairment, partial impairment, and no impairment rating, respectively. Score 0 is given when the patient cannot execute the pinching at all. Score 1 indicates that the pincer object is not held firmly against the gentle pull such that slippage occurs, while score 2 is given when the pincer object is held firmly without slippage against a gentle pull.

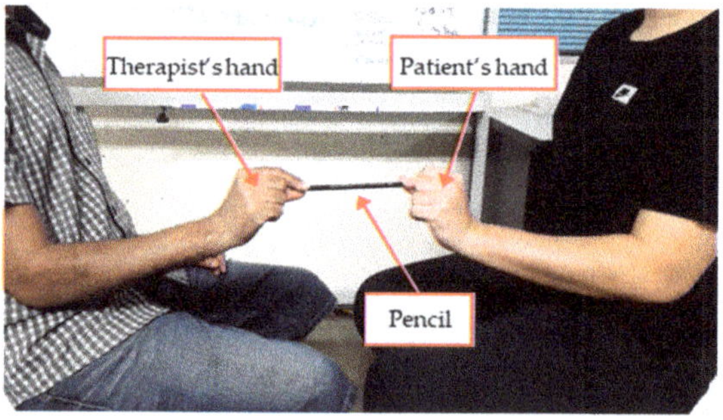

Figure 1. Pinch test as in the UE-FMA protocol.

The motivation of carrying out this study stemmed from the lack of studies related to a complete automated assessment of pinch impairment. In addition, the pinch assessment using UE-FMA is subjective due to the different interpretations of using the original guideline of UE-FMA [15]. These difference in interpretations may result in the following variations in clinical practice:

1. The orientation of the pincer object is not standardized. Some therapists pull the object horizontally, while others pull it vertically.
2. The posture of the patients' shoulder, elbow, forearm, and hand may differ between each clinic resulting in different pinch force exerted at different postures.
3. The amount of pulling force actuated by the therapist is subjective [16]. This opens the possibility for low intra-rater and inter-rater reliability of pinch evaluation.
4. The occurrence of slippage, which results in score of either 1 or 2 is visually observed and sensitive to the therapist experience, skills, and appreciation [17–23].

Due to the previous drawbacks of using the original protocol of UE-FMA assessment, several studies have been conducted to improve the original UE-FMA protocol. In addressing the orientation of the pincer object, Page et al. [24] suggested that the therapist should horizontally pull the pincer object outwards away from the patient's pinch grasp. Horizontal orientation choice is supported by other studies [25–27], which report that the brain response leading to pinch has higher latency responding to the pulling being applied vertically compared to horizontally. Page et al. [24] also proposed the upper limb posture and concurred with an independent study by Sullivan et al. [28]. Both studies recommend that the patient should be sitting on a chair in a 90°. Then, the patient should maintain shoulder at 0°, elbow at 90°, and forearm in resting position. The therapist, if necessary, may support the patient's elbow at 90°. Sullivan has suggested that the therapist can use a bedside table to support the patient's arm and elbow. The use of bedside table is supported by other studies, which reported that the arm weight leads to abnormal coupling between shoulder abduction and elbow flexion due to abnormal muscle co-activation required to compensate the weight of the arm [29]. Therefore, Ellis and colleagues used arm weight support to reduce the activation of those muscles (e.g., biceps), which otherwise would be involved in pinching [30,31].

Recently, the research community [21,32–39] have been working on automating UE-FMA, but few of them involved the pinch grip. Otten and his colleagues [40,41] implemented a glove with built-in Force Sensing Resistor (FSR) sensors as the data acquisition system to measure the patient's pinch force during the evaluation process. Nevertheless, some post-stroke patients with severe to moderate impairment have muscle contracture and spasticity that lead to difficulties in wearing the glove. In later studies [42,43], the FSR sensor was attached to the pincer object such that the patient is no longer required to wear a glove. The pinch force threshold values that distinguish between slip (score 1) and non-slip (score 2) onsets are useful in determining the baseline measurements to evaluate the effectiveness of pinch rehabilitation for both right and left hands. In both studies, the discrimination between scores 1 and 2 was based on the ability of the subject to exert pinch force up to a threshold value. However, this threshold pinch force value was selected by the therapist which may open the possibility for uncertainty. Moreover, these studies still have not addressed the issues related to subjectivity of the manual pull and the limitations of the FSR sensors used in the data acquisition system.

Generally, FSR sensors (also known as a thin-film sensor) have been used for many applications [44] but they suffer from performance variation as well as low performance [45,46]. The variation in performance is attributed to the conductive material on the sensor layers being sensitive to temperature and deformation on the surface [47]. In addition, the variation in performance can be attributed to the inability to follow the standard calibration procedures resulting in different calibration results compared to those in the manufacturer's datasheet. On the other hand, low performance can be attributed to variation in voltage gain, high hysteresis, nonlinearity, and inaccurate measurements in real-time pinching [48]. Likitlersuang et al. [49] investigated the performance of the thin-film sensor when attached directly to human skin. The results show a large measurement error of 23% using the standard calibration techniques. In addition, the pressure distribution on the sensing area may not be uniform during slippage due to different pinching contact location.

The main objective was to develop an automated data acquisition system for pinch assessment based on the standard UE-FMA protocol. In this study, the following research questions were addressed: (1) Does the slip onset occur at the maximum pinch force? (2) Is there any significant difference between right and left hands in pinch and pulling forces at the slip onset? In this study, the authors hypothesized that the maximum pinch force exerted by the volunteer against the automatic pulling would be located at the slip onset. The authors also hypothesized that the pinch and pulling force measurements of the right hands are significantly higher than pinch and pulling force measurements of left hands at the slip onset for right-handed volunteers.

To address the first question, an automated data acquisition system was developed. The automated system must be able to measure and acquire pinch and pulling forces as well as slip displacement accurately and automatically. The automated system uses linear electric actuator applying automatic pull, customized pinch force load cell measuring the pinch force, pulling force load cell measuring the pulling force, and Linear Variable Differential Transformer (LVDT) displacement sensor detecting the slip onset. The sensors and actuator were calibrated and then integrated together with Arduino® Due board. Subsequently, the data from 50 right-handed male volunteers within the age of 18–24 years were collected according to the UE-FMA protocol to establish the relationship between pulling and pinch forces in order to observe the slip onset. The paired *t*-test was performed on collected data to address the second question.

2. Materials and Methods

2.1. Pinch Data Acquisition System

The experimental setup of the data acquisition system is described in Figure 2. It consists of:

1. Displacement sensor: LVDT sensor (1) with Low Pass Filter (9).
2. Linear actuator system: including linear electric actuator (2) and servo motor driver (8).
3. Customized Pinch force load cell: pincer object (4), Wheatstone bridge (5), and an amplifier (6).
4. Pulling force load cell: load cell (3) and an amplifier (11).
5. Data acquisition card: Arduino® Due board (Arduino LLC, Torino, Italy) (11) and Arduino® IDE 1.8.5 software (arduino.cc) [50].
6. DC power supply (7).

In our previous work [51], a customized pinch force load cell based on strain gauge was designed, fabricated, and calibrated. The customized pinch force load cell is composed of three parts: pincer object, strain gauge, and signal conditioning circuit. The pincer object was fabricated to mimic a pencil, as used in UE-FMA, which represents a cylindrical object with a diameter of 12 mm and 150 mm length. In order for a strain gauge to measure the bending at the pinching location near the free edge of the pincer object, a 4 mm slot cut was created, as shown in Figure 3. The slot cut starts from the free end to the middle of the pincer object, similar to the shape of a tuning fork. The pinching location is within 20 mm from the free edge and is adequate to cover the fingertip length. A 350-ohm strain gauge (SGK-L1-D-K350P-PC11-E) was placed at the maximum stress point. The maximum stress point was determined based on the stress analysis performed using ANSYS® finite element software (Version 15.0, ANSYS Inc, Canonsburg, PA, USA) as shown in Figure 4. The conditioning circuit consists of a quarter Wheatstone bridge and an amplifier. The quarter Wheatstone bridge was implemented to convert the strain gauge resistance change to voltage change when pinching occurs. However, the voltage change was too small. Thus, a commercial amplifier (Wachendorff Strain Gauge Converter®, SENECA s.r.l., Padova, Italy) with a gain of 1000 was used.

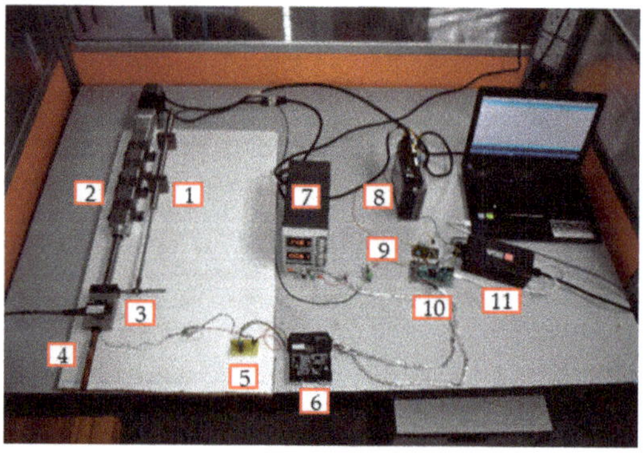

Figure 2. Experimental Setup of data acquisition system for pinch assessment.

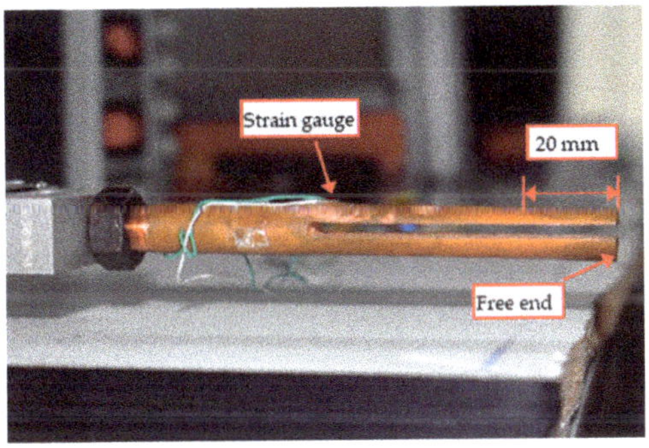

Figure 3. Pinch force load cell.

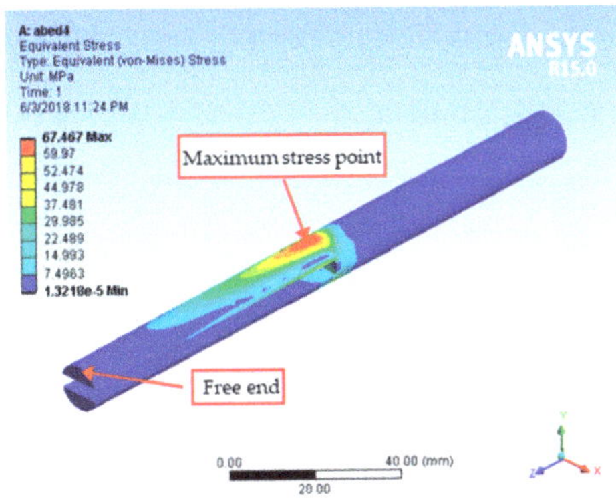

Figure 4. Stress analysis of the pincer object where an external force is applied at the free end.

The standard strain gauge calibration of force–voltage is usually performed for one contact location only. However, in the pinch test of UE-FMA, the contact location may vary due to the different initial pinching location of each volunteer. Furthermore, the contact location may also vary during the automatic pull, which causes slipping. Therefore, a continuous relationship of force–voltage for all possible contact points along the pinching length is required. This can be achieved by performing a multipoint calibration at five contact points away from the free edge (4.1, 5.8, 11.5, 16.5, and 20.2 mm). The multipoint calibration was performed using a mechanical testing system (INSTRON® 3366, Illinois Tool Works Inc, Norwood, MA, USA) by applying dynamic loading (0–50 N), as shown in Figure 5. The pincer object was held stationary on the INSTRON® 3366 base during calibration by applying glue on both sides of the pincer object to prevent any movement, as shown in Figure 5. As a result, five force–voltage curves were generated, as depicted in Figure 6a. The voltage and force data were recorded from the customized pinch force load cell and the INSTRON® 3366 machine. Subsequently, the continuous relationship of force as a function of voltage and position was modeled using linear regression function, as depicted in Figure 6b. The linear correlation coefficient R^2 of 0.9896 indicates high linearity. Using the linear plane model equation (Equation (1)), the force measurements can be estimated at any contact point within pinching length given the measurements of voltage and position, which were collected from the pinch force load cell and the LVDT sensor, respectively.

$$\text{Force} = -0.9447 + 0.02371 \times \text{Voltage} + 0.3632 \times \text{Position}, \quad (1)$$

In performing the hysteresis test to the customized pinch force load cell, dynamic loading (0–50 N) and unloading (from 50 to 0 N) were applied at the contact point 11.5 mm, which is approximately at the midpoint of the pinching length. Figure A1 (Appendix A) shows the loading and unloading curves of voltage and force measurements recorded from the pinch force load cell and INSTRON® 3366 machine, respectively. The solid line is the force–voltage relationship for dynamic loading, while the dotted line is the force–voltage relationship for dynamic unloading. The hysteresis is defined as the difference in voltage offset between loading and unloading lines at the midpoint on the force axis. The following formula gives the percentage of hysteresis:

$$\text{Hysteresis} = 100\% \times (V_{mu} - V_{ml})/(V_{max} - V_{min}) = 0.287\%. \quad (2)$$

where V_{mu} (892.46 mV) is the voltage value on the unloading line in Figure A1 at midpoint force F_m (23.86 N), V_{ml} (898.1 mV) is the voltage value on the loading line at F_m, V_{max} (1960.86 mV) is the maximum voltage value on the loading line, and V_{min} (0 mV) is the minimum voltage value on the unloading line. The hysteresis of 0.287% can be considered negligible [52]. Furthermore, it is much lower than the hysteresis of thin-film pressure sensors (5%) [53]. Similar hysteresis tests were performed at the other contact points resulting in similarly negligible hysteresis.

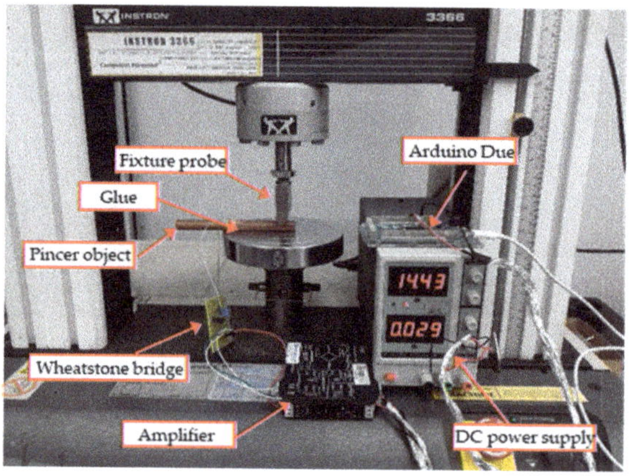

Figure 5. Calibration setup of the pinch force load cell.

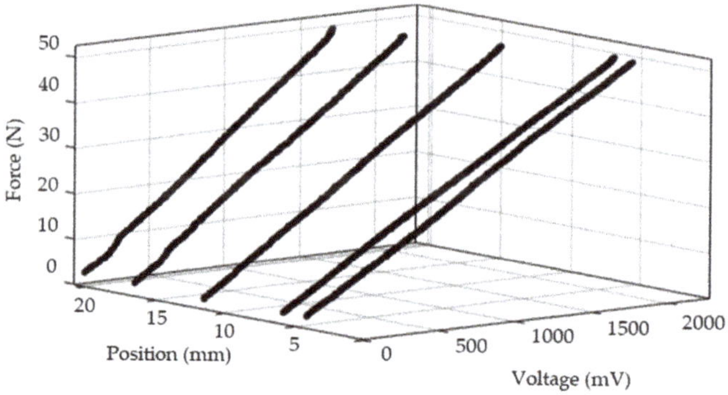

(**a**)

Figure 6. *Cont.*

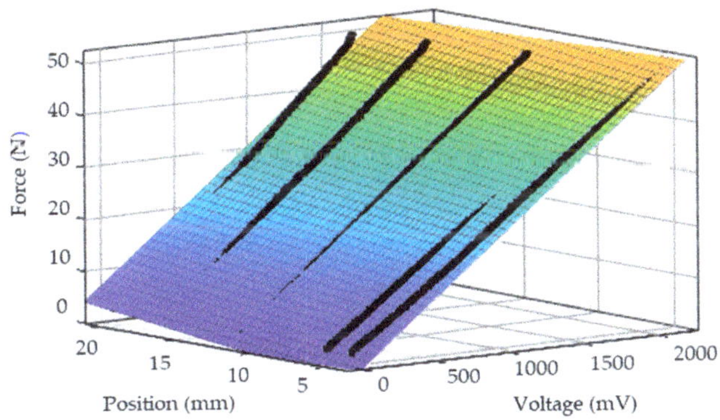

(b)

Figure 6. (a) Force–voltage curves at five contact points; and (b) estimated plane of force as a function of voltage and position.

In performing the repeatability test, a dynamic loading (0–50 N) was applied three times at the contact point 11.5 mm. The Coefficient of Variation (CV) was used to calculate the repeatability. CV is given by the ratio between standard deviation and mean values of three repeated voltage measurements at each force point ranging from F_{min} (2.28 N) to F_{max} (50 N). Then, all CVs are averaged to provide a single CV [49,54]. The "Cfvar" function in SPSS Statistics® V21 was used to calculate the CV. The CV value of 1.43 ± 0.447% was obtained which indicates high repeatability [55]. In addition, it is smaller than the CV value of thin-film pressure sensors (2.1 ± 2.3%) [49].

In replacing the manual gentle pull by the therapist, a linear electric system was used. The linear actuator system consists of a linear electric actuator, AC servo motor, AC servo driver, and Arduino® Due. To run the linear electric actuator, an AC servo motor (SM0602—400 Watt) was coupled in-line to the drive cap of the linear electric actuator. The output analog signal of Arduino® Due controls the AC servo motor through the AC servo driver (M2DV-3D02R). The linear actuator system can exert automatic pulling force ranging from 0 to 63.25 N with a resolution of 0.5 N.

A pulling force is produced when the volunteer pinches the pincer object and then resists the automatic pull generated by the linear actuator system. An off-the-shelf pulling force load cell (STA-1-50 Aluminium S Type Tension) is located between the pincer object and the linear electric actuator to measure the pulling force. The voltage–force relationship of the pulling force load cell was established by the manufacturer, as shown in Figure 7. The pulling force load cell has 0.03% accuracy, 0.01% repeatability, 0.05 N resolution, and $R^2 = 1$ linearity.

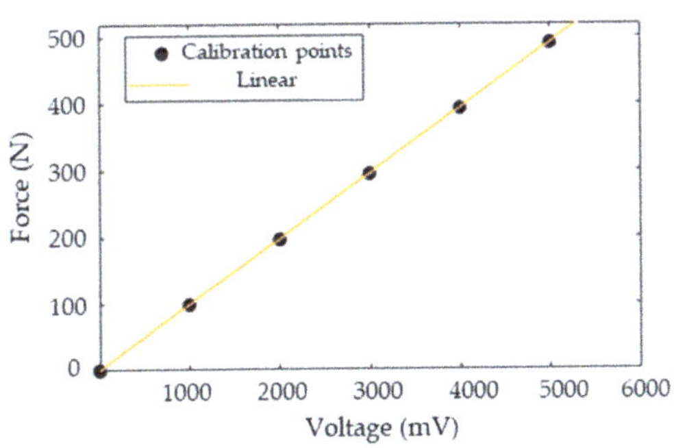

Figure 7. Voltage–force relationship for pulling force load cell.

In detecting the slip onset, the LVDT displacement sensor was used. The LVDT sensor and linear electric actuator were clamped together such that they move in parallel. The output signal of the LVDT sensor was filtered with an analog Low Pass Filter (LPF) of 20 Hz cut off frequency to remove the high-frequency noise. Since the pinching length is only 20 mm, the LVDT sensor was calibrated at 0-, 5-, 10-, 15-, and 20-mm displacement points using a standard digital caliper with 0.05 mm resolution. The calibration process was repeated three times to perform the repeatability test, as mentioned previously for the pinch force load cell. The three voltage–displacement relationships show high linearity of $R^2 = 0.9999$, as shown in Figure 8. The repeatability of the LVDT sensor is 9.57×10^{-5}%, which indicates high repeatability. The analog signals of the customized pinch force load cell, pulling force load cell, and LVDT sensors were converted to digital inside the Arduino® Due and then recorded in a PC installed with Arduino® IDE software. The diagram that illustrates the data acquisition process is shown in Figure 9.

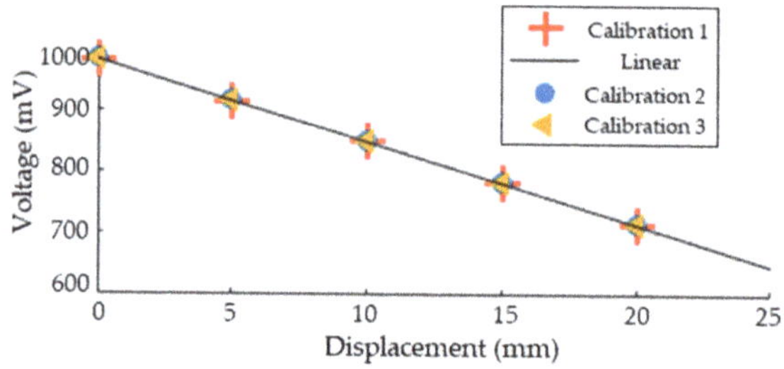

Figure 8. Displacement–voltage relationships of three times calibration.

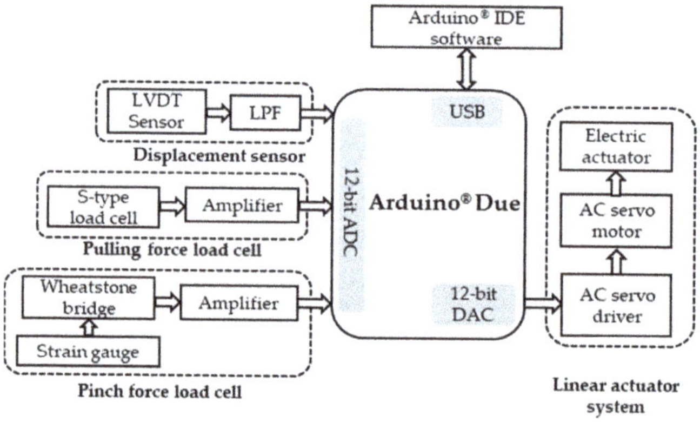

Figure 9. Block diagram of the data acquisition process.

2.2. *Volunteers Recruitment and Experimental Protocol*

Fifty right-handed healthy male students participated. All volunteers had no prior history of upper limb injury and passed the health status questionnaire (SF-36). To avoid bias in the force measurements, the study was conducted on a single-gender because males and females have different pinch strength [56]. Furthermore, the study excluded construction workers and professional sportsmen as they have stronger pinch force than the average male population [57]. The mean age, height, and weight for volunteers were 21.86 ± 1.41 years, 170.41 ± 7.03 cm, and 69.49 ± 17.9 kg, respectively. Ethical approval (Reference number: JKEUPM-2017-248) for this study was obtained from The Ethics Committee for Research Involving Human Subjects of Universiti Putra Malaysia.

In this study, the improved wrist/hand UE-FMA protocol by Page et al. [24] was adopted. The volunteers were instructed to sit at 90° on a chair, shoulder in the neutral position, elbow at 90°, and forearm in resting position. A bedside table was positioned to support the volunteer's arm and elbow, as shown in Figure 10a. The volunteers were also instructed to clean and dry their hands before commencing the experiment to avoid any influence of moisture through sweat. Then, they were instructed to hold the pincer object comfortably using their thumb and index fingertips, as in Figure 10b. Once the linear actuator started to pull the pincer object, the volunteers attempted to resist this pulling as much as possible. Since no visual feedback was provided, the volunteer had to react and adjust accordingly to the perceived amount of pulling force. To achieve pinch stability, the pinch force had to be adjusted in proportion to the changes in the pulling force. The linear actuator stopped exerting pulling force the moment the pincer object had totally slipped away from the volunteer's fingers, and this was indicated by a 20-mm displacement slip. Figure A2 (Appendix A) demonstrates the flowchart to execute the experiment. The data collection involved both right and left hands. The duration of one trial began when the linear actuator received the control signal from the Arduino® Due, and ended the instant the pincer object slipped totally away. The analysis of the recorded data showed the pinch and pulling force values at the slip onset. The slip onset was defined as the moment when the pincer object started to shift with respect to the starting position, indicating that the object slipped away from the volunteer's fingers. During the trial, one set of measurements, which included the pinch force, pulling force, and displacement, were recorded, resulting in 300 datasets (50 volunteers × 2 hands × 3 trials per hand). To reduce the effect of the physical and mental fatigue, an adequate resting period of 3 min was given between each trial. Cronbach's alpha (α) was used to measure the reliability among the three trials of the hand, which as computed using the following formula [58]:

$$\alpha = \frac{N.\bar{c}}{v + (N-1).\bar{c}} \quad (3)$$

where N is the number of trials, $\bar{c}$ is the average inter-item covariance among the trials, and v is the average variance.

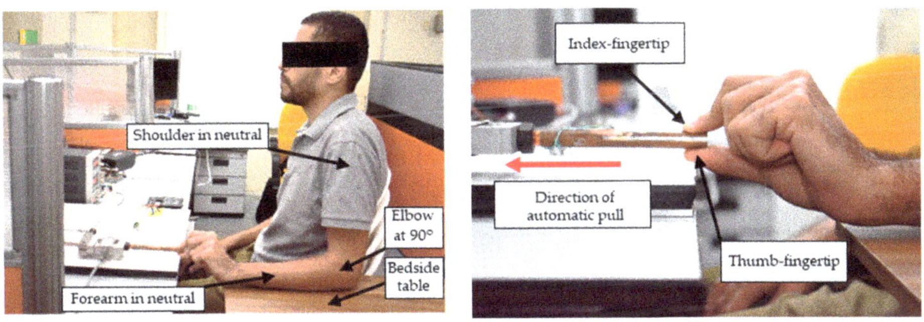

(a) (b)

Figure 10. (**a**) Body and upper limb positioning during the experiment; and (**b**) thumb and index finger opposed onto the pincer object. Note that the volunteer should resist the automatic pull applied by the linear actuator system.

3. Results

Figure 11a shows data recorded from one trial of one volunteer. The pinch force, pulling force, and displacement signals were recorded from the customized pinch force load cell, pulling force load cell, and the LVDT sensor, respectively. The linear electric actuator started to pull the pincer object, as indicated by a sudden increase in the pinch and pulling force signals at 3 s. The volunteer exerted pinch force increased as the pulling force increased gradually. From this point, the pincer object totally slipped away from volunteer's fingers, as indicated by a sudden increase in displacement measurements at 20 s. For each trial, it was important to extract the pinch and pulling force values at the slip onset. The slip onset is the point that differentiates between slip and non-slip where the pincer object begins to slip and is the first instance where the displacement started to increase from a relatively flat level. In Figure 11a, it is difficult to determine the slip onset visually due to the small magnitude of the increase in the displacement absolute scale (mm). In addition, the increase in displacement is nonlinear, thus a nonlinear scaling is required. Subsequently, the nonlinear decibel scale (dB) was used such that the logarithm base 10 of the displacement absolute value was multiplied by 20 to provide the necessary scaling. Upon that, the slip onset was precisely determined at 12 s, as shown in Figure 11b. The pinch and pulling force values at the slip onset (12 s) were 15.39 and 7.05 N, respectively.

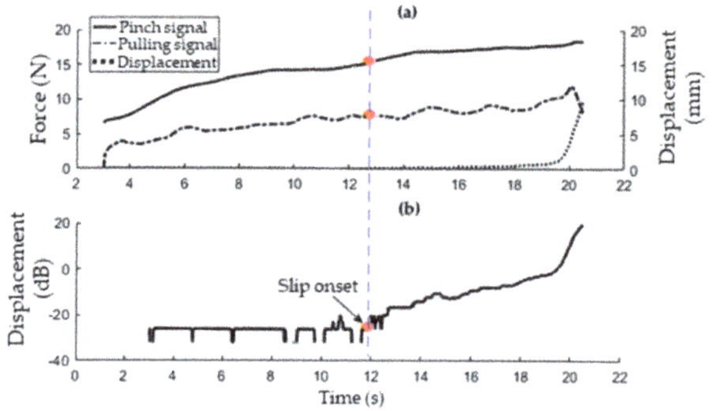

Figure 11. (**a**) One trial recorded from one volunteer; and (**b**) displacement in decibel scale.

Since the static coefficient of friction of copper alloy is constant, it is expected that the relationship between pinch and pulling force prior to slip onset is linear. Figure 12 shows the pinch–pulling force relationship before and after the slip onset. It was observed that the volunteer initially exerted excessive resistance (indicated by a sudden increase in pulling force), because he had to react and adjust the pinch force according to the perceived automatic pull generated by the linear actuator. There was also some nonlinearity observed at 12 and 14.5 N pinch forces that were probably due to the absence of visual feedback as well as the continuous increase in the amount of automatic pull. Prior to the slip onset, the pinch–force relationship had high linearity of $R^2 = 0.8971$, while the linearity decreased to $R^2 = 0.5845$ after the slip onset. This decrease in linearity as due to the inability of the volunteer to exert pinch force according to the continuous increasing of pulling force, which was indicated by the continuous slipping. The COF was computed as the ratio of pulling force to the pinch force at the slip onset, which was equal to 0.45 (7.05/15.39).

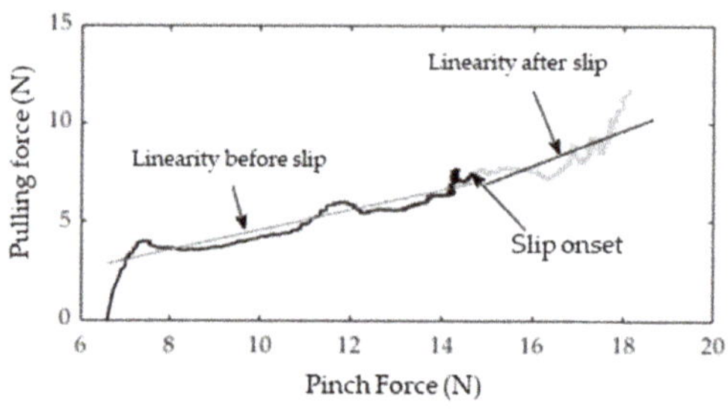

Figure 12. Pinch–pulling force relationship.

For each hand, three trials were recorded, and then the pinch and pulling force values were extracted at the slip onsets using the same criteria involved in the above trial. Figure A3a (Appendix A) shows the three trials collected from the right hand of a volunteer. It can be observed that the pinch force signals were close to each other during the three trials, and they had excellent reliability (Cronbach's alpha = 0.989). This was also observed for pulling force signals (Cronbach's alpha = 0.969). Based on the displacement measurements after being converted to decibels, the slip onsets were at 17, 14, and 12 s for Trials 1–3, respectively, as shown in Figure A3b. The pinch and pulling force values at the slip onsets as well as the static COFs are summarized in Table 1. The interpretation is that this volunteer could exert an average pinch force of 15.93 N against the maximum automatic pull of 7.65 N without slipping.

Table 1. Measurements recorded from the right hand of a volunteer.

Variables	Trial 1	Trial 2	Trial 3	Average
Pinch force (N)	14.61	17.81	15.39	15.93
Pulling force (N)	7.29	8.61	7.05	7.65
Static COF	0.498	0.483	0.458	0.48

For each hand of the 50 volunteers, the pinch and pulling force values as well as the static COFs of the three trials were averaged, resulting in three averaged values (as demonstrated in Table 1). Figure 13 shows the distribution of averaged values collected from the right and left hands of the 50 volunteers. It is clear that the distribution has an approximately normal bell-shape and no skewness exists. The mean, standard deviation, and range of pinch and pulling force values as well as static COF of right and left hands are summarized in Table 2. It was found that the mean pinch and pulling force values of right hands were slightly higher than those of lefts hands by 4.1% and 5.28%, respectively. However, there were no significant differences between right and left hands based on the paired *t*-test at the confidence level of $p = 0.05$. The paired *t*-test shows a *p*-value of 0.298 and 0.275 for pinch and pulling force, respectively. The average time duration for each hand was around 7 min, which included three trials (around 20 s for each trial depending on the amount of time for the pincer object to slip away from the fingers) with two resting periods in between (3 min for each resting period). Considering each volunteer performed the same experiment with both hands, a total time duration of around 14 min was involved.

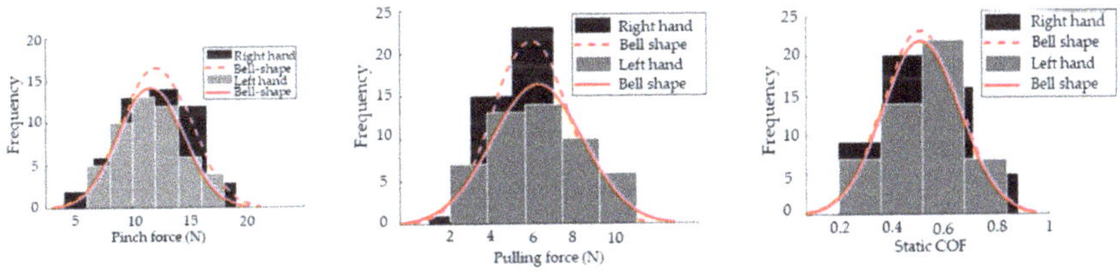

(a) (b) (c)

Figure 13. Distribution of: (**a**) pinch force; (**b**) pulling force; and (**c**) static COF for 50 volunteers.

Table 2. Summary of pinch force, pulling force, and static COF values at slip onset for 50 volunteers.

Variable		Mean	Standard Deviation	Range
Pinch force (N)	Right hand	12.17	3.02	5.36–18.48
	Left hand	11.67	2.82	6.79–17.67
Pulling force (N)	Right hand	6.25	2.19	2.37–10.77
	Left hand	5.92	1.86	1.88–10.67
Static COF	Right hand	0.518	0.146	0.27–0.85
	Left hand	0.517	0.145	0.23–0.81

4. Discussion

In 274 trials (91.3% of total trials), the volunteers could adjust the pinch force in proportion to the changes in pulling force. However, the remaining trials were excluded prior to analysis due to improper adjustment of pinch force to the pulling force. This is considered as human errors such that the volunteers might not follow the experiment's procedures and instructions probably due to delay in reaction, missing the instructions, or absence of visual feedback. According to the UE-FMA protocol, the subject should resist a gentle pull to avoid slip. This means that the pincer object slips when the subject is no longer able to exert enough pinch force. In this study, it was initially hypothesized that the slip onset was expected at the maximum pinch force just before the point where the pincer object totally slipped. Interestingly, it was found that the slip onset occurred much earlier and was subjected to a sub-maximal pinch force. This result opens up the possibility of using the slip onset as a measurement of pinch improvement such that the patient would develop a stronger pinch force over rehabilitation time at the slip onset. Furthermore, the continuous changes of slip onset would improve the responsiveness of the pinch assessment such that the scoring is not subjected only to score 1 and 2 as in the current UE-FMA scoring criteria.

The pinch–force relationship was found to be linear prior to the slip onset as the static COF between human skin and copper alloy was constant. However, there was some variability in pinch and pulling force, as indicated by peaks approximately at 7.5, 12, and 14.5 N, as depicted in Figure 12. The highest variability was observed at 7.5 N pinch force at the loading phase in which the linear electric actuator started to pull. This is consistent with previous studies stated that the subjects generally show stronger pinch forces at the loading phase [59]. Besides, the pinch force rate (N/s) reached the highest at the loading phase when the volunteer was subjected to unpredictable loading [27]. The study conducted by Takamuku et al. [60] showed that pinch force variability is lower when visual feedback is provided to the subject. In addition, the pinch force control is greatly influenced by the safety margin factor during the unpredictable loading [61]. The safety margin is an additional amount of pinch force applied to guard against slip, which is different from volunteer to volunteer [62]. The other two peaks at 12 and 14.5 N can be interpreted such that the volunteer should adapt his pinch force against the continuous automatic pull without visual feedback; hence, the safety margin is continuously changed to adapt the continuous increase in pulling force. Therefore, it is possible that the volunteer may unconsciously fail to adapt his pinch force to the increase in pulling force at some events. The safety margin is also influenced by the static COF [63], such that it will be high in the case of slippery object as in copper alloy used in our experiment. The pinch–pulling force relationship can be used as an assessment for pinch force control against unpredictable loading for stroke patients.

The CV value (represented by standard deviation divided by the mean) indicates how well the mean values of pinch and pulling forces summarizes the whole dataset. Interpretation of CV may vary; the rule of thumb can be used, which states the CV value of under 1 shows a small variance [64]. In this study, the CV values were 0.248 and 0.35 for pinch and pulling forces of the right hands, respectively. The CV values were 0.241 and 0.31 for pinch and pulling forces of left hands, respectively. Hence, the mean pinch and pulling force values are a reasonable representation of healthy adults in Malaysia. The results also indicate that the pinch and pulling force vary among the volunteers at the slip onset, which is expected as each volunteer has a different body size, hand size, and amount of safety margin. For instance, a bigger fingertip size leads to a larger contact area between the fingertip and the object's surface. Consequently, larger contact area would lead to larger pinch force, as reported in a related study by Derler et al. [65], who showed that there is a positive association between contact area and force exerted by the index finger. In addition, each volunteer has different skin characteristics such as oiliness, moisture, roughness, and age that may affect the friction between fingertip skin and pincer object's surface. In this study, the mean static COFs for right and left hands were computed as 0.518 ± 0.146 and 0.517 ± 0.145, respectively. This is relatively consistent with previous findings related to measuring COF between human skin and copper material, reported by Sivamani et al. [66] (mean COF = 0.55) and Li et al. [67] (mean COF = 0.58).

In the literature, many studies have reported a significant difference in maximum pinch force between right and left hands [68]. Although our results show a slight increase in the means of pinch and pulling force for right hand compared to the left hand, there was no significant difference ($p > 0.05$). Thus, it is not required to distinguish between right and left hands when a single-gender group is recruited for further research. Furthermore, the unaffected hand can be used as a reference for the affected hand during the rehabilitation of post-stroke patients. The explanation of this result can be that the volunteers did not exert their maximum pinch force at the slip onset. This is consistent with a study conducted by Li and Yu [69], who reported that there is no significant difference between left and right hands when the grip force is at 25%, 50%, and 75% of the maximum force. The results of their study indicate that there is a significant difference only at maximum grip force.

Recently, three main components of the assessment system are required for an assessment system to be fully automated: administration, data acquisition, and rating [19]. Administration involves the instructions and procedures to evaluate the patient, while the data acquisition involves obtaining the outcome measurements from patients. Rating is the criteria to evaluate the outcome measurements of the patients, which are usually designed as an ordinal scale. In this study, only the pinch data acquisition system is automated based on UE-FMA protocol. The limitation of this study is that the administration and rating components are not yet automated. In the future, this limitation will be addressed, such that stroke patients with pinch deficits at different levels of impairment (severe, moderate, and mild) will be recruited to determine the threshold values for each level so that an automated rating process can be performed. In addition, the administration of the pinch test will be automated by including a graphical user interface to visualize the instructions and procedures.

To use the automated data acquisition system in clinical practice based on UE-FMA protocol, the design can be improved to be more compact, less wires, and portable by: (1) replacing the DC power supply with a portable DC battery to power the two amplifiers and LVDT sensor; (2) using a customized small size linear electric actuator as the with pulling force of smaller than 15 N; and (3) using a wireless connection (e.g., ZigBee module) to transmit/receive digitalized signals to/from the PC instead of signal wires. This improvement will be addressed in the future work.

5. Conclusions

In this paper, an automated data acquisition system for pinch assessment based on UE-FMA protocol is presented. The automated system is able to provide objective measurements of pinch and pulling forces and detect slip onset rather than the subjective manual gentle pull and visual observation of slipping occurrence. In addition, the therapist's gentle pull has been replaced with a linear actuator sub-system exerting a consistent amount of pulling force. Right and lefts hands of 50 volunteers were recruited to investigate the pinch and pulling force measurements at the slip onset using the developed system. The pinch–pulling force relationship is linear, which is indicated by a proportional increase of pinch force against the continuous increase of pulling force prior to slip onset. In addition, the volunteers were subjected to submaximal pinch force at the slip onset. The mean pinch force values at the slip onset were 12.17 and 11.67 N for right and left hands, respectively. The mean pulling force values at the slip onset were 6.25 and 5.92 N for right and left hands, respectively. It was found that there is no significant difference in force measurements between right and left hands. A further study can be conducted to investigate the hypothesis of considering slip onset and pinch–pulling force relationship as a pinch assessment for stroke patients.

Author Contributions: A.A. (Abdallah Alsayed) and R.K. designed and developed the pinch assessment system; H.R. performed the integration of sensors and actuator; A.A. (Azizan As'arry) performed CAD and finite element analysis as well as provided access to volunteers; A.A. (Abdallah Alsayed) performed data collection and results analysis; and all authors contributed to writing and editing the paper. All authors have read and agreed to the published version of the manuscript

Funding: This research was funded by Universiti Putra Malaysia, grant number IPS No. 9574400.

Acknowledgments: We would like to thank Yayasan Khazanah Berhad, Malaysia for their technical and financial support.

Conflicts of Interest: The authors declare no conflict of interest.

Data Availability: The data and codes used in this study are available from the corresponding author upon request.

Appendix A

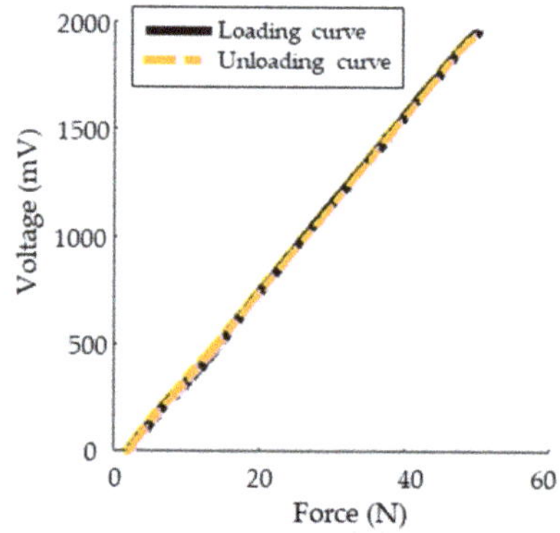

Figure A1. Force–voltage relationship during loading and unloading.

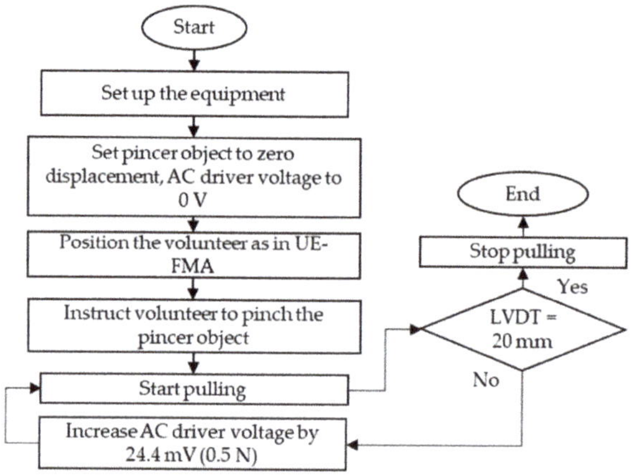

Figure A2. Flowchart of experiment execution.

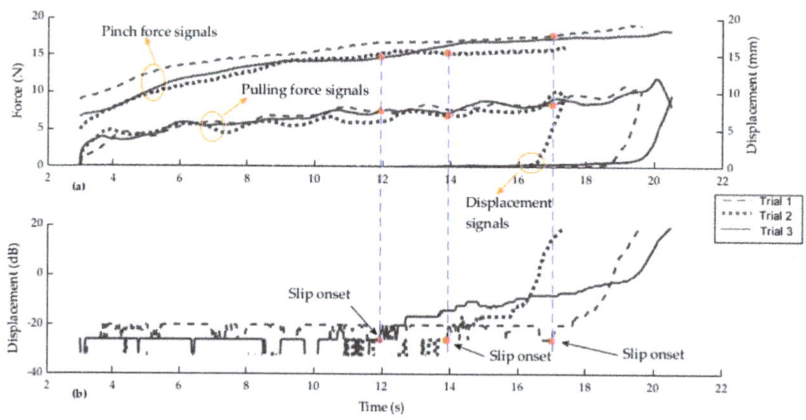

Figure A3. (**a**) Three trials recorded from one volunteer; and (**b**) displacement in decibel scale.

References

1. Miyasaka, H.; Takeda, K.; Ohnishi, H.; Orand, A.; Sonoda, S. Effect of Sensory Loss on Improvements of Upper-Limb Paralysis Through Robot-Assisted Training: A Preliminary Case Series Study. *Appl. Sci.* **2019**, *9*, 3925. [CrossRef]
2. Li, K.; Nataraj, R.; Marquardt, T.L.; Li, Z.-M. Directional coordination of thumb and finger forces during precision pinch. *PLoS ONE* **2013**, *8*, e79400. [CrossRef] [PubMed]
3. Pessina, M.A.; Bowley, B.G.; Rosene, D.L.; Moore, T.L. A method for assessing recovery of fine motor function of the hand in a rhesus monkey model of cortical injury: An adaptation of the Fugl-Meyer Scale and Eshkol-Wachman Movement Notation. *Somatosens. Mot. Res.* **2019**, *36*, 69–77. [CrossRef]
4. El-Katab, S.; Omichi, Y.; Srivareerat, M.; Davenport, A. Pinch grip strength as an alternative assessment to hand grip strength for assessing muscle strength in patients with chronic kidney disease treated by haemodialysis: A prospective audit. *J. Hum. Nutr. Diet.* **2016**, *29*, 48–51. [CrossRef] [PubMed]
5. Murphy, M.A.; Resteghini, C.; Feys, P.; Lamers, I. An overview of systematic reviews on upper extremity outcome measures after stroke. *BMC Neurol.* **2015**, *15*, 29. [CrossRef]
6. Valero-Cuevas, F.J.; Smaby, N.; Venkadesan, M.; Peterson, M.; Wright, T. The strength-dexterity test as a measure of dynamic pinch performance. *J. Biomech.* **2003**, *36*, 265–270. [CrossRef]
7. Harrison, J.K.; McArthur, K.S.; Quinn, T.J. Assessment scales in stroke: Clinimetric and clinical considerations. *Clin. Interv. Aging* **2013**, *8*, 201.
8. Yozbatiran, N.; Der-Yeghiaian, L.; Cramer, S.C. A standardized approach to performing the action research arm test. *Neurorehabilit. Neural Repair* **2008**, *22*, 78–90. [CrossRef]
9. Johansson, G.M.; Häger, C.K. A modified standardized nine hole peg test for valid and reliable kinematic assessment of dexterity post-stroke. *J. Neuroeng. Rehabil.* **2019**, *16*, 8. [CrossRef]
10. Sears, E.D.; Chung, K.C. Validity and responsiveness of the jebsen–taylor hand function test. *J. Hand Surg.* **2010**, *35*, 30–37. [CrossRef]
11. Wolf, S.L.; Catlin, P.A.; Ellis, M.; Archer, A.L.; Morgan, B.; Piacentino, A. Assessing Wolf motor function test as outcome measure for research in patients after stroke. *Stroke* **2001**, *32*, 1635–1639. [CrossRef] [PubMed]
12. Raghavan, P. Upper limb motor impairment after stroke. *Phys. Med. Rehabil. Clin.* **2015**, *26*, 599–610. [CrossRef] [PubMed]
13. Page, S.J.; Fulk, G.D.; Boyne, P. Clinically important differences for the upper-extremity Fugl-Meyer Scale in people with minimal to moderate impairment due to chronic stroke. *Phys. Ther.* **2012**, *92*, 791–798. [CrossRef] [PubMed]
14. Fugl-Meyer, A.R.; Jääskö, L.; Leyman, I.; Olsson, S.; Steglind, S. The post-stroke hemiplegic patient. 1. A method for evaluation of physical performance. *Scand. J. Rehabil. Med.* **1975**, *7*, 13–31. [PubMed]
15. Deakin, A.; Hill, H.; Pomeroy, V.M. Rough guide to the Fugl-Meyer Assessment: Upper limb section. *Physiotherapy* **2003**, *89*, 751–763. [CrossRef]
16. Paternostro-Sluga, T.; Grim-Stieger, M.; Posch, M.; Schuhfried, O.; Vacariu, G.; Mittermaier, C.; Bittner, C.; Fialka-Moser, V. Reliability and validity of the Medical Research Council (MRC) scale and a modified scale for testing muscle strength in patients with radial palsy. *J. Rehabil. Med.* **2008**, *40*, 665–671. [CrossRef]
17. Van Beers, R.J.; Haggard, P.; Wolpert, D.M. The role of execution noise in movement variability. *J. Neurophysiol.* **2004**, *91*, 1050–1063. [CrossRef]
18. Tanaka, K.; Yano, H. Errors of visual judgement in precision measurements. *Ergonomics* **1984**, *27*, 767–780. [CrossRef]
19. Simbaña, E.D.O.; Baeza, P.S.-H.; Jardón, A.; Balaguer, C. Review of Automated Systems for Upper Limbs Functional Assessment in Neurorehabilitation. *IEEE Access* **2019**, *7*, 32352–32367. [CrossRef]
20. de los Reyes-Guzmán, A.; Dimbwadyo-Terrer, I.; Trincado-Alonso, F.; Monasterio-Huelin, F.; Torricelli, D.; Gil-Agudo, A. Quantitative assessment based on kinematic measures of functional impairments during upper extremity movements: A review. *Clin. Biomech.* **2014**, *29*, 719–727. [CrossRef]
21. Yu, L.; Xiong, D.; Guo, L.; Wang, J. A remote quantitative Fugl-Meyer assessment framework for stroke patients based on wearable sensor networks. *Comput. Methods Programs Biomed.* **2016**, *128*, 100–110. [CrossRef] [PubMed]

22. Woytowicz, E.J.; Rietschel, J.C.; Goodman, R.N.; Conroy, S.S.; Sorkin, J.D.; Whitall, J.; Waller, S.M. Determining levels of upper extremity movement impairment by applying a cluster analysis to the Fugl-Meyer assessment of the upper extremity in chronic stroke. *Arch. Phys. Med. Rehabil.* **2017**, *98*, 456–462. [CrossRef] [PubMed]
23. Yu, L.; Wang, J.; Guo, L.; Zhang, Q.; Li, P.; Li, Y.; Yu, X.; Huang, Y.; Wu, Z. Transfer learning based quantitative assessment model of upper limb movement ability for stroke survivors. In Proceedings of the 2017 2nd International Conference on Information Technology (INCIT), Nakhon Pathom, Thailand, 2–3 November 2017; pp. 1–4.
24. Page, S.J.; Levine, P.; Hade, E. Psychometric properties and administration of the wrist/hand subscales of the Fugl-Meyer Assessment in minimally impaired upper extremity hemiparesis in stroke. *Arch. Phys. Med. Rehabil.* **2012**, *93*, 2373–2376. [CrossRef] [PubMed]
25. Westling, G.; Johansson, R. Factors influencing the force control during precision grip. *Exp. Brain Res.* **1984**, *53*, 277–284. [CrossRef]
26. Hu, W.; Wei, N.; Li, Z.-M.; Li, K. Effects of muscle fatigue on directional coordination of fingertip forces during precision grip. *PLoS ONE* **2018**, *13*, e0208740. [CrossRef]
27. Häger-Ross, C.; Cole, K.J.; Johansson, R.S. Grip-force responses to unanticipated object loading: Load direction reveals body-and gravity-referenced intrinsic task variables. *Exp. Brain Res.* **1996**, *110*, 142–150. [CrossRef]
28. Sullivan, K.J.; Tilson, J.K.; Cen, S.Y.; Rose, D.K.; Hershberg, J.; Correa, A.; Gallichio, J.; McLeod, M.; Moore, C.; Wu, S.S. Fugl-Meyer assessment of sensorimotor function after stroke: Standardized training procedure for clinical practice and clinical trials. *Stroke* **2011**, *42*, 427–432. [CrossRef]
29. Frisoli, A.; Procopio, C.; Chisari, C.; Creatini, I.; Bonfiglio, L.; Bergamasco, M.; Rossi, B.; Carboncini, M.C. Positive effects of robotic exoskeleton training of upper limb reaching movements after stroke. *J. Neuroeng. Rehabil.* **2012**, *9*, 36. [CrossRef]
30. Ellis, M.D.; Schut, I.; Dewald, J.P. Flexion synergy overshadows flexor spasticity during reaching in chronic moderate to severe hemiparetic stroke. *Clin. Neurophysiol.* **2017**, *128*, 1308–1314. [CrossRef]
31. Straathof, P.T.; Lobo-Prat, J.; Schilder, F.; Kooren, P.N.; Paalman, M.I.; Stienen, A.H.; Koopman, B.F. Design and control of the A-Arm: An active planar arm support for adults with Duchenne muscular dystrophy. In Proceedings of the 2016 6th IEEE International Conference on Biomedical Robotics and Biomechatronics (BioRob), Singapore, 26–29 June 2016; pp. 1242–1247.
32. Heyer, P.; Orihuela-Espina, F.; Castrejón, L.R.; Hernández-Franco, J.; Sucar, L.E. Sensor Abstracted Extremity Representation for Automatic Fugl-Meyer Assessment. In *Applications for Future Internet*; Springer: Cham, Switzerland, 2017; pp. 152–163.
33. Wang, J.; Yu, L.; Wang, J.; Guo, L.; Gu, X.; Fang, Q. Automated Fugl-Meyer assessment using SVR model. In Proceedings of the 2014 IEEE International Symposium on Bioelectronics and Bioinformatics (IEEE ISBB 2014), Chung Li, Taiwan, 11–14 April 2014; pp. 1–4.
34. Villán-Villán, M.A.; Pérez-Rodríguez, R.; Gómez, C.; Opisso, E.; Tormos, J.M.; Medina, J.; Gómez, E.J. A First Step for the Automation of Fugl-Meyer Assessment Scale for Stroke Subjects in Upper Limb Physical Neurorehabilitation. In Proceedings of the International Conference on Informatics, Management and Technology in Healthcare, Athens, Greece, 9–11 July 2015; pp. 45–48.
35. Oña, E.D.; Jardón, A.; Monge, E.; Molina, F.; Cano, R.; Balaguer, C. Towards Automated Assessment of Upper Limbs Motor Function Based on Fugl-Meyer Test and Virtual Environment. In Proceedings of the International Conference on NeuroRehabilitation, Pisa, Italy, 16–20 October 2018; pp. 297–301.
36. Eichler, N.; Hel-Or, H.; Shimshoni, I.; Itah, D.; Gross, B.; Raz, S. 3D motion capture system for assessing patient motion during Fugl-Meyer stroke rehabilitation testing. *IET Comput. Vis.* **2018**, *12*, 963–975. [CrossRef]
37. Kim, W.-S.; Cho, S.; Baek, D.; Bang, H.; Paik, N.-J. Upper extremity functional evaluation by Fugl-Meyer assessment scoring using depth-sensing camera in hemiplegic stroke patients. *PLoS ONE* **2016**, *11*, e0158640. [CrossRef] [PubMed]
38. Lee, W.W.; Yen, S.-C.; Tay, E.B.A.; Zhao, Z.; Xu, T.M.; Ling, K.K.M.; Ng, Y.-S.; Chew, E.; Cheong, A.L.K.; Huat, G.K.C. A smartphone-centric system for the range of motion assessment in stroke patients. *IEEE J. Biomed. Health Inform.* **2014**, *18*, 1839–1847. [CrossRef] [PubMed]
39. Olesh, E.V.; Yakovenko, S.; Gritsenko, V. Automated assessment of upper extremity movement impairment due to stroke. *PLoS ONE* **2014**, *9*, e104487. [CrossRef] [PubMed]

40. Otten, P.; Kim, J.; Son, S. A framework to automate assessment of upper-limb motor function impairment: A feasibility study. *Sensors* **2015**, *15*, 20097–20114. [CrossRef]
41. Otten, P.; Son, S.H.; Kim, J. Automating stroke patient evaluation using sensor data and SVM. In Proceedings of the 2014 IEEE 7th International Conference on Service-Oriented Computing and Applications, Matsue, Japan, 17–19 November 2014; pp. 223–229.
42. Lee, S.; Lee, Y.-S.; Kim, J. Automated evaluation of upper-limb motor function impairment using Fugl-Meyer assessment. *IEEE Trans. Neural Syst. Rehabil. Eng.* **2017**, *26*, 125–134. [CrossRef]
43. Lee, S.-H.; Song, M.; Kim, J. Towards clinically relevant automatic assessment of upper-limb motor function impairment. In Proceedings of the 2016 IEEE-EMBS International Conference on Biomedical and Health Informatics (BHI), Las Vegas, NV, USA, 24–27 February 2016; pp. 148–151.
44. Kim, Y.M.; Son, Y.; Kim, W.; Jin, B.; Yun, M.H. Classification of children's sitting postures using machine learning algorithms. *Appl. Sci.* **2018**, *8*, 1280. [CrossRef]
45. Paredes-Madrid, L.; Emmi, L.; Garcia, E.; De Santos, P.G. Detailed study of amplitude nonlinearity in piezoresistive force sensors. *Sensors* **2011**, *11*, 8836–8854. [CrossRef]
46. Parmar, S.; Khodasevych, I.; Troynikov, O. Evaluation of flexible force sensors for pressure monitoring in treatment of chronic venous disorders. *Sensors* **2017**, *17*, 1923. [CrossRef]
47. Rivera, J.; Carrillo, M.; Chacón, M.; Herrera, G.; Bojorquez, G. Self-calibration and optimal response in intelligent sensors design based on artificial neural networks. *Sensors* **2007**, *7*, 1509–1529. [CrossRef]
48. Almassri, A.; Wan Hasan, W.; Ahmad, S.; Shafie, S.; Wada, C.; Horio, K. Self-calibration algorithm for a pressure sensor with a real-time approach based on an artificial neural network. *Sensors* **2018**, *18*, 2561. [CrossRef]
49. Likitlersuang, J.; Leineweber, M.J.; Andrysek, J. Evaluating and improving the performance of thin film force sensors within body and device interfaces. *Med. Eng. Phys.* **2017**, *48*, 206–211. [CrossRef] [PubMed]
50. Ide, D.T.A. Available online: https://www.arduino.cc/en/Main/Software (accessed on 1 January 2019).
51. Alsayed, A.; Kamil, R.; Ramli, H.R.; As'arry, A. Design and Calibration of Pinch Force Measurement Using Strain Gauge for Post-Stroke Patients. *Int. J. Integr. Eng.* **2019**, 11. [CrossRef]
52. Oh, J.; Kim, J.O.; Kim, Y.; Choi, H.B.; Yang, J.C.; Lee, S.; Pyatykh, M.; Kim, J.; Sim, J.Y.; Park, S. Highly Uniform and Low Hysteresis Piezoresistive Pressure Sensors Based on Chemical Grafting of Polypyrrole on Elastomer Template with Uniform Pore Size. *Small* **2019**, *15*, 1901744. [CrossRef] [PubMed]
53. Matute, A.; Paredes-Madrid, L.; Gutierrez, E.; Vargas, C.A.P. Characterization of drift and hysteresis errors in force sensing resistors considering their piezocapacitive effect. In Proceedings of the 2017 IEEE SENSORS, Glasgow, UK, 29 October–1 November 2017; pp. 1–3.
54. Reed, G.F.; Lynn, F.; Meade, B.D. Use of coefficient of variation in assessing variability of quantitative assays. *Clin. Diagn. Lab. Immunol.* **2002**, *9*, 1235–1239. [CrossRef]
55. Zheng, Y.; Zhao, M.; Jiang, J.; Song, L. Dynamic Force Transducer Calibration Based on Electrostatic Force. *IEEE Access* **2019**, *7*, 48998–49003. [CrossRef]
56. Maleki-Ghahfarokhi, A.; Dianat, I.; Feizi, H.; Asghari-Jafarabadi, M. Influences of gender, hand dominance, and anthropometric characteristics on different types of pinch strength: A partial least squares (PLS) approach. *Appl. Ergon.* **2019**, *79*, 9–16. [CrossRef]
57. Angst, F.; Drerup, S.; Werle, S.; Herren, D.B.; Simmen, B.R.; Goldhahn, J. Prediction of grip and key pinch strength in 978 healthy subjects. *BMC Musculoskelet. Disord.* **2010**, *11*, 94. [CrossRef]
58. Cronbach, L.J. Coefficient alpha and the internal structure of tests. *Psychometrika* **1951**, *16*, 297–334. [CrossRef]
59. Blennerhassett, J.M.; Carey, L.M.; Matyas, T.A. Grip force regulation during pinch grip lifts under somatosensory guidance: Comparison between people with stroke and healthy controls. *Arch. Phys. Med. Rehabil.* **2006**, *87*, 418–429. [CrossRef] [PubMed]
60. Takamuku, S.; Gomi, H. Better grip force control by attending to the controlled object: Evidence for direct force estimation from visual motion. *Sci. Rep.* **2019**, *9*, 13114. [CrossRef]
61. Roberge, J.-P.; Ruotolo, W.; Duchaine, V.; Cutkosky, M. Improving industrial grippers with adhesion-controlled friction. *IEEE Robot. Autom. Lett.* **2018**, *3*, 1041–1048. [CrossRef]
62. Miall, R.C.; Rosenthal, O.; Ørstavik, K.; Cole, J.D.; Sarlegna, F.R. Loss of haptic feedback impairs control of hand posture. *Exp. Brain Res.* **2019**, *237*. [CrossRef] [PubMed]
63. Barrea, A.; Bulens, D.C.; Lefèvre, P.; Thonnard, J.-L. Simple and reliable method to estimate the fingertip static coefficient of friction in precision grip. *IEEE Trans. Haptics* **2016**, *9*, 492–498. [CrossRef] [PubMed]

64. Hall, D. (e) in Normandy: The sociolinguistics, phonology and phonetics of the Loi de Position. *J. Fr. Lang. Stud.* **2019**, *29*, 1–33. [CrossRef]
65. Derler, S.; Gerhardt, L.-C.; Lenz, A.; Bertaux, E.; Hadad, M. Friction of human skin against smooth and rough glass as a function of the contact pressure. *Tribol. Int.* **2009**, *42*, 1565–1574. [CrossRef]
66. Sivamani, R.K.; Wu, G.C.; Gitis, N.V.; Maibach, H.I. Tribological testing of skin products: Gender, age, and ethnicity on the volar forearm. *Ski. Res. Technol.* **2003**, *9*, 299–305. [CrossRef] [PubMed]
67. Li, W.; Qu, S.; Zhou, Z. Reciprocating sliding behaviour of human skin in vivo at lower number of cycles. *Tribol. Lett.* **2006**, *23*, 165. [CrossRef]
68. Petersen, P.; Petrick, M.; Connor, H.; Conklin, D. Grip strength and hand dominance: Challenging the 10% rule. *Am. J. Occup. Ther.* **1989**, *43*, 444–447. [CrossRef]
69. Li, K.W.; Yu, R. Assessment of grip force and subjective hand force exertion under handedness and postural conditions. *Appl. Ergon.* **2011**, *42*, 929–933. [CrossRef]

applied sciences

MDPI

Article

Investigation of the Functional Decline in Proprioceptors for Low Back Pain Using the Sweep Frequency Method

Reiya Nishio [1,*], Yohei Ito [1], Yoshifumi Morita [1], Tadashi Ito [2], Kazunori Yamazaki [3] and Yoshihito Sakai [4]

1 Department of Electrical and Mechanical Engineering, Graduate School of Engineering, Nagoya Institute of Technology, Nagoya 466-8555, Japan; y.ito.359@nitech.jp (Y.I.); morita@nitech.ac.jp (Y.M.)
2 Three-Dimensional Motion Analysis Room, Aichi Prefectural Mikawa Aoitori Medical and Rehabilitation Center for Developmental Disabilities, Okazaki 444-0002, Japan; sanjigen@mikawa-aoitori.jp
3 Faculty of Clinical Engineering, School of Medical Sciences, Fujita Health University, Toyoake 470-1192, Japan; ymzkk@fujita-hu.ac.jp
4 Department of Orthopedic Surgery, National Center for Geriatrics and Gerontology, Obu 474-8511, Japan; jsakai@ncgg.go.jp
* Correspondence: r.nishio.115@nitech.jp; Tel.: +81-52-735-5412

Received: 24 October 2019; Accepted: 18 November 2019; Published: 20 November 2019

Abstract: The purpose of this study was to investigate the functional decline in proprioceptors in patients with low back pain (LBP) by evaluating the entire range of response frequencies of proprioceptors. In previous studies, the function of proprioceptors was only evaluated by single frequency vibrations. However, because it is assumed that individual differences exist in response frequencies of proprioceptors, we developed a method using vibration with sweep frequency covering the entire range of response frequencies of proprioceptors. The center of pressure was determined in 35 elderly patients with non-specific LBP (NSLBP) and 28 elderly individuals with non-LBP (NLBP) during upright stances on a balance board without vision. Local vibratory stimulations to lower leg or trunk muscles were continuously varied between 27 and 272 Hz over 60 s to evaluate the proprioceptive predominance of a body location using the relative proprioceptive weighting ratio (RPW). Compared with the NLBP group, the NSLBP group exhibited a lower RPW value for the Vater-Pacini corpuscles. Thus, the NSLBP group relied more on the input of Vater-Pacini corpuscles in the trunk. A reduction in lower leg sensitivity at the Vater-Pacini corpuscles in the NSLBP group was observed.

Keywords: vibratory stimulation device; local muscle vibration; proprioceptors; low back pain; response frequency; postural control; Vater-Pacini corpuscles

1. Introduction

Worldwide, low back pain (LBP) has become the main cause of disability [1]. However, its causes are unidentifiable in 85% of the cases [2]. Postural control impairment is considered one of the main causes of LBP [3]. Postural control is affected by sensory inputs from visual and vestibular senses and somatosensory receptors which provide superficial and proprioception. A previous study indicated that the feedback control on the trunk posture is mainly influenced by the visual and proprioception [4]. Also, another previous study suggested that the ability to select and reweigh sensory inputs to adapt to situations is a critical factor for postural control [5]. In addition, non-specific LBP (NSLBP) may alter postural control by refocusing proprioception from the trunk to the ankles [6]. Furthermore, reduced precision of trunk movement in NSLBP might be explained by proprioceptive deficits [7]. Therefore, NSLBP could be caused by a decline in proprioception leading to problems in postural control and LBP.

Proprioception in the lower leg and trunk muscle plays an important role in maintaining postural stability [8]. Each proprioceptor has a response frequency of vibration. Meissner corpuscles are sensitive to lower vibration frequencies (peaks around 30 Hz), whereas the Vater-Pacini corpuscles are sensitive to higher frequencies (peaks around 240 Hz) [9]. Muscle spindles have a response frequency range of 40–100 Hz [10]. Mechanical vibration is commonly used to test externally induced balance control. This method has been widely used to investigate the role of muscle spindle in research focusing on postural control [11–14]. However, the role of other proprioceptors in NSLBP has not been evaluated. Thus, comparing postural sway of NSLBP during various local frequency vibrations with that of non-LBP (NLBP) will allow determination of the cause of LBP and frequencies of proprioception deficit.

Postural sway of NSLBP during local frequency vibrations of 30, 60 and 240 Hz has been compared with that of NLBP in our previous study. From the result, we found that not only does NSLBP involve the functional impairment of muscle spindles that other groups reported [11–13] but also Vater-Pacini corpuscles were functionally declined because of LBP [15,16]. However, in previous studies, only one frequency of vibration stimulation was applied within the range of response frequencies for each proprioceptor [11–13,15,16]. In addition, other studies investigating muscle spindle response to multiple vibration frequencies reported different results. A previous study using 90 Hz vibration frequency suggested that the weighting of proprioceptive feedback from lumbar muscle spindles did not differ between NSLBP and NLBP [17]. On the other hand, another previous study using 60 Hz vibration frequency suggested that LBP patients rely less on proprioceptive feedback from lumbar muscle spindles [18]. From these results, it is implied that the postural response induced by mechanical vibration varies with the applied vibration frequency even in one proprioceptor. Also, it is assumed that there are individual differences in response frequencies of proprioceptors. To the best of our knowledge, a protocol for measuring postural control while applying local vibration stimulation in the entire response frequency band has not been developed. Thus, we considered that the inspection device and a protocol for evaluating postural response in the range of the entire response frequencies of the proprioceptors are necessary.

Therefore, we developed a new experimental protocol to measure the center of pressure (CoP) of an individual on a gravicorder during local mechanical vibration with sweep frequencies, using our developed variable frequency vibratory stimulation device. This method can be analyzed by applying a vibration stimulus once in the entire frequency range covering the response frequencies of each proprioceptor. For this reason, there is a possibility that this method can identify proprioceptors related to NSLBP and the functionally declined frequency bands of proprioceptors. In addition, the method might be capable of more detailed analyses because the vibration is stimulated with the entire frequency range covering the response frequencies of each proprioceptor. Furthermore, this method is considered to be the shortest for measuring all the response frequency bands of the three proprioceptors. The physical burden on subjects during measurement can also be reduced because the measurement time can be shortened by about 4 min and 30 s compared with the time required for the conventional method with a constant frequency. Our final goal is to develop an inspection device for evaluating the functional decline in proprioceptors. In this study, we aimed to investigate the relationship between proprioception and LBP. Thus, postural sway response in NSLBP and NLBP was measured with our new sweep frequency method and compared with the responses obtained from past studies using the method of constant frequency.

2. Materials and Methods

2.1. Subjects

This study was carried out during a period of 30 months (August 2016–July 2019) in tandem with general clinical practice. Written informed consent was obtained from all participants before inclusion in the study. All investigations were conducted according to the principles expressed in the

Declaration of Helsinki. The Ethics Committee of the National Center for Geriatrics and Gerontology approved the study (IRB approval number: 586).

Sixty-three elderly individuals (age ≧ 65 years: age range 65–90; sex ratio: 36 women/27 men) were recruited, including 35 patients (22 women, 13 men) with NSLBP which lasted over 3 months who visited the National Center for Geriatric and Gerontology for orthopedic treatment. Among the recruited individuals, 28 individuals with NLBP (14 women and 14 men) acted as controls with no history of disabling NSLBP which is defined as LBP when the pathoanatomical cause of the pain cannot be determined [19]. Patients with the following characteristics were excluded—paralysis, astasia, spinal cord tumor, spinal infection and history of spinal surgery. Demographics of all individuals are presented in Table 1.

Table 1. Demographic characteristics and functional outcome of the patients.

Variables	NLBP (n = 28)	NSLBP (n = 35)	p
Age, years	75.9 ± 5.4	75.1 ± 5.2	0.552
Height, cm	154.8 ± 7.7	153.2 ± 7.3	0.410
Weight, kg	59.8 ± 11.4	57.7 ± 10.5	0.609
BMI, kg/m^2	24.8 ± 3.3	24.5 ± 3.7	0.774
VAS, cm	2.4 ± 2.5	6.2 ± 2.4	<0.001
RDQ (score)	9.1 ± 5.8	11.8 ± 4.0	<0.05

BMI: body mass index; NLBP: non-low back pain; NSLBP: non-specific low back pain; VAS: visual analog scale; RDQ: Roland-Morris disability questionnaire; SD: standard deviation. Data are presented as the mean ± SD. All p values were generated using the independent t-test, Mann-Whitney u test.

2.2. Device

Figure 1a,b shows an inspection system for evaluating the functional decline in proprioceptors. The system consists of a PC, an amplifier, four vibrators (NSW1-205-8A, Aurasound, Inc., CA, USA), three hook-and-loop fasteners and a Wii Balance Board (Nintendo Co., Ltd., Kyoto, Japan). Two hook-and-loop fasteners with one vibrator were used for the left and right lower legs and one hook-and-loop fastener with two vibrators was used for the left and right parts of the trunk.

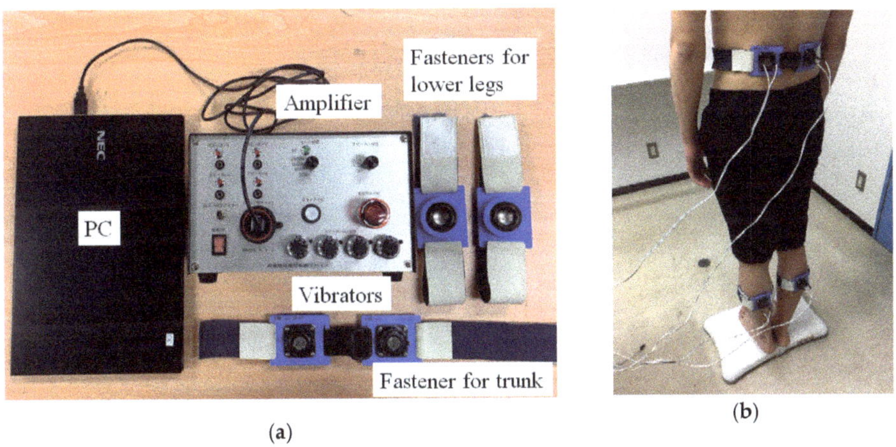

Figure 1. Inspection system for proprioceptive function: (**a**) Variable frequency vibratory stimulation device; and (**b**) experimental setup for a participant.

Vibration signals were generated using the PC. Then, vibrations were output from the vibrators as a mechanical vibration stimulus through the amplifier. In other previous studies, the mechanical vibration stimulus with an amplitude of 0.4 to 1.0 mm was used as the proprioceptive input [10,11,13].

Also, the maximum amplitude of our developed device which can be output without distortion of mechanical vibration at each frequency from 20 to 300 Hz is 0.8 mm. Thus, the amplitude of the vibration was selected at 0.8 mm of a sinusoidal wave. Vibrators were attached to the hook-and-loop fasteners using holders and fixed to the muscle belly of the gastrocnemius (GS) and the lumbar multifidus (LM) muscles. The contact pressure to the body locations of the vibrators was regulated by changing the length of the hook-and-loop fasteners according to the circumference of the fixed body locations. A Wii Balance Board was used as a gravicorder. The CoP was measured instead of the center of gravity. A Wii Balance Board has been reported as a reliable and valid tool for assessing standing balance [20]. It can acquire the time series data of CoP coordinates with a sampling frequency of 100 Hz. Moreover, it is possible to save the CoP as CSV data by running the self-made CoP data collection software on a PC connected with the Wii Balance Board via Bluetooth.

2.3. Experimental Method

The CoP in postural sway while the participant stood barefoot on the Wii Balance Board with his/her eyes closed and feet together was measured. Participants were instructed to remain still and relaxed in the standing position with their arms hanging loosely at their sides. In order to prevent injuries due to falls during measurements, two researchers stood on both sides of a subject with eyes closed. They paid attention to whether or not the subject was likely to fall and gave support if the subject was falling. As the vibration frequency, a sweep frequency (Figure 2) was used. The vibration was continuously changed from 27 Hz to 272 Hz (frequency ascend mode) or 272 Hz to 27 Hz (frequency descend mode) for 60 s, thus covering the response frequency ranges of the three proprioceptors. The relationships of sweep frequency between frequency $f(t)$ and measurement time t are shown in Equations (1) and (2). The coefficient a was determined so that the frequencies at $t = 15$ s and 75 s were set to 27 Hz and 272 Hz in the frequency ascend mode, respectively. Thus, the coefficient a was found to be 0.03851. Furthermore, the frequency descend mode had the same frequencies at $t = 15$ s and 75 s (272 Hz and 27 Hz, respectively).

$$f(t) = \begin{cases} 0\ (0 \le t \le 15) \\ 27e^{a(t-15)}\ (15 \le t \le 75) \end{cases} \quad \text{(frequency ascend mode)} \tag{1}$$

$$f(t) = \begin{cases} 0\ (0 \le t \le 15) \\ 27e^{a(75-t)}\ (15 \le t \le 75) \end{cases} \quad \text{(frequency descend mode)} \tag{2}$$

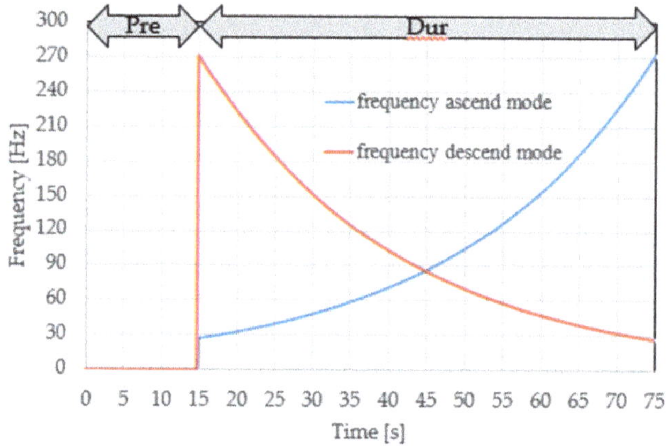

Figure 2. Sweep frequency of vibratory stimulation.

The measurement procedure consisted of two conditions that provided vibration stimulus—GS and LM. The time needed for one condition was 75 s and was divided into two sections. The first 15 s and last 60 s sections are referred to as Pre- and Dur-sections, respectively. In the Pre-section, a participant closed his/her eyes. Then, in the Dur-section, the vibratory stimulation was applied to the GS or LM of the participant with his/her eyes closed. The ascend or descend sweep frequency mode was randomly determined for each subject. A rest interval of 60 s was kept between the conditions. During this time, each participant sat in a chair and remained at rest.

2.4. Data Analysis Method

The postural control performance was evaluated from the CoP data while vibration stimulation was applied. Furthermore, the CoP data divided the Dur-section into three evaluation sections (ES_i, i = 1, 2, 3) determined according to the frequency of vibration stimulation and the response frequencies of the proprioceptors. Table 2 lists the names of each ES, the frequency of the vibration stimulus and the corresponding proprioceptors.

Table 2. Lists of names of each ES, frequency of vibration stimulus and corresponding proprioceptors.

ES_i	Frequency [Hz]	Corresponding Proprioceptors
ES_1	30–43	Meissner corpuscle
ES_2	49–71	Muscle spindles
ES_3	168–247	Vater-Pacini corpuscle

ES: Evaluation section.

Because the previous studies showed that a relationship existed between CoP displacement in an anteroposterior direction and the response of proprioceptors by vibration stimulus, the CoP displacement in the anteroposterior direction was only considered [10]. Previous studies used the relative proprioceptive weighting ratio (RPW) and the root mean square (RMS) of GS and LM as evaluation indexes to evaluate the proprioceptive predominance of body location and the magnitude of CoP in an anteroposterior direction [11,15,18]. The evaluation indexes applicable to the sweep frequency method were introduced as follows.

To obtain information regarding the proprioceptive predominance of body locations, modified RPW was calculated using the following equation.

$$RPW^i = \frac{RMS^i_{GS}}{RMS^i_{GS} + RMS^i_{LM}} \times 100\ [\%] \quad (i = 1,\ 2,\ 3) \tag{3}$$

where the subscript i is used to distinguish the ESs. The higher is the value of RPW^i from 50% to 100%, the higher is the reliability of GS input. On the other hand, the lower is the value of RPW^i from 50% to 0%, the higher is the reliability of LM input [11,15]. RMS^i_* means the effective value of the magnitude of CoP in an anteroposterior direction. In each ES in the Dur-section, the larger the RMS, the higher the amount of transition of postural sway in the anteroposterior direction compared to those in the Pre-section. RMS^i_* was calculated using the following equations.

$$RMS^1_* = \sqrt{\frac{1}{N}\sum_{n=n_1}^{n_2}\left\{Y_{Dur(*)}(n) - \overline{Y}_{pre(*)}\right\}^2} \tag{4}$$

$$RMS^2_* = \sqrt{\frac{1}{N}\sum_{n=n_3}^{n_4}\left\{Y_{Dur(*)}(n) - \overline{Y}_{pre(*)}\right\}^2} \tag{5}$$

$$RMS_*^3 = \sqrt{\frac{1}{N}\sum_{n=n_5}^{n_6}\left\{Y_{Dur(*)}(n) - \overline{Y}_{pre(*)}\right\}^2} \quad (6)$$

where N is the total sampling number of each ES; n is the number of data series; $Y_{Dur(*)}(n)$ is the CoP in the anteroposterior direction in the Dur-section; $\overline{Y}_{pre(*)}$ is the mean value of CoP in the anteroposterior direction in the Pre-section; the superscript "*" is used to distinguish stimulation body locations, namely, the GS or LM; and subscript numbers are used to distinguish the ESs. In this case, N was equal to 1000 because all ESs were analyzed in 10 s and the sampling frequency was 100 Hz. Table 3 shows the numbers of data series corresponding to the start frequency and the end frequency of ES_i in the case of frequency ascend mode and descend mode. In each of the ascend and descend modes, the numbers of sampling series n corresponding to measurement time t were calculated by Equations (1) and (2) in order to obtain the start and end frequencies of each ES, respectively, as shown in Table 2. To correct the CoP in the Dur-section based on the CoP in the Pre-section, the CoP in the Dur-section was subtracted from the mean value of the CoP in the Pre-section. These values were calculated using MATLAB (MathWorks, Inc., Natick, MA, USA).

Table 3. Numbers of data series corresponding to the start and end frequencies of ES_i.

ES_i	Number of Data Series	Frequency Ascend Mode	Frequency Descend Mode	Corresponding Frequency [Hz]
ES_1	n_1	1750	7250	30
	n_2	2750	6250	43
ES_2	n_3	3000	6000	49
	n_4	4000	5000	71
ES_3	n_5	6250	2750	168
	n_6	7250	1750	247

ES: evaluation section; n: the number of data series; The subscript numbers of ES are indicated as follows: "1: Meissner corpuscle; from 30 to 43 Hz; 2: muscle spindles; from 49 to 71 Hz, 3: Vater-Pacini corpuscle; from 168 to 247 Hz."

2.5. Statistical Analysis

All analyses were performed using IBM SPSS statistical software (Version 25.0; IBM Corp., Armonk, NY, USA). The frequency ascend and descend modes groups were combined into a population of RPW^i and RMS_*^i for statistical analysis to eliminate the order effects due to vibration application. Data were expressed as mean values and standard deviations for the NLBP and NSLBP. First, these data were analyzed to investigate whether they were normally distributed using the Shapiro-Wilk test. Variable data of RPW^i were compared between NLBP and NSLBP using the independent t-test if both the compared data were normally distributed. In statistical analysis, $p < 0.05(*)$ were considered statistically significant. In addition, a two-way mixed analysis of variance (ANOVA) was performed with RMS as the dependent variable. The independent variables were group (NSLBP group or NLBP group) and the stimulated body locations (GS or LM). Where significant interaction effects were identified, paired *t*-tests and univariate analysis with Bonferroni correction were conducted.

3. Results

3.1. Proprioceptive Postural Control

Table 4 shows the results of RPW^i. For each evaluation index, the mean values and SDs analyzed in the 3 sections are shown. The RPW values of NSLBP were smaller than 50%. NSLBP had a significantly smaller RPW compared with NLBP in Vater-Pacini corpuscles ($p < 0.05$). Meanwhile, there were no significant differences in the RPW values of Meissner corpuscles and muscle spindle between NSLBP and NLBP groups (Table 4).

Table 4. Mean values of RPW^i with standard deviations.

Variable	NLBP	NSLBP	p
RPW^1 [%]	52.2 ± 13.2	48.2 ± 9.3	0.184
RPW^2 [%]	53.8 ± 11.8	49.1 ± 10.2	0.092
RPW^3 [%]	53.3 ± 12.5	46.4 ± 9.8	<0.05

NLBP: non-low back pain; NSLBP: non-specific low back pain; RPW: relative proprioceptive weighting ratio. Data are presented as the mean ± standard deviation or n [%]. All p values were generated using an independent t-test. The superscript numbers indicated the following: "1: Meissner corpuscle; from 30 to 43 Hz; 2: muscle spindles; from 49 to 71 Hz, 3: Vater-Pacini corpuscle; from 168 to 247 Hz."

3.2. Postural Stability

Table 5 shows a result of a two-way mixed ANOVA in RMS^i_*. There was no significant interaction between the NLBP and NSLPB group and the stimulated body locations in the Meissner corpuscle and muscle spindle. In contrast, the interaction between the NLBP and NSLPB groups and the stimulated body locations in Vater-Pacini corpuscles were significantly observed ($p < 0.05$).

Figure 3 depicts the results of RMS in Vater-Pacini corpuscle. Compared with NLBP group, univariate analysis with Bonferroni correction showed that the NSLBP group exhibited significantly higher RMS values at LM in the Vater-Pacini corpuscle ($F(1,61) = 4.817$, $p < 0.05$), whereas there was no significant difference in the RMS values at GS in the Vater-Pacini corpuscle ($F(1,61) = 0.115$, $p = 0.735$). In addition, the paired t-test showed that there was a tendency of significantly higher RMS values at LM than at GS in the NSLBP group ($p = 0.055$). There was no significant difference between GS and LM in the NLBP group ($p = 0.189$).

Table 5. Results of RMS^i_* and two-way mixed analysis of variance.

Variable	NLBP	NSLBP	F(1, 61)	p
RMS^1_{GS} [mm]	10.7 ± 7.8	8.9 ± 4.3	2.874	0.095
RMS^1_{LM} [mm]	8.7 ± 4.0	9.5 ± 4.3		
RMS^2_{GS} [mm]	10.9 ± 7.4	9.6 ± 4.6	2.624	0.110
RMS^2_{LM} [mm]	8.8 ± 4.3	9.7 ± 4.5		
RMS^3_{GS} [mm]	9.6 ± 4.0	9.2 ± 3.9	5.255	<0.05
RMS^3_{LM} [mm]	8.4 ± 3.6	10.8 ± 4.9		

NLBP: non-low back pain; NSLBP: non-specific low back pain; RMS: root mean square of the center of pressure data. Data are presented as the mean ± standard deviation or n [%]. All F value and p values were generated using the two-way mixed analysis of variance. Independent variables: the NLBP and NSLPB group and the stimulated body locations. The subscripts and superscript numbers indicated the following: "GS: gastrocnemius; LM: lumbar multifidus; 1: Meissner corpuscle; from 30 to 43 Hz; 2: muscle spindles; from 49 to 71 Hz, 3: Vater-Pacini corpuscle; from 168 to 247 Hz."

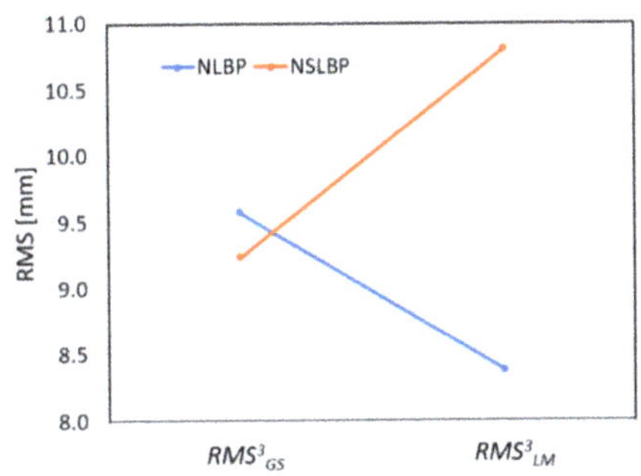

Figure 3. Result of RMS in Vater-Pacini corpuscle. RMS indicates root mean square of the center of pressure data; GS: gastrocnemius; LM: lumbar multifidus.

4. Discussion

To our knowledge, this is the first study to evaluate the postural response when vibration was applied in the entire response frequency range of proprioceptors using sweep frequency. Our main finding was that NSLBP relied more on LM input in the Vater-Pacini corpuscles during vibratory stimulus with a sweep frequency of 168–247 Hz to the lower leg in the standing position.

Past studies have already demonstrated that elderly patients with LBP relied on trunk strategies (Vater-Pacini corpuscles) for postural control during vibratory stimulation [15,16]. Compared with NLBP, lower RPW values in NSLBP during vibratory stimulations of 168 to 247 Hz indicate that NSLBP relies on trunk-muscle proprioceptive inputs. Moreover, the reduction in lower leg sensitivity at the Vater-Pacini corpuscles in the NSLBP group agrees well with the result of previous studies which reported that the NSLBP group decreased its reliance on proprioceptive signals from the lower leg during standing at 240 Hz of constant vibratory stimulus [15,16]. However, statistically significant differences were not observed between the NLBP and NSLBP groups in the Meissner corpuscles and muscle spindle. The decreased reliability on LM inputs in the Meissner corpuscles and muscle spindle may not have been recognized in the NSLBP group because this study was performed under different conditions from measurement methods and subjects in the previous studies [10–12,21–24]. In addition, the differences in age and LBP degree may have affected even the same cases of disease [15,16]. Therefore, the NSLBP group may depend more on the Vater-Pacini corpuscles of their trunk than on the Meissner corpuscles and muscle spindle.

Second, the resulting RMS values were considered. A significant interaction between the NLBP and NSLPB groups and the stimulated body location in Vater-Pacini corpuscles was confirmed. From Figure 3, the NSLBP group exhibited higher RMS values when vibration in the frequency band of the Vater-Pacini corpuscles was applied to LM. This result supports that the NSLBP group relied more on LM input in the Vater-Pacini corpuscles for postural control. These results agree well with those of previous studies where significant differences in postural sways were found between elderly individuals and healthy young adults during quiet standing conditions [25]. Moreover, a decreasing trend in the proprioceptive function in lower leg was observed in the NSLBP group. These findings clearly show the differences in postural control strategies of proprioceptive inputs with respect to NSLBP, suggesting that LBP may influence the postural control strategy of proprioceptive inputs derived from the Vater-Pacini corpuscles of the LM.

For the above reasons, we confirmed that the function of the Vater-Pacini corpuscles was decreased in relation to LBP using the sweep frequency method. Ito et al. reported that patients with NSLBP may have even greater postural instability because the sensitivity of GS at 240 Hz continues to decline [15]. Thus, the reduction possibly causes postural instability because the NSLBP group cannot switch to a more appropriate proprioceptive postural strategy. In addition, the effect of RPW on postural control under proprioceptive stimulation with the response frequency of Vater-Pacini corpuscles may be a good indicator of evaluating the degree of LBP under the sweep frequency method, same as the constant frequency method used in our previous study [15]. These findings could be used to diagnose NSLBP due to the reduced function of proprioceptors. The advantages of using the sweep frequency are as follows. Firstly, the sweep frequency method may eliminate individual differences in the response frequency of proprioceptors. Secondly, the resolution is high because the vibration is applied to the entire frequency band of the response frequency of each proprioceptor. In the future, it may be possible to conduct more detailed proprioceptive examination by analyzing finer frequency bands. Additionally, there is a possibility that the inspection using this method can identify the detailed frequency band of proprioceptors related to NSLBP. Thirdly, the examination time could be shortened. The measurement time of the sweep frequency method can be shortened by 4 min and 30 s compared with the time required for the constant frequency method used in our previous study. Thus, this method can reduce the burden on patients. Therefore, the method using sweep frequency could replace the conventional constant frequency method because it can measure the entire frequency band of proprioceptors in a short time.

Nevertheless, this study has a limitation in that only elderly persons with lumbar spondylosis were surveyed. Therefore, additional studies with healthy older persons and those with more severe disability must be conducted. Finally, the evaluation of the entire range of response frequencies of proprioceptors using sweep frequency could benefit from future studies investigating more optimal stimulation of response frequencies.

5. Conclusions

In this study, we aimed to investigate the functional decline in proprioceptors related to LBP by evaluating the entire range of response frequencies of proprioceptors using sweep frequency. Thus, we developed a variable frequency vibratory stimulation device and measured the CoP of NSLBP and NLBP on a Wii Balance Board during local vibration with sweep frequencies. As a result, compared with the NLBP group, the NSLBP group showed a reduction in lower leg sensitivity at Vater-Pacini corpuscles. Moreover, we believe that the sweep frequency method can replace the conventional constant frequency method because it can measure the entire frequency band of proprioceptors in a short time. Our future work will include investigating the treatment effect of vibration with the frequency evaluated by this device.

Author Contributions: Conceptualization, Y.M. and Y.S.; methodology, Y.M.; software, R.N., Y.I. and K.Y.; validation, T.I.; formal analysis, R.N.; investigation, R.N., Y.I., T.I. and K.Y.; resources, Y.M. and Y.S.; data curation, R.N. and Y.I.; writing—original draft preparation, R.N.; writing—review and editing, Y.I., Y.M., T.I., K.Y. and Y.S.; visualization, R.N.; supervision, Y.M.; project administration, Y.M. and Y.S.; funding acquisition, Y.S.

Funding: This research was funded by Geriatrics and Gerontology research and development cost, grant number 29-6. The APC was funded by Geriatrics and Gerontology research and development cost.

Acknowledgments: The authors thank the staff in the Department of Orthopedic Surgery and the National Center for Geriatric and Gerontology for their help in participant recruitment. This study was supported in part by a grant from the Geriatrics and Gerontology research and development cost.

Conflicts of Interest: The authors declare no conflicts of interest.

References

1. Hoy, D.; March, L.; Brooks, P.; Blyth, F.; Woolf, A.; Bain, C.; Williams, G.; Smith, E.; Vos, T.; Barendregt, J.; et al. Global burden of low back pain: Estimates from the Global Burden of Disease 2010 study. *Ann. Rheum. Dis.* **2014**, *76*, 968–974. [CrossRef] [PubMed]
2. Andrea, R.; Cholewicki, J.; Polzhofer, G.K.; Greene, H.S. Impaired postural control of the lumbar spine is associated with delayed muscle response times in patients with chronic idiopathic low back pain. *Spine* **2001**, *26*, 724–730. [CrossRef]
3. Henry, S.; Hitt, J.; Jones, S.; Bunn, J. Decreased limits of stability in response to postural perturbations in subjects with low back pain. *Clin. Biomech.* **2006**, *21*, 881–892. [CrossRef] [PubMed]
4. Willigenburg, N.W.; Kingma, I.; van Dieën, J.H. How is precision regulated in maintaining trunk posture? *Exp. Brain Res.* **2010**, *203*, 39–49. [CrossRef] [PubMed]
5. Sean, C.; Tim, K.; John, J.J. Modeling the dynamics of sensory reweighting. *Biol. Cybern.* **2006**, *95*, 123–134. [CrossRef]
6. Johansson, R.; Magnusson, M. Human postural dynamics. *Crit. Rev. Biomed. Eng.* **1991**, *18*, 413–437.
7. Willigenburg, N.W.; Kingma, I.; Hoozemans, M.J.; van Dieën, J.H. Precision control of trunk movement in low back pain patients. *Hum. Mov. Sci.* **2013**, *32*, 228–239. [CrossRef]
8. Bloem, B.R.; Allum, J.M.; Carpenter, M.G.; Honegger, F. Is lower leg proprioception essential for triggering human automatic postural responses? *Exp. Brain Res.* **2000**, *130*, 375–391. [CrossRef]
9. Kuroki, S.; Hagura, N.; Nishida, S.; Haggard, P.; Watanabe, J. Sanshool on The Fingertip Interferes with Vibration Detection in a Rapidly-Adapting (RA) Tactile Channel. *PLoS ONE* **2016**, *11*. [CrossRef]
10. Pyykko, I.; Jantti, P.; Aalto, H. Postural control in elderly subjects. *Age Ageing* **1990**, *19*, 215–221. [CrossRef]
11. Brumagne, S.; Janssens, L.; Janssens, E.; Goddyn, L. Altered postural control in anticipation of postural instability in persons with recurrent low back pain. *Gait Posture* **2008**, *28*, 657–662. [CrossRef] [PubMed]

12. Goossens, N.; Janssens, L.; Pijnenburg, M.; Severijns, P.; Van, R.C.; Meugens, P.; Brumagne, S. Test-Retest Reliability and Concurrent Validity of an fMRI-Compatible Pneumatic Vibrator to Stimulate Muscle Proprioceptors. *Multisens. Res.* **2016**, *29*, 465–492. [CrossRef] [PubMed]
13. Capicíková, N.; Rocchi, L.; Hlavacka, F.; Chiari, L.; Cappello, A. Human postural response to lower leg muscle vibration of different duration. *Physiol. Res.* **2006**, *55*, 129–134.
14. Nima, T.; Hossein, E.; Marco, M.; Jane, M. Proprioceptive impairments in high fall risk older adults: The effect of mechanical calf vibration on postural balance. *BioMed. Eng. Online* **2018**, *17*, 51. [CrossRef]
15. Ito, T.; Sakai, Y.; Morita, Y.; Yamazaki, K.; Igarashi, K.; Nishio, R.; Sato, N. Proprioceptive Weighting Ratio for Balance Control in Static Standing Is Reduced in Elderly Patients with Non–Specific Low Back Pain. *Spine* **2018**, *43*, 1704–1709. [CrossRef]
16. Ito, T.; Sakai, Y.; Yamazaki, K.; Igarashi, K.; Sato, N.; Yokoyama, K.; Morita, Y. Proprioceptive change impairs balance control in older patients with low back pain. *J. Phys. Ther. Sci.* **2017**, *29*, 1788–1792. [CrossRef]
17. Willigenburg, N.W.; Kingma, I.; van Dieën, J.H. Precision control of an upright trunk posture in low back pain patients. *Clin. Biomech.* **2012**, *27*, 866–871. [CrossRef]
18. Brumagne, S.; Cordo, P.; Verschueren, S. Proprioceptive weighting changes in persons with low back pain and elderly persons during upright standing. *Neurosci. Lett.* **2004**, *366*, 63–66. [CrossRef]
19. Chris, M.; Martin, U.; Rachelle, B. Non-specific low back pain. *Lancet* **2017**, *389*, 736–747. [CrossRef]
20. Clark, R.A.; Mentiplay, B.F.; Pua, Y.H.; Bower, K.J. Reliability and validity of the Wii Balance Board for assessment of standing balance: A systematic review. *Gait Posture* **2018**, *61*, 40–54. [CrossRef]
21. Claeys, K.; Brumagne, S.; Dankaerts, W.; Kiers, H.; Janssens, L. Decreased variability in postural control strategies in young people with non-specific low back pain is associated with altered proprioceptive reweighting. *Eur. J. Appl. Physiol.* **2011**, *111*, 115–123. [CrossRef] [PubMed]
22. Brumagne, S.; Janssens, L.; Knapen, S.; Claeys, K.; Suuden-Johanson, E. Persons with recurrent low back pain exhibit a rigid postural control strategy. *Eur. Spine J.* **2008**, *17*, 1177–1184. [CrossRef] [PubMed]
23. Mientjes, M.I.; Frank, J.S. Balance in chronic low back pain patients compared to healthy people under various conditions in upright standing. *Clin. Biomech.* **1999**, *14*, 710–716. [CrossRef]
24. Ito, T.; Sakai, Y.; Nakamura, E.; Yamazaki, K.; Yamada, A.; Sato, N.; Morita, Y. Relationship between paraspinal muscle cross-sectional area and relative proprioceptive weighting ratio of older persons with lumbar spondylosis. *J. Phys. Ther. Sci.* **2015**, *27*, 2247–2251. [CrossRef] [PubMed]
25. Ito, T.; Sakai, Y.; Yamazaki, K.; Nishio, R.; Ito, Y.; Morita, Y. Postural Strategy in Elderly, Middle-Aged, and Young People during Local Vibratory Stimulation for Proprioceptive Inputs. *Geriatrics* **2018**, *3*, 93. [CrossRef]

applied sciences

MDPI

Article

Temporal EEG Imaging for Drowsy Driving Prediction

Eric Juwei Cheng [1,*], Ku-Young Young [1] and Chin-Teng Lin [2]

1 Department of Electrical Engineering, National Chiao Tung University, Hsinchu 300, Taiwan; kyoung@mail.nctu.edu.tw
2 Centre for Artificial Intelligence, University of Technology Sydney, Sydney 2007, Australia; chintenglin@gmail.com
* Correspondence: fred7018@gmail.com

Received: 21 October 2019; Accepted: 21 November 2019; Published: 25 November 2019

Abstract: As a major cause of vehicle accidents, the prevention of drowsy driving has received increasing public attention. Precisely identifying the drowsy state of drivers is difficult since it is an ambiguous event that does not occur at a single point in time. In this paper, we use an electroencephalography (EEG) image-based method to estimate the drowsiness state of drivers. The driver's EEG measurement is transformed into an RGB image that contains the spatial knowledge of the EEG. Moreover, for considering the temporal behavior of the data, we generate these images using the EEG data over a sequence of time points. The generated EEG images are passed into a convolutional neural network (CNN) to perform the prediction task. In the experiment, the proposed method is compared with an EEG image generated from a single data time point, and the results indicate that the approach of combining EEG images in multiple time points is able to improve the performance for drowsiness prediction.

Keywords: electroencephalography; deep learning; driving fatigue; feature extraction; convolutional neural network

1. Introduction

The prevention of drowsy driving has become a major challenge in safety driving issues. Many drivers experience driving in drowsy conditions, especially in long-term driving. Continuous, unexciting driving reduces the vigilance of drivers and increases the risk of traffic accidents. To address this problem, the development of brain-computer interfaces (BCIs) to investigate the human's cognitive state is an urgent necessity. Electroencephalography (EEG) is one of the most direct and effective physiological measures for the estimation of brain dynamics. Recent EEG studies have demonstrated that changes in alertness during driving are related to changes in global brain dynamics [1,2]. It has also been shown that EEG is a robust measurement for the estimation of a driver's cognitive state [3–6]. In addition, EEG provides abilities of convenient measurement in real timeand is therefore widely used in real applications [3,7–9].

Although EEG has many advantages for the analysis of brain dynamics, the use of EEG-based BCIs in real applications remains challenging. The raw EEG signals acquired from the electrodes are often obscured by physiological artifacts such as eye movement and muscle movement, which is undesirable in the BCI system [10]. Therefore, removing these unwanted artifactsto capture brain activity has become a crucial issue in EEG-based BCI applications. Many studies have shown that independent component analysis (ICA) can effectively separate the artifacts from raw EEG data [11–14]. The mixture signal is decomposed into many statistically independent components by ICA. A non-artifact signal is obtained by excluding the components that are associated with artifacts. Although ICA is a powerful tool for extracting brain activity from raw EEGsignal, it cannot support real-time applications because the separated artifacts need to be removed manually. This drawback limits the utility of ICA for real-world BCI applications. An automatic processing BCI is strongly required for drowsy driving prediction since traffic accidents always occur in a very short time. Therefore, this study does not apply any artifact removal process to the raw EEG data during the experiment, ensuring that the proposed method does not use manual processes for the drowsy driving prediction task.

For EEG signals without artifact removal, how to correctly extract the informative features of EEG signals becomes a major challenge in BCI applications. A popular approach for feature extraction is transforming the EEG signals into a frequency domain [15,16]. Fast Fourier transform (FFT) is applied to compute the power spectra of the multi-channel time-series EEG signals into the frequency domain; then, the average of the power spectra value for each frequency is collected to obtain a feature vector for classification [3,17]. The main disadvantage of such an approach is that it only considers the frequency information. EEG is measured over the scalp in a three-dimensional space. It is obvious that the spatial information of EEG cannot be well described by a feature vector. Instead of the 1D feature vector, there is an increasing trend to use 2D feature maps for the analysis of EEG, which have achieved good performance in their application areas [18,19].

As the most popular machine learning technique in recent years, deep learning has achieved significant success in a variety of research fields, such as speech [20], images [21–23] and video [24]. The ability of deep learning techniques to learn unknown features from incoming data has gained considerable attention in EEG studies [25–27]. There is an increasing trend to use convolutional neural networks (CNNs) to analyze EEGs due to their state-of-the-art performance in the computer vision field. A popular approach is transforming the EEG measurement into a 2D feature map and then passing it into a CNN model for classification [28–31].

For drowsy driving prediction, it is difficult to identify the drowsy state using the single time point of EEG data because drowsiness is an ambiguous event. The driving performance may not immediately decrease with increasing drowsiness levels, which means that drivers maintain normal driving performance even though their vigilance level has started to decrease. To overcome these difficulties, this study proposes a new EEG image method that combines multiple frames of EEG images to examine the temporal activity of EEG data. Such approach not only focus on the current EEG data, but also considers the brain activity of the previous time period. The evaluation results show that the proposed method can improve the performance of EEG image-based BCI systems in drowsiness prediction.

2. Experimental Setup

2.1. Virtual Reality (VR)-Based Driving Environment

In our previous studies, to observe the subject's drowsy state during the driving task, a virtual reality (VR)-based realistic driving environment was developed to simulate a long-term driving situation [2,13,32–35]. As shown in Figure 1, the surrounding scenes were projected from six projectors to constitute a surrounding vision. A night-time driving scene at a fixed velocity of 100 km/h on a four-lane highway is set up in the VR experiment. Before the experiment was started, participants were directed to enter the real car mounted on a motion platform and then steer the vehicle according to the instructions. All participants were required to take a 5-min pre-test session to ensure that they clearly understood the instruction and did not suffer from simulator sickness. The highway scene was connected to a physiological measuring system, where the EEG and participants' performance were continuously and simultaneously measured and recorded.

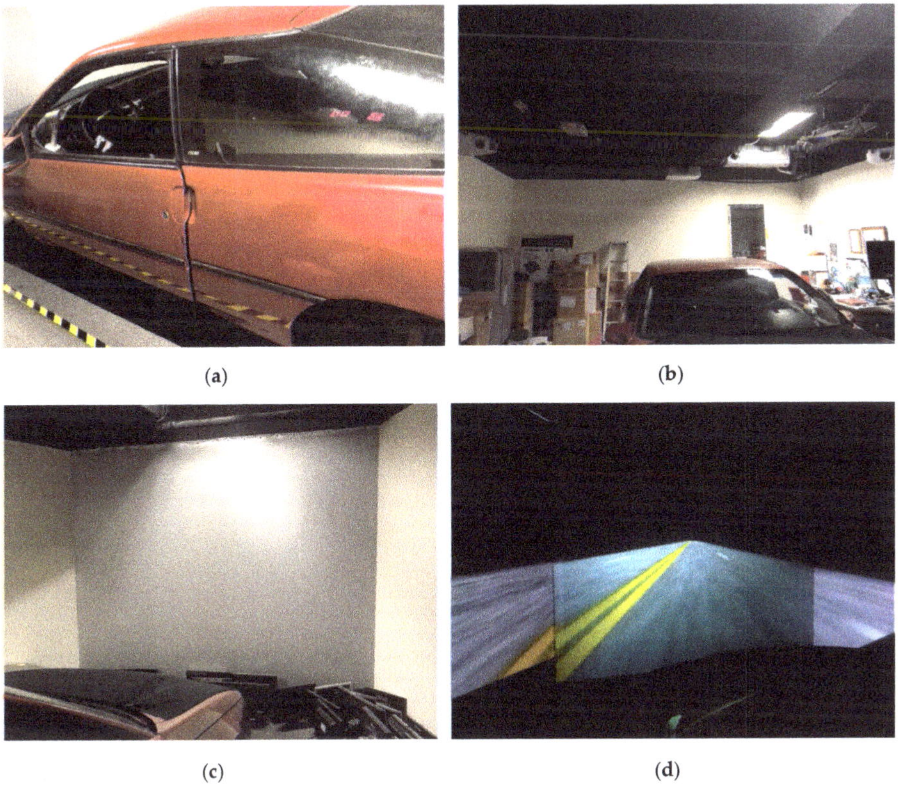

(a) (b) (c) (d)

Figure 1. The virtual reality (VR)-based driving environment. (**a**) Vehicle used for the driving task. (**b**) Projector platform. (**c**) The surrounding scene. (**d**) Front view of the VR scene.

2.2. Driving Fatigue Paradigm

The event-related lane-keeping driving task was adopted in this study for the evaluation of the brain dynamics occurring during the driving task, as illustrated in Figure 2. The participants were instructed to perform a 90-min driving task without breaking or resting in the VR driving environment. The driving experiment began in the early afternoon (13:00–14:00) after lunch because people often feel sleepy during this time [36]. During the sustained attention driving task, the VR paradigm randomly simulated a lane-departure event that caused the car to drift away from the center of the cruising lane. The participants were required to quickly steer the car back whenever the car started to deviate from the original cruising lane. There was no feedback to wake the participants even if they did not respond to the lane-departure event. The car continued to move along the curb until the participants steered it to return to the center of the cruising lane. Figure 2 describes a complete trial in the driving paradigm that includes the one-second baseline recording, deviation onset, response onset, and response offset. The time interval between the random lane-departure event was set to 5–10 s.

2.3. Participants

Thirty-eight right-handed, healthy young adults aged 20–30 years participated in the driving experiment. All subjects were required to have a driving license and sufficient sleep in the two preceding weeks. According to self-reporting, no subject had a history of psychological disorders. Before the experiment, the subjects were asked to answer a questionnaire about their sleep patterns to ensure that they had a normal cognitive state during the driving task, and they needed to complete a consent form explaining the experimental protocol that was approved by the Institutional Review Board of the Taipei Veterans General Hospital, Taiwan. The EEG signals of the subjects were captured from a Quik-Cap (CompumedicalNeuroScan) with 32 Ag/AgCl electrodes, including 30 EEG electrodes and two reference electrodes. The EEG electrodes were placed in accordance with a modified international 10–20 system. The impedance of all electrodes was kept under 5 kΩ during the experiments.

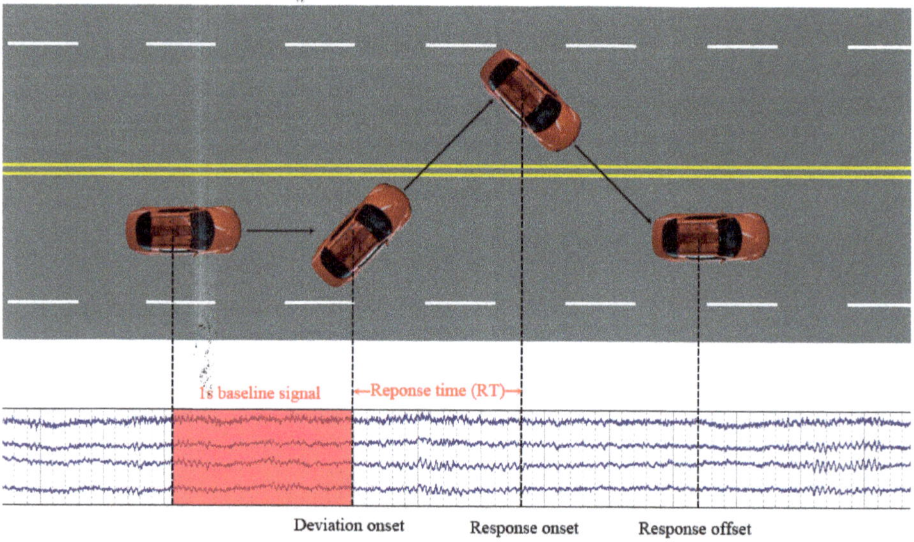

Figure 2. The event-related lane departure paradigm [2]. A complete trial consists of three steps: (**1**) deviation onset—the lane-departure event occurs, (**2**) response onset—the driver starts to steer the wheel back, and (**3**) the response offset—the car returns to the original cruising lane. Response time (RT) is defined as the time period between the deviation onset and the response onset. This study uses the 1 s baseline signal before the lane-departure event for predicting drowsiness according to the value of RT.

2.4. *Drowsiness Measurement*

The driving performance was defined based on the response time (RT), which represented the time between the deviation onset and the response onset. As the lane-departure event occurred, it was expected that the participant would take a long time to steer the car back to the center of the cruising lane if he/she was in a drowsy state; then, the response time (RT) in the trial could be very long. By contrast, when the participant was alert, he/she could respond to the lane-departure event in a short time. Previous studies have shown that baseline EEG activity is strongly correlated with changes in RT [34]. In this study, the 1s baseline signal (red region shown in Figure 2) was used to perform the drowsy prediction based on the trial's RT.

3. Approach

The general flowchart of our method is presented in Figure 3. Before data analysis, the acquired EEG records were processed using a 1-Hz high-pass and 50-Hz low-pass infinite impulse response filter to remove the noise and then down-sampled to 250-Hz to reduce the dimensions of the data. The power spectral activities of EEG signals were computed using FFT. To transform the EEG signal to a 2D image, we needed to address the following issues: (1) transforming the power spectrum of EEG signals to image values and (2) interpolating the points of the image data to a color image. The detailed approach is explained in the following sections.

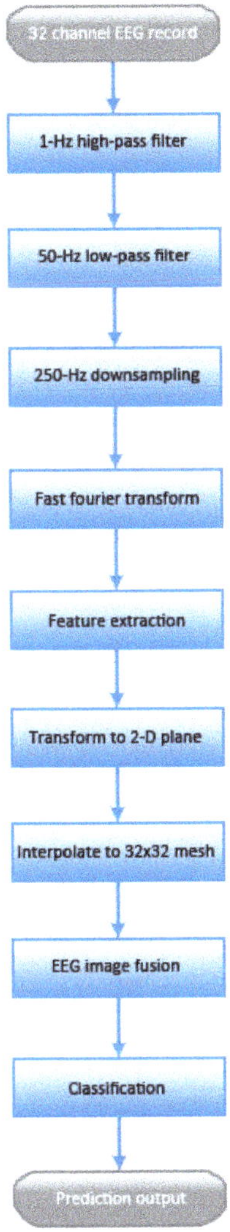

Figure 3. Flowchart of the proposed drowsiness prediction system.

3.1. Feature Extraction

To extract the physiological features, the 30-channel time-series EEG signal was transformed into a frequency domain via a 256-point FFT. Based on the findings in previous studies [14,37,38], the frequency band in the, theta (4–8 Hz), alpha (8–13 Hz), and beta (13–20 Hz) was suitable for estimating the driver's vigilance level. Our past studies also observed that the increasing power of theta band and alpha band had positive correlation with RT, and beta band had high correlation to kinesthetic stimuli which can affect the prediction performance [13,17]. The mean power of these frequency bands of interest was combined to form a feature vector. As depicted in Figure 4, this feature vector was considered a pixel value of the RGB image. Each channel of the colour image corresponds to a frequency band of interest.

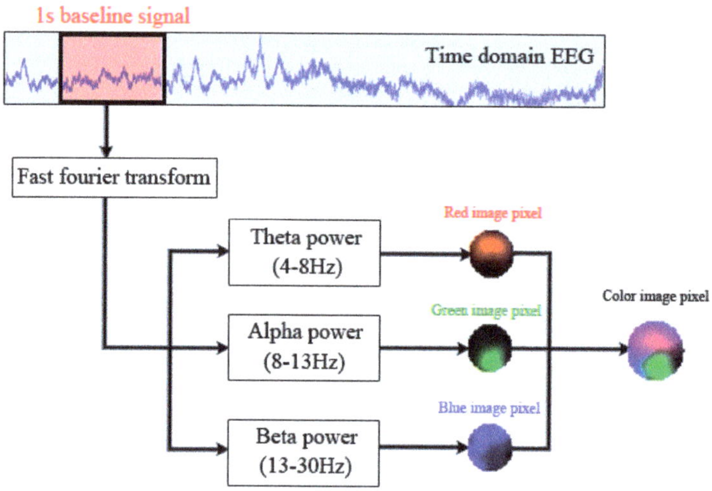

Figure 4. Schematic illustration used to transform the time domain electroencephalography (EEG) signals into a pixels in a color image.

3.2. Interpolation of the EEG Measurement to Image Pixels

As described in the previous section, we obtained 30 data points corresponding to the location of the EEG electrodes. First, we converted the magnitude of the power spectrum of the EEG signals into an image pixel value. Equation (1) shows the sigmoid function utilized to normalize the value of the EEG power spectrum to [0,1]:

$$P_t = \frac{1}{1 + e^{-0.5t}}, \quad (1)$$

where P_t is the normalized image pixel value, and t is the magnitude of the frequency response in the dB. Next, we needed to interpolate the scattered image data points to a color image. Figure 5 illustrates the interpolation scheme of the EEG image. The finite element method, a numerical technique that is usually applied for the approximate solution of engineering problems that are difficult to solve analytically, was adopted to perform the interpolation task. The Clough–Tocher scheme was used to interpolate a 32 × 32 mesh from the 30 image data points [39]. In this study, the EEG electrodes were placed in accordance with a modified international 10–20 system, which means that the location corresponding to each image point is known. Three topographical maps corresponding to the three frequency bands of interest were acquired by the Clough–Tocher scheme. The three spatial maps were then merged to create a 32 × 32 color image. Figure 6 demonstrates several samples of the EEG image.

3.3. Temporal EEG Image

One of the challenges in drowsy driving prediction is that some drowsy trials may have similar patterns of the alert trials. The driving performance might not degrade immediately, even if the alertness level of the drivers begins falling, which means that drivers can respond well to lane-departure events before they fall asleep (but the drowsy pattern of the EEG has appeared). In that case, the generated EEG images between the drowsy trial and its previous alert trials can be very similar. As the drivers wake up by themselves, their vigilance level dramatically recovers, and the RT returns to the alert state; then, the EEG pattern becomes completely different from the drowsy trials. Based on these findings, the drowsy state should be estimated not only using the current trial but also by examining the previous trials. This study proposes a temporal EEG image that is generated by a linear combination of a sequence of EEG images, as shown in Equation (2):

$$I'_t = \sum_{t=1}^{N} c_t I_t$$
$$\sum_{t=1}^{N} c_l = 1 \quad (2)$$

where I'_t is the generated temporal EEG image, I_t is the array of the EEG image data, c is a scalar coefficient and $c_i < c_j$ when $i < j$. N in this study is set to five. Figure 7 illustrates the schematic diagram of the temporal EEG image. Instead of asingle frame EEG image I_N, this approach estimates the drowsiness state using I'_N, which includes the information of the brain activity from multiple time points.

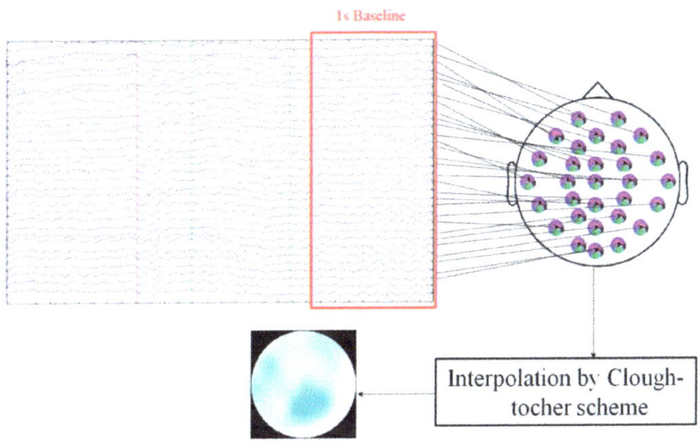

Figure 5. Interpolation scheme of the EEG image. The transformed image data points described in Figure 4 are arranged corresponding to the location of the EEG electrodes. These image pixels are interpolated over a 32 × 32 color image via Clough–Tocher scheme.

Figure 6. Examples of the generated EEG images. (**a**) The alert trials. (**b**) The drowsy trials.

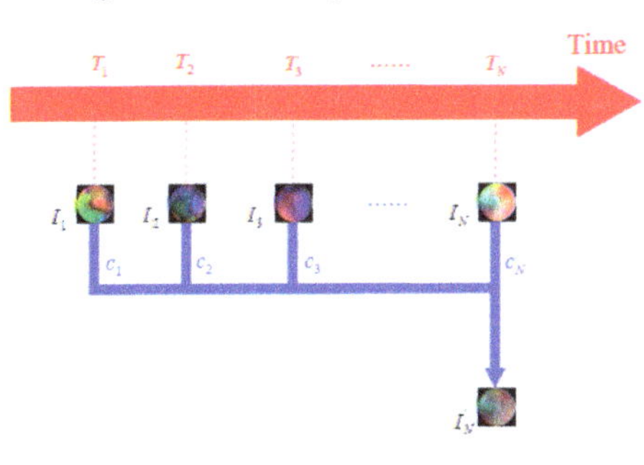

Figure 7. The schematic diagram of the temporal EEG image. I_N is the single frame EEG image generated from the current trial. Temporal EEG image I'_N is acquired by linearly combining I_N and its previous trials.

3.4. *Classification Using the CNN Model*

This study applies a CNN including six convolution layers, three max-pooling layers, and a layer to the classification of the input EEG image, as shown in Figure 8. A popular open source deep learning framework named Caffe is employed for implementing the CNN model [40]. The parameter setting of the overall CNN architecture is presented in Table 1. A set of filters is used to convolve the input EEG images for feature extraction. The convolved images are then subsampled by the max-pooling layer to derive compacted features. The convolution and pooling progress are repeated several times through CNN layers. The lower-level features of the input data are extracted via the early layers, and those features are collected in the later layers to hierarchically learn the higher-level features. Finally, the acquired high-level features are concatenated and passed into the fully-connected layer for the classification. The final prediction result is determined according to the output of the fully-connected layer. We only use the alert and drowsy classes in this study, so the output size of the fully-connected layer is 2 × 1.

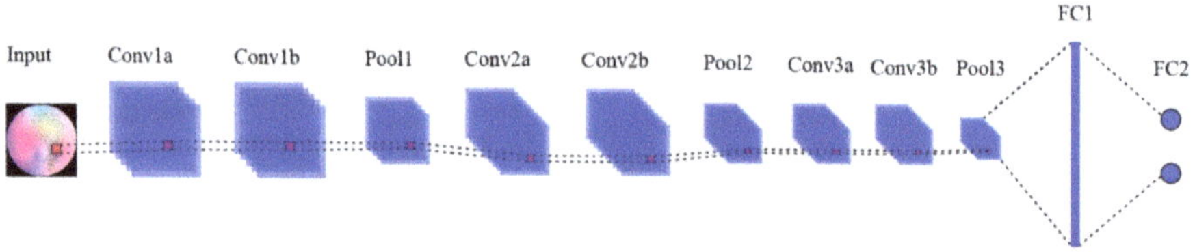

Figure 8. The convolutional neural network (CNN) architecture used in the experiment.

Table 1. Parameter setting of the CNN model.

Layer Name	Kernel Size	Stride	Num of Filters
Conv1a	3 × 3	1	16
Conv1b	3 × 3	1	16
Pool1	2 × 2	1	-
Conv2a	3 × 3	1	32
Conv2b	3 × 3	1	32
Pool2	2 × 2	2	-
Conv3a	3 × 3	1	32
Conv3b	3 × 3	1	32
Pool3	2 × 2	2	-
FC1	500	-	-
FC2	2	-	-

4. Experiment

The EEG dataset used in this study includes 10,395 alert trials and 3080 drowsy trials collected from 38 subjects. According to the suggestion of the previous research, if the drivers are fully aware of the driving situation, the average time for them to respond to the lane-departure event is approximately 0.7 s [41]. According to the previous studies [42–44], drivers provide poor performance when they don't respond the lane-departure event in three times the mean alert RT. Therefore, this study adopts three times the average response time as the classification boundary of the drowsy prediction task. The EEG trials are considered as alert trials as long as their RT are less than 2.1 s. In addition, the EEG trials with an RT larger than 2.5 s are labelled as drowsy trials, and the EEG trials with an RT between 2.1 s and 2.5 s are not used in this experiment. The evaluation of our approach is performed using leave-one-subject-out cross-validation. We select the data from one subject for testing and the data from the remaining subjects for training. This process is repeated for each of the 38 subjects. To evaluate the predictive performance of the proposed method, the temporal EEG image method is compared with the single frame EEG image method—directly using the current trial of the EEG image for drowsiness prediction.

Table 2 shows the comparison result of the temporal EEG image method and the single frame EEG image method. The average accuracy is an average of the accuracy of individual categories. It is apparent that the temporal EEG image method outperforms the single frame EEG image methods. In both methods, a similar accuracy of the alert class is given, but our approach achieves significant improvement in the accuracy of the drowsy class. Based on the aim of drowsiness prediction, the prediction rate of the drowsy class is more important than the alert class. Furthermore, the results also demonstrate that our approach has better prediction performance than the single frame EEG image method in most subjects, which proves that the improvement of our approach has universality in general users. Table 3 shows the evaluation result of EEGnet and hierarchical convolutional neural network (HCNN), which are CNN-based approach for EEG analysis and achieve good performance in their applications [45]. The results demonstrate that the proposed method yields superior prediction performance than EEGnet and HCNN.

Table 2. The prediction result of the single frame EEG image and temporal EEG image.

Subject	Single Frame EEG Image			Temporal EEG Image		
	Alert (%)	Drowsy (%)	Total Accuracy (%)	Alert (%)	Drowsy (%)	Total Accuracy (%)
S1	97.30	23.06	79.12	97.94	19.17	78.11
S2	88.21	20.13	52.25	82.72	29.78	53.95
S3	93.50	27.60	72.47	97.85	21.47	73.07
S4	96.76	9.30	71.85	97.79	14.19	73.33
S5	96.61	80.69	92.22	97.13	86.67	94.20
S6	96.72	84.03	93.22	98.21	96.25	97.66
S7	95.14	50.26	87.21	95.46	60.54	89.33
S8	98.30	27.69	87.31	96.42	66.15	91.60
S9	96.67	32.55	81.06	95.15	58.91	86.17
S10	96.50	12.10	80.25	96.21	8.39	79.09
S11	95.31	14.67	68.21	96.27	14.21	68.44
S12	95.60	15.80	66.30	94.50	25.71	68.90
S13	95.76	9.66	74.94	98.02	7.82	76.02
S14	92.04	17.44	80.25	96.24	16.46	83.53
S15	91.04	19.47	83.18	93.56	18.07	85.20
S16	94.10	18.25	80.62	95.06	20.00	81.62
S17	93.87	16.77	81.96	95.27	18.60	84.12
S18	92.30	16.13	85.02	92.79	16.45	85.46
S19	85.97	33.50	65.05	87.51	42.92	69.49
S20	92.40	21.44	76.24	95.15	20.17	77.94
S21	94.30	34.61	79.34	94.19	54.41	84.12
S22	92.98	21.86	76.20	94.49	21.86	77.07
S23	87.12	32.38	67.30	90.21	28.57	67.72
S24	95.68	20.28	76.46	94.83	30.35	78.27
S25	95.08	40.58	83.78	94.53	55.34	86.35
S26	92.02	37.11	85.08	91.68	48.67	86.22
S27	91.65	52.52	82.21	91.89	62.82	84.24
S28	97.60	26.50	85.55	97.29	30.25	85.73
S29	96.24	16.12	80.08	97.63	12.65	80.21
S30	94.79	12.10	74.83	95.24	13.45	75.34
S31	93.38	18.96	82.42	94.32	17.76	82.95
S32	87.57	23.57	73.37	87.92	29.76	74.78
S33	89.67	19.24	65.37	92.60	15.19	65.42
S34	89.21	25.00	84.49	90.36	22.00	85.26
S35	90.14	15.85	78.29	91.04	18.54	79.29
S36	88.33	23.57	52.43	79.65	35.63	54.75
S37	95.76	12.37	80.15	94.91	16.67	80.75
S38	69.29	40.23	53.27	52.73	53.60	53.22
Average	92.76	26.93	76.83	92.76	32.35	78.39

Table 3. The prediction result of EEGnet and HCNN.

Subject	EEGnet [45]			HCNN [28]		
	Alert (%)	Drowsy (%)	Total Accuracy (%)	Alert (%)	Drowsy (%)	Total Accuracy (%)
S1	94.39	13.33	73.99	99.07	11.11	76.92
S2	84.92	24.53	52.1	90.48	17.33	50.72
S3	93.84	16.4	68.7	95.51	21.33	71.43
S4	92.79	14.19	69.8	94.23	13.95	70.75
S5	90.81	18.89	70.66	99.45	86.11	95.72
S6	88.65	20.28	69.49	98.37	93.55	96.89
S7	91.26	23.24	79.29	94.82	32.43	83.88
S8	93.21	17.69	81.17	97.08	15.38	84.05
S9	92.46	16.73	73.69	98.2	12.73	77.03
S10	89.77	20.16	76.19	92.97	16.13	77.98
S11	91.8	12.12	64.8	92.54	12.72	65.5
S12	91.75	14.91	63.16	93.12	12.5	63.12
S13	96.4	9.08	75.1	94.31	11.76	74.18
S14	90.62	15.12	78.6	96.77	14.63	83.69
S15	91.18	16.67	82.93	89.96	21.05	82.33
S16	93.46	13.81	79.21	89.44	27.84	78.41
S17	91.69	15.91	80.66	84.43	17.2	74.65
S18	92.65	17.26	81.56	88.52	11.29	81.14
S19	91.41	14.75	60.44	94.92	17.5	63.63
S20	91.52	15.25	74.01	92.42	16.95	75.1
S21	94.09	19.31	75.16	97.34	20.59	77.92
S22	86.52	19.15	70.37	89.3	18.64	72.36
S23	85.65	19.63	61.56	96.35	7.94	64.09
S24	90.76	15.25	71.37	92.16	16.31	72.68
S25	87.33	20.58	73.38	97.69	22.33	81.95
S26	87.91	17.23	78.93	90.18	36.14	83.31
S27	89.78	15.95	71.84	96.08	28.24	79.59
S28	92.19	16.5	79.14	95.83	20	82.76
S29	92.37	13.27	76.15	95.79	12.24	78.66
S30	91.32	14.03	72.52	92.7	15.13	73.82
S31	91.95	16.27	80.71	92.19	17.91	81.15
S32	91.32	21.9	75.65	90.28	16.67	73.66
S33	100	0	64.89	90.41	11.39	62.67
S34	58.14	35	56.42	86.29	20	81.34
S35	83.3	29.76	74.62	88.68	17.07	77.08
S36	100	0	43.43	88.37	22.32	51.001
S37	83.62	17.22	71.61	92.02	22.22	79.4
S38	80	20.12	46.12	87.88	13.95	46.05
Average	90.02184	16.88132	71.03737	93.05658	21.64684	74.91082

5. Discussion and Conclusions

It is challenging to classify EEG data without an artifact removal process because drivers' brain activity can change over time due to many factors, such as their mental state and body movement, which result in the temporal fluctuations of the EEG signals. However, there still contains important information associated with the drowsiness level of drivers, and thus, the temporal analysis of the EEG signals becomes a crucial issue. This study proposes a temporal EEG image algorithm that combines a sequence of EEG images to form a new EEG image that contains brain dynamics from multiple time points. Our experimental results show that the proposed method achieves good performance in the drowsiness prediction.

Support vector machines (SVMs) are also employed for comparison with our approach because they are popular classifiers for EEG analysis. In our experiment, the computational cost is expensive and a bad prediction result is obtained if we use the EEG image as the input of SVM. Thus, the power spectrum of EEG is selected as the input of the SVM. Similar to the experiment described in the previous sections, the input data do not apply any artifact removal process. The experimental results indicate that SVM provides a biased prediction result towards the alert class. That is, it always predicts an alert regardless of the input and results in a perfect detection of alert trials but no detection of drowsy trials. We found that SVM only provides meaningful prediction results in a balanced training dataset that has a similar number of alert trials and drowsy trials. In real-world applications, BCI systems usually have to perform drowsiness prediction under imbalanced-datasets, which means that SVM cannot provide reasonable reliability in real-world BCI applications.

To find a suitable CNN model for the drowsiness prediction task, this study introduces two CNN architectures for further evaluation: (1) AlexNet—a very popular CNN architecture that is larger than the CNN architecture used in this study [18], (2) 3D CNN—by performing 3D convolutions, which is capable of learning features from both spatial and temporal dimensions. For AlexNet, the EEG measurement is transformed into 227×227 to fit the input size of AlexNet. For 3D CNN, different from our approach using a linear combination of a sequence of EEG images, these EEG images are directly fed into the 3D CNN model since they are 3D input data. Our results show that both AlexNet and 3D CNN cannot achieve better performance than the proposed CNN architecture in the present study. That is, high-complexity CNN is not required for the drowsiness prediction task.

The detection of drowsy trials remains a challenge because different subjects have different drowsy EEG patterns. A further investigation of brain dynamics is the key to improving the prediction performance of the BCI system.

Author Contributions: E.J.C. conceived the main idea and wrote the paper. K.-Y.Y. provided the academic support and checked the manuscript, C.-T.L. provided the funding support and performed the experiment.

Funding: This work was supported in part by the Australian Research Council under Grant DP180100670 and Grant DP180100656, Taiwan Ministry of Science and Technology under AQ1 Grant MOST 106-2218-E-009-027-MY3 and MOST 108-2221-E-009-120-MY2 and W911NF-10-2-0022.

Conflicts of Interest: The authors declare no conflict of interests.

References

1. Chuang, S.-W.; Ko, L.-W.; Lin, Y.-P.; Huang, R.-S.; Jung, T.-P.; Lin, C.-T. Co-modulatory spectral changes in independent brain processes are correlated with task performance. *Neuroimage* **2012**, *62*, 1469–1477. [CrossRef] [PubMed]
2. Huang, C.S.; Pal, N.R.; Chuang, C.H.; Lin, C.T. Identifying changes in EEG information transfer during drowsy driving by transfer entropy. *Front. Hum. Neurosci.* **2015**, *9*, 570. [CrossRef] [PubMed]
3. Liu, Y.T.; Lin, Y.Y.; Wu, S.L.; Chuang, C.H.; Lin, C.T. Brain Dynamics in Predicting Driving Fatigue Using a Recurrent Self-Evolving Fuzzy Neural Network. *IEEE Trans. Neural Netw. Learn. Syst.* **2015**, *27*, 347–360. [CrossRef] [PubMed]
4. Baulk, S.D.; Reyner, L.A.; Horne, J.A. Driver sleepiness—Evaluation of reaction time measurement as a secondary task. *Sleep* **2001**, *24*, 695–698. [CrossRef]
5. Banks, S.; Catcheside, P.; Lack, L.; Grunstein, R.R.; McEvoy, R.D. Low levels of alcohol impair driving simulator performance and reduce perception of crash risk in partially sleep deprived subjects. *Sleep* **2004**, *27*, 1063–1067. [CrossRef]
6. Campagne, A.; Pebayle, T.; Muzet, A. Correlation between driving errors and vigilance level: Influence of the driver's age. *Physiol. Behav.* **2004**, *80*, 515–524. [CrossRef]
7. Arunkumar, N.; Ramkumar, K.; Venkatraman, V.; Abdulhay, E.; Fernandes, S.L.; Kadry, S.; Segal, S. Classification of focal and non-focal EEG using entropies. *Pattern Recognit. Lett.* **2017**, *94*, 112–117.
8. Long, J.; Li, Y.; Wang, H.; Yu, T.; Pan, J.; Li, F. A hybrid brain computer interface to control the direction and speed of a simulated or real wheelchair. *IEEE Trans. Neural Syst. Rehabil. Eng.* **2012**, *20*, 720–729. [CrossRef] [PubMed]
9. Xiong, Y.; Gao, J.; Yang, Y.; Yu, X.; Huang, W. Classifying driving fatigue based on combined entropy measure using EEG signals. *Int. J. Control Autom.* **2016**, *9*, 329–338. [CrossRef]
10. Winkler, I.; Haufe, S.; Tangermann, M. Automatic classification of artifactual ICA-Components for artifact removal in EEG signals. *Behav. Brain Funct.* **2011**, *7*, 30. [CrossRef]

11. Hyvärinen, A.; Karhunen, J.; Oja, E. *Independent Component Analysis*; Wiley-Interscience: New York, NY, USA, 2001.
12. Tran, Y.; Craig, A.; Boord, P.; Craig, D. Using independent component analysis to remove artifact from electroencephalographic measured during stuttered speech. *Med. Biol. Eng. Comput.* **2004**, *42*, 627–633. [CrossRef] [PubMed]
13. Lin, C.-T.; Wu, R.-C.; Liang, S.-F.; Chao, W.-H.; Chen, Y.-J.; Jung, T.-P. EEG-based drowsiness estimation for safety driving using independent component analysis. *IEEE Trans. Circuits Syst. I Regul. Pap.* **2005**, *52*, 2726–2738.
14. Jung, T.P.; Humphries, C.; Lee, T.W.; Makeig, S.; McKeown, M.J.; Iragui, V.; Sejnowski, T.J. Extended ICA removes artifacts from electroencephalographic recordings. *Adv. Neural Inf. Process. Syst.* **1998**, *10*, 894–900.
15. Garipelli, G.; Chavarriaga, R.; R Millán, J. Single trial analysis of slow cortical potentials: A study on anticipation related potentials. *J. Neural Eng.* **2013**, *10*, 036014. [CrossRef]
16. Pfurtscheller, G.; Neuper, C. Motor imagery and direct brain- computer communication. *Proc. IEEE* **2001**, *89*, 1123–1134. [CrossRef]
17. Chuang, C.H.; Lai, P.C.; Ko, L.W.; Kuo, B.C.; Lin, C.T. Driver's cognitive state classification toward brain computer interface via using a generalized and supervised technology. In Proceedings of the 2010 International Joint Conference on Neural Networks (IJCNN), Barcelona, Spain, 18–23 July 2010; pp. 1–7.
18. Tabar, Y.R.; Halici, U. A novel deep learning approach for classification of EEG motor imagery signals. *J. Neural Eng.* **2016**, *14*, 016003. [CrossRef]
19. Carvalho, S.R.; Filho, I.C.; Resende, D.O.; Siravenha, A.C.; De Souza, C.R.B.H.; Debarba, B.D.; Gomes, R.; Boulic, A. Deep Learning Approach for Classification of Reaching Targets from EEG Images. In Proceedings of the 2017 30th SIBGRAPI Conference on Graphics, Patterns and Images (SIBGRAPI), Niteroi, Brazil, 17–20 October 2017.
20. Hinton, G.; Deng, L.; Yu, D.; Dahl, G.A.; Mohamed, N.; Jaitly, A.; Senior, V.; Vanhoucke, P.; Nguyen, T.; Sainath, B. Kingsbury, Deep neural networks for acoustic modeling in speech recognition. *IEEE Signal. Process. Mag.* **2012**, *29*, 82–97. [CrossRef]
21. Krizhevsky, A.; Sutskever, I.; Hinton, G. Imagenet classification with deep convolutional neural networks. In Proceedings of the NIPS, Doha, Qatar, 12–15 November 2012.
22. Taigman, Y.; Yang, M.; Ranzato, M.; Wolf, L. DeepFace: Closing the Gap to Human-Level Performance in Face Verification. In Proceedings of the IEEE Conference on Computer Vision and Pattern Recognition (CVPR), Columbus, OH, USA, 24–27 June 2014.
23. Sun, Y.; Wang, X.; Tang, X. Deep Learning Face Representation from Predicting 10,000 Classes. In Proceedings of the IEEE Conference on Computer Vision and Pattern Recognition (CVPR), Columbus, OH, USA, 24–27 June 2014.
24. Karpathy, A.; Toderici, G.; Shetty, S.; Leung, T.; Sukthankar, S.; Li, F.-F. Large-scale video classication with convolutional neural networks. In Proceedings of the IEEE Conference on Computer Vision and Pattern Recognition (CVPR), Columbus, OH, USA, 24–27 June 2014.
25. Jirayucharoensak, S.; Pan-Ngum, S.; Israsena, P. Eeg-based emotion recognition using deep learning network with principal component based covariate shift adaptation. *Sci. World J.* **2014**, *2014*, 627892. [CrossRef]
26. Zheng, W.-L.; Zhu, J.-Y.; Peng, Y.; Lu, B.-L. EEG-based emotion classification using deep belief networks. In Proceedings of the 2014 IEEE International Conference on Multimedia and Expo (ICME), Chengdu, China, 14–18 July 2014.
27. Hosseini, M.-P.; Soltanian-Zadeh, H.; Elisevich, K.; Pompili, D. Cloud-based deep learning of big eeg data for epileptic seizure prediction. In Proceedings of the 2016 IEEE Global Conference on Signal and Information Processing (GlobalSIP), Washington, DC, USA, 7–9 December 2016.
28. Li, J.; Zhang, Z.; He, H. Hierarchical convolutional neural networks for EEG-based emotion recognition. *Cogn. Comput.* **2017**, *10*, 368–380. [CrossRef]
29. Bashivan, P.; Rish, I.; Yeasin, M.; Codella, N. Learning representations from EEG with deep recurrent-convolutional neural networks. *Int. Conf. Learn. Represent.* **2016**, *1511*, 06448.
30. Emami, A.; Kunii, N.; Matsuo, T.; Shinozaki, T.; Kawai, K.; Takahashi, H. Seizure detection by convolutional neural network-based analysis of scalp electroencephalography plot images. *NeuroImage Clin.* **2019**, *22*, 101684. [CrossRef]

31. Comaniciu, A.; Najafizadeh, L. Enabling Communication for Locked-in Syndrome Patients using Deep Learning and an Emoji-based Brain Computer Interface. In Proceedings of the 2018 IEEE Biomedical Circuits and Systems Conference (BioCAS), Cleveland, OH, USA, 24 December 2018.
32. Lin, F.-C.; Ko, L.-W.; Chuang, C.-H.; Su, T.-P.; Lin, C.-T. Generalized EEG-based drowsiness prediction system by using a self-organizing neural fuzzy system. *IEEE Trans. Circuits Syst. I Reg. Pap.* **2012**, *59*, 2044–2055. [CrossRef]
33. Huang, R.-S.; Jung, T.-P.; Delorme, A.; Makeig, S. Tonic and phasic electroencephalographic dynamics during continuous compensatory tracking. *NeuroImage* **2008**, *39*, 1896–1909. [CrossRef] [PubMed]
34. Chuang, C.H.; Ko, L.W.; Jung, T.P.; Lin, C.T. Kinesthesia in a sustained-attention driving task. *Neuroimage* **2014**, *91*, 187–202. [CrossRef] [PubMed]
35. Lin, C.T.; Wu, R.C.; Jung, T.P.; Liang, S.F.; Huang, T.Y. Estimating Driving Performance Based on EEG Spectrum Analysis. *EURASIP J. Appl. Signal. Process.* **2005**, *19*, 3165–3174. [CrossRef]
36. Ueno, H.; Kaneda, M.; Tsukino, M. Development of drowsiness detection system. In Proceedings of the VNIS'94-1994 Vehicle Navigation and Information Systems Conference, Yokohama, Japan, 6 August 2002.
37. Jap, B.T.; Lal, S.; Fischer, P.; Bekiaris, E. Using EEG spectral components to assess algorithms for detecting fatigue. *Expert Syst.* **2009**, *36*, 2352–2359. [CrossRef]
38. Bashivan, P.; Bidelman, G.M.; Yeasin, M. Spectrotemporal dynamics of the EEG during working memory encoding and maintenance predicts individual behavioral capacity. *Eur. J. Neurosci.* **2014**, *40*, 3774–3784. [CrossRef]
39. Alfeld, P. A trivariate clough-tocher scheme for tetrahedral data. *Comput. Aided Geom. Des.* **1984**, *1*, 169–181. [CrossRef]
40. Jia, Y.; Shelhamer, E.; Donahue, J.; Karayev, S.; Long, J.; Girshick, R.; Guadarrama, S.; Darrell, T. Caffe: Convolutional architecture for fast feature embedding. *arXiv* **2014**, arXiv:1408.5093.
41. Marc, G. How Long Does It Take to Stop? Methodological Analysis of Driver Perception-Brake Times. *Transp. Hum. Factors* **2000**, *2*, 195–216.
42. Lin, C.T.; Huang, K.C.; Chuang, C.H.; Ko, L.W.; Jung, T.P. Can arousing feedback rectify lapses in driving? Prediction from EEG power spectra. *J. Neural Eng.* **2013**, *10*, 056024. [CrossRef] [PubMed]
43. Huang, K.C.; Jung, T.P.; Chuang, C.H.; Ko, L.W.; Lin, C.T. Preventing lapse in performance using a drowsiness monitoring and management system. In Proceedings of the 2012 Annual International Conference of the IEEE Engineering in Medicine and Biology Society, San Diego, CA, USA, 12 November 2012.
44. Lin, C.-T.; Huang, K.-C.; Chao, C.-F.; Chen, J.-A.; Chiu, T.-W.; Ko, L.-W.; Jung, T.-P. Tonic and phasic EEG and behavioral changes induced by arousing feedback. *NeuroImage* **2010**, *52*, 633–642. [CrossRef] [PubMed]
45. Lawhern, V.J.; Solon, A.J.; Waytowich, N.R.; Gordon, S.M.; Hung, C.P.; Lance, B.J. EEGNet: A compact convolutional neural network for EEG-based brain computer interfaces. *J. Neural Eng.* **2018**, *15*, 056013. [CrossRef] [PubMed]

applied sciences

MDPI

Article

Interval-Based LDA Algorithm for Electrocardiograms for Individual Verification

Chulseung Yang [1], Gi Won Ku [1], Jeong-Gi Lee [1] and Sang-Hyun Lee [2,*]

[1] Korea Electronics Technology Institute, Buk-gu, Gwangju 61011, Korea; yangcs@keti.re.kr (C.Y.); giwon9@keti.re.kr (G.W.K.); jklee@keti.re.kr (J.-G.L.)

[2] Department of Computer Engineering, Honam University, Gwangsan-gu, Gwangju 62399, Korea

* Correspondence: leesang64@honam.ac.kr

Received: 8 July 2020; Accepted: 27 August 2020; Published: 31 August 2020

Abstract: This paper presents an interval-based LDA (Linear Discriminant Analysis) algorithm for individual verification using ECG (Electrocardiogram). In this algorithm, at first, unwanted noise and power-line interference are removed from the ECG signal. Then, the autocorrelation profile (ACP) of the ECG signal, which is a mathematical representation of the degree of similarity between a given time series and a lagged version of itself over successive time intervals, is calculated. Finally, the interval-based LDA algorithm is applied to extract unique individual feature vectors that represent distance and angle characteristics on short ACP segments. These feature vectors are used during the processes of enrollment and verification of individual identification. To validate our algorithm, we conducted experiments using the MIT-BIH ECG and achieved EERs (Equal Error Rate) of 0.143%, showing that the proposed algorithm is practically effective and robust in verifying the individual's identity.

Keywords: Individual verification; Electrocardiogram (ECG); Interval based LDA; biometrics

---

1. Introduction

Since the invention of electrocardiogram (ECG) recording in 1903 [1], a considerable number of studies using the electrocardiogram have been performed in the fields of diagnosing cardiac-related diseases [2–5], detecting sleep apnea [6], monitoring driver drowsiness [7], and measuring blood pressure [8], because of its advantage of noninvasive convenience. Recently, researchers started to use the ECG biometrics to verify individual identity [9]. The ECG is a suitable tool because the signal itself constantly changes depending on the sympathetic/parasympathetic activities and it is resistant to attacks and duplication [10–12].

According to the latest US Federal Trade Commission report published in March 2019 [13], identity theft and cybercrime ranked third among the issues most complained about. Such kinds of identity fraud and hacking are becoming seriously dangerous. Unfortunately, they are unavoidable as long as we use password-based identity authentication systems. Therefore, new types of biometrics such as facial characteristics [14], gait [15], fingerprints and iris scans [16], vein pattern [17], etc., have been developed. Nevertheless, References [18,19] show that they are still vulnerable to a replay/falsification attack even though they perform better with very low identification errors. To avoid problems related to such attacks, the biometrics should be a variable, so that they cannot be stolen, duplicated, or shared. Thus, many studies are showing a great interest in the ECG signal as a promising alternative biometric [20–23].

The ECG signal is non-stationary and thus continuously changes over time. The two limbs of the autonomic nervous system (i.e., sympathetic and parasympathetic nerves) are the most important factors determining the changes in the heart rate. This type of variability is called intra-subject variability [24–26], that is, the changes occur for the same individual over time. There is also inter-subject variability found in the ECG's waveform that depends on the position, size, and anatomy of the heart, age, gender, relative body weight, chest configuration, and various other factors [27,28]. Therefore, the ECG signal varies from person to person, implying that we can use it as a biometric for the verification of an individual's identity.

There have been numerous attempts at applying ECG inter-subject variation for individual verification/identification [20–23,29–32], which can be classified into two categories: the fiducial dependent approach [20–23] and the fiducial independent approach [29–32]. The fiducial dependent approach relies on local features, such as time duration and amplitude differences between specific points of interest in an ECG, so it is less affected by heart rate variations. However, it has the possibility of missing the overall morphological information that might be useful in identifying individuals. Many studies have focused on fiducial dependent features such as time intervals, angles, and amplitudes between specific fiducial points on the QRS complex [20,33–35].

On the contrary, the fiducial independent approach is to find the feature of the ECG signal in the frequency domain rather than the time domain, or to extract the whole morphological features that appear to offer excellent discrimination information among individuals. Many algorithms based on frequency transformation have been proposed for extracting fiducial independent features from ECG signals, in which they use wavelet transform or DCT (Discrete Cosine Transform) to convert ECG signals in the time domain to the frequency domain [36–40]. The autocorrelation method also has been gaining considerable attention because of its excellent performances reported in many studies. It eliminates the need for fiducial point localization in the ECG signal by using an autocorrelation (AC) of the segmented ECG signal followed by DCT or LDA [41–43]. More recently, Safie [44] introduced a new feature extraction method, known as Pulse Active Ratio (PAR), that is implemented on electrocardiogram signals for biometric authentication. The Pulse Active Ratio uses pulse width modulation (PWM) to generate new ECG feature vectors and thereby can adapt to changes in heart rate. However, such fiducial independent approaches require tremendous computational effort because the parameters should be generated and compared for every person listed in the database during the verification stage. Furthermore, the database is often updated with intra-subject variability, in which the computational cost increases exponentially as the size of the database increases. This computational cost problem can be mitigated significantly by combining the above-mentioned two approaches.

In this paper, we propose an interval-based Linear Discriminant Analysis (LDA) algorithm, a type of hybrid algorithm that combines the complementary strengths of both fiducial dependent and fiducial independent features, and thereby is able to achieve higher accuracy of identification. The main characteristics of the proposed algorithm are as follows: (i) the use of an autocorrelation profile (ACP) of ECG signal; (ii) the use of short segments of the ACP to extract the interval-based feature vectors, which can be updated every 5 s, showing higher adaptability to intra-subject variation; (iii) the enrollment and verification of the individual's interval-based feature vectors; and (iv) the update of interval-based feature vectors when the user's identity claim is accepted.

The remainder of the paper is organized as follows. The proposed LDA algorithm based on the short ACP segments to determine the feature vectors on an interval basis as enrollment process, the verification process, and the algorithm to update the feature vectors are described in Section 2. In Section 3, the experimental results are compared with the existing ECG identification algorithms based on the same public database. Finally, the conclusion is drawn in Section 4.

2. Materials and Methods

The ECG signal reflects the electrical activity of the heart. A basic ECG signal cycle is shown in Figure 1 with a single lead ECG sensor. The signal cycle consists of the P wave, followed by the QRS complex, and the T wave [1,33].

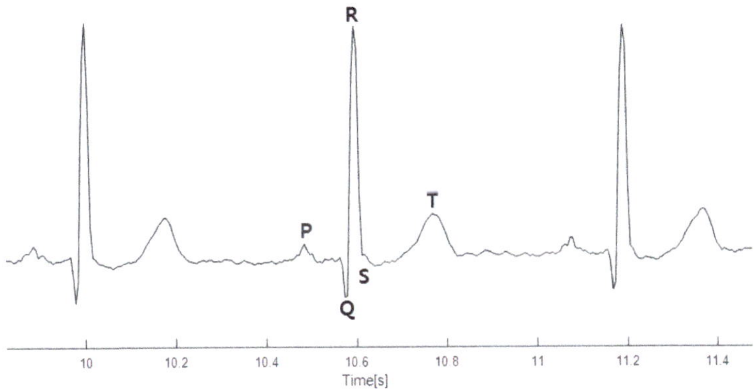

Figure 1. Basic ECG signal cycle.

Figure 2 shows a simplified block diagram of the proposed interval-based LDA algorithm. This algorithm mainly consists of signal preprocessing, enrollment, and verification.

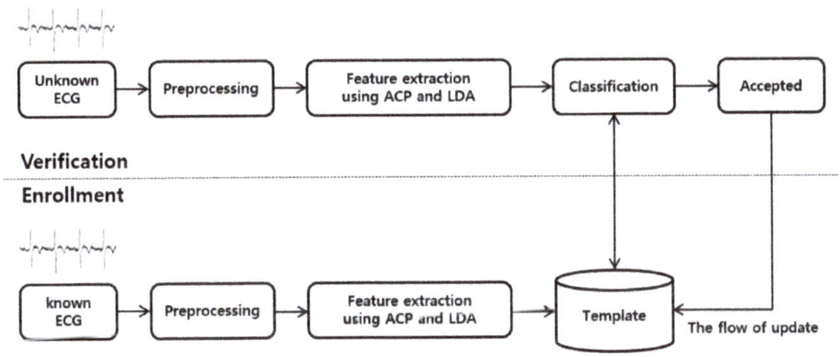

Figure 2. Simplified block diagram of the proposed interval-based LDA algorithm.

In the preprocessing stage, the raw ECG signal's noise is removed or suppressed. For biometric authentication, every user can use his/her ECG sensor in one of the two operation modes, that is, enrollment or verification mode.

In the enrollment mode/stage, a distinctive feature set is extracted from the ACP (autocorrelation profile) of the ECG on an interval basis to create an association between the user and his/her biometric characteristics. Subsequently, the extracted feature set is stored in a template. This process is called training in the enrollment mode. The template is updated over time, mainly to handle the intra-subject variation of cardiac signals. Longer training of the ECG signal leads to lower chances of false rejection (FR). In this study, the enrollment is performed for 2 min to provide a stable signature.

Similarly, during the verification mode/stage, the newly acquired unknown input is compared with the template stored in the database to verify a claimed identity. The verification decision is made through a matching process based on a feature vector threshold. The obtained matching score is finally compared with the threshold to either accept or reject the target user's identity claim.

2.1. Preprocessing

We used the ECG data from the MIT-BIH public database to test the proposed algorithm. The MIT-BIH database (MITDB) [34] is the standard ECG database universally used for ECG analyses that contains 48 records of heartbeats collected from 47 subjects; each record was sampled at 360 Hz and is approximately 30 min long. The subjects were 25 men with ages ranging from 32 to 89 years and 22 women with ages ranging from 23 to 89 years. The data were obtained with a two-channel ambulatory Holter monitor at the Bath Israel Deaconess Medical Center from 1975 to 1979. In most database records, the upper signal is a modified limb lead II(MLII) obtained by placing the electrode on the chest, and the lower signal is usually a V1(occasionally V2 or V5 or V4). Normal QRS complexes are usually prominent in the upper signal, thus, we only used the single lead data (the upper signal, MLII) in the experiment.

To minimize the negative effect of random noises, we applied the scheme proposed by Sornmo [35] in our study, in which noisy ECG signals are passed through the bandpass filters (BPF) spanning from 0.5 to 200 Hz to reject the DC components and high-frequency noise. Subsequently, we implemented a notch filter and a short-time Fourier transform (STFT) to eliminate the power line interference and its harmonics. Figure 3 shows the preprocessing procedures.

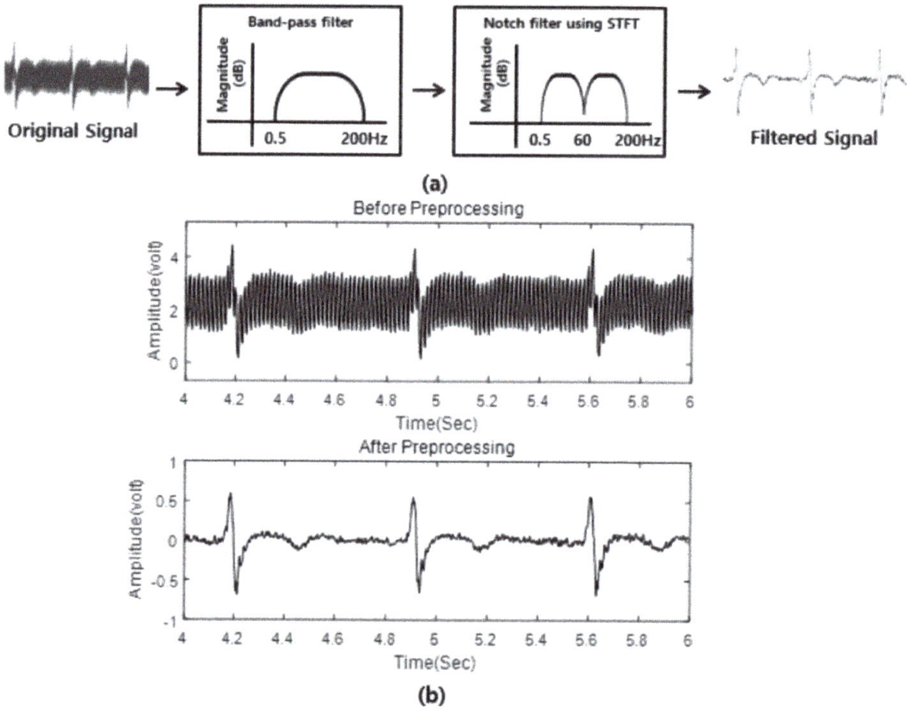

Figure 3. ECG signal preprocessing. (**a**): The simplified block diagram of signal preprocessing, where the BPF and notch filter are applied to remove the noise and power line interference related harmonics on ECG signals. (**b**): An example of an ECG signal before and after the preprocessing.

2.2. Feature Extraction

Let the noise-filtered ECG signals be $x(n)$, with the time index n. The proposed method strongly relies on the small pieces of $x(n)$, called classes, and calculates the ACP (autocorrelation profile) of each class. Then, each calculated ACP is sliced into short-time ACP segments; each segment is called ACPlet. By dividing each ACP into multiple ACPlets, we may easily capture the temporal pattern of the ACP in detail. The features we used are mainly based on the amplitude and angle patterns of the ACPlets to identify the similarities between an unknown ACP input and the reference profile.

2.2.1. Calculating the Reference Profile, $\overline{\Phi}(m)$

As mentioned in the previous section, during the enrollment mode, 2 min long ECG signals received from the ECG sensor are considered to be the training set $x(n)$. The training set $x(n)$ is then divided into several classes with a length of N. Each class of $x(n)$ is labeled $x_c(n)$ with the class "c", as follows

$$x_c(n) = x(n)\cdot R_N(n - N\cdot c) \text{ for a given class } c = 0, 1, \cdots, C-1 \tag{1a}$$

where

$$R_N(n) = \begin{cases} 1, & 0 \leq n \leq N-1 \\ 0, & \text{otherwise} \end{cases} \tag{1b}$$

Subsequently, the ACP $\Phi_c(m)$ of $x_c(n)$ can be computed by the following Equation (2a).

$$\Phi_c(m) = \sum_{n=0}^{N-1} x_c(n)\cdot x_c(n+m) \text{ for } 0 \leq m \leq Q-1 \text{ and a given class c} \tag{2a}$$

where $x(n+m)$ is the time-shifted version of ECG with a time lag m and Q is the length of the QRS complex. Figure 4 shows how the signals $x_c(n)$ and $x_c(n+m)$ are defined to obtain the ACP, that is, $\Phi_c(m)$.

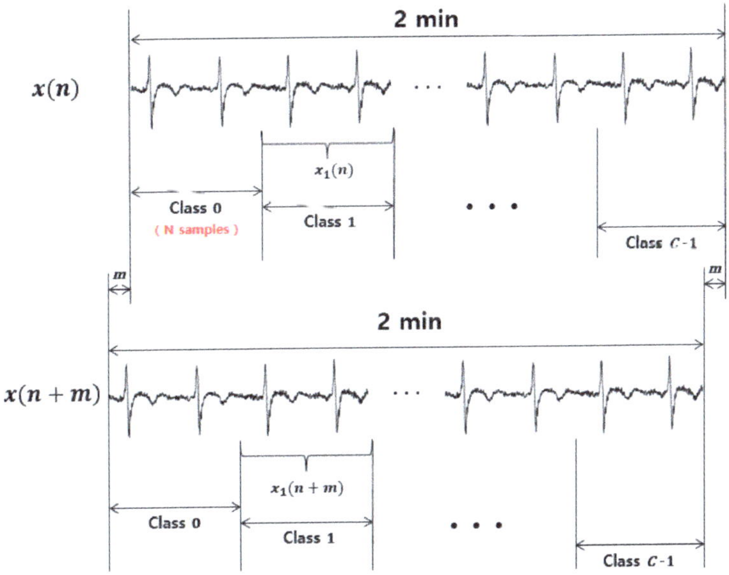

Figure 4. The two-minute-long training set $x(n)$ and its time-shifted version $x(n+m)$. The training set $x(n)$ is divided into C classes, each of which is N samples long. For example, $x_1(n)$ and $x_1(n+m)$ are the 1st class of $x(n)$ and $x(n+m)$, respectively.

Based on the previous research, the QRS complex is considered to be the most prominent waveform of ECG for individual verification [41,45,46]. Hence, the time shift parameter m of $\Phi_c(m)$ was chosen between 0 to $Q-1$. In this paper, the length Q, which depends on the individual, was calculated using the QRS detection method based on the Pan-Tompkins algorithm [47]. Depending on the individual, the different length of Q helps in achieving a high identification accuracy and robustness of the proposed algorithm.

Based on the assumption that the training set $x(n)$ includes C classes, we define a set of ACPs for C classes as follows:

$$\Phi_0(m), \Phi_1(m), \cdots, \Phi_c(m), \cdots, \Phi_{C-1}(m) \tag{2b}$$

Subsequently, the reference profile $\overline{\Phi}(m)$ is defined as the average of all ACPs contained in the training set $x(n)$, given by:

$$\overline{\Phi}(m) = \frac{1}{C}\sum_{c=0}^{C-1}\Phi_c(m) \tag{3}$$

where C is the total number of classes in the training set $x(n)$ for 2 min.

2.2.2. Interval-Based Feature Extraction

Figure 5 shows an example of ACPs for two different subjects (subject A and subject B). We assumed that each subject has C classes and those C classes are displayed together within the same time period Q.

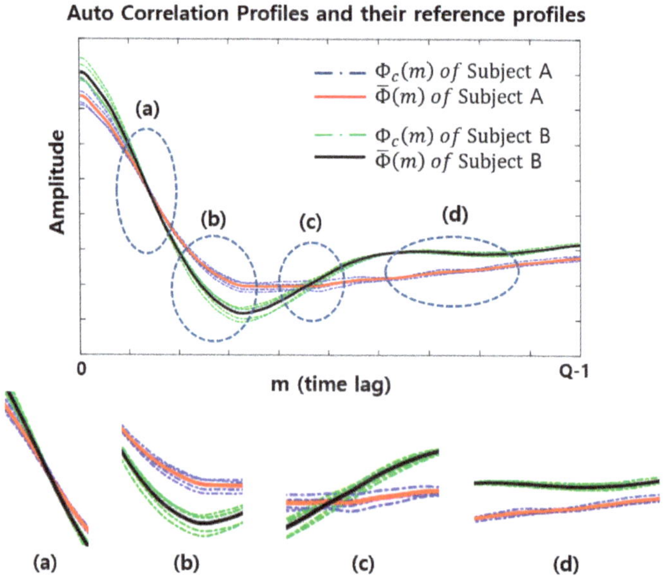

Figure 5. An example of a set of ACPs, displayed together within the same time period of Q, for two different subjects (subject A and subject B) when C = 5. (**a**–**d**) illustrate the differences in amplitude and angle patterns of ACPlet generated when the ACPs are segmented into six ACPlets, even though the two subjects have a very similar ACP shape.

From Figure 5, the intervals (a) and (c) are when the two subjects are difficult to distinguish by using only the amplitude patterns of $\Phi_c(m)$, but these two subjects are more easily distinguished using the angle patterns of $\Phi_c(m)$. It is obvious from the two profiles of (a) and (c) that their angle patterns are significantly different, in which the angle pattern is defined as the angles between the horizontal x-axis and the ACPlet. On the other hand, (b) and (d) are the intervals when the two subjects can be easily distinguished using the amplitude patterns of $\Phi_c(m)$.

These results lead to the idea of an interval-based feature vector. Accordingly, to facilitate the identification between different subjects, each class needs to be sliced into short-time ACP segments, i.e., the ACPlets. Therefore, in this work, the two features of amplitude and angle patterns of ACP produced for every ACPlet are used to distinguish the ECG signals.

2.2.3. Deciding the Optimal Interval Size for Interval-Based Feature Vector

In Section 2.2.2, we realized that dividing ACPs with a short interval and applying different feature vectors to each ACPlet yields a better identification; however, the interval size should be decided appropriately. The optimal interval size is determined based on the maximum interval that satisfies a mean square error (MSE) criterion between the ACPlets and their linearly fitted lines, such that each ACPlet provides clear angle information.

Based on the assumption of the optimal interval size L, the reference profile $\overline{\Phi}(m)$ can be equally divided into Q/L profiles $f_i(v)$, defined as follows:

$$\overline{\Phi}(m) = \{\ f_1(v), f_2(v), \cdots, f_i(v), \cdots, f_{NI}(v)\ \} \text{ for a given } L \tag{4}$$

where $i \left(= \frac{Q}{L}\right)$ is the number of intervals and $0 \leq v \leq L - 1$.

Here, let define $\hat{\Phi}(m)$ as the set of $g_i(v)$ that is the $f_i(v)$ linearly fitted, as follows:

$$\hat{\Phi}(m) = \{\ g_1(v), g_2(v), \cdots, g_i(v), \cdots, g_{NI}(v)\ \} \tag{5}$$

where $0 \leq v \leq L - 1$.

Thus, the optimal interval L is the maximum length that satisfies the following MSE (ζ) criterion between the two profiles $\overline{\Phi}(m)$ and $\hat{\Phi}(m)$.

$$L = \max_{l} \left[\sum_{i=1}^{NI} \sum_{v=0}^{l-1} (\ f_i(v) - g_i(v)\ )^2 \leq \varepsilon \right] \text{for } 2 \leq l \leq \frac{Q}{2} \tag{6}$$

where ε is the minimum tolerance of the MSE and NI (=Q/L) is the number of intervals. The symbol represented in the above process is illustrated in Figure 6.

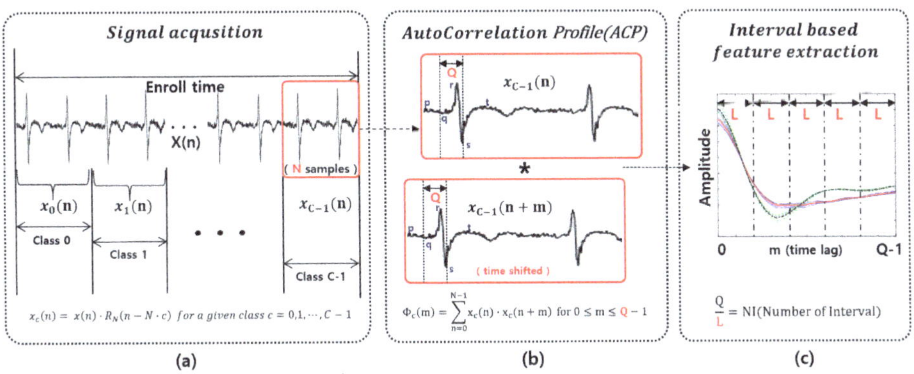

Figure 6. Symbol descriptions: (**a**) N is the update cycle of the verification process; (**b**) Q is the time-shifting size of the ACP process; (**c**) L is the size of an interval and the maximum length that satisfies the MSE (ζ) criterion.

Figure 7 shows typical examples of the MSE ζ and identification accuracy η as a function of NI (Number of Interval). In Figure 7, MSE ζ represents the ratio based on the assumption that MSE ζ is 100% when NI is set to 2, where (a) represents the minimum tolerance ε, which is experimentally set to 20%; (b) is the NI value satisfying (a), which corresponds to Q/L; and (c) is the identification accuracy when the NI value corresponds to (b).

The Figure 7 shows that the MSE ζ decreases as NI increases, where a higher NI is proportional to identification accuracy η but requires more computations for the verification process. Accordingly, there is a trade-off between the NI and computational cost. We also observed that the IDA improves with increasing NI; however, the improvement is not significant. For instance, as the NI increases from Q/L to 50, only an improvement of less than 0.2% is obtained. In conclusion, the performance of the proposed algorithm is not highly sensitive to the NI as long as it is greater than Q/L. Finally, the NI is set to Q/L when MSE ζ corresponds to ε, since there is no big difference in the final performance as long as the NI is greater than Q/L.

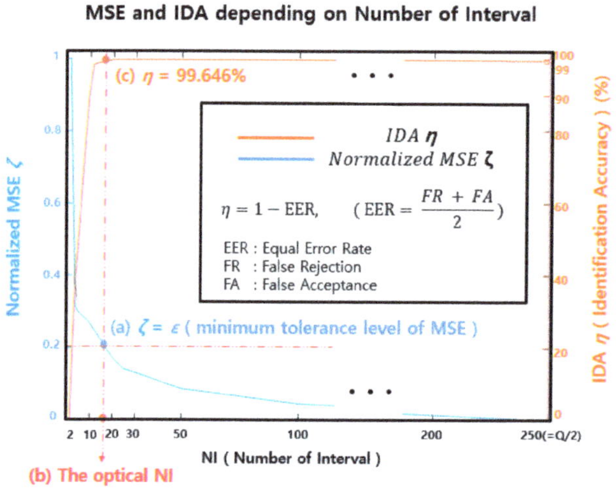

Figure 7. The relationship between the MSE (Mean Squared Error) ζ and the IDA (Identification Accuracy) η according to NI (Number of Interval) values to show the decision of the optimal NI. In this experiment, the optimal NI is set as the value when $\eta = 99.646\%$ and $\zeta = \varepsilon$

2.3. Threshold Values for Classification

Our algorithm requires two types of interval-based thresholds for the robust verification of intra-subject variation. One threshold is for the amplitude feature vector and another is for the angle feature vector, denoted $TH_{\phi}(i,v)$ and $TH_{\theta}(i)$, respectively, where the variable i indicates the interval index. These thresholds are used to check the similarity between the amplitude and angle feature vectors of the newly incoming ACPlets to those stored in the template.

2.3.1. Threshold of Amplitude Similarity: $TH_{\phi}(i,v)$

Step 1: Consider a subject A for the training set and let define the ACP as $\phi_c^A(m)$ that consists of the ACPlets $\phi_c^A(i,v)$ for $c \in \{0,\ldots,C-1\}$:

$$\phi_c^A(m) = \left\{\phi_c^A(1,v), \phi_c^A(2,v), \cdots, \phi_c^A(i,v), \cdots, \phi_c^A(NI,v)\right\} \tag{7}$$

In matrix form,

$$\begin{bmatrix} \phi_0^A(m) \\ \phi_1^A(m) \\ \vdots \\ \phi_c^A(m) \\ \vdots \\ \phi_{C-1}^A(m) \end{bmatrix} = \begin{bmatrix} \phi_0^A(1,v) & \phi_0^A(2,v) & \cdots & \phi_0^A(i,v) & \cdots & \phi_0^A(NI,v) \\ \phi_1^A(1,v) & \phi_1^A(2,v) & \cdots & \phi_1^A(i,v) & \cdots & \phi_1^A(NI,v) \\ \vdots & \vdots & & \vdots & & \vdots \\ \phi_c^A(1,v) & \phi_c^A(2,v) & \vdots & \phi_c^A(i,v) & \vdots & \phi_c^A(NI,v) \\ \vdots & \vdots & \cdots & \vdots & \cdots & \vdots \\ \phi_{C-1}^A(1,v) & \phi_{C-1}^A(2,v) & \cdots & \phi_{C-1}^A(i,v) & \cdots & \phi_{C-1}^A(NI,v) \end{bmatrix} \tag{8}$$

Step 2: Calculate the average in a vertical way to obtain the reference profile, $\overline{\phi}^{A}(m)$:

$$\overline{\phi}^{A}(m) = \left[\overline{\phi}^{A}(1,v)\ \overline{\phi}^{A}(2,v)\ \cdots \overline{\phi}^{A}(i,v)\ \cdots \overline{\phi}^{A}(NI,v)\right] \tag{9}$$

where

$$\overline{\phi}^{A}(i,v) = \frac{1}{C}\sum_{c=0}^{C-1} \phi_c^A(i,v) \text{ for a given } i \text{ and } v \tag{10}$$

Step 3: Define $\Delta\phi_c^A(i,v)$ as the amplitude distance between $\overline{\phi}^{A}(i,v)$ and $\phi_c^A(i,v)$:

$$\Delta\phi_c^A(i,v) = \left|\overline{\phi}^{A}(i,v) - \phi_c^A(i,v)\right| \tag{11}$$

where the distance $\Delta\phi_c^A(i,v)$ is large if the descriptors are further apart in the descriptor space and small if they are closer together.

We obtain:

$$\begin{gathered} TH_{\phi_{max}}(i,v) = max\left\{\Delta\varnothing_c^A(i,v)\middle|\ \forall c \in \{0,1,\cdots C-1\}\right\}, \\ \text{and } TH_{\phi}(i,v) = \mu_{\phi}^A(i,v) + \sigma_{\phi}^A(i,v) \text{ for a given } i \text{ and } v \end{gathered} \tag{12}$$

where

$$\begin{gathered} \mu_{\phi}^A(i,v) = mean\left\{\Delta\phi_c^A(i,v)\middle|\ \forall c \in \{0,1,\cdots C-1\}\right\} \\ \sigma_{\phi}^A(i,v) = std\left\{\Delta\phi_c^A(i,v)\middle|\ \forall c \in \{0,1,\cdots C-1\}\right\} \end{gathered} \tag{13}$$

Note that the amplitude distance $\Delta\phi_c^A(i,v)$ is small and thus $TH_{\phi}(i,v)$ is most likely smaller than $TH_{\phi_{max}}(i,v)$, since this parameter reflects the distance between intra-subjects.

2.3.2. Threshold of Angle Similarity: $TH_{\theta}(i)$

We calculated the scatter within class representing the degree of dispersion in a class which is called intra-subject variability, and the scatter between classes representing the degree of dispersion between different classes, which is called inter-subject variability, to obtain $TH_{\theta}(i)$ [43].

Step 1: Consider a data set of (x,y) and (x,r) for a given $i_o \in i$, where

$$\begin{gathered} x = \{\forall v\}, \\ y = \left\{\phi_c^A(i_o,v)\ \middle|\ i_o \in i,\ \forall v\right\} \text{ for a given class } c \\ r = \left\{\overline{\phi}^{A}(i_o,v)\ \middle|\ i_o \in i, \forall v\right\} \end{gathered} \tag{14}$$

Step 2: Calculate the scatter within class and scatter between classes of the dataset (x,y) and (x,r).

$$\begin{gathered} S_W = S_{(x,y)} + S_{(x,r)} \\ S_{(x,y)} = \Sigma\left(y-u_y\right)\left(y-u_y\right)^T \\ S_{(x,r)} = \Sigma\left(r-u_r\right)\left(r-u_r\right)^T \\ S_B = \left(u_y-u_r\right)\left(u_y-u_r\right)^T \end{gathered} \tag{15}$$

where u_y and u_r indicates the mean of y and r, respectively.

Step 3: Calculate the weight vector matrix in terms of S_W and S_B. Subsequently, use the eigenvector with the highest eigenvalue as the most discriminant projection vector.

$$\begin{gathered} W = \frac{S_B}{S_W} \\ WV = \lambda V \end{gathered} \tag{16}$$

where V is the eigenvector when λ corresponds to the highest eigenvalue.

Let's define $\theta_c^A(i_o)$ as the angle θ of the x-axis to the most discriminant projection vector V of the dataset, (x, y) and (x, r).

$$\theta_c^A(i_o) = tan^{-1}(V) \tag{17}$$

Step 4: Define $\theta_c^A(i_o)$ as the angle distance per interval between dataset (x, y) and (x, r) for a given c and i_o.

We obtain:

$$\begin{aligned} TH_{\theta_{max}}(i_o) &= max\left\{ \theta_c^A(i_o) \middle| \forall c \in \{0, 1, \cdots C-1\} \right\} \\ TH_{\theta_{min}}(i_o) &= min\left\{ \theta_c^A(i_o) \middle| \forall c \in \{0, 1, \cdots C-1\} \right\} \end{aligned} \tag{18}$$

and

$$TH_{\theta_1}(i_o) = \mu_\theta^A(i_o) + \sigma_\theta^A(i_o), \ TH_{\theta_2}(i_o) = \mu_\theta^A(i_o) - \sigma_\theta^A(i_o) \tag{19}$$

where

$$\begin{aligned} \mu_\theta^A(i_o) &= mean\left\{ \theta_c^A(i_o) \middle| \forall c \in \{0, 1, \cdots C-1\} \right\} \\ \sigma_\theta^A(i_o) &= std\left\{ \theta_c^A(i_o) \middle| \forall c \in \{0, 1, \cdots C-1\} \right\} \end{aligned} \tag{20}$$

Similar to the amplitude distance $\Delta\phi_c^A(i, v)$, the angle distance $\theta_c^A(i_o)$ is mostly between $TH_{\theta_1}(i_o)$ and $TH_{\theta_2}(i_o)$ and almost always between $TH_{\theta_{min}}(i)$ and $TH_{\theta_{max}}(i)$. Based on this process and the subsequent updating process, the range of the intra-subject variation can be determined and the advantage of the constantly changing ECG signal can be reaped, which is beneficial for the verification using ECG.

2.4. Verification

In addition to the biometric data for subject A, the template database includes information about the identifier related to the subject A. The term "biometric data" is referred to as the $\left\{ \overline{\phi}^A(i, v) \middle| \forall i \in \text{training set} \right\}$ stored along with the $TH_\phi(i, v)$ and $TH_\theta(i)$ in the template database. During the verification stage, the system validates the identity claimed by an unknown person by checking the similarity of the person's biometric data to the enrolled data retrieved by the identifier input by subject A.

Consider an unknown subject B to be verified. If an unknown person B is detected by the biometric system during the verification, first, a biometric input is extracted from the person by calculating the feature vector, $\left\{ \phi_c^B(i, v) \middle| \forall i, c \in \text{verification mode} \right\}$. Subsequently, the system calculates the difference between the unknown input and templates, resulting in the amplitude and angle distances in the same manner as during enrollment.

Next, the system calculates the similarity to determine how close the subjects A and B are. The computation is carried out by applying a similarity threshold to the distances and normalizing the resulting value to the range [0,1], where 0 means "no match at all" and 1 means "perfect match". In this paper, two similarity measures are considered, and those are the similarities of the amplitude and angle distances, denoted as S_ϕ and S_θ, respectively.

Subsequently, we define a matching score S to quantify the match between the unknown and template subject by concatenating the two similarities into $S = 0.5S_\phi + 0.5S_\theta$, where both similarities are equally weighted. Given a matching score S, the system checks whether the score S is above a certain threshold S_{TH} for the signature classification. A detailed explanation of the verification is given below.

Step 1: First, consider an unknown subject B to be verified and let its ACP be $\Phi_c^B(m)$, which belongs to class c. It can be represented by the set of the ACPlets of

$$\Phi_c^B(m) : \ \Phi_c^B(m) = \left\{ \phi_c^B(1, v), \phi_c^B(2, v), \cdots, \phi_c^B(i, v), \cdots, \ \phi_c^B(NI, v) \right\} \tag{21}$$

Step 2: Compute the amplitude distance $\Delta\varnothing_c^B(i,v)$ between

$$\overline{\phi}^A(i,v) \text{ and } \phi_c^B(i,v), \text{ given by}: \ \Delta\phi_c^B(i,v) = \left|\overline{\phi}^A(i,v) - \phi_c^B(i,v)\right| \quad (22)$$

where this distance shows how close the amplitudes of $\overline{\phi}^A(i,v)$ and $\phi_c^B(i,v)$ are. Smaller $\Delta\varnothing_c^B(i,v)$ leads to a higher possibility that the unknown subject B is to be subject A.

Step 3: Calculate the similarity of the amplitude distance: S_ϕ

The similarity of the amplitude distance (S_ϕ) can be obtained by applying the threshold of the amplitude similarity to $\Delta\varnothing_c^B(i,v)$ as follows:

$$\begin{array}{l} for\ i = 1\ : NI \\ \quad for\ v = 0\ : L-1 \\ \quad\quad \Omega(\mathrm{i,v}) = \frac{TH_\phi(i,v)}{\Delta\varnothing_c^B(i,v)} \text{ for a given class} \\ \quad\quad \Omega(\mathrm{i,v}) = 0 \text{ if } \Delta\varnothing_c^B(i,v) > TH_{\phi_{max}}(i,v) \\ \quad\quad \Omega(\mathrm{i,v}) = 1 \text{ if } \Delta\varnothing_c^B(i,v) \le TH_\phi(i,v) \\ \quad end \\ end \\ \mathrm{S}_\varnothing = \frac{1}{(NI \cdot L)} \sum_{i=1}^{NI} \sum_{v=0}^{L-1} \Omega(i,v) \text{ for a given class} \end{array} \quad (23)$$

Given $\Delta\varnothing_c^B(i,v)$, if $\Delta\varnothing_c^B(i,v) \le TH_\phi(i,v)$, the corresponding ACP pixel is counted as "accepted", i.e., the similarity $\Omega(i,v) = 1$. Else if $\Delta\varnothing_c^B(i,v) > TH_{\phi_{max}}(i,v)$, the corresponding ACP pixel is counted as "declined", that is, the similarity score $\Omega(i,v) = 0$. If $TH_\phi(i,v) < \Delta\varnothing_c^B(i,v) \le TH_{\phi_{max}}(i,v)$, the corresponding ACP pixel is counted as "partially accepted" and the similarity score is given by $\Omega(\mathrm{i,v}) = \frac{\mathrm{TH}_\phi(\mathrm{i,v})}{\Delta\varnothing_c^B(\mathrm{i,v})}$, which ranges from 0 to 1. Finally, the similarity of the amplitude distance (S_ϕ) is determined as the average of the similarity score calculated on a pixel-by-pixel basis.

Step 4: Compute the angle distance $\theta_c^B(i)$

Define $\theta_c^B(i_o)$ as the angle θ of the x-axis to the most discriminant projection vector of the datasets, (x, z) and (x, r) where

$$\begin{array}{c} x = \{\forall v\}, \\ z = \left\{\varnothing_c^B(i_o,v)\middle| i_o \in i, \forall v\right\} \text{ for a given class } c \\ r = \left\{\overline{\varnothing}^A(i_o,v)\middle| i_o \in i, \forall v\right\} \end{array} \quad (24)$$

The method of finding the most discriminant projection vector is the same as the method used to calculate the most discriminant vector in the enrollment mode. Subsequently, define $\theta_c^B(i_o)$ as the angle distance per interval between datasets (x, z) and (x, r) for a given c and i_o.

Step 5: Calculate the similarity of angle distance: S_θ

The similarity of angle distance (S_θ) can be obtained by applying the threshold of the angle similarity to $\theta_c^B(i)$ as follows:

$$\begin{array}{l} \text{for } i = 1\ : NI \\ \quad if\ \theta_c^B(i) \le TH_{\theta_1}(i),\ \Lambda(i) = \frac{TH_{\theta_1}(i)}{\theta_c^B(i)} \\ \quad if\ \theta_c^B(i) \ge TH_{\theta_2}(i),\ \Lambda(i) = \frac{TH_{\theta_2}(i)}{\theta_c^B(i)} \\ \Lambda(i) = 0 \text{ if } \left(\theta_c^B(i) < TH_{\theta_{min}}(i) \text{ or } \theta_c^B(i) > TH_{\theta_{max}}(i)\right) \\ \Lambda(i) = 1 \text{ if } TH_{\theta_1}(i) \le \theta_c^B(i) \le TH_{\theta_2}(i) \text{ for a given class} \\ end \\ S_\theta = \frac{1}{NI} \sum_{i=1}^{NI} \Lambda(i) \text{ for a given class} \end{array} \quad (25)$$

Given $\theta_c^B(i)$, if $TH\theta_{min}(i) \leq \theta_c^B(i) < TH\theta_1(i)$ or $TH\theta_2(i) < \theta_c^B(i) \leq TH\theta_{max}(i)$, the corresponding ACP interval is counted as "partially accepted", that is, the similarity score is $\Lambda(i) = \frac{TH_\theta(i)}{\theta_c^B(i)}$, which has the range [0,1], or else if $\theta_c^B(i) < TH\theta_{min}(i)$ or $\theta_c^B(i) > TH\theta_{max}(i)$, the corresponding ACP interval is counted as "declined", that is, the similarity score $\Lambda(i) = 0$, or else if $TH\theta_1(i) \leq \theta_c^B(i) \leq TH\theta_2(i)$, the corresponding ACP interval is counted as "accepted" and the similarity score is given by $\Lambda(i) = 1$. The similarity of the angle (S_θ) is determined as the average of the similarity scores calculated on an interval-by-interval basis.

Step 6: Calculate the matching score

For every incoming class, we define a matching score $S = 0.5S_\phi + 0.5S_\theta$ with a range from 0 to 1, where 1 means a perfect match. Given a matching score S, the system finally determines whether the score S is above a certain threshold S_{TH}, called as the matching threshold, for signature identification.

2.5. Updating Template

As mentioned in Section 1, an individual ECG signal changes over time due to intra-subject variability. Therefore, if a subject uses a fixed template for a long time, the subject may be falsely rejected during the verification. To solve this problem, we applied an updating process by continuously performing the training process for every class. For instance, for a new C^{th} class incoming shortly after the training process of class $c \in \{0, 1, \ldots C-1\}$, if the matching score S is larger than the threshold S_{TH}, the new C^{th} class is accepted as a new member for the template update. Subsequently, the newly accepted class is appended to the training set, and thus the latest accepted C classes $c \in \{1, 2, \ldots C-1\}$ are considered to be the new training set for the update. The template update for subject A is described below.

2.5.1. Calculating the Amplitude and Angle Distances

The latest accepted C classes are re-indexed using $c \in \{0, 1, \ldots C-1\}$.

(i) Use the C classes as the new training set for subject A and find the ACP as $\Phi_c^A(m)$ which consists of the ACPlets $\varnothing_c^A(i, v)$:

$$\Phi_c^A(m) = \left\{\phi_c^A(1, v), \phi_c^A(2, v), \cdots, \phi_c^A(i, v), \cdots, \phi_c^A(NI, v)\right\} \tag{26}$$

(ii) Next, find the profile $\overline{\Phi}^A(m)$:

$$\overline{\Phi}^A(m) = \left[\overline{\phi}^A(1, v)\ \ \overline{\phi}^A(2, v)\ \cdots \overline{\phi}^A(i, v)\ \cdots \overline{\phi}^A(NI, v)\right] \tag{27}$$

where

$$\overline{\phi}^A(i, v) = \frac{1}{C}\sum_{c=0}^{C-1} \phi_c^A(i, v) \text{ for a given } i \text{ and } v \tag{28}$$

(iii) Finally, compute the amplitude distance $\Delta\varnothing_c^A(i, v)$ and angle distance $\theta_c^A(i)$:

$$\begin{gathered}\Delta\phi_c^A(i, v) = \left|\overline{\phi}^A(i, v) - \phi_c^A(i, v)\right| \text{ and} \\ \theta_c^A(i) = tan^{-1}(V)\end{gathered} \tag{29}$$

where V is the most discriminant projection vector between $\overline{\phi}^A(i, v)$ *and* $\phi_c^A(i, v)$.

2.5.2. Updating the Threshold of Amplitude Similarity

$$
\begin{aligned}
\mu_{\phi}(i,v) &= mean\Big\{\ \Delta\varnothing_c^A(i,v)|\ \forall c \in \{0,1,\cdots C-1\ \}\Big\} \\
\sigma_{\phi}(i,v) &= std\Big\{\ \Delta\varnothing_c^A(i,v)|\ \forall c \in \{0,1,\cdots C-1\ \}\Big\} \\
TH_{\phi_{max}}(i,v) &= max\Big\{\ \Delta\varnothing_c^A(i,v)|\ \forall c \in \{0,1,\cdots C-1\ \}\Big\} \\
TH_{\phi}(i,v) &= \mu_{\phi}(i,v) + \sigma_{\phi}(i,v)
\end{aligned}
\tag{30}
$$

2.5.3. Updating the Threshold of Angle Similarity

$$
\begin{aligned}
\mu_{\theta}(i) &= mean\{\ \theta_c^A(i)|\ \forall c \in \{0,1,\cdots C-1\ \}\Big\} \\
\sigma_{\theta}(i) &= std\{\ \theta_c^A(i)|\ \forall c \in \{0,1,\cdots C-1\ \}\Big\} \\
TH_{\theta_{max}}(i) &= max\{\ \theta_c^A(i)|\ \forall c \in \{0,1,\cdots C-1\ \}\Big\} \\
TH_{\theta_{min}}(i) &= min\{\ \theta_c^A(i)|\ \forall c \in \{0,1,\cdots C-1\ \}\Big\} \\
TH_{\theta_1}(i) = \mu_{\theta}(i) + \sigma_{\theta}(i)\ \ & TH_{\theta_2}(i) = \mu_{\theta}(i) - \sigma_{\theta}(i)
\end{aligned}
\tag{31}
$$

By updating the threshold of the amplitude/angle similarity using the above process, the ECG changes due to intra-subject variation can be continuously reflected in the template. In this study, the verification/updating template is repeatedly performed for every class and each class is specified to be 5 s long, which is fast enough for real-life applications.

3. Experiment and Results

An experiment was carried out using a public open database, available on the PhysioNet website [34], to evaluate the proposed algorithm.

The MIT-BIT arrhythmia database (MITDB) contains 48 records of heartbeats collected from 47 subjects (records 201 and 202 are from the same male subject). Each record is approximately 30 min long and was sampled at 360Hz. The subjects are 25 men with ages between 32 and 89 years old and 22 women with ages between 23 to 89 years old. The data were obtained using a two-channel ambulatory Holter monitor at the Beth Israel Deaconess Medical Center from 1975 until 1979. This database is the most referred database related to arrhythmic studies. In this study, this database was used to evaluate the efficiency of the proposed algorithm. Two 2-min intervals were randomly selected for each data. The first 2-min interval of each data was used for training and the second data was used for verification. Only single lead data was used in the experiment.

The error rate can be calculated for each subject, as shown in Figure 8. The FR (False Rejection) means that the subject was rejected despite being himself. The FA (False Acceptance) means that the subject was accepted despite being someone else. The distance threshold was chosen to be the intersection of FR (False Rejection) and FA (False Acceptance), depending on the subject. The threshold and EER for each subject are calculated based on the intersection of FR and FA. If there is only one intersection point, the threshold and EER are determined based on the intersection point, whereas if there are more than two points, the threshold is determined by the average value of the largest and smallest value, and certainly, EER is calculated as 0, as shown in Figure 8.

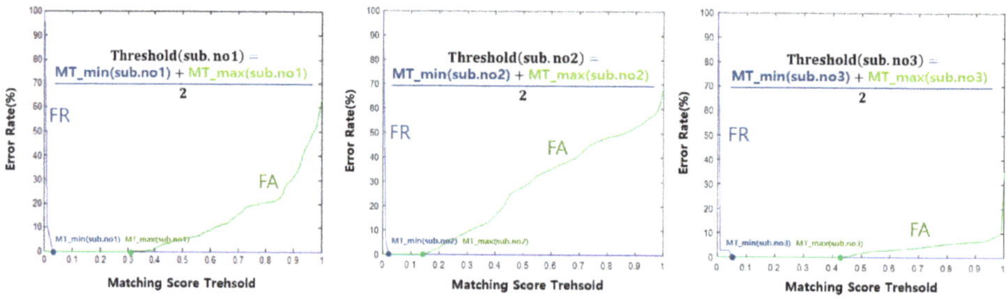

Figure 8. Examples of false rejection and acceptance rates for Subject no. 1, no. 2, and no. 3.

Figure 9a shows an example of obtaining the EER for each individual. A histogram of the EER (Equal Error Rates) is shown in Figure 9b. The average EER is 0.143%, which indicates the efficiency of the overall algorithm, as shown in Equation (32). Only a single user was enrolled at any point in time. In other words, at any point in time, we had one client and 46 imposters.

$$\text{Total efficiency of verification} = \frac{1}{\text{No. of subject}} \sum_{\text{Sub}=1}^{\text{No. of subject}} \text{EER}_{\text{Sub}} \quad (32)$$

where Sub indicates the subject number.

To demonstrate the efficiency of our proposed algorithm, we compared it with related works that used the same database (MIT-BIH), as shown in Table 1. Table 1 shows that the EER (Equal Error Rate) in Reference [48] is higher than the proposed algorithm despite using ECG data from two leads. Recently, most ECG wearable devices, such as smart watches, only use one lead; thus, the proposed algorithm is more suitable for real-life applications. In Reference [48], a fixed shifting size was used for the autocorrelation, while in the proposed algorithm, we used a flexible shifting size for the autocorrelation as the QRS width for the individual, which requires less computational effort. Although fused information from two ECG leads was used in Reference [49], their IDA (Identification Accuracy) is lower than the proposed algorithm; however, the enroll time is 5 min, which is longer than our algorithm. A fiducial point dependent method was used in References [28,50–52] which is vulnerable to noise. Only 10–20 s data were used for enrollment and verification in References [52,53]. However, we used two 2-min data for enrollment and verification to verify the stability of the proposed algorithm. Only a direct contact sensor was used in References [28,48–53] to verify the algorithms. In contrast, we used a non-contact single lead senor for real-life applications. In this paper, in the case of a fiducial dependent approach, although it has the advantage of being able to identify a subject in a relatively short time with less computational effort, it is easily affected by external noise. On the other hand, a fiducial independent approach has the advantage of being less affected by external noise, but it takes higher computational effort and longer authentication time. To compensate for the advantages and disadvantages of the above two methods, the amount of calculation was reduced through the amplitude feature using ACP and the identification accuracy was improved through the interval-based angle feature.

However, in this study, the matching score was calculated using the same weight for the two features, resulting in a disadvantage in terms of the weakness in various environmental variables. Therefore, we intend to explore the application of variable weights for future research to design a robust algorithm for environment variables.

Table 1. Comparison of the proposed algorithm with related research using the same public database.

Ref.	NS	ST	DB	Method	EER	IDA
[48]	56	2-lead Direct Contact type	MITDB NSRDB PTB database	Normalized AC and LDA	3.8 %	96.2 %
[49]	145	2-lead Direct Contact type	(a) MITDB (b) NSRDB (c) LTSTDB	WT and ICA	(a) 0.43 % (b) 0.67 % (c) 1.89 %	(a) 99.57 % (b) 99.33 % (c) 98.11 %
[28]	50	2-lead Direct Contact type (Only single lead of data used)	MITDB NSRDB	Delineation of P and T wave, Correlation	1.01 %	98.99 %
[50]	73	2-lead Direct Contact type (Only single lead of data used)	(a) MITDB (b) IIT(BHU) database (laboratory experiment)	Delineation of P and T wave, Eigenbeat feature extraction	(a) 8.58 % (b) 4.45 %	(a) 91.42 % (b) 95.55 %
[51]	47	2-lead Direct Contact type (Only single lead of data used)	MITDB	Sum of Gaussian, QDA	3 %	97 %
[52]	83	2-lead Direct Contact type (Only single lead of data used), AliveCor single lead Contact type [54]	(a) MITDB NSRDB (b) Laboratory experiment database	FP (Time and amplitude), Proposed hierarchical algorithm	(a) 9.795 % (b) 8.18 %	(a) 90.205 % (b) 91.82 %
[53]	65	2-lead Direct Contact type (Only single lead of data used)	(a) MITDB (b) NSRDB	DWT	(a) 6.9 % (b) 0.6 %	(a) 93.1 % (b) 99.4 %
Proposed method	72	2-lead Direct Contact type (Only single lead of data used)	MITDB	Unnormalized AC and Interval based LDA algorithm	0.143 %	99.857 %

NS number of subject, ST sensor type, DB database, EER equal error rate, IDA identification accuracy, AC autocorrelation, LDA linear discriminant analysis, WT wavelet transform, ICA independent component analysis, QDA quadratic discriminant analysis, DWT discrete wavelet transform.

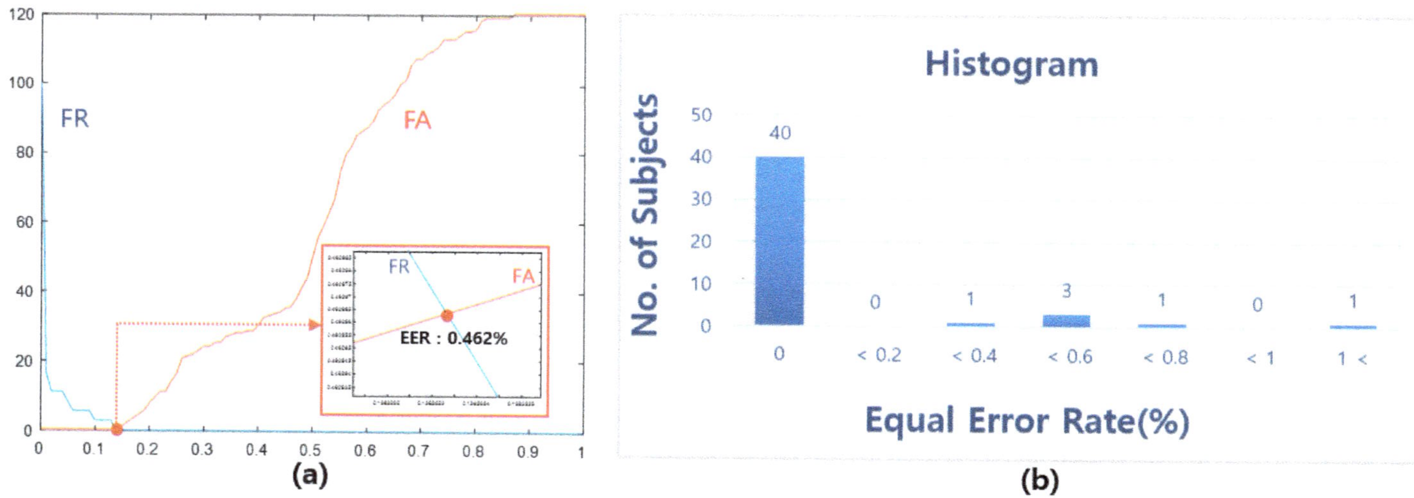

Figure 9. (**a**) Example of the EER depending on individuals (subject no. 41); (**b**) a histogram of the EER (Equal Error Rates) (average EER is 0.143%).

4. Conclusions

In this paper, a novel interval-based LDA algorithm is proposed. The implementation experiments of the proposed algorithm show higher identification accuracy compared to the existing ECG-based identification algorithm using the fiducial point dependent method or fiducial independent method. Based on this algorithm, we can complement each method's disadvantages by using both a fiducial point dependent method (amplitude distance by pixel) and fiducial point independent method (angle distance by feature vector). In addition, our proposed algorithm is able to maintain identification accuracy by tracking the ECG that changes in real-time using a template updating process.

This algorithm was tested with a public ECG database (MIT-BIH) and we compared its performance with related studies using the same database. The results show that the proposed algorithm is more reliable than other algorithms in terms of EER (Equal Error Rate). However, this algorithm has the disadvantage that it requires 5 s for each identification. Therefore, as future research, we intend to reduce the identification time and proceed to commercialize this as a product through multimodal identification combined with other biometrics.

Author Contributions: Conceptualization, C.Y. and G.W.K.; methodology, C.Y., J.-G.L. and G.W.K.; software, G.W.K.; validation, C.Y., J.-G.L. and S.-H.L.; formal analysis, C.Y. and S.-H.L.; investigation, C.Y. and G.W.K.; resources, C.Y. and G.W.K.; data curation, G.W.K.; writing-original draft preparation, C.Y.; writing-review and editing, S.-H.L.; visualization, C.Y. and G.W.K.; supervision, S.-H.L.; project administration, C.Y. and S.-H.L. All authors have read and all agree upon the published version of the manuscript.

Funding: This research was funded by the Technological Innovation R&D Program (S2827786) funded by the Ministry of SMEs and Startups (MSS, Korea).

Conflicts of Interest: The authors declare no conflict of interest.

References

1. Geddes, L.A.; Roeder, R.A. The first electronic electrocardiograph. *Cardiovasc. Eng. Int. J.* **2002**, *2*, 73–79. [CrossRef]
2. Myers, G.B.; Klein, H.A.; Stofer, B.E. The electrocardiographic diagnosis of right ventricular hypertrophy. *Am. Heart J.* **1948**, *35*, 1–40. [CrossRef]
3. Casale, P.N.; Devereux, R.B.; Alonso, D.R.; Campo, E.; Kligfield, P. Improved sex-specific criteria of left ventricular hypertrophy for clinical and computer interpretation of electrocardiograms: Validation with autopsy findings. *Circulation* **1987**, *75*, 565–572. [CrossRef] [PubMed]
4. Addison, P.S.; Watson, J.N.; Clegg, G.R.; Holzer, M.; Sterz, F.; Robertson, C.E. Evaluating arrhythmias in ECG signals using wavelet transforms. *IEEE Eng. Med. Biol. Mag.* **2000**, *19*, 104–109. [CrossRef] [PubMed]
5. Acharya, R.; Kannathal, N.; Krishnan, S.M. Comprehensive analysis of cardiac health using heart rate signals. *Physiol. Meas.* **2004**, *25*, 1139.
6. Bsoul, M.; Minn, H.; Tamil, L. Apnea MedAssist: Real-time Sleep Apnea Monitor Using Single-Lead ECG. *IEEE Trans. Inf. Technol. Biomed.* **2011**, *15*, 416–427. [CrossRef] [PubMed]
7. Karthikeyan, P.; Murugappan, M.; Yaacob, S. Detection of human stress using short-term ECG and HRV signals. *J. Mech. Med. Biol.* **2013**, *13*, 1350038. [CrossRef]
8. Kaniusas, E.; Pfutzner, H.; Mehnen, L.; Kosel, J.; Tellez-Blanco, C.; Varoneckas, G.; Alonderis, A.; Meydan, T.; Vazquez, M.; Rohn, M.; et al. Method for continuous nondisturbing monitoring of blood pressure by magnetoelastic skin curvature sensor and ECG. *IEEE Sens. J.* **2006**, *6*, 819–828. [CrossRef]
9. Saul, J.P.; Rea, R.F.; Eckberg, D.L.; Berger, R.D.; Cohen, R.J. Heart rate and muscle sympathetic nerve variability during reflex changes of autonomic activity. *Am. J. Physiol. Circ. Physiol.* **1990**, *258*, H713–H721. [CrossRef]
10. Pomeranz, B.; Macaulay, R.J.; Caudill, M.A.; Kutz, I.; Adam, D.; Gordon, D.; Kilborn, K.M.; Barger, A.C.; Shannon, D.C.; Cohen, R.J.; et al. Assessment of autonomic function in humans by heart rate spectral analysis. *Am. J. Physiol. Circ. Physiol.* **1985**, *248*, H151–H153. [CrossRef]
11. Agrafioti, F.; Gao, J.; Hatzinakos, D.; Yang, J. Heart biometrics: Theory, methods and applications. In *Biometrics*; InTech: Shanghai, China, 2011; pp. 199–216.

12. Biel, L.; Pettersson, O.; Philipson, L.; Wide, P. ECG analysis: A new approach in human identification. In Proceedings of the 16th IEEE Instrumentation and Measurement Technology Conference, Venice, Italy, 24–26 May 1999.
13. US Federal Trade Commission. Consumer Sentinel Network Data Book for January–December 2018. Available online: https://www.ftc.gov/system/files/documents/reports/consumer-sentinel-network-data-book-2018/consumer_sentinel_network_data_book_2018_0.pdf (accessed on 26 August 2020).
14. Burge, M.; Burger, W. Ear Biometrics. In *Biometrics*; Springer US: Boston, MA, USA, 1996; pp. 273–285.
15. Hossain, S.M.E.; Chetty, G. Next generation identity verification based on face-gait Biometrics. In Proceedings of the International Conference on Biomedical Engineering and Technology, Kuala Lumpur, Malaysia, 4–5 June 2011.
16. Prabhakar, S.; Pankanti, S.; Jain, A.K. Biometric recognition: Security and privacy concerns. *IEEE Secur. Priv.* **2003**, *1*, 33–42. [CrossRef]
17. Sanchit; Ramalho, M.; Correia, P.L.; Soares, L.D. Biometric identification through palm and dorsal hand vein patterns. In Proceedings of the 2011 IEEE EUROCON—International Conference on Computer as a Tool, Lisbon, Portugal, 27–29 April 2011; pp. 1–4.
18. Boulgouris, N.V.; Plataniotis, K.N.; Micheli-Tzanakou, E. (Eds.) *Biometrics: Theory, Methods, and Applications*; John Wiley & Sons, Inc.: Hoboken, NJ, USA, 2009; Volume 9, ISBN 9780470522356.
19. Jain, A.K.; Arun, R.; Salil, P. An introduction to biometric recognition. *IEEE Trans. Circuits Syst. Video Technol.* **2004**, *14*, 4–20. [CrossRef]
20. Israel, S.A.; Irvine, J.M.; Cheng, A.; Wiederhold, M.D.; Wiederhold, B.K. ECG to identify individuals. *Pattern Recognit.* **2005**, *38*, 133–142. [CrossRef]
21. Irvine, J.; Israel, S.; Wiederhold, M.; Wiederhold, B. A new biometric: Human identification from circulatory function. In Proceedings of the Joint Statistical Meetings of the American Statistical Association, San Francisco, CA, USA, 3–7 August 2003.
22. Zhang, Z.; Wei, D. A New ECG Identification Method Using Bayes' Teorem. In Proceedings of the Tencon 2006 IEEE Region 10 Conference, Hong Kong, China, 14–17 November 2006.
23. Singh, Y.N.; Gupta, P. ECG to Individual Identification. In Proceedings of the 2008 IEEE Second International Conference on Biometrics: Theory, Applications and Systems, Arlington, VA, USA, 29 September–1 October 2008.
24. Simon, B.P.; Eswaran, C. An ECG Classifier Designed Using Modified Decision Based Neural Networks. *Comput. Biomed. Res.* **1997**, *30*, 257–272. [CrossRef] [PubMed]
25. Hoekema, R.; Uijen, G.J.H.; van Oosterom, A. Geometrical aspects of the interindividual variability of multilead ECG recordings. *IEEE Trans. Biomed. Eng.* **2001**, *48*, 551–559. [CrossRef] [PubMed]
26. Tan, R.; Perkowski, M. ECG Biometric Identification Using Wavelet Analysis Coupled with Probabilistic Random Forest. In Proceedings of the 2016 15th IEEE International Conference on Machine Learning and Applications (ICMLA), Anaheim, CA, USA, 18–20 December 2016.
27. Agrafioti, F.; Hatzinakos, D.; Bui, F.M. Cardiac biometric security in welfare monitoring. In Proceedings of the 2012 IEEE International Conference on Acoustics, Speech and Signal Processing (ICASSP), Kyoto, Japan, 25–30 March 2012.
28. Singh, Y.N.; Gupta, P. Correlation-based classification of heartbeats for individual identification. *Soft Comput.* **2011**, *15*, 449–460. [CrossRef]
29. Fratini, A.; Sansone, M.; Bifulco, P.; Cesarelli, M. Individual identification via electrocardiogram analysis. *Biomed. Eng. Online* **2015**, *14*, 78. [CrossRef]
30. Singh, Y.N.; Singh, S.K. Evaluation of Electrocardiogram for Biometric Authentication. *J. Inf. Secur.* **2011**, *03*, 39–48. [CrossRef]
31. Guennoun, M.; Abbad, N.; Talom, J.; Rahman, S.M.M.; El-Khatib, K. Continuous authentication by electrocardiogram data. In Proceedings of the 2009 IEEE Toronto International Conference Science and Technology for Humanity (TIC-STH), Toronto, ON, Canada, 26–27 September 2009.
32. Kyoso, M.; Uchiyama, A. Development of an ECG identification system. In Proceedings of the 2001 Conference Proceedings of the 23rd Annual International Conference of the IEEE Engineering in Medicine and Biology Society, Istanbul, Turkey, 25–28 October 2001.
33. Kass, R.E.; Colleen, E.C. (Eds.) *Basis and Treatment of Cardiac Arrhythmias*; Springer Science & Business Media: Berlin/Heidelberg, Germany, 2005; Volume 171.

34. Physionet, Physiobank Archives, Massachusetts Institute of Technology Cambridge, January 2011. Available online: https://www.physionet.org/physiobank/database/#ecg (accessed on 26 August 2020).
35. Sörnmo, L.; Laguna, P. *Bioelectrical Signal Processing in Cardiac and Neurological Applications*; Academic Press: New York, NY, USA, 2005; Volume 8.
36. Chan, A.D.C.; Hamdy, M.M.; Badre, A.; Badee, V. Wavelet Distance Measure for Person Identification Using Electrocardiograms. *IEEE Trans. Instrum. Meas.* **2008**, *57*, 248–253. [CrossRef]
37. Ye, C.; Coimbra, M.T.; Vijaya Kumar, B.V.K. Arrhythmia detection and classification using morphological and dynamic features of ECG signals. In Proceedings of the 2010 Annual International Conference of the IEEE Engineering in Medicine and Biology, Buenos Aires, Argentina, 31 August–4 September 2010.
38. Belgacem, N. ECG Based Human Authentication using Wavelets and Random Forests. *Int. J. Cryptogr. Inf. Secur.* **2012**, *2*, 1–11. [CrossRef]
39. Choudhary, T.; Manikandan, M.S. A novel unified framework for noise-robust ECG-based biometric authentication. In Proceedings of the 2015 2nd International Conference on Signal Processing and Integrated Networks (SPIN), Noida, India, 19–20 February 2015.
40. Gutta, S. Biomedical Signal Processing and Inference in Wearable Sensing Applications. Ph.D. Thesis, Oklahoma State University, Stillwater, OK, USA, 1 December 2016.
41. Plataniotis, K.N.; Hatzinakos, D.; Lee, J.K.M. ECG Biometric Recognition Without Fiducial Detection. In Proceedings of the 2006 Biometrics Symposium: Special Session on Research at the Biometric Consortium Conference, Baltimore, MD, USA, 19–21 September 2006.
42. Matta, R.; Lau, J.K.H.; Agrafioti, F.; Hatzinakos, D. Real-Time Continuous Identification System Using ECG Signals. In Proceedings of the 2011 Canadian Conference on Electrical and Computer Engineering, Niagara Falls, ON, Canada, 8–11 May 2011; pp. 1313–1316.
43. Wang, Y.; Agrafioti, F.; Hatzinakos, D.; Plataniotis, K.N. Analysis of Human Electrocardiogram for Biometric Recognition. *EURASIP J. Adv. Signal Process.* **2007**, *2008*, 148658. [CrossRef]
44. Safie, S.I.; Soraghan, J.J.; Petropoulakis, L. Pulse Active Ratio (PAR): A new feature extraction technique for ECG biometric authentication. In Proceedings of the 2011 IEEE International Conference on Signal and Image Processing Applications (ICSIPA), Kuala Lumpur, Malaysia, 16–18 November 2011.
45. Mücke, D. *Elektrokardiographie Systematisch*; UNI-MED-Verlag: Bremen, Germany, 1996.
46. Khalil, I.; Sufi, F. Legendre Polynomials based biometric authentication using QRS complex of ECG. In Proceedings of the 2008 International Conference on Intelligent Sensors, Sensor Networks and Information Processing, Sydney, Australia, 15–18 December 2008.
47. Pan, J.; Tompkins, W.J. A Real-Time QRS Detection Algorithm. *IEEE Trans. Biomed. Eng.* **1985**, *BME-32*, 230–236. [CrossRef] [PubMed]
48. Agrafioti, F.; Hatzinakos, D. ECG biometric analysis in cardiac irregularity conditions. *Signal, Image Video Process.* **2009**, *3*, 329–343. [CrossRef]
49. Ye, C.; Coimbra, M.T.; Kumar, B.V.K.V. Investigation of human identification using two-lead Electrocardiogram (ECG) signals. In Proceedings of the 2010 Fourth IEEE International Conference on Biometrics: Theory, Applications and Systems (BTAS), Washington, DC, USA, 27–29 September 2010.
50. Singh, Y.N.; Singh, S.K. Identifying Individuals Using Eigenbeat Features of Electrocardiogram. *J. Eng.* **2013**, *2013*, 539284. [CrossRef]
51. Sarkar, A.; Abbott, A.L.; Doerzaph, Z. ECG biometric authentication using a dynamical model. In Proceedings of the 2015 IEEE 7th International Conference on Biometrics Theory, Applications and Systems (BTAS), Arlington, VA, USA, 8–11 September 2015.
52. Arteaga-Falconi, J.S.; Al Osman, H.; El Saddik, A. ECG Authentication for Mobile Devices. *IEEE Trans. Instrum. Meas.* **2016**, *65*, 591–600. [CrossRef]
53. Dar, M.N.; Akram, M.U.; Usman, A.; Khan, S.A. ECG biometric identification for general population using multiresolution analysis of DWT based features. In Proceedings of the 2015 Second International Conference on Information Security and Cyber Forensics (InfoSec), Cape Town, South Africa, 15–17 November 2015.
54. *Heart Monitor AC-002 User Manual*; AliveCor, Inc.: San Francisco, CA, USA, 2015.

applied sciences

MDPI

Article

Spectral and Nonlinear Analysis of Electrodermal Activity in Adolescent Anorexia Nervosa

Zuzana Visnovcova [1,2], Lucia Bona Olexova [1,2], Nikola Sekaninova [1,2], Igor Ondrejka [3], Igor Hrtanek [3], Dana Cesnekova [3], Simona Kelcikova [4], Ivan Farsky [5] and Ingrid Tonhajzerova [1,2,*]

1 Biomedical Center Martin, Jessenius Faculty of Medicine in Martin, Comenius University in Bratislava, 03601 Martin, Slovakia; zuzana.visnovcova@uniba.sk (Z.V.); olexova10@uniba.sk (L.B.O.); sekaninova1@uniba.sk (N.S.)

2 Department of Physiology, Jessenius Faculty of Medicine in Martin, Comenius University in Bratislava, 03601 Martin, Slovakia

3 Psychiatric Clinic, Jessenius Faculty of Medicine in Martin, Comenius University in Bratislava, University Hospital Martin, Kollarova 2, 03659 Martin, Slovakia; igor.ondrejka@uniba.sk (I.O.); igor.hrtanek@uniba.sk (I.H.); cesnekova.danka@gmail.com (D.C.)

4 Department of Midwifery, Jessenius Faculty of Medicine in Martin, Comenius University in Bratislava, 03601 Martin, Slovakia; simona.kelcikova@uniba.sk

5 Department of Nursing, Jessenius Faculty of Medicine in Martin, Comenius University in Bratislava, 03601 Martin, Slovakia; ivan.farsky@uniba.sk

* Correspondence: ingrid.tonhajzerova@uniba.sk; Tel.: +421-2633425

Received: 8 June 2020; Accepted: 25 June 2020; Published: 29 June 2020

Abstract: Anorexia nervosa (AN) is an eating disorder with increasing prevalence in childhood and adolescence. Sympathetic dysregulation is supposed to be the underlying mechanism of increased cardiovascular risk in AN. Thus, we assess the electrodermal activity (EDA) as a non-invasive index of sympathetic cholinergic activity using linear and nonlinear analysis in adolescent AN with the aim of detecting potential biomarkers for AN-linked cardiovascular risk. We examined 25 adolescent girls with AN and 25 age-matched controls. EDA was continuously recorded during a 5-min resting phase. Evaluated parameters were: time-domain (skin conductance level, non-specific skin conductance responses), frequency-domain (EDA in very low, low, sympathetic, high and very high frequency bands) and nonlinear (approximate, sample, symbolic information entropies, detrended fluctuation analysis (DFA)) parameters of EDA and peripheral skin temperature. Our findings revealed lower EDA values indicating a decrease in the sympathetic nervous activity in female adolescents with the acute phase of AN. Further, we found higher nonlinear index DFA in AN vs. controls. We assumed that nonlinear index DFA could provide novel and independent information on the complex sympathetic regulatory network. We conclude that the parameters of complex EDA analysis could be used as sensitive biomarkers for the assessment of sympathetic cholinergic dysregulation as a risk factor for AN-linked cardiovascular morbidity.

Keywords: anorexia nervosa; electrodermal activity; time-domain analysis; frequency-domain analysis; nonlinear analysis

1. Introduction

Anorexia nervosa (AN) is one of the most frequent eating disorders with the highest morbidity and mortality among intake malfunctions [1,2]. Patients with AN are characterised by low weight as a result of the restriction of food intake, fear of weight gain, and distortion of self-image, including fear of being or becoming fat [3,4]. Rates of lifetime prevalence of AN range from 1.4% to 3.3% in women [5]. AN mainly presents in mid- to late adolescence, although its occurrence in childhood and early adolescence is increasing [6].

In general, the autonomic nervous system (ANS) plays a crucial role in the maintenance of the homeostasis, flexibility and adaptability of the human body. It is assumed that disturbances of the ANS may participate in underlying mechanisms of cardiovascular disorders associated with AN [7]. For example, several studies revealed lower heart rate associated with higher heart rate variability indicating cardiovagal dominance, and sympathetic vascular underactivity indexed by lower blood pressure variability [7–11]. However, studies regarding other biosignals in AN are scarce.

Electrodermal activity (EDA) represents a non-invasive index of sympathetic cholinergic activity. Specifically, changes in sympathetic activity lead to altered activity of eccrine sweat glands, which is related to skin´s ability to conduct electricity and thus to EDA variations [12,13]. Moreover, the eccrine sweat glands are involved in thermoregulation of organisms, and cover the majority of the organism, but are the densest on palms and soles. Indeed, these localizations are mainly used in psychophysiological research, because it is assumed that they are associated with emotion-evoked sweating [13].

From the physiological aspect, the skin is relatively non-conductive, due to its most upper layer of derma—the *stratum corneum*. A variable amount of sweat is released from the ducts of the eccrine sweat glands, which open onto the skin surface. The degree of sympathetic activation affects the amount of filled sweat ducts, and the amount of produced sweat; in turn, this activation influences the degree of skin conduction. Variations of the amount of sweat in the ducts relates to the conductivity of the skin and its measurable changes in EDA [14]. Additionally, Posada-Quintero et al. (2016) revealed that variation in the dynamics of the sympathetic activity evaluated by EDA represent one of the important prognostic and diagnostic tools in diseases associated with autonomic dysfunction [15].

With respect to methodology, Lanata et al. (2012) evaluated changes of EDA by standard time and frequency indices and nonlinear parameters, such as deterministic chaos, recurrence plot, and detrended fluctuation analysis in the healthy group during exposure of sets of images from the International Affective Picture System (IAPS) [16]. This study revealed that nonlinear indices of EDA contribute to better extraction of EDA features, such as characteristics of sympathetic activity without significant differences in two types of used measuring devices (special textile gloves vs. bipolar electrodes) [16]. Our previous study evaluated the difference between standard and nonlinear parameters of EDA (approximate and symbolic information entropy) after cognitive stressors (recovery phase) in healthy young adults. Specifically, whereas skin conductance amplitude remained increased in the recovery phase compared to the baseline period, nonlinear indices decreased under the baseline value. It is assumed that this could be caused by complex mechanisms consisting of feedback loops forming a chaotic system on the central level [17].

Recently, evaluation of EDA in AN has been attracting more attention from scientists. Several studies found a lower amplitude of skin conductance response (SCR) in adult females suffering from AN versus age-matched healthy control (HC) in resting conditions [18–20]. Similarly, Crifaci et al. (2013) revealed a decrease of SCR, but a higher level of frequency of SCR in AN group with respect to healthy volunteers during baseline [10]. Additionally, Soussignan et al. (2011) observed significantly larger SCR during exposure to pictures of food and odorous foods in HC versus AN women [21]. Finally, AN women had lower SCR during rest and the Iowa Gambling Task compared to the HC group [22]. According to these results, Tchanturia et al. (2007) assumed that SCR represents a sensitive indicator of response to stimuli in AN patients [22]. While the variations of SCR in adult AN patients are relatively obvious [18,20], the changes of spectral components and nonlinear features of EDA in adolescent females suffering from AN remain unclear. Moreover, studies evaluating variables of EDA by time, spectral and nonlinear indices in AN adolescents are rare. There exist only a few studies concerning changes of these parameters, and only in healthy adults [15–17,23,24]. For example, time and spectral indices of EDA for frequency bands (range from 0.0 Hz to 0.5 Hz) increased during applications of various stressors (changes of position, cold pressor, and Stroop test) assuming that the

EDA evaluation by different parameters could provide an important quantitative tool for evaluating sudomotor activity regulated by the sympathetic nervous system [15].

Based on these studies, we hypothesized that complex EDA analysis (time-domain, spectral and non-linear) could provide important information related to sympathetic nervous activity in adolescent anorexic patients. Thus, we aimed to evaluate EDA by different parameters of the spectral and nonlinear analysis indicating complex sympathetic cholinergic regulatory network in adolescent anorexia nervosa. To the best of our knowledge, this is the first study to use complex EDA analysis with the aim of detecting potential biomarkers for complex sympathetic cholinergic dysregulation as a risk factor for AN-linked cardiovascular morbidity already present in adolescent age.

2. Materials and Methods

The study was approved by the Ethics Committee of Jessenius Faculty of Medicine in Martin, Comenius University in Bratislava in accordance with the 1964 Helsinki declaration and its later amendments. All participants and their parents/legal representatives were carefully instructed about the study protocol and gave written informed consent to participate in the study before the examination.

2.1. Study Population

The studied group consisted of 25 girls suffering from newly diagnosed AN recruited from inpatients admitted to the Psychiatric Clinic of Jessenius Faculty of Medicine and University Hospital in Martin (mean age: 14.8 ± 0.4 years), and control group of 25 healthy age-matched controls (mean age: 15.1 ± 0.3 years). Exclusion criteria were the following for both groups: history of recent acute illness or chronic cardiovascular, respiratory, endocrine, neurological, metabolic, and infectious diseases, smoking, medication/psychoactive substances, or dietary supplementation potentially affecting the ANS. Moreover, mental disorders and weight abnormalities (underweight/overweight/obesity) were considered as exclusion criteria in control group. The weight was defined according to extended International Obesity Task Force body mass index (BMI) cut-offs for thinness, overweight, and obesity using age- and sex- specific BMI cut-offs, which correspond to the adult BMI range between 18.5 and 25 kg/m^2 for normal weight, and the threshold of 30 kg/m^2 for obesity [25].

2.2. Diagnosis of Anorexia Nervosa

The diagnosis of anorexia nervosa without other comorbid mental disorders (e.g., depressive disorder) was assessed by a thorough clinical investigation based on a unstructured diagnostic interview by a staff child/adolescent psychiatrist according to the Diagnostic and Statistical Manual of Mental Disorders, DSM-5 (American Psychiatric Association, 2013). The AN patients were examined before pharmacotherapy and starting of nonpharmacological regimen interventions, including re-feeding, during the first days of hospitalization.

2.3. The Study Protocol and Evaluated Parameters

The examinations were performed in the Psychophysiological laboratory (Biomedical Center Martin, Jessenius Faculty of Medicine in Martin) under standard conditions with the minimalization of stimuli: a quiet room, 22 °C, in the morning between 8:00 and 11:30 a.m., after a normal breakfast.

Firstly, anthropometric parameters were assessed by body composition analyser InBody 120 (Biospace Co. Ltd, Seoul, Korea). Then, participants were instructed to sit comfortably and rest in a special armchair for 10 min to avoid potential effects of stress (laboratory environment and examining person). After the introductory 10 min, EDA was recorded using FlexComp Infinity Biofeedback (Thought Technology, Canada) with a sampling frequency of 256 Hz. The EDA signal was monitored by two bipolar electrodes composed of 10-mm diameter Ag-AgCl placed on the middle phalanges of two fingers on the non-dominant hand. In accordance with some other studies [26,27], the special EDA conductive gel was not used. Special FlexComp biofeedback electrodes containing a metallic disk set were used without conductive gel as recommended by Thought Technology, Ltd. [28]. These special electrodes were carefully cleaned with an alcohol wipe after each examination of the subject. In addition, peripheral skin temperature was recorded using FlexComp Infinity Biofeedback (Thought Technology, Canada). The temperature of the periphery was monitored by a sensor attached to the second phalange of one finger on the non-dominant hand. The baseline phase of the study lasted for 5 min.

2.4. *Anthropometric Measures*

Body composition was analysed using InBody 120 (Biospace Co. Ltd, Seoul, Korea) with multi-segmental multi-frequency (20/100 kHz) bioimpedance analysis, using eight tactile electrodes (thumbs and fingers of the hands, balls of the feet and heels) in a standing position. Thereafter, software Lookin'Body120 (Biospace Co. Ltd, Korea) processed the obtained data. The following body composition parameters were evaluated: weight (kg) and body mass index (BMI, kg/m^2) calculated as the weight (kg) divided by the square of height (m), waist circumference (WC, cm). Body surface area (BSA, m^2) was calculated using the Mosteller equation as:

$$BSA = \sqrt{h} \times w/3600, \quad (1)$$

where h is height (cm) and w is weight (kg) [29].

3. Data Analysis

3.1. *Time-Domain Parameters of EDA*

The tonic EDA, needed for analysis, was extracted by 10th order low-pass finite impulse response filter (cut off 0.0004 Hz) [15]. Next, the skin conductance level (SCL) was evaluated as a mean amplitude of tonic EDA. SCL informs about quantitative changes in the cholinergic sympathetic nervous system. The physiological values of SCL depend on the size of the sensors used. For 10 mm sensors, the range is from 0 to 30 micro Siemens [14,30].

Non-specific skin conductance responses (NS.SCRs) were obtained as the rate of spontaneous skin conductance responses that occur without external stimuli during examination [31]. Typical values ranged from 1 to 3 per min during baseline. The higher values of NS.SCRs indicate intensified arousal conditions [14]. Thus, the time-domain parameters provide information about changes associated with the sympathetic nervous system [15].

3.2. *Power Spectral Analysis of EDA*

At first, the data was filtered with an 8th-order Chebyshev Type I low-pass filter. Consequently, the signals were down-sampled to 2 Hz. Then, to eliminate any trend the data was high-pass filtered with an 8th-order Butterworth filter. The power spectra of signals were evaluated using Welch´s periodogram method with a 50% overlap. The mean power spectrum of the analysed unit was computed by fast Fourier transform (using a Blackman window length of 128 samples), and spectral powers [μS^2] in the appropriate frequency bands (VLF: 0.000–0.045 Hz; EDA-Symp: 0.045–0.25 Hz (sum of low (0.045–0.15 Hz) and high 1 (0.15–0.25 Hz) frequencies); HF: 0.25–0.40 Hz and VHF: 0.40–0.50 Hz) were obtained according to Posada-Quintero et al. [15].

From a physiological perspective, the frequency-domain parameters express the spectral distribution of sympathetic arousal in the skin surface [15].

3.3. *Nonlinear Indices of EDA*

3.3.1. Entropy

Entropy describes the unpredictability, randomness and uncertainty of a system. Entropy represents random and asymmetrical fluctuations in a given time series, and thus the complexity of the sympathetic cholinergic regulatory system [32]. The fundament of entropy indices calculation is the evaluation of probabilities distribution of the expected value in the time series [32,33].

3.3.2. Approximate and Sample Entropy of EDA

Approximate (ApEn) and Sample (SampEn) entropies describe the probability that the vectors of length m chosen from a time-series of length N that are related within a given tolerance range r, remain similar for the vector of length m + 1.

$$\text{ApEn}\ (m, r, N) = \Phi^m(r) - \Phi^{m+1}(r), \quad (2)$$

$$\text{SampEn}\ (m, r, N) = \log\left(\Phi^m(r) / \Phi^{m+1}(r)\right), \quad (3)$$

where N—number of points, m—length of sequence (for N = 300 length of sequence m = 2), r—tolerance of similarity (r ϵ< 0.1 SD; 0.25 SD>). Minimum values of ApEn and SampEn (around zero) indicate regular system. In contrast, the increasing ApEn and SampEn refer to a random and more complex system [32,34].

3.3.3. Symbolic Information Entropy

The analysis of symbolic information entropy (SIE) has been demonstrated to be sufficient for the investigation of complex systems and is able to describe nonlinear aspects within a time series. The concept of SIE is based on coarse-graining of the time series into symbolic sequences with a certainly given alphabet. This data was transformed by four numbers (0, 1, 2, 3) to classify the dynamic changes of the time series, then these symbols were used to generate SIE [17,24]. The numbers characterize changes of the time series, i.e., 0 and 1 indicate slow and fast increasing of the data, respectively; 2 and 3 specify fast and slowly decreasing signal, respectively. The patterns are divided into vectors of length L = 2 and they are paired into the class according to letters of the vector. For instance, vector 13 means the change of SCL from fast elevation to a slow decrease in the waveform. Index SIE characterizes the randomness and messiness of the system. Symbolic information entropy can be defined as

$$\text{SIE}(m) = -\Sigma^{M}{}_{i=1}\ (p_i * P_i), \quad (4)$$

where if $p_i > 0$, then $P_i = \log_2(p_i)$, else $P_i = 0$. The higher values of SIE represent more irregular, messy and unpredictable signals, whereas the values around zero characterize higher regularity [17,24].

3.3.4. Detrended Fluctuation Analysis of EDA

Detrended fluctuation analysis (DFA) is a scaling analysis method for evaluating the statistical self-similarity properties of the time series. DFA is used to confirm the existence of persistent long-term range associations in EDA data. Exponent α represents the slope of the trend line relating log F(n) to log (n) in the range of time-scales. It is estimated by linear regression. Generally, two scaling exponents—α1 and α2 are evaluated, which represent relationships over short (≤30 s) and long (>30 s) timescales, respectively [35]. At first, the analyzed data (length of N) was integrated and divided into boxes of equal length n. In individual boxes of n, a least square line fits the signals. The y coordinate of the straight stretches is marked by $y_n(k)$. Subsequently, we subtracted the local trend $y_n(k)$ from y(k) in each box, i.e., we detrended them. Finally, the root-mean-square fluctuation of this combined and detrended signals was computed according to Lanata et al. (2012) [16]:

$$F\ (n) = \sqrt{\frac{1}{N} \sum_{k=1}^{N} [y(k) - y_n(k)]^2}, \quad (5)$$

Figure 1 summarizes the EDA evaluated parameters using linear and nonlinear analysis.

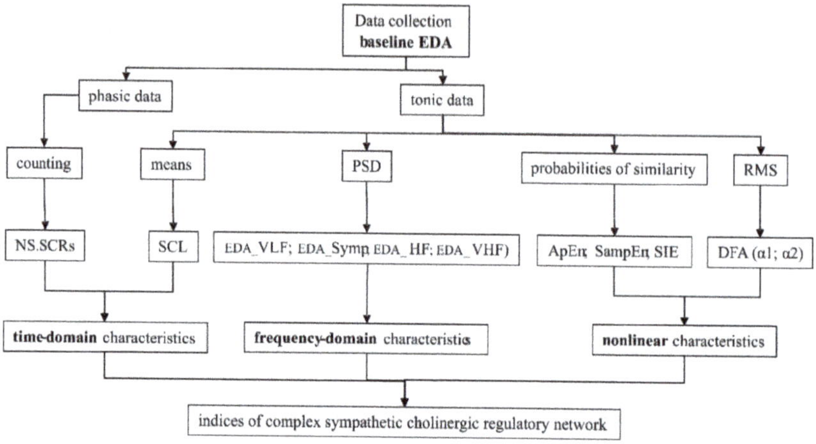

Figure 1. Evaluation of electrodermal activity (EDA) parameters. SCL—skin conductance level; NS.SCRs—nonspecific skin conductance responses; PSD—power spectral density; EDA-VLF—electrodermal activity in very low frequency band; EDA-Symp—EDA in the sum of low and high 1 frequency bands; EDA-HF—EDA in high 2 frequency band; EDA-VHF—EDA in very high frequency band; ApEn—approximate entropy; SampEn—sample entropy; SIE—symbolic information entropy; RMS—root mean square; DFA—detrended fluctuation analysis; $\alpha 1$, $\alpha 2$—indices of detrended fluctuation analysis short-term and long-term correlations, respectively.

4. Statistical Analysis

Statistical analysis was performed by SYSTAT (SSI, Richmond, CA, USA). The Lilliefors´ normality statistical test was used for evaluation data distributions (Gaussian/non-Gaussian). The Mann–Whitney test was used for between-group comparison of EDA indices and the peripheral temperature (PT) because data were not normally distributed. Anthropometric parameters were normally distributed and analyzed by unpaired Student´s *t*-test. The associations between the parameters of EDA and the peripheral temperature, and correlations between anthropometric characteristics (BMI, BSA) and EDA parameters were analyzed using Spearman´s rank-order correlation test. A value of $p < 0.05$ (two-tailed) was considered statistically significant. Anthropometric parameters were expressed as a mean ± standard error of the mean (SEM). Values of EDA and PT were expressed as median (interquartile range).

5. Results

Anthropometric characteristics of both groups (AN and controls) are summarized in Table 1.

Table 1. Anthropometric characteristics of groups.

	Anorexia Nervosa (N = 25)	Controls (N = 25)	*p* Value
Age (years)	14.76 ± 0.38	15.12 ± 0.32	0.580
Weight (kg)	46.04 ± 1.56	57.76 ± 2.58	0.0001
Height (m)	1.65 ± 1.28	1.66 ± 1.90	0.816
BMI (kg/m^2)	16.88 ± 0.48	20.88 ± 0.66	0.0001
BSA (m^2)	1.45 ± 0.03	1.62 ± 0.04	0.001
WC (cm)	64.79 ± 1.41	75.80 ± 1.99	0.0001

Values are expressed as mean ± SEM. BMI—body mass index; BSA—body surface area; WC—waist circumference. *p* values express the comparison between groups. Probabilities $p < 0.05$ were considered as significant.

5.1. Comparison between Groups of Time-Domain EDA Parameters

Index SCL was significantly lower in the AN group compared to controls ($p = 0.041$). No significant differences were found in index NS.SCRs (Table 2).

Table 2. Parameters of electrodermal activity and peripheral temperature.

Parameter	Anorexia Nervosa (n = 25)	Controls (n = 25)	*p* Value
	Time-domain indices of EDA		
SCL (μS)	0.67 (0.32–2.21)	2.10 (0.97–3.23)	0.041
NS.SCRs	0.50 (0.00–1.00)	1.00 (0.00–2.00)	0.257
	Frequency-domain indices of EDA		
EDA-VLF (μS^2)	18.85 (0.28–56.66)	230.84 (41.39–798.81)	0.001
EDA-Symp (μS^2)	16.33 (3.09–103.79)	815.60 (125.69–1427.03)	0.0001
EDA-HF2 (μS^2)	1.45 (0.37–5.94)	1.73 (0.44–5.07)	0.684
EDA-VHF (μS^2)	1.45 (0.37–5.94)	0.21 (0.03–1.07)	0.765
	Nonlinear indices of EDA		
ApEn	0.59 (0.41–0.82)	0.67 (0.45–0.82)	0.674
SampEn	0.48 (0.28–1.18)	0.48 (0.27–0.81)	0.562
SIE	2.14 (2.06–2.65)	2.35 (1.51–2.76)	0.734
α1	1.01 (0.65–1.20)	0.94 (0.77–1.17)	0.779
α2	0.49 (0.07–0.73)	0.11 (0.05–0.31)	0.032
	Index of peripheral temperature		
PT (°C)	30.53 (25.93–33.46)	31.43 (28.94–32.54)	0.487

5.2. Comparison between Groups of Spectral-Domain EDA Parameters

Index EDA in the very low frequency band was significantly lower for AN group compared to HC ($p = 0.001$). EDA-Symp was significantly lower during baseline for AN vs. HC groups ($p = 0.0001$). No significant differences were found in indices EDA in the high and very high frequency bands (Table 2).

5.3. Comparison between Groups of Nonlinear EDA Parameters

Only nonlinear index α2 was significantly higher for AN group compared to HC ($p = 0.032$). No significant differences were found in remaining nonlinear parameters—ApEn, SampEn, SIE and α1 between groups (Table 2).

5.4. Comparison between Groups of Peripheral Temperature

Peripheral temperature was without significant differences between groups (Table 2).

Values are expressed as median (interquartile range). SCL—skin conductance level; NS.SCRs—nonspecific skin conductance responses; EDA-VLF—electrodermal activity in very low frequency band; EDA-Symp—EDA in the sum of low and high 1 frequency bands; EDA-HF—EDA in high 2 frequency band; EDA-VHF—EDA in very high frequency band; ApEn—approximate entropy; SampEn—sample entropy; SIE—symbolic information entropy; α1, α2—indices of detrended fluctuation analysis short-term and long-term correlations, respectively; PT—peripheral temperature. The *p* value expresses the comparison between groups. Probabilities $p < 0.05$ were considered to be significant.

5.5. Correlation Analysis

Correlation analysis in the total group revealed significant positive correlations between BMI and SCL, and between BMI and PT ($r = 0.292$, $p = 0.042$; $r = 0.302$, $p = 0.041$, respectively). Index SCL was significantly positively correlated with indices of spectral-domain EDA-VLF and EDA-Symp ($r = 0.556$, $p = 0.0001$; $r = 0.277$, $p = 0.050$, respectively). No significant correlations were found in the remaining parameters.

6. Discussion

To the best of our knowledge, this study was the first to describe features of the sympathetic cholinergic regulatory system using spectral and nonlinear characteristics of EDA in AN adolescents before starting a treatment. The results can be summarized as follows: (1) several indices of time (SCL) and frequency (EDA-VLF and EDA-Symp) analysis were significantly lower, indicating insufficient sympathetic cholinergic activity already in adolescent anorexia nervosa; (2) in contrast, the index of DFA–α2 showed a different pattern compared to the time and spectral EDA characteristics—it was increased in anorectic patients thus providing independent and novel information related to altered complexity in sympathetic cholinergic regulatory network in anorexia; (3) correlation analysis revealed that BMI is positively associated with SCL and PT indicating possible alteration of sympathetically mediated and thermoregulatory mechanisms in AN, which could be induced by extensive reduction in body weight. Several mechanisms are proposed.

First, reduced SCL is in accordance with other studies that revealed a lower mean of tonic EDA in AN adults [18,20] and extend these findings into the adolescent age group. Additionally, our study revealed statistically important decrease of EDA in spectral-domain (EDA-VLF, EDA-LF and EDA-Symp) in AN compared to controls. Moreover, we found a positive correlation between SCL and EDA-VLF, and between SCL and EDA-Symp. Thus, we assume that EDA evaluation by SCL, EDA-VLF and EDA-Symp indices represents a sufficient non-invasive tool for assessment of sympathetic cholinergic underactivity linked to AN. Our previous study revealed reduced sympathetic activity based on blood pressure variability analysis in newly diagnosed AN girls, indicating nutritional trajectory as an adaptation process associated with bodyweight control from "normal weight to underweight" [8]. Thus, we assume that acute phase of adolescent AN is characterized by sympathetically mediated autonomic hypoactivation which could only partially be ascribed to the reduced body weight measured by BMI associated with reduced basal metabolism. Specifically, neurobiological features such as hormonal deficiencies (e.g., decreased level of leptin and increased level of ghrelin), neurotransmitters (e.g., serotonin and dopamine) [36–38] and electrolyte dysfunctions (e.g., mainly potassium) [39] could play an important role in AN-linked sympathetic underactivity. Notably, leptin as an important molecular signal of energy abundance is closely linked to sympathetic regulation [40,41], and reduced leptin concentrations associated with sympathetic hypoactivity were found in anorexia nervosa [42].

Further, psychopathological features linked to AN could also importantly contribute to sympathetic hypoactivity in AN. A recent study found an inverse correlation between the negative beliefs concerning the consequences of not controlling thoughts and SCL, suggesting that metacognitive variables (i.e., the inescapability of negative thoughts and the need for thought control) could have a key role in modulating patients' sympathetic hypoactivation. Moreover, this worry could be considered a cognitive strategy to avoid intense emotions. From this perspective the dysfunctional metacognitions about worry might yield also a reduced sympathetic activity in anorexia nervosa [20]. Therefore, it is questionable whether AN-linked sympathetic hypoactivity is predominantly a result of neurobiological changes or is more a reflection of psychopathological features associated with AN.

Due to the fact that biological signals are very difficult to interpret and with various timing and diverse structure, we used nonlinear methods of analysis, which better characterize complex qualitative features of the system. Contrary to time and frequency domain indices, we revealed that the index α2 was significantly higher in AN group compared to controls. Thus, we assume that the α2 index could provide independent information about the complexity of the sympathetic cholinergic regulatory network compared to other time and frequency parameters of EDA. With respect to another biosignal (heart rate), Ishizawa et al. (2008) revealed increased complexity of heart rate variability based on DFA in AN group [9]. Additionally, Jelinek et al. (2017) found an increase of α2-HRV during orthostasis in AN compared to controls [43]. It seems that this important issue remains for further research to elucidate pathways linking sympathetic complexity and anorexia nervosa.

In light of these findings, we could speculate about the complex mechanisms. The excitatory and inhibitory influences of the sympathetic regulation derive from different structures of the central nervous system. We assume that the control mechanisms and neural pathways' influence on the central regulation of EDA are complex. Boucsein et al. (1992) define two potential independent paths affecting sweat production and thus skin conductance: A) cortical level—central control of EDA applies to inclination of cortical centers and basal ganglia as brain structures encompassed in motor control [44]. Notably, cortical pathways encompass excitatory control via pyramidal pathways and the premotor cortex [26,44,45]; B) subcortical level—EDA control encompasses effects from the hypothalamus and the limbic system related to thermoregulatory sweating. Additionally, the inhibitory effects of EDA are mediated by the hippocampus, whereas the excitatory effects are mediated by the amygdala [26,44,45]. Furthermore, the EDA is changed by control effects of reticular formation in the brainstem [26]. Therefore, we assume that the complexity of the brain regulatory regions (prefrontal cortex, hypothalamus) and the existence of feedback loops create a chaotic system influencing EDA.

Conversely, decreased prefrontal cortex (PFC) activity has often been observed among AN patients compared to healthy controls using functional neuroimaging techniques [46–49]. The PFC is a key area for executive functioning, which includes inhibitory control, working memory, and planning. Behavioural studies have revealed that AN patients often have complications with tasks requiring inhibitory control, including response inhibition and task switching [50]. Thus, inhibitory control might be a key element influencing the salient characteristics of AN. Functional imbalance between reward and inhibitory control in the cerebral cortex, especially the PFC, might be related to AN susceptibility and pathology. However, regional structural relevance regarding functional abnormalities in AN remains largely unknown [51–53]. Moreover, the PFC is one of the last brain regions to fully develop relative to subcortical areas [53]. PFC organization undergoes dynamic variations in terms of synaptic density throughout prepubescence and adolescence [54]. This might represent one of key pathomechanisms leading to both AN onset in adolescence and abnormalities in complex feedback regulatory loops resulting in AN-linked sympathetic disruption.

Taken together, the conventional and nonlinear EDA parameters reveal different features of the complex sympathetic cholinergic system consisting of multiple feedback regulatory loops in adolescent AN. While time and spectral parameters of EDA characterize quantitative features of sympathetic cholinergic regulation indicating sympathetic hypoarousal in adolescent AN, the parameter of DFA–$\alpha 2$ could provide novel and distinct information about the qualitative feature complexity of the sympathetic regulatory network, as well.

7. Conclusions

Our study revealed decreased time and spectral indices of EDA indicating sympathetic underactivity in never-treated AN presenting at adolescent age. However, the nonlinear parameter $\alpha 2$ showed a different pattern, indicating abnormal complexity of the sympathetic cholinergic regulatory network in adolescent AN. We suggest that the complex EDA analysis seems to be a sensitive tool for evaluating the activity of the sympathetic cholinergic modulation in AN adolescents. Further, the discrete alteration of the complex sympathetic regulatory network indexed by EDA might represent a risk factor for greater cardiovascular complications already present in adolescent AN.

Future studies should be directed to clarify pathomechanisms linking AN and increased cardiovascular risk, and reveal a complex profile of the sympathetic nervous activity in anorexic patients, particularly in adolescent age. Further clinical research is necessary to identify biomarkers for the assessment of sympathetic complex dysfunction, which could be used for early prevention and personalized treatment of cardiovascular diseases in AN patients.

8. Limitations of the Study

In present study, the cohort consisted of a relatively small homogenous sample of female adolescents with newly diagnosed AN; therefore, it is needs to be validated in a larger cohort with respect to gender. Further, EDA represents a noninvasive index of the sympathetic cholinergic system; therefore, it is not possible to apply these results in terms of general sympathetic dysregulation associated with anorexia nervosa. In this aspect, future research based on continual analysis of additional parameters evaluating the sympathetic control and exploring complex relations between EDA signal and other sympathetically-mediated biosignals (e.g., pre-ejection period, blood pressure variability) under physiological and pathological conditions is needed.

Author Contributions: Z.V., researcher in the field of normal and pathological physiology. She wrote and contributed to the preparation of all parts of the original and revised manuscript and participated in final version of the manuscript. L.B.O., N.S., young researchers in the field of normal and pathological physiology. A significant contribution of all named authors was searching and selection of the scientific studies according to the topic of manuscript, writing and final editing of the original and revised manuscript. I.O., I.H., D.C., clinicians in pediatric psychiatry. They significantly participated in writing and final supervision of the manuscript parts related to the pedopsychiatric topic. S.K., researcher in the field of midwifery and I.F. researcher in the field of nursing. They were involved in searching and selection of the scientific studies according to the topic of manuscript. I.T., professor of normal and pathological physiology, and head of research team. She contributed significantly to the preparation of all parts of the manuscript, final version, supervision and approval of the original and revised manuscript. All authors have read and agreed to the published version of the manuscript.

Funding: This work was funded by the Slovak Scientific Grant Agency under grants VEGA 1/0044/18 and VEGA 1/0190/20, Ministry of Health of the Slovak Republic under the project registration number 2018/20-UKMT-16.

Conflicts of Interest: The authors declare no conflict of interest.

Abbreviations

AN	anorexia nervosa
EDA	electrodermal activity
DFA	detrended fluctuation analysis
ANS	autonomic nervous system
IAPS	International Affective Picture System
SCR	skin conductance response
HC	healthy controls
BMI	body mass index
DSM-5	Diagnostic and Statistical Manual of Mental Disorders
Ag-AgCl	silver-silver chloride
WC	waist circumference
BSA	body surface area
h	height
w	weight
SCL	skin conductance level
NS.SCRs	non-specific skin conductance responses
PSD	power spectral density
EDA-VLF	electrodermal activity in very low frequency band
EDA-Symp	electrodermal activity in the sum of low and high 1 frequency bands
EDA-HF	electrodermal activity in high 2 frequency band
EDA-VHF	electrodermal activity in very high frequency band
ApEn	approximate entropy
SampEn	sample entropy
SIE	symbolic information entropy
RMS	root mean square
$\alpha 1$	indices of detrended fluctuation analysis short-term correlation
$\alpha 2$	indices of detrended fluctuation analysis long-term correlation
VLF	very low frequency band
HF	high frequency band
VHF	very high frequency band
N	number of points
r	tolerance range
m	length of sequence
L	length of vector
p	probability
y	coordinate of the strait stretches
F	fluctuation of combined and detrended signal
SEM	standard mean error
PT	peripheral temperature
PFC	prefrontal cortex

References

1. Arcelus, J.; Mitchell, A.J.; Wales, J.; Nielsen, S. Mortality rates in patients with anorexia nervosa and other eating disorders: A meta-analysis of 36 studies. *Arch. Gen. Psychiatry* **2011**, *68*, 724–731. [CrossRef] [PubMed]
2. Lavender, J.M.; Wonderlich, S.A.; Engel, S.G.; Gordon, K.H.; Kaye, W.H.; Mitchell, J.E. Dimensions of emotion dysregulation in anorexia nervosa and bulimia nervosa: A conceptual review of the empirical literature. *Clin. Psychol. Rev.* **2015**, *40*, 111–122. [CrossRef] [PubMed]
3. American Psychiatric Association. *Diagnostic and Statistical Manual of Mental Disorders*, 5th ed.; American Psychiatric Publishing: Arlington, VA, USA, 2013.
4. Attia, E. Anorexia Nervosa: Current Status and Future Directions. *Annu. Rev. Med.* **2010**, *61*, 425–435. [CrossRef]
5. Favaro, A.; Ferrara, S.; Santonastaso, P. The spectrum of eating disorders in young women: A prevalence study in a general population sample. *Psychosom. Med.* **2003**, *65*, 701–708. [CrossRef] [PubMed]
6. Madden, S.; Morris, A.; Zurynski, Y.A.; Kohn, M.; Elliot, E.J. Burden of eating disorders in 5–13-year-old children in Australia. *Med. J. Aust.* **2009**, *190*, 410–414. [CrossRef]
7. Mazurak, N.; Enck, P.; Muth, E.; Teufel, M.; Zipfel, S. Heart rate variability as a measure of cardiac autonomic function in anorexia nervosa: A review of the literature. *Eur. Eat. Disord. Rev.* **2011**, *19*, 87–99. [CrossRef]
8. Tonhajzerova, I.; Mestanikova, A.; Jurko, A.; Grendar, M.; Langer, P.; Ondrejka, I.; Jurko, T.; Hrtanek, I.; Cesnekova, D.; Mestanik, M. Arterial stiffness and haemodynamic regulation in adolescent anorexia nervosa versus obesity. *Appl. Physiol. Nutr. Metab.* **2020**, *45*, 81–90. [CrossRef]
9. Ishizawa, T.; Yoshiuchi, K.; Takimoto, Y.; Yamamoto, Y.; Akabayashi, A. Heart Rate and Blood Pressure Variability and Baroreflex Sensitivity in Patients With Anorexia Nervosa. *Psychosom. Med.* **2008**, *70*, 695–700. [CrossRef]
10. Crifaci, G.; Billeci, L.; Tartarisco, G.; Balocchi, R.; Pioggia, G.; Brunori, E.; Maestro, S.; Morales, M.A. ECG and GSR measure and analysis using wearable systems: Application in anorexia nervosa adolescents. In Proceedings of the International Symposium on Image and Signal Processing and Analysis (ISPA), Trieste, Italy, 4–6 September 2013; pp. 499–504. [CrossRef]
11. Sachs, K.V.; Harnke, B.; Mehler, P.S.; Krantz, M.J. Cardiovascular complications of anorexia nervosa: A systematic review. *Int. J. Eat. Disord.* **2016**, *49*, 238–248. [CrossRef]
12. Figner, B.; Murphy, R.O. Using Skin conductance in judgment and decision making research. In *A Handbook of Process Tracing Methods for Decision Research*; Schutle-Mecklenbeck, M., Kuehberger, A., Ranyard, R., Eds.; Psychology Press: New York, NY, USA, 2010; pp. 163–184.
13. Dawson, M.E.; Schell, A.M.; Filion, D.L. The Electrodermal System. In *Handbook of Psychophysiology*; Cacioppo, J.T., Tassinary, L.G., Berntson, G.G., Eds.; Cambridge University Press: Cambridge, MA, USA, 2000; pp. 200–223.
14. Dawson, M.E.; Schell, A.M.; Filion, D.L. The Electrodermal System. In *Handbook of Psychophysiology*; Cacioppo, J.T., Tassinary, L.G., Berntson, G.G., Eds.; Cambridge University Press: New York, NY, USA, 2007; pp. 159–181.
15. Posada-Quintero, H.F.; Florian, J.P.; Orjuela-Cañón, A.D.; Aljama-Corrales, T.; Charleston-Villalobos, S.; Chon, K.H. Power Spectral Density Analysis of Electrodermal Activity for Sympathetic Function Assessment. *Ann. Biomed. Eng.* **2016**, *44*, 3124–3135. [CrossRef]
16. Lanatà, A.; Valenza, G.; Scilingo, E.P. A novel EDA glove based on textile-integrated electrodes for affective computing. *Med. Biol. Eng. Comput.* **2012**, *50*, 1163–1172. [CrossRef]
17. Visnovcova, Z.; Mestanik, M.; Gala, M.; Mestanikova, A.; Tonhajzerova, I. The complexity of electrodermal activity is altered in mental cognitive stressors. *Comput. Biol. Med.* **2016**, *79*, 123–129. [CrossRef] [PubMed]
18. Bellodi, L.; Martoni, R.M.; Galimberti, E.; Cavallini, M.C. Low baseline startle in anorexia nervosa patients. *Prog. Neuro-Psychopharmacol. Biol. Psychiatry* **2013**, *40*, 26–29. [CrossRef]
19. Clarke, J.; Ramoz, N.; Fladung, A.K.; Gorwood, P. Higher reward value of starvation imagery in anorexia nervosa and association with the Val66Met BDNF polymorphism. *Transl. Psychiatry* **2016**, *6*, e829. [CrossRef]
20. Palomba, D.; Venturini, M.; Rausa, M.; Contin, S.A.; Penolazzi, B.; Schumann, R.; Ballardini, D. Reduced sympathetic activity and dysfunctional metacognition in patients with anorexia nervosa: A preliminary study. *J. Evid. Based Psychother.* **2017**, *17*, 1–20. [CrossRef]

21. Soussignan, R.; Schaal, B.; Rigaud, D.; Royet, J.P.; Jiang, T. Hedonic reactivity to visual and olfactory cues: Rapid facial electromyographic reactions are altered in anorexia nervosa. *Biol. Psychol.* **2011**, *86*, 265–272. [CrossRef]
22. Tchanturia, K.; Liao, P.C.; Uher, R.; Lawrence, N.; Treasure, J.; Campbell, I.C. An investigation of decision making in anorexia nervosa using the Iowa Gambling Task and skin conductance measurements. *J. Int. Neuropsychol. Soc.* **2007**, *13*, 635–641. [CrossRef]
23. Piacentini, R. *Emotions at Fingertips: Revealing Individual Features in Galvanic Skin Response Signals—Roberto Piacentini—Google Books*; Sapienza Università di Roma: Roma, Italy, 2004.
24. Yang, Z.; Liu, G. An entropy measure of emotional arousal via skin conductance response. *J. Fiber Bioeng. Inform.* **2014**, *7*, 67–80.
25. Cole, T.J.; Lobstein, T. Extended international (IOTF) body mass index cut-offs for thinness, overweight and obesity. *Pediatr. Obes.* **2012**, *7*, 284–294. [CrossRef] [PubMed]
26. Blain, S.; Power, S.D.; Sejdic, E.; Mihailidis, A.; Chau, T. A cardiorespiratory classifier of voluntary and involuntary electrodermal activity. *Biomed. Eng. Online* **2010**, *9*, 11. [CrossRef] [PubMed]
27. Poh, M.Z.; Swenson, N.C.; Picard, R.W. A wearable sensor for unobtrusive, long-term assessment of electrodermal activity. *IEEE Trans. Biomed. Eng.* **2010**, *57*, 1243–1252. [PubMed]
28. FlexComp Infinity Hardware Manual. Available online: https://bio-medical.com/media/support/t7555m.pdf (accessed on 23 June 2020).
29. Mosteller, R. Simplified calculation of body-surface area. *N. Engl. J. Med.* **1987**, *317*, 1098.
30. Venables, P.H.; Gartshore, S.A.; O'Riordan, P.W. The function of skin conductance response recovery and rise time. *Biol. Psychol.* **1980**, *10*, 1–6. [CrossRef]
31. Boucsein, W.; Fowles, D.C.; Grimnes, S.; Ben-Shakhar, G.; Roth, W.T.; Dawson, M.E.; Filion, D.L. Publication recommendations for electrodermal measurements. *Psychophysiology* **2012**, *49*, 1017–1034. [PubMed]
32. Pincus, S.M. Approximate entropy as a measure of system complexity. *Proc. Natl. Acad. Sci. USA* **1991**, *88*, 2297–2301. [CrossRef] [PubMed]
33. Richman, J.S.; Moorman, J.R. Physiological time-series analysis using approximate and sample entropy. *Am. J. Physiol. Hear. Circ. Physiol.* **2000**, *278*, H2039–2049. [CrossRef] [PubMed]
34. Pincus, S.M.; Goldberger, A.L. Physiological time-series analysis: What does regularity quantify? *Am. J. Physiol. Hear. Circ. Physiol.* **1994**, *266*, H1643–1656. [CrossRef]
35. Peng, C.K.; Havlin, S.; Stanley, H.E.; Goldberger, A.L. Quantification of scaling exponents and crossover phenomena in nonstationary heartbeat time series. *Chaos* **1995**, *5*, 82–87. [CrossRef]
36. Ravaldi, C.; Vannacci, A.; Ricca, V. Cardiac complications of anorexia nervosa. *Recenti Prog. Med.* **2003**, *94*, 267–270.
37. Vanderdonckt, O.; Lambert, M.; Cornejo Montero, M.; Boland, B.; Brohet, C. The 12-lead electrocardiogram in anorexia nervosa: A report of 2 cases followed by a retrospective study. *J. Electrocardiol* **2001**, *34*, 233–242. [CrossRef]
38. Tanaka, M.; Nakahara, T.; Muranaga, T.; Kojima, S.; Yasuhara, D.; Ueno, H.; Nakazato, M.; Inui, A. Ghrelin concentrations and cardiac vagal tone are decreased after pharmacologic and cognitive-behavioral treatment in patients with bulimia nervosa. *Horm. Behav.* **2006**, *50*, 261–265. [CrossRef] [PubMed]
39. Ballardini, D.; Schumann, R. *La Riabilitazione Psiconutrizionale nei Disturbi Dell'alimentazione*; Carocci Faber: Roma, Italy, 2011; ISBN 9788874666218.
40. Zigman, J.M.; Elmquist, J.K. Minireview: From anorexia to obesity—The Yin and Yang of body weight control. *Endocrinology* **2003**, *144*, 3749–3756. [CrossRef] [PubMed]
41. Rahmouni, K. Leptin-induced sympathetic nerve activation: Signaling mechanisms and cardiovascular consequences in obesity. *Curr. Hypertens. Rev.* **2010**, *6*, 104–109. [CrossRef]
42. Murialdo, G.; Casu, M.; Falchero, M.; Brugnolo, A.; Patrone, V.; Cerro, P.F.; Ameri, P.; Andraghetti, G.; Briatore, L.; Copello, F.; et al. Alterations in the autonomic control of heart rate variability in patients with anorexia or bulimia nervosa: Correlations between sympathovagal activity, clinical features, and leptin levels. *J. Endocrinol. Investig.* **2007**, *30*, 356–362. [CrossRef] [PubMed]
43. Jelinek, H.; Cornforth, D.; Tarvainen, M.; Spence, I.; Russell, J. Decreased sample entropy to orthostatic challenge in anorexia nervosa. *J. Metab. Syndr.* **2017**, *6*. [CrossRef]
44. Sequeira, H.; Roy, J.-C. Cortical and hypothalamo-limbic control of electrodermal responses. In *Progress in Electrodermal Research*; Springer: Plenum Press: New York, NY, USA, 1993; pp. 93–114.

45. Boucsein, W. *Electrodermal Activity*; Plenum University Press: New York, NY, USA, 1992.
46. Lock, J.; Garrett, A.; Beenhakker, J.; Reiss, A.L. Aberrant brain activation during a response inhibition task in adolescent eating disorder subtypes. *Am. J. Psychiatry* **2011**, *168*, 55–64. [CrossRef]
47. Nagamitsu, S.; Araki, Y.; Ioji, T.; Yamashita, F.; Ozono, S.; Kouno, M.; Iizuka, C.; Hara, M.; Shibuya, I.; Ohya, T.; et al. Prefrontal brain function in children with anorexia nervosa: A near-infrared spectroscopy study. *Brain Dev.* **2011**, *33*, 35–44. [CrossRef]
48. Oberndorfer, T.A.; Kaye, W.H.; Simmons, A.N.; Strigo, I.A.; Matthews, S.C. Demand-specific alteration of medial prefrontal cortex response during an inhibition task in recovered anorexic women. *Int. J. Eat. Disord.* **2011**, *44*, 1–8. [CrossRef]
49. Suda, M.; Brooks, S.J.; Giampietro, V.; Friederich, H.C.; Uher, R.; Brammer, M.J.; Williams, S.C.R.; Campbell, I.C.; Treasure, J. Functional neuroanatomy of body checking in people with anorexia nervosa. *Int. J. Eat. Disord.* **2013**, *46*, 653–662. [CrossRef]
50. Galimberti, E.; Martoni, R.M.; Cavallini, M.C.; Erzegovesi, S.; Bellodi, L. Motor inhibition and cognitive flexibility in eating disorder subtypes. *Prog. Neuropsychopharmacol. Biol. Psychiatry* **2012**, *36*, 307–312. [CrossRef]
51. Castro-Fornieles, J.; Bargalló, N.; Lázaro, L.; Andrés, S.; Falcon, C.; Plana, M.T.; Junqué, C. A cross-sectional and follow-up voxel-based morphometric MRI study in adolescent anorexia nervosa. *J. Psychiatr. Res.* **2009**, *43*, 331–340. [CrossRef] [PubMed]
52. Chui, H.T.; Christensen, B.K.; Zipursky, R.B.; Richards, B.A.; Hanratty, M.K.; Kabani, N.J.; Mikulis, D.J.; Katzman, D.K. Cognitive function and brain structure in females with a history of adolescent-onset anorexia nervosa. *Pediatrics* **2008**, *122*, e426–e437. [CrossRef] [PubMed]
53. Somerville, L.H.; Casey, B.J. Developmental neurobiology of cognitive control and motivational systems. *Curr. Opin. Neurobiol.* **2010**, *20*, 236–241. [CrossRef] [PubMed]
54. Chowdhury, U.; Gordon, I.; Lask, B.; Watkins, B.; Watt, H.; Christie, D. Early-onset anorexia nervosa: Is there evidence of limbic system imbalance? *Int. J. Eat. Disord.* **2003**, *33*, 388–396. [CrossRef] [PubMed]

applied sciences

MDPI

Article

Feature Assessment of Toe Area Activity during Walking of Elderly People with Stumbling Experiences through Wearable Clog-Integrated Plantar Visualization System

Yingjie Jin [1,*], Miho Shogenji [2] and Tetsuyou Watanabe [3,*]

1 Graduated School of Natural Science and Technology, Kanazawa University, Kakuma-machi, Kanazawa 9201192, Japan
2 Faculty of Health Sciences, Institute of Medical, Pharmaceutical and Health Sciences, Kanazawa University, 5-11-80 Kodatsuno, Kanazawa 9200942, Japan; shogen@mhs.mp.kanazawa-u.ac.jp
3 Faculty of Frontier Engineering, Institute of Science and Engineering, Kanazawa University, Kakuma-machi, Kanazawa 9201192, Japan
* Correspondence: e.k-kin@stu.kanazawa-u.ac.jp (Y.J.); twata@se.kanazawa-u.ac.jp (T.W.)

Received: 21 January 2020; Accepted: 12 February 2020; Published: 17 February 2020

Abstract: In this study, we investigated the relationship between toe-area activity and stumbling experiences utilizing our developed sensing system, in order to assess toe-area activity in elderly people with stumbling experiences. The sensing system enables the visualization of the plantar aspect while walking on any surface and under any condition. An image of the plantar aspect is received at a reflecting surface and captured by a camera attached to a clog. The toe-area activity was evaluated by comparing the difference between the toe contact areas at heel-strike and push-off. Thirteen young individuals (nine men and four women, age 22.4 ± 2 years) and nine elderly individuals (five men and four women, age 65.3 ± 2 years) participated in the experiment by walking along a straight line wearing the plantar sensing system on their feet. The analysis found that a low value of the mean toe activity for multiple walking cycles was associated with high stumbling risk, irrespective of age, whereas large variations in toe activity was associated with aging. These results indicate that toe activity can predict stumbling risk irrespective of age. We also found that a large value of the maximum toe activity during multiple walking cycles indicates aging, whereas a low value is associated with high stumbling risk.

Keywords: wearable sensors; gait analysis; stumbling; plantar visualization

1. Introduction

The Cabinet Office of the Government of Japan [1] has reported falls as the fourth leading reason for elderly individuals above 65 years requiring support from caregivers. The demand for risk assessment to prevent falls increases with age. Fall risk assessment considers different types of factors [2]. This study focused on the motion of the plantar region, especially the toe area. Examination of the sole of the foot, or the plantar aspect, which is often conducted as a self-health check in Oriental medicine, is expected to reveal significant health-related information, such as irregularity in toe-area shape corresponding to stumbling risks. However, the number of observational methods used to obtain quantitative data is limited. Measurement systems for observing parts of the foot involved in walking are classified into two types.

The first type comprises pressure-sensor-based systems. Pressure profiles during walking can be obtained by embedding pressure sensors into the insoles found in every shoe. There are several commercially available pressure sensors for investigating the relationship between plantar pressure and walking speed [3,4], age [5], and foot type [6]. Majumder et al. [7] predicted falls using sensory data from a smartphone and four pressure sensors on a shoe. Ayena et al. [8] assessed the risk of falls at home using a system similar to that of Majumder et al. [7]. Yu et al. [9] developed a shoe system that monitored walking with shoe-integrated force sensors. Crea et al. [10] developed a pressure-sensitive insole based on optoelectronic technology.

The second type of measurement system comprises the inertial measurement unit (IMU)-sensor-based systems [11]. Hung and Suh [12] reduced the position error of shoes estimated by IMU sensors by utilizing additional camera information. Do and Suh [13] developed a walking measurement system that employed inertial sensors and a camera with floor markers to estimate step length and foot angle. An optical sensor was utilized in this case, although the targets did not correspond to plantar observation. Foxlin [14] tracked pedestrians with shoe-mounted IMU sensors. Ojeda and Borenstein proposed a personal dead-reckoning navigation system with a shoe-mounted IMU sensor. Sim et al. [15] proposed an algorithm to detect falls with acceleration sensors on a shoe. Rampp et al. [16] estimated a stride parameter that utilized IMU sensors on a shoe. Mariani et al. [17] used IMU sensors on a shoe to assess 3D spatial parameters of gait.

There are also systems that have both pressure and IMU sensors. Scheoers et al. [18] proposed a shoe fitted with force and IMU sensors. Bamberg et al. [19] proposed a wireless wearable sensor system with an accelerometer, a gyroscope, and a ground force reaction sensor embedded in a shoe. Bebek et al. [20] developed a personal navigation system with pressure and IMU sensors, while Chen et al. [21] detected an abnormality in shoes with four force and IMU sensors. Hegde et al. [22] developed the SmartStep shoe with embedded pressure and acceleration sensors. Kawsar et al. [23] presented an activity detection system that utilized pressure data from a shoe and accelerometers and gyroscope data from a smartphone.

Several studies have focused on toe plantar flexor strength, which is closely related to toe activity. Tsuyuguchi et al. [24] and Kim et al. [25] showed that toe grip strength is an independent risk factor for falling. Mickle et al. [26] showed that the reduction of toe flexor strength and the presence of toe deformities increased the risk of falling among older people. Hiura et al. [27] investigated the relationship between walking ability and falls in elderly Japanese people. Their results showed that lower toe-gap force was significantly associated with falling. Uritani et al. [28] presented a study showing that timed up-and-go test (a measure of functional ability) results were significantly correlated with toe grip strength. Branthwaite et al. [29] assessed the effect of toe flexor exercises on apical plantar pressure, and showed that such exercises can improve forefoot stability and gait efficiency. Takatori et al. [30] showed that the toe elevation angle of participants who had experienced a fall was significantly smaller compared with that of participants who did not fall.

Plantar pressure is another important indicator in walking evaluation, such as for the diagnosis and determination of the course of a disease. Bus et al. [31] showed that the claw/hammer toe deformity is associated with a distal-to-proximal transfer of load in the forefoot, and elevated the plantar pressure at the sub-metatarsal heads in neuropathic diabetic patients. Morag et al. [32] developed predictive models to identify the potential etiological factors associated with elevated plantar pressure. Payne et al. [33] investigated the factors associated with increased plantar pressure on the diabetic foot. Buldt et al. [34] compared plantar pressure among healthy individuals with normal, planus, or cavus feet. Mickle et al. showed that high plantar pressure may increase the risk of falls.

However, the above observation methods can provide only specific data, including pressure or IMU data; thus, they yield a limited amount of information. Examples of unobservable information obtained by conventional measurements include accurate contact area, skin deformation, and skin color. It is difficult for pressure sensors to detect or identify the exact contact area when the load is low (for example, during the swing phase). Thus, we propose a novel wearable sensing system for direct observation of the plantar aspect while walking on any surface and condition [35]. The plantar image is reflected by a reflective film and captured by a camera embedded in a clog (shoe). Passive joints are embedded in the camera holder and around the tarsometatarsal joints to facilitate walking while wearing the system. However, the main contribution of this sensing system is the investigation of fall or stumbling risks, which were not considered previously. As a first step, we focus on toe activity, which is closely related to walking issues such as stumbling, which in turn is directly connected to falling [36,37]. We did not find research that collected and analyzed visual data on toe movements in the above literature, because of the difficulty in directly obtaining visual data from the sensor. The utilization of the developed system is expected to provide clearer information on toe movements, especially in the swing phase. The movements in the swing phase are difficult to observe by conventional IMU or pressure sheet sensors.

In this study, we investigated the relationship between toe-area activity and stumbling utilizing the developed sensing system. Specifically, we identify the features of toe-area activity while walking in elderly people with a history of stumbling. The main difference of this study from the previous one [35] is the utilization of the sensing system. The previous study provided only primary results among young male participants, whereas this study investigated the effects of aging to identify new features of walking in elderly people with a stumbling history. We also improved the sensing system. The difference between the actual sensing system and previous system [35] is the data processing method: the partially manual data processing used in the previous study was changed to fully automatic data processing to determine the contact area in the toe and the corresponding criterion for evaluation. The remainder of this paper is organized as follows. Section 2 presents the developed clog-integrated plantar visualization system. The experimental investigation and data analysis method for evaluating toe activity are presented in Section 3. The results are provided in Section 4 and discussed in Section 5. A summary of the study is presented in Section 6.

2. Clog-Integrated Plantar Visualization System

Overview of the System

Figure 1a shows an overview of the sensing system. The system was constructed by covering the sensing area, shown in Figure 1b, with the upper part of the clog (Crocs, Niwot, CO, USA). A schematic view of the lower part (sensing area) of the system, which consists of a reflecting surface and camera, is shown in Figure 1b. The image of the plantar aspect was reflected by the reflecting surface and captured by the camera fixed at the holder. Transparent silicone and acrylic plates were utilized as the foot sole to enable visualization of the planta. The sensing area was surrounded by a frame manufactured by a 3D printer (Raise 3D N2) to support the weight of the wearer (exerted through the acrylic plates) and minimize the total weight of the clog.

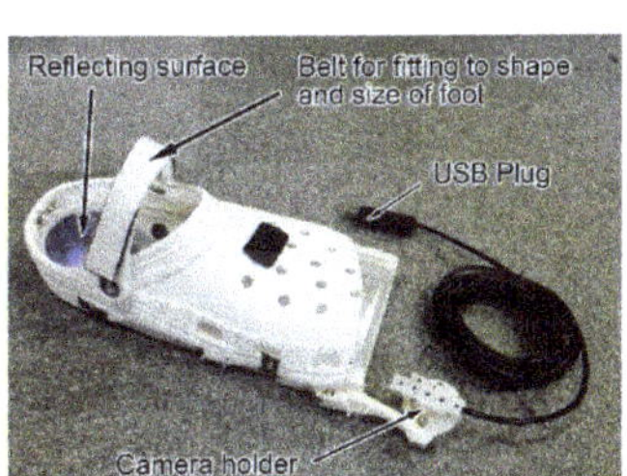

(a)

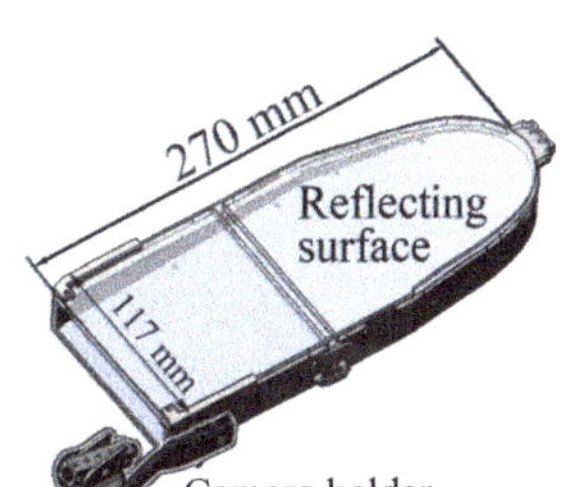

(b)

Figure 1. Overview of the system: (**a**) photo of the clog-integrated plantar visualization system and (**b**) dimensions of the underpart of the plantar observation system.

A schematic illustration of the sensing system is shown in Figure 2. The embedding of the camera to capture the visual information from the plantar aspect was a crucial step. Positioning the camera at the bottom of the clog for direct visualization of the planta is difficult because the space under the insole is very narrow. If the area under the plantar aspect is too thick, it becomes difficult to walk, which makes the observation of walking difficult. Multiple cameras or a camera with an extremely wide field of view and short focal length could be utilized in this setting. However, the use of multiple cameras is unsuitable because they should be sufficiently small to be embedded into or attached to the bottom space of the clog; the processing costs would also increase. Thus, the use of only one camera is ideal to reduce costs and simplify the system. The reflecting surface was placed at the insole of the clog while the camera was set outside the clog to capture the reflected plantar image (Figure 2). The camera was positioned in such a way that the main target areas could be clearly observed. Usually, the heel area of the foot is utilized only during heel-strike, and not during push-off or other swing-phase parts (see Figure 3 for the definition of phases and their components). Therefore, the primary observation target is the toe area, particularly the big toe used during push-off motion.

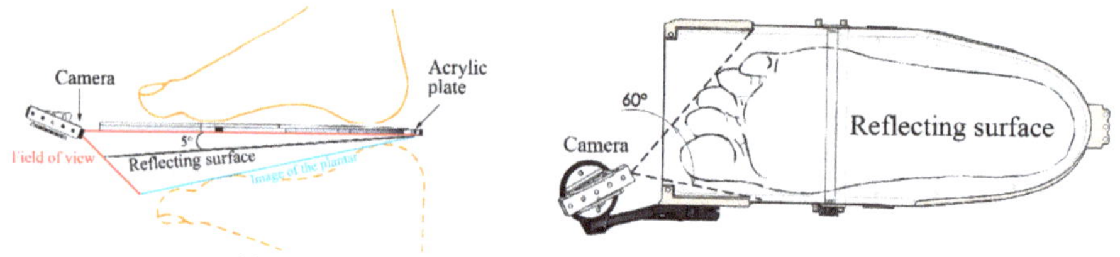

Figure 2. Schematic illustration of the plantar observation system: (**a**) side view and (**b**) bottom view.

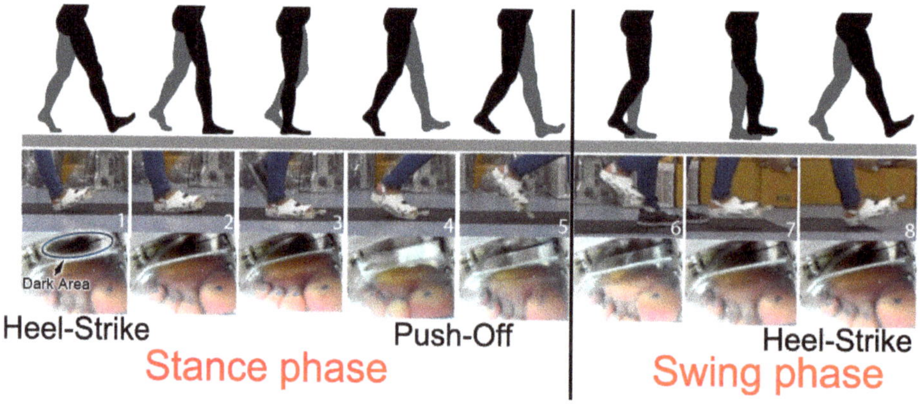

Figure 3. Phases and phase components of a walking cycle and representative images captured by the developed plantar observation system.

An endoscopic camera (GIWOX; resolution: 640 × 480 pixels, focal length: 6 cm to infinity, angle of view: 60°) with a short focal length was utilized to obtain a wide field of view. A 0.6 mm thick reflecting film was utilized to ensure that the thickness of the clog sole was close to its original thickness. Sequential images showing a single walking cycle are presented in Figure 3. The toe area is visible in all phases, while the heel area (dark area in the figure) is visible at heel-strike.

Two passive joints were embedded into the clog (Figure 4). One was installed in the clog near the tarsometatarsal joints to minimize the effect of the 3D-printed solid part on walking and enable the foot to bend naturally. The maximum allowable joint angle was set to 40°. The heel area cannot be observed during push-off because of the passive joints. The other passive joint was installed where the holder is attached to the clog to minimize the disturbance on walking by the camera holder. The camera holder rotates during push-off and the heel area is not observable.

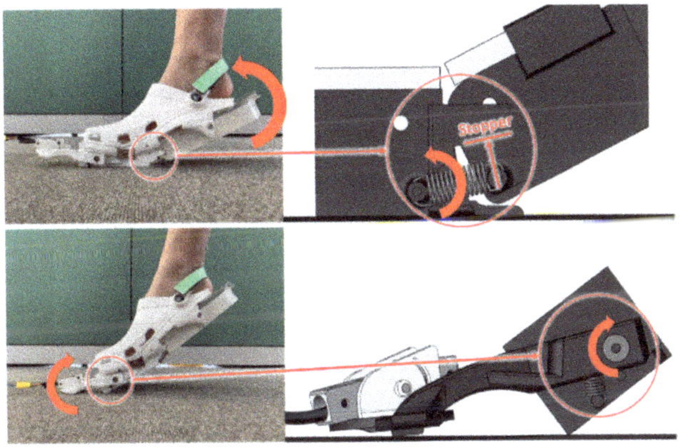

Figure 4. Passive joints to allow the foot to bend at the tarsometatarsal joints and toward the camera.

The system currently costs less than $70, and the sensor for one foot weighs 448 g (the weight of the original clog is 190 g). The camera cable was connected to a Wi-Fi transmitter (F99 WIFI; weight: 27.5 g, size: 63 × 35 × 18 mm^3) with a built-in battery, which was attached to the shank of the wearer to wirelessly transmit the captured images. The electric power reserve of the sensor was sufficient to complete a test lasting approximately 1 h. Hence, the proposed system has the potential for use in a wide variety of experimental environments.

3. Methods

3.1. Participants

The participants were 13 young and healthy individuals (nine men and four women, age 22.4 ± 2 years, weight = 58.3 ± 10.6 kg, height = 165.6 ± 8.9 cm, and foot size = 24–27 cm), and 9 elderly individuals (five men and four women, age 65.3 ± 2 years, weight = 61.4 ± 10.9 kg, height = 163.6 ± 10.8 cm, and foot size = 24–27 cm) without any diseases, impediments, or history of walking problems. The procedure and purpose of the experiments were approved by the Medical Ethics Committee of Kanazawa University (No. 33). Information on the stumbling experiences of participants were obtained via a questionnaire; individuals who had stumbled within a period of three months were designated as individuals with stumbling experience. The participants were separated into two groups according to their responses: those without stumbling experiences (eight young and five elderly individuals) and those with stumbling experiences (five young and four elderly individuals). The toe-area activities of the two groups were compared. The information on participants is summarized in Table 1.

Table 1. Information of participants.

	Young	Elderly
Number of participants	13 (M:9 F:4)	9 (M:5 F:4)
Without stumbling experience	8 (M:5 F:3)	5 (M:3 F:2)
With stumbling experience	5 (M:4 F:1)	4 (M:2 F:2)
Information of participants		
Age	22.4 ± 2 years	65.3 ± 2 years
Weight	58.3 ± 10.6 kg	61.4 ± 10.9 kg
Height	165.6 ± 8.9 cm	163.6 ± 10.8 cm
Foot Size	24–27 cm	24–27 cm

M: Male; F: Female.

3.2. Procedure

The participants walked a straight line along a 6.5 m long flat road wearing the developed plantar sensing system on their right foot and a normal shoe on their left foot. The thickness of the sole of the left shoe was the same as that of the sensing system so that there was no impact on walking. There was initially a start and stop zone, measuring 1 m each; and the walking data obtained in the remaining 4.5 m, which corresponds to three walking cycles, were analyzed. The camera cable was attached to the leg of participants to avoid disturbing the walking motion. Before the experiment, we conducted a practice session enabling participants to adapt to the plantar sensing system. This minimized the effects of differences between the left and right footwear on the results. The participants practiced walking while wearing the experimental system until they could walk normally without the feeling of strangeness, and the motion of the left foot (wearing a normal shoe) was close to that of the right foot (wearing the developed sensing system). The practice session was also aimed at minimizing the effect of the camera holder on walking. We determined the position and angle of the camera based on the principle of specular reflection and the horizontal angle of view of the camera. Before each experiment, we controlled the camera position and angle so that the image captured includes all the areas of the forepart of the foot and was focused. We also checked whether the average image intensity was in the range of the desired reference values: 0.53 ± 0.02, which were determined through trial and error based on the results of preliminary experiments. If not, we adjusted the camera's auxiliary brightness so that the average image intensity was within the desired range. With these methods, we obtained the plantar images for all the participants under almost the same illumination conditions and foot positions. The participants were instructed to walk while facing forward as much as possible to avoid looking at their feet during the experiment. Each participant performed the experiment three times; thus, we analyzed nine walking cycles for each participant.

3.3. Data Analysis: Toe-Area Activity

First, each frame containing the toe area was extracted from the captured video streams. Second, the toe area was extracted from the image frame and evaluated. Third, the contact area in the toe area was derived. The information about the contact area was obtained using the skin color of the bare foot. The skin turns white when in contact with something because the blood flow inside capillary vessels is blocked. Thus, the white area in the image corresponds to the contact area. The RGB (red, green, blue) image was transformed into the hue, saturation, value (HSV) image, and the extracted area that satisfied $0.055 < H < 0.167$, $0.05 < S < 0.75$, $0.4 < V$ corresponded to the white area. It should be noted that the contact area can be detected by the naked eye. Therefore, the range of H, S, V values for detecting the contact area was manually determined through trial and error, utilizing sample images obtained at preliminary experiments. The same range of HSV values was applied to all the data for all participants. Fourth, using binarization, the extracted area was assigned a value of one, whereas the other areas were assigned a value of zero. Finally, the small noise area was removed. Figure 5 outlines the overall procedure. Figure 6 shows the time series for the contact area in the toe area for the two young study groups over one walking cycle in a single trial. The first peak resulted from pushing-off before taking-off, while the second peak indicates the heel-strike. The difference in the values of the contact areas at the two peaks was large for the group with no stumbling experiences and small for the group with stumbling experiences. We assumed that the former group moved or rotated their toes while pushing off, thus increasing the contact area, whereas the latter group did not. We examined the difference in the two peaks corresponding to toe-area activity, called the difference of two peak-contact areas (DPCA). The DPCA was normalized for variations in the width of the contact area (Figure 5) to compare the value for each group considering individual differences. The normalized DPCA (NDPCA) is defined as follows:

$$\mathrm{NDPCA} = \frac{A_{\mathrm{peak1}} - A_{\mathrm{peak2}}}{A_{\mathrm{peak1}} - A_{\mathrm{min}}} \tag{1}$$

where A_{peak1} denotes the contact area at the first peak, A_{peak2} denotes the contact area at the second peak, and A_{min} denotes the minimum contact area. In short, we used the DPCA as an indicator of toe movement, and the min–max normalization of the DPCA was performed using the maximum and minimum difference between the contact area in each cycle. This eliminated data from individual differences, and finally, the NDPCA was obtained. The NDPCA was automatically derived by the method presented in the Appendix A.

Algorithm: Extraction of the contact area of the toe area	
Given	Image captured by the camera at each frame
1	Select the toe area from the captured image.
2	An RGB image is transformed into an HSV image.
3	Extracted area satisfies $0.055 < H < 0.167$, $0.05 < S < 0.75$, $0.4 < V$.
4	Binarization: Extracted area value = 1, the other area value = 0.
5	Remove the small area.
6	Count the area with a value = 1.
End	

Figure 5. Procedure to derive the toe contact area.

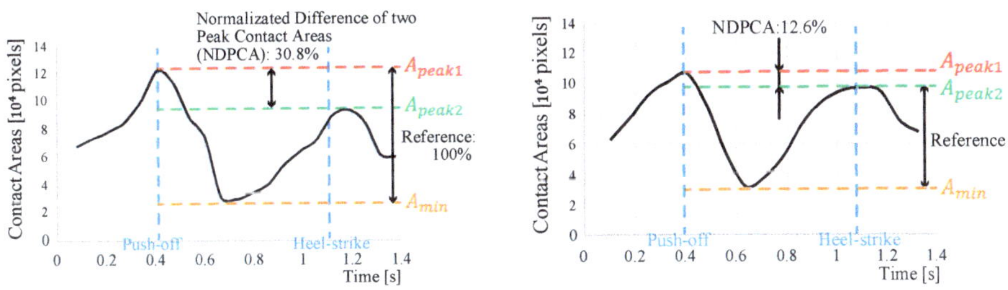

(a) (b)

Figure 6. Time series of the toe contact area for (**a**) the group without stumbling experiences and (**b**) the group with stumbling experiences. Both subfigures display one example obtained from a single trial for a single young individual. NDPCA: normalized difference of two peak-contact areas.

4. Results

The mean, maximum, minimum, and standard deviation values of NDPCAs for nine walking cycles were used as evaluation values for each participant. There were no significant differences between the groups with regard to gender proportion, as shown in Table 2; thus, we evaluated the results according to stumbling experiences and aging effect.

Table 2. Effect of gender on NDPCA.

NDPCA (Mean ± Std.)	Male	Female	*p*-Value *
Young	0.318 ± 0.073	0.300 ± 0.024	0.536
Elderly	0.317 ± 0.047	0.301 ± 0.059	0.690

* *p* is probability value.

Figure 7 and Table 3 show the calculated NDPCAs. Assessment of the mean NDPCAs showed statistically significant differences between the groups with and without stumbling experiences, regardless of the difference in age between groups. Note that the bars shown in Figure 7 represent not standard error but standard deviation, and thus there could be an overlap in the bars for the groups even if there was a statistically significant difference between them. Toe-area activity (NDPCA) was high for the group with no stumbling history, and low for the group with a stumbling history. The results also showed no significant difference in the NDPCA between the elderly and young people in both groups. We evaluated the maximum and minimum NDPCAs to observe extreme cases in toe activity. The order of the mean value of the maximum NDPCA was highest for elderly individuals without stumbling experience, followed by elderly individuals with stumbling experience, and then young individuals without and with stumbling experience. A statistically significant difference was observed between young and elderly groups ($p < 0.01$), between young and elderly groups with stumbling experiences, and between groups with and without stumbling experiences irrespective of age. A difference was also observed between the young and elderly groups with no stumbling experiences, although it was not statistically significant. No statistically significant difference was observed for any pairs in the minimum NDPCAs. The variations in toe activity were evaluated via the standard deviation of NDPCA. A difference between the young and elderly groups without stumbling experiences was observed, although it was not statistically significant.

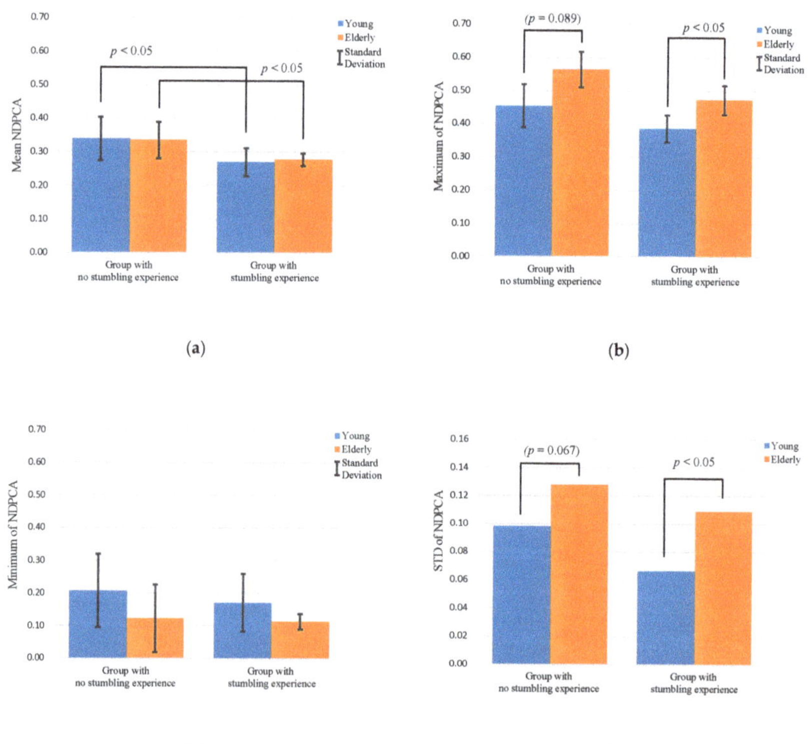

Figure 7. Main results of NDPCA: (**a**) Mean NDPCA; (**b**) Maximum NDPCA; (**c**) Minimum NDPCA; (**d**) Standard deviation of NDPCA.

Table 3. Mean, maximum, minimum, and standard deviation of the NDPCA for groups with and without stumbling histories.

	Young	Elderly	p-Value
Mean NDPCA			
Without stumbling experience	0.339 ± 0.064	0.335 ± 0.054	0.451
With stumbling experience	0.269 ± 0.041	0.278 ± 0.019	0.348
***p*-value**	**0.020** *	**0.039** *	/
Maximum NDPCA			
Without stumbling experience	0.454 ± 0.065	0.564 ± 0.053	**0.012** *
With stumbling experience	0.385 ± 0.041	0.471 ± 0.043	**0.038** *
***p*-value**	**0.089**	**0.017** *	/
Minimum NDPCA			
Without stumbling experience	0.207 ± 0.112	0.123 ± 0.104	0.101
With stumbling experience	0.171 ± 0.087	0.113 ± 0.023	0.110
***p*-value**	0.268	0.572	/
NDPCA Standard Deviation			
Without stumbling experience	0.098	0.128	**0.067**
With stumbling experience	0.063	0.109	**0.048** *
***p*-value**	**0.096**	0.151	/

* Significant at $p < 0.05$.

5. Discussion

Toe-activity analysis: Using videos, various toe movements were observed during walking. Some participants without stumbling history showed noticeable toe movements in the swing phase. Preservation of posture in the swing phase by participants with stumbling experiences was observed. These factors monitored by our proposed system are difficult to measure with conventional pressure sheet sensors. These factors also explain why we focused on the first and second peaks (contact areas at the start and end of the swing phase) and the difference between them to evaluate toe activity.

Effects of toe activity on stumbling: All the evaluation values shown in Figure 7 and Table 3 exhibit the same tendency—that is, higher toe activity decreased the stumbling risk. This tendency was particularly clear at the mean and maximum NDPCAs, as the statistically significant differences were evident. Zhang et al. [38] assessed the effects of restricting the flexion–extension motion of the first metatarsophalangeal (MTP) joint on the walking gait of humans, and found a high risk of slipping and falling when the MTP joint is constrained. They also found that the flexion–extension motion of the MTP joint is important for normal walking. This supports our finding that low toe activity was associated with high stumbling risk. Tsuyuguchi et al. [24] and Kim et al. [25] investigated the relationship between the risk of falling and toe flexor strength of middle-aged and elderly individuals. Their results showed that toe flexor strength is related to the incidence of falls. Given that high toe activity corresponds to high flexion–extension ability, it was assumed that the risk of stumbling and falling could be estimated by the NDPCA. Kurihara et al. [39] reported a strong correlation between maximum isometric toe flexor strength and the cross-sectional area of intrinsic and extrinsic plantar muscles. Guillén-Rogel et al. [40] reported a moderate correlation between toe flexor strength and ankle dorsiflexion range of motion (DF ROM). Mecagni et al. [41] reported correlations between ankle ROM and balance among community-dwelling elderly women. The aforementioned studies show that toe flexor strength is closely related to balance in walking because the ankle DF ROM can reflect toe flexor strength. We believe that high NDPCA corresponds to high ankle mobility, which means that individuals with high NDPCA have better walking dynamics than those with low NDPCA. Furthermore, low NDPCA indicates that the flexion–extension motion of the MTP joint is not working properly, and that the DF ROM is low, which increases the risk of falling and stumbling. A well-known criterion for evaluating stumbling is the minimum toe clearance (MTC), which is the minimum distance between the toe and the floor. Moosabhoy et al. [42] investigated the sensitivity of MTC to the hip, knee, and ankle joint angles of the swing leg, and found that MTC was most sensitive to the ankle joint angle. Therefore, we speculate that toe activity may be related to MTC. The relationship between toe activity and MTC needs to be clarified in future studies.

Aging effect: The mean NDPCA evaluates the toe activity for nine walking cycles. The difference in mean NDPCAs between healthy elderly and young individuals was very small when the groups with and without stumbling histories were compared. This indicates that the mean NDPCA is a valid parameter for evaluating stumbling risk, irrespective of age. With regard to the maximum NDPCA, a high value is associated with aging (whereas a low value indicates high stumbling risk). We speculate that the high value of the maximum NDPCA resulted from a temporary action that compensated for the decline in muscle strength—in other words, a temporary increase in kicking force to raise the foot high and prevent stumbling. A low value of the minimum NDPCA was also associated with aging, although statistically significant differences were not observed. We speculate that the low value of the minimum NDPCA indicates low toe activity, which is caused by aging and leads to high stumbling risk. A large value obtained by subtracting the minimum NDPCA from the maximum NDPCA indicates considerable fluctuations in toe activity, which indicates aging. In the same context, we also evaluated the standard deviation of NDPCA. For elderly individuals with stumbling experience, the standard deviation of NDPCA was higher than that of young people. This highlights the fact that their toe activities were more volatile and may increase their risk of stumbling. The considerable variations in NDPCA could be attributed to the decline in muscle strength due to aging.

Limitations and further applications: The developed sensing system has many potential applications in future research. These include the analysis of the plantar surface during walking by participants with conditions such as toe deformity and weakness [43] or plantar calluses [44], and the exploration of the plantar region to investigate plantar flexion [26] or diabetes [45]. However, the developed sensing system still has some deficiencies that need to be addressed. We provided only one size of sensor. For a few participants in certain testing environments, foot motion around the MTP joints was affected by the rigidity of the clog. Given that this study is based on image analysis, the sensor was sensitive to the illumination of the experimental environment. Illumination may affect the validity of the image data. Color-based image extraction was also affected by the camera settings. This issue can be resolved by locking the white balance of the camera and adjusting the average brightness. The effect of ambient lighting can be reduced by using the auxiliary lighting provided by the endoscope. Further improvements and application of the sensing system to other studies will be part of our future research. The experimental setup for the left and right footwear was different, but the disparity between results was minimized by conducting a practice walking session with participants. The practice session also minimized the difference in weight and toe length (due to the camera holder) between normal shoes and the developed system. However, the influence of these disparities cannot be completely eliminated. In the future, the investigation could be performed using the same experimental setup for the right and left feet, optimizing materials and topology for reducing the weight, and shortening of the length of the camera holder. Consolidating the validity of the obtained results by increasing the number of participants and its application to the assessment of stumbling or fall risk might also be part of our future work. In this study, we investigated only the case in which participants walked straight ahead. The investigation of the plantar aspects of several types of walking, including turning and speed change [46], will be considered in future works. The developed plantar visualization system has potential applications in different types of clinical measurements of fall risk, including the single-leg-stance test and postural sway test. Further investigation of other measurements will be part of future work.

6. Conclusions

This study investigated the relationship between toe activity during walking and stumbling experiences using a clog-integrated plantar visualization system [35]. The system can accurately observe features such as the contact area, skin deformation, and skin color. We focused on the contact area in the toe area and evaluated toe-area activity as participants walked in a straight line.

The participants were separated into four groups: young individuals with and without stumbling experiences and elderly individuals with and without stumbling experiences. The difference between the two peaks at the toe contact area was normalized and calculated to evaluate the toe-area activity. The two peaks correspond to heel-strike and push-off. We evaluated the mean, maximum, minimum, and standard deviation of the toe-area activity for multiple walking cycles. We found that the mean toe-area activity was high for the group with no stumbling experiences, and low for the group with stumbling experiences, irrespective of age. In contrast, the deviations in toe-area activity were high in the elderly group and low in the young group, irrespective of stumbling histories. Therefore, the mean toe-area activity is useful for evaluating stumbling risk, irrespective of age. We also found that a large value of the maximum toe-activity indicates aging, whereas a low value indicates stumbling risk.

Author Contributions: Conceptualization, M.S. and T.W.; Data curation, Y.J.; Formal analysis, Y.J. and T.W.; Funding acquisition, T.W.; Investigation, Y.J. and T.W.; Methodology, Y.J., M.S. and T.W.; Project administration, T.W.; Software, Y.J.; Supervision, M.S. and T.W.; Validation, M.S. and T.W.; Writing—original draft, Y.J.; Writing—review & editing, T.W. All authors have read and agreed to the published version of the manuscript.

Funding: This research was partly funded by Mitutoyo Association for Science and Technology, grant number R1810.

Conflicts of Interest: The authors declare no conflict of interest.

Appendix A

Automatic Deviation of NDPCA

To automatically determine the NDPCA, we developed a method to separate the continuous time-series data of the contact area in the toe area into wave data for one walking cycle; detect the extreme points; and identify the first and second peaks as well as the troughs.

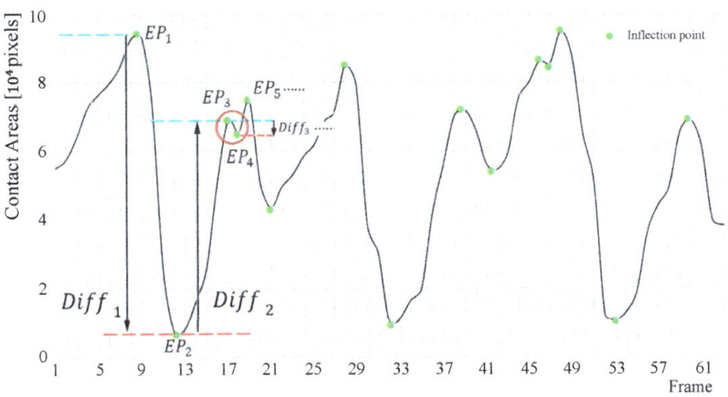

Figure A1. Extreme points extracted by determining the change in the toe contact area.

As shown in Figure A1, the calculation of the DPCA requires the two peaks (A_{peak1}, A_{peak2}) and the minimum point (A_{min}) of each walking cycle from the time-series data. We derived these extreme points by determining where the time derivative of the pixel time-series data was zero. The local peak or maximum is where the time derivative changes from positive to negative. Conversely, the point where the time derivative changes from negative to positive is the local trough or minimum. Figure A1 shows the extreme points (in green) derived from three walking cycles. Some of the extreme points were unnecessary for the calculation of the NDPCA (e.g., EP_2 and EP_3, where EP_i denotes the ith extreme point) and were deleted before calculation. The unnecessary extreme points are characterized by a small difference between contact areas of adjacent points, as seen in the red circle in Figure A1. Thus, we checked the distance between two adjacent extreme points, $Diff_i$, defined by

$$Diff_i = \left|EP_i - EP_{i+1}\right| \text{ for } i = 1,\ 2,\ 3\ldots \tag{A1}$$

We then removed the unnecessary extreme points. If $Diff_i$ was less than the threshold θ, then EP_i was removed. The threshold was determined through trial and error, based on the preliminary experiments. Here, the difference between $Diff_{\max}$ and $Diff_{\min}$ was used as the reference for considering the individual difference and normalizing the range of time-series data. In addition, if both EP_i and EP_{i+2} were the same type of extreme points (i.e., peak or trough), then EP_i was removed. The condition for removing EP_i was defined as

$$Diff_i < \theta(i = 1,\ 2,\ 3\ldots) \tag{A2}$$

where

$$\begin{aligned} \theta &= 0.25(Diff_{max} - Diff_{min}) \\ Diff_{max} &= \mathrm{Max}(Diff_i)\ (i = 1,\ 2,\ 3\ldots) \\ Diff_{min} &= \mathrm{Min}(Diff_i)\ (i = 1,\ 2,\ 3\ldots) \end{aligned} \tag{A3}$$

Figure A2 shows time-series data after removing the unnecessary extreme points in Figure A2. Each walking cycle has three extreme points associated with two peaks (A_{peak1} and A_{peak2}) and one trough between the two peaks. The trough between A_{peak1} and A_{peak2} is the minimum point (A_{min}). A_{peak1} and A_{peak2} correspond to the values of the contact areas at push-off and heel-strike, respectively, and A_{min} corresponds to the moment when the contact area during the swing phase is the smallest. The swing phase is detected upon detection of A_{min}. Given that the order of appearance of A_{peak1}, A_{min}, and A_{peak2} is known, the peaks corresponding to A_{peak1} and A_{peak2} are recognized by the location of A_{min}. One walking cycle ends at the trough that occurs after A_{peak2}. The NDPCA is calculated using A_{peak1}, A_{peak2}, and A_{min} in Equation (A2). The overall automatic gait recognition procedure is summarized in Figure A3.

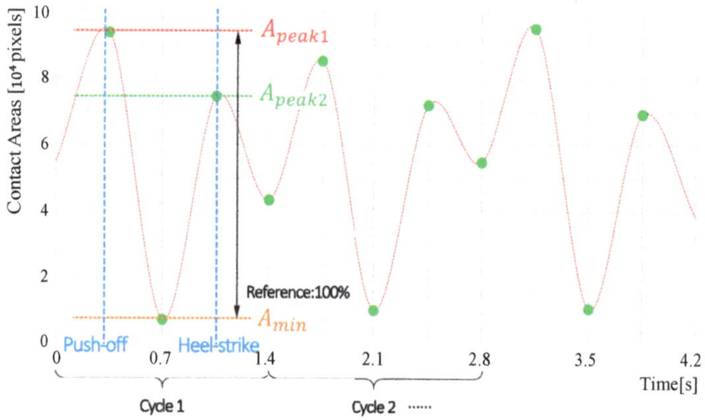

Figure A2. First and second peaks and troughs extracted from the calculation for three cycles.

Algorithm 2: Automatic NDPCA derivation	
Given	Time-series data from Algorithm 1
1	Find all extreme points.
2	Remove the extreme points that are not needed to identify the walking cycle and the NDPCA.
3	Identify the walking cycle and the first and second peaks as well as the troughs.
4	Derive the NDPCA for each cycle.
End	

Figure A3. Procedure for deriving the NDPCA automatically.

References

1. Cabinet Office, Government of Japan. Annual Report on the Aging Society. 2016. Available online: https://www8.cao.go.jp/kourei/english/annualreport/2016/2016pdf_e.html (accessed on 15 January 2020).
2. Oliver, D.; Daly, F.; Martin, F.C.; McMurdo, M.E.T. Risk factors and risk assessment tools for falls in hospital in-patients: A systematic review. *Age Ageing* **2004**, *33*, 122–130. [CrossRef]
3. Taylor, A.J.; Menz, H.B.; Keenan, A.-M. The influence of walking speed on plantar pressure measurements using the two-step gait initiation protocol. *Foot* **2004**, *14*, 49–55. [CrossRef]
4. Burnfield, J.M.; Few, C.D.; Mohamed, O.S.; Perry, J. The influence of walking speed and footwear on plantar pressures in older adults. *Clin. Biomech.* **2004**, *19*, 78–84. [CrossRef] [PubMed]
5. Preis, S.; Klemms, A.; Müller, K. Gait analysis by measuring ground reaction forces in children: Changes to an adaptive gait pattern between the ages of one and five years. *Dev. Med. Child Neurol.* **2008**, *39*, 228–233. [CrossRef] [PubMed]
6. Chuckpaiwong, B.; Nunley, J.A.; Mall, N.A.; Queen, R.M. The effect of foot type on in-shoe plantar pressure during walking and running. *Gait Posture* **2008**, *28*, 405–411. [CrossRef]
7. Majumder, A.J.A.; Zerin, I.; Uddin, M.; Ahamed, S.I.; Smith, R.O. Smartprediction: A Real-time Smartphone-based Fall Risk Prediction and Prevention System. In Proceedings of the 2013 Research in Adaptive and Convergent Systems, Montreal, OC, Canada, 1–4 October 2013; pp. 434–439.
8. Ayena, J.C.; Otis, M.J.D.; Menelas, B.A.J. An efficient home-based risk of falling assessment test based on Smartphone and instrumented insole. In Proceedings of the 2015 IEEE International Symposium on Medical Measurements and Applications (MeMeA) Proceedings, Turin, Italy, 7–9 May 2015; pp. 416–421.
9. Yu, H.; Wang, D.; Yang, C.-J.; Lee, K.-M. A walking monitoring shoe system for simultaneous plantar-force measurement and gait-phase detection. In Proceedings of the 2010 IEEE/ASME International Conference on Advanced Intelligent Mechatronics, Montreal, ON, Canada, 6–9 July 2010; pp. 207–212.
10. Crea, S.; Donati, M.; De Rossi, S.M.M.; Oddo, C.M.; Vitiello, N. A wireless flexible sensorized insole for gait analysis. *Sensors* **2014**, *14*, 1073–1093. [CrossRef]
11. Yang, S.; Li, Q. Inertial sensor-based methods in walking speed estimation: A systematic review. *Sensors* **2012**, *12*, 6102–6116. [CrossRef]
12. Hung, T.N.; Suh, Y.S. Inertial sensor-based two feet motion tracking for gait analysis. *Sensors* **2013**, *13*, 5614–5629. [CrossRef]
13. Do, T.N.; Suh, Y.S. Gait Analysis Using Floor Markers and Inertial Sensors. *Sensors* **2012**, *12*, 1594–1611. [CrossRef]
14. Foxlin, E. Pedestrian Tracking with Shoe-Mounted Inertial Sensors. *IEEE Comput. Graph. Appl.* **2005**, *25*, 38–46. [CrossRef]
15. Sim, S.Y.; Jeon, H.S.; Chung, G.S.; Kim, S.K.; Kwon, S.J.; Lee, W.K.; Park, K.S. Fall detection algorithm for the elderly using acceleration sensors on the shoes. In Proceedings of the Annual International Conference of the IEEE Engineering in Medicine and Biology Society, EMBS, Boston, MA, USA, 30 August–3 September 2011; pp. 4935–4938.
16. Rampp, A.; Barth, J.; Schülein, S.; Gaßmann, K.G.; Klucken, J.; Eskofier, B.M. Inertial Sensor-Based Stride Parameter Calculation from Gait Sequences in Geriatric Patients. *IEEE Trans. Biomed. Eng.* **2015**, *62*, 1089–1097. [CrossRef]

17. Mariani, B.; Hoskovec, C.; Rochat, S.; Büla, C.; Penders, J.; Aminian, K. 3D gait assessment in young and elderly subjects using foot-worn inertial sensors. *J. Biomech.* **2010**, *43*, 2999–3006. [CrossRef]
18. Schepers, H.M.; Koopman, H.F.J.M.; Veltink, P.H. Ambulatory assessment of ankle and foot dynamics. *IEEE Trans. Biomed. Eng.* **2007**, *54*, 895–902. [CrossRef] [PubMed]
19. Bamberg, S.J.M.; Benbasat, A.Y.; Scarborough, D.M.; Krebs, D.E.; Paradiso, J. A Gait analysis using a shoe-integrated wireless sensor system. *IEEE Trans. Inf. Technol. Biomed.* **2008**, *12*, 413–423. [CrossRef] [PubMed]
20. Bebek, Ö.; Suster, M.A.; Rajgopal, S.; Fu, M.J.; Huang, X.; Cavusoglu, M.C.; Young, D.J.; Mehregany, M.; van den Bogert, A.J.; Mastrangelo, C.H. Personal Navigation via High-Resolution Gait-Corrected Inertial Measurement Units. *IEEE Trans. Instrum. Meas.* **2010**, *59*, 3018–3027. [CrossRef]
21. Chen, M.; Huang, B.; Xu, Y. Intelligent shoes for abnormal gait detection. In Proceedings of the IEEE International Conference on Robotics and Automation, Pasadena, CA, USA, 19–23 May 2008; pp. 2019–2024.
22. Hegde, N.; Sazonov, E.S. SmartStep 2.0—A completely wireless, versatile insole monitoring system. In Proceedings of the 2015 IEEE International Conference on Bioinformatics and Biomedicine (BIBM), Washington, DC, USA, 9–12 November 2015; pp. 746–749.
23. Kawsar, F.; Hasan, M.K.; Love, R.; Ahamed, S.I. A Novel Activity Detection System Using Plantar Pressure Sensors and Smartphone. In Proceedings of the IEEE 39th Annual Computer Software and Applications Conference, Taichung, Taiwan, 1–5 July 2015; pp. 44–49.
24. Tsuyuguchi, R.; Kurose, S.; Seto, T.; Takao, N.; Tagashira, S.; Tsutsumi, H.; Otsuki, S.; Kimura, Y. Toe grip strength in middle-aged individuals as a risk factor for falls. *J. Sports Med. Phys. Fit.* **2018**, *58*, 1325–1330.
25. Kim, Y.-W.; Kwon, O.-Y.; Cynn, H.-S.; Weon, J.-H.; Yi, C.-H.; Kim, T.-H. Comparison of Toe Plantar Flexors Strength and Balancing Ability between Elderly Fallers and Non-fallers. *J. Phys. Ther. Sci.* **2011**, *23*, 127–132. [CrossRef]
26. Mickle, K.J.; Munro, B.J.; Lord, S.R.; Menz, H.B.; Steele, J.R. ISB Clinical Biomechanics Award 2009. *Clin. Biomech.* **2009**, *24*, 787–791. [CrossRef] [PubMed]
27. Hiura, M.; Nemoto, H.; Nishisaka, K.; Higashi, K.; Katoh, T. The Association Between Walking Ability and Falls in Elderly Japanese Living in the Community Using a Path Analysis. *J. Community Health* **2012**, *37*, 957–962. [CrossRef] [PubMed]
28. Uritani, D.; Fukumoto, T.; Matsumoto, D.; Shima, M. The Relationship Between Toe Grip Strength and Dynamic Balance or Functional Mobility Among Community-Dwelling Japanese Older Adults: A Cross-Sectional Study. *J. Aging Phys. Act.* **2016**, *24*, 459–464. [CrossRef] [PubMed]
29. Branthwaite, H.; Grabtree, G.; Chockalingam, N.; Greenhalgh, A. The Effect of Toe Flexion Exercises on Grip. *J. Am. Podiatr. Med. Assoc.* **2018**, *108*, 355–361. [CrossRef]
30. Takatori, K.; Matsumoto, D. Relationships Between Simple Toe Elevation Angle in the Standing Position and Dynamic Balance and Fall Risk Among Community-Dwelling Older Adults. *PM&R* **2015**, *7*, 1059–1063.
31. Bus, S.A.; Maas, M.; de Lange, A.; Michels, R.P.J.; Levi, M. Elevated plantar pressures in neuropathic diabetic patients with claw/hammer toe deformity. *J. Biomech.* **2005**, *38*, 1918–1925. [CrossRef] [PubMed]
32. Morag, E.; Cavanagh, P.R. Structural and functional predictors of regional peak pressures under the foot during walking. *J. Biomech.* **1999**, *32*, 359–370. [CrossRef]
33. Payne, C.; Turner, D.; Miller, K. Determinants of plantar pressures in the diabetic foot. *J. Diabetes Complicat.* **2002**, *16*, 277–283. [CrossRef]
34. Buldt, A.K.; Forghany, S.; Landorf, K.B.; Levinger, P.; Murley, G.S.; Menz, H.B. Foot posture is associated with plantar pressure during gait: A comparison of normal, planus and cavus feet. *Gait Posture* **2018**, *62*, 235–240. [CrossRef]
35. Jin, Y.; Shogenji, M.; Watanabe, T. Clog-integrated plantar visualization system for evaluating activity during walking. In Proceedings of the 2017 IEEE International Conference on Advanced Intelligent Mechatronics (AIM), Munich, Germany, 3–7 July 2017; pp. 862–867.
36. Byju, A.G.; Nussbaum, M.A.; Madigan, M.L. Alternative measures of toe trajectory more accurately predict the probability of tripping than minimum toe clearance. *J. Biomech.* **2016**, *49*, 4016–4021. [CrossRef]
37. Santhiranayagam, B.K.; Sparrow, W.A.; Lai, D.T.H.; Begg, R.K. Non-MTC gait cycles: An adaptive toe trajectory control strategy in older adults. *Gait Posture* **2017**, *53*, 73–79. [CrossRef]
38. Zhang, J.; Si, Y.; Zhang, Y.; Liu, Y. The effects of restricting the flexion–extension motion of the first metatarsophalangeal joint on human walking gait. *Biomed. Mater. Eng.* **2014**, *24*, 2577–2584. [CrossRef]

39. Kurihara, T.; Yamauchi, J.; Otsuka, M.; Tottori, N.; Hashimoto, T.; Isaka, T. Maximum toe flexor muscle strength and quantitative analysis of human plantar intrinsic and extrinsic muscles by a magnetic resonance imaging technique. *J. Foot Ankle Res.* **2014**, *7*, 26. [CrossRef]
40. Guillén-Rogel, P.; San Emeterio, C.; Marín, P.J. Associations between ankle dorsiflexion range of motion and foot and ankle strength in young adults. *J. Phys. Ther. Sci.* **2017**, *29*, 1363–1367. [CrossRef] [PubMed]
41. Mecagni, C.; Smith, J.P.; Roberts, K.E.; O'Sullivan, S.B. Balance and ankle range of motion in community-dwelling women aged 64 to 87 years: A correlational study. *Phys. Ther.* **2000**, *80*, 1004–1011. [CrossRef] [PubMed]
42. Moosabhoy, M.A.; Gard, S.A. Methodology for determining the sensitivity of swing leg toe clearance and leg length to swing leg joint angles during gait. *Gait Posture* **2006**, *24*, 493–501. [CrossRef] [PubMed]
43. Menz, H.B.; Zammit, G.V.; Munteanu, S.E. Plantar pressures are higher under callused regions of the foot in older people. *Clin. Exp. Dermatol.* **2007**, *32*, 375–380. [CrossRef]
44. Fox, J.; Docherty, C.L.; Schrader, J.; Applegate, T. Eccentric plantar-flexor torque deficits in participants with functional ankle instability. *J. Athl. Train.* **2008**, *43*, 51–54. [CrossRef]
45. Mueller, M.J.; Hastings, M.; Commean, P.K.; Smith, K.E.; Pilgram, T.K.; Robertson, D.; Johnson, J. Forefoot structural predictors of plantar pressures during walking in people with diabetes and peripheral neuropathy. *J. Biomech.* **2003**, *36*, 1009–1017. [CrossRef]
46. Verlinden, V.J.A.; van der Geest, J.N.; Hoogendam, Y.Y.; Hofman, A.; Breteler, M.M.B.; Ikram, M.A. Gait patterns in a community-dwelling population aged 50 years and older. *Gait Posture* **2013**, *37*, 500–505. [CrossRef]

applied sciences

MDPI

Article

A Novel Artificial Pancreas: Energy Efficient Valveless Piezoelectric Actuated Closed-Loop Insulin Pump for T1DM

Alp Kaçar [1,*], Mehmet Bülent Özer [2] and Yiğit Taşcıoğlu [3]

1 Department of Mechanical Engineering, TOBB University of Economics and Technology, Ankara 06530, Turkey
2 Department of Mechanical Engineering, Middle East Technical University, Ankara 06800, Turkey; ozerb@metu.edu.tr
3 Department of Mechanical Engineering, TED University, Ankara 06420, Turkey; yigit.tascioglu@tedu.edu.tr
* Correspondence: a.kacar@etu.edu.tr

Received: 16 June 2020; Accepted: 29 July 2020; Published: 30 July 2020

Abstract: The objective of this work is to develop a closed-loop controlled insulin pump to keep the blood glucose level of Type 1 diabetes mellitus (T1DM) patients in the desired range. In contrast to the existing artificial pancreas systems with syringe pumps, an energy-efficient, valveless piezoelectric pump is designed and simulated with different types of controllers and glucose-insulin models. COMSOL Multiphysics is used for piezoelectric-fluid-structural coupled 3D finite element simulations of the pump. Then, a reduced-order model (ROM) is simulated in MATLAB/Simulink together with optimal and proportional-integral-derivative (PID) controllers and glucose–insulin models of Ackerman, Bergman, and Sorensen. Divergence angle, nozzle/diffuser diameters, lengths, chamber height, excitation voltage, and frequency are optimized with dimensional constraints to achieve a high net flow rate and low power consumption. A prototype is manufactured and experimented with different excitation frequencies. It is shown that the proposed system successfully controls the delivered insulin for all three glucose–insulin models.

Keywords: piezo-fluid-structural coupled simulation; APS; valveless micropump; closed-loop insulin pump

1. Introduction

Piezoelectric pumps are a compact and lightweight alternative in microfluidic flow control. However, artificial pancreas systems (APSs) typically use syringe pumps for insulin delivery. In this paper, we propose a closed-loop controlled valveless insulin pump to keep the blood glucose level of Type 1 diabetes mellitus (T1DM) patients in the desired range. Our objective is to develop a piezoelectric pump that delivers the right amount of insulin while maintaining energy efficiency. Design and optimization of the pump are accomplished by using multi-physics finite element simulations, and the resulting performance is evaluated by numerical tests on mathematical glucose–insulin models and also by laboratory experiments with a preliminary prototype. The rest of this section reviews the related work in the subject matter. Section 2 presents a mathematical model of the pump and three glucose–insulin models with different complexity. Section 3 explains piezoelectric-fluid-structural coupled finite element implementation and simulation-based design optimization in COMSOL to achieve a high flow rate and low power consumption. The designed closed-loop controllers are outlined in Section 4. Sections 5 and 6 present model-based simulations in MATLAB/Simulink and the experiments, respectively. Finally, in Section 7, we discuss current and future work.

1.1. Diabetes Mellitus

Insufficient insulin production in islet cells in the pancreas or tissue resistance against secreted insulin leads to excessive glucose concentration in the blood, which is called diabetes mellitus [1]. Insulin and glucagon, the two necessary secretions of the pancreas, regulate the glucose level in the body. Insulin is secreted to stimulate the cells to absorb the extra glucose for the energy needed, and glucagon is secreted to increase the blood glucose level up to its desired level required for the regulation of the system [2].

According to the findings of the IDF (International Diabetes Federation) in 2017, approximately 425 million adult people (20–79 years) were living with diabetes, and this will rise to 629 million by 2045 [3]. Furthermore, diabetes caused 4 million deaths and at least 727 billion dollars in health expenditure in 2017 and 12% of total spending on adults [3]. Furthermore, statistics of the WHO (World Health Organization) predict a dramatic increase of the adult diabetes population from 4% (in 2000, 171 million people) to 5.4% (366 million worldwide) by the year 2030 [4].

1.2. Closed-Loop Insulin Pumps

Manual injection with an insulin pen or pump causes patients to appear twice in a discrete control loop: as the glucose metabolic system which must be controlled and as the controller itself (cognitive system). Since it is difficult to take all external disturbances and internal systems into consideration, the discrete control method is generally followed by hypoglycemic or hyperglycemic events. Thus, to avoid patients having to determine each insulin dose manually, and to limit the variation in blood glucose concentration, an artificial pancreas needs to be developed as an essential scientific research aim. The required insulin dose is calculated using a control algorithm based on continuous glucose measurements obtained via a sensor without human input, and the precise insulin dose is automatically administered via a pump that continuously delivers insulin [5]. Improvement in glycemic control, in terms of glycated HbA1c (hemoglobin A1c) concentration and a reduction in hypoglycemia, and automatic changes in the basal insulin infusion rate at the time of exercise without any extra high-carbohydrate intake are essential advantages of APSs [6].

A T1DM patient needs to calculate his/her meal intake and required exercise interval in order to inject insulin, and glucose level also needs to be measured several times during a loop [4]. In this manual loop, the whole responsibility is on the patient, and if the patient cannot balance the glucose level correctly, it may cause many different diseases. Consequently, using a closed-loop system controlled by an artificial pancreas is quite beneficial for T1DM patients [6]. Artificial pancreases should be convenient to use, accessible, long-lasting, secure, safe, and affordable. Furthermore, they should also consume low energy. In this study, being low-cost and disposable, and consuming low energy are the primary focused features of an artificial pancreas.

Most of the APSs in the market includes a syringe pump such as Medtronic's MiniMed 670G [7], Beta Bionics' iLet [8], and Tandem's t:slim X2 [9]. However, the clinical use of syringe pumps is associated with several risk factors, including overdose or underdose due to lack of control of the infusion speed, bleeding from backflow, and introduction of air. Overdose is the most problematic and significant one among all. It can lead to severe complications such as hypersecretion, respiratory depression, and hypoxia [10]. Additionally, the use of syringe pumps may not be ideal due to their large size and high-power demands for APS applications, where portability and mobility are important requirements. The novel artificial pancreas, which we propose in this study, functions with a piezoelectric actuated pump and can eliminate the accuracy, portability, and power efficiency issues of the old methods and serve as a solution.

1.3. Piezoelectric Actuated Pumps

A piezoelectric material is a smart mechanical material that can deform when the voltage is applied to it [11]. Piezoelectric actuation is advantageous from many aspects such as high actuation force, fast mechanical response, ease of integration, compact design, and low power consumption compared to the power required for the actuation of syringe pumps. Piezoelectric actuators can be used in different configurations such as in cylinder shape encircling a tube and moving the fluid in the tube under compressions and expansions along the radial direction [12] or a disc shape piezoelectric actuator moving the fluid with a dome deflected shape [13–17]. Moreover, even valveless designs of the disc configuration are possible for improved reliability due to the elimination of valves [13]. To manipulate, transport, and control the exact amount of fluid in the order of microliters and milliliters, micropumps are of great importance in the microfluidic systems [14]. In order to predict the flow rate and working of the micropump, analytical and numerical results are compared to each other [15].

Furthermore, Laser and Santiago [16] have categorized the micropumps in terms of valves, chamber, thickness, voltage, frequency, pressure, and flow values. Optimization of diffuser dimensions such as diffuser length, diffuser neck width, and diffuser angle is necessary in order to enhance the performance of the micropump on the net flow [17]. Finally, in order to design and manufacture a low-cost micropump that is useful in a variety of applications, the overall micropump design should enable the use of mass fabrication methods [14].

1.4. Insulin/Glucose Models

Modeling of glucose–insulin control systems has gained significant importance in the last 50 years [18], and various glucose–insulin kinetic models have been found in order to capture the glucose–insulin mechanism [19]. Even though there is a wide range of models from simple expressions to complex mathematical models, there are three general groups for mathematical models including linear models (Ackerman [18], Ceresa [20]), nonlinear models (Bergman [21,22], Furler [23]), and comprehensive models (Sorensen [24], Cobelli [25], and Hovorka [26]) [19]. In this work, the Ackerman model is used for optimal controller design, the Bergman model is used for PID controller parameters, and finally, the Sorensen model is used for validation and reliability of models.

2. Mathematical Modeling

2.1. The Piezoelectric Pump Analytical Modeling

The insulin rate that needs to be pumped for an adult is defined as a minimum of 1.67 μL/min [27]. To pump this amount with piezoelectric disk deflection, the valve selection of the pump is one of the most sensitive parts. Neither active nor passive valves were used in this study because the opening-closing speed and back-pressure constraints of the active valves were inadequate for the designed pump, and passive valves are not only difficult to manufacture but also inapplicable in terms of numerical implementation. Instead, a valveless design was desired to be used. In this valveless design, literature reviews of 3D conical, 2D cone-shaped, or planar cone projected options were examined [16]. When the voltage sent to piezoelectric disk is positive, and it does an upward displacement, insulin is pumped from both inlet and outlet, but more from the outlet. When the voltage is negative, the piezoelectric disk will be in downward displacement, and it will pump insulin from the reservoir from both the inlet and outlet, but more from the inlet. According to the difference in pumping frequency, the speed difference between inlet and outlet can be defined. This is illustrated in Figure 1 below. Nozzle and diffuser cone angle, in other words, divergence angle ($2\theta°$), differs from 5° to 70°. Optimum angle needs to be calculated for each system, but studies have searched for angle-length variances as well [28].

The deformation of the piezoelectric pump d_z was calculated using Timoshenko's theory of thin circular plates with small deflection [29].

$$d_z(r,t) = A_0(t)\left[1-\left(\frac{r}{R}\right)^2\right]^2 \tag{1}$$

where r is the radial coordinate, R is the radius of the piezoelectric disk, and A_0 is the time-dependent deflection of the piezoelectric disc. Maximum of d_z is used to calculate the volume. The elliptic conical form is used to calculate the approximate delivered insulin volume analytically. This volume is considered a quarter of one period.

$$V = \frac{\pi 4R^2 h}{6}, \tag{2}$$

where h is the maximum deflection (d_z) of the piezoelectric disk. The total flow rate from both inlet and outlet can be calculated as follows:

$$Q_{total} = \frac{V}{t_{quarter}}, \text{ where } Q_{total} = Q_{inlet} + Q_{outlet}, \tag{3}$$

and $t_{quarter}$ represents time (e.g., 1/400 s for 100 Hz). In order to validate numerical results easily, Q_{total} equation is used to compare total flow rates in a quarter of one period. Therefore, Equation (3) does not calculate the net flow. The net flow occurs from the difference of Q_{inlet} and Q_{outlet} instead. The following net flow rate equation is calculated in Section 3 using summation of inlet and outlet flow rate difference [30].

$$Q_{net} = \frac{1}{2}\sum_{0}^{1/FR} Q_{inlet} - Q_{outlet}, \tag{4}$$

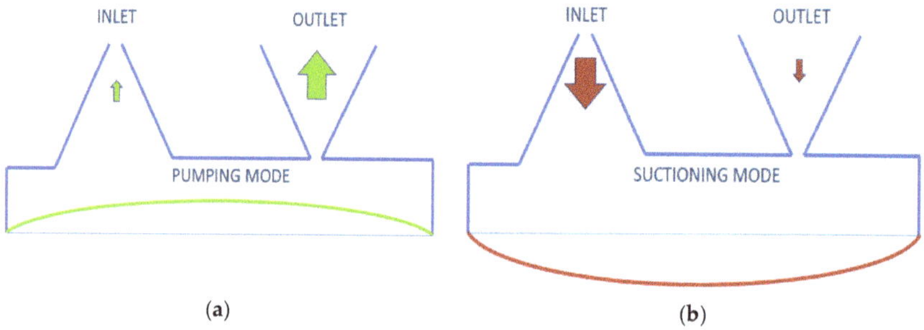

Figure 1. Working principle of valveless piezoelectric pump illustration: (**a**) pumping mode; (**b**) suctioning mode.

2.2. *Insulin/Glucose Plant Modeling*

Plant models were developed to see T1DM patients' blood glucose levels in the simulation environment and to calculate the controller parameters. These models differ in terms of complexity and accuracy. One linear, one nonlinear, and one comprehensive model were chosen among a great many models in the literature. The Ackerman model was used in the design of an optimum controller as a linear model; the Bergman model was selected as the nonlinear model with medium-level detail applied on all of the controllers. Finally, the Sorensen model was used as the model which shows the insulin glucose level in each part of the body in a detailed way [6].

Ackerman [18,31,32], developed a model utilizing four first-order rate constants for the glucose–insulin interactions. The model is described on two poles; the first one is G = blood glucose level; another one is H = hormone level. g and h parameters in the linear model are defined by considering the differences in fasting values for $g \equiv G - G_0$ and $h \equiv H - H_0$.

$$\frac{dg}{dt} = m_1 g - m_2 h + J(t), \tag{5}$$

$$\frac{dh}{dt} = -m_3 h + m_4 g + K(t); \tag{6}$$

m_1 represents glucose activity, m_2 is a parameter for the removal of glucose due to blood hormone concentration, m_3 is defined for the duration of insulin action, and m_4 is for the removal of hormone for high glucose concentration (typically 0 for T1DM). $J(t)$ and $K(t)$ are rates of exogenous glucose and hormone, respectively.

Although the model of insulin and glucose dynamics adapted from Bergman et al. [21] define a similar model with Ackerman [18] for glucose–insulin interactions, it has three compartments for plasma insulin $I(t)$, remote insulin $X(t)$, and plasma glucose $G(t)$. p_1 is a parameter at which glucose is removed from the plasma space independent of the influence of insulin, and the rate of insulin appearance and remote insulin disappearance are governed by p_3 and p_2, respectively. The time constant for insulin disappearance is defined with n. Subscript B is defined for basal plasma concentrations, and (0) is the level of the parameter at $t = 0$, and external insulin infusion is defined as $u(t)$, and meal intake is defined with $p(t)$ [23,24].

$$\frac{dG(t)}{dt} = -(p_1 + X(t))\, G(t) + p_1 G_B + p(t), \text{ where } G(0) = G_b, \tag{7}$$

$$\frac{dX(t)}{dt} = -p_2 X(t) + p_3[I(t) - I_B], \text{ where } X(0) = 0, \tag{8}$$

$$\frac{dI(t)}{dt} = -n[I(t) - I_B] + u(t), \text{ where } I(0) = I_b, \tag{9}$$

Finally, the Sorensen model is used for advanced glucose–insulin metabolism modeling. Sorensen [24] used the compartment modeling technique. Six compartments are used: brain, heart/lungs, gut (stomach–intestine), liver, kidney, and periphery (muscle–adipose). Mass balance for glucose of brain is $\frac{dG_{BV}}{dt}$ and $\frac{dG_{BI}}{dt}$, heart and lungs: $\frac{dG_H}{dt}$, gut: $\frac{dG_G}{dt}$, liver: $\frac{dG_L}{dt}$, kidney: $\frac{dG_K}{dt}$. The periphery is defined, and periphery equations are given below. First subscripts are defined as follows: B is the brain, G is the gut, H is the heart, L is the liver, K is the kidney, and P is the periphery. Second subscripts are as follows: I is interstitial fluid space, and V is vascular blood space; also B is defined as basal value, and N is defined as normalized value by using basal values [33].

$$\frac{dG_{PI}}{dt} = \frac{V_{PI}}{T_P^G V_{PI}} \cdot (G_{PV} - G_{PI}) - \frac{\Gamma_{PGU}}{V_{PI}}, \tag{10}$$

On the other hand, mass balance for insulin of the brain is $\frac{dI_B}{dt}$, and for heart and lungs $\frac{dI_H}{dt}$. In terms of insulin intake, $i(t)$, gut $\frac{dI_G}{dt}$, liver $\frac{dI_L}{dt}$, and kidney $\frac{dI_K}{dt}$. The periphery is represented as follows:

$$\frac{dI_{PI}}{dt} = \frac{V_{PI}}{T_P^I V_{PI}^I} \cdot (I_{PV} - I_{PI}) - \frac{\Gamma_{PIC}}{V_{PI}}, \tag{11}$$

Another part of the model is metabolic sources and sinks for glucose compartment such as peripheral glucose uptake, hepatic glucose production and uptake, and kidney glucose excretion. Additionally, for insulin compartment, liver insulin clearance, peripheral insulin clearance, and kidney insulin clearance variables are used while defining the glucagon model [19]. Sorensen model parameters are defined with delivered insulin $i(t)$, blood glucose concentration G, blood insulin concentration I, vascular blood flow rate Q, metabolic secreting rate Γ, volume V, and trans-capillary diffusion rate T [24]. All glucose–insulin mathematical models were designed with T1DM parameters in MATLAB-Simulink, and minute-based open-loop simulations and step responses were identified. The model parameters of Ackerman can be found in Section 4, the Bergman Model was taken from [22], and the Sorensen model was taken from [19] Appendix B5.

3. Numerical Implementation

3.1. Piezoelectric Pump Finite Element Implementation

Solid mechanics and electrostatics equations were used to calculate piezoelectric dynamics numerically in the stress-charge form.

$$\{T\} = \left[c^E\right]\{S\} - [e]^t\{E\}, \tag{12}$$

$$\{D\} = [e]\{S\} + \left[\varepsilon^S\right]\{E\}, \tag{13}$$

where T is stress vector, S is elastic strain vector, $[c^E]$ is elastic stiffness under constant electric field, $[e]$ is a coupling matrix, and superscript t on $[e]$ means transpose of the matrix $[e]$. D is electric flux density of piezoelectric material, E is electric field intensity, and $\left[\varepsilon^S\right]$ is dielectric permittivity under constant strain [34]. The values of these parameters are presented in Table 1 with other material properties. The dimensions of the pump are illustrated in the same table as well. Some of dimensions are optimized in Section 3.2, so initial values are given.

Table 1. Material properties [34,35].

Material/Dimensions	Parameter	Value	Unit
PZT-5J	Density	7800	kg/m^3
Diameter (D_p): 25 mm *Thickness* (t_p): 0.4 mm	Elasticity matrix $\left[c^E\right]$	$\begin{bmatrix} 8.2 & 3.4 & 3.0 & 0 & 0 & 0 \\ 3.4 & 8.2 & 3.0 & 0 & 0 & 0 \\ 3.0 & 3.0 & 5.9 & 0 & 0 & 0 \\ 0 & 0 & 0 & 2.1 & 0 & 0 \\ 0 & 0 & 0 & 0 & 2.1 & 0 \\ 0 & 0 & 0 & 0 & 0 & 2.4 \end{bmatrix} 10^{10}$	N/m^2
Chamber Dimensions: *Inlet/Outlet Diameter* (D_{out}): 3 mm *Nozzle Small Radius* (r_d): 50 µm	Piezoelectric Coupling Matrix $[e]^t$	$\begin{bmatrix} 0 & 0 & -10.48 \\ 0 & 0 & -10.48 \\ 0 & 0 & 16.58 \\ 0 & 0 & 0 \\ 0 & 14.25 & 0 \\ 14.25 & 0 & 0 \end{bmatrix}$	C/m^2
Nozzle Big Radius (R_d): 0.25 mm (Initial) *Chamber Height* (mu): 0.5 mm (Initial)	Permittivity $\left[\varepsilon^S\right]/\varepsilon_0$, where $\varepsilon_0 = 8.85 * 10^{-12}\ F/m$	$\begin{bmatrix} 1641 & 0 & 0 \\ 0 & 1641 & 0 \\ 0 & 0 & 1143 \end{bmatrix}$	1
Nozzle Length (L): 0.5 mm (Initial)	Damping (Isotropic Loss Factor) η_s	0.02	1
Divergence Angle (θ): 11° (Initial)	Dielectric Loss (Isotropic Loss Factor) for Electrical Permittivity εS	0.02	1
Brass *Diameter* (D_c): 27 mm *Thickness*(t_b): 0.2 mm	Density Young modulus Poisson's ratio	8500 90,000 0.32	kg/m^3 MPa 1
Insulin	Density Kinematic viscosity	997 0.0014	kg/m^3 kg/m.s

For numerical simulation of the piezoelectric actuator, solid mechanics and two-way electrostatic coupling were used, and appropriate dimensions were determined through iterative design and simulations. Piezoelectric material used in the simulations were decided to be the 27 × 0.5 mm piezoelectric disk of the STEMinc (Davenport, FL, USA) [35] company. Last but not least, the optimum design was found by conducting a three-way analysis (e.g., electrostatics, structural mechanics, and laminar flow) to the system. The requirements for a valveless design are as follows: the flow velocity needs to reach to 1 m/s but should be less than 8 m/s, the piezoelectric disk diameter should be 27 mm, the maximum piezoelectric disk displacement should be 50 µm and minimum displacement should be 2.5 µm, net flows should be at least 2.5 mL/h (~0.1 µL/s), the size of the pump should fit the palm of a hand, the maximum voltage supply should be 100 Vp-p, and, finally, 0.1 A continuous, the instantaneous current (transient) value of one ampere, should be the maximum value needed for current.

Planar/horizontal design was implemented, as shown in Figure 2, and when this design was finished and after defining a starting point, parameter ranges such as diffuser/nozzle angle, nozzle's wide diameter, nozzle length, and length of the chamber were decided accordingly. Between the range of minimum 1 degree and maximum 30 degrees, 10-degree nozzle angle; between the range of minimum 0.1 mm and maximum 1 mm, 1 mm nozzle length; and, finally, between the range of minimum 0.5 mm and maximum 2 mm, 0.5 mm nozzle diameter were used for simulation iterations.

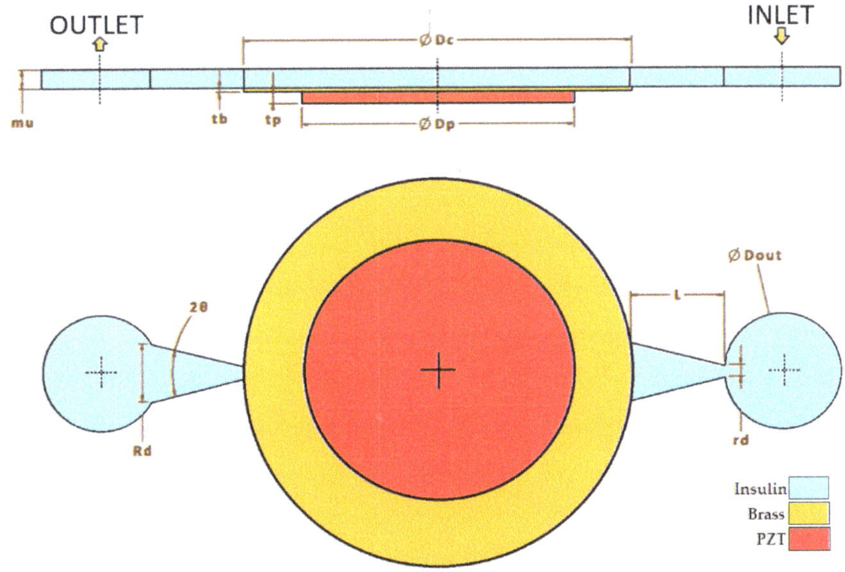

Figure 2. Geometry and dimensions.

The input signal to the piezoelectric pump is defined as follows, where V is voltage, FR is the frequency (in Hz) of applied voltage and t is time:

$$V(t) = V \sin(2\pi \cdot FR \cdot t), \tag{14}$$

By using COMSOL Multiphysics 5.4 [34], analyses were made in different domains both in a single and a coupled way, and with different time steps, mesh sizes, and discretization values. Different relative tolerance values were tried as well. According to the results of Eigenfrequency study analysis, the piezoelectric actuator's first natural frequency was calculated as 6373 Hz without fluid. Then the transient analysis setup was started. Damping was defined to the linear elastic material, dielectric loss was defined to the piezoelectric disk, a fixed constraint was defined to the outer surface of the brass located under the piezo, and lastly, the ground was defined to the one surface of the piezoelectric material and terminal to the other surface. Force, displacement, and voltage were assigned as 0 for initial conditions. In the piezoelectric actuator's first mode shape, it was seen that it had displacements as shown in Figure 3. According to the piezoelectric actuator data [35], expected displacement could reach up to 50 μm at 155 Vp-p; hence, arrangements that provide this displacement were determined. The applied voltage was 100 Vp-p, and the frequency was 100 Hz. How voltage and frequency affected the net flow in the optimization part can also be seen in Section 3.2.

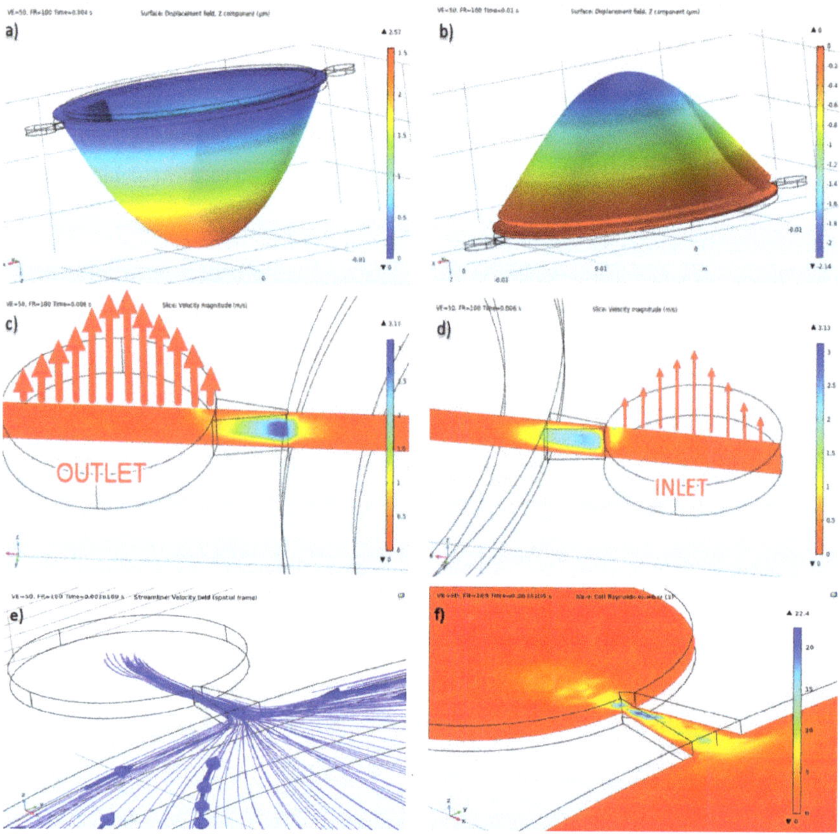

Figure 3. (**a**) Displacement of the piezoelectric actuator at pumping mode, (**b**) suction mode, (**c**) inlet velocity illustration, (**d**) outlet velocity illustration at the same time, (**e**) velocity field streamlines, (**f**) cell Reynold's number.

Insulin flow is defined as an incompressible flow, at 293.15 K temperature, under 1 atm reference pressure, and without a turbulence model. Cell Reynold's numbers shown in Figure 3f verified that there was no need for a turbulence model. As an initial condition, velocity and pressure were given as 0 (zero). As a boundary condition, after defining the stable wall, the wall condition was given as no-slip, and it was defined with inlet and outlet, as seen in Figure 3. For inlet and outlet, 0 Pa pressure condition was given, and no constraint was defined for flow direction. Absolute tolerance was given as 0.001, and the PARDISO solver was used in a coupled analysis study. The results of displacement of the piezoelectric actuator under insulin pressure, the pressure inside the pump chamber, the current passing through the terminal in the electrical circuit, and velocity values in inlet and outlet were calculated and presented in different plots of this study.

3.2. Piezoelectric Pump Optimization

Analyses were done between 100 K to 500 K mesh elements for coupled analysis to prove mesh independency. After mesh convergence analyses, <1% for structure-based error and ~8% flow-based error were calculated for 200 K. The analysis time measured a minimum 2.5 h to 262 h for 100 K to 500 K grid analysis with 1% relative tolerance and 1/40 step size (40 time steps per cycle); as a result, the 200 K grid had an acceptable error, so it was used. There were many analyses done, as well, for different time-steps from 20 steps per cycle to 1600 steps per cycle. These analyses took a minimum of 1.5 h to 48 h for 100 K grid size and 1% relative tolerance. Error was calculated with respect to 1600 steps per cycle. The error was, respectively, >41% for 20 step, 8.5% for 40 step, 7.5% for 100 step, and <1.5% for 400 step per one cycle. i7-7700-32GB hardware was used for this analysis and time optimization.

Nozzle/diffuser divergence angle, nozzle/diffuser wide radius, nozzle/diffuser length, and chamber height were optimized. According to [13] and [30], when the chamber diameter increases, the efficiency of nozzle/diffuser and the net flow increases. When diameter of the piezoelectric disk increases, the expected deformation of the piezoelectric disk increases as well. These parameters were optimized for best efficiency, maximum deflection, maximum velocity, and net flow, as well as manufacturability considerations. Analyses were performed to achieve the most efficient nozzle and diffuser design. Since chamber diameter was fixed by piezoelectric actuator selection, only chamber height (*mu*) was optimized in a range of 0.2 mm to 2.5 mm, and the one with the highest net flow rate using Equation (4) was 1 mm, shown in Figure 4a.

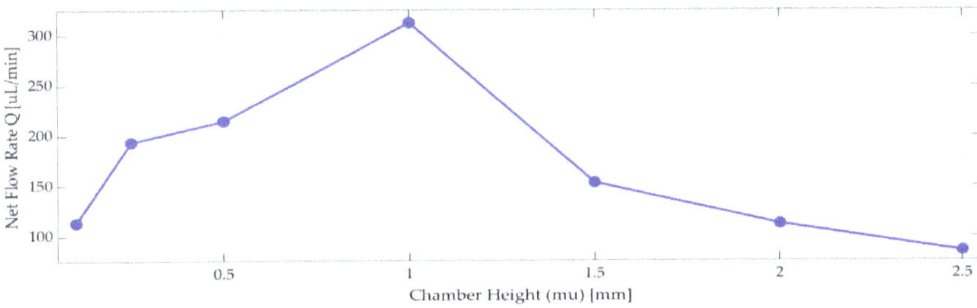

(**a**)

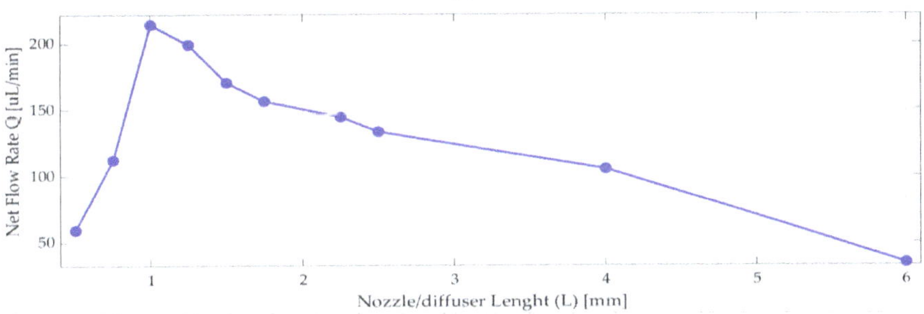

(**b**)

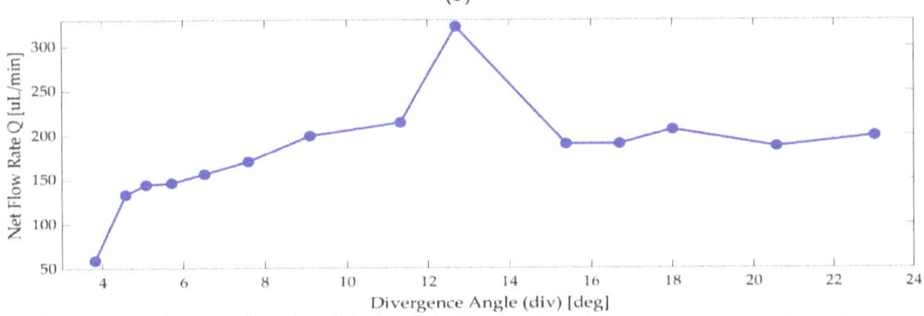

(**c**)

Figure 4. Numerical simulation results for net flow rate versus (**a**) chamber height; (**b**) length of nozzle/diffuser, (**c**) divergence angle.

As seen in Figure 4b, nozzle/diffuser length affects the net flow rate, so more than 10 different lengths were analyzed. The net flow rate significantly changed around 1 mm. There are many published pieces of research about nozzle and diffuser design [13,30,36,37] that are typically above 4 degrees half angle. When the nozzle and diffuser angle increases, the net flow rate increases too. However, there are some velocity values and venturi effects that may constrict the flow, so analyses were done to have the most efficient planar nozzle/diffuser. As seen in Figure 4c above, the most efficient nozzle and diffuser selected half-angle value θ was in between 12 and 14 degrees.

In addition to that, voltage and frequency parameters sweep was done. The piezoelectric actuator terminal was connected to an electrical circuit, and there was a voltage source in the circuit. Different voltages and frequencies were applied to the voltage source (see Figure 5). When the voltage increased, displacement and net flow increased explicitly. It was observed that the flow rate peaked at 100 Hz for the frequency range between 20 Hz and 200 Hz. (see Figure 5).

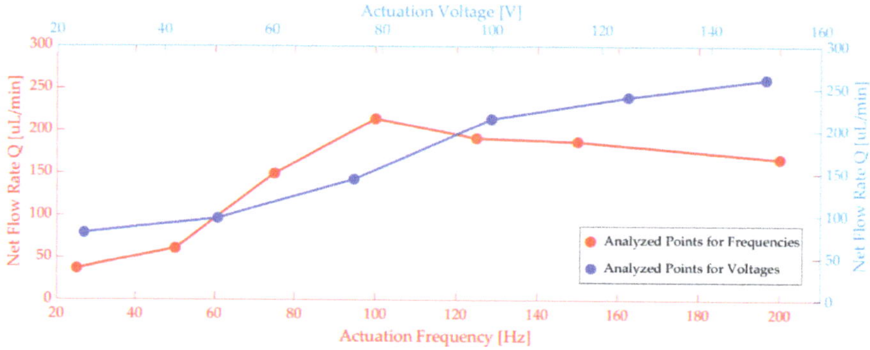

Figure 5. Numerical simulations of net flow rate versus voltage and frequency.

Another case considered was the calculation of the current drawn by the piezoelectric disk. There is an electrical circuit defined in COMSOL workspace, and the cable connected between the electrical source and the piezoelectric actuator was used as a resistor; the current through the cable was calculated. As expected, it increased when the voltage increased. Using the voltage and the current values, power consumption was calculated for different voltage and frequencies applied to the system. It is shown in Figure 6 below.

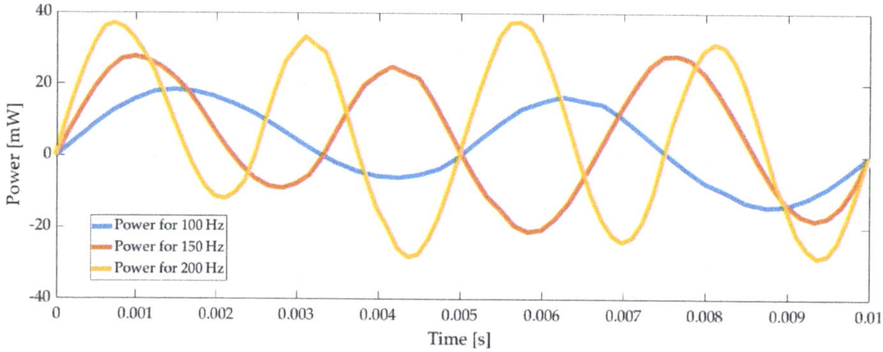

Figure 6. Power consumption of the pump with time for 100 Vp-p, 100 Hz, 150 Hz, and 200 Hz.

Last but not least, outlet and inlet flow rates were recorded, as illustrated in Figure 7 for 3 µm displacement amplitude of the piezoelectric disc. The net flow was calculated with Equation (4) as 220 µL/min in average.

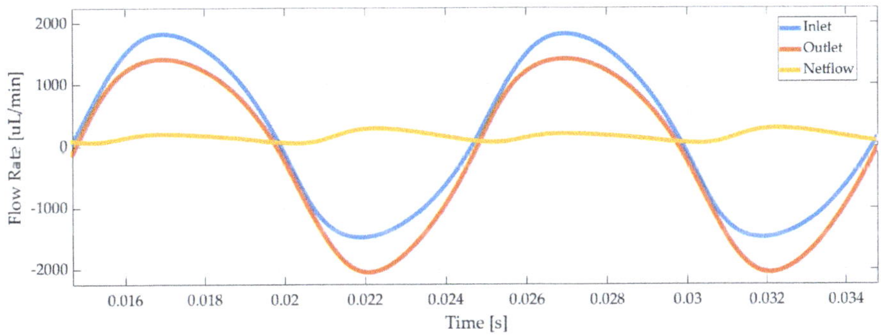

Figure 7. Numerical simulation results of inlet, outlet, and net flow rate of the piezoelectric pump.

4. Controller Design

First, Ackerman [18], Sorensen [24], and Bergman [21] glucose–insulin models were implemented in MATLAB-Simulink. Additionally, for each of these three models, a PID controller was designed. Additionally, the optimum controller gains, calculated by using Ackerman model, were implemented on all these models. It is known that daily insulin delivery and human glucose levels should be in the desired range [3] of 70–130 mg/dl. The three models used in this study have parameters for different patients. It is easy to observe that Ackerman's model can be used as a state-feedback controller design. Bergman's minimal model is easier to adopt, and using this simpler non-linear model can be less time-consuming than a more complicated non-linear model. The Sorensen model is comprehensive and detailed, so insulin and glucose levels all-around the body can be simulated with it. It is also very complex and contains several parameters.

After the glucose–insulin models are established, control of glucose level is the next step. For those controllers, there are two types of control reference methods: target tracking (e.g., 90 mg/dl) and reference tracking (e.g., 80–110mg/dl) [19]. For this article, PID with bang–bang coupling, and optimal controllers were selected from control algorithms, and the target tracking method was used [6].

To design the optimal controller, state–space forms had to be established. Since Sorensen and Bergman's models are nonlinear, the optimal controller was designed regarding Ackerman's linear model. Ackerman equations are formed in the state–space form as shown using Equations (5) and (6).

$$\begin{bmatrix} \dot{x}_1 \\ \dot{x}_2 \end{bmatrix} = \begin{bmatrix} -m_1 & -m_2 \\ m_4 & -m_3 \end{bmatrix} \begin{bmatrix} x_1 \\ x_2 \end{bmatrix} + \begin{bmatrix} 0 \\ 1 \end{bmatrix} u. \tag{15}$$

After the state–space model is defined, a performance criterion is defined considering the deviation from the target of blood glucose level and delivered insulin; also, δ is taken as the balance between them. δ defines the coefficiency of the effect of input, which typically describes the delivered insulin.

$$P.I. = \int_{\circ}^{\infty} \left(x_1 - x_{ref}\right)^2 + \delta\, u^2\,, \tag{16}$$

While Q is defined as 1, and P is as δ, Ricatti's equation is applied for A and B from (15) as follows:

$$A^T R + RA - RBP^{-1} B^T R + Q = 0 \text{ where } R = \begin{bmatrix} r_{11} & r_{12} \\ r_{12} & r_{22} \end{bmatrix}. \tag{17}$$

For calculated R, $m_1 = 0.005$, $m_2 = 0.001$, $m_3 = 0.002$, $m_4 = 0$, for $\delta = 0.25, 0.5, 0.75, 1$ parameters applied, and from the gain equation, K is observed and u_0 input is derived from following equation. On the other hand, u is defined as pump working time, which represents the power consumption of the pump as well, so $\delta = 0.25$ represents a less power efficient pump, and $\delta = 1$ represents a more power-efficient pump.

$$u_0 = -K^T x \text{ where } K = RBP^{-1}, \tag{18}$$

Finally, the PID algorithm is established, and the reference glucose and actual glucose measurement information is transmitted to the PID controller and delivered to the bang–bang controller and ROM of the piezoelectric actuated pump at the input of the plant model. The flowing insulin is transmitted to the insulin/glucose model. PID controller parameters *Kp, Ki, Kd*, is optimized for the best performance for overshoot *Me* = *40%* rising time *tr* = *5* mins and settling time *ts* = *15* mins for all glucose–insulin models [38].

5. Model-Based Simulation and Results

Based on the results of COMSOL simulations, a reduced-order model (ROM) of the pump was generated in the form of a look-up table. This model output time-dependent net flow and electrical current from excitation voltage and frequency inputs. Figure 8 shows an overview of the closed-loop control simulations in MATLAB/Simulink, where the pump supplied insulin to Ackerman, Bergman, and Sorensen glucose–insulin models with T1DM parameters. The R2019b version of MATLAB was used, and the solver settings were configured as 6×10^{-7} minutes minimum step size, auto solver selection, 1×10^{-4} relative tolerance, adaptive zero-crossing, 1440 min. (i.e., one day) stop time.

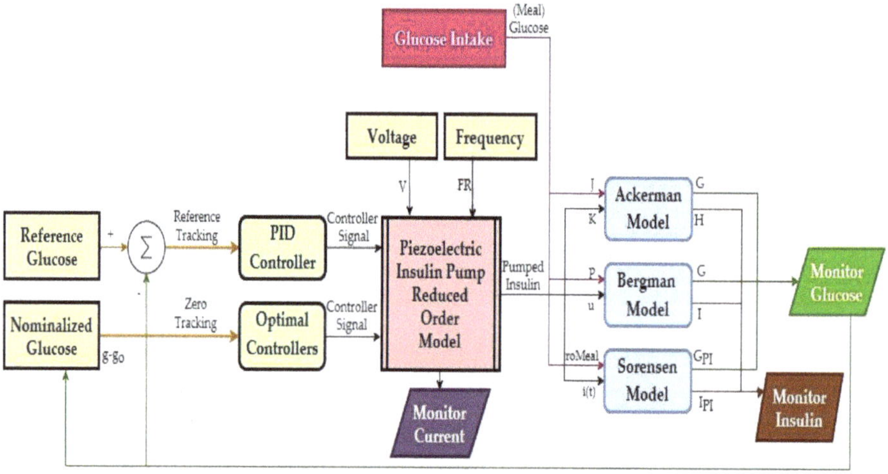

Figure 8. Overview of the closed-loop control simulations.

The glucose intake model supplied glucose input to the glucose–insulin models. On top of the 100 mg/dL basal glucose, a 150 mg/dL increase lasted for 18 min after each meal. This increase was supplied via a 1/(2s + 1) delay transfer function. Voltage and frequency of the pump excitation signal were constant, and the controllers switched the pump on/off based on the difference between the reference and measured glucose. Pump displacement and the electrical current were monitored as well as the inlet and outlet flow rates. The amount of delivered insulin was calculated from the net flow rate and the volume integral of the outlet.

As explained in Section 1.4, the simpler models, Ackerman and Bergman, are used for controller design, whereas the comprehensive Sorensen model is used for validation. Figure 9 shows the results of model-based simulations with three meal disturbances at 400, 750, and 1150 minutes. The PID controller successfully maintained the blood glucose level of the Sorensen model at around 90 mg/dL (Figure 9) and kept the insulin level at 18 mU/dL (Figure 10a). On the other hand, the δ value of the optimal controller had a negligible effect on the insulin level. This was shown with the Bergman model, as seen in Figure 10b, where the insulin level decreased minimally with increasing δ. The corresponding difference between the glucose levels was smaller than 0.05 mg/dL.

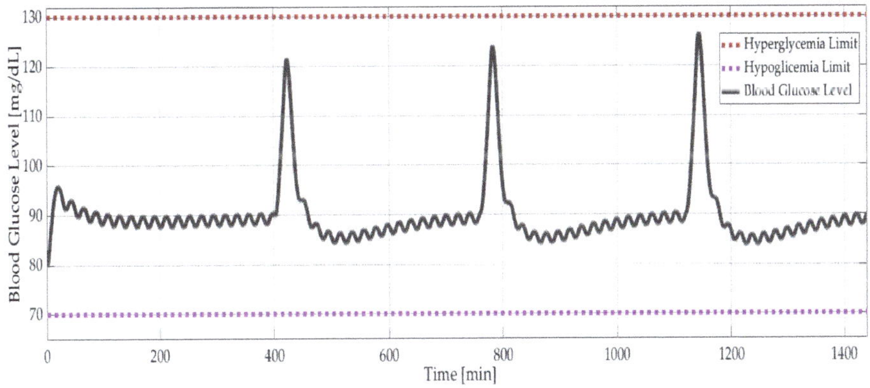

Figure 9. Model-based simulation results: Blood glucose level versus time with PID controlled Sorensen model.

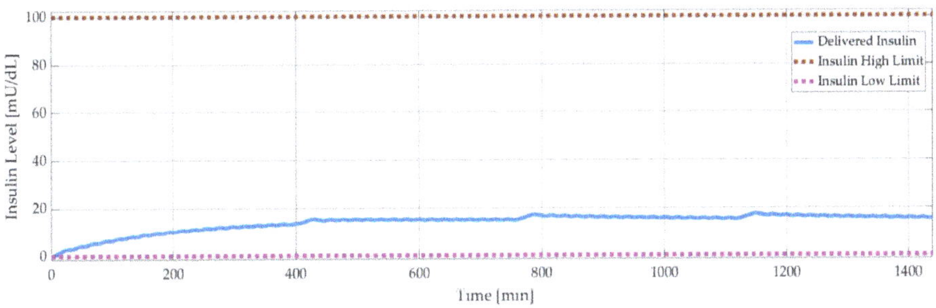

(**a**)

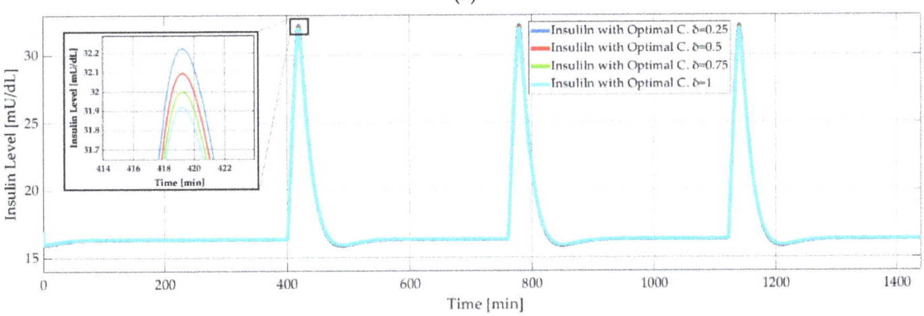

(**b**)

Figure 10. Model-based simulation results: (**a**) insulin level versus time with PID controlled Sorensen model; (**b**) insulin level versus time with optimal controlled Bergman model.

6. Experimental Verification and Results

The prototype pump consisted of two 3D printed closures (Figure 11a) and the piezoelectric actuator (STEMinc SMBA27T05PP), as seen in Figure 11b. Wires of the actuator were soldered and then it was glued at the brass substrate and sealed between the closures. The sealing was tested with compressed air at 100 kPa. Plastic hoses with a 3 mm inner diameter were attached to the inlet and outlet of the pump, as shown in Figure 11c.

(a) (b) (c)

Figure 11. Pump prototype: (**a**) 3D printed bottom closure, (**b**) the piezoelectric actuator, (**c**) assembled and sealed pump.

The test stand is shown in Figure 12. The signal generator could provide a 2 Vp-p sinus signal with an adjustable frequency between 1 Hz and 3 kHz. The amplitude could be increased up to 100 Vp-p at the Falco Systems WMA-300 amplifier. An amperemeter (Keithley 2000 Series Digital Multimeter) was connected in series between the amplifier and the pump to measure the electrical current. A Polytec CLV-2534 laser vibrometer measured the velocity of the piezoelectric actuator and displays on the oscilloscope.

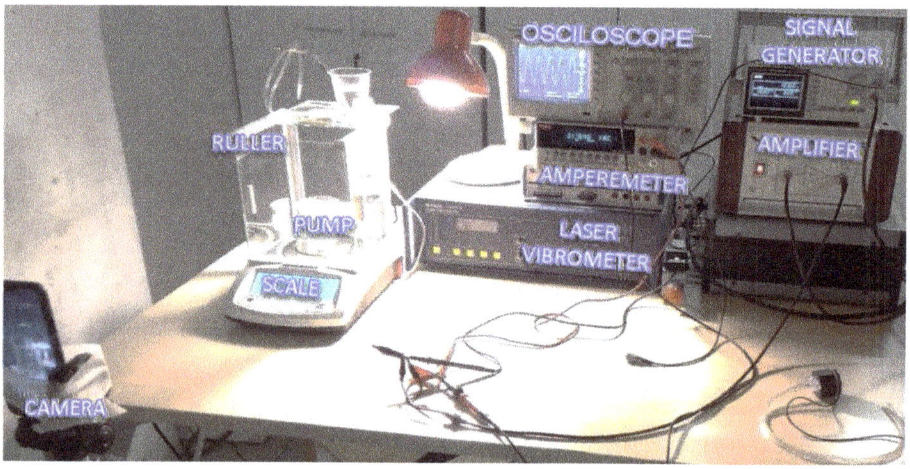

Figure 12. Experiment setup and architecture.

Water was used to simulate insulin in the experiments. The hoses were attached side-by-side to a vertical ruler to ensure equal water level at the initial condition (i.e., 0 Pa boundary condition). A 100 Vp-p sinusoidal voltage with different frequencies was applied to the pump until the head difference reached 1 mm. The change in boundary conditions was ignored in this interval when calculating net flow rates.

Figure 13 shows the measured maximum displacement of the actuator at different frequencies alongside the corresponding numerical results, since the differences between numerical simulation results and experiments were less than 10%.

Measured RMS electrical current versus excitation frequency is plotted in Figure 14 with the corresponding numerical results. On average, the experiments showed 22% higher current draw than the simulations. At 100 Hz with 100 Vp-p 0.7 mA RMS current corresponded to 18 mW RMS power consumption.

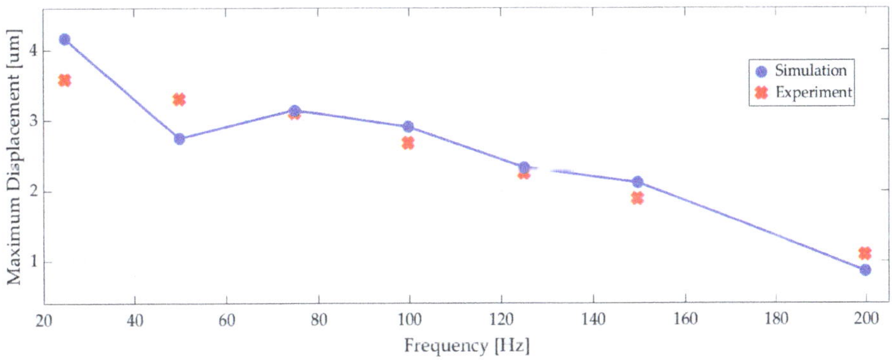

Figure 13. The experimental and simulated maximum displacement of the piezoelectric actuator at different excitation frequencies.

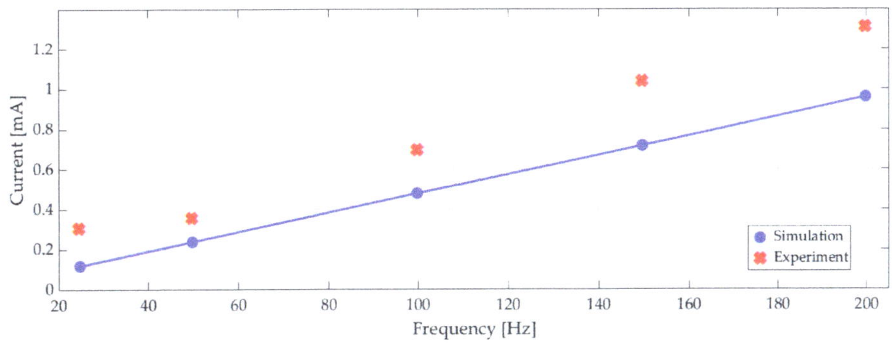

Figure 14. Experimental and simulated electrical current (RMS) at different excitation frequencies.

7. Discussion and Future Work

In this paper, we demonstrated a novel artificial pancreas system with a valveless piezoelectric pump. Although the pump is mechanically simple, its design requires detailed analysis to ensure efficient operation. We optimized our design, based on high net flow rate and low power consumption objectives, by using numerical multi-physics simulations in COMSOL. Once satisfied, we derived a reduced-order mathematical model of the pump in MATLAB/Simulink for model-based closed-loop control of blood glucose level in T1DM. In these simulations, the pump supplies insulin to three commonly used insulin–glucose models, namely, Ackerman, Bergman, and Sorensen. It must be noted that these models have parameters of different T1DM patients. The input to the controllers is the difference between the target glucose level and measured blood glucose level of the patient. The reference value is given and the instantaneous value is the output of the insulin–glucose model. A standalone APS gets the instantaneous value from a sensor known as a continuous glucose monitor. Integrating a continuous glucose monitor model in our simulations is a part of future work.

The produced prototype includes two closures and a piezoelectric actuator. A multi-jet fusion 3D printer successfully printed the closures with embedded inlet and outlet nozzles, which require high precision. The piezoelectric actuator is selected due to availability and it consists of a PZT disk attached to a thin brass substrate. Although the PZT is lead-based, and potentially toxic, it is not exposed to insulin in our design since the brass side faces the chamber. Lead-free piezoelectric materials, such as alkaline niobates or bismuth titanate [39], are interesting alternatives for future work. Another option is to use a separating layer, like a non-toxic coating or diaphragm, between the actuator and the chamber.

The prototype pump does not include embedded electronics at this stage because it is experimented in a laboratory test setup. This enables us to test the pump with a range of inputs and to observe a multitude of outputs during development phase. A fully coupled simulation model of the APS is one of the major contributions of this work. Hence, the experiments serve as a basis and validation for the development of the simulation model as well as the pump. The main drawback of the existing test setup is the lack of an accurate flow sensor. Because of this, instead of measuring the net flow rate we measured the maximum displacement of the actuator and compared with the simulations to find that the values are within 10% of each other at the experimented frequencies. This discrepancy can be extended to net flow rate, but supporting this with experimental proof is our first priority in future work.

In the literature, there are several studies [13,30] that investigate the dynamics of valveless piezoelectric pumps. Our contribution here is that we coupled the piezoelectric pump dynamics with three types of well-known glucose–insulin models. Furthermore, we applied PID and optimal controllers on the blood glucose level control problem. For blood glucose level control with an APS, it is crucial to show that the proposed pump can deliver insulin with the flow rates required by the blood glucose controller. The Ackerman's and Bergman's glucose–insulin models are relatively simple so that they are used for the design of optimal and PID controllers, respectively. The comprehensive Sorensen model is used for performance assessment in model-based simulations. Because of its complexity, the Sorensen model requires more computational power and time. This issue must be considered when electing an embedded controller for a standalone APS.

We showed with analysis and experiment that a valveless piezoelectric pump is significantly more energy-efficient compared to a syringe pump, which is typically used in APS applications. Both the numerical and experimental studies of the proposed pump predicted a power requirement of nearly 20 mW for a flow rate in the range of 200–300 μL/min. A commercial syringe pump's technical specification document states that almost 5 W is required for pumping (additional power requirement on top of the standby condition of the pump) [40]. Such orders of magnitude decreases in the power requirement are essential for APS applications where portability and battery life considerations are crucial.

Author Contributions: Conceptualization, A.K. and M.B.Ö.; methodology, A.K., Y.T., and M.B.Ö.; mathematical modeling, A.K.; numerical implementation, A.K.; writing—original draft preparation, A.K.; writing—review and editing, M.B.Ö. and Y.T. All authors have read and agreed to the published version of the manuscript.

Funding: This research received no external funding.

Conflicts of Interest: The authors declare no conflict of interest.

References

1. Ntaganda, J.M.; Minani, F.; Banzi, W.; Mpinganzima, L.; Niyobuhungiro, J.; Gahutu, J.B.; Rutaganda, E.; Kambutse, I.; Dusabejambo, V. Simplified mathematical model of glucose-insulin system. *Am. J. Comput. Math.* **2018**, *8*, 233–244. [CrossRef]
2. Saleem, M.U.; Farman, M.; Meraj, M.A. Stability analysis of sorensen's model for controllability and observability. *Proc. Pak. Acad. Sci. Part B* **2017**, *54*, 133–145.
3. IDF. *IDF Diabetes Atlas*, 9th ed.; IDF: Brussels, Belgium, 2019.
4. Kovacs, L.; Kulcsar, B.; Bokor, J.; Benyo, Z. Model-based nonlinear optimal blood glucose control of Type I diabetes patients. In Proceedings of the 2008 30th Annual International Conference of the IEEE Engineering in Medicine and Biology Society, Vancouver, BC, Canada, 20–25 August 2008; pp. 1607–1610. [CrossRef]
5. Lunze, -I.K.; Singh, T.; Walter, M.; Brendel, M.D.; Leonhardt, S. Blood glucose control algorithms for type 1 diabetic patients: A methodological review. *Biomed. Signal Process. Control* **2013**, *8*, 107–119. [CrossRef]
6. Nath, A.; Biradar, S.; Balan, A.; Dey, R.; Padhi, R. Physiological models and control for type 1 diabetes mellitus: A brief review. *IFAC PapersOnLine* **2018**, *51*, 289–294. [CrossRef]
7. MiniMed^{TM} 670G Insulin Pump System. Available online: https://www.medtronicdiabetes.com/products/minimed-670g-insulin-pump-system (accessed on 1 June 2020).

8. Beta Bionics: Introducing the iLet. Available online: http://www.betabionics.com/ (accessed on 1 June 2020).
9. T:slim X2 Insulin Pump w/ Dexcom G6 CGM. Available online: https://www.tandemdiabetes.com/products/t-slim-x2-insulin-pump (accessed on 1 June 2020).
10. Jung, B.; Seo, K.-S.; Kwon, S.J.; Lee, K.; Hong, S.; Seo, H.; Kim, G.-Y.; Park, G.-M.; Jeong, J.; Seo, S.W. Efficacy evaluation of syringe pump developed for continuous drug infusion. *J. Dent. Anesthesia Pain Med.* **2016**, *16*, 303–307. [CrossRef]
11. Yang, J. *An Introduction to the Theory of Piezoelectricity*; Kluwer Academic Publishers: Boston, MA, USA, 2005; Volume 9.
12. Lumentut, M.; Friswell, M.I. A smart pipe energy harvester excited by fluid flow and base excitation. *Acta Mech.* **2018**, *229*, 4431–4458. [CrossRef]
13. Singh, S.; Kumar, N.; George, D.; Sen, A.K. Analytical modeling, simulations and experimental studies of a PZT actuated planar valveless PDMS micropump. *Sens. Actuators A Phys.* **2015**, *225*, 81–94. [CrossRef]
14. Khalilian, S. Development of a Low Cost Micro-Pump. Master's Thesis, University of Calgary, Calgary, AB, Canada, 2012; pp. 1–185. [CrossRef]
15. Atul, S.T.; Lenin Babu, M.C. Characterization of valveless micropump for drug delivery by using piezoelectric effect. In Proceedings of the 2016 International Conference on Advances in Computing, Communications and Informatics (ICACCI), Jaipur, India, 21–24 September 2016; pp. 2138–2144. [CrossRef]
16. Laser, D.J.; Santiago, J.G. A review of micropumps. *J. Micromech. Microeng.* **2004**, *14*, R35–R64. [CrossRef]
17. Nisar, A.; Afzulpurkar, N.; Tuantranont, A.; Mahaisavariya, B. Three dimensional transient multifield analysis of a piezoelectric micropump for drug delivery system for treatment of hemodynamic dysfunctions. *Cardiovasc. Eng.* **2008**, *8*, 203–218. [CrossRef]
18. Yipintsoi, T.; Gatewood, L.C.; Ackerman, E.; Spivak, P.L.; Molnar, G.D.; Rosevear, J.W.; Service, F.J. Mathematical analysis of blood glucose and plasma insulin responses to insulin infusion in healthy and diabetic subjects. *Comput. Boil. Med.* **1973**, *3*, 71–78. [CrossRef]
19. Chee, F.; Fernando, T. *Closed-Loop Control of Blood Glucose*; Springer: Berlin/Heidelberg, Germany, 2007; Volume 368.
20. Ceresa, F.; Ghemi, F.; Martini, P.F.; Martino, P.; Segre, G.; Vitelli, A. Control of blood glucose in normal and in diabetic subjects: Studies by compartmental analysis and digital computer technics. *Diabetes* **1968**, *17*, 570–578. [CrossRef]
21. Bergman, R.N.; Phillips, L.S.; Cobelli, C. Physiologic evaluation of factors controlling glucose tolerance in man: Measurement of insulin sensitivity and beta-cell glucose sensitivity from the response to Intravenous glucose. *J. Clin. Investig.* **1981**, *68*, 1456–1467. [CrossRef] [PubMed]
22. Roy, A.; Parker, R.S. Dynamic modeling of exercise effects on plasma glucose and insulin levels. *J. Diabetes Sci. Technol.* **2007**, *1*, 338–347. [CrossRef] [PubMed]
23. Furler, S.M.; Kraegen, E.W.; Smallwood, R.H.; Chisholm, D. Blood glucose control by intermittent loop closure in the basal mode: Computer simulation studies with a diabetic model. *Diabetes Care* **1985**, *8*, 553–561. [CrossRef] [PubMed]
24. Sorensen, J.T. A Physiologic Model of Glucose Metabolism in Man and Its Use to Design and Assess Improved Insulin Therapies for Diabetes. Ph.D. Thesis, University of California, Berkeley, CA, USA, 1985; p. 557.
25. Cobelli, C.; Mari, A. Control of diabetes with artificial systems for insulin delivery algorithm independent limitations revealed by a modeling study. *IEEE Trans. Biomed. Eng.* **1985**, *32*, 840–845. [CrossRef]
26. Hovorka, R.; Shojaee-Moradie, F.; Carroll, P.V.; Chassin, L.J.; Gowrie, I.J.; Jackson, N.C.; Tudor, R.S.; Umpleby, A.M.; Jones, R.H. Partitioning glucose distribution/transport, disposal, and endogenous production during IVGTT. *Am. J. Physiol. Metab.* **2002**, *282*, E992–E1007. [CrossRef]
27. Kaçar, A.; Özer, M.B.; Taşcıoğlu, Y. Piezoelectric actuated insulin pump simulation model and controller design. In Proceedings of the 8th International Conference, ICAT'19, Sarajevo, Bosnia and Herzegovina, 26–30 August 2019; pp. 351–356.
28. Chandrasekaran, A.; Packirisamy, M. Improved efficiency of microdiffuser through geometry tuning for valveless micropumps. *J. Fluids Eng.* **2015**, *138*, 031101. [CrossRef]
29. Nguyen, A.V.; Huang, X. Miniature valveless pumps based on printed circuit board technique. *Sens. Actuators A Phys.* **2001**, *88*, 104–111. [CrossRef]
30. Gidde, R.R.; Pawar, P.M.; Dhamgaye, V.P. Fully coupled modeling and design of a piezoelectric actuation based valveless micropump for drug delivery application. *Microsyst. Technol.* **2019**, *26*, 633–645. [CrossRef]

31. Hariri, A.; Wang, L.Y. Observer-based state feedback for enhanced insulin control of type 'I' diabetic patients. *Open Biomed. Eng. J.* **2011**, *5*, 98–109. [CrossRef]
32. Swan, G.W. An optimal control model of diabetes mellitus. *Bull. Math. Biol.* **1982**, *44*, 793–808. [CrossRef]
33. Acikgoz, S.U.; Diwekar, U.M. Blood glucose regulation with stochastic optimal control for insulin-dependent diabetic patients. *Chem. Eng. Sci.* **2010**, *65*, 1227–1236. [CrossRef]
34. COMSOL: Multiphysics Software for Optimizing Designs. Available online: https://www.comsol.com/ (accessed on 1 June 2020).
35. Piezo Round Bimorph Actuator for Pumps 27 × 0.5mm. Available online: https://www.steminc.com/PZT/en/piezo-round-bimorph-actuator-for-pumps-27x05mm (accessed on 1 June 2020).
36. Schmandt, B.; Herwig, H. Diffuser and nozzle design optimization by entropy generation minimization. *Entropy* **2011**, *13*, 1380–1402. [CrossRef]
37. Ben Mansour, R.; Sahin, A.Z. Optimization of conical micro-diffusers and micro-nozzles considering entropy generation. *Arab. J. Sci. Eng.* **2012**, *38*, 3161–3170. [CrossRef]
38. Laxminarayan, S.; Reifman, J.; Steil, G.M. Use of a food and drug administration-approved type 1 Diabetes mellitus simulator to evaluate and optimize a proportional-integral-derivative controller. *J. Diabetes Sci. Technol.* **2012**, *6*, 1401–1412. [CrossRef]
39. Maeder, M.D.; Damjanovic, D.; Setter, N. Lead free piezoelectric materials. *J. Electroceram.* **2004**, *13*, 385–392. [CrossRef]
40. SPETEC—Syringe Pump Symax. Available online: https://www.spetec.de/en/products/laboratory-equipment/syringe-pumps/syringe-pump-symax (accessed on 1 June 2020).

applied sciences

MDPI

Article

Virtual Reality Exposure Therapy for Driving Phobia Disorder: System Design and Development

Amy Trappey [1,*], Charles V. Trappey [2], Chia-Ming Chang [3], Routine R.T. Kuo [1], Aislyn P.C. Lin [1] and C.H. Nieh [3,4]

1 Department of Industrial Engineering and Engineering Management, National Tsing Hua University, Hsinchu 300, Taiwan; s108034518@m108.nthu.edu.tw (R.R.T.K.); s105034013@m105.nthu.edu.tw (A.P.C.L.)
2 Department of Management Science, National Chiao Tung University, Hsinchu 300, Taiwan; trappey@faculty.nctu.edu.tw
3 Psychiatry Department, Chang Gung Memorial Hospital, Taipei 333, Taiwan; cmchang58@cgmh.org.tw (C.-M.C.); 408456055@gapp.fju.edu.tw (C.H.N.)
4 Department of Clinical Psychology, Fu Jen Catholic University, Taipei 242, Taiwan
* Correspondence: trappey@ie.nthu.edu.tw; Tel.: +886-3572-7686

Received: 25 June 2020; Accepted: 13 July 2020; Published: 15 July 2020

Abstract: Driving phobia is an anxiety disorder. People are greatly impaired in their daily lives when suffering from driving phobia disorders. The anxieties can be triggered under various conditions, such as driving over bridges, driving at high speeds, or driving in close proximity to large trucks. Traditional cognitive behavioral therapy (CBT) and exposure therapy are the most common approaches used in the treatment of psychological disorders, such as anxiety disorder (AD) and panic disorder (PD). This research focuses on virtual reality (VR)-based exposure therapy, called VRET, and describes the design and development of a system which uses alternating levels of fear-based driving scenarios that can be recorded and automatically adjusted to maximize exposure effectiveness without causing the subjects to panic. The proposed VRET integrates an advanced feedback database module for tracing and analyzing the system, along with the user's bio-data to show the valid data collection of the system and its effectiveness for future use in clinical trials. The research conducts a system's pre-test analysis using 31 subjects to demonstrate the effectiveness of the system. This research demonstrates the systematic development of the VRET for driving phobia disorder by depicting the system framework, key system modules, system integration, bio-database management, and pre-test data analysis to support our next research efforts in hospital-based clinical trials and for additional VRET development applications for clinical psychology.

Keywords: virtual reality exposure therapy; driving phobia; post-traumatic stress disorder; physiological signal

1. Introduction

Post-traumatic stress disorder (PTSD) is a general term for any mental health issue that is triggered by a traumatic event experienced or witnessed by the patient. Many types of assaults can result in PTSD, such as physical or mental abuse, sexual assault, and serious physical accidents. People that suffer from PTSD have problems with daily, societal, and work-related activities. The condition interferes with one's ability to deal with normal life tasks. In 1980, PTSD was officially recognized as a mental illness by the American Psychiatric Association after the analysis of soldiers injured and exposed to traumatic experiences during the Vietnam war [1].

Cognitive behavioral therapy (CBT) and exposure therapy (ET) are the most common approaches for treating PTSD, anxiety disorder (AD), and panic disorder (PD) [2]. The goal of therapies is to change patterns of thinking or behavior that cause mental disorders. CBT works by changing people's attitudes and their behaviors by focusing on the thoughts, images, beliefs and attitudes. Because there is no invasive treatment or psychoactive drugs used while conducting CBT, it is generally recommended by medical institutions [3]. The challenges of CBT are that it requires patient cooperation, is time consuming, and is difficult to execute as a standard procedure.

An emerging treatment approach for PTSD takes advantage of immersive-based information technology to advance exposure therapy using virtual reality (VR) and more sophisticated physical measuring devices (called VR-based exposure therapy (VRET)). VRET allows patients to slowly expose themselves to different levels of traumatic stimuli using immersions of vision, sound and tactile feedback, which matches the principle of systematic desensitization (SD). SD is evaluated by researchers to be as effective as CBT [2,4]. Using VRET, patients interact and are exposed to a simulation of the phobic surroundings, but do not physically encounter the feared situations and the immersive condition can be immediately adjusted to reduce additional harm that can be caused by stress and panic attacks resulting from overexposure. The curative effects of VRET are more significant (and more realistic) then CBT [5]. There are some issues to be considered when designing VRET as a therapeutic treatment [6]. First, VRET is an evolving technology, psychiatrists and psychotherapists have to experiment with designs and configurations of therapy, develop base scenes, changing scenarios, and integrate complex monitoring equipment while demonstrating that the treatment is effective, reliable, and valid. The development costs and design thresholds of VRET are considerably more challenging than traditional CBTs. Further, researchers must consider the realism of the immersive environments. If the immersive environment is not realistic, the patients will lack exposure and the curative effect (or an effective meta-analytic effect size for clinical trials) will be compromised. If the immersive environment is too realistic, it could trigger severe phobic reactions during the experiments and endanger the subject's safety. Finally, some patients may not be familiar with VR hardware and software settings. Thus, severe dizziness during the experiment may hinder the VRET results. Thus, patients' physical conditions and their suitability for VRET need to be carefully investigated to ensure the effectiveness of VRET implementations and applications [6].

Section 2 presents a literature review of immersive applications in PTSD, including traditional ET, CBT, and VRET. Section 3 introduces the methodology framework and the approaches used in this research. Section 4 demonstrates the system architecture and the hardware and software components that are adopted and integrated to build the immersive environment. Section 5 performs statistical analysis to the pre-test data, using the correlation coefficients between system and bio-data, nonparametric testing for gender difference and anxiety group difference, and an ANOVA test for ET level effects. A Shewhart control chart was applied to monitor the subjects' emotional arousal. Section 6 summarizes the VRET systematic research, describes the research contributions, and highlights the next phase VRET research for controlled clinical trials.

In principle, this research focuses on VRET design and development using seven increasingly frightening driving scenario levels. The proposed VRET integrates with an advanced system and bio-database module for tracing and analyzing the system and user's bio-data. The research conducts the system's pre-test analysis to demonstrate the effectiveness of the system. The main objectives of this research are to present the systematic development of the VRET for driving phobia disorder, including the VRET system framework, key system modules, system design theory, bio-feedback database management, and pre-test data analysis. The research outcomes will support our next efforts in using VRET for hospital based clinical trials for driving phobia disorders, and potentially extending VRET development to other applications in clinical psychology.

2. Literature Review

In this section, the literature review focuses on the latest state of immersive technologies and applications in PTSD-related psychotherapies. Publications from the American Psychological Association (APA), the Institute of Electrical and Electronics Engineers (IEEE) Xplore, and the Web of Science (WoS) digital library are searched using keywords Cognitive Behavior Therapy, Post Traumatic

Stress Disorder, and Exposure Therapy. The results were further refined to include Virtual Reality Exposure Therapy and Driving Phobia. The two groups of literature are simply divided into VRET and traditional treatments. The traditional treatment types are further divided into CBT and ET.

CBT and ET are commonly used to treat PTSD, depression, anxiety, and panic disorders. CBT reduces cognitive anxiety by letting patients understand the causes of their reactions and how to manage and change reactions that may exasperate their mental condition. The efficacy of CBT on anxiety-related mental disorders, especially PTSD, has been confirmed by many studies [7–10].

Exposure therapy exposes patients to various levels of stimulating situations and continuously increases the intensity of the exposure to gradually induce a tolerance toward the exposure. Subjects are taught to self-analyze their reactions and change their behavioral response to better adapt to the changing conditions of exposure. Exposure therapy includes both imaginary and in vivo exposure. For imagined exposure, the therapist may ask the patient to imagine what they are afraid of. On the other hand, in vivo exposure allows patients to face the phobia directly. Some clinical trials have confirmed the effectiveness of exposure for treating PTSD. For example, Bryant et al. [11] divided 45 patients with PTSD into three groups—waiting list, exposure or counseling. The results showed that only 14% and 20% of the group members that received prolonged exposure and prolonged exposure plus anxiety management still suffered from symptoms of PTSD after treatment. Fifty-six percent of the supportive counseling group members were not affected by the PTSD treatment. The findings indicated that prolonged exposure may be a critical approach in the treatment of PTSD [11].

Aside from treating phobias, one of the successful applications of virtual reality technology is in the field of education. In one study, researchers have conducted a bibliometric study of the last twenty years from the Scopus database, and conclude that the use of virtual reality in education can improve the quality of the teaching and learning processes [12]. In Jesús López Belmonte's research, he asserted that digital technology has become a part most of people's lives and entered various fields of society. For example, he applies VR to education and reports that VR yields many benefits such as optimizing teaching and learning processes, timely access to new information, greater mobility, ubiquitous access, and adaptable to the uniqueness of each student [13].

At present, there are many studies on the applications of VRET for various kinds of phobias such as acrophobia, arachnophobia, and aviophobia [14]. In one study, a 50-year-old officer who suffered from PTSD caused by the Vietnam War underwent VRET. After 14 treatments of up to 90 min over seven weeks, his PTSD symptoms fell 34% under standard clinical measures. Self-assessment decreased by more than 45% [15]. Another study of flying phobia ET confirmed that VRET and traditional ET have no difference in statistical effect size. Seventy-five patients with fear of flying (FOF) were divided into three groups (25 people in each group): VRET, traditional ET and the waitlist (do nothing). After treatment, the results of VRET and traditional ET were almost equivalent while both groups were superior to those on the waitlist [16]. Another study used meta-analysis to integrate effects from 21 clinical studies with a total of 300 subjects. The analysis results (effect size) demonstrated that VRET effectively reduces anxiety [17]. Finally, our research studied the application of VRET in arachnophobia treatment, incorporating Arduino bio-sensors to collect real-time bio-data from subjects. Bio-data are integrated with the VRET system data and stored in SQL database for data mining and real-time VRET level adjustment is used to minimize over-exposer to the treatment [18].

Driving is an essential skill that facilitates independence and mobility of individuals in modern society and is often a requirement of employment. Being diagnosed with a driving phobia limits an individual's work options and creates issues associated with social interaction when mass transportation or transportation for hire is not available or unaffordable. According to emotional processing theory, successful exposure therapy leads to new and more neutral memory structures that overrule old memories that provoke and heighten anxiety [19]. In a VRET for driving phobia pilot study, 14 subjects with serious driving phobia were placed into a series of experimental conditions, including psychotherapeutic, medical examination, psychotherapy sessions, VRET sessions, and a final behavioral avoidance test (BAT) when driving in traffic. The therapy session for each subject required 10 days to complete the experiment. After 6 to 12 weeks, follow-up tracking of the subjects was performed to access the subjects' driving performance and evaluate the effectiveness of the treatment for each subject. The follow-up tracking indicated that 13 out of the 14 patients maintained successful treatment results [20]. Another VRET experiment for driving phobia used eight females who were willing to try VRET but unwilling participants used in vivo exposure. The VRET exposure environment consisted of eight sessions which gradually exposed participants to increasingly uncomfortable exposure over 50-minute intervals. Factors, such as mean mood, anxiety, cognitive evaluation, quality of life, sense of presence, subjective discomfort, and heart rate were recorded during the experiment. The result showed that the sense of presence decreased as the session progressed while subjects' subjective discomfort scores and heart rate increased. After the experiment, subjects were determined to be more confident with their driving ability. Six out of eight subjects were willing to conduct in vivo exposure after VRET [21]. A VRET experiment for driving phobia related to fear of driving in tunnels was conducted. In the virtual reality environment, the car passes through a tunnel with lighting adjusted to resemble natural daylight on entry and exit, with increasing darkness and increasing lightness upon exit. The sound of traffic was also included to enhance the realism of the simulations [22]. The above driving phobia VRET studies provide a solid basis for continued research. The numbers of subjects in these studies are generally too small to draw a decisive conclusion or provide suggestions for future VRET refinement. Moreover, most of the literature evaluated their VRET system based on subjects' driving performance, and few focused on the subjects' biological senses and system recorded data analysis. Therefore, this pre-test experiment will examine the feasibility of using biological sensors and system-data collection and analyses to measure subjects' driving performance to ensure the safety and effectiveness of VRET toward phobia treatment. Further, to ensure the VRET is scientifically conducted, collecting real-time bio-data of patients is vital during the ET sessions. Matured bio-sensor solutions, such as the Arduino micro-controller, can transmit bio-data to a remote computer or cloud database using an XBee wireless network [23].

3. Methodology Applied in This Research

Figure 1 presents the research flow of this study for the driving phobia VRET design and development. The project formulation made at the beginning was to decide how to implement the research, perform risk assessment, and hypothesize expected results. Literature related to VRET, driving phobia, CBT, and PTSD were reviewed and referenced before designing the exposure treatment scenario. The virtual reality immersive environments were constructed to model different fear levels of scenarios. The constructed VRET prototype has been examined by psychologists and an iterative modification process was executed until the psychologists were confident that the system matched the needs of ethical and medically acceptable driving phobia treatments.

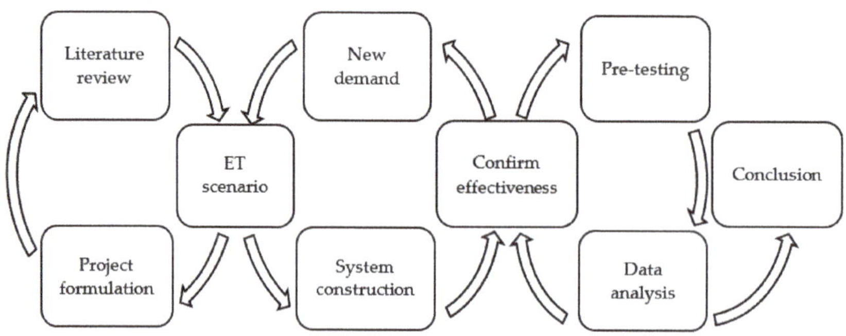

Figure 1. The research flow of the virtual reality-based exposure therapy (VRET) system design and development.

The VRET system is constructed using a seven-level exposure therapy design as shown in Figure 2. Since some patients are extremely afraid of car crashes, pedestrians and other driving vehicles do not appear in the immersive environment for the initial test runs. All of the simulation scenes follow a single route map. Subjects drive a specified distance to reach a destination before advancing to a higher level. Each level has its unique simulation environment, time, and treatment objectives. The details of each level are highlighted as follows.

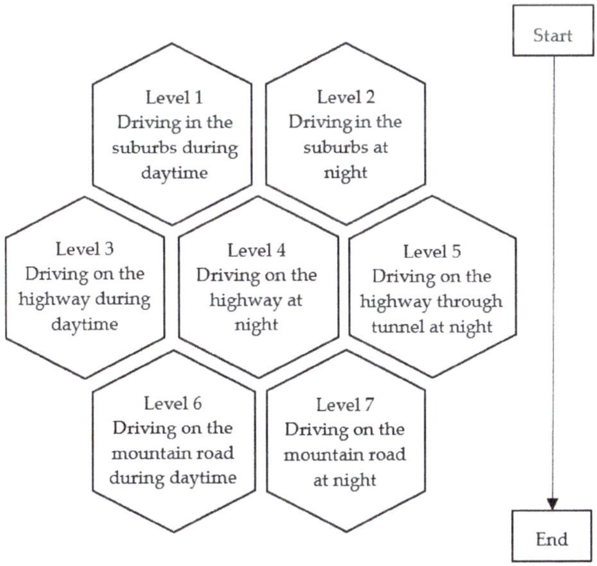

Figure 2. Scenario design level chart.

Level 1 Driving in the suburbs during the daytime, the speed limit is 40 km/h.

Level 2 Driving in the suburbs at night, the speed limit is 40 km/h.

Level 3 Driving on the highway during the daytime, the speed limit is 110 km/h, and the system will ask subjects to drive faster if the driving speed is less than 90 km/h.

Level 4 Driving on the highway at night, the speed limit is 110 km/h, and the system will ask subjects to drive faster if the driving speed is less than 90 km/h.

Level 5 Driving on the highway at night, the speed limit is 110 km/h, and the system will ask subjects to drive faster if the driving speed is less than 90 km/h. Subjects will be expected to traverse a tunnel.

Level 6 Driving on a mountain road during the daytime, the speed limit is 50 km/h.

Level 7 Driving on the mountain road at night, the speed limit is 50 km/h.

The VRET experiment flow is illustrated in Figure 3. The treatment begins with a detailed introduction about the experimental objective, method, procedure, expected risk, and relevant disclaimer information. A pre-test questionnaire is completed by the subject before the experiment to measure the subject's mental state. The questionnaire asks about the subject's behavior, thoughts, and emotions in regard to previous driving experiences. During the experiment, bio-data including heart rate, skin conductance, body temperature, respiration, and head movement are recorded in the computer database for statistical analysis. If virtual reality vertigo or other motion sickness symptoms occur to the subject during the VRET session, an emergency stop is triggered and the experiment is terminated. After the immersive treatment is finished, subjects are asked to complete a post-test feedback form.

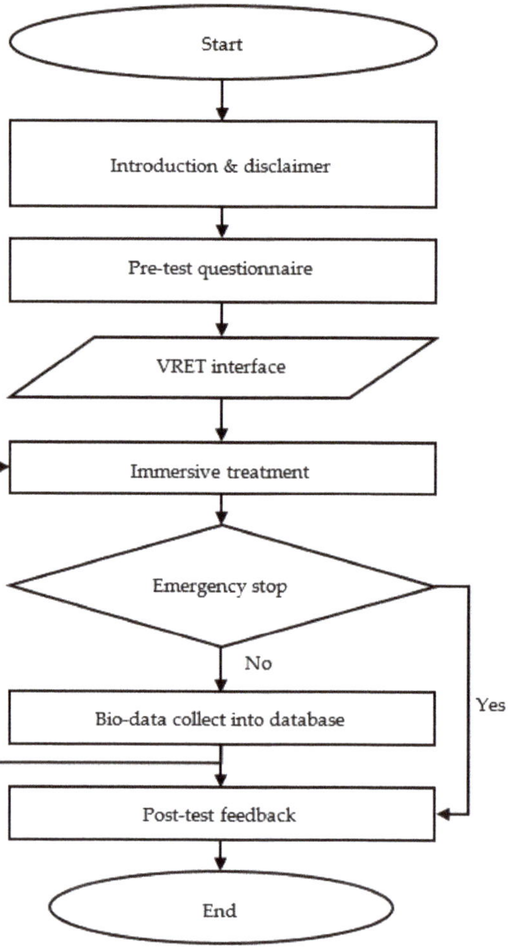

Figure 3. The driving phobia VRET experiment flow.

A total of thirty-one subjects including 14 males and 17 females at the university participated in the pre-test of the VRET prototype. The inclusion criteria are aged between 20 and 25 years old, possession of a valid driving license, and described a fear of at least one type of driving phobias such as driving on highways, mountain roads, or congested traffic.

4. System Architecture

The overview system architecture of this VRET research is presented in Figure 4. The VRET system (software) consists of four modules: the exposure therapy (ET) scenario module (7 scenario levels), the VR environment module, the data collection module, and the data analysis module. The VR–User interface (hardware) includes a VR headset, steering wheel, and bio-sensors. A database management system is developed to collect and integrate system and bio-data from the VRET system and the VR-User interface. The system's hardware, software, and database designs are described in the following paragraphs. For more technical details, please refer to [16].

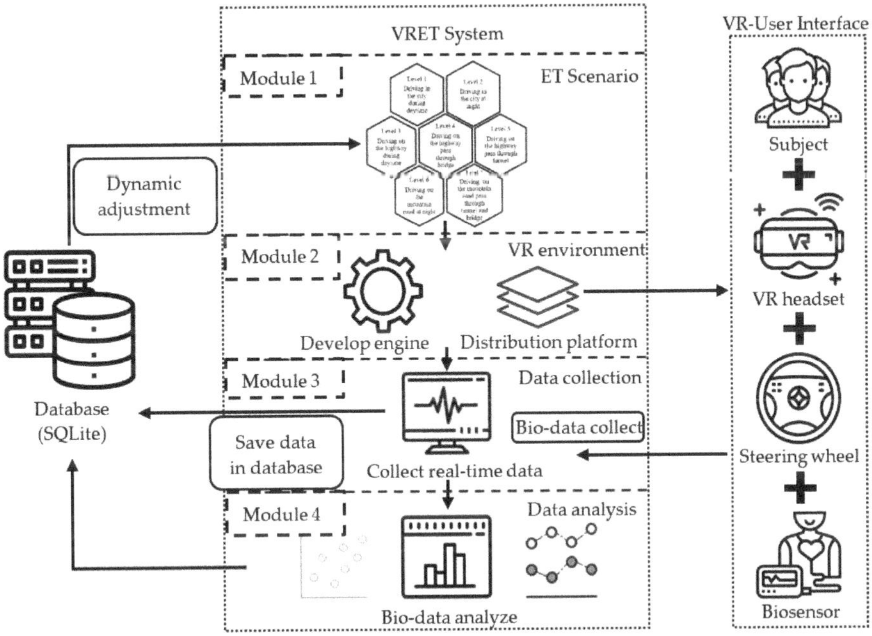

Figure 4. The system architecture for the driving phobia VRET prototype (refer to Figure 2 for Module 1 details).

4.1. *Hardware*

The hardware components used by the system include the VR headset (HTC, VIVE Pro, Taipei, Taiwan), the steering wheel (Guillemot Co., Thrustmeter T300 RS GT, Carentoir, France), and bio-sensors (Thought Technology Ltd., ProComp Infiniti, Montreal, Canada). The HTC VIVE Pro is a popular virtual reality immersive headset. The kit headset includes VR HMD, two base stations for movement tracking, and two motion controllers for virtual world interaction. The two base stations are able to connect the immersive environment with the subject's reactions to scene objects via motion controllers. The Thrustmeter T300 RS GT Edition is a high-end racing steering wheel equipped with a feedback mechanism, a specialized T3PA acceleration pedal that is integrated with steering wheel motion to provide realistic driving feedback to the subjects. The ProComp Infiniti is an eight-channel biofeedback system for real-time data acquisition that can be used with combinations of sensors including EEG, EKG, RMS EMG, skin conductance, heart rate, blood volume, pulse, respiration, temperature, force, and voltage.

4.2. *Software*

The software components adopted into the system include the development engine (Unity Technologies, Unity, San Francisco (HQ), CA, USA) and the distribution platform (Valve Corp., Steam, Bellevue, Washington, USA). Unity is a well-known real-time 3D development platform which enables users to develop both 2D and 3D games and models in Windows, iOS, Android, and over twenty types of systems. Microsoft C# is used as the programming language; users are allowed to control actions of virtual objects using specific C# scripts. Unity is compatible with other design software and gaming platforms that have external operating devices also compatible with Unity. Steam is one of the largest video game digital distribution service platforms in the world and also provides a virtual reality service called SteamVR to support the HTC VIVE virtual reality headset. Data collection and data management use SQLite (SQLite.org) as the database engine and BioGraph Infinite software to record bio-data. The VRET database management system (DBMS)'s entity-relation diagram is shown in Figure 5.

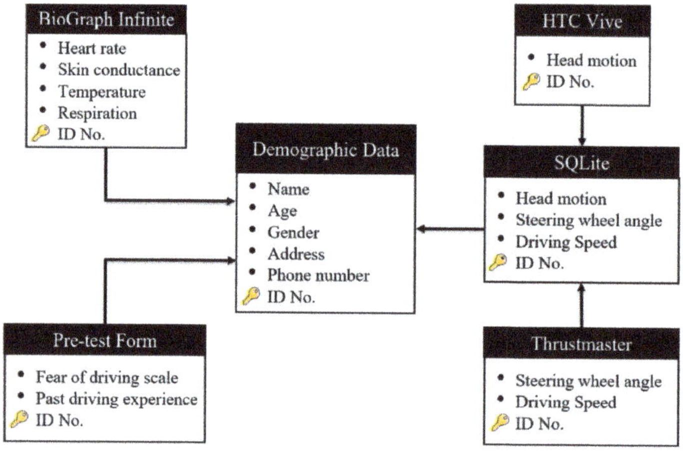

Figure 5. The entity-relation (ER) diagram for the VRET database.

4.3. Database

SQLite is a relational database management system (DBMS) contained in the C library. The reason for choosing SQLite over MySQL is because SQLite is a popular choice as an DBMS for client storage. SQLite is used to collect head motion, steering wheel rotation angle, and driving speed data in real time. These data are accessed from the database using the Python programming script. Th database storage model is shown in Figure 6. BioGraph Infinite is a software platform that is specifically used for ProComp Infiniti data access, allowing high end users to create their own screens, scripts and channel sets. The purpose of BioGraph Infinite is to collect heart rate, skin conductance, body temperature, and respiration rate data during ET experiments. The biological and system data are integrated for analysis.

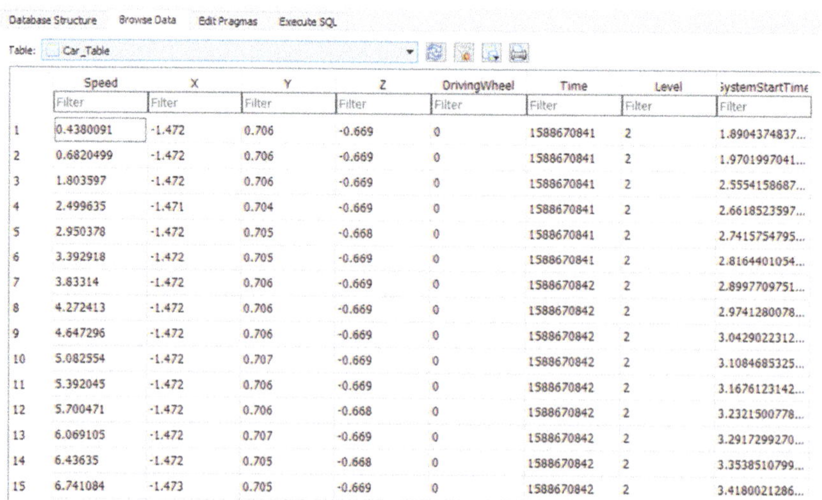

	Speed	X	Y	Z	DrivingWheel	Time	Level	SystemStartTime
	Filter	Filter	Filter	Filter	Filter	Filter	Filter	Filter
1	0.4380091	-1.472	0.706	-0.669	0	1588670841	2	1.8904374837...
2	0.6820499	-1.472	0.706	-0.669	0	1588670841	2	1.9701997041...
3	1.803597	-1.472	0.706	-0.669	0	1588670841	2	2.5554158687...
4	2.499635	-1.471	0.704	-0.669	0	1588670841	2	2.6618523597...
5	2.950378	-1.472	0.705	-0.668	0	1588670841	2	2.7415754795...
6	3.392918	-1.472	0.705	-0.669	0	1588670841	2	2.8164401054...
7	3.83314	-1.472	0.706	-0.669	0	1588670842	2	2.8997709751...
8	4.272413	-1.472	0.706	-0.669	0	1588670842	2	2.9741280078...
9	4.647296	-1.472	0.706	-0.669	0	1588670842	2	3.0429022312...
10	5.082554	-1.472	0.707	-0.669	0	1588670842	2	3.1084685325...
11	5.392045	-1.472	0.706	-0.669	0	1588670842	2	3.1676123142...
12	5.700471	-1.472	0.706	-0.668	0	1588670842	2	3.2321500778...
13	6.069105	-1.472	0.707	-0.669	0	1588670842	2	3.2917299270...
14	6.43635	-1.472	0.705	-0.668	0	1588670842	2	3.3538510799...
15	6.741084	-1.473	0.705	-0.669	0	1588670842	2	3.4180021286...

Figure 6. Example of VRET's system database storage.

5. Pre-Test and Pre-Test Data Analysis

In order to examine the effectiveness of the VRET system and to improve the VRET system to facilitate future clinical trials, this study recruited 31 subjects for pre-testing. Before the pre-test experiment, subjects were asked to fill out the "Self-rating fear of driving scale," for the purpose of measuring their degree of fear of driving. The rating scale was designed according to emerging measures offered by American Psychiatric Association [24]. During the pre-test experiment, the system data, bio-data, and subjects' suggestions for refinement of the VRET system were recorded. This study was approved by the Chang Gung Medical Foundation Institution Review Board (IRB), the institution's ethical commission, approval number 201901262B0.

5.1. Pre-Test Subjects' Demographic Data

This study recruited 14 males and 17 females at the university participated in the pre-test of the VRET prototype. Pre-testing subjects' demographic data including gender, height, weight, age, and the years of driving experience are presented in Table 1. 14 males with a mean height of 172.5 cm (SD = 5.21 cm) and a mean weight of 64.29 kg (SD = 7.61 kg), 17 males with a mean height of 160.43 cm (SD = 4.75 cm) and a mean weight of 52.71 kg (SD = 5.44 kg). All the subjects with a mean age of 22.39 years (SD = 0.95 year) and the mean duration of driving experience is 2.10 years (SD = 1.13 years).

Table 1. Pre-test subjects' demographic data.

Subject No.	Gender	Height (cm)	Weight (kg)	Age	Driving Experience (Years)
1	Female	164	52	23	4
2	Female	164	62	21	2
3	Male	173	57	24	3
4	Male	178	74	24	2
5	Male	160	48	22	1
6	Female	156	48	22	1
7	Male	174	75	22	2
8	Female	153	52	23	1
9	Female	152	47	22	1
10	Female	162	52	23	3
11	Male	167	69	22	3
12	Male	172	58	22	2
13	Female	164	49	22	1
14	Female	161	55	23	4
15	Female	167	49	22	2
16	Female	166	62	22	2
17	Female	158	50	22	1
18	Male	175	65	22	1
19	Male	170	65	22	2
20	Female	165	50	22	1
21	Male	171	60	22	3
22	Male	170	65	22	1
23	Female	162	53	22	3
24	Female	160	65	21	2
25	Female	164	52	22	1
26	Male	174	71	24	1
27	Male	177	71	24	2
28	Female	162	46	22	2
29	Male	172	65	22	3
30	Male	182	57	25	6
31	Female	162	52	21	2

5.2. Pre-Test Procedures

1. Explain the purpose and precautions of the experiment and ask subjects to sign the consent form.

2. Subjects filled out the "Self-rating fear of driving scale."
3. Subjects were equipped with bio-sensors for the purpose of detecting the real-time bio-data. Bio-sensors including heart rate, temperature, respiration, and skin conductance.
4. Conducting the pre-test VRET experiment and collect subjects' real-time bio- and system data.
5. After the pre-test experiment, subjects were asked to fill out the feedback form and provide some suggestions to the refinement of VRET system.

5.3. Pre-Test Data Analysis

The real-time VRET system data, shown in Figures 7 and 8, and the pre-test subject's bio-data, are shown in Figures 9 and 10. For the current prototype of VRET for driving phobia, the bio-data of each subject are monitored and analyzed for trends and correlations between bio-data and system data. The data spike abnormalities are monitored during the pre-testing experiment sessions. The goal is to improve the effectiveness of the therapy design of the VRET immersive system based on data analytic outcomes to avoid levels of driving conditions showing insignificant bio-effects or causing overly dramatic effects (i.e., lacking the feature of gradual systematic desensitization). The following discussion describes the statistical analysis results of the system's pre-test experiments with 31 subjects.

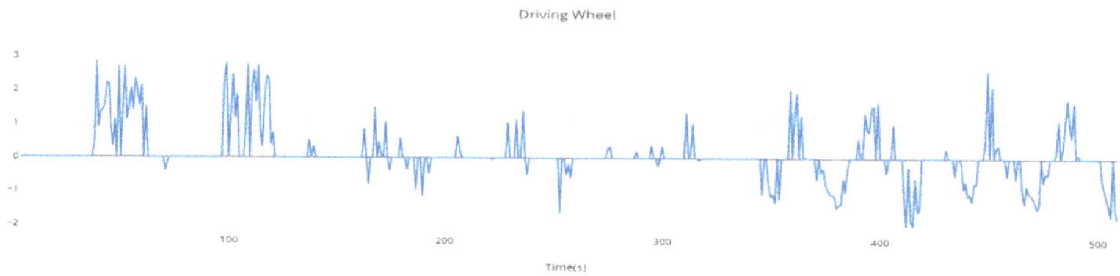

Figure 7. System data—Steering wheel rotation data tracking.

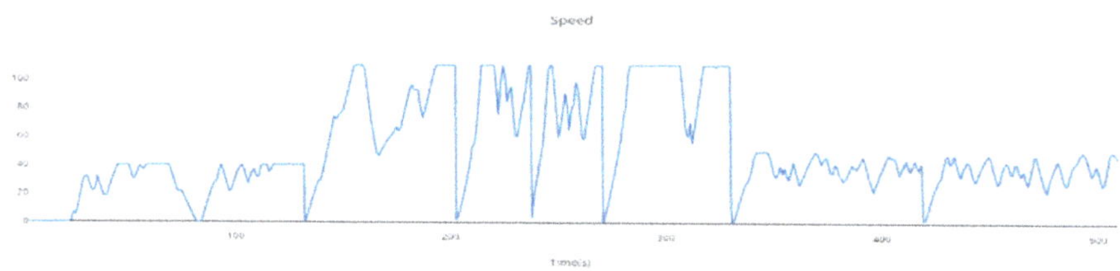

Figure 8. System data—Tracking driving speeds over time.

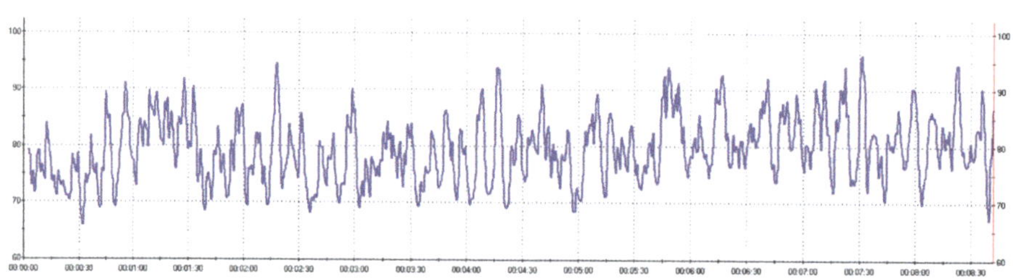

Figure 9. Bio-data—Tracking subject's heart rate during experiment session.

Figure 10. Bio-data—Subject's skin conductance during the experiment session.

The Pearson correlation coefficient has been applied to measure the strength of linear association between the VRET system data and the subject's bio-data. The result indicates that the correlations are relatively small with most coefficients falling between −0.3 and 0.3. Nonparametric test for gender differences compared the 14 males and 17 female subjects that participated in the pre-test experiment. The Mann–Whitney U test was used to test if there was a statistically significant difference between the overall satisfaction medians of the male and female groups. The null hypothesis is $\eta_{male} = \eta_{female}$, using a significance level of $\alpha = 0.05$, the p-values related to four kinds of bio-data are all higher than alpha value (0.05), so there is insufficient evidence to conclude a statistically significant difference between the overall satisfaction medians of the male and female groups.

The Mann–Whitney U test was also used to test if there is a statistically significant difference between the overall satisfaction medians of the high anxiety and low anxiety groups. Subjects were divided into two groups, high anxiety groups and low anxiety groups, according to the rating of "Self-rating fear of driving scale" filled out before the pre-test experiment. Eleven subjects were in the high anxiety group and 20 subjects were in the low anxiety group. The null hypothesis is $\eta_{high} = \eta_{low}$, using a significance level of $\alpha = 0.05$, three out of the four p-values related to bio-data (heart rate, respiration, and skin conductance) are lower than alpha value (0.05), so there is statistical evidence that there is a difference between the overall satisfaction medians of the two anxiety types. The VRET system does affect the subject's bio-data according to their degree of fear of driving.

One-way ANOVA tests comparing means of bio-data in seven scenarios are used to estimate the 95% confidence interval and determine whether the associated population means in different scenarios are significantly different. The results are presented in Figure 11a,c,e,g. Fisher's least significant difference method for multiple comparisons was applied to examine which means at different levels are different, as presented in Figure 11b,d,f,h.

The bio-feedback data of the one-way ANOVA interval plots, presented in Figure 11, indicate that pre-test subjects become increasingly anxious as the fear level of the scenarios increases (shown by heart rate, temperature, and skin conductivity trends). For detailed information from the ANOVA results, multiple comparisons analysis between different levels was also conducted, as higher fear levels tend to be assigned to the same groups and lower fear levels tend to be assigned to the same groups, a significant difference in the sample means between higher levels and lower levels was demonstrated. This trend supports the hypothesis that increases in the severity of pre-existing phobias increase the subject's bio-feedback responses.

An interesting observation is that the bio-feedback data in level 1 seems to be significantly higher than level 2 data as shown in heart rate and temperature plots and the Fisher's LSD method supports the same result. A possible cause of this phenomenon may be related to the experimental design where subjects never exposed to a VR environment may become more anxious during the initial ET session. As the experiment progresses, the subjects become acclimated to the VR environment so there is less tension and greater ease of use of the system. In order to prevent this problem from happening, a ten-minute static VR experience can be added to the experiment flow between pre-test questionnaire and VRET interface to stabilize the subject's bio-response before the experiment begins.

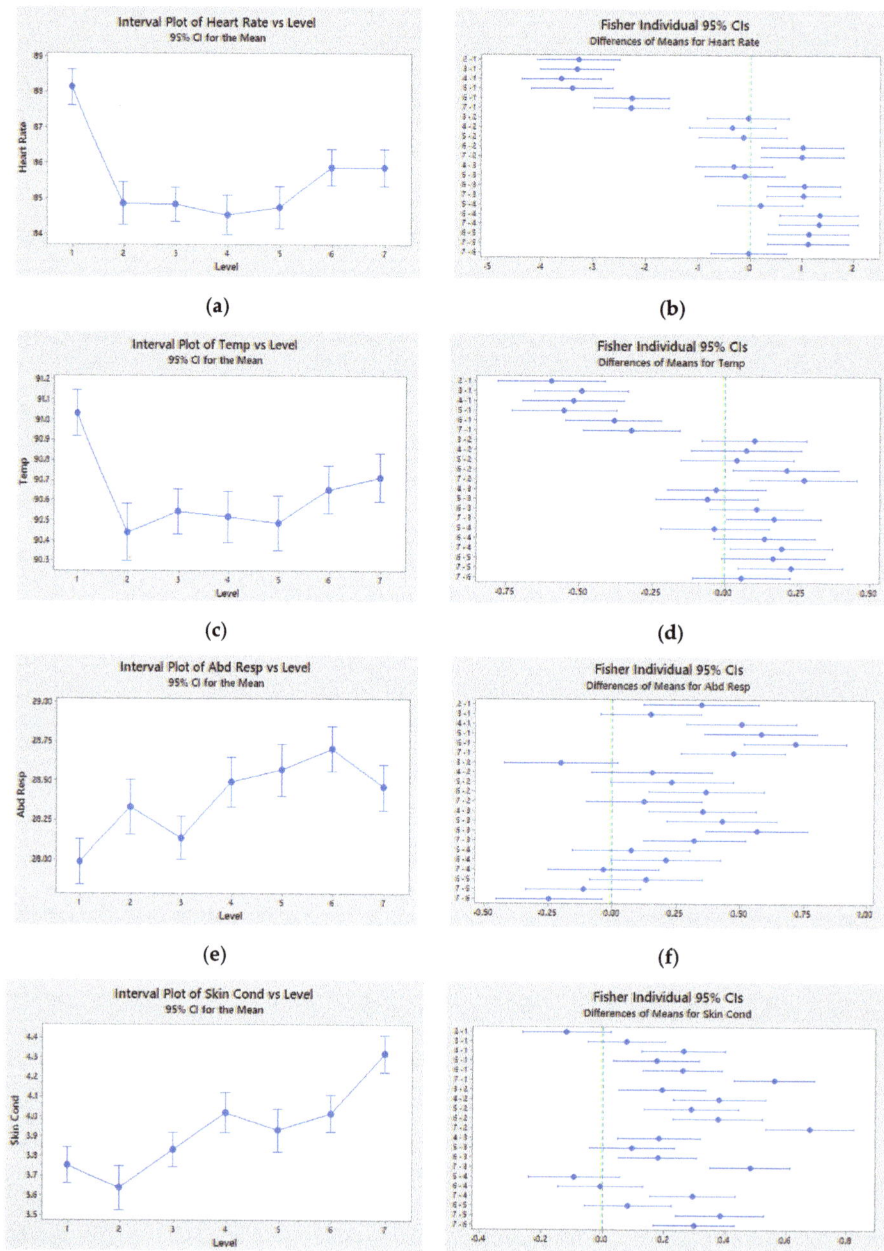

Figure 11. (**a**) One-way ANOVA interval plot of heart rate vs. level; (**b**) Fisher's least significant difference (LSD) method for multiple comparisons of heart rate vs. level; (**c**) One-way ANOVA interval plot of temperature vs. level; (**d**) Fisher's LSD method for multiple comparisons of temperature vs. level; (**e**) One-way ANOVA interval plot of respiration vs. level; (**f**) Fisher's LSD method for multiple comparisons of respiration vs. level; (**g**) One-way ANOVA interval plot of skin conductance vs. level; (**h**) Fisher's LSD method for multiple comparisons of skin conductance vs. level.

According to Dawson's research, if the amplitude of skin conductance is larger than 0.1 µs, then the skin conductance indicates a larger emotional response [25]. In this research, skin conductance data are converted into the form of amplitude data. A Shewhart control chart of amplitude data is presented in Figure 12. If four out of five data points exceed 0.1µs, it is marked in red, which means the subject has a larger emotional response at this point of time. Sixteen of the thirty-one subjects meet the above conditions and have data points marked in red. These data points frequently appeared for the 6th, 21st, 25th, and 32nd subjects.

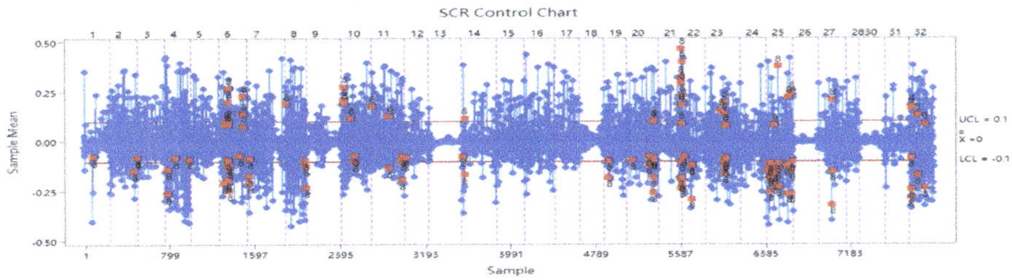

Figure 12. Shewhart control chart of skin conductance amplitude data.

The heart rate data are standardized according to individual data distribution and a Shewhart control chart of the standardized data is presented in Figure 13. If four out of five points exceed two standard deviations, the data points are marked in red indicating the subject has experienced a larger emotional response at this point of time. Twenty-one of the thirty-one subjects met the above condition with their heart rate data points marked in red.

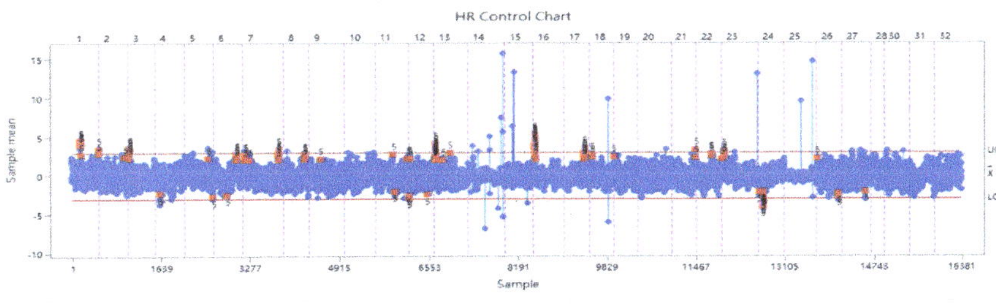

Figure 13. Shewhart control chart of standardized heart rate data.

5.4. Discussion of Pre-Test

The proposed VRET system has integrated system and bio-data collection for real-time data analytics compared to previous VRET research. In the Kaussner et al. VRET pilot study, anxiety and habituation during the VRET sessions were assessed using ratings on subjective units of a distress scale as well as the heart rate [20]. Costa et al. used subjective discomfort scores and heart rate to measure subjects' physiological response and sense of presence [21]. Additionally, in Claudio et al.'s fear of driving in tunnels research, questionnaires measured the outcome of the experiment and no real-time data was collected [22]. Most pervious VRET literature uses heart rate for real-time data tracking. Temperature, respiration, and skin conductance are for real-time data tracking in our VRET system. These preliminary design and development efforts are for the construction of a self-adaptive VRET system that will automatically adjust the levels of exposure according to real-time data tracking. The proposed future system will be tested for improvements to the immersive experience and to better ensure subjects' safety.

Compared to previous VRET research which only demonstrates the development of VRET without verification, this research has uniquely combined VRET design and development with the verification of its efficacy through pre-test. According to the results of Mann–Whitney U test, there is a significant difference between the overall satisfaction medians of high anxiety groups and low anxiety groups. Moreover, the results of ANOVA analysis show that bio-data and level are positively correlated except for Level 1. Therefore, through pre-test data analysis, VR is demonstrated to provide the subject with a sense of authenticity and can indeed cause anxiety similar to actual driving.

The problem of high level 1 bio-data is also emerging within pre-test data analysis. For this problem, follow-up studies may consider adding a ten minutes of VR experience before the experiment starts to assist the subjects to adapt to the VR environment and to improve the VRET system for official clinical trials.

6. Conclusions

The immersive technologies, most typically referred to as virtual reality (VR) and lately also to include augmented reality (AR) enabling technologies, are popularly researched and developed to improve healthcare and treatments for physical, psychological and social wellbeing with respect to different populations and their health issues (e.g., elderly, autism spectrum disorder, PTSD, phobia, etc.) [26,27]. For example, Lee et al. [24] have comprehensively reviewed VR and AR technologies developed for physical, psychological and social wellbeing of an elderly population. The review article discovers that plenty of R&D has been deployed for improving physical healthcare and treatments. Nonetheless, considering the rising psychological (mental and social) health issues, there are great needs to pursue immersive technologies and design immersive solutions for psychological therapies and treatments. The research presents the systematic design and development of virtual reality exposure therapy applied to fear of driving. Therapeutic treatments for this phobia remain as a major clinical challenge for psychiatrists, psychotherapists, and their patients. A VR headset, a high specification steering wheel, bio-sensors, and software are integrated to develop a realistic driving experience in an immersive environment for effective exposure therapy. The subjects are exposed to varying levels of fear inducing driving levels for treatment. Thirty subjects provided pre-test data for the system. The bio-feedback data and suggestions for improvement were derived from the experiments. Through pre-test data analysis, VR convincingly causes anxiety like actual driving. Furthermore, some systemic problems are discovered, so the follow-up research can improve the VRET system and improve the system design. Carefully crafted experimental design is required for collecting valid and reliable data for the VRET analytical models and system modification. The clinical psychiatrists' feedback is critical to improve the design before the VRET can be used for clinical trials.

In summary, this research has presented the VRET technical components, the system framework, key modules for implementation, and the experiment database design and development for the research. Specifically, the bio- and system-data analysis demonstrates the efficacy of the VRET for use in initial hospital-based clinical trials with continued improvement as the trails are analyzed. In the future, the research team hopes to extend the VRET treatments to other phobias, such as acrophobia, arachnophobia, and claustrophobia. The future research direction also includes the intelligent development of self-adaptive VRET systems incorporating progressive brain neurology and brain computer interface (BCI) technologies. When the degree of a subject's fears are detected through advanced BCI sensors, the self-adaptive VRET system will have sufficient knowledge and intelligence to automatically adjust the levels of exposure to ensure safe and effective gradual systematic desensitization.

Author Contributions: A.T.: Constructed the research framework and incorporated VR and IoT technologies for the system; C.V.T.: CBT and ET literature overview for anxiety disorders and panic disorders, interpretation of meta-analysis studies; C.-M.C.: PTSD, OCD, and PD theories and clinical ET design; R.R.T.K.: VRET system development and implementation, data analysis; A.P.C.L.: VR and VRET literature search, VRET prototype testing, data validation; C.H.N.: CBT and ET literature review and data validation. All authors have read and agreed on the published version of the manuscript.

Funding: This research is partially supported by the joint research grant of National Tsing Hua University and Chang Gong Memorial Hospital and the individual research grants of Ministry of Science and Technology, Taiwan (Grant numbers: MOST-108-2221-E-007-075-MY3 and MOST-108-2410-H-009-025-MY2).

Conflicts of Interest: The authors declare no conflict of interest.

References

1. American Psychiatric Association. *Diagnostic and Statistical Manual of Mental Disorders*; American Psychiatric Pub.: Washington, DC, USA, 2013.
2. Ougrin, D. Efficacy of exposure versus cognitive therapy in anxiety disorders: Systematic review and meta-analysis. *BMC Psychiatry* **2011**, *11*, 1–13.
3. Foa, E.B.; Rothbaum, B.O. *Treating the Trauma of Rape: Cognitive-Behavioral Therapy for PTSD*; Guilford Press: New York, NY, USA, 2001.
4. Wolpe, J. Psychotherapy by reciprocal inhibition. *Cond. Reflex* **1968**, *3*, 234–240. [PubMed]
5. Scozzari, S.; Gamberini, L. Virtual reality as a tool for cognitive behavioral therapy: A review. In *Advanced Computational Intelligence Paradigms in Healthcare 6. Virtual Reality in Psychotherapy, Rehabilitation, and Assessment*; Springer: Berlin/Heidelberg, Germany, 2011; pp. 63–108.
6. Meldrum, D.; Herdman, S.; Moloney, R.; Murray, D.; Duffy, D.; Malone, K.; French, H.; Hone, S.; Conroy, R.; McConn-Walsh, R. Effectiveness of conventional versus virtual reality based vestibular rehabilitation in the treatment of dizziness, gait and balance impairment in adults with unilateral peripheral vestibular loss: A randomised controlled trial. *BMC Ear Nose Throat Disord.* **2012**, *12*, 1–8. [CrossRef] [PubMed]
7. Matsuoka, Y.; Nishi, D.; Nakajima, S.; Yonemoto, N.; Noguchi, H.; Otomo, Y.; Kim, Y. Impact of psychiatric morbidity on quality of life after motor vehicle accident at 1-month follow up. *Psychiatry Clin. Neurosci.* **2009**, *63*, 235–237. [CrossRef] [PubMed]
8. Norris, F.H. Epidemiology of trauma: Frequency and impact of different potentially traumatic events on different demographic groups. *J. Consult. Clin. Psychol.* **1992**, *60*, 409. [CrossRef] [PubMed]
9. Hofmann, S.G.; Asnaani, A.; Vonk, I.J.; Sawyer, A.T.; Fang, A. The efficacy of cognitive behavioral therapy: A review of meta-analyses. *Cogn. Ther. Res.* **2012**, *36*, 427–440. [CrossRef] [PubMed]
10. Bisson, J.; Andrew, M. Psychological treatment of post-traumatic stress disorder (PTSD). *Cochrane Database Syst. Rev.* **2007**, *2007*, CD003388.
11. Bryant, R.A.; Sackville, T.; Dang, S.T.; Moulds, M.; Guthrie, R. Treating acute stress disorder: An evaluation of cognitive behavior therapy and supportive counseling techniques. *Am. J. Psychiatry* **1999**, *156*, 1780–1786. [PubMed]
12. Campos Soto, M.N.; Navas-Parejo, M.R.; Moreno Guerrero, A.J. Realidad virtual y motivación en el contexto educativo: Estudio bibliométrico de los últimos veinte años de Scopus. *ALTERIDAD Rev. Educ.* **2020**, *15*, 47–60. [CrossRef]
13. Belmonte, J.L.; Sánchez, S.P.; Cevallos, M.B.M.; Meneses, E.L. Competencia digital de futuros docentes para efectuar un proceso de enseñanza y aprendizaje mediante realidad virtual. *Edutec Rev. Electrón. Tecnol. Educ.* **2019**, 1–15. [CrossRef]
14. Gerardi, M.; Cukor, J.; Difede, J.; Rizzo, A.; Rothbaum, B.O. Virtual reality exposure therapy for post-traumatic stress disorder and other anxiety disorders. *Curr. Psychiatry Rep.* **2010**, *12*, 298–305. [CrossRef] [PubMed]
15. Rothbaum, B.O.; Hodges, L.; Alarcon, R.; Ready, D.; Shahar, F.; Graap, K.; Pair, J.; Hebert, P.; Gotz, D.; Wills, B. Virtual reality exposure therapy for PTSD Vietnam veterans: A case study. *J. Trauma. Stress* **1999**, *12*, 263–271. [CrossRef] [PubMed]
16. Rothbaum, B.O.; Anderson, P.; Zimand, E.; Hodges, L.; Lang, D.; Wilson, J. Virtual reality exposure therapy and standard (in vivo) exposure therapy in the treatment of fear of flying. *Behav. Ther.* **2006**, *37*, 80–90. [CrossRef] [PubMed]
17. Parsons, T.D.; Rizzo, A.A. Affective outcomes of virtual reality exposure therapy for anxiety and specific phobias: A meta-analysis. *J. Behav. Ther. Exp. Psychiatry* **2008**, *39*, 250–261. [CrossRef]
18. Trappey, A.J.; Trappey, C.V.; Chang, C.-M.; Govindarajan, U.H.; Shih, X.-Y.; Su, I.-A.; Gupta, N. Behavioral Therapy for Phobias Using Immersive Virtual Reality Technology. *J. Chin. Soc. Mech. Eng.* **2020**, *41*, 131–140.
19. Costa, R.T.D.; Carvalho, M.R.D.; Nardi, A.E. Virtual reality exposure therapy in the treatment of driving phobia. *Psicol. Teor. Pesqui.* **2010**, *26*, 131–137. [CrossRef]

20. Kaussner, Y.; Kuraszkiewicz, A.; Schoch, S.; Markel, P.; Hoffmann, S.; Baur-Streubel, R.; Kenntner-Mabiala, R.; Pauli, P. Treating patients with driving phobia by virtual reality exposure therapy–A pilot study. *PLoS ONE* **2020**, *15*, e0226937. [CrossRef] [PubMed]
21. Costa, R.T.D.; Carvalho, M.R.D.; Ribeiro, P.; Nardi, A.E. Virtual reality exposure therapy for fear of driving: Analysis of clinical characteristics, physiological response, and sense of presence. *Rev. Bras. Psiquiatr.* **2018**, *40*, 192–199. [CrossRef] [PubMed]
22. Claudio, A.P.; Carmo, M.B.; Gomes, I.L.; Esteves, F.; Gaspar, A. Sense of presence inside a feared (virtual) tunnel. In Proceedings of the 2015 10th Iberian Conference on Information Systems and Technologies (CISTI), Aveiro, Portugal, 30 July 2015; pp. 1–7.
23. Kioumars, A.H.; Tang, L. ATmega and XBee-based wireless sensing. In Proceedings of the 5th International Conference on Automation, Robotics and Applications, Wellington, New Zealand, 6–8 December 2011; pp. 351–356.
24. Craske, M.; Wittchen, U.; Bogels, S.; Stein, M.; Andrews, G.; Lebeu, R. *Severity Measure for Specific Phobia—Adult*; Lake St. Louis: American Psychiatric Association: Washington, DC, USA, 2013.
25. Dawson, M.E.; Schell, A.M.; Filion, D.L. The electrodermal system. In *Handbook of Psychophysiology*; Cacioppo, J.T., Tassinary, L.G., Berntson, G.G., Eds.; Cambridge University Press: Cambridge, UK, 2017. [CrossRef]
26. Liu, X.; Wu, Q.; Zhao, W.; Luo, X. Technology-facilitated diagnosis and treatment of individuals with autism spectrum disorder: An engineering perspective. *Appl. Sci.* **2017**, *7*, 1051. [CrossRef]
27. Lee, L.N.; Kim, M.J.; Hwang, W.J. Potential of augmented reality and virtual reality technologies to promote wellbeing in older adults. *Appl. Sci.* **2019**, *9*, 3556. [CrossRef]

applied sciences

MDPI

Article

Serious Game Platform with Haptic Feedback and EMG Monitoring for Upper Limb Rehabilitation and Smoothness Quantification on Spinal Cord Injury Patients

Álvaro Gutiérrez [1,*], Delia Sepúlveda-Muñoz [1], Ángel Gil-Agudo [2] and Ana de los Reyes Guzmán [2]

1 Escuela Técnica Superior de Ingenieros de Telecomunicación, Universidad Politécnica de Madrid, Av. Complutense 30, 28040 Madrid, Spain; delia.sepulvedam@alumnos.upm.es

2 Biomechanics and Technical Aids Unit, National Hospital for Paraplegics, Finca La Peraleda, 45071 Toledo, Spain; amgila@sescam.jccm.es (Á.G.-A.); adlos@sescam.jccm.es (A.d.l.R.G.)

* Correspondence: a.gutierrez@upm.es; Tel.: +34-91-06-72452

Received: 4 October 2019; Accepted: 24 January 2020; Published: 2 February 2020

Abstract: Cervical Spinal Cord injury (SCI) is a neurological disease that produces, as a consequence, impairments of the upper limb function. This paper illustrates a virtual reality platform based on three serious games for upper limb rehabilitation with electromyography monitoring, providing force feedback to the patient. In the rehabilitation process proposed, haptic feedback was provided to the patients to strength the arm muscles by means of the Novint Falcon device. This end-effector device was used to manipulate the serious games. During the therapy performance, the system recorded electromyography signals from the patient's arm muscles, which may be used to monitor muscle contraction. The work presented a virtual reality system developed for spinal cord-injured patients. Each virtual reality environment could be modified in strength and duration according to the patients' needs and was implemented for recording quantitative data about the motor performance. The platform was validated as a proof of concept in cervical spinal cord-injured patients. Results showed that this rehabilitation platform could be used for obtaining objective information in relation to motor control characteristics.

Keywords: upper limb rehabilitation; serious games; haptic feedback; electromyography sensors; virtual reality; smoothness

1. Introduction

Spinal Cord Injury (SCI) is one of the most important neurological diseases that produce deficiencies in the field of physical disability [1]. It is often a catastrophic condition requiring chronic care. Despite an enormous amount of research in SCI, the neurological prognosis for a patient with severe SCI remains dismal. Life expectancy after such an injury is markedly reduced due to complications proportional to the severity of injury or remaining neurological functions. Therefore, SCI is associated with a risk of developing secondary conditions that can be debilitating and even life-threatening, such as deep vein thrombosis, urinary tract infections, muscle spasms, osteoporosis, pressure ulcers, chronic pain, and respiratory complications [2]. Acute care, rehabilitation services and ongoing health maintenance are essential for the prevention and management of these conditions [3].

The worldwide incidence of SCI lies between 10 and 83 per million inhabitants per year [4]. In Spain, approximately 1000 new cases of SCI per year occur due to trauma, which is the main cause; half due to traffic accidents, the rest due to falls, blows, sport accidents or other injuries [5]. Most of these people have permanent damages. It is observed that people with SCI have difficulties and limitations in mobility (96.9% of cases), in self-care (81.1% of cases) and in carrying out tasks of domestic life (84.3% of cases) [6]. This situation means that the majority of the population with SCI requires personal and technical support to carry out these activities. The treatment for SCI is often expensive: it can reach up to 55,000€ the first 60 days for a patient in acute phase, with an additional 10,000€ for the intensive care stay. Hence, the economic and societal impact is enormous, both to the immediate family and the society at large [7]. The inclusion of lower cost technologies is a key aspect for reducing the therapy cost. Therefore, there is a great need for new systems that improve the quality of life for SCI injured patients [8–10].

The impairment of the Upper Limb (UL) is one of the most common sequelae following neurological injuries [11]. The loss of arm and hand function is one of the most devastating consequences in tetraplegia [12]. One important aspect of this lost function is smoothness. Smoothness is a characteristic of a skilled and well-coordinated human movement [13] and it is an independent measure with respect to speed and distance [14]. When moving the hand between pairs of targets, people tend to generate roughly straight hand trajectories with a single-peaked, bell-shaped speed profile [15]. However, in presence of neurological pathologies, the UL movement is characterized by the lack of smoothness [16], the performance of more curved trajectories with directional changes and a higher peaks number in the speed profile. Smoothness has been used as a measure of motor performance in healthy people and neurological pathologies, mainly stroke patients [11,14,17,18], being the neurological injury with the highest incidence and prevalence in the population. This control motor characteristic has been quantified by means of different variables, extracted of the movement analysis as the peaks number in the speed profile, the ratio between the maximal and mean velocity and the jerk metric computed as the time derivative of acceleration [13,19]. As our expertise in this field, after the developmental work of kinematic indices applied to spinal cord injured patients, smoothness metric computed from the peaks number in the speed profile is discriminative between healthy and neurological patients [20,21].

Small progresses in arm and hand function may lead to increase autonomy in activities of daily living [22], improving independence and quality of life. For this reason, rehabilitation for the improvement in UL function after cervical SCI is a top priority in individuals with tetraplegia [23]. In rehabilitation, considerable amounts of practice are required to induce neuroplastic changes and functional recovery of neurological motor deficits [24]. SCI rehabilitation works through mechanisms of experience-dependent plasticity [25]. The process of plasticity consists of changes in the activation pattern, either of structure or function of neural connections. These changes involve the modification of the strength of existing connections or the creation of new neural connections. Although it is likely that some of the reorganization after central nervous system injuries take place in the cortex, plasticity changes may also occur in subcortical structures, such as the spinal cord [26]. Therefore, neuroplasticity is essential in the SCI rehabilitation process.

Nowadays, technology is typically involved in rehabilitation. It makes use of high cost devices, such as exoskeletons or robotic devices [27], or low cost devices, such as the Leap Motion, Kinect, or Novint Falcon, among others [28]. In this paper, we make use of the Novint Falcon device, a desktop haptic robot with 3 degrees of freedom (see Section 2). Several studies found in the literature have investigated the use of the Novint Falcon in healthy subjects: Scalona et al. analyzed the motor performance during six reaching movement sessions varying the force levels [29]. The subjects had to reach eight different objectives placed on a circumference (vertical and horizontal line and in each diagonal). Cappa et al. analyzed the response to different force feedback in healthy population with an educational purpose [28]. Palsbo et al. used the Novint Falcon to train fine movements in a group of healthy developing children and noted significant improvements in handwriting performance [30]. However, there is no evidence of research studies with the Novint Falcon in spinal cord injured patients. Nonetheless, some studies have been performed with the Novint Falcon to analyze the UL function in patients who have suffered stroke [31,32]. These studies have been conducted with a rehabilitation purpose. Chortis et al. evaluated the effects of repetitive arm movements with a group of eight post-stroke subjects by using the commercial virtual games included in Novint

Falcon [33]. These games were oriented to reach UL functional tasks, following the rehabilitative principles. The patient performance was compared between pre and post condition by using only clinical tests. Unfortunately, in those studies, smoothness was not analyzed.

Additionally, the use of technology in rehabilitation in combination with virtual reality (VR) environments has increased in the last years [34–37], mainly because the quantity of therapy that patients receive increases. VR has many advantages for intervention, such as enabling the grading of activities, obtaining precise performance measures, providing a safe and ecologically valid environment, and being enjoyable and motivating [24,38]. There is evidence about the acceleration in the rehabilitation process by the combination of these VR treatments with conventional rehabilitation [39]. Moreover, the use of VR environments can increase the intensity of the therapy through the adequate repetitions number [27]. VR reality games increase motivation in patients both in occupational and physical therapy [40]. Therefore, the health care sector is showing steadily increasing interest in serious games [41].

The work presented in this paper is centered in reaching the patient's assessment in an objective way by means of the smoothness movement characteristic. Therefore, the aim of the present paper, an extension of [42], is to develop three serious games based on UL functional tasks, manipulated by the Novint Falcon device, and to collect quantitative data about SCI patients for analyzing movement smoothness. Moreover, this research involves the novelty of applying force feedback to SCI patients while analyzing the movement smoothness. In addition, electromyographic signals are obtained, integrating a Myo Gesture armband into the platform, which could be useful in the future for analyzing impairments in the muscles activation pattern.

The rest of the paper is organized as follows: Section 2 describes the methodology in relation to the integration of a commercial haptic device (Novint Falcon) and a commercial electromyography (EMG) sensor (Myo Gesture armband) into a virtual reality platform; the development of the serious games and the validation proof. Section 3 presents the results of the implemented proof of concept. Section 4 includes the discussion and Section 5 concludes the paper.

2. Materials and Methods

2.1. UL Rehabilitation Platform

The platform was composed of a PC, including the serious games developed in Unity3D, and two hardware devices: Novint Falcon for providing force feedback and manipulating the serious games; and Myo Gesture Arm Band for providing information about muscle contraction (see Figure 1).

Figure 1. Scheduled illustration of the upper limb rehabilitation platform: Myo Gesture armband (left object in image) and Novint Falcon device (right object in image).

2.1.1. Novint Falcon

The Novint Falcon (see Figure 1, right) is a desktop haptic robot device with 3 degrees of freedom (DOF). It provides haptic feedback where the users feel virtual objects as they virtually touch them [43]. The device consists of three motorized arms attached to an interchangeable end-effector providing kinesthetic feedback. It uses a USB 2.0 port, has a capacity of 9 N of force, and provides a $10 \times 10 \times 10$ cm^3 working space. The device has a refresh rate up to 1000 times per second [44].

In this paper, the Novint Falcon was selected because of its low cost, at least one order of magnitude, compared to similar devices, such as the Delta.3 [45] or Omega-3 [46], and because it can generate more force feedback than other low cost haptic devices [28]. It was used as an end-effector robotic device for upper limb rehabilitation. However, the Novint Falcon comes with a standard ball grip which is difficult to use by patients with SCI, as a fine pinch is necessary to perform the grasp [47]. The grip selected for SCI patients was of a cylinder shape for a better hand-grasp. This was a novelty in comparison to previous studies and it was designed and printed on a 3D printer.

In previous studies, tests were performed with an EMG recording system located in the main arm muscles [48]. The tests showed that using the Novint Falcon to give force feedback activates the Anterior Deltoid muscle, which flexes and medially rotates the arm, and the Extensor Carpi Ulnaris muscle, which extends and adducts the hand at the wrist joint. This was demonstrated by applying different degrees of haptic feedback in an exercise where participants would press down on a vibrating cube [48]. Therefore, the tests concluded that, by using the Novint Falcon, the shoulder and the forearm are activated.

2.1.2. Myo Gesture Arm Band

The Myo Gesture armband (see Figure 1, left) is a wearable gesture and motion control device working with EMG sensors to measure electrical pulses in the arm [49]. While muscles expand and contract, the armband sends signals wirelessly to other devices. It uses a Bluetooth 4.0 low energy connection, it has a proprietary muscle activity EMG sensor, a 9-axis inertial measurement unit (IMU), a rechargeable lithium-ion battery with one full day use out of single charge and haptic feedback which provides short, medium or long vibrations.

Deeping into the sensors, the device has 8 EMG sensors and an IMU, containing a three-axis gyroscope, a three-axis accelerometer and a three-axis magnetometer. Thanks to this 9-axis IMU, Myo Gesture armband senses the motion, orientation and rotation of the forearm, used to recognize hand and wrist movements.

In this paper, the device was used for obtaining a correct EMG monitoring during the performance of the rehabilitation games. By using this armband, the muscle contraction and the arm kinematics were stored for a posterior data analysis. Because of the Myo Gesture armband design, the 8 EMG sensors are located equally distributed in the perimeter of the forearm. The EMG data was streamed at 200 Hz by Bluetooth communication to the PC.

Moreover, the rehabilitation platform implemented provided the opportunity to visualize in real time the data obtained from the EMG sensors. This functionality allowed to measure the muscle strength that the patient performs during each game.

2.1.3. Communication Protocol

The communication between the virtual environment and the devices were implemented via sockets. The communication frequency between these devices was very important. Because the user needed to perceive continuity in the images, the frequency of 30 Hz was selected for the graphics. However, for haptic stability without oscillations a sampling rate greater than 500 Hz was mandatory. In this paper, the frequency selected was around 650 Hz.

2.2. *Serious Games Development*

The first step to create the rehabilitation platform was to implement the virtual end-effector. A software created the socket between the Unity 3D platform and the Novint Falcon device, enabling the bidirectional transmission and reception of information.

To study the movement of the virtual object, it was necessary to establish a relationship between the Cartesian coordinates of the end of the haptic device and the coordinates of the virtual object. This module updated every frame in the scene with the position of the end-effector. The axes reference of the Novint Falcon and the games are shown in Figure 2, where the x and y axes are horizontal and vertical movements on the screen respectively, and the z axis comes out of the screen.

(a) **(b)**

Figure 2. (**a**) Novint Falcon and (**b**) serious games reference axes.

Then, the serious games have been implemented with a therapeutic sense. To create a complete platform, different menus focused on the patients were developed. The configuration of the games attributes was essential to fit the games according to the needs of each patient in relation to the functional capabilities. Within each serious game proposed, patients had to reach the number of objectives selected by the therapist. To achieve this goal, the time of execution was restricted. If the patient did not achieve the objectives in the available time, the game would finish. To help patients in developing this task, the movement made with the end-effector was visualized in the virtual environment.

A summary of the main characteristics analyzed in each game is shown in Table 1. The behavior and characteristics of each serious game are explained in subsequent sections. These features were measured in terms of movement smoothness. Moreover, each serious game included elements of visual feedback, recommended to boost the rehabilitation. Therefore, each serious game sought to check if it was able to positively modify neural mechanisms and improve motor performance by tuning the control structure of a patient. This effect should be analyzed in a future work, where a clinical validation should be carried on. This effect has been mostly analyzed in stroke patients [50] but these feedback sources can be extended to patients who have suffered other neurological diseases. Nevertheless, more studies focused exclusively on SCI upper limb impairments and their recovery are needed to elucidate if there are any difference in the residual motor control after an SCI compared to post-stroke patients.

Table 1. Summary with the main characteristics of each game.

Game	Exercise Type	Force Type	Therapeutic Objective
Following the path Picking bananas Collecting stars	Path guidance Resistive Resistive	Hooke's law Viscosity Hooke's law	Accuracy Accuracy, Strength Accuracy, Strength

2.2.1. Following the Path

The end-effector in the scene was represented by a pencil which emulates a path drawn by the patient. In this game, there was not a selection of the number of the objectives. The path was the same in all configurations. Users had to travel inside the path without going out of the edges and when they got out of it, an opposed force was exerted. When the user completed a segment, it changed its color to a darker one (see Figure 3a) until the user completed all segments or the time was exhausted.

This game was implemented in 2D (x and y axes) because the depth of the scene could cause trajectory disturbance in patients. The range of motion was of 5 cm in x axis and of 7.5 cm in y axis. The path of the trajectory had a width of 0.5 cm.

Therapeutic Objective

This rehabilitation game was based on a path guidance exercise. The therapeutic objective was to improve the precision and smoothness in the UL movement made by the patient and to recover fine motor control. This training modality using robot-mediated therapy was a combination of assistive and active exercise. When the patient was moving the end-effector inside the path, the active mode was working and the patient felt no force. However, when the patient went out the path, the assistive modality started to push the patient's hand inside. Therefore, the haptic robot assisted by providing force feedback opposing to the movement of the patient.

In this training, the trajectory performed by the patient acquires relevant importance. For this reason, the trajectory made by the patient was collected, so the movement smoothness could be analyzed.

The force feedback ($\boldsymbol{F_V}$) was implemented by using Hooke's law (see Equation (1)).

$$\boldsymbol{F_V}\,(kT) = -k_V\,\Delta\boldsymbol{X}(kT), \tag{1}$$

where T was the sampling time, kT a specific instant in the discrete domain ($k \in [0,1,2,\ldots]$), $\Delta\boldsymbol{X}(kT)$ the distance to the edge of the path, and k_V the stiffness or elasticity simulated in the edge of the path. The force was implemented in the edges of the path, so patients felt these edges like walls that they should not cross.

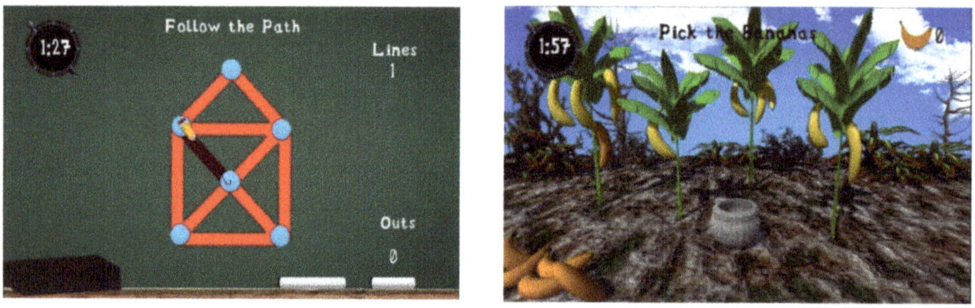

(a) (b)

Figure 3. (**a**) Serious game "Following the path"; (**b**) Serious game "Picking bananas".

2.2.2. Picking Bananas

The end-effector was a basket which must collect different bananas that were located in different trees (see Figure 3b). The patient moved into the environment with the basket and when a banana fell from the tree, he had to get it with the basket before it touched the floor.

The therapist selected the number of bananas in the scene and the time for the duration of the task. Then, the game selected the frequency of falling and it randomized the falls for every session. The game notified the user when one banana was going to fall by changing its color and producing a sound. If the banana fell into the basket, the score increased. If it did not fall into it, it simply disappeared. When all the bananas fell, the game ended and the final result screen was visualized.

The force feedback provided by the haptic device was proportional to the number of bananas inside the basket, as a concept of weight and friction. Depending on the user, a high number of bananas could be tiring. Therefore, users could empty the basket in the left corner of the screen when they passed by.

The trees were disposed in the floor using a distance of 6 cm. With this distribution, users had enough space to move the end-effector in the total range of motion offered by the haptic device.

Therapeutic Objective

The rehabilitation exercise implemented in this serious game was resistive, so the robot opposed to the movement that the patient performed. In this kind of exercises, the number of repetitions should be small to avoid fatigue. A rest between exercises should be sufficient for recovery, and the resistance should be modulated to increase it as the patient ability grows.

The force feedback implemented was a viscosity resistance. This force was proportional to the velocity and opposed to the end-effector movement (see Equation (2)). This force (F_V) should increase according to the number of objectives collected.

$$F_V(kT) = -b_V \dot{X}(kT), \tag{2}$$

where T was the sampling time, kT a specific instant in the discrete domain ($k \in [0,1,2,\ldots]$), $\dot{X}(kT)$ the velocity of the end-effector and b_V a constant referred to the viscosity. In this exercise, an isotonic muscle contraction was produced. This muscle contraction implied the shortening and lengthening of muscle fibers. Depending on the amount of force working against an individual's body, one of two kinds of isotonic contractions took place: concentric and eccentric contractions. Concentric contractions occur when muscles shorten while its tension is greater than the force opposing it. On the other hand, eccentric contractions occur when muscles extends in length [51].

2.2.3. Collecting Stars

The virtual visualization of the end-effector was a rocket and the objectives were stars which must be eliminated (see Figure 4). To destroy the stars, the patient moved the rocket to the star position. The game finished when the total number of objectives were achieved or the time was up. The stars had different sizes: the bigger ones were easier to catch than the others.

Figure 4. Collecting stars interface; one star has been collected.

A star was eliminated when the rocket collided with it. Once the collision was done, the star disappeared from the scene, the score was incremented and a sound was produced. After collecting a star, the user had to come back to the center of the screen before performing other destruction. To achieve this behavior, the collision detection module was disabled during this period.

This game was designed in 2D, because it focused on following a straight trajectory, rejecting the depth of the scene. Therefore, the allowed movements were in the y and x axes, but not in z axis.

The center of the stars was disposed in a circumference with a radius of 4.5 cm, centered in x and y equal to zero. Consequently, the range of motion of the haptic device was fully covered, allowing a range of motion of −5 cm to 5 cm in both axes.

Therapeutic Objective

The exercise implemented in this game was resistive. The robot provided force opposing the movement. Thus, its goal was to obtain the strengthening of the UL muscles. In addition, with this game the linearity of the movement could be checked. Patients with UL lesion usually tried to compensate movements performing curved trajectories when they tried to catch an objective which was in a straight direction.

In this case, there was a zone without force in the middle of the game circle, but when the patient was close to the targets, the force applied increased. The force began to act only in a radius of 2.5 cm from the objective. The force applied followed the Hooke's law (see Equation (1)). Therefore, the force scaled linearly with respect to the distance when the end-effector was entering into the force area.

As in the picking bananas game, the muscle contraction was isotonic, because the force depended on the movement and the position. A shortening and lengthening of the muscle fibers occurred during the game performance.

2.3. Participants

A total of 8 people participated in the study divided into two groups: a group of healthy people (n = 4) and a group of cervical SCI patients (n = 4). All patients fulfilled the following inclusion criteria: age 16 to 65 years, at least 6 months from the injury onset, and cervical injury of the metameric levels between C1 (first cervical vertebra) and C7 (7th cervical vertebra), neurologically classified according to the American Spinal Injury Association (ASIA) scale into grades C or D [52]. Patients who presented any vertebral deformity, joint constraint, surgery on any of the UL, balance disorders, dysmetria due to associated neurologic disorders, visual acuity defects, cognitive deficit, or head injury associated with the SCI were excluded. Background data of patients are provided in Table 2. The healthy subjects were chosen according to reach two groups of similar demographic characteristics (3 males of 19, 21 and 23 years old and a female of 44). The guidelines of the declaration of Helsinki were followed in every case. Informed consent was obtained from all individual participants included in the study, which was approved by the Local Ethics Committee, Toledo, Spain.

Table 2. Spinal Cord Injury (SCI) patients' characteristics.

Patient	Age	Sex	Injury Level	ASIA
1	19	M	C4	C
2	18	M	C4	C
3	45	F	C5	D
4	20	M	C6	C

2.4. Experimental Setup

The study was performed during two days, with two sessions per day. Therefore, the study was formed by data obtained from four different sessions. In all sessions, participants performed the test before starting with the serious games. This methodology was used to obtain clear results. In this way, participants had a previous knowledge of the games and the results obtained did not show the variability of the learning process.

The three games were performed in all sessions, always in the same order and with the same characteristics to avoid including variability into the study. First, the "Picking bananas" game was performed for 2 min and 15 objectives were selected. Second, the "Collecting stars" game was performed for a maximum of 2 min and 16 objectives were selected. Finally, the "Following the path" game was also performed for a maximum of 2 min. The maximum force selected in all games was at a value of 4.5 N, half of the maximum available force at the Novint Falcon. This parameter selection was chosen because in all tests participants showed skills to perform the task with this configuration.

All participants were right handed. The end-effector of the Novint Falcon was held with the right hand and the Myo Gesture armband placed in the same arm, specifically in the forearm, with the electrodes located on the flexor and extensor wrist muscles. The tasks were performed in the frontal plane of the patients (see Figure 5).

Figure 5. Example of a healthy trial with the Novint Falcon held on the right hand and the Myo Gesture armband on the right forearm.

2.5. Data Processing

In this study, the UL motor performance of each participant was measured in terms of movement smoothness, computing the peaks number from the velocity profile during the task execution proposed by the serious games.

A processing script was developed by using the software MATLAB (MathWorks®). The speed profile was obtained from the position data obtained by the haptic robot. These data were stored into text files during the performance of each serious game. The trajectories were analyzed in 2D.

The statistical non-parametric U Mann-Whitney test was applied to analyze the possible differences between both populations (healthy and patients groups) with a significance level of 5%. The results were expressed as mean and standard deviation for the following variables: mean and peak speed (m/s), number of peaks on the trajectory, and duration of the task.

3. Results

The results in relation to the motor performance in each variable within the serious games are shown in Tables 3 and 4.

Table 3. Results about duration, as well as peak and mean velocity in each serious game for both experimental groups (a, statistical differences $p < 0.05$).

Following the Path	Healthy Group (n = 4)	SCI Patients (n = 4)
Duration (s)	**28.55 (7.91)** [a]	**49.52 (13.33)** [a]
Peak velocity x (m/s)	0.255 (0.121)	0.406 (0.133)
Peak velocity y (m/s)	0.556 (0.203)	0.270 (0.125)
Mean velocity x (m/s)	0.063 (0.008)	0.058 (0.015)
Mean velocity y (m/s)	0.070 (0.018)	0.058 (0.020)
Picking bananas		
Duration (s)	120.00 (0)	120.00 (0)
Peak velocity x (m/s)	1.600 (0.441)	1.752 (0.591)
Peak velocity y (m/s)	0.799 (0.813)	0.818 (0.902)
Mean velocity x (m/s)	**0.060 (0.013)** [a]	**0.141 (0.065)** [a]
Mean velocity y (m/s)	**0.047 (0.013)** [a]	**0.091 (0.022)** [a]
Collecting stars		
Duration (s)	34.66 (7.91)	41.78 (15.83)
Peak velocity x (m/s)	1.232 (0.223)	0.878 (0.231)
Peak velocity y (m/s)	1.317 (0.387)	0.898 (0.145)
Mean velocity x (m/s)	0.160 (0.082)	0.140 (0.056)
Mean velocity y (m/s)	0.158 (0.054)	0.137 (0.054)

Table 4. Peaks number in each serious game for both experimental groups (a, statistical differences $p < 0.05$).

Peaks Number	Healthy Group (n = 4)	SCI Patients (n = 4)
Following the path	51.12 (11.53) [a]	81.32 (22.93) [a]
Picking bananas	154.75 (12.63)	164.72 (17.22)
Collecting stars	54.48 (10.25)	67.42 (24.58)

3.1. Following the Path

The time spent in the performance of the task was significantly different: healthy people spent on average 28.55 s and SCI patients 49.52 s. This indicates that patients needed to perform the movement slowly for avoiding going out of the edges. During the task execution, a counter quantified the number of times that a participant went out of the path. The analysis shows that healthy subjects went out of the path an average of 2.75 times and SCI patients 13.75 times. This fact increases the corrections number of the hand due to the force feedback applied. So, the number of peaks in the speed profile is statistically different between both groups.

3.2. Picking Bananas

In this serious game, all the participants spent 2 min. The average of the speed profile is slightly higher in SCI patients than in healthy subjects. This is because SCI patients made quick movements trying to catch the objectives.

The number of peaks obtained in the analysis shows that SCI patients perform more accelerations and decelerations in the speed profile. However, the number of peaks was very high in both groups, not showing statistical differences. Nonetheless, by observing the speed profile we observe that the speed profile of SCI group is more abrupt than the healthy participants. An example of average speed profile on axis x during the execution of the Picking bananas game for one healthy and one SCI participant is shown in Figure 6. This profile shows a maximum speed peak of 1.3 m/s for the healthy participant and 1.9 m/s for the SCI one. Two peaks higher than 0.5 m/s can be observed in the healthy subject, while this number rises up to fourteen for the SCI participant within the same time frame. Similar results are observed in all axes for all games and participants. However, a larger clinical study should be carried on to better analyze those differences and obtain statistical differences.

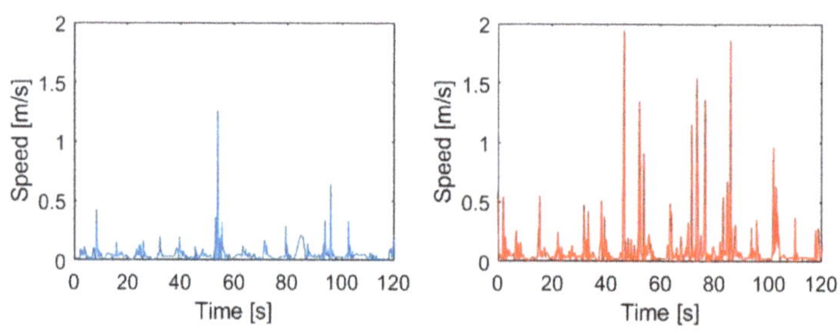

Figure 6. Example of a Healthy (blue) and SCI (red) participant speed profile of x axis measured from the end-effector of Novint Falcon device in the Picking bananas game.

3.3. Collecting Stars

During the execution of this serious game, SCI patients spent more time for reaching the 16 objectives (41.78 s) than healthy subjects (34.66 s). The duration of the exercise could affect the smoothness of the movement. This was because carrying out the task more slowly produced segmentation of the movement and more peaks appeared in the speed profile. In addition, SCI patients performed longer trajectories with a higher number of peaks than the healthy subjects (see Figure 7 for an example). This longer trajectories were due to the difficulty of accurately reaching the objectives, both the starts and the center position. For the examples shown in Figure 7, it can be observed how the healthy performed more accurately trajectories both on the center and the start positions. It can also be observed that just one approximation, or two at maximum in some points, was performed to reach every star. On the contrary, the SCI participant performed multiple passes through the same point. Some stars, especially shown at the left-bottom part of the image, needed more than five approximations to be reached. Moreover, the lines accumulation on the center of the image is clearly stronger in the SCI patient plot.

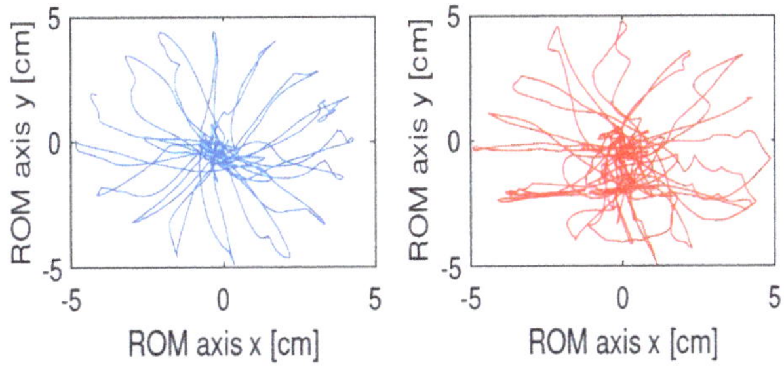

Figure 7. Example of a Healthy (blue) and SCI (red) participant range of motion (ROM) measured from the end-effector of Novint Falcon device in the Collecting stars game.

3.4. Electromyography Recording

An example of the activation pattern in flexors and extensors muscles measured from the Myo Gesture armband placed on the forearm of a healthy subject and an SCI patient during the execution of the serious games is shown in Figure 8. These data were analyzed taking into account the potential values obtained from the Myo Gesture armband. The EMG data provided by the Myo Gesture armband was unitless, 8 bit value. Hence, it provided a potential range between 0 and 255. To represent the muscle contraction, the value of the data collected from the sensor placed in the muscles of interest had been calculated. In all trials, SCI patients showed a stronger activation than healthy people. However, for all users, the muscle contraction was low, obtaining an average of 38.13 in "Following the path" game, 39.64 in "Picking bananas" games, and, finally, 53.35 in "Collecting stars".

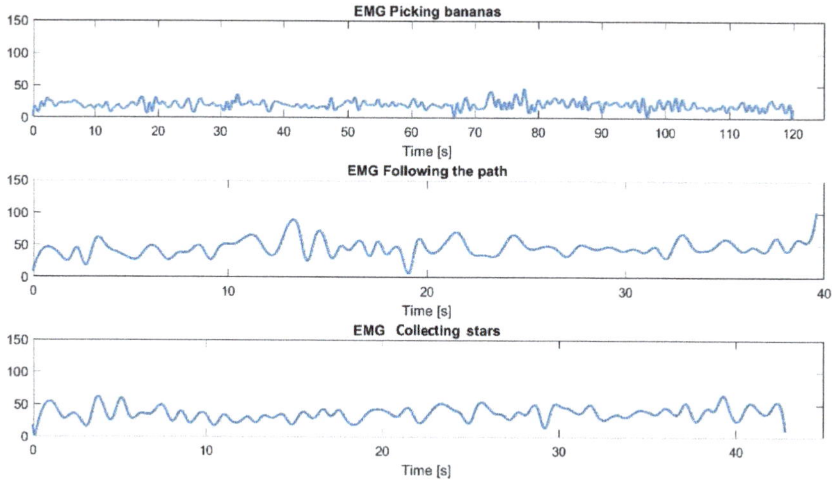

(a)

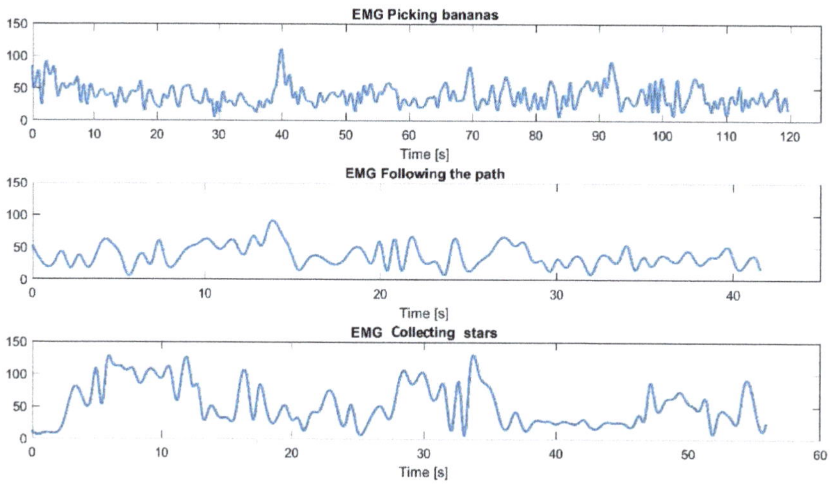

(b)

Figure 8. Activation pattern in the flexors and extensors muscles from Novint Falcon device placed on the forearm for the three serious games for (**a**) a healthy subject and (**b**) an SCI patient. The information provided by the Myo Gesture armband is unitless, with a potential value in the range of [0,255]. EMG = electromyography.

4. Discussion

In the present study, all virtual environments developed focus on reaching UL functional tasks. For the performance, they require a combination of analytical movements in each involved joint. The force feedback delivered by means of Novint Falcon provides sensitive information to the patient that modifies the kinematic patterns performed and, as a consequence, the kinematic variables measured.

An important contribution in this research with respect to previous studies is the integration of the Novint Falcon device and the Myo Gesture armband into a unique platform, obtaining EMG signals from flexors and extensors muscles of the wrist joint. The monitoring of EMG signals could be useful in the future for analyzing impairments in the muscles activation pattern during the serious games performance. The EMG values obtained are low, far from the maximum value provided by the Myo Gesture armband. This effect occurs because the load of the therapy proposed is also moderate. Therefore, the muscle activity that patients need to overcome the force feedback is moderate. It can be appreciated that, in the "Collecting star" game, where the force feedback applied is higher, the muscle contraction is also higher. To represent the muscle contraction, the value of the data collected has been calculated. For that reason, it is necessary to validate this EMG signal with those activation patterns measured from a clinical EMG system. In previous studies, differences in EMG patterns between healthy subjects and patients with cervical SCI have been reported to match with the differences in kinematic results [53]. Therefore, the EMG signal is essential for monitoring the muscle contraction during the games performance. In a posterior analysis, these data could be used to check the changes in the patterns of motor activity during the rehabilitation process. The data provided by the EMG could also be useful for controlling the appearance of peripheral fatigue [54]. Moreover, its supervision is important to avoid damaging the patient's muscle with an overload of work.

In relation to the content of the games, the "Collecting stars" game has been developed similar to those proposed by Scalona et al., where they provide 8 objectives to reach with different levels of force feedback [29]. In our proposal, the game involves 16 objectives around the circumference: four in the vertical and horizontal line, and the other 12 in the diagonal lines. The other two serious games were oriented to complete functional tasks, like the commercial games included in Novint Falcon and used by Chortis et al. [33]: one of them, the "Picking bananas" game, to pick up and carry objects; and the other one, the "Following the path" game, to follow a hand trajectory.

Nonetheless, the lack of movement smoothness has already been demonstrated in presence of neurological pathologies [16]. It is measured at the distal segment and shows as a consequence longer hand trajectories with a higher number of directional changes. This characteristic provides objective quantitative information of UL motor function and control. The ideal condition when moving the hand between pairs of targets is to generate roughly straight hand trajectories with a single-peaked, bell-shaped speed profile [15]. This behavior was proved to be independent of the part of the work-space in which the movement was performed. Moreover, the results obtained were strong indicators that movement planning takes place in terms of UL distal segment, by means of hand trajectories rather than joint rotations [55,56]. Nevertheless, smoothness metric has shown to be discriminative between healthy and neurological patients when smoothness is computed from the peaks number in the speed profile [20,21]. Therefore, in this paper, smoothness has been measured by computing the peaks number within the speed profile. Results from the proof of concept show that this behavior was maintained, and a higher peaks number was obtained in SCI patients than in healthy people for the three serious games. However, statistical differences were only obtained in the "Following the path" game.

5. Conclusions

The use of new technologies in the rehabilitation field has increased in the last years. Robot aided and VR rehabilitation provide higher motivational environments to the patient, facilitate the performance of the necessary movement repetitions for inducing the UL motor learning and increase the amount of therapy that patients receive. However, these technologies are expensive, and a need for low-cost devices exists. This work takes advantage of low-cost VR systems and robotic haptic devices to enhance involvement and engagement of patients to provide a congruent multisensory afferent feedback during motor exercises and benefit from the flexibility of virtual scenarios.

The low cost platform developed in the present work enables the UL rehabilitation with haptic feedback and EMG monitoring for SCI patients. First of all, three serious games have been designed and implemented to compute within them the smoothness movement. The design and implementation of each serious game required the definition of a unique interface and physical models related to haptic feedback. In all of them, the movement smoothness has been reported lower in patients than in healthy people, showing an inverse relation with the peaks number obtained. As a conclusion, the three serious games proposed are suitable for measuring movement smoothness, but the most suitable game for discriminating between healthy and SCI patients is the "Following the path" game. This is because, in this game, the order to reach the different nodes has been imposed, being always the same and allowing an identical comparison between participants. Nonetheless, these assumptions should be confirmed with a large clinical validation.

A proof of concept to check the functionality of the platform with SCI patients was made. Results obtained in terms of movement smoothness show differences in the movement patterns. Some benefits drawn from this research are exposed as follows:

- All the devices of the platform are low-cost and easily adaptable for different people, with a short time spent in preparation for the serious games performance.
- The configuration options used in this research allow to create different virtual scenarios, in order to personalize each patient rehabilitation.
- The final design generates a good haptic response and reliable monitoring of the muscle contractions.

Author Contributions: Conceptualization, Á.G., D.S.-M. and A.d.l.R.G.; methodology, Á.G., D.S.-M., A.d.l.R.G.; software, D.S.-M.; validation, D.S.-M., A.d.l.R.G., and Á.G.-A.; formal analysis, A.d.l.R.G., Á.G.; investigation, D.S.-M., A.d.l.R.G.; resources, Á.G. and Á.G.-A.; data curation, D.S.-M. and A.d.l.R.G.; writing—original draft preparation, Á.G., D.S.-M. and A.d.l.R.G.; writing—review and editing, Á.G., A.d.l.R.G. and Á.G.-A.; visualization, Á.G., A.d.l.R.G.; supervision, Á.G., A.d.l.R.G.; project administration, Á.G.-A. and A.d.l.R.G.; funding acquisition, Á.G. and A.d.l.R.G. All authors have read and agreed to the published version of the manuscript.

Funding: This research has been funded by grant from the Spanish Ministry of Economy and Competitivity and co-funded from FEDER, National Plan for Scientific and Technological Research and Innovation. Project RehabHand (Plataforma de bajo coste para rehabilitación del miembro superior basado en Realidad Virtual, ref. DPI2016-77167-R).

Acknowledgments: This work was supported by the National Hospital for Paraplegics in Toledo (Spain) and the Universidad Politécnica de Madrid (Spain). The authors acknowledge the collaboration of the participants in this study.

Conflicts of Interest: The authors declare no conflict of interest.

References

1. Chen, Y.; Tang, Y.; Vogel, L.C.; Devivo, M.J. Causes of spinal cord injury. *Top. Spinal Cord Inj. Rehabil.* **2013**, *1*, 1–8. [CrossRef]
2. World Health Organization. Spinal Cord Injury. Available online: http://www.who.int/news-room/fact-sheets/detail/spinal-cord-injury (accessed on 1 August 2019).
3. Humphreys, J.S. Key considerations in delivering appropriate and accessible health care for rural and remote populations: Discussant overview. *Aust. J. Rural Health* **2009**, *17*, 34–38. [CrossRef]
4. Wyndaele, M.; Wyndaele, J.J. Incidence, prevalence and epidemiology of spinal cord injury: What learns a worldwide literature survey? *Spinal Cord* **2006**, *44*, 523–529. [CrossRef] [PubMed]
5. DeVivo, M.; Biering-Sorensen, F.; Charlifue, S.; Noonan, V.; Post, M.; Stripling, T.; Wing, P. International Spinal Cord Injury Core Data Set. *Spinal Cord* **2006**, *44*, 535–540. [CrossRef] [PubMed]
6. Pérez, K.; Novoa, A.M.; Santamariña-Rubio, E.; Narvaez, Y.; Arrufat, V.; Borrell, C.; Cabeza, E.; Cirera, E.; Ferrando, J.; García-Altes, A.; et al. Incidence trends of traumatic spinal cord injury and traumatic brain injury in Spain. *Accid. Anal. Prev.* **2012**, *46*, 37–44. [CrossRef] [PubMed]
7. Goel, S.; Modi, H.; Dave, B.; Patel, P. Socio-Economic Impact of Cervical Spinal Cord Injury Operated in Patients with Lower Income Group. *Glob. Spine J.* **2016**, *6*, 264. [CrossRef]
8. Kemal, N.; Levent, Y.; Volkan, S.; Abdulkadir, A.; Kadriye, O. Rehabilitation of spinal cord injuries. *World J. Orthop.* **2015**, *6*, 8–16.
9. Oldrige, N.B. Outcomes Measurement: Health-Related Quality of Life. *Assist. Technol.* **1996**, *8*, 82–93. [CrossRef]

10. Steel, E.J. Content analysis to locate assistive technology in Queensland's motor injury insurance rehabilitation legislation and guidelines. *Assist. Technol.* **2018**, *13*, 1–5. [CrossRef]
11. Murphy, M.A.; Sunnerhagen, K.S.; Johnels, B.; Willen, C. Three-dimensional kinematic motion analysis of a daily activity drinking from a glass: A pilot study. *J. Neuroeng. Rehabil.* **2006**, *3*, 18. [CrossRef]
12. Lu, X.; M Zoghi, C.R.; Galea, M.P. Effects of training on upper limb function after cervical spinal cord injury: A systematic review. *Clin. Rehabil.* **2015**, *29*, 3–13. [CrossRef]
13. Rohrer, B.; Fasoli, S.; Krebs, H.I.; Hughes, R.; Volpe, B.; Frontera, W.R.; Hogan, N. Movement smoothness changes during stroke recovery. *J. Neurosci.* **2002**, *22*, 8297–8304. [CrossRef]
14. Bartolo, M.; De Nunzio, A.M.; Sebastiano, F.; Spicciato, F.; Tortola, P.; Nilsson, J.; Pierelli, F. Arm weight support training improves functional motor outcome and movement smoothness after stroke. *Funct. Neurol.* **2014**, *29*, 15. [CrossRef]
15. Morasso, P. Spatial control of arm movements. *Exp. Brain Res.* **1981**, *42*, 223–227. [CrossRef]
16. Krebs, H.I.; Aisen, M.L.; Volpe, B.T.; Hogan, N. Quantization of continuous arm movements in humans with brain injury. *Proc. Natl. Acad. Sci. USA* **1999**, *96*, 4645–4649. [CrossRef]
17. Platz, T.; Denzler, P.; Kaden, B.; Mauritz, K.-H. Motor learning after recovery from hemiparesis. *Neuropsychologia* **1994**, *32*, 1209–1223. [CrossRef]
18. Kahn, L.E.; Zygman, M.L.; Rymer, W.Z.; Reinkensmeyer, D.J. Robot-assisted reaching exercise promotes arm movement recovery in chronic hemiparetic stroke: A randomized controlled pilot study. *J. Neuroeng. Rehabil.* **2006**, *3*, 12. [CrossRef]
19. Casadio, M.; Pressman, A.; Fishbach, A.; Danziger, Z.; Acosta, S.; Chen, D.; Tseng, H.-Y.; Mussa-Ivaldi, F.A. Functional reorganization of upper-body movement after spinal cord injury. *Exp. Brain Res.* **2010**, *207*, 233–247. [CrossRef]
20. De los Reyes-Guzmán, A.; Dimbwadyo-Terrer, I.; Pérez-Nombela, S.; Monasterio-Huelin, F.; Torricelli, D.; Pons, J.L.; Gil-Agudo, A. Novel kinematic indices for quantifying upper limb ability and dexterity after cervical spinal cord injury. *Med. Biol. Eng. Comput.* **2017**, *55*, 833–844. [CrossRef]
21. De los Reyes-Guzman, A.; Dimbwadyo-Terrer, I.; Pérez-Nombela, S.; Monasterio-Huelin, F.; Torricelli, D.; Pons, J.L.; Gil-Agudo, A. Novel kinematic indices for quantifying movement agility and smoothness after cervical Spinal Cord Injury. *NeuroRehabilitation* **2016**, *38*, 199–209. [CrossRef]
22. Hoffman, H.G.; Patterson, D.R.; Carrougher, G. Use of Virtual Reality for Adjunctive Treatment of Adult Burn Pain During Physical Therapy: A Controlled Study. *Clin. J. Pain* **2000**, *16*, 244–250. [CrossRef] [PubMed]
23. Kalsi-Ryan, S.; Curt, A.; Verrier, M.C.; Fehlings, M.G. Development of the Graded Redefined Assessment of Strength, Sensibility and Prehension (GRASSP): Reviewing measurement specific to the upper limb in tetraplegia. *J. Neurosurg. Spine* **2012**, *17*, 65–76. [CrossRef] [PubMed]
24. Lohse, K.R.; Hilderman, C.G.E.; Cheung, K.L.; Tatla, S.; Van Der Loos, H.F.M. Virtual reality therapy for adults poststroke: A systematic review and meta-analysis exploring virtual environments and commercial games in therapy. *PLoS ONE* **2014**, 3, e93318.
25. Ding, Y.; Kastin, A.J.; Pan, W. Neural plasticity after spinal cord injury. *Curr. Pharm. Des.* **2005**, *11*, 41–50. [CrossRef] [PubMed]
26. Kaas, J.H. *International Encyclopedia of the Social & Behavioral Sciences*; Elsevier: Amsterdam, The Netherlands, 2001.
27. Yozbatiran, N.; Francisco, G.E. Robot-assisted Therapy for the Upper Limb after Cervical Spinal Cord Injury. *Phys. Med. Rehabil. Clin.* **2019**, *30*, 367–384. [CrossRef]
28. Cappa, P.; Clerico, A.; Nov, O.; Porfiri, M. Can Force Feedback and Science Learning Enhance the Effectiveness of Neuro-Rehabilitation? An Experimental Study on Using a Low-Cost 3D Joystick and a Virtual Visit to a Zoo. *PLoS ONE* **2013**, *8*, e83945. [CrossRef]
29. Scalona, E.; Hayes, D.; del Prete, Z.; Palermo, E.; Rossi, S. Perturbed Point-to-Point Reaching Tasks in a 3D Environment Using a Portable Haptic Device. *Electronics* **2019**, *8*, 32. [CrossRef]
30. Palsbo, S.E.; Marr, D.; Streng, T.; Bay, B.K.; Norblad, A.W. Towards a modified consumer haptic device for robotic-assisted fine-motor repetitive motion training. *Disabil. Rehabil. Assist. Technol.* **2011**, *6*, 546–551. [CrossRef]
31. Andaluz, V.H.; Salazar, P.J.; Silva, S.M.; Escudero, V.M.; Bustamante, D.C. Rehabilitation of upper limb with force feedback. In Proceedings of the 2016 IEEE International Conference on Automatica (ICA-ACCA), Curico, Chile, 19–21 October 2016; pp. 1–6.

32. Yeh, S.C.; Lee, S.H.; Chan, R.C.; Wu, Y.; Zheng, L.R.; Flynn, S. The efficacy of a haptic-enhanced virtual reality system for precision grasp acquisition in stroke rehabilitation. *J. Healthc. Eng.* **2017**, *2017*, 9840273. [CrossRef]
33. Chortis, A.; Standen, P.J.; Walker, M. Virtual reality system for upper extremity rehabilitation of chronic stroke patients living in the community. In Proceedings of the 7th ICDVRAT with ArtAbilitation, Maia, Portugal, 8–11 September 2008; pp. 221–228.
34. Alcover, E.A.; Jaume-i-Capó, A.; Moya-Alcover, B. PROGame: A process framework for serious game development for motor rehabilitation therapy. *PLoS ONE* **2018**, *13*, e0197383.
35. Pirovano, M.; Surer, E.; Mainetti, R.; Lanzi, P.L.; Borghese, N.A. Exergaming and rehabilitation: A methodology for the design of effective and safe therapeutic exergames. *Entertain. Comput.* **2016**, *14*, 55–65. [CrossRef]
36. Dimbwadyo-Terrer, I.; Trincado-Alonso, F.; los Reyes-Guzmán, A.D.; López-Monteagudo, P.; Polonio-López, B.; Gil-Agudo, A. Activities of daily living assessment in spinal cord injury using the virtual reality system toyra: Functional and kinematic correlations. *Virtual Real.* **2016**, *20*, 17–26. [CrossRef]
37. Eckert, M.; Gómez-Martinho, I.; Meneses, J.; Martínez, J.F. New approaches to exciting exergame-experiences for people with motor function impairments. *Sensors* **2017**, *17*, 354. [CrossRef] [PubMed]
38. Bart, O.; Agam, T.; Weiss, P.L.; Kizony, R. Using videocapture virtual reality for children with acquired brain injury. *Disabil. Rehabil.* **2011**, *33*, 1579–1586. [CrossRef]
39. Dimbwadyo-Terrer, I.; Gil-Agudo, A.; Segura-Fragoso, A.; de los Reyes-Guzman, A.; Trincado-Alonso, F.; Piazza, S.; Polonio-López, B. Effectiveness of the virtual reality system toyra on upper limb function in people with tetraplegia: A pilot randomized clinical trial. *Biomed. Res. Int.* **2016**, *2016*, 6397828. [CrossRef]
40. Omelina, L.; Jansen, B.; Bonnechère, B.; Van Sint Jan, S.; Cornelis, J. Serious games for physical rehabilitation: Designing highly configurable and adaptable games. In Proceedings of the International Conference on Disability, Virtual Reality and Associated Technologies, Laval, France, 10–12 September 2012; The University of Reading: Reading, UK, 2012; pp. 195–202.
41. Graafland, M.; Dankbaar, M.; Mert, A.; Lagro, J.; Wit-Zuurendonk, L.D.; Schuit, S.; Schaafstal, A.; Schijven, M. How to systematically assess serious games applied to health care. *JMIR Serious Games* **2014**, *2*, e11. [CrossRef]
42. Sepúlveda-Muñoz, D.; de los Reyes Guzmán, A.; Gil-Agudo, A.; Gutiérrez, A. Design and implementation of a Virtual Reality platform for Upper Limb rehabilitation. In Proceedings of the XXXVI Congreso Anual de la Sociedad Española de Ingeniería Biomèdica, Ciudad Real, Spain, 21–23 November 2018; Ediciones Visilab: Ciudad Real, Spain, 2018; pp. 203–206.
43. Virtual Reality Society. The Novint Falcon Haptic System—Virtual Reality Society. 2019. Available online: http://www.vrs.org.uk/virtual-reality-gear/haptic/novint-falcon.html (accessed on 1 August 2019).
44. Novint Falcon. Wide Variety of Games. 2019. Available online: http://www.robotshop.com/media/files/PDF/datasheet-nf1-s01.pdf (accessed on 1 August 2019).
45. Delta.3. Force Dimension. 2019. Available online: https://www.forcedimension.com/downloads/specs/specsheet-delta.3.pdf (accessed on 3 December 2019).
46. Omega.3. Force Dimension. 2019. Available online: https://www.forcedimension.com/downloads/specs/specsheet-omega.3.pdf (accessed on 3 December 2019).
47. Curtin, M. An Analysis of Tetraplegic Hand Grips. *Br. J. Occup. Ther.* **1999**, *62*, 444–450. [CrossRef]
48. Nagaraj, S.B.; Constantinescu, D. Effect of Haptic Force Feedback on Upper Limb. In Proceedings of the Second International Conference on Emerging Trends in Engineering & Technology, Nagpur, India, 16–18 December 2009; pp. 55–58.
49. ThalmicLabs. Myo Gesture Armband Product Specs—Welcome to Myo Support. 2019. Available online: http://support.getmyo.com/hc/en-us/articles/202648103-Myo-armband-product-specs (accessed on 1 August 2019).
50. Urra, O.; Casals, A.; Jané, R. The impact of visual feedback on the motor control of the upper-limb. In Proceedings of the 2015 37th Annual International Conference of the IEEE Engineering in Medicine and Biology Society (EMBC), Milan, Italy, 25–29 August 2015; pp. 3945–3948.
51. Muscle Physiology. Muscle Physiology—Types of Contractions. 2006. Available online: http://muscle.ucsd.edu/musintro/contractions.shtml (accessed on 1 August 2019).
52. Kirsblum, S.C.; Burns, S.P.; Biering-Sorensen, F.; Donovan, W.; Graves, D.E.; Jha, A. International standards for neurological classification of spinal cord injury (revised 2011). *J. Spinal Cord Med.* **2011**, *34*, 535–546. [CrossRef]

53. De los Reyes-Guzmán, A.; López-Dolado, E.; Lozano-Berrio, V.; Pérez-Nombela, S.; Torricelli, D.; Pons, J.L.; Gil-Agudo, A. Upper Limb Electromyographic Analysis Synchronized with Kinematics in Cervical Spinal Cord Injured Patients during the Activity of Daily Living of Drinking. *JSM Phys. Med. Rehabil.* **2017**, *1*, 1004.
54. Cifrek, M.; Medved, V.; Tonkovic´, S.; Ostojic´, S. Surface EMG based muscle fatigue evaluation in biomechanics. *Clin. Biomech.* **2009**, *24*, 327–340. [CrossRef]
55. Flash, T.; Hogan, N. The coordination of arm movements: An experimentally confirmed mathematical model. *J. Neurosci.* **1985**, *5*, 1688–1703. [CrossRef] [PubMed]
56. Krebs, H.I.; Ferraro, M.; Buerger, S.P.; Newbery, M.J.; Makiyama, A.; Sandmann, M.; Lynch, D.; Volpe, B.T.; Hogan, N. Rehabilitation robotics: Pilot trial of a spatial extension for MIT-Manus. *J. Neuroeng. Rehabil.* **2004**, *1*, 5. [CrossRef] [PubMed]

applied sciences

MDPI

Article

Improving the Efficiency of Robot-Mediated Rehabilitation by Using a New Organizational Model: An Observational Feasibility Study in an Italian Rehabilitation Center

Irene Aprile [1], Cristiano Pecchioli [1], Simona Loreti [1], Arianna Cruciani [1], Luca Padua [2,3] and Marco Germanotta [4,*]

1 IRCCS Fondazione Don Carlo Gnocchi, 20121 Milan, Italy; iaprile@dongnocchi.it (I.A.); cpecchioli@dongnocchi.it (C.P.); sim.lor@tiscali.it (S.L.); ariannacrux@gmail.com (A.C.)
2 Department of Geriatrics, Neurosciences and Orthopaedics, Università Cattolica del Sacro Cuore, 00168 Rome, Italy; luca.padua@unicatt.it
3 Fondazione Policlinico Universitario Agostino Gemelli IRCSS, 00168 Rome, Italy
4 IRCCS Fondazione Don Carlo Gnocchi, 50143 Florence, Italy
* Correspondence: mgermanotta@dongnocchi.it; Tel.: +39-06-3308-6448

Received: 28 October 2019; Accepted: 4 December 2019; Published: 8 December 2019

Featured Application: By using new organizational models, robotic rehabilitation can be sustainable; therefore, it is possible to provide high technological upper limb treatment to a high number of patients.

Abstract: Robotic rehabilitation is a promising way to restore upper limb function, but new organizational models are crucial in order to improve its sustainability. We aimed to analyze the feasibility of a robot-Assisted REhabilitation Area (robotAREA) equipped with four devices, using a new organizational model in which one physiotherapist supervises up to four patients. In 33 days, 60 patients were treated in the robotAREA. Two hundred fifty-five rehabilitation sessions were analyzed, each including two, three, or four patients supervised by one physiotherapist (the R2, R3, and R4 group, respectively). For each session, the duration of (a) setup, (b) training, (c) supervision, (d) session closure, and participant satisfaction were measured. For each patient, upper limb impairment, disability, mobility, and comorbidity were recorded. The time of training in the R4 group was lower by only 2 and 3 min compared to the R2 and R3 groups, respectively, while the supervision significantly changed between the R2, R3, and R4 groups. No differences were observed in satisfaction. Moreover, according to the logistic regression analysis, when the mean disability of the group is moderate, four patients can be supervised by one physiotherapist. These results suggest that the proposed model is feasible, and the mean disability of the group could influence the proper physiotherapist/patient ratio.

Keywords: rehabilitation; robotics; technological devices; upper limb impairment; organizational model

1. Introduction

Upper limb impairment is defined by the International Classification of Functioning, Disability and Health model [1] as (1) impairments of body function, such as a significant deviation or loss in neuromusculoskeletal and movement-related function related to joint mobility, muscle power, muscle tone, and/or involuntary movements, or (2) impairment of body structures, such as a significant deviation in structure of the nervous system or structures related to movement.

Robot-mediated therapy after upper limb impairment, because of neurological or orthopedic diseases, is emerging in the rehabilitation field. A recent Cochrane review highlighted a potential benefit of the electromechanical and robot-assisted upper limb training in stroke patients [2]. In addition, robotic therapy can increase the efficiency of therapy, and therefore the accessibility to treatment [3]. In fact, in an era of increasing focus on healthcare resource utilization, a considerable interest is devoted toward reducing costs while preserving a high-quality level. From this perspective, economic evaluations of rehabilitation interventions using cost-effectiveness and/or cost-utility analysis are needed [4]. However, few studies have been conducted on the real costs of robotic and technological rehabilitation when compared to traditional rehabilitation. Wagner et al. [5] found that the costs of an additional upper limb treatment (robotic or conventional) in stroke patients can be recuperated because of lower general healthcare costs. Moreover, a recent systematic review, including five studies, found that robotic therapy produces a better economic outcome when compared to conventional therapy [6].

In addition, Masiero et al. [7] suggested that the availability of a room equipped with more than one device can further increase the sustainability of the treatment, addressing an essential need of under-resourced healthcare systems [8]. In fact, although robotic devices are expensive, new organizational models where one physiotherapist supervises more than one patient could optimize the human resources. Some authors [9–11] suggested that this approach can be as effective as traditional therapy for stroke patients, presenting a more cost- and labor-efficient option. However, to the best of our knowledge, there have been no previous studies that analyzed either the rehabilitation dose provided to the patients or patient's satisfaction when different physiotherapist/patient ratios have been used. Moreover, the clinical characteristics of patients that could influence the proper physiotherapist/patient ratio remain uninvestigated.

In April 2016, to implement the robotic rehabilitation in a sustainable way, a set of four technological and robotic devices for the upper limb was identified, and a robotic rehabilitation area was built and equipped with the set [12]. This allowed experimenting with a new organizational model where more than one patient is supervised by one physiotherapist, in order to increase the efficiency of the robotic treatment.

Then, to critically evaluate the feasibility of this model, a study based on data related to the use of these devices in clinical practice was carried out.

Specifically, the aim of this study was twofold: (1) to verify the feasibility, in terms of rehabilitation dose (as measured by the time of training) and patient satisfaction (as measured by the Visual Analogue Scale for Satisfaction), of a new organizational model in which one physiotherapist supervises up to four patients using a set of four robotic-technological devices; and (2) to identify the optimal number of patients that can be supervised by one physiotherapist, in accordance to the patients' upper limb impairment (measured by the Motricity Index), disability (measured by the Barthel Index), mobility (measured by the Deambulation Index), and comorbidity (measured by the Cumulative Illness Rating Scale).

2. Materials and Methods

2.1. Study Design

An observational feasibility study examining the daily activity of the robot-Assisted REhabilitation Area (robotAREA) of a rehabilitation center in Italy.

2.2. Ethics

The study was carried out in compliance with the Helsinki Declaration. All the patients (both inpatients and outpatients) provided signed written informed consent to use collected data for clinical research purposes at the admission in the rehabilitation center. This study was approved by the Ethics and Experimental Research Committee of the "IRCCS Fondazione Don Carlo Gnocchi" (Prot. N.8/2016CE_FDG/FC/SA) on 6 April 2016.

2.3. Setting

The study was conducted over a period of 33 days (from February to March 2017). The RobotAREA is equipped with a set of four devices, three robots (MOTORE, Humanware, Italy, and Amadeo and Diego, Tyromotion, Austria), and one sensor-based device (Pablo, Tyromotion, Austria) [12]. MOTORE is a robotic device that allows passive, active, and active-assistive planar movements of the shoulder and elbow joints; Amadeo is a robotic device that allows passive, active, and active assistive flexion and extension finger movements (also separately for each finger); Diego is a robotic device that allows three-dimensional, unimanual, and bimanual movements of the shoulder joint, both in sitting and in standing, with gravity compensation; Pablo is a sensor-based device that requires patients to perform unassisted three-dimensional movements of the shoulder, elbow, and wrist joints, both unimanual and bimanual, both when sitting and standing. All physiotherapists who worked in the RobotAREA had an expertise in robotic rehabilitation and used these devices for more than one year. Figure 1 shows the distribution of the device stations inside the robotic-technological area.

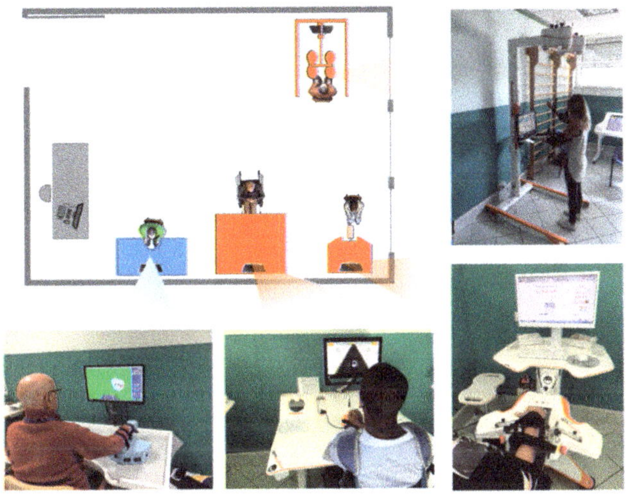

Figure 1. Distribution of the devices in the robot-Assisted REhabilitation Area (robotAREA). Beginning by the door, clockwise: Diego; Amadeo; Pablo; MOTORE.

2.4. Participants

Inclusion criteria: both inpatients and outpatients admitted to our rehabilitation facility with an upper limb impairment due to neurologic or orthopedic disorders. We excluded patients with a contraindication to robotic therapy, which was identified as: (1) inability to understand the physiotherapist's instructions and device exercises; (2) fixed contraction deformity/excessive spasticity; (3) any psychiatric disorder or severe cognitive disturbance that could reduce treatment compliance; and/or (4) severe visual impairment.

2.5. Robotic Rehabilitation Treatment

Each RS lasted 45 min, and the training frequency was 3 or 5 times/week, depending on the rehabilitation setting (outpatients or inpatients, respectively). During the treatment, both the distal and the proximal segments of the patients' upper arm were treated by means of the robotic and sensor-based devices. Motor and cognitive tasks were performed during the treatment, while visual and auditory feedback was provided to help the patients. During each session, the physical therapist uses one system for each patient to minimize the time required to move the patients from one system to another. The rehabilitation program starts with the robotic device for the shoulder and elbow joints, followed by the robotic device for the hand, the sensor-based device for the shoulder, elbow, and wrist and, finally, the robotic system for the shoulder. The adopted protocol followed the general indications, which were laid out in a flow chart, in order to ensure the homogeneity of treatment; however, for each patient, the physiotherapist selected the device and adapted the exercises to his/her residual abilities. More details on the rehabilitation protocol were previously reported [13]. During the rehabilitation session (RS), one physiotherapist supervised more patients at once with a physiotherapist/patient ratio of 1/2 (R2), 1/3 (R3), or 1/4 (R4). The number of patients treated in each RS was chosen by the physician depending on the upper limb impairment, disability, mobility, and comorbidity of the entire group, based on the experience done with a wide multicenter study [13].

2.6. Data Collection

A trained researcher monitored the patients during each robotic-technological rehabilitation session (RS). The data collected during each session for each patient was called "observation". During the study, for each patient, several observations were performed, according to the individual rehabilitation program. In order to limit bias due to researcher's presence, neither physiotherapists nor patients knew when data collection would take place.

To investigate whether and in which patients this new organizational model was applicable, we compared the different rehabilitation settings (R2, R3, and R4) in terms of the rehabilitation dose and clinical characteristics of the patients. For each session, the number of patients supervised by one physiotherapist was recorded, together with the clinical status of the patients, using validated clinical scales administered by trained operators. Specifically, we evaluated disability (according to the Barthel Index, BI [14]), upper limb impairment (using the Motricity Index, MI [15]), mobility (according to the Deambulation Index, DI [16]), comorbidities and severity (described according to the Cumulative Illness Rating Scale-Comorbidity Index, CIRS-CI, and the Cumulative Illness Rating Scale-Severity Index, CIRS-SI [17–19]). Complete demographic and clinical details were collected at the first observation for each patient.

Moreover, to evaluate the feasibility of the model, the following measures were recorded: (a) tSetup (minutes): time spent for the correct patient placement and for planning of the session exercises; (b) tTraining (minutes): time spent by patient to perform the training; (c) tSupervision (minutes): time spent by the physiotherapist in the individual interaction with the patient; and (d) tClosure (minutes): time spent by the physiotherapist to close the exercise session and remove the patient from the device. These measures were obtained in a subsample of observations. In addition, the Visual Analogue Scale for Satisfaction [20], a self-assessment scale ranging from 0 (no satisfaction) to 10 (extreme satisfaction), was used to measure patients' satisfaction. The exact question was "Are you satisfied with your rehabilitation session?"

2.7. Statistical Analysis

To evaluate the feasibility of the model, data obtained from the three groups were compared by using one-way ANOVA tests, followed by post-hoc analysis with Bonferroni correction, or Kruskal–Wallis tests, followed by post-hoc analysis, using Dunn's procedure with a Bonferroni correction, as appropriated.

To evaluate the impact of disability, upper limb impairment, mobility, and comorbidity on the number of patients in each group, for each session, the mean data of the group were computed and compared between groups (univariate analysis) using Kruskal–Wallis tests, followed by post-hoc analysis, using Dunn's procedure with a Bonferroni correction. Then, to identify the best physiotherapist/patient ratio, a multivariate logistic analysis was performed, considering the variables identified in the univariate analysis as covariates. Finally, a receiver operating characteristics ROC analysis was used to identify a possible cutoff.

For all statistical tests, the significance level was set to 0.05. Statistical analysis was performed using SPSS (IBM Corp. Released 2017. IBM SPSS Statistics for Windows, Version 25.0. Armonk, NY, USA: IBM Corp.).

3. Results

In a period of 33 days, 60 patients were observed in our rehabilitation department. Table 1 shows the demographic and clinical characteristics of the analyzed sample. Spinal disorders included spinal cord injury after trauma or surgery on the column for neoplasia/lumbar spinal stenosis or spinal disc herniation; orthopedic disease included patients with shoulder prosthesis, upper limb fractures/surgery, and polytrauma. Upper limb impairment, disability, mobility, and comorbidities are reported in Table 2. During the study, each patient performed a variable number of rehabilitation sessions depending on the hospitalization stay and frequency of treatment per week. No serious adverse events were observed, and the robotic therapy was well accepted by all patients, without changes in the initial plan. None of the patients declined to participate in the study, since it did not interfere with their rehabilitation plan.

Table 1. Demographic and clinical characteristics of the observed sample.

Age (Years)	**64 ± 16 (Range: 19–91)**	
Sex	Male	27 (45.0%)
	Female	33 (55.0%)
Setting	Outpatient	11 (18.3%)
	Inpatient	49 (81.7%)
Diagnosis	Stroke	41 (68.3%)
	Multiple sclerosis	3 (5.0%)
	Charcot Marie Tooth	1 (1.7%)
	Guillain Barrè syndrome	1 (1.7%)
	Motoneuron disease	1 (1.7%)
	Parkinson disease	1 (1.7%)
	Spinal disorder	7 (11.7%)
	Orthopedic disease/trauma	5 (8.3%)

Table 2. Disability, upper limb impairment, mobility, and comorbidities of the sample.

Variable	Mean ± SD	Range
Barthel Index	57.7 ± 23.1	(14–97)
Motricity Index	57.5 ± 23.2	(1–93)
Deambulation Index	2.9 ± 2.2	(0–7)
CIRS-SI	1.0 ± 0.4	(0.3–2)
CIRS-CI	1.9 ± 1.0	(0–6)

SD: standard deviation; CIRS-CI: Cumulative Illness Rating Scale-Comorbidity Index; CIRS-SI: Cumulative Illness Rating Scale-Severity Index.

The total observations (data collected at each session for each patient) were 782, with a mean of 24 observations/day (range: 17–31) and a mean of 13 observations/patient (range: 1–50) (see also the supporting document: Supplementary Table S1). Two hundred fifty-five rehabilitation sessions were recorded with a ratio of physiotherapist/patient as follows: 45 sessions (17.7%) with a ratio of 1/2 (R2); 148 sessions (58.0%) with a ratio of 1/3 (R3), and 62 sessions (24.3%) with a ratio of 1/4 (R4). The mean value of the physiotherapist/patient ratio was 1/3.1.

3.1. Feasibility of the Model

Data related to the rehabilitation dose, supervision, and patients' satisfaction were recorded in a subsample of 329 observations (supporting document: Supplementary Table S1) and reported in

Table 3. As expected, the tTraining was significantly different between groups ($p = 0.007$); post-hoc analysis showed that it was significantly lower in the R4 group, when compared with both R2 (mean difference: 3.2 min) and R3 groups (mean difference: 2.3 min). Instead, no differences in tTraining were found between R2 and R3 groups. The tSupervision was significantly different between all groups ($p < 0.001$), as confirmed also by post-hoc analysis. With respect to patient satisfaction, no significant differences were observed between groups ($p = 0.39$).

Table 3. Feasibility measurements: rehabilitation dose (tTraining) and supervision time (tSupervision) during a 45-min rehabilitation session, and patients' satisfaction in the whole sample and the three groups, separately, together with the statistical analysis. Data were obtained from 329 observations.

	Whole Group	R2	R3	R4	*p*	Post-Hoc Analysis		
	Mean ± SD	Mean ± SD	Mean ± SD	Mean ± SD		R2 vs. R3	R2 vs. R4	R3 vs. R4
tTraining (min)	31.3 ± 6.4	32.8 ± 5.0	31.9 ± 6.0	29.6 ± 7.2	**0.007**	>0.99	**0.04**	**0.01**
tSupervision (min)	9.7 ± 2.3	11.0 ± 3.2	9.7 ± 2.0	8.9 ± 2.1	**<0.001**	**0.005**	**<0.001**	**0.007**
VAS for satisfaction	8 ± 1	8 ± 2	8 ± 1	8 ± 1	0.39	–	–	–

R2, R3 and R4 indicate groups with 2, 3 or 4 patients supervised by one physiotherapist, respectively. Values in bold indicate statistical significance ($p < 0.05$). SD: standard deviation; VAS: Visual analogue scale.

In addition, we found that the mean time needed for the setup (tSetup) was 3.3 ± 1.3 min, while the mean time needed for the session closure (tClosure) was 1.4 ± 0.5 min.

3.2. Impact of Disability, Upper Limb Impairment, Mobility, and Comorbidity on the Model

A comparison of upper limb impairment, disability, mobility, and comorbidity mean score between the three groups was made and reported in Table 4. The results showed significant differences only in BI, DI, and CIRS-SI between the groups R2 versus R4, and R3 versus R4. Considering the lack of statistical significance between the R2 and R3 groups, a further analysis was made combining the R2 and R3 groups (R2 + R3 group). The logistic regression model was carried out on the three statistically significant variables obtained from the univariate analysis. We found that the model was statistically significant, $\chi2(3) = 36.096$, $p < 0.001$. As shown in Table 5, among the three predictor variables, only the BI was statistically significant. To obtain a reference value for the only significant variable (BI) in the logistic regression model, an ROC analysis was carried out. According to the Youden index, the obtained cut-off value was equal to 57.

Table 4. Disability, mobility, and comorbidity variables for the three groups, separately. Data were obtained from 782 observations.

	R2	R3	R4	*p*	Post-Hoc Analysis		
	Mean ± SD	Mean ± SD	Mean ± SD		R2 vs. R3	R2 vs. R4	R3 vs. R4
BI	49.0 ± 10	53.0 ± 12.6	63.0 ± 14.3	**<0.001**	0.24	**<0.001**	**<0.001**
MI	59.5 ± 14.7	57.2 ± 11.8	57.9 ± 8.8	0.42	-	-	-
DI	2.1 ± 1.2	2.4 ± 1.4	3.3 ± 1.4	**<0.001**	0.98	**<0.001**	**<0.001**
CIRS-SI	1.1 ± 0.3	1.0 ± 0.2	0.9 ± 0.2	**<0.001**	0.26	**<0.001**	**<0.001**
CIRS-CI	2.1 ± 0.8	2.0 ± 0.6	1.8 ± 0.4	0.06	-	-	-

Values in bold indicate statistical significance ($p < 0.05$). R2, R3 and R4 indicate groups with two, three, or four patients supervised by one physiotherapist, respectively. SD: standard deviation; BI: Barthel Index; MI: Motricity Index; DI: Deambulation Index; CIRS-CI: Cumulative Illness Rating Scale-Comorbidity Index; CIRS-SI: Cumulative Illness Rating Scale-Severity Index.

Table 5. Logistic regression model for the identification of the best physiotherapist/patient ratio (R2/R3 versus R4), considering the mean disability, mobility, and comorbidity of the group (identified in the univariate analysis). Data were obtained from 255 rehabilitation sessions (782 observations).

	β	Standard Error	p	Exp (β)	95% Confidence Interval for Exp (β)	
				Relative Risk	Lower Bound	Upper Bound
BI	0.099259	0.039318	**0.01**	1.1044	1.0224	1.1928
DI	−0.52213	0.37	0.16	0.5933	0.2873	1.2252
CIRS-SI	−1.41553	0.75418	0.06	0.2428	0.0554	1.0647

Values in bold indicate statistical significance ($p < 0.05$). BI: Barthel Index; DI: Deambulation Index; CIRS-SI: Cumulative Illness Rating Scale-Severity Index.

4. Discussion

Robotic tools are emerging in the rehabilitation field, showing promising expectations for an intensive and personalized treatment. Nonetheless, these devices are expensive. Therefore, it is necessary to find the best organizational models to increase the number of patients who may have access to the technological rehabilitation, optimizing human resources. In this perspective, it is crucial to know if and how patient's disability, mobility, or comorbidity can influence an organizational model where a new, non-traditional ratio of physiotherapist/patient would be adopted.

This study shows the feasibility, in terms of rehabilitation dose and patient's satisfaction, of an organizational model with one physiotherapist supervising more than one patient, using a set of four devices for a comprehensive upper limb rehabilitation. In a 45-min rehabilitation session, regardless of the physiotherapist/patient ratio, each patient performed about 30 min of training with an individual physiotherapist supervision of about 10 min. As expected, the time of training was significantly lower in the R4 group but with a gap of only 2 and 3 min compared to the R2 and R3 groups, respectively. Regarding the difference in the training time recorded between the R2/R3 and R4 groups, an increase of about 3 min of the rehabilitation session of the R4 group can be used to overcome this small gap.

Therefore, our results showed that the physiotherapists in this study were able to supervise up to four patients, without a reduction of the treatment dose provided to the patient. This is a crucial point: in fact, several studies report an intensity–effect relation between the amount of therapy that individuals receive and the motor recovery [21,22]; additionally, no ceiling effect for the intensity of the therapy has been observed [23–25]. Note that patient satisfaction was very high (8/10) without significant differences between groups, so patients were very satisfied even if treated together with one, two, or three other patients at the same time. Regarding the clinical characteristics of the patients, our data did not show differences when two or three patients were supervised by one physiotherapist. On the contrary, though as expected, the group in which four patients were supervised by one physiotherapist (R4) had a lower disability and comorbidity and higher mobility than the R2 and R3 groups. A remarkable result is that upper limb impairment did not influence the organizational model proposed, since the MI not significantly different between the groups (R2, R3, and R4). The multivariate analysis showed that only the BI can predict significantly the treatment group, meaning that the maximum number of patients supervised by one physiotherapist can be dependent on their level of autonomy. Therefore, the BI seems to be the most important variable that could influence the physiotherapist/patient ratio. Based on our results, a ratio of 1/4 could be used if the mean value BI of the group of patients is higher than 57. Note that patients with BI values lower than 60 are considered patients with a severe disability [14]. Our results showed a similar cut-off of disability, even if calculated as the mean of the group (BI mean score of group <57).

The analysis of both clinical data and feasibility measures suggests that patients with a severe disability (BI mean score of the group <57) can be equally treated in groups of two or three patients, providing that they had the same dose. Instead, when a moderate disability is present, up to four patients can be supervised by one physiotherapist. This study also confirms the possibility of optimizing resources allocation. In fact, in our rehabilitation planning, a mean of 24 patients were treated each day with a mean value of physiotherapist/patient ratio of 1/3.1, which means eight rehabilitation sessions/day. In a traditional rehabilitation model (with a physiotherapist/patient ratio of 1/1) to treat 24 patients/day, obviously 24 rehabilitation sessions/day are needed. As highlighted in the proposed organizational model, it is possible to reduce the number of rehabilitation sessions/day (needed to treat the same number of patients) by about 1/3, with better human resources allocation.

Moreover, it is worthy to note that we have specifically chosen a set of devices with a limited time required to start the rehabilitation session, as highlighted by the reported tSetup, and therefore, these have a limited impact on the provided rehabilitation dose. On the contrary, more complex devices, as exoskeletons, require a longer time to fit the robot for individual patients, and it is one of the main barriers to efficient use [26]. As in our approach, two studies on the same topic [8,9] used both robotic and non-robotic devices, supporting the idea that this approach can be effective and sustainable.

Anyway, robotic rehabilitation is not merely a matter of increasing the amount and intensity of therapy [27]. It provides extrinsic biofeedback that makes the patient aware of his movements and consequently allows him to modify his movement patterns, thus fostering his participation in the rehabilitation exercise [28–30]. Furthermore, it seems that the impact of technological rehabilitation on functional outcome could be optimized by offering more chances to the nervous system to experience "real" activity-related sensorimotor input during the training of upper limb movement [31].

From this knowledge and from our experience, it is very important to underline that physiotherapists have to perform an adequate training to acquire expertise with regard to these devices. With a correct use of them, the workload of the physiotherapists can be alleviated, their work time can be optimized, and they can focused toward personalizing rehabilitation treatments, according to each patient's residual abilities and progress in order to optimize functional recovery [27].

The organizational model herein discussed shows the feasibility of providing a comprehensive and a high technological level upper limb treatment that would be accessible to a high number of patients at once, while preserving an adequate training time and physiotherapist supervision. Finally, thanks to the sustainability in terms of optimizing human resources, this model can be applied to the usual care of the clinical practice considering the advantages that the robotic devices can provide (for example, to objectively measure the improvement during the therapy) [32,33].

Limitations of the Study

This study was conducted in a highly specialized rehabilitation center where all physiotherapists had an expertise in robotic rehabilitation and used these devices for more than one year. Therefore, this could limit the extension of our results to different rehabilitation units with lower levels of expertise. However, we believe that this limitation does not severely impact on the application of the results, since a high level of expertise is required to efficiently use these devices. Moreover, this study was not aimed to evaluate the validity and the reliability of the used outcome measures. Finally, this study did not compare the effect of different physiotherapist/patient ratios on the outcome of the rehabilitation process. These aspects should be addressed in future studies.

5. Conclusions

In conclusion, this study showed the feasibility, in terms of rehabilitation dose and patients' satisfaction, of a new organizational model, where a group of patients supervised by one physiotherapist is treated with a set of robotic and sensor-based devices. Moreover, it highlighted that the mean disability of the group, as measured by the Barthel Index, could influence the proper physiotherapist/patient ratio. Future studies should be addressed to analyze the impact of this new model on the rehabilitation outcomes.

Supplementary Materials: The following are available online at http://www.mdpi.com/2076-3417/9/24/5357/s1.

Author Contributions: Conceptualization, I.A. and M.G.; methodology, I.A., M.G.; formal analysis, M.G.; investigation, C.P. and A.C.; data curation, C.P. and M.G.; writing—original draft preparation, I.A., S.L. and M.G.; writing—review and editing, I.A. and M.G.; supervision, I.A. and L.P.

Funding: This research received no external funding.

Acknowledgments: The authors would like to thank Enrica Di Sipio for her help in making figures, Manuel Triolo for the support given from a linguistic standpoint and Claudia Loreti for administrative and technical support.

Conflicts of Interest: The authors declare no conflict of interest.

References

1. World Health Organization. *International Classification of Functioning, Disability and Health: ICF*; World Health Organization: Geneva, Switzerland, 2001.
2. Mehrholz, J.; Pohl, M.; Platz, T.; Kugler, J.; Elsner, B. Electromechanical and robot-assisted arm training for improving activities of daily living, arm function, and arm muscle strength after stroke. *Cochrane Database Syst. Rev.* **2015**. [CrossRef]
3. Laut, J.; Porfiri, M.; Raghavan, P. The Present and Future of Robotic Technology in Rehabilitation. *Curr. Phys. Med. Rehabil. Rep.* **2016**, *4*, 312–319. [CrossRef]
4. Hayward, K.S.; Kramer, S.F.; Thijs, V.; Ratcliffe, J.; Ward, N.S.; Churilov, L.; Jolliffe, L.; Corbett, D.; Cloud, G.; Kaffenberger, T.; et al. A systematic review protocol of timing, efficacy and cost effectiveness of upper limb therapy for motor recovery post-stroke. *Syst. Rev.* **2019**, *8*, 187. [CrossRef] [PubMed]
5. Wagner, T.H.; Lo, A.C.; Peduzzi, P.; Bravata, D.M.; Huang, G.D.; Krebs, H.I.; Ringer, R.J.; Federman, D.G.; Richards, L.G.; Haselkorn, J.K.; et al. An economic analysis of robot-assisted therapy for long-term upper-limb impairment after stroke. *Stroke* **2011**, *42*, 2630–2632. [CrossRef] [PubMed]
6. Lo, K.; Stephenson, M.; Lockwood, C. The economic cost of robotic rehabilitation for adult stroke patients: A systematic review. *JBI Database Syst. Rev. Implement. Rep.* **2019**, *17*, 520–547. [CrossRef] [PubMed]
7. Masiero, S.; Poli, P.; Armani, M.; Ferlini, G.; Rizzello, R.; Rosati, G. Robotic upper limb rehabilitation after acute stroke by NeReBot: Evaluation of treatment costs. *Biomed Res. Int.* **2014**, *2014*. [CrossRef]
8. Atashzar, S.F.; Shahbazi, M.; Tavakoli, M.; Patel, R.V. A Computational-Model-Based Study of Supervised Haptics-Enabled Therapist-in-the-Loop Training for Upper-Limb Poststroke Robotic Rehabilitation. *IEEE/ASME Trans. Mechatron.* **2018**, *23*, 563–574. [CrossRef]
9. Buschfort, R.; Brocke, J.; Heß, A.; Werner, C.; Waldner, A.; Hesse, S. The arm studio to intensify the upper limb rehabilitation after stroke: Concept, acceptance, utilization and preliminary clinical results. *J. Rehabil. Med.* **2010**, *42*, 310–314. [CrossRef] [PubMed]
10. Hesse, S.; Heß, A.; Werner, C.C.; Kabbert, N.; Buschfort, R. Effect on arm function and cost of robot-assisted group therapy in subacute patients with stroke and a moderately to severely affected arm: A randomized controlled trial. *Clin. Rehabil.* **2014**, *28*, 637–647. [CrossRef]
11. Bustamante Valles, K.; Montes, S.; de Madrigal, M.J.; Burciaga, A.; Martínez, M.E.; Johnson, M.J. Technology-assisted stroke rehabilitation in Mexico: A pilot randomized trial comparing traditional therapy to circuit training in a Robot/technology-assisted therapy gym. *J. Neuroeng. Rehabil.* **2016**, *13*, 83. [CrossRef]
12. Aprile, I.; Cruciani, A.; Germanotta, M.; Gower, V.; Pecchioli, C.; Cattaneo, D.; Vannetti, F.; Padua, L.; Gramatica, F. Upper Limb Robotics in Rehabilitation: An Approach to Select the Devices, Based on Rehabilitation Aims, and Their Evaluation in a Feasibility Study. *Appl. Sci.* **2019**, *9*, 3920. [CrossRef]
13. Aprile, I.; Germanotta, M.; Cruciani, A.; Loreti, S.; Pecchioli, C.; Cecchi, F.; Montesano, A.; Galeri, S.; Diverio, M.; Falsini, C.; et al. Upper Limb Robotic Rehabilitation after Stroke: A Multicenter, Randomized Clinical Trial. *J. Neurol. Phys. Ther.* **2019**, in press. [CrossRef]
14. Shah, S.; Vanclay, F.; Cooper, B. Improving the sensitivity of the Barthel Index for stroke rehabilitation. *J. Clin. Epidemiol.* **1989**, *42*, 703–709. [CrossRef]
15. Demeurisse, G.; Demol, O.; Robaye, E. Motor evaluation in vascular hemiplegia. *Eur. Neurol.* **1980**, *19*, 382–389. [CrossRef] [PubMed]
16. Korner-Bitensky, N.; Mayo, N.; Cabot, R.; Becker, R.; Coopersmith, H. Motor and functional recovery after stroke: Accuracy of physical therapists' predictions. *Arch. Phys. Med. Rehabil.* **1989**, *70*, 95–99. [PubMed]
17. Miller, M.D.; Paradis, C.F.; Houck, P.R.; Mazumdar, S.; Stack, J.A.; Rifai, A.H.; Mulsant, B.; Reynolds, C.F. Rating chronic medical illness burden in geropsychiatric practice and research: Application of the Cumulative Illness Rating Scale. *Psychiatry Res.* **1992**, *41*, 237–248. [CrossRef]

18. Conwell, Y.; Forbes, N.T.; Cox, C.; Caine, E.D. Validation of a measure of physical illness burden at autopsy: The Cumulative Illness Rating Scale. *J. Am. Geriatr. Soc.* **1993**, *41*, 38–41. [CrossRef]
19. Salvi, F.; Miller, M.D.; Grilli, A.; Giorgi, R.; Towers, A.L.; Morichi, V.; Spazzafumo, L.; Mancinelli, L.; Espinosa, E.; Rappelli, A.; et al. A manual of guidelines to score the modified cumulative illness rating scale and its validation in acute hospitalized elderly patients. *J. Am. Geriatr. Soc.* **2008**, *56*, 1926–1931. [CrossRef]
20. Brokelman, R.B.G.; Haverkamp, D.; van Loon, C.; Hol, A.; van Kampen, A.; Veth, R. The validation of the visual analogue scale for patient satisfaction after total hip arthroplasty. *Eur. Orthop. Traumatol.* **2012**, *3*, 101–105. [CrossRef]
21. Kwakkel, G.; van Peppen, R.; Wagenaar, R.C.; Wood Dauphinee, S.; Richards, C.; Ashburn, A.; Miller, K.; Lincoln, N.; Partridge, C.; Wellwood, I.; et al. Effects of augmented exercise therapy time after stroke: A meta-analysis. *Stroke* **2004**, *35*, 2529–2539. [CrossRef]
22. Byl, N.N.; Pitsch, E.A.; Abrams, G.M. Functional outcomes can vary by dose: Learning-based sensorimotor training for patients stable poststroke. *Neurorehabil. Neural Repair* **2008**, *22*, 494–504. [CrossRef] [PubMed]
23. Langhorne, P.; Wagenaar, R.; Partridge, C. Physiotherapy after stroke: More is better? *Physiother. Res. Int.* **1996**, *1*, 75–88. [CrossRef] [PubMed]
24. Kwakkel, G. Impact of intensity of practice after stroke: Issues for consideration. *Disabil. Rehabil.* **2006**, *28*, 823–830. [CrossRef] [PubMed]
25. Huang, H.-C.; Chung, K.-C.; Lai, D.-C.; Sung, S.-F. The impact of timing and dose of rehabilitation delivery on functional recovery of stroke patients. *J. Chin. Med. Assoc.* **2009**, *72*, 257–264. [CrossRef]
26. Swank, C.; Sikka, S.; Driver, S.; Bennett, M.; Callender, L. Feasibility of integrating robotic exoskeleton gait training in inpatient rehabilitation. *Disabil. Rehabil. Assist. Technol.* **2019**. [CrossRef]
27. Poli, P.; Morone, G.; Rosati, G.; Masiero, S. Robotic Technologies and Rehabilitation: New Tools for Stroke Patients' Therapy. *Biomed Res. Int.* **2013**, *2013*. [CrossRef]
28. van Vliet, P.M.; Wulf, G. Extrinsic feedback for motor learning after stroke: What is the evidence? *Disabil. Rehabil.* **2006**, *28*, 831–840. [CrossRef]
29. Masiero, S.; Poli, P.; Rosati, G.; Zanotto, D.; Iosa, M.; Paolucci, S.; Morone, G. The value of robotic systems in stroke rehabilitation. *Expert Rev. Med. Devices* **2014**. [CrossRef]
30. Gassert, R.; Dietz, V. Rehabilitation robots for the treatment of sensorimotor deficits: A neurophysiological perspective. *J. Neuroeng. Rehabil.* **2018**, *15*, 46. [CrossRef]
31. Nudo, R.J. Functional and structural plasticity in motor cortex: Implications for stroke recovery. *Phys. Med. Rehabil. Clin. N. Am.* **2003**, *14*, S57–S76. [CrossRef]
32. Germanotta, M.; Vasco, G.; Petrarca, M.; Rossi, S.; Carniel, S.; Bertini, E.; Cappa, P.; Castelli, E. Robotic and clinical evaluation of upper limb motor performance in patients with Friedreich's Ataxia: An observational study. *J. Neuroeng. Rehabil.* **2015**, *12*, 41. [CrossRef] [PubMed]
33. Germanotta, M.; Cruciani, A.; Pecchioli, C.; Loreti, S.; Spedicato, A.; Meotti, M.; Mosca, R.; Speranza, G.; Cecchi, F.; Giannarelli, G.; et al. Reliability, validity and discriminant ability of the instrumental indices provided by a novel planar robotic device for upper limb rehabilitation. *J. Neuroeng. Rehabil.* **2018**, *15*, 39. [CrossRef] [PubMed]

MDPI
St. Alban-Anlage 66
4052 Basel
Switzerland
Tel. +41 61 683 77 34
Fax +41 61 302 89 18
www.mdpi.com

Applied Sciences Editorial Office
E-mail: applsci@mdpi.com
www.mdpi.com/journal/applsci

www.ingramcontent.com/pod-product-compliance
Lightning Source LLC
Chambersburg PA
CBHW061522210926
51850CB00008B/2002
9783036523118